ADVANCES IN
FOOD BIOCHEMISTRY

ADVANCES IN FOOD BIOCHEMISTRY

Edited by

Fatih Yildiz

CRC Press
Taylor & Francis Group
Boca Raton London New York

CRC Press is an imprint of the
Taylor & Francis Group, an **informa** business

CRC Press
Taylor & Francis Group
6000 Broken Sound Parkway NW, Suite 300
Boca Raton, FL 33487-2742

First issued in paperback 2017

© 2010 by Taylor and Francis Group, LLC
CRC Press is an imprint of Taylor & Francis Group, an Informa business

No claim to original U.S. Government works

ISBN-13: 978-0-8493-7499-9 (hbk)
ISBN-13: 978-1-138-11548-4 (pbk)

Visit the Taylor & Francis Web site at
http://www.taylorandfrancis.com

and the CRC Press Web site at
http://www.crcpress.com

Contents

Foreword

Knowledge of food biochemistry is critical to the development and growth of major aspects of food science, including production, processing, preservation, distribution, safety, and engineering and technology. Each of these areas is directly related to the current effort of commercializing food products and adding new value to them. Putting this science into practice also requires an understanding of broader issues, such as national and global regulations, concerning the identity, manufacture, and transport of foods. Appreciating food biochemistry also offers insights into consumer perspectives and preferences regarding issues such as genetically modified foods, nanomaterials in foods, functional foods and nutraceuticals, and food safety. The successful application of this knowledge is ultimately essential to promoting health and wellness through food and nutrition.

Advances in Food Biochemistry provides updated information on fundamental topics such as food acids, additives, flavors, contaminants, and safety. However, it is unique in that it covers topics on the emerging science of food glycobiology, plant and animal hormones, nutrigenetics, and toxicogenomics. While traditional food biochemistry texts deal with the major macronutrients and micronutrients, this book addresses the modern application of phytochemicals from plant foods and herbal medicines to functional foods and dietary supplements. Few books in this field have recognized the influence of nutrigenomics on food science, and the implication of the impact of foods and nutrients on the genome, transciptome, proteome, and metabolome. Applying this knowledge about the interactions between our genes, nutrition, and lifestyle presents an opportunity to develop new food products directed to optimize our health based on an individual's unique needs. Similarly, toxicogenomics is covered here to describe not only its applications in deciphering mechanisms of toxicity, but also in aiding risk assessments. This emerging aspect of food biochemistry, including the use of microarray data, is now beginning to be used by government regulatory agencies to better predict harm and guarantee the safety of foods as well as drugs.

Professor Fatih Yildiz has edited this book to provide a practical overview of foods from a biochemical perspective. He received his undergraduate education in Turkey and did his graduate training in the United States, where he served on the faculty at the University of Maryland, College Park before joining the Middle East Technical University, Ankara, Turkey. He has published an extensive body of research and worked on food and nutrition science projects with the FAO, UNIDO, UNICEF, and NATO. Most recently, Dr. Yildiz received the Ambassador for Turkey Award from the European Federation of Food Science and Technology. He has worked to ensure that each chapter emphasizes the nutritional and health aspects of food and its components as well as associated issues such as technology and toxicology. Dr. Yildiz and the other chapter authors have brought together their renowned expertise and decades of experience to this book, designing it for advanced college students who can use it to learn about this field without the need of additional references. Food science and policy professionals will also find this book a useful resource in their field of work. If we are to feed the world, books like *Advances in Food Biochemistry* will play an important role in leading the way.

Jeffrey B. Blumberg, PhD, FACN, CNS
Tufts University
Boston, Massachusetts

Preface

This book is an outgrowth of my research and teaching to graduate and undergraduate students at the University of Maryland, College Park; the University of Minnesota, Minneapolis; the Institut National Recherches Agronomiques, Avignon, France; the Mediterranean Agronomic Institute of Chania, Crete, Greece; and the Middle East Technical University, Ankara, Turkey. The pace of discovery in biochemistry and its application to food biochemistry has been very rapid during the past decade. This increase in knowledge has enriched our understanding of the molecular basis of living systems as well as biomaterials and has opened many new areas of applications in food production, processing, storage, distribution, consumption, health, and toxicology of food and food components. Understanding the biochemistry of food and feed is basic to all the other research and development in the fields of food science, technology, and nutrition.

Chapter 1 covers water and its relation to food. This book covers classical as well as new information on the glycobiology of carbohydrates in Chapter 2 and amino acids, oligopeptides, polypeptides, and proteins in Chapter 3. Chapter 4 covers new and classical applications of enzymes in the food industry. Lipids, fats, and oils are covered in Chapter 5, along with the latest developments. Nucleic acid and DNA identification of foods are covered in Chapter 6, along with other applications. Chapters 2 through 6 include the most recent developments along with their applications in food and nutrition. There are several topics included in these chapters that are not covered in most food chemistry textbooks. The essential topic of plant, animal, and human hormones and their risks and benefits to producers and consumers are covered in Chapter 7. Chapter 8 covers the topic of functional foods, herbs, and dietary supplements. Most classical food chemistry/biochemistry books mention close to 20 carbohydrates (glucose, fructose, lactose, etc.), 22 basic amino acids, 20 or more fatty acids, 13 vitamins, and 30 minerals in food systems. But now, we know that foods may contain at least 100,000 different chemicals. Chapter 8 covers, classifies, and explains the functions of some these compounds. Herbs and dietary supplements are considered as food for all legal and practical purposes. The recent renewal of interest in herbal medicines has made it necessary to have a clearer understanding of their structure and function in human diet. Natural dietary supplements can include a variety of products derived from or by products of foods, can be high in certain ingredients, and can contain bioactive compounds that aid in maintaining the health and wellness of individuals. Chapter 9 covers flavor compounds in foods, while Chapter 10 deals with organic acids, including phenolic acids. Other topics covered include the interactions of the environment with food components, that is, oxidations and changes of foods during their life cycles. Chapters 11 and 12 cover biological and lipid oxidations in foods and their controls. Chapter 13 covers the issue of food safety—major contaminants and additives are discussed and their regulatory limits and status are explained.

Chapters 14 and 15 summarize the interactions of nutrition and the genetic makeup of the individual components. Chapter 14 covers an emerging field of food research and focuses on identifying and understanding, at the molecular level, interaction between nutrients and other dietary bioactives with the human genome during transcription, translation and expression, the processes during which proteins encoded by the genome are produced and expressed.

Chapter 15, which deals with toxicogenomics, discusses the interaction of food additives and contaminants with the genome of the individuals, and how this translates into human disease conditions.

The rapid growth of the food and dietary supplement industry into big business and the increase in the number of items on the shelves, the new controls on food additives and contaminants, and the attempts to standardize these articles all serve to indicate the growing importance of this timely book on advances in the biochemistry of foods.

Thus, the aim of this book is to provide a unified picture of foods from a biochemical point of view. The primary emphasis is on the technological, nutritional, health functionality, and toxicological properties of foods, and the changes that occur during the processing, storage, preparation, and consumption processes. The authors have written this book trying to keep in mind undergraduates and graduates who have had several courses in chemistry/biology; however, an effort has been made to provide the necessary background in chemistry/biology and sufficient explanations/illustrations for students to proceed without additional references.

I would like to thank the authors, reviewers, publishers, and all those involved in making this book a reality.

Fatih Yildiz
Middle East Technical University
Ankara, Turkey

Editor

Dr. Fatih Yildiz received his BS from Atatürk University in Erzurum, Turkey. He received his MS and PhD in food biochemistry from the University of Maryland, College Park. He received his assistant and associate professorships from the Middle East Technical University, Ankara, Turkey. He worked as a faculty member at the University of Maryland for five years. Dr. Yildiz worked as a professor in the Department of Food Science and Nutrition at the University of Minnesota, Minneapolis. He has also conducted research at the Institut National Recherches Agronomiques (INRA), Avignon, France as a visiting professor in 1997.

Currently, he teaches and conducts research in the biochemistry, biotechnology, and food engineering departments at the Middle East Technical University, Ankara, Turkey. He has worked on research projects with the Food and Agriculture Organization of the United Nations (FAO), United Nations Industrial Development Organization (UNIDO), the United Nations Children's Fund (UNICEF), and the North Atlantic Treaty Organization (NATO) as a project director.

Dr. Yildiz has published more than 130 research and review papers, mostly in English, in international and national journals as the major author. He has published in Turkish, French, and German. His research studies have been cited by Science Citation Index (SCI) more than 50 times. He has coauthored a book, *Minimally Processed and Refrigerated Fruits and Vegetables*, published by Chapman & Hall in 1994, which was then a new concept in the food industry.

His current research interests include health nutrition and safety attributes of Mediterranean diet. He has edited a book, *Phytoestrogens in Functional Foods*, published by CRC Press.

Professor Yildiz, is listed in the Who's Who in Turkey and Europe and serves on numerous advisory committees of the Ministry of Health and Agriculture. He is also a member of the National Codex Commission in Turkey, and a member of 10 scientific and academic organizations in the United States, France, and Turkey.

Scientific and professional society memberships that Professor Yildiz is a member of

1. Institute of Food Technologists, Chicago, IL (since 2001)
2. Turkish Food Technologist Association, Ankara, Turkey (since 1981)
3. American Society for Microbiology, Washington, DC (since 1975)
4. Global Food Traceability Forum, Halifax, West Yorkshire, United Kingdom (member since 2004)
5. European Federation of Food Science and Technology, Wageningen University, Wageningen, the Netherlands
6. Who's Who in Turkey and Who's Who in European Research and Development

Honors and awards

1. INRA Research Award (1997) on Post-Harvest Mushroom Technology, INRA, Avignon Research Station, France
2. Minnesota–South Dakota Dairy Foods Research Center Grant to conduct research in the University of Minnesota, Minneapolis (1991–1992)
3. Gaziantep Chamber of Commerce and Industry Award for outstanding service to the region (1998)
4. Ambassador for Turkey Award (2008) by EFFoST

Contributors

Dimitris G. Arapoglou
Laboratory of Biotechnology
National Agricultural Research Foundation
Athens, Greece

Ioannis S. Arvanitoyannis
School of Agricultural Sciences, Animal
Production and Agricultural Environment
University of Thessaly
Volos, Greece

Bruno Biavati
Department of Agroenvironmental Science
and Technology
University of Bologna
Bologna, Italy

Sara Bosi
Department of Agroenvironmental Science
and Technology
University of Bologna
Bologna, Italy

Dilek Boyacioglu
Department of Food Engineering
School of Chemical and Metallurgical
Engineering
Istanbul Technical University
Istanbul, Turkey

Esra Capanoglu
Department of Food Engineering
School of Chemical and Metallurgical
Engineering
Istanbul Technical University
Istanbul, Turkey

Pietro Catizone
Department of Agroenvironmental Science
and Technology
University of Bologna
Bologna, Italy

Muharrem Certel
Faculty of Engineering
Department of Food Engineering
Akdeniz University
Antalya, Turkey

Çağatay Ceylan
Department of Food Engineering
Izmir Institute of Technology
Urla, Turkey

Meltem Yalinay Cirak
Faculty of Medicine
Department of Microbiology and Clinical
Microbiology
Gazi University
Ankara, Turkey

Giovanni Dinelli
Department of Agroenvironmental Science
and Technology
University of Bologna
Bologna, Italy

Abdullah Ekmekci
Faculty of Medicine
Department of Medical Biology and Genetics
Gazi University
Ankara, Turkey

Sibel Fadıloğlu
Department of Food Engineering
University of Gaziantep
Gaziantep, Turkey

Diana Di Gioia
Department of Agroenvironmental Science
and Technology
University of Bologna
Bologna, Italy

Fahrettin Göğüş
Department of Food Engineering
University of Gaziantep
Gaziantep, Turkey

Bensu Karahalil
Faculty of Pharmacy
Department of Toxicology
Gazi University
Ankara, Turkey

Barçın Karakaş
Faculty of Engineering
Department of Food Engineering
Akdeniz University
Antalya, Turkey

Ayhan Karakoç
Department of Endocrinology
Gazi University
Ankara, Turkey

Sotirios Kiokias
Laboratory of Food Chemistry
    and Technology
School of Chemical Engineering
National Technical University of Athens
Athens, Greece

Işıl A. Kurnaz
Faculty of Engineering and Architecture
Genetics and Bioengineering Department
Yeditepe University
Istanbul, Turkey

Athanasios E. Labropoulos
Food Technology and Nutrition Department
Technological Educational Institute of Athens
Athens, Greece

Frank A. Manthey
Department of Plant Sciences
North Dakota State University
Fargo, North Dakota

Ilaria Marotti
Department of Agroenvironmental Science
    and Technology
University of Bologna
Bologna, Italy

Dilara Nilufer
Department of Food Engineering
School of Chemical and Metallurgical
    Engineering
Istanbul Technical University
Istanbul, Turkey

Çiğdem Soysal
Department of Food Engineering
University of Gaziantep
Gaziantep, Turkey

Theodoros H. Varzakas
Department of Food Technology
School of Agriculture
Technological Educational Institution of
    Kalamata
Kalamata, Greece

Y. Sedat Velioğlu
Faculty of Engineering
Department of Food Engineering
Ankara University
Ankara, Turkey

Yingying Xu
Department of Cereal and Food Sciences
North Dakota State University
Fargo, North Dakota

Fatih Yildiz
Department of Food Engineering and
    Biotechnology
Middle East Technical University
Ankara, Turkey

1 Water and Its Relation to Food

Barçın Karakaş and Muharrem Certel

CONTENTS

1.1 ABUNDANCE AND SIGNIFICANCE OF WATER IN FOODS

Water can universally be found in the solid, gaseous, and liquid states. Saltwater oceans contain about 96.5% of our global water supply. Ice, the solid form of water, is the most abundant form of freshwater and most of it, nearly 68.7%, is currently trapped in the polar ice caps and glaciers [1]. About 30% of the freshwater sources are present in aquifers as groundwater. The remaining freshwater is surface water in lakes and rivers, soils, wetlands, biota, and atmospheric water vapor.

Climate change due to global warming coupled with the ever-increasing demand for water by a growing human population have a surmounting effect on the present and future scarcity of this valuable resource. Besides the unavailability of water itself, shortages also result in problems, such as increased impurities and contaminants in the available water. In facing these situations, populations are obligated to manage and reuse water sources with awareness and diligence.

Water is the most abundant and surely the most frequently overlooked component in foods. It is estimated that over 35% of our total water intake comes from the moisture in the foods we consume [2]. The other contributors to our water intake are beverages and metabolic water, which is produced through chemical reactions in the body.

The water content of foods is very variable. It may be as low as 0% in vegetable oils and as high as 99% in some vegetables and fruit. Water by itself is free of calories and plain water does not contain nutritive substances, but it may be an ingredient itself in foods. Foods are described as dry or low-moisture foods if they have very low water content. These are most often solid food systems. Liquid food systems and tissue foods where water is the dominating constituent of the solution are high-moisture foods. Foods that contain moderate levels of water are intermediate-moisture foods (IMFs).

For the food industry, water is essential for processing, as a heating or a cooling medium. It may be employed in processes in the form of liquid water or in the other states of water such as ice or steam.

1

Almost all food-processing techniques involve the use or the modification of water in food. Freezing, drying, concentration, and emulsification processes all involve changes in the water fraction of the food. Without the presence of water, it would not be possible to achieve the physicochemical changes that occur during cooking such as the gelatinization of starch. Water is important as a solvent for dissolving small molecules to form solutions and as a dispersing medium for dispersing larger molecules to form colloidal solutions.

Water has historically been the primary solvent of choice for the extraction of apolar substances. Recently, the use of subcritical water, hot water at 100°C–374°C under high pressure, is being investigated as an environmentally friendly alternative to organic substances that are used for the extraction of polar substances [3].

The control of water activity in foods is an important tool for extending shelf life. It is responsible for the quality of foods affected by microbiological, chemical, and physical changes. The physical properties, quantity, and quality of water within the food have a strong impact on food effectiveness, quality attributes, shelf life, textural properties, and processing.

Food-preservation processes have a common goal of extending the shelf life of foods to allow for storage and convenient distribution. The activity of microorganisms is the first and most dangerous limitation of shelf life. Water is essential for microorganisms that may cause food spoilage if they are present in a food that offers them favorable conditions for growth. Hence, many food-preservation techniques were developed to reduce the availability or activity of water in order to eliminate the danger of microbial spoilage.

The presence or activity of water in foods may also enhance the rate at which deteriorative chemical reactions occur. Some products may become rancid through free radical oxidation even at low humidities and thus become unacceptable. Labile nutrients such as vitamins and natural color compounds are oxidized more rapidly when stored at low moisture levels. Enzyme-mediated hydrolytic reactions may reduce the quality of the food product. Other reactions such as the Maillard type of nonenzymatic browning may be enhanced by the presence of higher levels of water. On the other hand, water content is crucial for the textural characteristics and the sensory perception of foods. A food may be found unacceptable by consumers simply because it does not satisfy their textural (sensory) anticipation.

For the reasons mentioned above, it is important to have control over the quantity and quality of water in foods and in the processing thereof. In order to do so, it is essential to have a better understanding of the chemical and physical properties that determine the fundamental functions of water.

Controlling the growth of ice crystals is a primary concern for food technologists. Developments in the field of molecular biology have enabled recombinant production of many proteins. Two of these protein groups are the antifreeze proteins [4] and ice nucleation proteins [5,6], which have potential applications in the food industry. The application of these proteins and other techniques such as dehydro freezing, high-pressure freezing/thawing, and ohmic and microwave thawing are other options that are now available for achieving rapid freezing or thawing of foods [7].

1.2 CHEMICAL AND PHYSICAL PROPERTIES OF WATER

Water, with the chemical formula H_2O, has unique properties determining its physical and chemical nature. Water behaves unlike other compounds of similar molecular weight and atomic composition, which are mostly gasses at room temperature. Some physical properties of water are presented in Table 1.1. It has relatively low melting and boiling points, unusually high values for surface tension, permittivity (dielectric constant), and heat capacities of phase transition (heat of fusion, vaporization, and sublimation). Another unusual behavior of water is its expansion upon solidification.

The unexpected and surprising properties of water can be better understood after taking a closer look at its structure at a molecular level. The inter- and intramolecular forces involved are governed by the physical and chemical states and reactions of water.

TABLE 1.1
Some Physical Properties of H_2O

Property	Value
Freezing temperature at 101.3 kPa (1 atm)	0°C
Boiling temperature at 101.3 kPa (1 atm)	100°C
Heat of fusion at 0°C and 1 atm	6.01 kJ/mol
Heat of vaporization at 100°C and 1 atm	40.66 kJ/mol
Heat capacity of water at 20°C	75.33 J/mol K
Heat capacity of ice at 0°C	37.85 J/mol K
Density of water at 20°C	0.998 g/mL
Density of ice at 0°C	0.917 g/mL
Dielectric constant of water at 20°C	80.4

A molecule of water consists of two hydrogen atoms that are each bonded to an oxygen atom by a covalent bond, which is partially ionic in character. In this conformation where oxygen is at the center, there is an angle of 104.5° between the nuclei. Oxygen is strongly electronegative and can be visualized as partially drawing away the electrons from the hydrogen atoms leaving them with a partial positive charge. Due to the partially ionic nature of the bonds, the spatial localization of the electrons forms a tetrahedral quadruple where the oxygen is at the center and the two negative charges of oxygen's lone pair electron orbitals and the two positive charges of the hydrogen atoms form the four corners. The structure is polar, and due to this separation of charges the molecule behaves like an "electric dipole." This polarity represents the governing force between the water molecules rather than the weaker van der Waals forces.

The water molecule is attracted to neighboring water molecules by the affinity of the positive charged site of one molecule to a negative charged site of its neighbor and *vice versa*. This attractive force is termed as a "hydrogen bond." Each water molecule can therefore support four hydrogen bonds. Hydrogen bonds are stronger than that of van der Waals' and much weaker than covalent bonds, but the ability of a water molecule to form multiple hydrogen bonds with its neighbors in three dimensional space can help to visualize the effectiveness of these forces in establishing the strong association of water molecules. This strong association between the water molecules provides a logical explanation for its unusual properties.

Similar interactions occur between the OH and NH groups and between strongly electronegative atoms such as O and N. This is the reason for the strong association between alcohol, amino acids, and amines and their great affinity to water. Intra- and intermolecular hydrogen bonding occurs extensively in biological molecules. A large number of the hydrogen bonds and their directionality confer very precise three-dimensional structures upon the proteins and nucleic acids.

Water may influence the conformation of macromolecules if it has an effect on any of the non-covalent bonds that stabilize the conformation of the large molecule. These noncovalent bonds may be hydrogen bonds, ionic bonds or apolar bonds.

The arrangement of water molecules in ice is very close to a perfect tetrahedral configuration. The angle formed between the oxygen atoms of three adjacent water molecules is approximately 109°. The same angle for liquid water is about 105°. Because liquid water has a smaller bond angle than ice, molecules in water can be packed more closely and so water has a greater coordination number, which is basically the average number of the neighbors. Ice is therefore less dense and has a lower density than liquid water. Liquid water has the greatest density at about 4°C. Above this temperature as heat is introduced into the system, the hydrogen bonds weaken and the intramolecular space increases.

Although, for the sake of visualization, the explanation made above pictures a static molecular and intermolecular conformation of water and ice, in reality the atoms in the molecules and the molecules themselves are highly dynamic. In water the hydrogen bonds break and reform with high frequency. Even in ice, where the molecules are bound with stronger forces, water molecules are thought to vibrate and diffuse in the interstitial spaces within the crystal lattice.

Another point to be considered is the dissociation of water molecules to hydronium (H_3O^+) and hydroxyl ions (OH^-) that complicates the structural formations by providing altered potentials for the hydrogen bond formation.

$$2H_2O \leftrightarrow H_3O^+ + OH^-$$

Water ionizes reversibly as shown in the above equation and the dissociation reaction occurs in equilibrium. Therefore, in pure water, the number of ionized water molecules at any given time is the same. The concentration of the dissociated molecules is equal to the concentration of the hydronium ions, 10^{-7} mol/L that is equivalent to a pH of 7.

1.2.1 WATER ISOTOPES

Oxygen is known to exist in nature as six stable isotopes (^{14}O, ^{15}O, ^{16}O, ^{17}O, ^{18}O, and ^{19}O). Similarly, there are three known isotopes of hydrogen (^1H, ^2H (deuterium, D), and ^3H (tritium)). Of these isotopes, ^{14}O, ^{15}O, ^{19}O, and tritium are radioactive with relatively short half-lives and their presence in natural water is not significant. Water consists of the stable isotopes and the majority of water will consist of $H_2^{16}O$. The abundance of the other stable isotopes may vary depending on the origin of the water but within the limits of variation, the abundances of the remaining isotopes $H_2^{18}O$, $H_2^{17}O$, and HDO may be stated as 0.2%, 0.04%, and 0.03%, respectively.

Isotopes share the same chemical properties because they depend only on the number of protons. But the difference in the number of neutrons will result in different weights and this is particularly evident in the case of hydrogen. Deuterium is twice as heavy as ^1H. Isotopes of an element can have different weights and this will cause different physical properties. Due to its extra weight deuterium heavy water (D_2O) boils at 101.4°C and freezes at 3.8°C. The difference in weight affects the speed of the reactions involving water. For this reason, heavy water is not safe to drink [8].

Since the preparation of pure $H_2^{16}O$ is extremely difficult, virtually all existing experimental determinations on water were performed on the naturally occurring substance.

Techniques involving the use of D_2O and nuclear magnetic resonance (NMR) imaging were developed as noninvasive methods to observe the time course of water transport in microporous food materials [9].

The assessment of the stable isotope ratio of foods is a convenient method that was developed for the determination of fraud in the food products industry. Determination of D and ^{18}O content is a method utilized in the determination of watering of fruit juices and wines [10]. Stable isotope ratios of $^{13}C/^{12}C$, $^{15}N/^{14}N$, $^{18}O/^{16}O$, and $^2H/^1H$ are useful in the discrimination of other adulteration schemes and mislabeling of food products [11,12].

Enzymes may be sensitive to the presence of heavy isotopes and the isotope selection effects may be high for certain transformations, especially for hydrogen. The D/H ratio can thus be used to determine the mechanism of the action of enzymes and biosynthetic pathways [13].

1.2.2 TYPES OF WATER

Water in foods and biological materials can be grouped in three categories: free water, entrapped water, and bound water. Free water is easily removed from foods or tissues by cutting, pressing, or centrifugation. Entrapped water is immobilized within the lattices of large molecules, capillaries, or cells. Entrapped water, although not free flowing, does have the properties of free water. Free and entrapped water together may be considered bulk water. This water will behave almost like pure water during food processes. It is easily removed by drying, easily converted to ice during freezing, and available as a solvent. Bound water however, exhibits properties significantly altered than those of bulk water in the same system. Bound water is the layer of water molecules closest to solutes and other nonaqueous constituents and is more structurally bonded than free or entrapped water thus having hindered mobility.

Bound water is not free to act as a solvent for additional solutes, can be frozen only at very low temperatures below the freezing point of water, is denser than water, and exhibits no vapor pressure. In a medium-moisture food only a minute fraction of water that is directly in the vicinity of the solutes is bound water [14]. The fact that water molecules in solutions and in food systems are in constant motion applies to bound water as well as free water meaning that there is a constant exchange of individual water molecules at binding locations [15]. Although the mechanism of attraction allows this motion it is not practically possible to remove all bound water from foods by dehydration. This water is also resistant to freezing, is not available for chemical reactions and does not serve as a solvent [16]. The water mobility and its relation to water retention in foods is an area invoking interest. The effect of water mobility on the rate of chemical reactions is multidimensional and cannot be reduced to a single physicochemical parameter [17].

Water states and displacements can be investigated with thermogravimetry (TG) either in its classical or in the Knudsen version (where standard pans are replaced with Knudsen cells) [18].

The analysis of water content in foods is one of the most frequently and routinely performed analyses of foods. Methods used in the analysis of water are outlined in Figure 1.1. The great variability in water contents of food materials requires that the results of other analyses of foods are to be reported on a dry weight basis. This is one of the main reasons for the routine determination of water content in the food laboratory. While there is ongoing research to achieve improvements in the methods of water analysis [19,20], recent interest is also forwarded to the economic aspects of water determination in foodstuffs [21].

1.3 WATER AS A SOLVENT

Water is a good solvent for most biomolecules, which are generally charged or polar compounds. The form and function of biomolecules in an aqueous environment are governed by the chemistry of

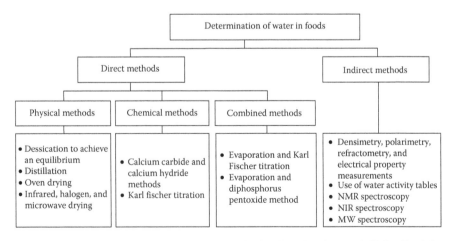

FIGURE 1.1 Methods for determining water content of foods. (Based on Isengard, H.-D., *Food Control*, 12, 395, 2001.)

their component atoms and the effect of water molecules surrounding them. The polar and cohesive properties of water are especially of importance for its solvent characteristics.

The introduction of any substance to water results in altered properties for the substance and water itself. The degree and type of alteration is dependent on the molecular and electronic structure, ionization characteristics, size, and stereochemistry of the solute.

The substance introduced into water, depending on these properties, will be dissolved, dispersed, or suspended. Water dissolves small molecules such as salts, sugars, or water-soluble vitamins to form "true solutions" which may either be ionic or molecular.

When different chemical and biochemical materials are introduced into water, there is a rise in the boiling point and viscosity and a decrease in the freezing point and surface tension. Solubility increases with increasing temperature as the heat introduced reduces water–water hydrogen bond attractions and facilitates solute hydration.

Ionic solutions are formed when the solute ionizes in water. The ions of the molecule separate in water and are surrounded or "hydrated" by the water molecules. Ionic molecules greatly influence the mobility of water molecules surrounding them and affect the colligative properties of solvent water. The degree to which the structure of bulk water is disrupted depends on the valence, size, and concentration of the ion in solution. In ice, the presence of ions interferes with intermolecular forces between water molecules and disrupts the crystal lattice structure. Hence the presence of salt decreases the melting point of water.

The great ionizing potential of water can be attributed to its high dielectric constant unmatched by other liquids [23]. Electromagnetic fields at radio and microwave frequencies interacts with food constituents, resulting in internal heat generation due to the dipole rotation of free water and the conductive migration of charged molecules [16].

Other molecules, such as sugars or alcohols dissolve in water to form molecular solutions. These hydrophilic molecules are also hydrated by water molecules clinging to them by hydrogen bonding but the solute molecules stay intact.

In solution glucose molecules have a strong stabilizing effect on the clusters of water molecules around them. This gelling effect is more pronounced in the polysaccharides [24]. The same effect is true of the fiber polysaccharides that are generally less soluble and the gums. The presence of OH groups on the carbon atoms and the compatible stereochemistry of monosaccharides render them highly soluble organic compounds. As for the oligosaccharides, the presence of some hydrophobic regions in the solute molecules affects their solubility. In polysaccharides, although the formation of the gel structure requires an aqueous environment, the gelling properties of the polysaccharides are determined primarily by polymer–polymer interactions [25].

Molecules that are too big to form true solutions can also be dispersed in water. Smaller molecules ranging in size of up to 100 nm are dispersed to form colloidal dispersions. Such molecules are generally organic products and polymers such as cellulose, pectin, starch, or proteins. Generally a two-phase colloid solution of relatively lower viscosity, where the solute is dispersed in a matrix of water is termed a sol. A gel is also a two-phase colloidal dispersion however in this case the solvent is dispersed in a solute matrix forming an elastic or semisolid. In a gel, the solute matrix does not necessarily have to constitute the greater mass portion of the total. "Water holding capacity" is a term that is frequently used to describe the ability of a matrix of molecules, usually macromolecules at low concentrations, to physically entrap large amounts of water. The stability of colloid solutions is particularly vulnerable to factors such as heating, freezing, and changes in pH. The conformation of macromolecules and stability of colloids are greatly affected by the kinds and concentrations of ions present in the medium.

Physical gels are formed when water is added to a lyophilic polymer but in insufficient amounts to completely dissolve the individual chains. Various polysaccharides such as pectin, carrageenan, and agarose, and proteins such as gelatin form physical gels in aqueous solution. These types of gels are usually reversible, i.e., they can be formed and disrupted by changing the pH, temperature, and other solvent properties.

When particles too large to form colloidal solutions, greater than 100 nm are dispersed in water they form suspensions. When the dispersion involves nonsoluble (nonpolar) liquids this is called an emulsion. Emulsions of water and oil may consist of droplets of oil in water (o/w emulsion) or droplets of water in oil (w/o emulsion) depending on which is the dispersed and continuous phase. Suspensions and emulsions have a tendency to separate when left to stand for long periods of time. Addition of stabilizers or emulsifiers to these solutions may facilitate dispersion and protect the solution from phase separation.

Some foods, like margarine, butter, and chocolate, consist of semisolid fat in a continuous phase. In these foods, formation of water bridges results in an increase in the yield stress value of semisolid fat upon the addition of water [26].

Water interacts strongly with polar groups and resists nonpolar structures, such as molecules containing carbon atom chains. In an aqueous environment, there is a tendency for water to be more ordered, forming lattice-like structures around these nonpolar molecules. Hence, there is a tendency for the nonpolar molecules to aggregate and release some of the water molecules, increasing the total entropy of the water molecules. This phenomenon is called the hydrophobic effect or hydrophobic interaction. The hydrophobic effect plays an important role in many biochemical processes.

Most biomolecules are amphiphilic, meaning they have both hydrophilic and hydrophobic groups. In an aqueous environment, these molecules self-assemble where the apolar parts of the molecules are grouped together. Some examples of structures formed as a result of this self-assembly are illustrated in Figure 1.2. The lipid bilayer is in fact the backbone of all biological membranes that makes compartmentalization possible in all cells. The assembly of these molecules would not be possible without the presence of water around them.

Protein–water interactions are vitally important in the application of proteins in model and food systems. Physicochemical properties of proteins such as dispersibility and viscosity are directly affected by solubility. Water associates with proteins in a progressive manner that may range from water molecules associated with specific groups to the more hydrodynamic hydration layers.

The functional properties of a protein are determined by its three dimensional conformation determined by the water molecules surrounding it [27,28]. Natural proteins are stable globular structures consisting of carbon rich amino acids that form the core, taking advantage of the hydrophobic effect. The charged and hydrogen bonding amino acids that tend to associate with water to give the protein its specific stereochemistry, tend to position themselves at the globule interface.

In this dynamic equilibrium state, charged groups of the protein are almost always located at the surface of the native protein whereas the interior contains hydrophobic groups tucked away from the surrounding water molecules (Figure 1.3). When the protein is denatured it loses functionality and the coil expands as all groups are hydrated and the charge distribution becomes more even [29].

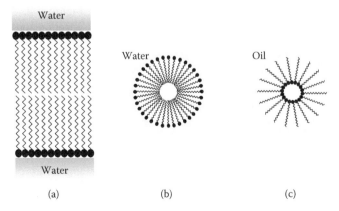

FIGURE 1.2   Schematic representation of structures formed as a result of lipid assembly in solution in biological and food systems. (a) Cross sections of a lipid bilayer, (b) a micelle, and (c) an inverted micelle.

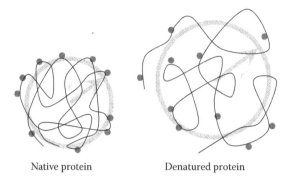

Native protein   Denatured protein

FIGURE 1.3 Schematic representation of protein conformation in an aqueous environment. (Adapted from Norde, W., *Colloids and Interfaces in Life Science*, Marcel Dekker Inc., New York, 2003, 1–433.)

1.4   CONCEPTS RELATED TO WATER MOBILITY IN FOOD

Biochemical constituents may partially immobilize water by stopping vaporization and lowering its chemical reactivity. The state and availability of water in food is just as important as the content of water. As a measurable quantity, water activity, expressed as a_w is used to express the availability of water in a food product. It is defined as the partial lowering in partial pressure created by the food and is the ratio of the partial pressure of water in the food to the partial pressure of pure water at the same temperature. Water activity is also equal to the equilibrium relative humidity (ERH) divided by 100.

$$a_w = \left[\frac{p_{food}}{p_{water}} \right]_{P,T} = \% \, ERH$$

Water activity in all foods is always a value below 1.0 because water is associated with the surfaces and solutes in the food.

Differences among water activities of food components, food domains, and the external environment outside the food induces a driving force of water transport from high a_w to low a_w until a common equilibrium value for the food system is reached.

Moisture sorption isotherm is a graphical representation of the variation in water activity or % ERH with change in moisture content of a sample at a specified temperature. Sorption isotherms of foods are generally nonlinear and very often sigmoid in shape (see Figure 1.4).

The difference in the equilibrium water content between the adsorption and desorption curves is called hysteresis. The hysteresis phenomenon is observed in highly hygroscopic materials [30]. The curve is divided into three regions. In the first region, the water in the food consists of water that is tightly bound to the product. In the second region, there is more water available and it is less tightly bound and present in small capillaries. In the third region, where the food contains higher levels of water and has a high a_w, there is plenty of water that is free or loosely held in capillaries. A detailed knowledge of water sorption isotherms are essential in food technology applications such as concentration, dehydration, and drying of foods, as well as in determining the quality, stability, and shelf life of foods.

Statistical models of the sorption behavior of foods have been applied for predicting the sorption behavior of foods. One of the most well-known and used model is the equation proposed by Brunauer, Emmett, and Teller known as the BET sorption isotherm model. This model is used extensively in food research. Another model extensively used for foods proposed by Guggenheim, Anderson, and de Boer is known as the GAB model or equation. The predicted values by either

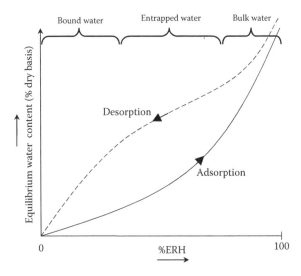

FIGURE 1.4 Sorption isotherm for a food product showing hysteresis.

equation are equivalent to the critical minimum moisture levels in foods above which sensory properties change and rehydration occurs.

The storage life of foods is influenced significantly by the water activity in the product. Intermediate-moisture foods, which have a_w values between 0.6 and 0.9, have drawn considerable attention [31]. Many of the intermediate-moisture foods have reduced moisture contents and are palatable without the need to rehydrate them [32]. Microbial proliferation is terminated at a_w values below 0.6. Even at higher moisture levels, the effect of lowered a_w has a synergistic effect in prolonging shelf life when combined with other methods of preservation. This technique of combining multiple preservation techniques to extend the shelf life stability of foods is known as the hurdle technology.

a_w and temperature are important factors influencing the state of amorphous food materials that are lacking in molecular order. These materials can be either in the rubbery or the glassy state. In the case of the latter, the material can be described as rigid, yet brittle, and has a high internal viscosity and a high internal mobility. The same amorphous material will become rubbery if the temperature is increased or if water is added into the system as a plasticizer. The temperature at which the critical conversion takes place is the glass transition temperature (T_g). Glass transition is a kinetic equilibrium process at temperatures below T_g. Mapping of food stability in relation to a_w and T_g is for the moment not extensive, however, a better understanding of the sorption properties of foods and T_g can be used in controlling stability and food packaging requirements [33]. The glass transition allows for the prediction of several physical evolutions in food products at low water content or in the frozen state which affect stability. The glass transition temperature is of importance especially for dry food products that are predominantly in a glassy amorphous form. When the products are stored at temperatures above T_g, the increased rate of physicochemical reactions that take place causes sticking, collapse, caking, agglomeration, crystallization, loss of volatiles, browning, and oxidation, which ultimately result in quality loss [34]. It is increasingly recognized that other parameters besides T_g, such as fragility, $\alpha\beta$ crossover temperature, and distribution of relaxation times, can also be used to study the time- and temperature-dependent distribution and mobility of water in foods [35].

The state diagram shown in Figure 1.5 is a plot of the states of a food as a function of the water or solids content and temperature. Most of the transitions of state can be measured by the differential scanning colorimetry (DSC) method, which detects the change in heat capacity occurring over the glass transition temperature range. Mechanical spectroscopy (or dynamic mechanical thermal

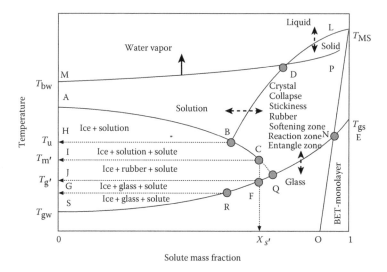

FIGURE 1.5   State diagram of foods (T_{bw}, boiling point; T_{u}, eutectic point; T'_{m}, end of freezing; T'_{g}, glass transition at endpoint of freezing; T_{gw}, glass transition of water; T_{ms}, melting point of dry solids; T_{gs}, glass transition of dry solids. (From Rahman, M.S., *Trends Food Sci. Technol.*, 17, 129, 2006. With permission.)

analysis, DMTS) and dielectric spectroscopy are alternatives, and some times more sensitive methods aimed at determining the change in the state of the food [35,36]. Physical and structural changes affecting quality of foods, such as collapse during drying, stickiness, and caking of dry powders, crystallization, change in viscosity or other textural changes, occur principally when the storage or processing temperature is above T_{g}. The rate of chemical reactions taking place in foods can be described using the Arrhenius relationship:

$$k = k_0 \exp\left(-E_a / RT\right)$$

In this model, the rate constant, k, is expressed as a function of the pre-exponential factor, k_0, the ideal gas constant, R, temperature, T, and the activation energy, E_a. However, the Arrhenius temperature model often falls short of explaining the physical behavior of foods, especially of macromolecular solutions at the temperatures above T_{g}. A better description of the physical properties is offered by the Williams–Landel–Ferry (WLF) model, which is an expression relating the change of the property to the $T - T_{g}$ difference [37,38]. That is,

$$\log a_T = \frac{-C_1(T - T_g)}{C_2 + T - T_g}$$

where
    a_T is the ratio of the relaxation phenomenon (i.e., η/η_g for viscosity)
    C_1 and C_2 are system-dependent constants

As a plasticizer or softener for macromolecular food structures, water reduces T_{g} due to the reduction in the inter- and intramolecular forces.

    Freezing, drying, and increasing the content of solutes in a food system are practical methods of decreasing a_w and increasing shelf life. The T_{g} curve can be used to determine the critical value for water activity ("critical a_w") equivalent to the "bound" or "monolayer" water activity value. Plots

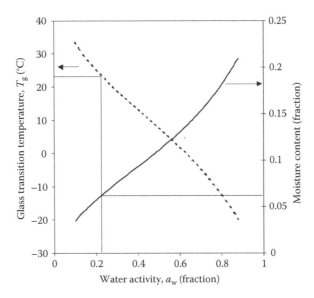

FIGURE 1.6 Use of the sorption isotherm and state diagram in determining the critical a_w. (From Sablani, S.S. et al., *Int. J. Food Prop.*, 10, 61, 2007. With permission.)

combining the glass transition and sorption curves as shown in Figure 1.6 can be useful in identifying the critical moisture content of a food at a certain processing or storage temperature.

1.5 DRINKING WATER

The individual requirement for the consumption of fluids may differ greatly depending on the physical activity and weight of the person as well as the ambient conditions. Nevertheless, a person is recommended to consume the equivalent of 1 mL/kcal energy expenditure [40] or at least 2–3 L of water daily, apart from the water ingested with foods and beverages. A potable water supply is essential for the survival of human life. Society, in general, requires water for the maintenance of public health, fire protection, cooling, electricity generation, use in industrial and agricultural processes, and navigation. Within the concept of food, water may be utilized for different purposes: direct consumption as drinking water, as a primary source during agricultural production of plant and animal products, and during the industrial and domestic processing of foods.

It is not a requirement for water to be extra pure in order for it to be potable. In fact many of the minerals necessary for our health are present as solutes in water. The presence of solutes in water not only contributes to the beneficial presence of minerals, but also to the desirable pleasant taste of the water. On the other hand water may be contaminated by chemical, physical, microbiological, or radiological factors of natural or anthropogenic origin. The levels and types of various contaminants actually form the basis for the definition of the quality of water. The adequacy of water for specific uses may be decided on by its quality.

The quality of water, as true for other commodities, depends on its origin or source, the factors which influence its composition during transport and storage, and on the processes which have been applied to improve its quality. The concerns for quality can, therefore, be source related if the source of water is contaminated, treatment related if the concern stems from the techniques used to treat water, or distribution related if it is the distribution system that is in question. The quality requirements for water depend on its intended use, and the criteria identified for monitoring quality are determined accordingly.

Potable drinking water can be defined as the water delivered to the consumer that can be safely used for drinking, cooking, and washing. Drinking water is intended for direct consumption for humans, and, thus, must be free of any hazards to health. Providing wholesome, clean, and safe water to the masses requires control that can only be achieved by setting standards, monitoring of quality parameters, and legislative directives which must be adhered to. Drinking water supplied to the consumer must meet the physical, chemical, microbial, and radionuclide parameters or standards.

Analysis and periodic monitoring are primary to achieve quality standards. However, a more total quality approach is required [39]. Periodic control and review of water treatment and transport facilities, evaluation and protection of the water source, and the evaluation of the quality and performance of the laboratory analyses are also essential to achieve better quality. Implementation of the hazard analysis critical control points (HACCP) approach to water supplies and use, specifically in the food industry, has been recommended [41,42]. The hazards encountered in water treatment in food factories must be identified and controlled. Possible hazards encountered have recently been presented as a report of the European Hygienic Engineering and Design Group [43].

The setting of standard limits and practicability of the water monitoring system of control depends on the identification and quantification or analysis of the contaminants present. Recent advances in the species of contaminants and the techniques used for their analyses have made it possible to better standardize and regulate water quality.

Freshwater sources generally consist of ground and surface water sources. In rare cases, water may be obtained from sea or ocean water by desalination processes that are relatively costly. Most municipal systems utilize surface water whereas most of the industrial consumers of water prefer to pump water from ground sources than to obtain it from municipal systems. Tap water provided by the municipal system to the community is treated to an extent which generally finds a balance between the economy and the practicality of the treatment and the safety of the water delivered. Nevertheless, the basic principle is that the quality of the water should be suitable for consumers to drink and use for domestic purposes without subsequent risk of adverse effects on their health throughout their lifetime. Also, special attention is necessary to protect vulnerable groups, such as pregnant women and children.

Agencies, international, state, or local, involved in the establishment of quality control of water have the common primary aim of protecting public health. Internet addresses of the U.S., the EU, and the WHO organizations concerned with water quality are presented in Table 1.2.

There are international agencies such as the World Health Organization (WHO) that set guidelines for the quality of drinking water [44,45]. Much of the WHO work is associated with water supply and sanitation in developing countries. The evaluation of risk, extrapolation of data, and the acceptance of risk are used to develop guidelines on drinking water quality. Many of these guidelines have subsequently been included in the EU directives and the legislation of other

TABLE 1.2
Internet Addresses where More Information on Standards, Guidelines, and Legislation Can Be Obtained

Area of Application	Organization	Address
International	WHO	http://www.who.int/water_sanitation_health/dwq/en/
United States	USEPA	http://www.epa.gov/safewater/creg.html
EU	Council of the EU	http://ec.europa.eu/environment/water/water-drink/index_en.html

governments in monitoring and sustaining drinking water quality. These WHO guidelines are applicable to packaged water and ice intended for human consumption.

The regulations determining water quality differ from state to state and from country to country; however, generally, it is the environmental agencies that are responsible for regulating the quality of municipal water as well as providing the legal grounds for the protection of freshwater sources from environmental pollution.

In the United States, the Safe Drinking Water Act (SDWA) authorizes the Environmental Protection Agency (EPA) to establish national health-based standards that reduce public exposure to the contaminants of concern. Each contaminant of concern has an unenforceable health goal, also called a maximum contaminant level goal (MCLG), and an enforceable limit referred to as maximum contaminant level (MCL). The EPA promulgates the National Primary Drinking Water Regulations (NPDWRs) that specify enforceable MCLs or treatment techniques for drinking water contaminants. The NPDWRs contain specific criteria and procedures, including the requirements for the monitoring, analysis, and quality control so that the drinking water system is in compliance with the MCL. The EPA sets MCLs as close to MCLGs as is technically and economically feasible. When the hazard of concern cannot be analyzed adequately, the EPA can require for a certain treatment technique to be employed during processing instead of setting a MCL. The EPA also sets secondary drinking water regulations that set monitoring recommendations or secondary MCLs for contaminants that affect the aesthetic, cosmetic, or technical qualities of drinking water.

The water policy of the EU is guided by directives set by the EU, which member states adopt by enacting laws in accordance with these directives. The management of water as a resource is controlled by the water framework directive (2000/60/EC), and the control of discharge of municipal and industrial wastewater is controlled by the urban waste water treatment directive (91/271/EEC). Drinking water in the EU is comprehensively regulated by the current drinking water directive 98/83/EC adopted in 1998. This directive is currently under revision for improvement. Monitoring and compliance to essential quality and health parameters are obligatory for the member states by the EU Drinking Water Directives (DWDs) [46]. The DWDs cover all water intended for human consumption except natural mineral waters, medicinal waters, and water used in the food industry not affecting the final product.

Bottled waters are alternatives to drinking water from the tap. The term "bottled water" is a generic term that describes all water sold in containers. The consumption of bottled water has risen phenomenally over the past years and the bottled water industry has grown to accommodate the demand [47]. Bottled waters are classified as a food material. They are monitored by the national food agencies and some international institutions. Many countries require the manufacturer to indicate the source of the water and the date of production on the label. The International Bottled Water Association (IBWA) also inspects its members annually to check for compliance to the IBWA standards. In the United States, bottled waters are regulated by the Food and Drug Administration (FDA) and state governments.

The laws and regulations prompting compliance are constantly under revision and development. With the advances in analytical techniques and better scientific knowledge of the impurities that may be present in water and their health implications necessitates modification of the current status. There is growing interest in the concept of a total quality approach toward water quality management to achieve a more comprehensive and preventive system of control rather than a system that relies on the compliance monitoring of treated waters [48]. The concept of risk assessment and risk management during the production and distribution of drinking water was introduced by WHO in the Guidelines for Drinking Water Quality in 2004. The methodology is considered for adoption by the European Commission as well as individual state governments.

Water, unless highly purified by distillation or membrane filtration technologies, is destined to contain differing amounts of various solutes such as minerals or suspended particles. Pure water at 25°C has a pH of 7 and it is a very strong solvent. Raindrops dissolve atmospheric gasses as they are being formed and collect dust particles and colloidal material in their fall. Carbon dioxide will readily dissolve in water and dissociate to form carbonic acid, reducing the pH. The ultimate source of water, meteoritic water, is thus acidic in nature. As the water accumulates in aquifers and open bodies of water, it picks up solutes that are present in the surrounding soil, sediment layers, and rocks. Calcium and magnesium dissolved in water are the two most common minerals that make water "hard." The degree of hardness becomes greater as the calcium and magnesium content increases and is related to the concentration of multivalent cations dissolved in the water.

Hard water may be a nuisance because it reacts with soap to produce soap curd and interferes with the cleaning processes. Where hot water is used the minerals in hard water may settle out of solution forming scales in pipes and equipment reducing efficiency and increasing the cost of maintenance. However, it has not been shown to be a health hazard. In fact, the National Research Council (National Academy of Sciences) states that hard drinking water generally contributes a small amount toward total calcium and magnesium in human dietary needs. They further state that in some instances, where dissolved calcium and magnesium are very high, water could be a major contributor of calcium and magnesium to the diet.

Water is not a strong contributor of dietary magnesium however the form of magnesium in water is thought to offer higher bioavailability than magnesium in foods. Epidemiological studies suggest that a negative correlation exists between the hardness of drinking water and cardiovascular mortality. Magnesium deficiency accelerates the development of atherosclerosis and the induction of thrombocyte aggregation, and, consequently, it is described as a risk factor for acute myocardial infarction and for cerebrovascular disease [49]. Because of this, magnesium supplementation of drinking water has been suggested.

Another mineral which is considered to be supplemented in drinking water is fluoride. Fluoride is added to drinking water to prevent the incidence of dental caries, especially seen in children with a fluoride-deficient diet. The subject is, however, a controversial one since excess fluoride is believed to have adverse effects such as dental fluorosis, which is the yellow staining of the teeth.

Mineral water is water from an underground source that contains at least 250 ppm of total soluble solids. Minerals and trace elements dissolved in it must come from the source and cannot be added later. Some mineral waters are naturally carbonated at the source, however most brands supplying fizzy mineral water add the CO_2 gas artificially.

With the exception of a few minerals which increase the health potential of drinking water, the presence of all other constituents is considered to be contaminants that pose potential health hazards.

1.5.1 WATER CONTAMINANTS

Contaminants polluting our water may be chemical or microbiological. Chemical hazards are caused by the chemical compounds which may be inorganic or organic. They may be present in the water due to the natural source, such as arsenic, which is a natural component of some soils and may be dissolved in groundwater. Alternatively, the contaminant may have been introduced into the ecosystem or the water source by human activity, such as pesticides, that trickle down sediment layers and streams and find their way into our freshwater. Some contaminants are by-products of chemicals used during the disinfection processes applied.

The types of chemical contaminants that are present in water are numerous. The chemical species emerging as contaminants is ever increasing. It is estimated that 1000 new chemicals are identified in water each year. Setting standards for and monitoring such a wide array of chemicals are impractical. Acute health problems are seldom associated with toxic chemical contaminants in water except on rare cases of massive accidents where a chemical may be introduced into the water supply in very large amounts. This type of contamination is also rarely dangerous, because often the toxic chemical makes the water unsuitable for consumption due to unacceptable taste, odor, or color. Microbial contaminants, on the other hand, have acute and widespread effects, and thus require a higher priority.

1.5.1.1   Microbiological Parameters

Infectious diseases caused by pathogenic bacteria, viruses, and protozoa or by parasites are the most common and widespread health risk associated with drinking water. Diseases such as cholera, typhoid fever, dysentery, diarrhea, enteritis, and infectious hepatitis are transmitted primarily through human and animal excreta. Human and animal wastes from sewage and farmyard runoff are principal sources for microbiological pollution. A list of the most important microbial pathogens in water is given in Table 1.3. Other pathogens which are not listed also exist. Bacteria such as *Pseudomonas*

TABLE 1.3
Orally Transmitted Waterborne Pathogens and Their Significance in Water Supplies

Pathogen	Health Significance	Persistence in Water	Resistance to Chlorine	Relative Infective Dose
Bacteria				
Burkholderia pseudomallei	Low	May multiply	Low	Low
Campylobacter jejuni, C. coli	High	Moderate	Low	Moderate
Enteropathogenic, enterotoxigenic and enteroinvasive *Escherichia coli*	High	Moderate	Low	Low
Enterohemorrhagic *E. coli*	High	Moderate	Low	High
Legionella spp.	High	Multiply	Low	Moderate
Nontuberculous mycobacteria	Low	Multiply	High	Low
Pseudomonas aeruginosa	Moderate	May multiply	Moderate	Low
Salmonella typhi	High	Moderate	Low	Low
Other *Salmonellae*	High	May multiply	Low	Low
Shigella spp.	High	Short	Low	Moderate
Vibrio cholerae	High	Short	Low	Low
Yersinia enterocolitica	High	Long	Low	Low
Viruses				
Adenoviruses	High	Long	Moderate	High
Enteroviruses	High	Long	Moderate	High
Hepatitis A virus	High	Long	Moderate	High
Hepatitis E virus	High	Long	Moderate	High
Norwalk virus	High	Long	Moderate	High
Noroviruses and sapoviruses	High	Long	Moderate	High
Rotaviruses	High	Long	Moderate	High
Protozoa				
Acanthamoeba spp.	High	Long	High	High
Cryptosporidium parvum	High	Long	High	High
Cyclospora cayetanensis	High	Long	High	High
Entamoeba histolytica	High	Moderate	High	High
Giardia intestinalis	High	Moderate	High	High
Naegleria fowleri	High	May multiply	High	High
Toxoplasma gondii	High	Long	High	High
Helminths				
Dracunculus medinensis	High	Moderate	Moderate	High
Schistosoma spp.	High	Short	Moderate	High

Source: World Health Organization (WHO), *Guidelines for Drinking-Water Quality*, Vol. 1, *Recommendations*, 3rd edn., World Health Organization, Albany, NY, 2006.

aeruginosa, and species of *Flavobacterium*, *Acetinobacter*, *Klebsiella*, *Serratia*, and *Aeromonas* can cause illness in people with impaired immunity. The microbiological quality of water is monitored throughout different stages of its distribution by testing for the coliform group or fecal coliforms (*E. coli*), which are indicators of fecal contamination. Most of the pathogens listed in Table 1.3 are so infective that the ingestion of even one viable organism could lead to the contraption of the disease, so the presence of any fecal contamination is unacceptable in drinking water. Therefore, the acceptable level of coliforms and fecal coliforms is zero. The tests are based on the presence or absence of the organisms; however, the regulatory limits allow for 5% of monthly analyzed samples to test positive. The sampling frequency is predetermined depending on the number of consumers.

The conventional methods for the analysis of microbes in water depend on culturing the organisms followed by application of biochemical tests for identification. In the past decade, rapid methods of detection of these organisms have been achieved by a number of newly developed methods. The surface plasmon resonance (SPR), amperometric, potentiometric, and acoustic wave sensors and their applications in biosensor systems are used for the detection of pathogens [51]. Biomolecular techniques based on the polymerase chain reaction to amplify genetic material have also been applied as rapid methods for detection. The PCR based methods have been successfully applied for virtually all types of pathogens listed in Table 1.1 and provide greater sensitivity, specificity, and accuracy with the drawbacks of being expensive and complicated [51,52].

Turbidity is a physical parameter monitored in relation to the microbiological quality of drinking water. The aim of monitoring turbidity is to reduce interference of particulate matter with disinfection by sheltering microorganisms, maintenance of chlorine residual, and problems that can arise in microbiological testing resulting from high bacterial populations. Suspended matter, even if biologically inert, can eventually settle out causing silting and anaerobic niches in waterways. Turbidity in water is also an indicator of poor treatment due to improper operations or inadequate facilities.

Blooms of algae in water are also undesirable because their proliferation results in technical problems as well as depreciation of the aesthetic qualities of the water. On the other hand, there are species of algae (i.e., *Cyanobacteria* spp.) which release hepatotoxic compounds into the water they grow in. The analysis of the toxins and the treatment of waters containing them are not practically possible, so suppliers must constantly monitor reservoirs for the development of algae. Waters containing excess amounts of nitrates and phosphates, which are nutrients for algae, are more susceptible for sustaining algal growth.

There are no general or standard methods employed in the disinfection of water. Water disinfection methods must be decided on after a thorough evaluation of the source water and the route of delivery. Water treatment is a multiple barrier system and the terminal disinfection step should be applied to water that is nearly completely free of pathogens, pollutants, and biodegradable products. The majority of pathogens (>99%) are successfully removed by the coagulation, flocculation, sedimentation, and filtration steps during treatment.

The primary purpose of disinfection is to kill or inactivate the remaining pathogens. The second purpose is to provide a disinfectant residual in the finished water and prevent microbial regrowth in the distribution systems. Chlorine, chloramines, ozone, chlorine dioxide, and ultraviolet radiation are the common disinfectants used in water treatment plants.

The use of electrolyzed water is also gaining popularity as a sanitizer in the food industry. The acidic and basic fractions obtained by the electrolysis of dilute NaCl solution are applied on food surfaces to reduce the microbial loads [53].

1.5.1.2 Chemical Parameters

Water resources may contain a wide range of harmful and toxic compounds. Maximum levels and recommended guide values for selected contaminants are shown in Table 1.4.

TABLE 1.4
Selected Primary Maximum Contaminant Levels in Potable Water Determined by the USEPA, the Equivalent Limits and Recommended Guideline Values Set by the EU Directive (98/83/EC), and the WHO

Contaminant	USEPA	EU	WHO
Inorganic Chemicals			
Antimony	0.006	0.005	0.020
Arsenic	0.05	0.01	0.01 (P)[a]
Asbestos	7 MFL[b]	n.l.[c]	n.l.
Barium	2	n.l.	0.7
Beryllium	0.005	n.l.	n.l.
Cadmium	0.005	0.005	0.003
Chromium (total)	0.1	0.05	0.05 (P)[a]
Cyanide	0.2	0.05	0.07
Fluoride	4	1.5	1.5
Lead	0.015	0.01	0.01
Mercury (inorganic)	0.002	0.001	0.006
Nitrate	10	50	50
Nitrite	1	0.5	0.2 (P)[a]
Selenium	0.05	0.01	0.01
Thallium	0.0005	n.l.	n.l.
Microbial Factors			
Total coliform	<5% +[d]	0 cells/100 mL	0 cells/100 mL
Turbidity	n.l.	1 NTU[e]	5 NTU
Organic Chemical Pesticides			
Alachlor	0.002		0.02
Atrazine	0.003		0.002
Carbofuran	0.04	Limits total pesticides as 0.0005 mg/L and pesticides by 0.00001 mg/L measured in supplies over 10,000 m³/day	0.007
Chlordane	0.002		0.0002
2,4-D	0.07		0.03
Dibromochloropropane	0.0002		0.001
1,2-Dichloropropane	0.005		0.04 (P)[a]
Endrin	0.002		0.0006
Lindane	0.0002		0.002
Methoxychlor	0.04		0.02
Silvex (Fenoprop; 2,4,5-TP)	0.05		0.009
Simazine	0.004		0.002
Other Organic Contaminants			
Arylamide	n.l.	0.0001	0.0005
1,2-Dichloroethane	0.005	0.003	0.03
Epichlorohydrin	n.l.	0.0001	0.0004 (P)[a]
Total trihalomethanes	0.1	0.1	0.01

(continued)

TABLE 1.4 (continued)
Selected Primary Maximum Contaminant Levels in Potable Water
Determined by the USEPA, the Equivalent Limits and Recommended
Guideline Values Set by the EU Directive (98/83/EC), and the WHO

Contaminant	USEPA	EU	WHO
Other Organic Contaminants			
PAHs (benzo(a)pyrene)	0.0002	0.00001	0.0007
Vinyl chloride	0.002	0.0005	0.0003
Benzene	0.005	0.001	0.01

Note: Numbers indicate levels in mg/L unless specified.

[a] Provisional guideline value.

[b] MFL, million fibers per liter.

[c] n.l., no limit was stated however treatment techniques may be required.

[d] Less than 5% of tested samples should be positive.

[e] NTU, nephelometric turbidity units.

The chemical characteristics of natural water reflect the soils and rocks with which the water has been in contact with. Water may also contain chemicals and suspended particles it picked up during its descent as rain, on its way to the catchments, during treatment, and distribution processes, on its route to the consumer. Groundwater is naturally filtered to some degree but may still contain significant amounts of particulate matter.

The chemicals in water may be inorganic elements or molecular species. Inorganic chemicals characteristic of groundwaters and surface waters are different. The importance given to the type of chemical contaminant present is dependent primarily on the potential health risk associated with it. Arsenic, barium, cadmium, and chromium are potentially carcinogenic elements. Mercury, lead, and selenium also have been proven to be toxic. These elements and their molecular species are high priority contaminants because their presence even in very small amounts could render the water toxic and nonpotable.

Nitrates and nitrites are naturally occurring ions that are part of the nitrogen cycle. Nitrites are a more reactive species formed as a result of reduction of nitrates by bacterial action or environmental conditions. Agricultural application of nitrogen rich fertilizers may contribute to the presence of nitrates in waters. Although nitrates and nitrites are not considered to be direct carcinogens, concern for their presence can be attributed to the endogenous or exogenous formation of N-nitroso compounds many of which are known to be carcinogenic. However, the main concern associated with high nitrate concentration in drinking water is the development of methemoglobinemia in infants. Nitrite binds to hemoglobin restricting its ability to carry oxygen from the lungs.

Aluminum, cyanide, molybdenum, nickel, silver, sodium, asbestos, and zinc are other inorganic chemicals that may contaminate drinking water supplies. The risks associated with these chemicals are relatively less either due to their rare occurrence in water or because the health risk associated with their presence is of limited concern. Nevertheless, government standards and the WHO guide values have been established for most of these chemicals.

Calcium, magnesium, potassium, phosphates (phosphorus), silica, and carbon dioxide may be present in water, and their presence generally does not necessitate control. Calcium and magnesium are contributors to the hardness of water. The control of their levels in water may be required mainly as a preventive measure for scale formation especially on heated surfaces or for taste appeal of the water.

There is no significant scientific basis to conclude that possible interactions of the above-mentioned contaminants with each other will produce other contaminant species. However, it is generally advised that contaminants that have similar health effects should be considered to have additive effects, and precaution should be taken to avoid elevated health risks. However, substances added to water in the disinfection stage of treatment are known to react with natural organic matter (humic and fulvic substances) and bromide in water to form various organic and inorganic by-products. These products are collectively referred to as disinfection by-products, or DBPs. Trihalomethanes and haloacetic acids are common DBPs in chlorinated and chloraminated waters. Chloride dioxide may be degraded to chlorite. Bromates are main DBPs associated with the ozonation of bromine rich waters. These are some of the many toxic compounds that may be present in water due to the treatment and a recent book on the subject is available [54].

Biodegradable organic substances originating from municipal, agricultural, and industrial wastes may contaminate water supplies. Human and animal wastes as well as food-processing wastes are mainly organic. Laboratory tests such as the biochemical oxygen demand (BOD) and the chemical oxygen demand (COD) have been developed to measure the amount of organic pollution especially significant for wastewater [55].

1.5.1.3 Radionuclides

Radionuclides in water can be classified according to the type of radiation they release: α-radiation is the release of positively charged He nuclei, β-radiation is the release of electrons, and γ-radiation is the release of electromagnetic energy [56]. Although there is a small amount of natural background radioactivity in waters, the main reason of concern is the amounts of radiation contributed as a result of human activity, such as the use of radiation in medicine, the testing of weapons, processing of nuclear fuel, and unforeseen accidents involving nuclear materials. The amount of health risk associated with drinking contaminated water would depend on the radioactivity concentration in the water (expressed in units of becquerel or curie) and the amount of contaminated water consumed [57].

REFERENCES

1. Mays, L. W., Water resources: An introduction. In: *Water Resources Handbook*, Mays, L. W., Ed., McGraw-Hill Companies, Inc., New York, pp. 1.1–1.34, 1996.
2. Thompson, J. and Manore, M., *Nutrition for Life*, Pearson Education, Inc., San Francisco, CA, 2007.
3. Kataoka, M., Wiboonsirikul, J., Kimura, Y., and Adachi, S., Properties of extracts from wheat bran by subcritical water treatment, *Food Science and Technology Research*, 14(6): 553–556, 2008.
4. Venketesh, S. and Dayananda, C., Properties, potentials, and prospects of antifreeze proteins, *Critical Reviews in Biotechnology*, 28: 57–82, 2008.
5. Palaiomylitou, M. A., Matis, K. A., Zubulis, A. I., and Kriakidis, D. A., A kinetic model describing cell growth and production of highly active recombinant ice nucleation protein in *Escherichia coli T*, *Biotechnology and Bioengineering*, 78(3): 321–332, 2002.
6. Hwang, W. Z., Coetzer, C., Tumer, N. E., and Lee, T. C., Expression of bacterial ice nucleation gene in a yeast *Saccharomyces cerevisiae* and its possible application in food freezing processes, *Journal of Agricultural and Food Chemistry*, 49: 4662–4666, 2001.
7. Li, B. and Sun, D. W., Novel methods for the rapid freezing and thawing of foods—A review, *Journal of Food Engineering*, 54: 175–182, 2002.
8. Conti, A., Isotopes. In: *Water Encyclopedia*, Lehr, J. H. and Keeley, J., Eds., John Wiley & Sons, Inc., New York, pp. 499–500, 2005.
9. Hills, B. P., NMR Micro-imaging studies of water diffusivity in saturated microporous systems. In: *Food Colloids and Polymers: Stability and Mechanical Properties*, Dickenson, E. and Walstra, P., Eds., The Royal Society of Chemistry, Cambridge, U.K., pp. 235–242, 1993.
10. Calderone, G. and Guillou, C., Analysis of isotopic ratios, for the detection of illegal watering of beverages, *Food Chemistry*, 106(4): 1399–1405, 2008.
11. Luykxs, D. M. A. M. and van Ruth, S. M., An overview of analytical methods for determining the geographical origin of food products, *Food Chemistry*, 107: 897–911, 2008.
12. Reid, L. M., O'Donnel, C. P., and Downey, G., Recent technological advances for the determination of food authenticity, *Trends in Food Science and Technology*, 17: 344–353, 2006.

13. Robins, R. J., Billault, I., Duan, J. R., Guiet, S., Pionnier, S., and Zhang, B. L., Measurement of ^{2}H distribution in natural products by quantitative ^{2}H NMR: An approach to understanding metabolism and enzyme mechanism, *Phytochemistry Reviews*, 2: 87–102, 2003.

14. Fennema, O. R., Water and ice. In: *Food Chemistry*, 3rd edn., Fennema, O. R., Ed., Marcel Dekker, New York, pp. 1–94, 1996.

15. Le Meste, M., Roudaut, G., Champion, D., Blond, G., and Simatos, D., Interaction of water with food components. In: *Ingredient Interactions Effects on Food Quality*, 2nd edn., Gaonkar, A. G. and McPherson A., Eds., Taylor & Francis Group, LLC, Boca Raton, FL, pp. 87–138, 2006.

16. Lewicki, P. P., Water as the determinant of food engineering properties, *Journal of Food Engineering*, 61: 483–495, 2004.

17. Bell, L. N., Bell, H. M., and Glass, T. E., Water mobility in the glassy and rubbery solids as determined by oxygen-17 nuclear magnetic resonance: Impact on chemical stability, *Lebensmittel Wissenchaft und Technologie*, 35: 108–113, 2002.

18. Schiraldi, A. and Fessas, D., Classical and Knudsen thermogravimetry to check states and displacements of water in food systems, *Journal of Thermal Analysis and Calorimetry*, 71(1): 225–235, 2003.

19. Schöffski, K., New Karl Fischer reagents for the water determination in food, *Food Control*, 12: 427–429, 2001.

20. Felgner, A., Schlink, R., Kirchenbühler, P., Faas, B., and Isengard, H. D., Automated Karl Fischer titration for liquid samples—Water determination in edible oils, *Food Chemistry*, 106: 1379–1384, 2008.

21. Isengard, H. D., Water determination—Scientific and economic dimensions, *Food Chemistry*, 106: 1393–1398, 2008.

22. Isengard, H.-D., Water content, one of the most important properties of food, *Food Control*, 12: 395–400, 2001.

23. Pande, A., *Handbook of Moisture Determination and Control-Principles, Techniques and Applications*, Vol. 1, Marcel Dekker, Inc., New York, 1974.

24. Coultate, T. P., Water. In: *Food: The Chemistry of Its Components*, 3rd edn., The Royal Society of Chemistry, Cambridge, U.K., pp. 320–338, 1996.

25. Suggett, A., Water—Carbohydrate interactions. In: *Water Relations of Foods*, Duckworth, R. B., Ed., Academic Press, Inc., London, U.K., pp. 23–36, 1975.

26. Johansson, D., Bergenståhl, B., and Lundgren, E., Sintering of fat crystal networks in oils. In: *Food Macromolecules and Colloids*, Dickinson, E. and Lorient D., Eds., Royal Society of Chemistry, Cambridge, U.K., pp. 418–430, 1995.

27. Scnepf, M. I., Protein–water interactions. In: *Biochemistry of Food Proteins*, Hudson, B. J. F., Ed., Elsevier Applied Science, New York, pp. 1–33, 1992.

28. Philips, L. G., Whitehead, D. M., and Kinsella, J., *Structure-Function Properties of Food Proteins*, Academic Press, Inc., New York, 1994.

29. Norde, W., *Colloids and Interfaces in Life Science*, Marcel Dekker, Inc., New York, pp. 1–433, 2003.

30. Rahman, S., Water activity and sorption properties of foods. In: *Food Properties Handbook*, CRC Press, Boca Raton, FL, pp. 1–84, 1995.

31. Bell, L. N. and Labuza, T. P., *Moisture Sorption Practical Aspects of Isotherm Management and Use*, 2nd edn., The American Association of Cereal Chemists, Inc, Eagan, MN, 2000.

32. Lewis, M. J., *Physical Properties of Foods and Food Processing Systems*, Woodhead Publishing, New York, 1990.

33. Sablani, S. S., Kasapis, S., and Rahman, M. S., Evaluating water activity and glass transition concepts for food stability, *Journal of Food Engineering*, 78: 266–271, 2007.

34. Bhandari, B. R. and Howes, T., Implication of glass transition for the drying and stability of dried foods, *Journal of Food Engineering*, 40(1–2): 71–79, 1999.

35. Champion, D., Le Meste, M., and Simatos, D., Towards an improved understanding of glass transition and relaxations in foods: Molecular mobility in the glass transition range, *Trends in Food Science & Technology*, 11: 41–55, 2000.

36. Rahman, M. S., State diagram of foods: Its potential use in food processing and product stability, *Trends in Food Science & Technology*, 17: 129–141, 2006.

37. Nelson, K. A. and Labuza, T. P., Water activity and food polymer science: Implications of state on Arrhenius and WLF models in predicting shelf life, *Journal of Food Engineering*, 22: 271–289, 1994.

38. Maltini, E., Torreggiani, D., Venir, E., and Bertolo, G., Water activity and the preservation of plant foods, *Food Chemistry*, 82: 79–86, 2003.

39. Sablani, S. S., Al-Belushi, K., Al-Marhubi, I., and Al-Belushi, R., Evaluating the stability of vitamin C in fortified formula using water activity and glass transition, *International Journal of Food Properties*, 10: 61–71, 2007.

40. Food and Nutrition Board Commission on Life Sciences, National Research Council (NRC) water and electrolytes. In: *Recommended Dietary Allowances*, 10th edn., National Academy Press, Washington, DC, pp. 247–261, 1989.

41. Casani, S. and Knøchel, S., Application of HACCP to water reuse in the food industry, *Food Control*, 13(4): 315–327, 2002.

42. Kirby, R. M., Bartram, J., and Carr, R., Water in food production and processing: Quantity and quality concerns, *Food Control*, 14: 283–299, 2003.

43. European Hygienic Engineering & Design Group (EHEDG), Safe and hygienic water treatment in food factories, *Trends in Food Science & Technology*, 18: S93–S100, 2006.

44. Casani, S., Rouhany, M., and Knøchel, S., A discussion paper on challenges and limitations to water reuse and hygiene in the food industry, *Water Research*, 39(4): 1134–1146, 2005.

45. World Health Organization (WHO), *Guidelines for Drinking-Water Quality*. Volume 1, *Recommendations*, 2nd edn., World Health Organization, Geneva, Switzerland, 1996.

46. Hecq, P., Hulsman, A., Hauchman, P., McLain, J. L., and Schmitz, F., Drinking water regulations. In *Analytical Methods for Drinking Water*, Quaver, P. and Thompson, K. C., Eds., John Wiley & Sons Ltd, West Sussex, U.K., pp. 1–37, 2006.

47. Gray, N. F., *Drinking Water Quality Problems and Solutions*, 2nd edn., Cambridge University Press, Cambridge, U.K., 2008.

48. Quevauviller, P., *Quality Assurance for Water Analysis, Water Quality Measurements Series*, John Wiley & Sons Ltd., West Sussex, U.K., pp. 1–252, 2002.

49. Sauvant, M.-P. and Pepin, D., Drinking water and cardiovascular disease, *Food and Chemical Toxicology*, 40: 1311–1325, 2002.

50. World Health Organization (WHO), *Guidelines for Drinking-Water Quality*. Volume 1, *Recommendations*, 3rd edn., World Health Organization, Albany, NY, 2006.

51. Leonard, P., Hearty, S., Brennan, J., Dunne, L., Quinn, J., Chakraborty, T., and O'Kennedy, R., Advances in biosensors for the detection of pathogens in food and water, *Enzyme and Microbial Technology*, 32: 3–13, 2003.

52. Hill, W. E. and Olsvik, Ø., Detection and identification of foodborne microbial pathogens by the polymerase chain reaction: Food safety applications. In: *Rapid Analysis Techniques in Food Microbiology*, Patel, P., Ed., Chapman & Hall, London, U.K., pp. 268–289, 1994.

53. Hricova, D., Stephan, R., and Zweifel, C., Electrolyzed water and its application in the food industry, *Journal of Food Protection*, 71(9): 1934–1947, 2008.

54. Xie, Y. F., *Disinfection Byproducts in Drinking Water, Formation, Analysis and Control*, CRC Press, Boca Raton, FL, 2004.

55. Vigil, K. M., *Clean Water, an Introduction to Water Quality and Water Pollution Control*, 2nd edn., OSU Press, Corvallis, OR, 2003.

56. Malina, J. F. Jr. and Dee, P. E. Water quality. In: *Water Resources Handbook*, Mays, L., Ed., McGraw Hill Companies, Inc., New York, pp. 8.3–8.49, 1996.

57. Wilkins, B. T., Incidents involving radionucleotides. In: *Water Contamination Emergencies*, Thompson, K. C. and Gray J., Eds., RSC Publishing, Dorset, U.K., pp. 240–250, 2006.

2 Glycobiology of Foods: Food Carbohydrates—Occurrence, Production, Food Uses, and Healthful Properties

Frank A. Manthey and Yingying Xu

CONTENTS

2.1 INTRODUCTION

Carbohydrates are ubiquitous. Every organism contains some carbohydrate. Carbohydrates can range from a simple monosaccharide to a large complex polysaccharide. Polysaccharides in combination with proteins, lipids, and nucleic acids play an important role in many plant and animal metabolic systems. Carbohydrates have many roles in food systems, where they function to provide flavor, structure, and texture to food and nutritional benefits to the consumer. This chapter attempts to address the role of common plant monosaccharides, oligosaccharides, and polysaccharides as ingredients in food systems by discussing their occurrence in plants, commercial processing, functionality, food uses, and healthful properties.

2.2 MONOSACCHARIDES

2.2.1 Occurrence in Plants

Monosaccharides are chemically grouped into two families: aldose sugars and ketose sugars (Figures 2.1 and 2.2). The most common sugars are pentoses and hexoses. The most common pentose sugars are D-xylose and L-arabinose, which are both aldoses. The most common hexose sugars are D-glucose, D-galactose, D-mannose, and D-fructose. D-Glucose, D-galactose, and D-mannose are aldoses. From a food ingredient use point of view, D-fructose is the only common ketose sugar in plants. The remaining five or six carbon sugars are unknown or rarely found in nature and are referred to as rare sugars [1]. Of the monosaccharides commonly found in plants, only glucose and fructose are found uncombined in plants.

2.2.2 Commercial Production

Glucose can be commercially produced by degradation of starch to glucose. Starch can be liquefied by hot acids. Commercially, hydrochloric acid is sprayed onto well-mixed starch, and the mixture is heated until the desired degree of hydrolysis is obtained. The acid is neutralized and the product is recovered by filtration or centrifugation, and washed and dried.

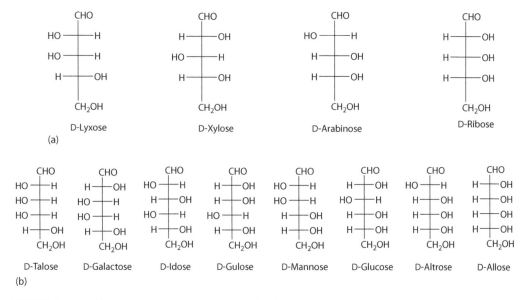

FIGURE 2.1 D-Aldose sugars: (a) D-pentose and (b) D-hexose.

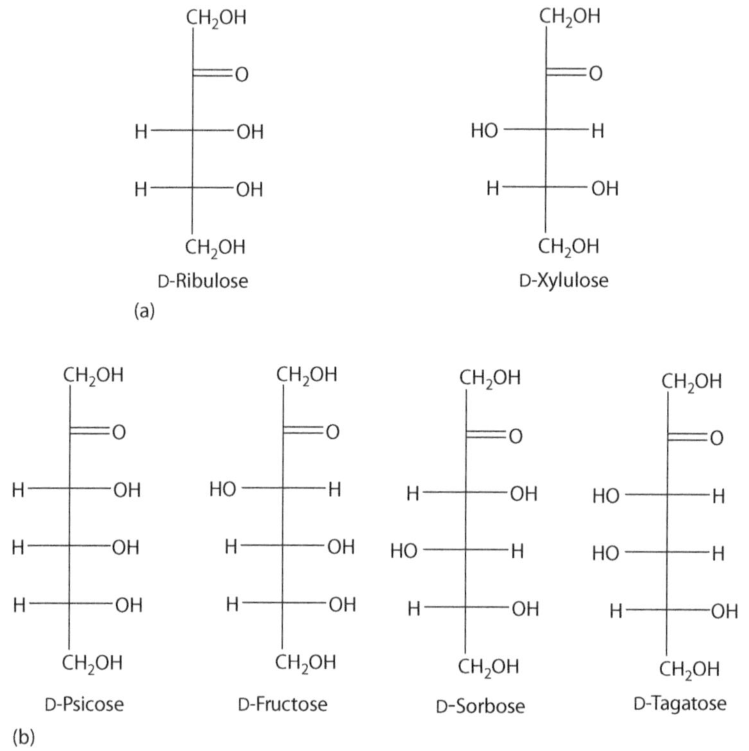

CH₂OH ... (chemical structures)

FIGURE 2.2   D-Ketose sugars: (a) pentulose and (b) hexulose.

Liquefied starch can be used to produce syrups with dextrose equivalence values from 50 to 98. (Dextrose equivalence = 100/average number of monosaccharide units in the oligosaccharides found in the syrup). These are usually produced by using an exoamylase (glucan 1,4-α-glucosidase) also known as amyloglucosidase. This enzyme can convert liquefied starch into 95%–98% glucose. The syrup is filtered and passed over activated charcoal and an ion-exchange resin.

Glucose can be converted to fructose by passing glucose syrup through a column containing glucose isomerase. This step typically converts about 40% of the glucose to fructose. The fructose is concentrated by passing the mixture through a chromatography column. High-fructose corn syrup contains ≈55% fructose and is produced by blending fructose back into the glucose syrup.

Rare sugars can be manufactured using microorganisms or by the use of enzymes [1]. *Agrobacterium tumefaciens* has been used to convert fructose to D-psicose [2]. Other researchers have reported on the manufacturing of rare sugars using immobilized enzymes [3]. A large-scale production of D-allose has been demonstrated using rhamnose isomerase from *Pseudomonas stutzeri* to convert D-psicose to D-allose [4].

2.2.3   FOOD USES

Glucose and fructose are generally used as syrups in food ingredients. They have film-forming and adhesive properties and are used as coatings for roasted nuts, candy, fillers, and spray-dried flavors. High-fructose corn syrups are used as sweeteners and seem almost ubiquitous where sweetening is desired.

Rare sugars such as D-allose, D-psicose, D-tagatose, and D-talose have been incorporated into health foods and drinks [5], where they are used as bulking, browning, and low-calorie sweetening agents in food preparation [6–8]. These rare sugars have a taste similar to sucrose.

2.2.4 Healthful Properties

Glucose is a quick source of energy in human beings. Glucose, galactose, mannose, fucose, xylose, *N*-acetylglucosamine, *N*-acetylgalactosamine, and *N*-acetylneuraminic acid have been identified as essential for glycoprotein synthesis in humans [9,10]. Cellular glycoproteins are important in cell receptor sites and cell recognition sites. Recent research has shown that all eight essential sugars can be readily absorbed into the body and directly incorporated into glycoproteins and glycolipids [9].

Rare sugars have been reported to have healthful properties [1]. D-Psicose, a ketose sugar, has been shown to inhibit intestinal sucrase and maltase and suppress plasma glucose increase after sucrose and maltose ingestion in rats [11]. D-Allose has been reported to improve blood flow in the liver [12] and inhibit cancer cell proliferation [13].

2.3 OLIGOSACCHARIDES

Oligosaccharides are composed of up to 20 monosaccharide units. Oligosaccharides often have colligative properties; which means that they can cause freezing point depression and boiling point elevation in food systems.

Digestible oligosaccharides represent a quick source of energy in human beings. Nondigestible oligosaccharides often have prebiotic properties. Prebiotics are food ingredients that selectively enhance a population of desirable bacteria (*Bifidobacteria* and *Lactobacillus*) in the large intestine [14]. Oligosaccharides can also be used to help stabilize probiotics in food systems. Probiotic foods contain a culture of microorganisms that improve the intestinal microbial balance of the consumer.

2.3.1 Sucrose

2.3.1.1 Occurrence in Plants

Sucrose, a disaccharide, is composed of α-D-glucose and β-D-fructose units that are linked (1→2) through their reducing ends (Figure 2.3). Thus, sucrose is a nonreducing sugar. Sucrose is made in the cytoplasm of cells in photosynthetically active tissue and is translocated via phloem to metabolically active sites, where it is usually cleaved into glucose and fructose and used as intermediates in various metabolic pathways. In plants such as sugar beet (*Beta vulgaris*) and sugarcane (*Saccharum* spp.), sucrose is a storage carbohydrate. In sugarcane, sucrose accumulates in the vacuoles of cells located in the internodal region of the stem, while in sugar beet, sucrose accumulates in the vacuoles of cells located in root tissue [15,16].

2.3.1.2 Commercial Production

Sucrose is commercially extracted from sugar beet and sugarcane. Extraction from sugar beet begins with the washing and slicing of the roots into strips. The strips are placed in a large tank where raw sugar is extracted in hot water. Sucrose extraction from sugarcane begins by breaking

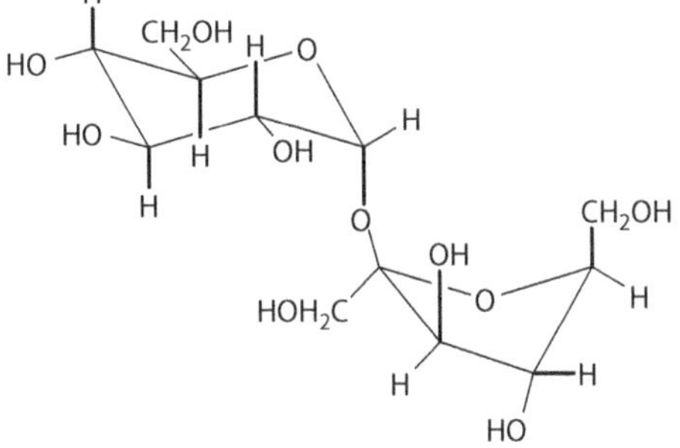

FIGURE 2.3   Sucrose.

and grinding the stems. The juice is squeezed out by passing the ground stems through a series of rolls. For both sugar beet and sugarcane, the raw sugar juice is clarified using a mixture of lime and carbon dioxide. The carbon dioxide reacts with the lime to produce calcium carbonate. Nonsugar particles attach to the calcium carbonate and precipitate. The juice is filtered and boiled under vacuum, which removes water and leaves a thick syrup. This material is filtered and centrifuged, and the sugar is washed with hot water and dried.

2.3.1.3   Food Uses

Sucrose affects colligative properties of water in food systems by depressing freezing point and elevating boiling point and can delay starch gelatinization [17]. Sucrose can interact with ingredients to enhance the retention of aroma and flavor of foods [18]. Its antioxidant properties have been used to prevent the deterioration of flavor in canned fruit. Its humectant properties have been used to prevent moisture loss from baked products. Sucrose can provide a yellow-brown color to food products, through thermal degradation, alkaline degradation, and Maillard products.

2.3.1.4   Healthful Properties

Sucrose is rapidly degraded into glucose and fructose, which are readily absorbed into the bloodstream in the small intestine. Sucrose provides a quick source of energy for the human body. However, overconsumption of sucrose can cause adverse health problems. It can contribute to obesity, dental caries, and can be problematic to people suffering from defects in glucose metabolism, e.g., hypoglycemia.

2.3.2   RAFFINOSE FAMILY OLIGOSACCHARIDES

2.3.2.1   Occurrence in Plants

Raffinose family oligosaccharides (RFO) are composed of sucrose and a varying number of α-D-galactosyl units. Raffinose and stachyose are two common members of RFO, found in plants. Raffinose is composed of glucose, fructose, and galactose (Figure 2.4a). Stachyose is similar to raffinose but contains an additional α-D-galactosyl unit (Figure 2.4b). These α-D-galactosides are formed by $\alpha(1–6)$ galactosides linked to the C-6 of the glucose unit of sucrose.

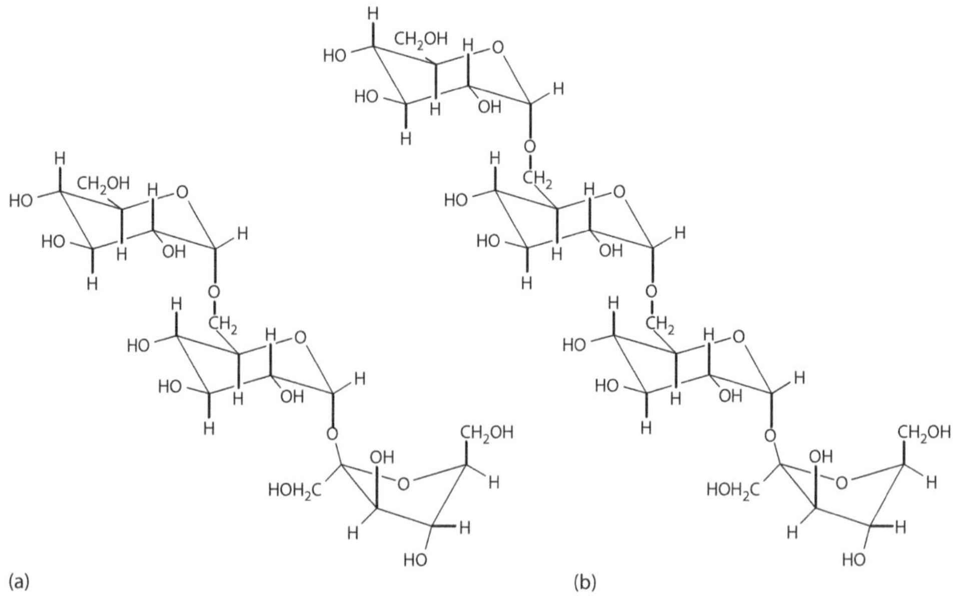

(a)   (b)

FIGURE 2.4   (a) Raffinose and (b) stachyose.

Stachyose is the principal transport carbohydrate in the phloem of some herbaceous and woody plants [19,20]. Raffinose and stachyose are found in leguminous seeds such as soybean (*Glycine max*) and lupine (*Lupinus* spp.) where they prevent desiccation of seeds after maturity [21] and serve as carbon reserves for use during germination [22].

2.3.2.2 Commercial Production

Sugar beet molasses contains about 18% raffinose. Commercially, raffinose is removed from beet molasses through chromatographic separation.

2.3.2.3 Food Uses

Raffinose and stachyose occur naturally in food containing legumes. They can be degraded during food processing since they are heat labile. Raffinose and stachyose, along with other RFOs, have potential use in aiding the survival of probiotics in food systems [23,24]. However, because of their tendency to produce flatulence in human beings, these sugars are generally not selectively added to food products.

2.3.2.4 Health Properties

Human beings lack α-galactosidase enzymes needed to digest these sugars. Raffinose and stachyose contain α-(1–6) linked galactose, which is cleaved only by bacterial α-galactosidases in the lower part of the gut. Galactose released from raffinose and stachyose can be a source of absorbed galactose. Raffinose and stachyose can promote the growth of *Bifidobacteria* population in the large intestine, and consequently have potential as prebiotic ingredients in functional foods [25].

2.3.3 FRUCTANS

Fructans can occur as oligosaccharides and polysaccharides. This means that their degree of polymerization can vary from below 20 sugar units to well above 20 sugar units. As a food ingredient, fructans often have a degree of polymerization of 25 or less. Thus, they are discussed as oligosaccharides.

2.3.3.1 Occurrence in Plants

There are five classes of fructans: inulin, levan, mixed levan, inulin neoseries, and levan neoseries [26]. Inulin is a linear polysaccharide composed of (2-1)-β-D-fructosyl units (Figure 2.5a). Levan is a linear polysaccharide composed of (2-6)-β-D-fructosyl units (Figure 2.5b). Mixed levan is a branched polysaccharide composed of (2-1) and (2-6)-β-D-fructosyl units. Inulin neoseries is a linear polysaccharide composed of two inulin polymers that are connected together by a sucrose molecule. Levan neoseries is a linear polysaccharide composed of two levan polymers linked together by the glucose unit of the sucrose molecule. The type of fructan produced varies with plant species. For example, plants such as chicory (*Cichorium intybus*) and Jerusalem artichoke (*Helianthus tuberosus*) in the Asteraceae family produce inulin. Plants in the Liliaceae family such as garlic (*Allium sativum*) produce inulin neoseries. Plants in the Poaceae family such as wheat (*Triticum* spp.), barley (*Hordeum vulgare*), and oats (*Avena sativa*) produce mixed levan or levan neoseries.

Plants that produce fructans grow in temperate climates where seasonal drought and/or frost are common. There is some evidence that fructans are involved in providing tolerance to drought and/or low temperature. In certain plants, such as chicory and Jerusalem artichoke, fructans serve as a carbohydrate reserve. Fructans are synthesized in the vacuoles of plants. Sucrose from chloroplast is translocated to a storage vacuole in the same cell or translocated to the primary storage organ (tuber, root). In the vacuole, the sucrose is converted into fructan by two or more different fructosyltransferases. The first enzyme initiates fructan synthesis by catalyzing the transfer of a fructosyl residue from sucrose to another sucrose molecule. The second enzyme transfers fructosyl residues from a fructan molecule to another fructan molecule or to a sucrose [26].

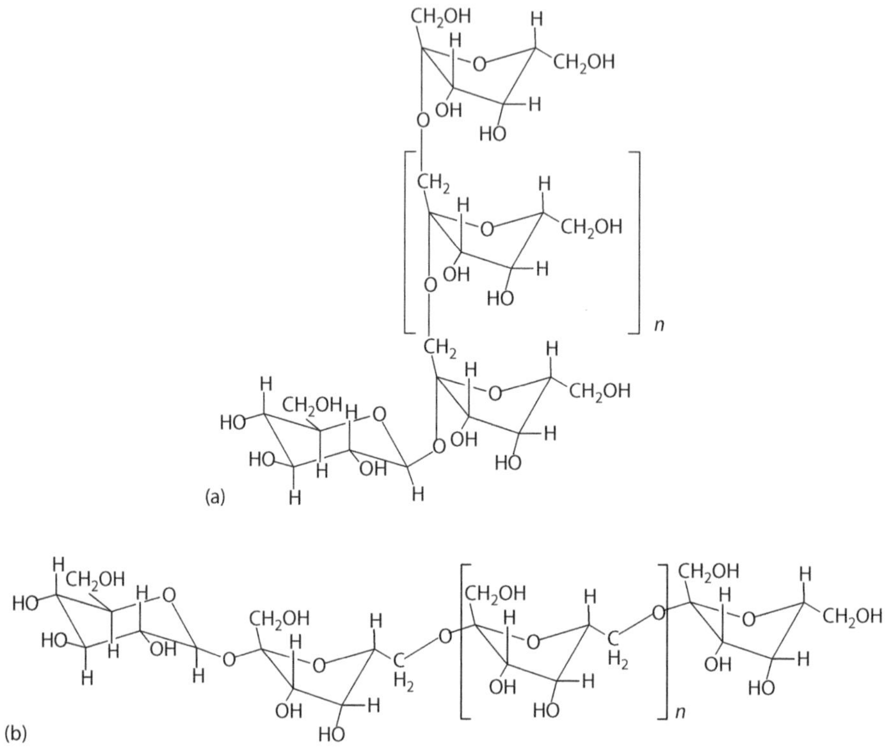

FIGURE 2.5   (a) Inulin and (b) levan.

2.3.3.2   Commercial Production

Roots of chicory and Jerusalem artichoke can contain more than 70% inulin on a dry basis [27,28]. Thus, much work has been conducted on commercializing fructan production using chicory and Jerusalem artichoke. Commercially, inulin is extracted by partial enzymatic hydrolysis using an endo-inulinase, followed by spray drying [27]. Fructo-oligosaccharides (FOS) can be synthesized from sucrose using fructosyltransferase [27]. Extracted inulin typically has a degree of polymerization of 12–25, while FOS have a degree of polymerization of 3–6 [27]. Ground Jerusalem artichoke (>70% inulin) has been used as an ingredient without further processing [28].

2.3.3.3   Food Uses

Fructans, inulin, and FOS have food ingredient status in most countries and are recognized as GRAS ingredients in the United States. In general, fructans have a bland neutral taste with no off-flavor or aftertaste [27,30]. Fructans are used in a wide range of food products such as bakery products, breakfast cereals, drinks, and dairy products. Fructans are heat labile and will break down under low pH [30,31].

Fructans have humectant properties that reduce water activity and improve microbiological stability. When mixed with water, inulin can form a particle gel network, resulting in a white creamy structure with a short spreadable texture [32]. Emulsion of long-chain fructan in water has organoleptic properties similar to fat and has been used as a fat replacement in food systems. Inulin works in synergy with most gelling agents.

FOS are very water soluble and have some sweetness (30%–35% of sucrose) [27]. FOS are used in beverages where they improve mouthfeel, enhance fruit flavor, and sustain flavor with less aftertaste when added to artificial sweeteners, aspartame or acesulfame K.

2.3.3.4 Healthful Properties

Fructans are a recognized form of dietary fiber. Digestive enzymes in the human small intestine are specific for α-glycosidic linkages. Thus the β-configuration of the anomeric carbon makes fructans resistant to hydrolysis in the human small intestine [14]. Kolida et al. [33] summarized several studies that document the prebiotic effect of fructans. Fructans are an efficient carbon source of beneficial *Bifidobacteria* in the colon, which enhances the growth of healthy gut microflora while suppressing the growth of pathogenic bacteria [25,32,33]. Fructans promote mineral absorption, particularly calcium [14], enhance immune functions and antitumor activity, and may modulate lipid and carbohydrate metabolism for good cardiovascular and diabetic health [30].

2.4 POLYSACCHARIDES

2.4.1 GENERAL

Polysaccharides are composed of more than 20 monosaccharide units. Polysaccharides often are classified as being starch or nonstarch. Starch polysaccharides represent a source of energy in human beings while nonstarch polysaccharides generally are nondigestible and are important in maintaining intestinal health.

2.4.2 STRUCTURE

Polysaccharides are a very diverse set of compounds. They can consist of one type of sugar (homoglycans) or several types of sugars (heteroglycans). They can be branched or nonbranched and can vary in type of linkage. The three dimensional structure or conformation of polysaccharide chains is determined by the monosaccharide units and position and type of glycosidic linkages. For example, polysaccharides with α-(1-4)-D-glucosyl linkages or β-(1-3)-D-glucosyl linkages form hollow helixes while those with β-(1-4)-D-glucosyl linkages form ribbonlike conformations. The ribbonlike and hollow helix conformations are the two basic chain conformations.

Polysaccharides associated with cell walls generally have a ribbonlike conformation. The ribbon conformation allows the chains to align parallel to each other. This promotes efficient packing and strong hydrogen bonds, which in turn allows them to form fibrous water insoluble aggregates. All ribbonlike conformations have some degree of zigzag structure. Hollow helices tend to be more flexible than ribbonlike conformations. The flexibility is reduced by formation of an inclusion complex with another molecule (e.g., amylose–lipid complex) or by combining to form multiple helixes or nesting helical conformations.

2.4.3 WATER SOLUBILITY

Unsubstituted glycosyl units contain three hydroxyl groups and one ring oxygen all of which can hydrogen bond with other polymers or with water. Hydrogen bonding has a major influence on the water solubility of polysaccharides. During hydration, the hydrogen bonding changes from hydroxyl groups of adjacent glycan molecules to hydrogen bonding with water, the reverse is true during dehydration.

Solubility of polysaccharides depends on intermolecular hydrogen bonding and whether steric hindrance (often in the form of branches) keeps the chains at a distance from each other so that water can penetrate and hydrate the polymers. Single, neutral monosaccharide units with one type of linkage and few or no branches are usually insoluble in water and are difficult to solubilize. They form orderly conformation within the chain and chain–chain interaction. Branched polysaccharides are more water soluble. Branches reduce chain–chain interaction, which allows easier hydration. Dried branched polysaccharides rehydrate more quickly than do dried nonbranched polysaccharides.

2.4.4 Viscosity

Most polysaccharides increase solution viscosity. Solution properties of a polysaccharide depend on its structure, molecular weight, and concentration. Linear molecules in solution sweep out a large space, which increases their probability of colliding with other molecules and forming a viscous solution. In contrast, highly branched molecules of the same molecular weight sweep less space and form less viscous solutions. Nonstarch polysaccharides generally have Newtonian flow properties at low concentrations and non-Newtonian flow behavior at moderate to high concentrations. Flow is based on the cohesiveness of the mixture of molecules with different shapes and sizes and how much force is required to move them.

2.4.5 Gel Formation

Gel formation requires a balance of chain homogeneity with chain irregularities. Thus, many polysaccharides do not form gels but form entanglement networks. A homoglycan polysaccharide will tend to associate greatly with other chains and precipitate out of solution. Chains that allow no chain–chain interaction cannot form a gel since they are unable to form junction zones necessary to build a three-dimensional network. Thus, chains that allow some chain–chain interaction are more likely to form a gel. The region of chain–chain interaction (junction zones) is often terminated by structural irregularities which prevent complete aggregation and precipitation of the polysaccharide. For this reason it is not uncommon for a mixture of polysaccharides to form a gel even though singly they are unable to. For example, locust bean gum (nongelling) and xanthan gum (weak gel) together form a strong gel. This is attributed to the interaction among different chain polymers and formation of mixed junction zones.

2.4.6 Starch

2.4.6.1 Occurrence in Plants

The primary function of starch is to serve as energy storage and as a carbon source for *de novo* biosynthesis of macromolecules. Starch can accumulate temporarily in the chloroplast of cells found in photosynthetic tissue. Most starch is found in the storage organs such as the endosperm of seeds or in roots and tubers.

Starch is composed of two polysaccharides: amylose and amylopectin. Amylose and amylopectin are both polymolecular [contain α-(1-4) and α-(1-6)-D-glucosyl linkages] and polydisperse (vary in degree of polymerization). Amylopectin is a large, highly branched molecule with a degree of polymerization of 10^4–10^7 with 4%–5% of the linkages involved with branch points (Figure 2.6) [34]. Amylopectin branches are categorized as A-chains, B-chains, and C-chains.

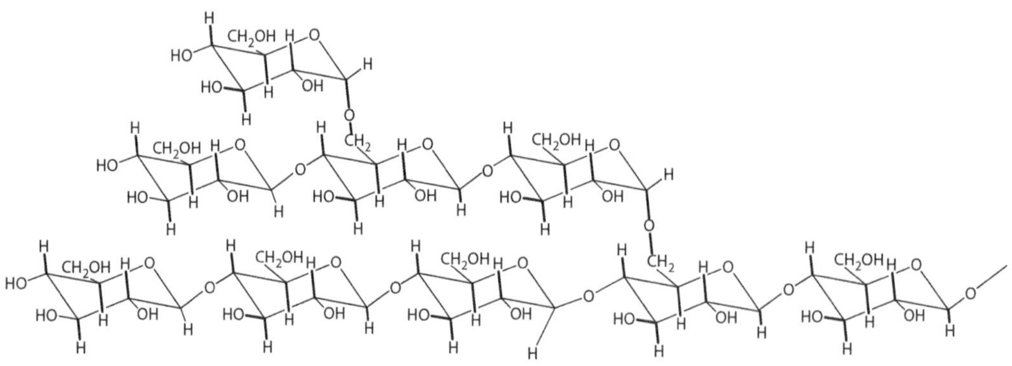

FIGURE 2.6   Amylopectin.

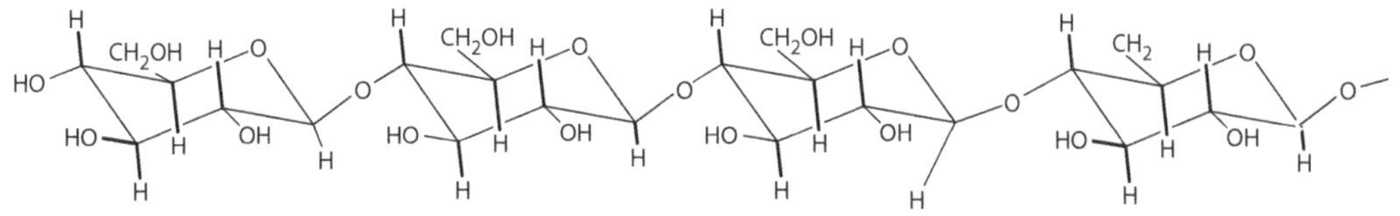

FIGURE 2.7 Amylose.

The C-chain is the original chain. B-chains are branch chains that have branches. A-chains are branch chains that are not branched; hence, they are the outermost chains. Amylose is smaller than amylopectin with a typical degree of polymerization of 10^3–10^4 and is nearly nonbranched with only 0.3%–0.6% of the linkages involved with branch points (Figure 2.7) [34]. In general, amylose comprises 20%–30% of the starch, although high (≈65%) and low (<1%) amylose starches exist [35].

Amylopectin forms the principal structure of the starch granule. Vandeputte and Delcour [36] reviewed three models that have been proposed for amylopectin synthesis. All three models include soluble starch synthase to increase chain length, branching enzymes to incorporate branch points, and debranching enzymes to remove excess or ill-placed branches. There are several isozymes of starch synthase and branching enzymes. Each isozyme has a function [34,37,38]. For example, rice (*Oryza sativa*) has four soluble starch synthases (SSI, SSII, SSIIIa, SSIIIb) and three branching enzymes (BEI, BEIIa, BEIIb) [36]. SSI catalyzes short amylopectin chains, SSII catalyzes the elongation of A-chains and B-chains, and SSIIIa and SSIIIb catalyze the elongation of B-chains. BEI adds branches to amylose, and BEIIa and BEIIb add branches to amylopectin. High-amylose starches are formed in plants that lack branching enzymes [34].

Amylopectin synthesis involves chain elongation followed by branching. Branching occurs in clusters. Twenty to twenty-five chains may be included in a cluster, which is called the crystalline lamellae. About 80%–90% of the chains are involved in clusters [39]. Some of the chains in a cluster may exist as left-handed double stranded helices, which increase the stability of the clusters. Clusters are linked by long B-chains. Most B-chains extend into two to three clusters [39]. Between the clusters is a region with few branches. This region is called the amorphous lamellae. Thus, there is an alternating sequence of crystalline and amorphous lamellae within the amylopectin molecule.

Three models also have been proposed for amylose synthesis [40,41]. Regardless of the model, amylose synthesis occurs within the amylopectin molecule, behind the zone of amylopectin synthesis. Amylose is synthesized by Granule Bound Starch Synthase (GBSS). GBSS is located within the crystalline lamellae of the amylopectin molecule. As an amylose chain elongates it moves into the amorphous lamellae of amylopectin. Most amylose is found in the amorphous regions of the starch granule. Starch from plants lacking the genes for GBSS contains only amylopectin [34]. These starches are referred to as waxy starches.

Starch synthesis occurs in amyloplasts. Amyloplasts contain either simple or compound starch granules. Simple starch granules occur when only one granule forms per amyloplast. Compound starch granules occur when many small starch granules form per amyloplast. The occurrence of simple and compound starch granules is determined by the plant species. For example wheat, barley, and rye (*Secale cereale*) form simple starch granules while rice and oat form compound starch granules. Granule size and shape varies with the botanical source and growth environment. Starch granules within a plant source are not uniform. For example, cereals with simple starch granules will have a bimodal or trimodal size distribution of starch granules. These granule sizes are often referred to as A-granules, B-granules, and C-granules. Size differences are often related to when the granule synthesis was initiated [42]. In wheat, A-granules are large (≥14 μm) and are initiated soon after fertilization; B-granules have medium size (5–16 μm) and form about 10 days after fertilization; and C-granules are quite small (≤5 μm) and form about 21 days after fertilization.

Starch granules have several levels of structure. The first level consists of crystalline and amorphous lamellae. Lamellae consist of highly branched region (crystalline lamellae) alternating with a region containing few branches (amorphous lamellae). A second level of structure consists of growth rings. A cross section of the granule shows alternating zones of crystalline and amorphous zones that together constitutes growth rings. These growth rings are concentric rings of zones that have different molecular packing density.

Crystalline characteristics of a starch granule can be studied using x-ray diffraction. Through x-ray diffraction techniques, it has been estimated that native starch granules have crystallinity ranging from 15% to 45% [35,43]. Native starches have one of three x-ray diffraction patterns. A-crystalline pattern, common of starch found in cereal grains, indicates densely packed crystals arranged in a monoclinic array, with few water molecules associated with the starch. B-crystalline pattern indicates less densely packed crystals arranged in a hexagonal array, with more water associated with the starch. C-crystalline pattern indicates that the granule contains both A-crystalline and B-crystalline types [35,44], where the B-crystalline type is found in the center of the granule and is surrounded by the A-crystalline type.

Evidence is accruing that indicates a third level of granule structure called blocklets [45]. Atomic force microscopy of starch granules that have been lintnerized (acid etching of starch granules) has shown a regularity of block-like structures [46]. These blocklets appear to be groupings of amylopectin lamellae into spheres. The blocklet structure contains several repeating amorphous and crystalline lamellae. These blocklets differ in size depending on the botanical source and location in the granule. Blocklets are larger for B-crystalline types than for A-crystalline types.

2.4.6.2 Commercial Processing

Starch isolation from cereal grains involves steeping, coarse milling, degerming, fine milling, screening, centrifugal separation, and drying. During steeping, the grains are soaked in an aqueous solution of sulfur dioxide and lactic acid to loosen the granules in the packed endosperm cell structure. The softened grain is coarsely ground and a hydrocyclone separator separates the germ which is less dense because it contains a high level of lipid. The remaining material is finely ground to release the starch. The starch is isolated from other cellular components by centrifugation.

Starch isolation from tubers/roots involves washing and peeling the tubers. The tubers are soaked in an aqueous solution of sodium bisulfite to prevent discoloration. The tubers are ground by a cylindrical drum containing rotary blades. Starch is isolated from other cellular components by centrifugation.

2.4.6.3 Functionality

Starch functionality in food systems is primarily related to its gelatinization, retrogradation, and pasting properties. The functionality of native starch varies with botanical source, amylopectin fine structure, and amylose:amylopectin ratio.

2.4.6.4 Gelatinization

Gelatinization is the irreversible swelling of starch granules in excess hot water. Gelatinization involves plasticization of the amorphous and crystalline lamellae by water and heat. Water moves more quickly into the amorphous zone and lamellae than into the crystalline zone and lamellae. Hydration of the amorphous lamellae induces a transition from a glassy state to a rubbery state. This allows the amorphous lamellae to swell, which facilitates hydration and dissociation of the double helices in the crystalline lamellae. At room temperature, there is strong hydrogen bonding among amylose and amylopectin chains. Energy in the form of heat is required to break these hydrogen bonds. As the temperature increases, there is a change over from intermolecular hydrogen bonding to forming hydrogen bonds with water, which increases the mobility of the amylose and amylopectin chains and promotes the dissociation of the double helices in the crystalline lamellae. Eventually, amylose leaches from the granule and the amylose and amylopectin chains are unable to return to their original position upon cooling and dehydration. The temperature at which gelatinization occurs is lower for granules with the B-crystalline than the A-crystalline order.

2.4.6.5 Pasting/Retrogradation

If a shear force is applied to gelatinized starch granules, they will disrupt and a paste is formed. Starch paste is a viscous mass consisting of a continuous phase of solubilized entangled amylose molecules and a discontinuous phase of granule remnants. As the hot starch paste cools, the amylose molecules re-associate into aggregates and form a gel. Gel firmness is greater for starches with high than with low amylose contents. Gelatinized starch molecules re-associate and form an ordered structure with double helices during storage. This process is called retrogradation.

Retrogradation occurs in two stages. The first stage is the rapid formation of the crystalline region from the retrograded amylose. The second stage is the slow formation of an ordered structure within the amylopectin. Amylose forms double helical associations. Amylopectin crystallization occurs through the association of the outer most starch chains. The absence of amylose, e.g., waxy starches, delays retrogradation. Additionally, short outer chains reduce the retrogradation of the amylopectin.

2.4.6.6 Physical and Chemical Modifications

Starch functionality can be changed through physical and chemical modifications. Retrogradation can be modified by enzymatically reducing the length of the amylopectin outer chains. Annealing can change the functionality of starch. Annealing is a physical reorganization of starch granules when heated in excess water at or below its glass transition temperature. Annealed starch retains its granular structure. Annealing optimizes crystalline order and increases rigidity of the amorphous regions. It does not form new double helices but optimizes existing ones. Annealing results in reduced starch swelling, elevated starch gelatinization temperature, diminished gelatinization temperature range, and restricted amylose leaching [47].

Heat-water treatment involves exposing starch with low moisture content (18%–27%) to high temperature. Heat-water treatment physically modifies the starch. The low moisture content limits molecular mobility and gelatinization during treatment. Starch that has been heat-water treated has increased gelatinization temperature.

Common chemical modifications include derivatization of hydroxyl groups, cross-linking chains, and depolymerization of chains. Starch derivatives involve very few of the hydroxyl groups. The degree of substitution with ester or ether groups generally ranges from 0.002 to 0.2 [48]. Adding branches reduces the interaction between the chains that delays retrogradation. Phosphate ester cross-linking is a common linkage. Cross-linking strengthens the granule, reduces the rate and amount of swelling, and the subsequent disintegration.

2.4.6.7 Presence of Other Compounds

Starch functionality can also be affected by the presence of other compounds that are often found in food systems [49]. Salts and sugars lower water activity and delay gelatinization and increase gelatinization temperature by binding water in competition with starch [50]. Sugars also decrease the peak viscosity and gel strength, with disaccharides being more potent than monosaccharides. Acids disrupt hydrogen bonding, which results in a more rapid swelling of the granule. Soluble solids interfere by tying up water necessary for hydration. Fats and proteins tend to coat starch, which delays granule hydration and lowers the rate of swelling. Amylose is able to encapsulate lipids. This complex is relatively insoluble and resists gelatinization.

2.4.6.8 Food Uses

Starch and/or starch derivatives are nearly ubiquitous in food systems. Starch affects the sensory and textural properties of food. Starch is used to stabilize structure by acting as a bulking agent.

Starch can function as a thickening agent for sauces and pie fillings and as a colloidal stabilizer for salad dressings. Starch is used as coating, glazing, and gel-forming agents for gum confections.

High temperature and shear makes native starch granules fragile and susceptible to rupture during processing. Thus, native starches have limited use in the food industry. Starches are often modified to improve the desired properties [51]. Starch suitability depends on the processing temperature, length of time at temperature, shear forces, and pH. Cross-linked starches are more tolerant to harsh food processing conditions. In general, increased cross-linking results in more tolerance to acidic conditions and less breakdown of starch.

2.4.6.9 Healthful Properties

Starch can be degraded into glucose which is readily absorbed into the bloodstream in the small intestine. Glucose provides a source of energy for the body. However, overconsumption of starch can cause adverse health problems. Starch in cooked starchy foods typically is readily digested. However, not all starch is easily digested. These starches are referred to as resistant starch.

Resistant starch is not digested in the small intestine and undergoes some fermentation in the large intestine similar to that of traditional dietary fiber [51]. There are four types of resistant starch (RS). RS-I is physically protected starch. This starch is generally entrapped within the plant cell or a very dense food matrix. RS-I would include whole or partially ground grains. RS-II comes from certain plants whose native (nongelatinized) starch limits accessibility of digestive enzymes. These include nongelatinized granules with B-type crystallinity, such as raw potatoes and green bananas. It also includes high-amylose starch, which is a major commercial source of resistant starch [52]. RS-III is retrograded starch. Annealing and heat moisture treatments can increase the amount of RS-III in a food product. RS-IV is chemically modified starch.

Resistant starch represents a fiber source in food systems. Unlike more traditional fiber sources that have high water-binding capacity, resistant starch does not have elevated water-binding properties. Thus, resistant starch can often replace native starch 1:1 without any adverse effect on processing or end-use quality [39].

2.4.7 Nonstarch Polysaccharides

2.4.7.1 Occurrence in Plants

Nonstarch polysaccharides have three major functions in plants. Nonstarch polysaccharides can serve as primary carbohydrate storage. Galactomannans in endospermic legumes are a storage source of carbohydrate. Similarly glucomannan is the primary storage carbohydrate found in tubers from konjac plant (*Amorphophallus konjac*). Cellulose and hemicellulosic polysaccharides such as arabinoxylan and pectin, function as major components of cell walls. Lastly, polysaccharides can function to protect the plant from invasion, injury, and unfavorable water-relations. For example, the exudate gums such as gum arabic and gum ghatti are exuded from the plant in response to injury. Mucilage gums are found in the epidermal layer of the seed coat of certain plant species such as flaxseed (*Linum usitatissimum*), yellow mustard seed (*Sinapis alba*), and plants in the *Plantago* genus [53]. These gums protect the seed from dehydration and ensure that there is sufficient moisture for germination.

2.4.7.2 Food Uses

Meer et al. [54] listed five primary functions of nonstarch polysaccharides in food systems: retention of water, reduction in moisture evaporation, alteration of freezing rate, modification of ice crystal formation, and regulation of rheological properties or viscosity.

2.4.7.3 Healthful Properties

In general, nonstarch polysaccharides are resistant to digestion in the small intestine and are fermentable in the large intestine of humans. Thus, nonstarch polysaccharides are a source of dietary fiber. The ability to ferment in the large intestine seems to be related to water solubility. Highly fermentable polysaccharides are water soluble while poorly fermented are water insoluble [30]. Dietary fiber has been shown to reduce cholesterol and attenuate blood glucose, maintain gastrointestinal health, and positively affect calcium bioavailability and immune function [30]. Certain nonstarch polysaccharides can be used as fermentable substrate for the growth of probiotic microorganisms in food systems and can be used as prebiotics that promote growth of *Bifidobacteria* and *Lactobacilli* in the large intestine.

2.4.7.4 Commercial Nonstarch Polysaccharides

There are many important nonstarch polysaccharides commonly used in food systems. However, given the limitations inherent to a single chapter, only a select few are discussed below.

2.4.8 CELLULOSE

2.4.8.1 Occurrence in Plants

Cellulose is a linear, nonbranched polysaccharide that can contain up to 10,000 β-D-(1-4) linked glucosyl units (Figure 2.8a) [56]. Cellulose has a flat ribbonlike conformation. Cellulose molecules can associate to form fibrous crystalline bundles that are highly insoluble and impermeable to water.

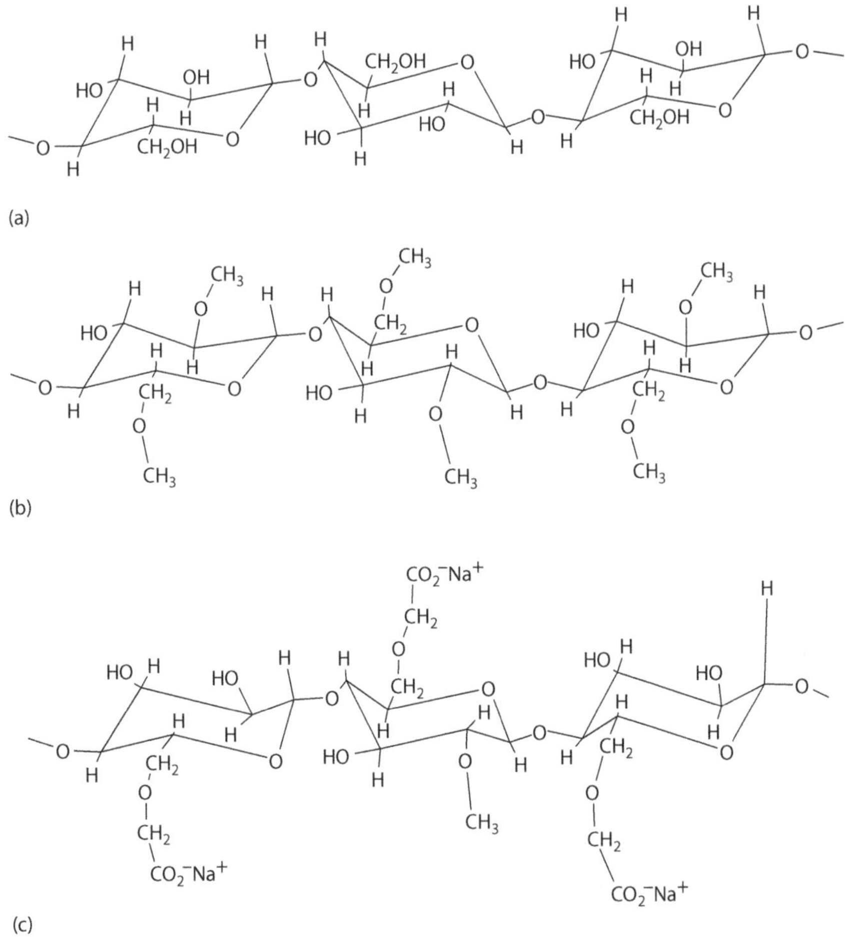

FIGURE 2.8   (a) Cellulose, (b) methylcellulose, and (c) sodium carboxymethylcellulose.

These bundles are the primary component of cell walls in plants. Cellulose provides the cell wall with the strength to resist the turgor pressure in plant cells, maintain size and shape, and influence the direction of plant growth [57].

Cellulose is synthesized by cellulose synthase, a large enzyme complex that appears as a rosette with a sixfold symmetry and is located in the plasma membrane. A large number of glucan chains are synthesized simultaneously [57]. Six glucan chains are directed into the exit channel of the complex where they are then hydrogen bonded into crystalline cellulose I microfibril. The parallel arrangement of the glucan chains in the cellulose microfibril requires that the newly synthesized glucan chains align with each other and lock into a specific crystalline arrangement.

There are two classes of cellulose: cellulose I and cellulose II. Most cellulose is produced as crystalline cellulose and is defined as cellulose I. Glucan chains in cellulose I are parallel to each other and are packed side by side to form microfibrils. There are three suballomorphs of cellulose I. These are cellulose Iα, Iβ, and Iγ. α-Cellulose is insoluble in 18% alkali; β-cellulose is soluble in alkali but precipitates when solution is neutralized; and γ-cellulose is what remains in solution even after neutralization. Cellulose Iα, Iβ, and Iγ differ in their crystalline packing, molecular conformation, and hydrogen bonding which affects their physical properties. Cellulose II is formed naturally in a few organisms and by mutants. The glucan chain arrangement in cellulose II is antiparallel, and occurs as a result of chain folding during synthesis.

2.4.8.2   Commercial Processing

Commercial sources of cellulose include wood pulp and cotton linters. Cotton linters are the short fibers remaining on cottonseeds after the long fibers have been removed. Cotton fibers are about 98% cellulose; while, wood is 40%–50% cellulose, 30% hemicellulose, and 20% lignin. Wood requires extensive processing to solubilize and remove the hemicellulose and lignins.

Microcrystalline cellulose is purified and partially depolymerized cellulose. Commercial processing of microcrystalline cellulose involves the hydrolysis of the purified wood pulp using hydrochloric acid to reduce the degree of polymerization, resulting in small acid resistance crystalline regions. Microcrystalline cellulose has a particle size range of 20–90 µm. Microcrystalline cellulose can be subjected to mechanical energy after hydrolysis, which tears apart the microfibrils and provides colloidal sized aggregates.

Cellulose quality is measured by the content of α-cellulose. Powdered food grade cellulose is not required to reach 99% purity because all cellulosic cell wall materials occur naturally in fruits, vegetables, and cereals. Highly purified forms (over 99% α-cellulose) are used to make cellulose derivatives such as methyl cellulose (Figure 2.8b) and sodium carboxymethyl cellulose (Figure 2.8c). Methyl cellulose is made by reacting cellulose with methyl chloride. Carboxymethyl cellulose is made by treating cellulose with alkali and then reacting with monochloroacetic acid. Carboxymethyl cellulose generally has a degree of substitution of 0.6–0.9. Substitutions occur on $O2$ and/or $O6$, and are occasionally found on $O3$.

2.4.8.3   Food Uses

Cellulose is used as a processing aid in the filtration of juices, as an anticaking agent for shredded cheese, as a fat substitute and a bulking agent in low calorie foods [56,58]. Cellulose needs to be physically or chemically modified to be used in food systems. The use of powdered cellulose is limited since it can result in poor mouthfeel [59]. Powdered cellulose is used in cake batters to help foam stability. It can be used to reduce fat and increase the moisture content of fried foods. Cellulose restricts water displacement by fat during frying due to strong hydrogen bonding with water. Temperature and pH do not significantly affect the water retention of powdered cellulose.

Microcrystalline cellulose is water dispersible and has properties similar to water soluble gums. Microcrystalline cellulose is used as a bulking agent, as a fat replacer in emulsion based food products, and is used to add creaminess to drinks, e.g. chocolate drinks. Decreasing particle size of microcrystalline cellulose increases its water retention capacity.

Methyl cellulose is nonionic, but can act as an emulsifier, because it contains both hydrophilic and hydrophobic groups. Methyl cellulose functions as an emulsifier/stabilizer in low oil/no oil salad dressings and can be applied to fried foods to reduce oil absorption. Methyl cellulose is used in baked goods to prevent boil over of pastry fillings and to aid gas retention during baking. Methyl cellulose can form gel at elevated temperatures due to hydrophobic interactions between highly substituted regions that stabilize intermolecular hydrogen bonding.

Carboxymethyl cellulose is an anionic water soluble polymer capable of forming very viscous solutions. Carboxymethyl cellulose is soluble in basic conditions and insoluble in acidic conditions. It is used as a thickener, stabilizer, and suspending agent. Carboxymethyl cellulose inhibits ice crystal growth in frozen desserts and soft serve ice cream. It is also used to improve loaf volume during baking by stabilizing gas bubbles.

2.4.9   Mixed Linkaged β-Glucan

2.4.9.1   Occurrence in Plants

Mixed linkage β-glucan, commonly referred to as β-glucan, is a major structural component of cell walls of several common cereal grains, e.g., barley, oat, and rye. Total β-glucan content ranges from 3%–8% in barley, 4%–6% in oat, and 1%–3% in rye [60,61]. β-Glucan content in grain from wheat generally does not exceed 1% and is concentrated in the crease of the kernel.

Mixed linked β-glucans are linear chains of β-D-glucosyl units, where about 70% are linked β-(1-4) and 30% are linked β-(1-3) [62]. The β-(1-3) linkages occur singly and are separated by cellotriosyl or cellotetrasyl units (Figure 2.9). In oat, the amount of cellotriosyl units is about twice as much as that of cellotetrasyl [63]. In barley, the amount of cellotriosyl units is about 3–3.5 times more than that of cellotetrasyl units [62,63]. The β-(1-3) linkage disrupts the uniform ribbon structure that is typical of β-(1-4) glucans (cellulose) and allows for increased flexibility and water solubility. Gomez et al. [64] described the molecular conformation of β-glucan as a partially stiff, wormlike cylinder.

Mixed linked β-glucan is synthesized by (1-3),(1-4)-β-glucan synthase which is located in the Golgi apparatus and consists of two glycotransferases. One transferase primarily synthesizes cellotetrasyl units and the other synthesizes cellotriosyl units [65].

2.4.9.2   Commercial Processing

Enzymatic hydrolysis represents a major problem in the extraction and utilization of β-glucan. Endo-(1-3),(1-4)-β-D-glucanase hydrolyzes a (1-4) linkage in mixed linkage of (1-3),(1-4) β-D-glucans where the D-glucosyl residue is substituted at the 3 position. Glucanase enzymes may be native to the grain or originate from microorganisms. Thus, extraction of β-glucan from cereal grains requires that the hydrolytic enzymes be deactivated.

Enzymes are deactivated by refluxing cereal grain flour in aqueous ethanol or treating the flour with dilute aqueous acid. Next, the β-glucan is extracted with hot water or a sodium carbonate solution. Extraction with sodium carbonate solution at pH 10 allows the extraction to occur at relatively low temperatures. Water temperature is usually kept below 60°C to avoid starch gelatinization. The extract is centrifuged to remove insoluble material and dried on heated rolls or by spray drying.

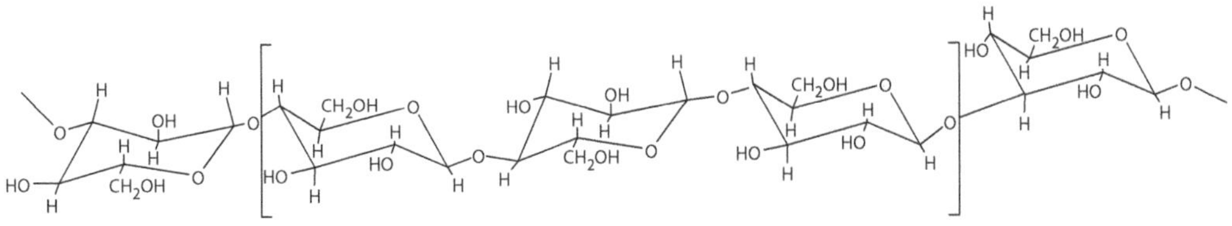

FIGURE 2.9   Mixed linkage β-D-glucan.

Solubility, extractability, and yield of β-glucan are influenced by the particle size and pretreatment of cereal materials as well as by the extraction conditions (temperature, pH, and ionic strength) [66,67].

2.4.9.3 Functional Properties

Solutions of β-glucans have high viscosity at relatively low concentrations. The viscosity is stable over a wide range of pH 2–10 but decreases with increased temperature. Compared with other thickeners at a concentration of 0.5%, oat β-glucan is less viscous than xanthan and guar gum but more viscous than locust bean gum and gum arabic [68]. Viscosity properties vary with the botanical source. For example, viscosity is greater with an equivalent solution of oat β-glucan than with barley β-glucan. This probably reflects the differences in cellotriosyl and cellotetrasyl contents in barley and oats as mentioned above.

At high concentrations (10% β-glucan), β-glucan molecules can associate and form aggregates. Formation of aggregates occurs upon heating samples of different molecular weights and forming a weak gel structure by physical links between the short sequences of the chains [69]. Acid hydrolysis of β-glucan chains can result in more gel-like behavior by promoting aggregation and formation of three-dimensional network [68].

2.4.9.4 Food Uses

β-Glucan is neutral and nonionic, making it suitable for a wide range of food products. Good viscosity forming properties make β-glucan a potential thickening agent in different food applications such as ice creams, sauces, and salad dressings. β-Glucan is used in beverages and has the potential to be used as a fat replacer in meat and dairy systems [69,70]. β-Glucan can be used to make edible films [71].

2.4.9.5 Healthful Properties

Consumption of β-glucan has been reported to reduce serum cholesterol and reduce blood glucose levels [72,73]. High viscosity is crucial for achieving the positive effect on peak blood glucose. Physiological effects are probably due to increased viscosity of the intestinal contents.

2.4.10 GALACTOMANNAN

2.4.10.1 Occurrence in Plants

Galactomannans are embedded in the primary cell walls of a wide assortment of plants such as purple morning glory (*Ipomoea purpurea*), bird's-foot trefoil (*Lotus japonicus*), pagoda tree (*Sophora japonica*), lettuce (*Lactuce sativa*), and mesquite pods (*Prosopis* spp.). Galactomannans in cell walls help maintain water relations before and during germination [74]. Galactomannans are also a major carbohydrate (energy) reserve in the endosperm of endospermic legumes such as *Cyamopsis tetragonoloba* (guar gum), *Ceratonia siliqua* (carob or locust bean gum), *Caesalpinia spinosa* (tara gum), and *Cassia tora/obtusifolia* (cassia gum). Galactomannans accumulate in the vacuole of endosperm cells generally as an amorphous structure. Galactomannans in the endosperm are thought to contribute to the stabilization of enzymes in the dry endosperm by preventing dehydration [75].

Galactomannans are composed of β-(1-4)-D-mannosyl backbone with substituted α-(1-6)-D-galactosyl side chains (Figure 2.10). Galactomannan biosynthesis involves two Golgi membrane bound glycosyltransferases, mannan synthase, and galactomannan galactosyltransferase. The galactomannan galactosyltransferase specificity is important in regulating the distribution and amount of α-(1-6)-galactosyl substitution on the β-(1-4) linked mannan backbone [76]. Galactose substitution keeps the molecules from self-associating. Species differ in the galactose/mannose ratio. For example, the galactose/mannose ratio is 1:2 for guar gum, 1:3 for tara gum, 1:4 for locust bean gum, and 1:5 for cassia gum.

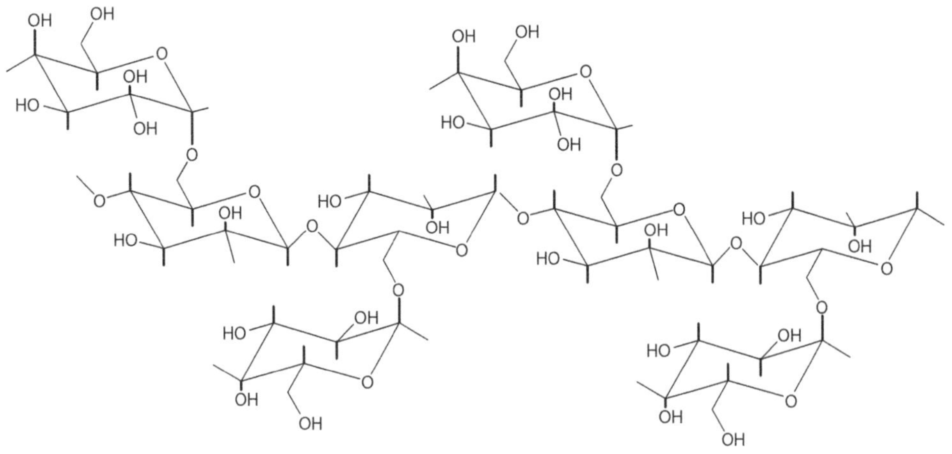

FIGURE 2.10   Galactomannan.

2.4.10.2   Commercial Processing

Galactomannan is isolated from seeds of endospermic legumes such as *Cyamopsis tetragonoloba*, *Ceratonia siliqua*, *Caesalpinia spinosa*, and *Cassia tora/obtusifolia*. The seeds are removed from dried pods and the seed hull is removed by mechanical abrasion or by roasting. Roasting loosens the hull allowing it to be removed from the seed. The seeds are broken and the germ is separated from the endosperm. Then the endosperm is ground to the desired particle size.

2.4.10.3   Functional Properties

Galactomannans are neutral polymers that have a ribbonlike structure and take on a random coil conformation in aqueous solutions. Varying amounts of galactose affects water solubility and gelling properties of these gums [77]. Gel formation occurs with the association of galactose-free regions of the polymer, where mannan segments of the chain can interact together. Solubility increases with increased number of galactosyl branches. Thus, thickening power decreases in order of increasing galactose content. At low concentration or shear rate, galactomannans exhibit Newtonian flow behavior. At high concentration/high shear, they exhibit pseudoplastic viscosity, where viscosity decreases with increased shear rate.

2.4.10.4   Food Uses

Guar gum is soluble in cold and hot water and can form highly viscous solutions. Guar gum is used in food systems as a film, emulsifier, stabilizer, suspending agent, thickening agent, and water-binding agent. Guar gum is used as a stabilizing agent at <1% of food weight to prevent settling in salad dressings, soft drinks, and fruit juices. It is also used in baked products and pastries to reduce starch retrogradation and improve texture and shelf life.

Locust bean gum is only partially soluble in cold water. It is generally synergistic with other gums particularly xanthan gum and is used to improve freeze thaw behavior of frozen products, e.g., ice cream and frozen desserts. Locust bean gum typically is used at ≤0.5%. Its main role is to provide viscosity, specific mouthfeel, and texture for foods such as ice cream, sherbet, cream cheese, yogurts, pie filings, sauces, and dressings.

2.4.10.5   Healthful Properties

Galactomannans such as guar gum and locust bean gum might have some therapeutic applications in treating obesity, hypercholesterolemia, gastritis, gastroduodenal ulcer, and constipation [78].

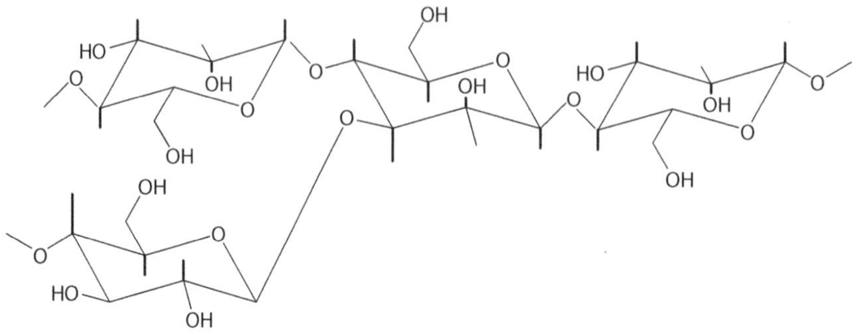

FIGURE 2.11 Glucomannan.

2.4.11 Glucomannan

2.4.11.1 Occurrence in Plants

Glucomannan accumulates in the tubers of the konjac plant (*Amorphophallus konjac*) where it is the primary storage carbohydrate. Konjac glucomannan is a neutral heteropolysaccharide consisting of a linear polymer of β-(1-4)-D-mannosyl units and β-(1-4)-D-glucosyl units in the ratio of 1.6:1 [79], with an occasional acetylated sugar (Figure 2.11) [80]. Konjac glucomannan is a semiflexible linear chain with some branching at the C-3 of both D-mannose and D-glucose [80,81].

2.4.11.2 Commercial Processing

Konjac tubers are washed, peeled, sliced, and dried. The dried slices are milled into flour and air classified (Konjac flour). Konjac flour contains about 70% glucomannan. Konjac flour can be further purified by alcohol to remove soluble starch to produce Konjac gum.

2.4.11.3 Functional Properties

Glucomannan is soluble in cold and hot water and will form a viscous solution. Glucomannan forms a gel when heated in an alkaline solution. Konjac glucomannan will interact with starch and is synergistic with kappa carrageenan, xanthan gum, and locust bean gum.

2.4.11.4 Food Uses

Konjac flour is used to make noodles and other food products. Konjac glucomannan has a very high water-binding capacity and is used in fiber drinks and as a thickening agent in food such as sauces and pie fillings.

2.4.11.5 Health Properties

Konjac glucomannan is regarded as a no calorie food in China and Japan. It is considered to provide high-quality dietary fiber that has been demonstrated to aid in weight reduction, glucose control, and cholesterol reduction.

2.4.12 Pectin

2.4.12.1 Occurrence in Plants

Pectin is a complex class of plant cell wall polysaccharides, composed of galacturonans and rhamnogalacturonans. Pectin has a backbone of β-(1-4)-D-galacturonic acid with α-(1-2)-L-rhamnose units interspersed (Figure 2.12). The galacturonic acid units are partially acetylated and methylated, and the L-rhamnose units contain side chains containing arabinose and/or galactose [82].

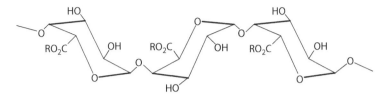

FIGURE 2.12 Pectin.

2.4.12.2 Commercial Processing

Pectin is commercially derived from citrus pomace and apple pomace. Canteri-Schemin et al. [83] described a method for pectin extraction from apple pomace. In their procedure, pectin was extracted in a hot (97°C) citric acidic solution and then cooled quickly to minimize depolymerization. The cooled extract was treated with alcohol. The pectin floats at the surface of an alcohol/water mixture. The pectin was filtered, dried, and ground.

2.4.12.3 Functional Properties

Pectins are water soluble and adopt a random coil conformation in solution. There are two major types of pectin. High methoxy pectins (HM-pectin) have >50% methylation, are soluble in water, and form gels in low pH and in the presence of a high concentration of sugar. Low methoxy pectins (LM-pectin) swell and hydrate in cold water and form gels in the presence of calcium. LM-pectin does not need sugar to form a gel.

2.4.12.4 Food Uses

Pectins are used as gelling agents either through the use of calcium cross-linking in the case of LM-pectins or sugar-induced dehydration with HM-pectin. Pectins are used as thickeners and stabilizers in jams, jelly, and fruit products [82]. Pectin can act as a protein dispersion stabilizer in acidified dairy products like yogurt and milk-based fruit drinks.

2.4.12.5 Healthful Properties

Pectin is a soluble dietary fiber and exerts physiological effects on the gastrointestinal tract such as delayed gastric emptying, reduced glucose absorption, and reduced serum cholesterol, all of which probably relate to its gel-forming and water-holding capacity. Kelley and Tsai [84] proposed that pectin lowered cholesterol levels by interfering with cholesterol absorption and by increasing cholesterol turnover. Pectin that has been partially hydrolyzed (pectin oligosaccharides) has some prebiotic activity. Pectin oligosaccharides have been shown to increase *Bifidobacteria* and *Eubacterium* numbers, but not as much as other prebiotic oligosaccharides such as FOS [85,86].

Degree of methylation, molecular weight, and pH appear to have a significant impact on the potential role of pectin in human health. Greater fermentation occurs with low than high methylation. LM pectins are better prebiotics than HM pectins [86]. Interestingly, LM pectin might be useful in inhibiting the spread of tumor cells associated with prostate cancer. Methylation influences the number of receptor binding sites that inhibit the spread of tumor cells associated with prostate cancer, the more methylation the fewer binding sites. Citrus pectin can be modified to have very low methylation (5%–10%) and reduced molecular weight (1,000–15,000) compared to unmodified pectin (20,000–400,000).

2.4.13 XYLOGLUCAN

2.4.13.1 Occurrence in Plants

Xyloglucans are the predominant hemicelluloses in the primary walls of edible vegetables and fruits of dicotyledonous plants. They are formed in the Golgi apparatus and secreted into the cell wall

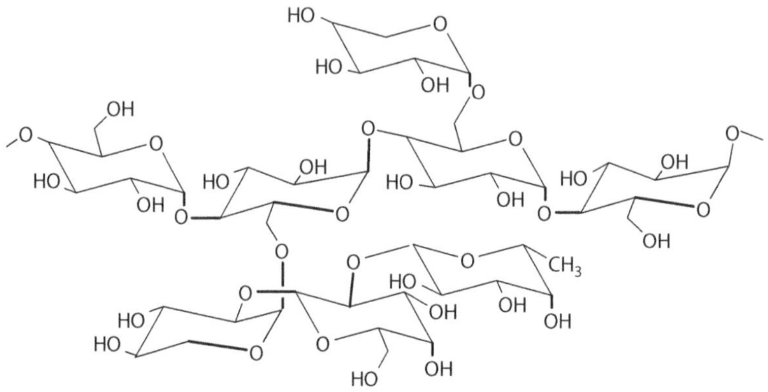

FIGURE 2.13 Xyloglucan.

via secretory vesicles [87]. Xyloglucans cross-link cellulose microfibrils and provide the flexibility needed for the microfibrils to slide during cell growth and enlargement. Xyloglucans are also found in the endosperm cell walls of seeds, for example, tamarind seeds [82].

Xyloglucans consist of a cellulose backbone of β-(1-4)-D-glucosyl units with side chains of α-(1-6) xylanose (Figure 2.13). Other sugars such as β-D-galactose, and disaccharides of fucose and galactose can be found attached at *O*-2. Xyloglucans have a linear flexible coil conformation. The chains are fairly stiff, being stiffer than galactomannan chains but less stiff than xanthan chains [88].

2.4.13.2 Commercial Isolation

Xyloglucans are generally not isolated but are used as the ground flour of tamarind (*Tamarindus indica*) seeds or the seed of *Detarium senegalense*, an African leguminous plant [88].

2.4.13.3 Functional Properties

Xyloglucans in tamarind has some random coil behavior. The cellulose-like backbone promotes interchain interactions. Xyloglucans are water soluble, but individual macromolecules tend not to fully hydrate, and thus supramolecular aggregated species remain present even in very dilute solutions. Tamarind is a gelling gum [89]. It can form a gel with sugar concentrates even in cold water. Tamarind and other exudate gums are high molecular weight macromolecules that can be easily dissolved and dispersed in water under appropriate conditions. They can modulate rheological properties of foods, and are generally used as food thickeners, texture modifiers, stabilizers, and emulsifiers for various applications [90].

2.4.13.4 Food Uses

Tamarind flour is used as food thickeners, stabilizers, and gelling agents. Polysaccharides from tamarind seeds can be used to replace pectin in the manufacture of jellies and jams, can be used in fruit preserving with or without acids, and can be used as a stabilizer in ice cream, mayonnaise, and cheese.

2.4.14 Exudate Gums

2.4.14.1 Occurrence in Plants

Gum arabic, gum ghatti, gum karaya, and gum tragacanth are the major commercial exudate gums. Gum arabic originates from *Acacia senegal* or *Acacia seyal* trees; gum ghatti comes from *Anogeissus latifolia*; gum karaya comes from *Sterculia* spp.; and gum tragacanth comes from

Astragalus gummifer. Exudate gums are a plants defense mechanism. These polysaccharides are exuded from plants as a result of injury or fungal attack.

Exudate gums are very diverse and complex in structure and composition [91,92]. Their composition vary with species and with seasonal and geographical variations. Exudate gums can be grouped based on their backbone as arabinogalactans, e.g., gum arabic, substituted glucuronomannans, e.g., gum ghatti, and substituted rhamnogalacturonans, e.g., gum karaya. Gum tragacanth is a mixture of arabinogalactan and galacturonan. Each gum type can come from a range of species and growing environments, all of which result in variation in composition and functionality [92–94].

2.4.14.2 Commercial Processing

Gums arabic, ghatti, karaya are collected by tapping branches and trunk of trees. Gum tragacanth is collected by tapping the root. Tapping involves making a superficial incision. Gum is manually collected as partially dried tears, hand cleaned to remove dirt, bark, and other foreign material [92]. The gum is dried. Dried gums are hard and glass-like and can be milled on a hammer mill. The milled gums are sieved to a uniform particle size.

2.4.14.3 Functional Properties

Gums arabic, ghatti, karaya, and tragacanth are nongelling. Gum arabic is a nongelling and low viscosity gum. It is soluble in hot and cold water and exhibits Newtonian flow even at high concentrations. Gum ghatti is somewhat soluble in hot and cold water. Gum ghatti does not completely dissolve in water but forms dispersion. Aqueous solutions of gum ghatti exhibit a non-Newtonian flow behavior. Gum karaya does not dissolve in water but swells to give dispersion. Gum karaya can swell to more than 60 times its original volume [92]. Aqueous solutions of gum karaya exhibit a Newtonian flow behavior at low concentrations <0.5%, shear thinning at 0.5%–2%, and become a paste or spreadable gels above 2%. Gum tragacanth swells rapidly in cold and hot water to form a viscous colloidal suspension. The viscosity of the suspension reaches a maximum after 24 h at room temperature. Gum tragacanth exhibits shear thinning behavior.

2.4.14.4 Food Uses

Gum arabic is used to stabilize oil-in-water emulsions, acts as an emulsifier in citrus juice and cola-flavored soft drinks, and is used in the confectionery industry where it prevents sucrose crystallization, provides controlled flavor release, and slows down melting in the mouth [92]. Gum ghatti is used as an emulsifier and stabilizer in beverages. It is used to encapsulate and stabilize oil-soluble vitamins. Gum karaya is used as a stabilizer in dairy products such as whipped cream and cheese spreads, and as a water-binding agent in processed meats and pasta. Gum karaya can control ice crystal formation in frozen desserts. The acid stability, high viscosity, and suspension properties of gum karaya make it suited for stabilizing low pH emulsions such as sauces and dressings [92]. Gum tragacanth is an excellent oil-in-water emulsifying agent with good stability to heat, acidity, and aging. It is used in salad dressings, oil and flavor emulsions, ice creams, bakery fillings, icings, and confectionaries. Gum tragacanth is used to create a creamy mouthfeel. It is used in lozenges and gum drops as a thickener and provides texture.

REFERENCES

1. Ahmed, Z., Production of natural and rare pentoses using microorganisms and their enzymes, *Electron. J. Biotechnol.*, 4, 103, 2001. Available online: www.ejb.org/content/vol4/issue2/full/7
2. Kim, H.-J. et al., Characterization of an *Agrobacterium tumefaciens* D-psicose 3-epimerase that converts D-fructose to D-psicose, *Appl. Environ. Microbiol.*, 72, 981, 2006.
3. Takeshita, K. et al., Mass production of D-psicose from D-fructose by a continuous bioreactor system using immobilized D-tagatose 3-epimerase, *J. Biosci. Bioeng.*, 90, 453, 2000.
4. Morimoto, K. et al., Large scale production of D-allose from D-psicose using continuous bioreactor and separation system, *Enzyme Microb. Technol.*, 38, 855, 2006.

5. Nagata, M., D-Psicose-containing hypoglycemic compositions, foods, and feeds and method for suppression of rapid increase of blood sugar level, Japanese Patent JP 2005213227 A2 20050811, *Jpn. Kokai Tokkyo Koho*, 2005, English abstract.

6. Arena, B.J. and Arnold, E.C., Low calorie D-aldohexose monosaccharides and their use, U.S. Patent Appl. 89-369985 19890622, U.S. patent 4963382 A 19901016, 1990.

7. Kim, P., Current studies on biological tagatose production using L-arabinose isomerase: A review and future perspective, *Appl. Microbiol. Biotechnol.*, 65, 243, 2004.

8. Petersen-Skytte, U., D-Tagatose as a flavor creator in various food systems plus examples of preparing these food systems, *Res. Disclosure*, 491, P232, 2005.

9. Martin, A. et al., Availability of specific sugars for glycoconjugate biosynthesis: A need for further investigations in man, *Biochimie,* 80, 75, 1998.

10. McAnalley, B.H. and Vennum, E., Introduction to glyconutritionals, *GlycoSci. Nutr.*, 1, 1, 2009. Available online: www.usa.glycoscience.com

11. Matsuo, T. and Izumori, K., D-psicose inhibits intestinal α-glucosidase and suppresses glycemic response after carbohydrate ingestion in rats, *Kagawa Daigaku Nogakubu Gakujutsu Hokoku,* 58, 27, 2006, English abstract.

12. Hossain, M.A. et al., Improved microcirculatory effect of D-allose on hepatic ischemia reperfusion following partial hepatectomy in cirrhotic rat liver, *J. Biosci. Bioeng.*, 101, 369, 2006.

13. Sui, L. et al., The inhibitory effect and possible mechanisms of D-allose on cancer cell proliferation, *Int. J. Oncol.*, 27, 907, 2005.

14. Roberfroid, M.B., Functional foods: Concepts and application to inulin and oligofructose, *Br. J. Nutr.*, 87, Suppl. 2, S139, 2002.

15. Kenter, C. and Hoffmann, C.M., Seasonal patterns of sucrose concentration in relation to other quality parameters of sugar beet (*Beta vulgaris* L.), *J. Sci. Food Agric.*, 86, 62, 2006.

16. Zhu, Y.J., Komor, E., and Morre, P.H., Sucrose accumulation in the sugarcane stem is regulated by the difference between the activities of soluble acid invertase and sucrose phosphate synthase, *Plant Physiol.*, 115, 609, 1997.

17. Clarke, M.A., Technological value of sucrose in food products, in *Sucrose Properties and Applications,* Mathlouthi, M. and Reiser, P., Eds., Blackie Academic Professional, New York, 1995, pp. 223–247.

18. Godshall, M.A., Role of sucrose in retention of aroma and enhancing the flavor of foods, in *Sucrose Properties and Applications*, Mathlouthi, M. and Reiser, P., Eds., Blackie Academic Professional, New York, 1995, pp. 248–263.

19. Knop, C., Voitsekhovskaja, O., and Lohaus, G., Sucrose transporters in two members of the Scrophulariaceae with different types of transport sugar, *Planta*, 213, 80, 2001.

20. Sprenger, N. and Keller, F., Allocation of raffinose family oligosaccharides to transport and storage pools in *Ajuga reptans*: The roles of two distinct galactinol synthases, *Plant J.*, 21, 249, 2000.

21. Martinez-Villaluenga, C., Frias, J., and Vidal-Valverde, C., Raffinose family oligosaccharides and sucrose contents in 13 Spanish lupin cultivars, *Food Chem.*, 91, 645, 2005.

22. Brenac, P. et al., Raffinose accumulation related to desiccation tolerance during maize (*Zea mays*) seed development and maturation, *J. Plant Physiol.*, 150, 481, 1997.

23. Gibson, G.R. and Roberfroid, M.D., Dietary modulation of the human colonic microbiota: Introducing the concept of prebiotics, *J. Nutr.*, 125, 1401, 1995.

24. Martinez-Villaluenga, C. et al., Influence of addition of raffinose family oligosaccharides on probiotic survival in fermented milk during refrigerated storage, *Int. Dairy J.*, 16, 768, 2006.

25. Bouhnik, Y. et al., The capacity of nondigestible carbohydrates to stimulate fecal bifidobacteria in healthy humans: A double-blind, placebo-controlled, parallel-group, dose-response relation study, *Am. J. Clin. Nutr.*, 80, 1658, 2004.

26. Vinj, I. and Smeekens, S., Fructan: More than a reserve carbohydrate, *Plant Physiol.*, 120, 351, 1999.

27. Franck, A., Technological functionality of inulin and oligofructose, *Br. J. Nutr.*, 87, Suppl. 2, S287, 2002.

28. Praznik, W., Cieslik, E., and Filipiak-Florkiewicz, A., Soluble dietary fibres in Jerusalem artichoke powders: composition and application in bread, *Nahrung*, 46, 151, 2002.

29. Tungland, B.C. and Meyer, D., Nondigestible oligo- and polysaccharides (dietary fiber): Their physiology and role in human health and food, *Comprehen. Rev. Food Sci. Food Safety*, 1, 73, 2002.

30. Shene, C., Cabezas, M.J., and Bravo, S., Effect of drying air temperature on drying kinetics parameters and fructan content in *Helianthus tuberosus* and *Cichorium intybus*, *Drying Technol.*, 21, 945, 2003.

31. Ritsema, T. and Smeekens, S., Fructans: Beneficial for plants and humans, *Curr. Opin. Plant Biol.*, 6, 223, 2003.

32. Kolida, S., Tuohy, K., and Gibson, G.R., Prebiotic effects of inulin and oligofructose, *Br. J. Nutr.*, 87, Suppl. 2, S193, 2002.
33. Rahman, S. et al., Genetic alteration of starch functionality in wheat, *J. Cereal Sci.*, 31, 91, 2000.
34. Parker, R. and Ring, S.G., Aspects of the physical chemistry of starch, *J. Cereal Sci.*, 34, 1, 2001.
35. Vandeputte, G.E. and Delcour, J.A., From sucrose to starch granule to starch physical behaviour: A focus on rice starch, *Carbohydr. Polym.*, 58, 245, 2004.
36. Morell, M.K. et al., Wheat starch biosynthesis, *Euphytica*, 119, 55, 2001.
37. Imparl-Radosevich, J.M. et al., Understanding catalytic properties and functions of maize starch synthase isozymes, *J. Appl. Glycosci.*, 50, 177, 2003.
38. Sajilata, M.G., Singhal, R.S., and Kulmarni, P.R., Resistant starch—A review, *Comprehen. Rev. Food Sci. Food Safety*, 5, 1, 2006.
39. van de Wal, M. et al., Amylose is synthesized in vitro by extension of and cleavage from amylopectin, *J. Biol. Chem.*, 273, 22232, 1998.
40. Zeeman, S.C., Smith, S.M., and Smith, A.M., The priming of amylose synthesis in *Arabidopsis* leaves, *Plant Physiol.*, 128, 1069, 2002.
41. Bechtel, D.B. et al., Size-distribution of wheat starch granules during endosperm development, *Cereal Chem.*, 67, 59, 1990.
42. Zobel, H.F., Molecules to granules: A comprehensive review, *Starch/Stärke*, 40, 1, 1988.
43. Bogracheva, T.Y. et al., The granular structure of C-type pea starch and its role in gelatinization, *Biopolymers*, 45, 323, 1998.
44. Tang, H., Mitsunaga, T., and Kawamura, Y., Molecular arrangement in blocklets and starch granule architecture, *Carbohydr. Polym.*, 63, 555, 2006.
45. Baker, A.A., Miles, M.J., and Helbert, W., Internal structure of the starch granule revealed by AFM, *Carbohydr. Res.*, 330, 249, 2001.
46. Tester, R.F. and Debon, S.J.J., Annealing of starch—A review, *Int. J. Biol. Macromol.*, 27, 1, 2000.
47. Whistler, R.L. and BeMiller, J.N., *Carbohydrate Chemistry for Food Scientists*, 1st edn., Eagan Press, St. Paul, MN, 1997, p. 143.
48. Kim, S.S. and Setser, C.S., Wheat starch gelatinization in the presence of polydextrose or hydrolyzed barley β-glucan, *Cereal Chem.*, 69, 447, 1992.
49. Crochet, P. et al., Starch crystal solubility and starch granule gelatinisation, *Carbohydr. Res.*, 340, 107, 2005.
50. Sajilata, M.G. and Singhal, R.S., Specialty starches for snack foods, *Carbohydr. Polym.*, 59, 131, 2005.
51. Liu, Q., Understanding starches and their role in foods, in *Food Carbohydrates*, Cui, S.W., Eds., Taylor & Francis, New York, 2005, pp. 309–355.
52. Cui, S.W., *Polysaccharide Gums from Agricultural Products: Processing, Structures and Functionality*, 1st ed., Technomic Publishing, Lancaster, U.K., 2001, Chaps. 1, 2, and 5.
53. Meer, W.A., Meer, G., and Gerard, T., Natural plant hydrocolloids in bakery application, *The Bakers Digest*, June 1973, pp. 45–47,67.
54. Theander, O., Westerlund, E., and Aman, P., Structure and components of dietary fiber, *Cereal Foods World*, 38, 135, 1993.
55. Saxena, I.M. and Brown Jr, R.M., Cellulose biosynthesis: Current views and evolving concepts, *Ann. Bot.*, 96, 9, 2005.
56. Gu, L. et al., Structure-function relationships of highly refined cellulose, *Trans. ASAE*, 44, 1707, 2001.
57. Ang, J.F. and Miller, W.B., Multiple functions of powdered cellulose as a food ingredient, *Cereal Foods World*, 36, 558, 1991.
58. Beresford, G. and Stone, BA., (1-3), (1-4)-β-D-Glucan content of *Triticum* grains, *J. Cereal Sci.*, 1, 111, 1983.
59. Saastamoinen, M., Plaami, S., and Kumpulainen, J., Genetic and environmental variation in β-glucan content of oats cultivated or tested in Finland, *J. Cereal Sci.*, 16, 279, 1992.
60. Wood, P.J., Weisz, J., and Blackwell, B.A., Molecular characterization of cereal β-D-glucans. Structural analysis of oat β-D glucans from different sources by high-performance liquid chromatography of oligosaccharides released by lichenase, *Cereal Chem.*, 68, 31, 1991.
61. Papgeorgiou, M. et al., Water extractable (1-3,1-4)-β-D-glucans from barley and oats: An intervarietal study on their structural features and rheological behaviour, *J. Cereal Sci.*, 42, 213, 2005.
62. Gomez, C. et al., Physical and structural properties of barley (1-3), (1-4)-β-D-glucan. Part II. Viscosity, chain stiffness and macromolecular dimensions, *Carbohydr. Polym.*, 32, 17, 1997.
63. Urbanowicz, B., Rayon, C., and Carpita N.C., Topology of the maize mixed linkage (1-3),(1-4)-β-D-glucan synthase at the Golgi membrane, *Plant Physiol.*, 134, 758, 2004.

64. Gaosong, J. and Vasanthan, T., Effect of extrusion cooking on the primary structure and water solubility of β-glucans from regular and waxy barley, *Cereal Chem.*, 77, 396, 2000.

65. Zhang, D., Doehlert, D.C., and Moore, W.R., Rheological properties of (1-3), (1-4)-β-D-glucans from raw, roasted, and steamed oat groats, *Cereal Chem.*, 75, 433, 1998.

66. Dawkins, N.L. and Nnanna, L.A., Studies on oat gum [(1-3, 1-4)-β-D-glucan]: Composition, molecular weight estimation and rheological properties, *Food Hydrocolloids*, 9, 1, 1995.

67. Gomez, C. et al., Physical and structural properties of barley (1-3), (1-4)-β-D-glucans III. Formation of aggregates analysed through its viscoelastic and flow behaviour, *Carbohydr. Polym.*, 34, 141, 1997.

68. Doublier, J-L. and Wood, P.J., Rheological properties of aqueous solutions of (1-3), (1-4)-β-D-glucan from oats (*Avena sativa* L.), *Cereal Chem.*, 72, 335, 1995.

69. Morin, L.A., Temelli, F., and McMullen, L., Interactions between meat proteins and barley (*Hordeum* spp.) β-glucan within a reduced-fat breakfast sausage system, *Meat Sci.*, 68, 419, 2004.

70. Tudorica, C.M. et al., The effects of refined barley β-glucan on the physico-structural properties of low-fat dairy products: Curd yield microstructure, texture and rheology, *J. Sci. Food Agric.*, 84, 1159, 2004.

71. Tejinder, S., Preparation and characterization of films using barley and oat β-glucan extracts, *Cereal Chem.*, 80, 728, 2003.

72. Braaten, T.J. et al., Oat β-glucan reduces blood cholesterol concentration in hypercholesterolemic subjects, *Eur. J. Clin. Nutr.*, 48, 465, 1994.

73. Wood, P.J. et al., Effect of dose and modification of viscous oat gum on plasma glucose and insulin following an oral glucose load, *Br. J. Nutr.*, 72, 731, 1994.

74. Reid, J.S.G. and Bewley, J.D., A dual role for the endosperm and its galactomannan reserves in the germinative physiology of fenugreek (*Trigonella foenum-graecum*), an endospermic leguminous seed, *Planta*, 147, 145, 1979.

75. Cappelletti, C. I. et al., Weak bond-linked components of the reserve and fibrillar cell walls of the endosperm of *Gleditsia triacanthos* (*Leguminosae*), *ARKIVOC*, 12, 62, 2005. Available online: http://www.arkat-usa.org/ark/journal/2005/l12_Lederkremer/1565/1565.pdf

76. Edwards, M.E. et al., The seeds of *Lotus japonicus* lines transformed with sense, antisense, and sense/antisense galactomannan galactosyltransferase constructs have structurally altered galactomannans in their endosperm cell walls, *Plant Physiol.*, 134, 1153, 2004.

77. Richardson, P.H. and Norton, I.T., Gelation behavior of concentrated locust bean gum solutions, *Macromolecules*, 31, 1575, 1998.

78. Castillo, G.E., Lopez, C.G., and Lopez, C.A., Guar gum characteristics and applications, *Ciencia Tecnologia Pharmaceutica*, 15, 3, 2005, English abstract.

79. Kato, K. and Matsuda, K., Studies of the chemical structure of konjac mannan: I. Isolation and characterization of oligosaccharides from the partial hydrolyzate of the mannan, *Agric. Biol. Chem.*, 33, 1446, 1969.

80. Maeda, M., Shimahara, H., and Sugiyama, N., Detailed examination of the branched structure of konjac glucomannan, *Agric. Biol. Chem.*, 44, 245, 1980.

81. Li, B., Xie, B., and Kennedy, J.F., Studies on the molecular chain morphology of konjac glucomannan, *Carbohydr. Polym.* 64, 510, 2006.

82. Waldron, K.W., Parker, M.L., and Smith, A.C., Plant cell walls and food quality, *Comprehen. Rev. Food Sci. Food Safety*, 2, 101, 2003.

83. Canteri-Schemin, M.H. et al., Extraction of pectin from apple pomace, *Brazilian Arch. Biol. Technol.*, 48, 259, 2005.

84. Kelley, J.J. and Tsai, A.C., Effect of pectin, gum Arabic and agar on cholesterol absorption, synthesis, and turnover in rats, *J. Nutr.*, 108, 630, 1978.

85. Manderson, K. et al., In vitro determination of prebiotic properties of oligosaccharides derived from an orange juice manufacturing by-product stream, *Appl. Environ. Microbiol.*, 71, 8383, 2005.

86. Olano-Martin, E., Gibson, G.R., and Rastall, R.A., Comparison of the in vitro bifidogenic properties of pectins and pectic-oligosaccharides, *J. Appl. Microbiol.*, 93, 505, 2002.

87. Lerouxel, O. et al., Biosynthesis of plant cell wall polysaccharides—A complex process, *Curr. Opin. Plant Biol.*, 9, 621, 2006.

88. Picout, D.R. et al., Pressure cell assisted solubilization of xyloglucans: Tamarind seed polysaccharide and detarium gum, *Biomacromolecules*, 4, 799, 2003.

89. Mathur, N.K. and Mathur, V., Polysaccharide gums as stabilizers and thickeners in food systems, *Trend Carbohydr. Chem.*, 7, 193, 2001.

90. Lin H.-Y. and Lai, L.-S., Isolation and viscometric characterization of hydrocolloids from mulberry (*Morus alba* L.) leaves, *Food Hydrocolloids,* 23(3), 840–848, May 2009.

91. Tischer, C.A., Iacomini, M., Wagner, R., and Gorin, P.A.J., New structural features of the polysaccharide from gum ghatti (*Anogeissus latifola*). *Carbohydr. Res.* 337, 2205–2210, 2002.
92. Verbeken, D., Dierckx, S., and Dewettinck, K., Exudate gums: Occurrence, production, and applications, *Appl. Microbiol. Biotechnol.*, 63, 10, 2003.
93. Baldwin, T.C., Quah, P.E., and Menzies, A.R., A serotaxonomic study of Acacia gum exudates, *Phytochemistry*, 50, 599, 1999.
94. Yadav, M.P., Johnston, D.B., and Hicks, K.B., Corn fiber gum: New structure/function relationships for this potential beverage flavor stabilizer, *Food Hydrocolloids*, 23(6), 1488–1493, August 2009.

3 Amino Acids, Oligopeptides, Polypeptides, and Proteins

Fatih Yildiz

CONTENTS

3.1 INTRODUCTION

Proteins are among the fundamental molecules of biology. They are common to all life, are present in every cell on Earth today, and are responsible for most of the complex functions that make life possible. There are an estimated 100,000 different proteins in the human body alone [1]. In plants, an inventory of more than 13,000 plant proteins has been cataloged [2].

They are also the major structural constituent of living systems. According to the central dogma of molecular biology [3], information is transferred from DNA to RNA to proteins. DNA functions as a storage medium for the information necessary to synthesize proteins, and RNA is responsible for (among other things) the translation of this information into protein molecules.

Virtually all the complex chemical functions of the living cell are performed by protein-based catalysts called enzymes, hormones, signals, and material carriers in biological fluids. Specifically, enzymes either make or break chemical bonds. Protein enzymes should not be confused with RNA-based enzymes (also called ribozymes), a group of macromolecules that perform functions similar to protein enzymes [4]. Further, most of the structural and scaffolding molecules that hold cells and organelles together are made of proteins. In addition to their catalytic functions, proteins can transmit and communicate signals from the extracellular environment, duplicate genetic information, assist in transforming the energy in light and chemicals with high efficiency, convert chemical energy into mechanical work, and carry molecules between cell compartments. A significant number of proteins, especially large proteins, have a structure divided into several independent domains. These domains can often perform specific functions in a protein. For example, a cell membrane receptor might have an extracellular domain to bind a target molecule and an intracellular domain that binds other proteins inside the cell, thereby transducing a signal across the cell membrane.

The domain of a protein is determined by its secondary structure. There are four main types of domain structures: alpha-helix, beta sheet, beta-turn, and random coil.

Storage proteins [5] are a group that comprises of proteins generated mainly during seed production and stored in (vacuoles) the seed serve as nitrogen sources for the developing embryo during germination. It is obviously more effective for the plant to use proteins instead of secondary plant products for this purpose. The average protein content of cereal grains is 10%–15% of their dry weight and that of leguminose seeds 20%–25%, while it is only 3%–5% in normal leaves. Besides seeds, storage proteins can also be found in root and shoot tubers, for example, potatoes.

No clear definition of what a storage protein is exists. The term is operational and was coined for all those proteins whose share in the total protein amount of the cell is greater than five percent. Usually, the following properties can be found, too: (1) the storage proteins have no enzymatic activities, (2) they serve as nitrogen sources for the germinating seed, (3) they occur normally in an aggregated state within a membrane surrounded vesicle (protein bodies, aleuron grains), and (4) they are often built from a number of different polypeptide chains.

Storage proteins are important for human nutrition (plant proteins) and numerous studies concerning their structure and biosynthesis have therefore been published during the last few years. Furthermore, there is considerable interest in the production of mutants with increased protein content or increased amount of essential amino acids [6].

3.1.1 Amino Acids

Proteins consist of polymers of amino acids (AA) that are characterized by the following general chemical structure:

$$R-CH(NH_2)COOH$$

Two hydrogens and a nitrogen comprise the *amino* group, $-NH_2$, and the *acid* entity is the carboxyl group, $-COOH$. Amino acids link to each other when the carboxyl group of one molecule reacts with the amino group of another molecule, creating a peptide bond $-C(=O)NH-$ and releasing a molecule of water (H_2O). Amino acids (AA) are the basic building blocks of enzymes, hormones, proteins, and body tissues. A *peptide* is a compound consisting of two or more amino acids. *Oligopeptides* have 10 or fewer amino acids. *Polypeptides* and *proteins* are chains of 10 or more amino acids, but peptides consisting of more than 50 amino acids are classified as proteins. The chemical structure of amino acids is given in Table 3.1 [7].

3.1.1.1 Stereochemistry

In all the 20 amino acids, except glycine, the carbon atom with the amino group is attached to four different substituents. The tetrahedral bond angles of carbon and the asymmetry of the attachments make it possible for the amino acids to have two non-superimposable structures, the L and R forms, which are mirror images of each other. Only L-amino acids are found in natural proteins. L-Amino acids have the amino group to the left when the carboxyl group is at the top, as illustrated here. The wedge bonds are above the display plane and the dotted bonds are below the display plane. The rotating molecular model represents oxygen atoms as light, nitrogen as light, carbon as dark, and hydrogen as dark colored for the dipolar ion form of L-alanine ($CH_3CH(NH_3^+)COO^-$):

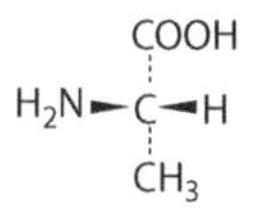

$$\begin{array}{c} COOH \\ | \\ H_2N \blacktriangleright C \blacktriangleleft H \\ | \\ CH_3 \end{array}$$

L-Alanine

TABLE 3.1
Naturally Occurring Amino Acids, Their Abbreviations, and Structural Formulas

- Ala = alanine
  $CH_3CH(NH_2)COOH$

- Asn = asparagine
  $H_2N-C(=O)CH_2CH(NH_2)COOH$

- Cys = cysteine (2 Cys = Cystine)
  $HS-CH_2CH(NH_2)COOH$

- Glu = glutamic acid
  $HOOC-CH_2CH_2CH(NH_2)COOH$

- His = histidine[1] (Babies only)

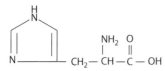

- Leu = leucine[3]
  $CH_3CH(CH_3)CH_2CH(NH_2)COOH$

- Met = methionine[5]
  $CH_3-S-CH_2CH_2CH(NH_2)COOH$

- Pro = proline (Hydroxyproline)

- Thr = threonine[7]
  $CH_3CH(OH)CH(NH_2)COOH$

- Tyr = tyrosine

- Arg = arginine
  $H_2N-C(=NH)NHCH_2CH_2CH_2CH(NH_2)COOH$

- Asp = aspartic acid
  $HOOC-CH_2CH(NH_2)COOH$

- Gln = glutamine
  $H_2N-C(=O)CH_2CH_2CH(NH_2)COOH$

- Gly = glycine
  H_2N-CH_2COOH

- Ile = isoleucine[2]
  $CH_3CH_2CH(CH_3)CH(NH_2)COOH$

- Lys = lysine[4]
  $H_2N-CH_2CH_2CH_2CH_2CH(NH_2)COOH$

- Phe = phenylalanine[6]

- Ser = serine
  $HOCH_2CH(NH_2)COOH$

- Trp = tryptophan[8]

- Val = valine[9]
  $CH_3CH(CH_3)CH(NH_2)COOH$

Notes:
- Essential amino acids are indicated with superscript numbers.
- The term "essential amino acid" refers to an amino acid that is required to meet physiological needs and must be supplied in the diet. Arginine is synthesized by the body, but at a rate that is insufficient to meet growth needs. Methionine is required in large amounts to produce cysteine if the latter amino acid is not adequately supplied in the diet. Similarly, phenylalanine can be converted to tyrosine, but is required in large quantities when the diet is deficient in tyrosine. Tyrosine is essential for people with the disease phenylketonuria (PKU) whose metabolism cannot convert phenylalanine to tyrosine. Isoleucine, leucine, and valine are sometimes called "branched-chain amino acids" (BCAA) because their carbon chains are branched. Amino acids are grouped according to the characteristics of the side chains:
- Aliphatic—alanine, glycine, isoleucine, leucine, proline, and valine.
- Aromatic—phenylalanine, tryptophan, and tyrosine.
- Acidic—aspartic acid, and glutamic acid.
- Basic—arginine, histidine, and lysine.
- Hydroxylic—serine and threonine.
- Sulfur-containing—cysteine and methionine.
- Amidic (containing the amide group)—asparagine and glutamine.

TABLE 3.2
General Chemical Properties of Amino Acids

Amino Acid	Short	Abbrev.	Avg. Mass (Da)	pI	pK₁ (α-COOH)	pK₂ (α-⁺NH₃)
Alanine	A	Ala	89.09404	6.01	2.35	9.87
Cysteine	C	Cys	121.15404	5.05	1.92	10.70
Aspartic acid	D	Asp	133.10384	2.85	1.99	9.90
Glutamic acid	E	Glu	147.13074	3.15	2.10	9.47
Phenylalanine	F	Phe	165.19184	5.49	2.20	9.31
Glycine	G	Gly	75.06714	6.06	2.35	9.78
Histidine	H	His	155.15634	7.60	1.80	9.33
Isoleucine	I	Ile	131.17464	6.05	2.32	9.76
Lysine	K	Lys	146.18934	9.60	2.16	9.06
Leucine	L	Leu	131.17464	6.01	2.33	9.74
Methionine	M	Met	149.20784	5.74	2.13	9.28
Asparagine	N	Asn	132.11904	5.41	2.14	8.72
Pyrrolysine	O	Pyl				
Proline	P	Pro	115.13194	6.30	1.95	10.64
Glutamine	Q	Gln	146.14594	5.65	2.17	9.13
Arginine	R	Arg	174.20274	10.76	1.82	8.99
Serine	S	Ser	105.09344	5.68	2.19	9.21
Threonine	T	Thr	119.12034	5.60	2.09	9.10
Selenocysteine	U	Sec	169.06			
Valine	V	Val	117.14784	6.00	2.39	9.74
Tryptophan	W	Trp	204.22844	5.89	2.46	9.41
Tyrosine	Y	Tyr	181.19124	5.64	2.20	9.21

Note: The pK_a values of amino acids are slightly different when the amino acid is inside of a protein [10].

3.1.2 STRUCTURE AND BIOCHEMICAL PROPERTIES OF AMINO ACIDS

Table 3.2 is a listing of the one-letter symbols, the three-letter symbols, and the chemical properties of the side chains of the standard amino acids (Table 3.3) [8]. The masses listed are based on weighted averages of the elemental isotopes at their natural abundances. Note that forming a peptide bond results in the elimination of a molecule of water, so the mass of an amino acid unit within a protein chain is reduced by 18.01524 Da. The one-letter symbol for an undetermined amino acid is X. The three-letter symbol *Asx* or one-letter symbol *B* means the amino acid is either asparagine or aspartic acid; *Glx* or *Z* means either glutamic acid or glutamine; and *Xle* or *J* means either leucine or isoleucine. The IUPAC and IUBMB [9] now also recommend that *Sec* or *U* refers to selenocysteine, and *Pyl* or *O* refers to pyrrolysine.

3.2 PEPTIDES

3.2.1 FORMATION OF A PEPTIDE FROM TWO AMINO ACIDS

Figure 3.1 shows the reaction of two amino acids, where R and R' are any functional groups from the table above. The small circle shows the water (H_2O) that is released, and the larger circle shows the resulting peptide bond (−C(=O)NH−).

The reverse reaction, i.e., the breakdown of peptide bonds into the component amino acids, is achieved by *hydrolysis*. Many commercial food products use hydrolyzed vegetable proteins as

TABLE 3.3
Remarks about the Amino Acids

Amino Acid	Remarks
Alanine	Very abundant, very versatile. More stiff than glycine, but small enough to pose only small steric limits for the protein conformation. It behaves fairly neutrally, and can be located in both hydrophilic regions on the protein outside and the hydrophobic areas inside.
Cysteine	The sulfur atom binds readily to *heavy metal* ions. Under oxidizing conditions, two cysteines can join together in a *disulfide bond* to form the amino acid *cystine*. When cystines are part of a protein, *insulin* for example, this stabilizes *tertiary structure* and makes the protein more resistant to *denaturation*; disulfide bridges are therefore common in proteins that have to function in harsh environments including digestive enzymes (e.g., *pepsin* and *chymotrypsin*) and structural proteins (e.g., *keratin*). Disulfides are also found in peptides too small to hold a stable shape on their own (e.g., *insulin*).
Aspartic acid	Behaves similarly to glutamic acid. Carries a hydrophilic acidic group with strong negative charge. Usually is located on the outer surface of the protein, making it water soluble. Binds to positively charged molecules and ions, often used in enzymes to fix the metal ion. When located inside of the protein, aspartate and glutamate are usually paired with arginine and lysine.
Glutamic acid	Behaves similar to aspartic acid. Has longer, slightly more flexible side chain.
Phenylalanine	*Essential* for humans. Phenylalanine, tyrosine, and tryptophan contain large rigid *aromatic* groups on the side chain. These are the biggest amino acids. Like isoleucine, leucine, and valine, these are hydrophobic and tend to orient toward the interior of the folded protein molecule.
Glycine	Because of the two hydrogen atoms at α carbon, glycine is not *optically active*. It is the smallest amino acid, rotates easily, and adds flexibility to the protein chain. It is able to fit into the tightest spaces, e.g., the triple helix of *collagen*. As too much flexibility is usually not desired, as a structural component it is less common than alanine.
Histidine	In even slightly acidic conditions *protonation* of the nitrogen occurs, changing the properties of histidine and the polypeptide as a whole. It is used by many proteins as a regulatory mechanism, changing the conformation and behavior of the polypeptide in acidic regions such as the late *endosome* or *lysosome*, enforcing conformation change in enzymes. However only a few histidines are needed for this, so it is comparatively scarce.
Isoleucine	*Essential* for humans. Isoleucine, leucine, and valine have large aliphatic hydrophobic side chains. Their molecules are rigid, and their mutual hydrophobic interactions are important for the correct folding of proteins, as these chains tend to be located inside of the protein molecule.
Lysine	*Essential* for humans. Behaves similarly to arginine. Contains a long flexible side chain with a positively charged end. The flexibility of the chain makes lysine and arginine suitable for binding to molecules with many negative charges on their surfaces. For example, *DNA*-binding proteins have their active regions rich with arginine and lysine. The strong charge makes these two amino acids prone to be located on the outer hydrophilic surfaces of the proteins; when they are found inside, they are usually paired with a corresponding negatively charged amino acid, e.g., aspartate or glutamate.
Leucine	*Essential* for humans. Behaves similar to isoleucine and valine. See isoleucine.
Methionine	*Essential* for humans. Always the first amino acid to be incorporated into a protein; sometimes removed after translation. Like cysteine, contains sulfur, but with a *methyl* group instead of hydrogen. This methyl group can be activated, and is used in many reactions where a new carbon atom is being added to another molecule.
Asparagine	Similar to aspartic acid. Asn contains an *amide* group where Asp has a *carboxyl*.
Proline	Contains an unusual ring to the N-end amine group, which forces the CO-NH amide sequence into a fixed conformation. Can disrupt protein folding structures like α-*helix* or β *sheet*, forcing the desired kink in the protein chain. Common in *collagen*, where it often undergoes a *posttranslational modification* to *hydroxyproline*. Uncommon elsewhere.
Glutamine	Similar to glutamic acid. Gln contains an amide group where Glu has a *carboxyl*. Used in proteins and as a storage for *ammonia*.
Arginine	Functionally similar to lysine.

TABLE 3.3 (continued)
Remarks about the Amino Acids

Amino Acid	Remarks
Serine	Serine and threonine have a short group ended with a *hydroxyl* group. Its hydrogen is easy to remove, so serine and threonine often act as hydrogen donors in enzymes. Both are very hydrophilic, therefore the outer regions of soluble proteins tend to be rich with them.
Threonine	*Essential* for humans. Behaves similarly to serine.
Valine	*Essential* for humans. Behaves similarly to isoleucine and leucine. See isoleucine.
Tryptophan	*Essential* for humans. Behaves similarly to phenylalanine and tyrosine (see phenylalanine). Precursor of *serotonin.*
Tyrosine	Behaves similarly to phenylalanine and tryptophan (see phenylalanine). Precursor of *melanin, epinephrine,* and *thyroid hormones.*

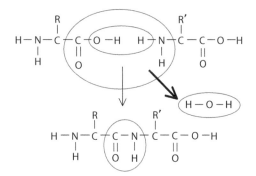

FIGURE 3.1   Peptide formation and a peptide bond.

flavoring agents. Soy sauce is produced by hydrolyzing soybean and wheat protein by fungal fermentation or by boiling with acid solutions. Monosodium glutamate (MSG), a flavor enhancer, is a sodium salt of glutamic acid that is found naturally in seaweed and fermented soy products. All amino acids are naturally occurring.

3.2.2 PEPTIDES, OLIGOPEPTIDES, POLYPEPTIDES, AND PROTEINS

Peptides consist of two or more amino acids, oligopeptides (2–10 amino acids), *polypeptides* (10–50 amino acids or more), and *proteins* both contain 10 or more amino acids. Those polypeptides consisting of more than 50 amino acids are classified as proteins. A *protein* is a complex, high molecular weight organic compound that consists of amino acids joined by peptide bonds.

Peptide hormones are produced by the endocrine glands (pituitary, thyroid, pineal, adrenal, and pancreas) or by various organs such as the kidney, stomach, intestine, placenta, or liver (Table 3.4). Peptide hormones can have complex, convoluted structures with hundreds of amino acids. Figure 3.2 illustrates the chemical structure of human insulin and its three-dimensional shape. Insulin is made of two amino acid sequences. The *A-Chain* has 21-amino acids, and the *B-Chain* has 30-amino acids. The chains are linked together through the sulfur atoms of cysteine (Cys). Peptide hormones are generally different for every species, but they may have similarities [11]. Human insulin is identical to pig insulin, except that the last amino acid of the B-Chain for the pig is alanine (Ala) instead of threonine (Thr) (IUPAC and IUBMB) [9] and [11].

TABLE 3.4
Some Important Peptide Hormones

Hormone	Number of Amino Acids	Function
Insulin	51	Lowers blood glucose level, promotes glucose storage as glycogen and fat. Fasting decreases insulin production.
Glucagon	29	Increases blood glucose level. Fasting increases glucagon production.
Ghrelin	28	Stimulates release of growth hormone, increases feeling of hunger.
Leptin	167	Its presence suppresses the feeling of hunger. Fasting decreases leptin levels.
Growth hormone	191	Human growth hormone (HGH), also called somatotropin, promotes amino acid uptake by cells and regulates development of the body. Growth hormone levels increase during fasting.
Prolactin	198	Initiates and maintains lactation in mammals.
Human placental lactogen	191	Produced by the placenta late in gestation.
Luteinizing hormone	204	Induces the secretion of testosterone.
Follicle-stimulating hormone	204	Induces the secretion of testosterone and dihydrotestosterone.
Chorionic gonadotropin	237	Produced after implantation of an egg in the placenta.
Thyroid-stimulating hormone	201	Stimulates secretion of thyroxin and triiodothyronine.
Adrenocorticotropic hormone	39	Stimulates production of adrenal cortex steroids (cortisol and costicosterone).
Vasopressin	9	Increases the reabsorption rate of water in kidney tubule cells (antidiuretic hormone).
Oxytocin	9	Causes contraction of mammary gland cells to produce milk and stimulation of uterine muscles during childbirth.
Angiotensin II	8	Regulates blood pressure through vasoconstriction.
Parathyroid hormone	84	Increases calcium ion levels in extracellular fluids.
Gastrin	14	Regulates secretion of gastric acid and pepsin, a digestive enzyme consisting of 326 amino acids.

Note: See also Chapter 10.

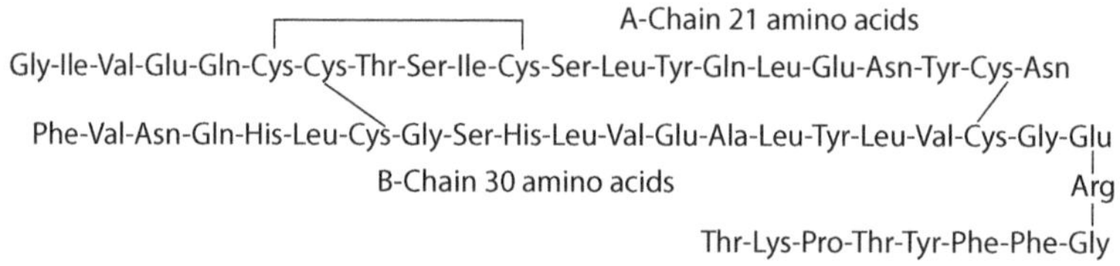

FIGURE 3.2 Chemical structure of human insulin.

3.2.3 SOME EXAMPLES OF POLYPEPTIDES

Adenylate cyclase activating polypeptide 1 (*pituitary*), also known as *ADCYAP1*, is a human gene. They are also considered a type of paracrine and regulators of enteroendocrine cells (neuroendocrine cells) [12].

Neuropeptide Y (NPY) is a 36-amino acid peptide neurotransmitter found in the brain and autonomic nervous system. It is one of the pancreatic polypeptides [13].

Peptide YY, also known as *PYY* or peptide tyrosine tyrosine, is the name of a human gene and protein. Peptide YY_{3-36} (PYY) is a linear polypeptide consisting of 36 amino acids with structural homology to NPY and pancreatic polypeptide. PYY exerts its action through NPY receptors, inhibits gastric motility, and increases water and electrolyte absorption in the colon [14].

Pancreatic polypeptide is a 36-amino acid peptide that inhibits pancreatic exocrine function [15]. The pancreatic polypeptide (PP) affects the secretion of the pancreatic enzymes, water, and electrolytes. PP increases gastric emptying and gut motility [15].

Glucagon is a 29-amino acid polypeptide. Glucagon is an important hormone involved in carbohydrate metabolism. Produced by the pancreas, it is released when the glucose level in the blood is low (hypoglycemia), causing the liver to convert stored glycogen into glucose and release it into the bloodstream. The action of glucagon is thus opposite to that of insulin. Its primary structure in humans is NH_2-His-Ser-Gln-Gly-Thr-Phe-Thr-Ser-Asp-Tyr-Ser-Lys-Tyr-Leu-Asp-Ser-Arg-Arg-Ala-Gln-Asp-Phe-Val-Gln-Trp-Leu-Met-Asn-Thr-COOH. This consists of 20 amino acid residues [16].

Calcitonin is a 32-amino acid polypeptide hormone that is produced in humans primarily by the parafollicular (also known as C-cells) of the thyroid, and in many other animals in the "ultimobranchial body" (or gland is an embryological structure that gives rise to the calcitonin producing cells, which are called parafollicular cells of the thyroid gland). It acts to reduce blood calcium (Ca^{2+}), opposing the effects of the parathyroid hormone (PTH) [17].

Antifreeze polypeptides are found in the blood plasma of bony fishes from the polar and subpolar oceans. The major antifreeze polypeptide (AFP) from winter flounder (37 amino acid residues) is a single alpha-helix [18].

3.3 OVERALL CONFORMATION OF PROTEINS

Proteins have a covalently bonded backbone, as discussed before, in relation to amino acid sequence determination. But the 3-D shape or conformation is held together by weaker bonding of the noncovalent type. The linear form of the polypeptide backbone of the protein folds into a tightly held shape, which is chemically stabilized by weak bonds like hydrogen bonds, ionic bonds, and hydrophobic interactions among nonpolar amino acid side chains [19].

To reduce the complexity of the protein structure to a manageable level for our study and understanding, the protein is considered to have four levels of structure [20].

3.3.1 FOUR LEVELS OF PROTEIN STRUCTURE

1. Primary structure: polypeptide backbone
2. Secondary structure: local hydrogen bonds along the backbone
3. Tertiary structure: long distance bonding involving the AA side chains
4. Quaternary structure: protein–protein interactions leading to formation of dimers and tetramers

3.3.1.1 Primary Structure of Proteins

Protein characterization include primary structure which determines the following parameters of the protein [21]:

1. Native molecular weight (MW)
2. Subunit composition (i.e., if the protein is a dimer, trimer, tetramer, etc.) by finding subunit MW
3. Amino acid (AA) composition and sequence (sequencing)
4. N-Terminal end and C-Terminal end determinations

3.3.1.2 Secondary Structure of Proteins

In the 1950s, Linus Pauling [19] named the first structure he found by x-ray diffraction, the alpha-helix, and the second structure, beta sheet. These names are used even today for the two forms of the protein secondary structure and a third type of form—in the region where the protein bends back on itself to form its compact shape or conformation—has been added. Thus the types of protein secondary structures include [22]

1. Alpha-helix
2. Beta-sheet
3. Turns or bends (bends in the backbone to fold the polypeptide back on itself) where the local hydrogen bonding forms the secondary structure in proteins [19].

3.3.1.2.1   Alpha-Helix

The alpha-helix is held together by hydrogen bonds between the amide hydrogen on the nitrogen and another amide carbonyl oxygen of every fourth amino acid residue (approximately). These are intrachain H-bonds that are along the same region of the backbone of the polypeptide, or in other words, within the same region of the amino acid sequence (Figure 3.3). The side chains of the amino acids project out from the core of the alpha-helix. Water is excluded from the tight inner core of the alpha-helix, which is very hydrophobic [23].

3.3.1.2.1   Beta Sheet

Beta sheets are also held together by hydrogen bonds between the hydrogen on the nitrogen and another amide carbonyl oxygen of the peptide bonds, but between the chains of the backbone rather than along it, as was found for the alpha-helix [24]. These are called interchain H-bonds since they form between two parts of the polypeptide backbone separated from one another by some distance or length of the amino acid sequence of the polypeptide. In the beta sheet structure the amino acid side chains project above and below the plane of the sheet. The R groups (amino acid side chains) alternate above and below the sheet along the backbone chain (Figure 3.4).

The two types of backbone chain orders are

1. Parallel, where the chains run in the same direction
2. Antiparallel, where the chains run in the opposite direction

3.3.1.3   Tertiary Structure—The Third Level of Protein Structure

The tertiary protein structure is formed by long distance interactions of the amino acid side chains.

There are several types of bonds which make the tertiary structure possible [25]:

1. Hydrogen bonds
2. Ionic bonds
3. Hydrophobic interactions
4. Disulfide bonds—weak covalent bonds between 2 Cys (R-Cys-S-S-Cys-R) (Figure 3.5)

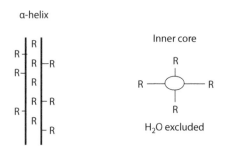

FIGURE 3.3   Side and top views of alpha-helix illustrating the position of the side chains.

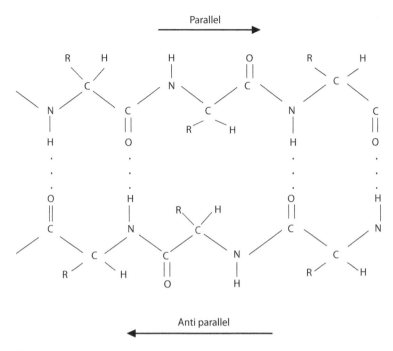

FIGURE 3.4 Hydrogen bonding between two polypeptide chains in a beta sheet.

3.3.1.3.1 Hydrogen Bonds in the Tertiary Structure

Hydrogen bonds in the tertiary structure involve polar non-charged amino acid side chains: (1) alcohols—Ser Thr Tyr and (2) amides—Asn Gln (as shown in Figure 3.6).

These AAs can serve as both H-donors and acceptors. These are weak bonds like all H-bonds. In some cases, the acids (Glu and Asp) and His—when not charged—form hydrogen bonds. Cys thiols (−S−H) do not form hydrogen bonds instead they form disulfide bonds [26].

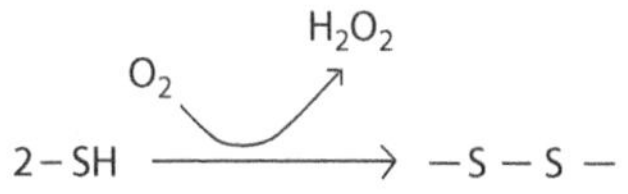

$$2-SH \xrightarrow{\quad O_2 \quad H_2O_2 \quad} -S-S-$$

FIGURE 3.5 Disulfide bond formation between two −SH containing amino acids.

3.3.1.4 Quaternary Structure—The Fourth Level of Protein Structure

The quaternary structure is formed between polypeptide chains and leads to the formation of dimers, trimers, tetramers, etc. The fourth level of structure depends on the noncovalent bonding between

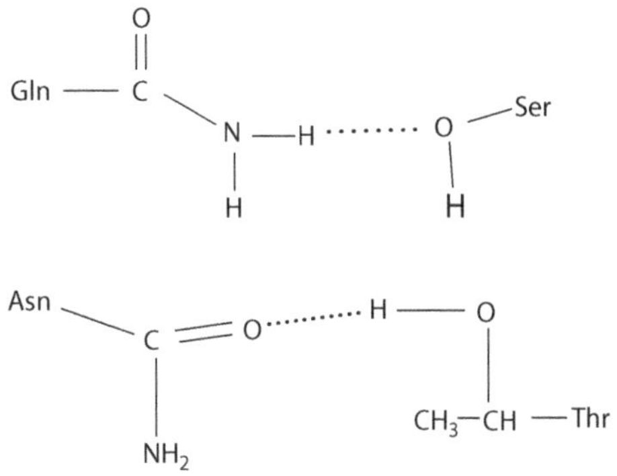

Gln — C (=O) N—H ⋯⋯ O—Ser
with H below N and H below O

Asn — C =O ⋯⋯ H — O
with NH₂ below C and CH₃—CH —Thr below O

FIGURE 3.6 Hydrogen bonds between AA side chains.

two polypeptide chains. The commonly found bonds in the quaternary structure are [27] (a) ionic bonds and (b) hydrogen bonds. The less commonly found bonds in the quaternary structure are (a) hydrophobic interfaces and (b) interchain disulfides.

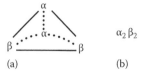

FIGURE 3.7 Quaternary structure of proteins: (a) tetrahedral shape and (b) subunit type.

3.3.1.4.1 Protein Subunit Composition

The composition of the protein subunit is determined by dividing the native molecular weight by the subunit molecular weight. The native molecular weight is determined by gel filtration as described above in the section on using gel filtration for purification. If the native molecular weight = 80,000 Da, and subunit molecular weight = 38,000 Da, then the protein is a dimer (i.e., 80,000/38,000 = ~2). If a protein is a monomer, its native molecular weight should be the same as its subunit molecular weight. Some proteins are composed of two different subunits, and therefore two polypeptides are found when they are analyzed by SDS-PAGE, and therefore, a pure protein must be used to do SDS-PAGE in order to avoid confusion [28].

3.3.1.4.2 Native Molecular Weight and Subunit Composition

Some proteins and enzymes are composed of a single polypeptide chain, which are called *monomers*. Many have multiple copies of the same polypeptide chain—homodimers and homotetramers. Others have more than one polypeptide chain; for example, hemoglobin has a and b chains with two of each to form a heterotetramer (Figure 3.7 and Table 3.5).

3.4 REQUIREMENTS FOR PROTEIN PURIFICATION

There is a need to keep track of the protein purification studies: if it is an enzyme then use its enzyme activity; or if it is a colored protein, then monitor its specific wavelength of absorbance. But it must maintain biological conditions of pH, salt concentration, and temperature to keep the protein from denaturing and losing its biological activity [26].

3.4.1 GENERAL PROTEIN PURIFICATION STRATEGY

To purify the protein of interest (i.e., separate it from other proteins in a mixture), one should take advantage of its general and specific properties: its native surface charge using ion exchange

TABLE 3.5
List of Native Molecular Weights and Subunit Compositions of Selected Proteins

Protein/Enzyme	MW (Da)	Subunits	Type
Glucagon	3,000	1	Monomer
Cytochrome c	13,000	1	Monomer
Ribonuclease A	13,700	1	Monomer
Lysozyme	13,900	1	Monomer
Myoglobin	16,900	1	Monomer
Chymotrypsin	21,600	1	Monomer
Carbonic anhydrase	30,000	1	Monomer
Hexokinase	102,000	2	Dimer
Glycogen phosphorylase	194,000	2	Dimer
Hemoglobin	64,500	4	Tetramer
Lactate dehydrogenase	140,000	4	Tetramer
Glutamine synthetase	600,000	12	Dodecamer

chromatography, its unique shape and size using gel filtration column chromatography, and its biological activity using affinity chromatography. These steps are sometimes applied in succession: first, ion exchange to separate other proteins that have a different charge from the protein of interest, next, gel filtration to separate all other proteins with a different size/shape than the protein of interest, and, finally, affinity chromatography to separate them based on biological activity, which is usually highly specific for the protein/enzyme of interest. In some cases, an enzyme may be purified to a homogeneous state (i.e., completely purified) using affinity chromatography alone since it is so effective at separating a specific enzyme from all other proteins in a mixture [28,29].

3.4.1.1 Details of the Three Common Protein Purification Methods

3.4.1.1.1 Surface Charge Properties Can Be Used in Ion Exchange Chromatography

The surface charge properties can be used in ion exchange chromatography, where the charge on the protein depends on its pI and the pH of the buffer used. Negatively charged proteins are separated using anion exchange chromatography on a support like (diethyl amino ethyl) DEAE Cellulose (positively charged groups extend out from the cellulose particles). Positively charged proteins are separated using cation exchange chromatography on a support like (microgranular cellulose) CM Cellulose (negatively charged groups extend out from the cellulose particles).

3.4.1.1.2 Gel Filtration Takes Advantage of Protein Size and Shape to Separate It from Other Proteins

The gel is a molecular sieve where large proteins exit from the column first, followed by medium size proteins and finally small proteins elute. If the gel filtration column is calibrated with proteins of known native molecular weight, then one can estimate the proteins native molecular weight by comparison to the standard proteins using a calibration curve.

3.4.1.1.3 Affinity Chromatography Takes Advantage of the Biological Activity of Proteins for Separation

Make a solid support (gel) by covalently linking the substrate of the enzyme (the substrate is the small molecule or metabolite on which the enzyme normally acts on to catalyze the reaction specific to the enzyme and binds in the enzymes' active site). Then put the substrate-gel (sometimes called the affinity material or affinity gel) in a column and allow the enzyme of interest to bind, and then wash away all unbound proteins by passing a lot of buffer over the column. Next, the enzyme is eluted by adding a buffer to the column containing the substrate, the free substrate in solution binds more tightly to the enzyme than the substrate covalently linked to the gel. The enzyme–substrate complex can then be dialyzed to remove the substrate or the complex can be passed over a gel filtration column where the substrate will elute after the enzyme [27].

3.4.2 Protein Purity Testing

To test the purity of a protein after each separation step use a Native PAGE gel (PAGE = polyacrylamide gel electrophoresis). This method separates proteins based on charge density while maintaining their biological activity. The proteins separated on the Native PAGE gel will retain their biological activity or enzyme activity and this can be detected by using an appropriate activity stain. The proteins will also bind dyes which recognize all proteins and so even nonenzyme proteins on the gel can be identified by a total protein stain. Thus, the specific enzyme of interest can be revealed by an activity stain, and by comparing its position on the gel to the all the proteins revealed by the total protein stain, it can be determined if the enzyme or protein of interest is pure or not. If you find you have only one protein on the gel and it stains for both enzyme activity and protein, then you have a homogeneous protein and do not need to do more purification [30].

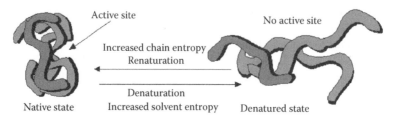

FIGURE 3.8 Protein denaturation and renaturation.

3.5 DENATURATION AND RENATURATION IN PROTEINS

1. When proteins are heated, or exposed to acids or bases, or high salt concentrations, the variety of weak bonds holding tertiary and quaternary structure together can be disrupted so that the protein unfolds. Unfolding = *denaturation*; resulting in loss of biological function (Figure 3.8).
2. Unfolding can proceed even to disrupt the secondary structure.
3. Denaturation is sometimes *reversible*; an unfolded protein can be restored to correct folding and regain its biological activity. This is called *renaturation*.
4. Denaturation can also occur *irreversibly* (as when egg white protein, albumin, is denatured by boiling to congeal as egg white). Renaturation is then no longer possible [31].

3.6 CONJUGATED PROTEINS

3.6.1 GLYCOPROTEINS

Glycoproteins are proteins that contain oligosaccharide chains (glycans) covalently attached to their polypeptide side chains. The carbohydrate is attached to the protein in a cotranslational or posttranslational modification. This process is known as glycosylation. In proteins that have segments extending extracellularly, the extracellular segments are often glycosylated. Glycoproteins are often important integral membrane proteins, where they play a role in cell–cell interactions [32] (Table 3.6).

There are two types of glycosylation:

- In N-glycosylation, the addition of sugar chains can happen at the amide nitrogen on the side chain of the asparagine.
- In O-glycosylation, the addition of sugar chains can happen on the hydroxyl oxygen on the side chain of hydroxylysine, hydroxyproline, serine, or threonine [33].

3.6.1.1 Functions of Glycoproteins

The functions of glycoproteins are summarized in Table 3.6.

3.6.2 LIPOPROTEINS

A *lipoprotein* is a biochemical assembly that contains both proteins and lipids. The lipids or their derivatives may be covalently or noncovalently bound to the proteins. Many enzymes, transporters, structural proteins, antigens, adhesins, and toxins are lipoproteins. Examples include the high density and low density lipoproteins of the blood, the transmembrane proteins of the mitochondrion and the chloroplast, and bacterial lipoproteins [34]. Lipoproteins in the blood, an aqueous medium, carry fats around the body. The protein particles have hydrophilic groups aimed outward so as to attract water molecules; this makes them soluble in the salt water based blood pool. Triglyceride-fats and cholesterol are carried internally, shielded from the water by the protein particle [35].

TABLE 3.6
Some Functions of Glycoproteins

Function	Glycoproteins
Structural molecule	Collagens
Lubricant and protective agent	Mucins
Transport molecule	Transferrin, ceruloplasmin
Immunologic molecule	Immunoglobins, histocompatibility antigens
Hormone	Chorionoic gonadotropin, thyroid-stimulating hormone (TSH)
Enzyme	Various, e.g., alkaline phosphatase
Cell attachment-recognition site	Various proteins involved in cell–cell (e.g., sperm–oocyte), virus–cell, bacterium–cell, and hormone–cell interactions
Antifreeze	Certain plasma proteins of coldwater fish
Interact with specific carbohydrates	Lectins, selectins (cell adhesion lectins), antibodies
Receptor	Various proteins involved in hormone and drug action
Affect folding of certain proteins	Calnexin, calreticulin
Regulation of development	Notch and its analogs, key proteins in development
Hemostasis (and thrombosis)	Specific glycoproteins on the surface membranes of platelets

3.6.2.1 Classification of Lipoproteins by Density

The general categories of lipoproteins are listed in the order from larger and less dense (more fat than protein) to smaller and denser (more protein, less fat):

- Chylomicrons—carry triacylglycerol (fat) from the intestines to the liver, skeletal muscle, and adipose tissue.
- Very low density lipoproteins (VLDL)—carry (newly synthesized) triacylglycerol from the liver to the adipose tissue.
- Intermediate density lipoproteins (IDL)—have densities between that of VLDL and LDL. They are not usually detectable in the blood.
- Low density lipoproteins (LDL)—carry cholesterol from the liver to cells of the body. This is sometimes referred to as the "bad cholesterol" lipoprotein.
- High density lipoproteins (HDL)—these collect cholesterol from the body's tissues, and bring it back to the liver. Sometimes referred to as the "good cholesterol" lipoprotein.

3.6.3 NUCLEOPROTEINS

A *nucleoprotein* is any protein which is structurally associated with nucleic acid (either DNA or RNA). The prototypical example is any of the histone class of proteins, which are identifiable on strands of chromatin. Telomerase, a RNP (RNA/protein complex), and protamines are also nucleoproteins [36]. *Ribonucleoprotein* (RNP) is a nucleoprotein that contains RNA, i.e., it is a compound that combines ribonucleic acid and protein together. It is one of the main components of the nucleolus. The *signal recognition particle* (SRP) is a ribonucleoprotein (protein–RNA complex) that recognizes and transports specific proteins to the endoplasmic reticulum in eukaryotes and the plasma membrane in prokaryotes [37]. Deoxyribonucleoproteins (complexes of DNA and proteins) constitute the genetic material of all organisms and of many viruses. They function as the chemical basis of heredity and are the primary means of its expression and control. Most of the mass of chromosomes are made up of DNA and proteins whose structural and enzymatic activities are required for the proper assembly and expression of the genetic information, encoded in the molecular structure of the nucleic acid.

Ribonucleoproteins (complexes of RNA and proteins) occur in all cells as part of the machinery for protein synthesis. This complex operation requires the participation of messenger RNAs (mRNAs), amino acid transfer RNAs (tRNAs), and ribosomal RNAs (rRNAs), each of which interacts with specific proteins to form functional complexes called polysomes, on which the synthesis of the new proteins occurs [38].

3.7 BIOSYNTHESIS OF PROTEINS

Amino acids are the monomers which are polymerized to produce proteins. Amino acid synthesis is the set of biochemical processes which build the amino acids from carbon sources like glucose. Not all amino acids may be synthesized by every organism, for example adult humans have to obtain 8 of the 20 amino acids from their diet.

To understand how proteins are biosynthesized, we have to divide the decoding process into two steps. The DNA only stores the genetic information and is not involved in the process by which the information is used. The first step in protein biosynthesis therefore has to involve *transcribing* the information in the DNA structure in a useful form. In a separate step, this information can be *translated* into a sequence of amino acids [39].

3.7.1 Transcription

Before the information in the DNA can be decoded, a small portion of the DNA double helix must be uncoiled. A strand of the RNA that is a complementary copy of one strand of the DNA is then synthesized [27]. Assume that the section of the DNA that is copied has the following sequence of nucleotides, starting from the 3′ end.

3′ T-A-C-A-A-G-C-A-G-T-T-G-G-T-C-G-T-G… 5′ DNA

When we predict the sequence of the nucleotides in the RNA complement, we have to remember that the RNA uses U where T would be found in the DNA. We also have to remember that base pairing occurs between two chains that run in *opposite directions*. The RNA complement of this DNA should therefore be written as follows:

3′ T-A-C-A-A-G-C-A-G-T-T-G-G-T-C-G-T-G…5′ *DNA*
5′ A-U-G-U-U-C-G-U-C-A-A-C-C-A-G-C-A-C…3′ *mRNA*

Since this RNA strand contains the message that was coded in the DNA, it is called the *messenger RNA (mRNA)*.

3.7.2 Translation

The messenger RNA now binds to a ribosome, where the message is translated into a sequence of amino acids. The amino acids that are incorporated into the protein that is synthesized are carried by relatively small RNA molecules known as *transfer RNA* or *tRNA*. There are at least 60 tRNAs that differ slightly in their structures, in each cell. At one end of each tRNA is a specific sequence of three nucleotides (anticodons) that can bind to the messenger RNA. At the other end is a specific amino acid. Thus, each three-nucleotide segment of the messenger RNA molecule codes for the incorporation of a particular amino acid. The relationship between the triplets, or codons, on the mRNA and the amino acids is shown in Table 3.7 [40].

3.8 FOOD PROTEINS

3.8.1 Wheat Proteins

Wheat proteins contain albumins, globulins, gliadins, and glutenins, these four basic proteins depending on their varied solubility in different solvents (Figure 3.9).

TABLE 3.7
Gene Expression and Biochemistry of Amino Acids

Amino Acid	Short	Abbrev.	Codon(s)	Occurrence in Proteins (%)	Essential[a] in Humans
Alanine	A	Ala	GCU, GCC, GCA, GCG	7.8	—
Cysteine	C	Cys	UGU, UGC	1.9	Conditionally
Aspartic acid	D	Asp	GAU, GAC	5.3	—
Glutamic acid	E	Glu	GAA, GAG	6.3	Conditionally
Phenylalanine	F	Phe	UUU, UUC	3.9	Yes
Glycine	G	Gly	GGU, GGC, GGA, GGG	7.2	Conditionally
Histidine	H	His	CAU, CAC	2.3	Yes
Isoleucine	I	Ile	AUU, AUC, AUA	5.3	Yes
Lysine	K	Lys	AAA, AAG	5.9	Yes
Leucine	L	Leu	UUA, UUG, CUU, CUC, CUA, CUG	9.1	Yes
Methionine	M	Met	AUG	2.3	Yes
Asparagine	N	Asn	AAU, AAC	4.3	—
Pyrrolysine	O	Pyl			
Proline	P	Pro	CCU, CCC, CCA, CCG	5.2	—
Glutamine	Q	Gln	CAA, CAG	4.2	—
Arginine	R	Arg	CGU, CGC, CGA, CGG, AGA, AGG	5.1	Conditionally
Serine	S	Ser	UCU, UCC, UCA, UCG, AGU, AGC	6.8	—
Threonine	T	Thr	ACU, ACC, ACA, ACG	5.9	Yes
Selenocysteine	U	Sec	UGA[b]		—
Valine	V	Val	GUU, GUC, GUA, GUG	6.6	Yes
Tryptophan	W	Trp	UGG	1.4	Yes
Tyrosine	Y	Tyr	UAU, UAC	3.2	Conditionally
Stop codon[c]	—	Term	UAA, UAG, UGA	—	—

Notes:

UGA is normally a stop codon, but encodes selenocysteine if a (selenocysteine insertion sequence) SECIS element is present.

SLC—the single-letter code—is used to represent the amino acids in the protein databases.

Example: The single-letter code for human glucagon is HSQGTFTSDYSKYLDSRRAQDFVQWLMNT

Codon letters: A = adenine, C = cytosine, G = guanine, U = uracil.

[a] An essential amino acid cannot be synthesized in the humans and must, therefore, be supplied in the diet. Conditionally, essential amino acids are not normally required in the diet, but must be supplied exogenously to specific populations that do not synthesize them in adequate amounts.

[b] AUG signals the "start" of translation when it occurs at the beginning of a gene.

[c] The stop codon is not an amino acid, but is included for completeness.

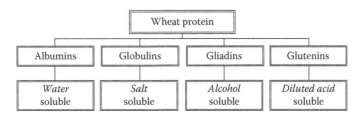

FIGURE 3.9 Wheat proteins.

As far as the practical utilization and commercial benefit for industries, two of these proteins are of maximum value in terms of food processing and food quality and, they are the water-soluble gliadins and glutenins.

FIGURE 3.10 Wheat flour fractions.

During the production of wheat starch, dough is firstly formed by mixing water and flour together, and then under the stream of water, the starch and the solubles are washed away and the gluten is left (Figure 3.10). Gluten generally contains 75%–80% protein which is mostly composed of the two proteins: gliadins and glutenins.

Wheat gluten has some properties specific to the baking quality due to its special amino acid composition and structure. Figure 3.11 can tell us the structures of the glutenins and gliadins contributing to the development of gluten. There is one important difference between gliadins and glutenins, that is, gliadins have an intramolecular disulfide linkage while glutenins have both inter- and intramolecular disulfide linkages. That is why the gliadins are compact and globular in shape and the glutenins are linear and have relatively higher molecular weight, 50,000 to millions in comparison with the molecular weight of the gliadins between 20,000 and 50,000. To some degree, refer to the left of the figure; you can imagine how tacky, sticky, and rubbery the gluten is. The subunits of the gliadins and glutenins are shown in Figure 3.12. Some researchers [41] have shown that the high molecular weight glutenin subunits are the exact parts that contribute to the development of the gluten, and that the α-gliadins are the real parts that cause the coeliac disease.

3.8.1.1 Functional Properties of Wheat Proteins

Wheat gluten proteins—gliadin and glutenin—determine viscoelastic properties of dough by forming a continuous network. These proteins are characterized by unusually high content of glutamine residues. Wheat gluten is a valuable source of plant protein. Functional properties of the wheat proteins include gelation, emulsification, solubility, taste, foaming, color, flavor formation after baking, water holding, and fat absorption capacities. Protein functionality is closely related to its

FIGURE 3.11 Major wheat proteins.

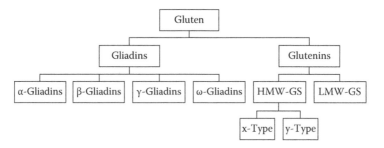

FIGURE 3.12 Subunits of gliadins and glutenins.

conformational state, which in turn is influenced by processing conditions. Wheat proteins are mostly used in their native form, and a small proportion is used in the form of protein concentrate and vital and nonvital wheat gluten isolates.

3.8.1.1.1 The Gluten: Water Solubilized Wheat Protein (SWP)

Although the insoluble nature of gluten is a desirable attribute in traditional applications of this protein in bread and baked products, its insolubility in water limits its usefulness in many other applications. Wheat gluten in modified forms (following modification by enzymes, chemicals, and physical treatments) becomes solubilized wheat protein (SWP) which has improved functional properties such as solubility, foaming properties, water holding capacity, fat absorption capacity, and emulsifying properties. Solubilized wheat protein could be used in cookies, muffins, and hamburger patties.

3.8.1.2 Gluten: Water Soluble Wheat Protein (SWP)

3.8.1.3 Gluten Proteins

The storage proteins of wheat are unique because they are also functional proteins. They do not have enzyme activity, but they are the only cereal proteins to form a strong and cohesive dough that will retain gas and produce light baked products. They can be easily isolated by removing starch and albumins/globulins by gently working dough under a small stream of water. After washing, a rubbery ball is left, which is called gluten. Traditionally, gluten proteins have been classified into four types according to their solubility [42] as follows:

- Albumins are soluble in water or dilute salt solutions and are coagulated by heat.
- Globulins are insoluble in pure water, but soluble in dilute salt concentrations and insoluble at high salt concentrations.
- Prolamins are soluble in aqueous alcohol.
- The glutelins are soluble in dilute acid or bases, detergents, or dissociating (urea) or reducing (beta-mercaptoethanol) agents.

An alternative classification to that described above has been proposed [41] based on composition and structure rather than solubility. Most of the physiologically active proteins (enzymes) are found in the albumin or globulin groups. Nutritionally, the albumins and globulins have a very good amino acid balance. They are relatively high in lysine, tryptophan, and methionine [43]. The prolamins were among the earliest proteins to be studied, with the first description of wheat gluten [44]. The prolamins have always been considered to be unique to the seeds of cereals and other grasses and unrelated to other proteins of seeds or other tissues. They have been given different names in different cereals, such as gliadin in wheat; avenins in oats; zeins in maize; secalins in rye; and hordein in barley. The glutelins are called glutenins in wheat. The gliadins and glutenins are the storage proteins of wheat endosperm and they tend to be rich in asparagine, glutamine, arginine, and proline [45–48], but very low in nutritionally important amino acids such as lysine, tryptophan, and methionine.

Cereals are an important protein source and are processed into bread, pasta and noodles, breakfast cereals, and fermented drink. For all these applications the quality is determined, to a greater or lesser extent, by the gluten proteins which account for about half of the total grain nitrogen. There are also opportunities to develop novel uses for cereal proteins in both the food and nonfood industries. It is important to study the wheat gluten proteins, particularly the deamidated SWP (soluble wheat protein) in relation to their structure and function.

3.8.1.4 Gliadin

The gliadins are divided into four groups, called alpha-, beta-, gamma-, and omega-gliadins, based on their electrophoretic mobility at low pH [49], but more than 30 components can be separated by two-dimensional electrophoretic procedures. The amino acid compositions of the alpha-, beta-, and gamma-gliadins are similar to each other and to that of the whole gliadin fraction [50]. The omega-gliadins contain little or no cysteine or methionine and only small amounts of basic amino acids. All gliadins are monomers with either no disulfide bonds (omega-gliadins) or intrachain disulfide bonds (alpha-, beta-, and gamma-gliadins). Although no complete sequences of omega-gliadins have been determined [50], purified a number of individual components from bread and pasta wheats and determined their relative molecular weights (MW), by sodium dodecyl sulfate polyacrylamide gel electrophoresis (SDS-PAGE); and the MW fell between 44,000 and 74,000, with most above 50,000. The alpha-, beta-, and gamma-gliadins have lower MWs, ranging between about 30,000 and 45,000 by SDS-PAGE and by amino acid sequencing. The latter approach has shown that the alpha- and beta-gliadins are closely related, and both are now usually referred to as alpha-type (in contrast to the gamma-type) gliadins. Although there is some overlap, the alpha-type gliadins generally have lower MWs than gamma-type gliadins [51].

In dough formation, the gliadins do not become covalently-linked into large elastic networks as the glutenins but act as a "plasticizer," promoting viscous flow and extensibility which are important rheological characteristics of dough. They may interact through hydrophobic interactions and hydrogen bonds [52]. Ultracentrifugation may be used to analyze the gliadin fraction extracted with aqueous ethanol. Although the protein was not homogeneous, a principal nonspherical component with a molecular weight of about 34,500 was present [53] using diffusion measurements showed gliadin to have a high degree of asymmetry with a calculated axial ratio of 8:1 while [53] using dielectric measurements showed an axial ratio of 13:1. Hydrodynamic studies showed that gliadins had a low intrinsic viscosity, indicating a compact and globular conformation [52]. Detailed x-ray scattering and electron microscopy studies of the two gluten fractions—the first gliadins and the second fraction a major glutenin subfraction called a_1-gluten (not the same as alpha-gliadin)—was reported by [53].

The gliadins appeared to have a doughnut shape, about 340 Å in diameter and 90 Å thickness, with a central hole of about 100 Å in diameter [53], suggesting that this shape resulted from the flattening of a hollow sphere during drying. Based on these studies wheat varieties were [52] divided into two types (Type 1 and Type 2) based on the major features of the gel electrophoretic patterns of their alpha-gliadins. A sub-fraction of the alpha-gliadins of Type 1 wheat variety was named A-gliadin which was found to be encoded by genes on chromosome 6A. The other alpha-gliadins were coded by genes on chromosomes 6A, 6B, and 6D. At pH 3, the A-gliadins were found to be partially unfolded and dissociated into monomers. As the pH was increased, the molecules became compactly folded and aggregated to form microfibrils [54]. At pH 5 and at an ionic strength of 0.005 M, electron microscopy revealed that the aggregates collected by ultracentrifugation had a microfibrillar structure about 80 Å in diameter and up to 3000 or 4000 Å long [54]. CD and ORD studies of A-gliadin dissolved in 10^{-5} M HCl at pH 5 showed that it was made up of 33%–34% alpha helix. The molecule was found to be more stable than most globular proteins. 65% of the helical structure present at 25°C remained when the temperature was raised to 90°C. The effects of heating were reversible when cooled [51].

The primary structure of alpha-type gliadins can be divided into five different domains [55]. Domain I consists of nonrepetitive N-terminal sequences and of repetitive sequences rich in glutamine, proline, and aromatic amino acids. Domain II contains a polyglutamine sequence with a maximum of 18 residues of glutamine. Domains III and V are homologous to the corresponding domains of gamma-type gliadins and low MW subunits of glutenin. Domain IV is unique to alpha-type gliadins and is rich in glutamine but poor in proline. Most alpha-type gliadins contain six cysteine residues, located in domains III (four residues) and V (two residues). Because of the monomeric character of alpha-type gliadins, and the absence of free sulfhydryl groups, it has been assumed that the cysteine residues are linked by three intramolecular disulfide bonds [55]. On the basis of the sequence homology to g-type gliadins and low MW subunits, definitive positions for disulfide bridges have been postulated for alpha-type gliadins [56].

The gamma-type gliadins are single monomeric proteins with only intrachain disulfide bonds and are considered to be the ancestral type of S-rich prolamin [41]. Complete amino acid sequences of several gamma-gliadins have been deduced from genomic and cDNA sequences [57]. These sequences showed a clear domain structure, with a nonrepetitive sequence of 14 residues at the N-terminus, an N-terminal repetitive domain based on a heptapeptide repeat motif (consensus Pro. Gln.Gln.Pro.Phe.Pro.Gln) and a nonrepetitive C-terminal domain which contained all the cysteine residues. Structural studies, using circular dichroism and structure prediction, indicated that the two domains adopt different conformations. Whereas the repetitive domain adopts a beta-reverse turn rich conformation, the nonrepetitive domain is rich in alpha-helix [56]. Scanning tunneling microscopy (STM) and small-angle x-ray scattering (SAXS) results indicated that the gamma-gliadins have a compact conformation, with axial ratios of approximately 1.5:1 [58]. A SAXS study of an S-poor prolamin, C hordein, with a repetitive motif (consensus Pro.Gln.Gln.Pro.Phe.Pro.Gln.Gln) similar to that of the repetitive domain of the gamma-gliadins (consensus Pro.Gln.Gln.Pro.Phe.Pro. Gln) has been reported [58]. The SAXS data indicated that in solution C hordein behaved as a worm-like chain (an extended conformation) with a high degree of flexibility. The similarity of the repeat motifs would indicate that the N-terminal domain of the g-gliadin may adopt a similar conformation in solution to that of C hordein.

Small deformation oscillatory measurements on gliadin between 50°C and 70°C showed the elastic modulus (G') to be roughly equal to the viscous modulus (G'') in magnitude, but a large increase in G' was observed at temperatures above 70°C [59]. The increase in the elastic component was attributed to cross-linking reactions occurring among the gliadin molecules, resulting in the formation of a network structure. The G' reached a peak at 120°C and the G'' increased to a plateau value in the temperature range 90°C–110°C. On further heating G'' fell to a minimum value at 120°C, whereas G' was at its maximum, indicating maximum structure build-up. At this point, the aggregation reaction appeared to have been completed and a highly crosslinked network formed. On increasing the temperature even further, a reduction in G' with a simultaneous peak in G'' was observed. This indicated a softening of the crosslinked gliadin at 130°C [59].

3.8.1.5 Glutenin

The glutenin proteins make up 35%–40% of flour protein and consist of subunits that form large polymers (MWs above 1×10^6 and possibly exceeding 1×10^7) stabilized by interchain disulfide bonds [60]. After reduction, both high molecular weight (HMW) and low molecular weight (LMW) subunits were observed by SDS-PAGE. The latter resembled the monomeric alpha-type gliadins in amino acid composition and had MWs in a similar range. The HMW subunit contains significantly less proline and had MWs value from 80,000 to 160,000 by PAGE. Extensive genetical analysis of the HMW subunits by Payne and coworkers [61], established that genes' coding for the HMW subunits are located on the long arms of chromosomes 1A, 1B, and 1D at complex *loci* designated Glu-A1, Glu-B1, and Glu-D1, respectively. By contrast, genes for the LMW subunits and the gliadins are located on the short arms of the same chromosomes (designated Gli-A1, Gli-B1, and Gli-D1, respectively). The HMW subunits consist of nonrepetitive domains of 88–104 and 42 residues at the N- and C-termini, respectively, separated by a longer repetitive domain (481–690 residues). Variation in the repetitive domain is responsible for most of the variation in the size of the whole protein, and it is based on random and interspersed repeats of hexapeptide and nonapeptide motifs, with tripeptides also present in x-type subunits only (Figure 3.12) [12]. Structure prediction indicated that the N- and C-terminal domains are predominantly alpha-helical, while the repetitive domains are rich in beta-turns.

Several models for the structure of wheat glutenin have been proposed. One of the earliest molecular models was that of Ewart [62]. He subsequently modified the model. Ewart's latest model shows one disulfide bond between two adjacent polypeptide chains of glutenins, which consist of linear polymers. Ewart pointed out that the rheological properties of dough are dependent on the presence of rheologically active disulfide bonds and thiol groups as well as on secondary forces in the concatenations [63].

Kasarda et al. [64] proposed an alternative model in which glutenin has only the intrachain disulfide bonds. They suggested that the intrachain disulfide bonds force glutenin molecules into specific conformations that facilitated interaction of adjacent glutenin molecules through noncovalent bonds, thereby causing aggregation.

A different model has been proposed [65] of functional glutenin complexes that contained both inter- and intrachain disulfide bonds. On the basis of results from SDS-PAGE, they proposed an aggregate of two types of glutenin complexes, I and II. In their model, glutenin I comprised subunits of molecular weight 6.8×10^4 and lower, held together through secondary forces, such as hydrogen bonds and hydrophobic interactions: glutenin II comprised cross-linked subunits of molecular weights above 6.8×10^4, linked by interchain disulfide bonds [66].

3.8.1.6 Deamidated Gluten

Wheat gluten is available as a by-product of the wheat starch industry and is used in food applications. The insolubility of gluten in aqueous solutions is one of the major limitations for its more extensive use in food processing for example in dairy products. Gluten insolubility is due to the high concentration of nonpolar amino acid residues such as proline and leucine and the polar but nonionizable residue glutamine, and to the low concentration of ionizable side chains such as lysine, arginine, glutamic acid, and aspartic acid. The interactions between glutamine and asparagine side chains through hydrogen bonds play an important role in promoting the association of gliadin and glutenin molecules [67]. Many researchers have developed methods for modifying the solubility and functional properties of gluten. Gluten modification via deamidation can be achieved in two ways, namely chemical deamidation (acid solubilization) under acidic conditions and high temperature [68] or enzymatic treatment [69]. Whether chemically or enzymatically induced, the deamidation of gluten proteins resulted in an increased charge density on the protein, causing changes in the protein conformation due to electrostatic repulsion. These charge-induced conformational changes resulted in enhanced surface hydrophobicity due to the exposure of hydrophobic residues (Figure 3.13). The increased surface hydrophobicity coupled with the presence of more negatively charged polar groups resulted in a modified protein with amphiphilic characteristics which made an ideal surface active agent for use as an emulsifier or foam stabilizer. Even though surface hydrophobicity increased, protein solubility was also enhanced due to decreased protein–protein interactions. Levels of deamidation, as low as 2%–6%, can enhance the functional properties of proteins [70]. Acid deamidation has been reported to leave behind an astringent mouthfeel, although this can be overcome by extraction with alkaline isopropanol and then isopropanol after deamidation [71].

Deamidation is a hydrolytic reaction, similar to the peptide-bond cleavage reaction, which is catalyzed by proteases. Deamidation is catalyzed by acids and bases (nucleophiles), and requires a water molecule. The general acid, HA, catalyzes the reaction by protonating the amido -NH- leaving group of the Asn side chain. A general base (the conjugate base, A- or hydroxide ion) can attack the carbonyl carbon of the amido group or activate another nucleophile by the abstraction of a proton for attack on the amide carbon. The transition state is inferred to be an oxyanion tetrahedral intermediate, whose stabilization by proton donors increases the rate of the reaction. The order of the acid- and base-catalyzed steps in Table 3.10 varies with reaction conditions, particularly pH. The pH of maximum stability of Asn and Gln in peptides is around pH 6. It has been shown [72] that how specific amino acid side chains are likely to function in catalyzing the deamidation of Asn and Gln in peptides and proteins. The Ser and Thr side chains can function as general acid groups, providing a proton to the leaving group or stabilizing the transition state. Asp, Glu, and His side chains are all nucleophiles at neutral pH, which can attack the carbonyl carbon of the amide side chain or function as general bases to activate nucleophiles. The gluten proteins lack Asp and do not have enough of His. Lys and Arg, which correlate with high deamidation rates when next to Asn and Gln in sequence, may stabilize the oxyanion intermediate. Gluten proteins lack Lys, although HMW glutenin has one Lys in the middle of Gln residues. However, there are numerous Arg residues next to Gln in HMW glutenins which may influence the reaction rate. The gamma-gliadins and the HMW glutenins, but not the alpha-gliadin

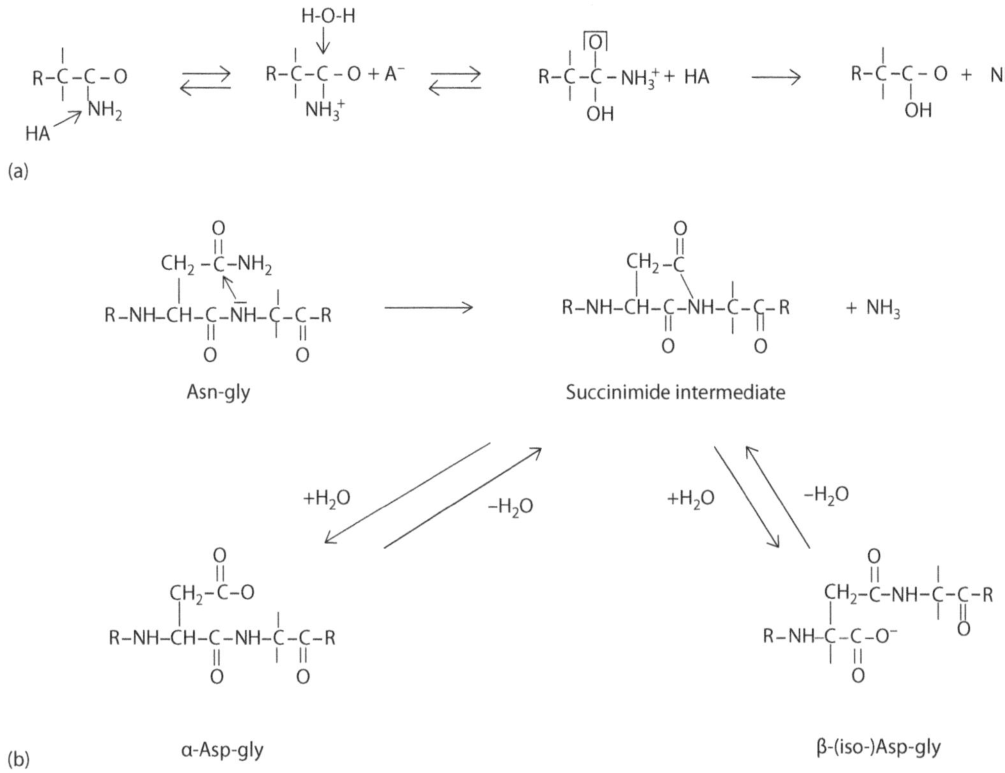

(a)

(b)

FIGURE 3.13 Deamidation according to the hydrolysis mechanism (a) and according to the succinimide intermediate mechanism (b). HA represents a general acid, R represents a protein chain. (From Metwalli, A.A.M. and van Boekel, M.A.J.S., *Food Chem.*, 61(1–2), 53, January 1998.)

and LMW glutenins have, terminal Gln which, may cyclize with a terminal carboxylate group to form an unstable anhydride which may break down to deamidated glutamate [73].

In studies on model tetra- and penta-peptides of different sequences, a broad range of deamidation rates was observed [74]. Several generalizations were made from the sequence dependence of these rate data:

- Polar residues preceding Asn and Gln increased deamidation rates.
- Neighboring Ser and Thr increased deamidation rates.
- Bulky, hydrophobic residues preceding Asn and Gln correlated with low deamidation rates.

Deamidation in peptides and proteins generally require the participation of a water molecule to go to completion. In peptides, there are minimal obstructions to water access to the labile amide. However, the more stable protein structures may limit access of water to the amide groups and so influence the rates of any deamidation reactions. Deamidation rates of Asn and Gln residues on the surface of the proteins will not be limited by water access, while those that occur in the interior of proteins may be. Such a limitation will be determined by the static protein structure and by the frequency with which buried Asn and Gln are exposed to solvent during rapid dynamic changes in the structure due to thermal motion.

3.8.1.7 Soluble Wheat Protein

SWP is the product of the deamidation (20%) of gluten produced commercially. The following properties were provided in the product specification information sheet.

3.8.1.7.1   Properties

SWP is a bland, white creamy low viscous sodium-salt of a soluble wheat protein isolate. By its unique combination of functional properties like emulsifying capacity, gelling, binding, and water-retention, the product can be used in several food applications like meat-preparations, soups, sauces, dressings, imitation dairy, etc. It can also be combined with other functional proteins like casein-ates, soy isolates, and may result in all types of synergies.

3.8.1.7.2   Chemical Data

Protein ($N \times 6.25$): 86%; moisture: 5%; fat: 8%; ash: 5%; carbohydrates: 1%; pH 6.5–7.5.

3.8.1.7.3   Solubility

A 1% protein solution was stirred for 30 min, the pH adjusted, and the solution was centrifuged at 5000 g for 20 min. The solubility (nitrogen solubility index) was expressed at different pH values in percent as

$$NSI = \frac{\text{Nitrogen content in the supernatant}}{\text{Total nitrogen content}}$$

which is used in making a solubility profile, for proteins.

3.8.1.7.4   Emulsifying Properties

Three characteristics may be distinguished as given in Table 3.8.

- **Emulsifying capacity (EC)**
  Soy-oil was continuously added to 100 mL of a 1% protein solution up to the reversion of the emulsion. The reversion is measured by conductivity.

$$E = \frac{(\text{g oil})}{(\text{g protein})} = \frac{\text{Amount of oil added (in grams)}}{1 \text{ g of protein}}$$

  A mixer and conductivity meter instruments may be used.
- **Emulsifying activity (EA)**
  An emulsion of 100 g soy-oil in 100 mL of a 1% protein solution was centrifuged at 1300 g for 5 min. The volume of emulsion phase after centrifugation was compared to the total volume.

TABLE 3.8
Emulsifying Properties at Different pH Values

pH	EC (g oil/g protein)	EA (%)	ES (%)
5	470	52.9	100
6	570	51.7	100
7	620	52.5	100

Note: The emulsifying capacity, in relation to sodium caseinate, is comparable at pH 7 and higher at pH 5 and 6.

TABLE 3.9
Effect of Temperature and pH on Gel Strength

Temperature (°C)	pH	Gel Strength
95	6	45
	7	48
105	6	180
	7	260
120	6	480
	7	370

$$EA\ (\%) = \frac{\text{Volume emulsion phase}}{\text{Total volume}}$$

- **Emulsifying stability (ES)**
  This is the ratio of the EA after 30 min of heating at 80°C to the EA.

$$ES(\%) = \frac{EA_{(30'-80°C)}}{EA} \times 100$$

3.8.1.7.5 Gelling Properties

A 15% protein solution was heated in 200 mL cans for 60 min. The gel strength, expressed in grams, was measured at room temperature with a Stevens Texture Analyzer [74] (cylindrical plunger 1") to a penetration depth of 40 mm at a speed of 2.0 mm/s (Table 3.9).

SWP (15%–20% w/v) formed a gel at 115°C. Use of reducing agents such as cysteine and bisodium sulfite (Na_2SO_3) reduced the gelling temperature to 65°C–70°C, accompanied by an increase in gel strength [74].

3.8.1.7.6 Viscosity

The solutions were prepared by mixing for 13 min (3 min at 2000 rpm, 10 min at 1500 rpm) 15–30 g of product per 100 mL (for, respectively, 15% and 20%). The pH was adjusted and the viscosity was measured within the first 3 min with a Brookfield viscometer at 20 rpm (Table 3.10) [74].

3.8.1.8 Intergenetic Comparisons of Cereal Prolamines

On the basis of information presented by several authors [75,76], a tree diagram summarizing currently accepted genetic relationships of cereal grains was constructed by Bietz [76].

Wheat, barley, and rye are classified in the same subfamily (*Festucoideae*) and tribe (*Triticeae*) of the grass family (*Gramineae*), and this close relationship is reflected in the structures of their prolamin storage proteins. Only in wheat, however, do these proteins form a cohesive mass (gluten). Barley and rye are diploids, each with seven pairs of chromosomes, while wheat species are diploid, tetraploid, or hexaploid (Figure 3.14).

The total number of chromosomes is 42 for hexaploids (used for bread making), 28 for tetraploids (durum wheat used for pasta), and 14 for diploids (primitive wheat).

The last two originated from the hybridization of related wild diploid species. Bread wheat (*Triticum aestivum*) is an

TABLE 3.10
Effect of Concentration and pH on Viscosity

Concentration (%)	pH	Viscosity (mPa s)
15	7	1000
20	6	6000
25	7	3300
30	8	1900

Chromosome Number in Each Genome	Genome		
	A	B	D
1	1A	1B	1D
2	2A	2B	2D
3	3A	3B	3D
4	4A	4B	4D
5	5A	5B	5D
6	6A	6B	6D
7	7A	7B	7D

FIGURE 3.14 Chromosome structure of wheat.

allohexaploid with three genomes (each having seven pairs of chromosomes called A, B, and D). The A and D genomes are thought to be derived from a wild diploid wheat (*Triticum* sp.) and the related wild grass *Aegilops squarrosa* (*T. tauschii*), respectively. The origins of the B genome are not known. Durum or pasta wheats (*T. durum*) are tetraploid, with only the A and B genomes. Because the A, B, and D genomes are derived from related diploid species, they have genes that encode related but not identical proteins. The interactions of proteins encoded by the different genomes are important in determining the precise physical and technological properties of dough produced from pasta and bread wheats. Each chromosome consists of a long arm and a short arm joined by a centromere.

The genes coding for the gliadin proteins are located on the short arms of groups 1 and 6 chromosomes [77,78]. The group 1 chromosomes control all the omega-gliadins, most of the gamma-gliadins, and a few of the beta-gliadins, whereas genes on the group 6 chromosomes code for all the alpha-gliadins, most of the beta-gliadins, and some of the gamma-gliadins. The genes coding for gliadin proteins occur as a single complex *locus* on each of the short arms of groups 1 and 6 chromosomes rather than at two or more *loci*. The HMW glutenin subunits are coded by genes (*loci*) on the arms of chromosomes 1A, 1B, and 1D. These *loci* are designated Glu-A1, Glu-B1, and Glu-D1, respectively. Electrophoretic studies have revealed appreciable polymorphism in the number and mobility of the HMW subunits in different wheat cultivars. That is, the genes on the chromosome 1 long arms show multiple-allelism. Based on electrophoretic studies, the authors of [79] proposed that there were two main types of subunits, the x type (of high relative molecular weight, MW) and the y type (low MW). This subdivision has been supported by chemical and genetic evidence (Figure 3.15).

A particular subunit is specified by noting its chromosome, followed by its classification as x or y, and finally a number (designating the protein subunit), this number increases with decreasing MWs. In contrast to the HMW subunits, the LMW glutenin subunits are encoded by genes on the short arms of the chromosomes 1A, 1B, and 1D, these *loci* are being designated Glu-A3, Glu-B3, and Glu-D3. The allocation of the genes coding for particular proteins has been achieved by the use of genetic variants in combination with electrophoresis. Examples of these variants are aneuploids—lines that are deficient in a single chromosome. Much of the pioneering work in this area was conducted by Sears [80]. In simple terms, when a particular chromosome is missing, this may coincide with the disappearance of a certain band or bands in the electrophoretic pattern of the

(a) Genes for HMW glutenins = Glu-A1, Glu-B1, and Glu-D1.
(b) Genes for LMW glutenins = Glu-A3, Glu-B3, and Glu-D3; Gli-A1, B1, and D1 = all omega gliadin, most of the gamma gliadin, and a few of the beta gliadin; Gli-A2, B2, and D2 = all alpha gliadin, most of beta gliadin, and some of gamma gliadin.

FIGURE 3.15 Chromosomal locations of major protein groups of hexaploid wheats.

wheat protein, thus identifying the chromosome that carries the gene(s) responsible for the synthesis of the missing protein(s).

Storage proteins are usually synthesized in very large amounts and consequently need to be stored in a highly concentrated form and in a subcellular compartment in which they are separate from other metabolic processes. This is achieved by a combination of specific solubility properties and deposition into protein bodies (1–20 mm). The protein bodies have been found both in the aleurone cells and the endosperm. The solubility properties are determined by the primary structures of the individual proteins and their interactions by noncovalent forces (notably hydrogen bonds and hydrophobic interactions) and by covalent disulfide bonds. Storage proteins are synthesized on the rough endoplasmic reticulum with a signal peptide that directs the nascent polypeptide into the lumen of the endoplasmic reticulum and is itself removed by proteolytic cleavage. The signal sequences of the storage proteins all display the characteristics of the transported proteins in other organisms [81] such as length, hydrophobicity, and the presence of an amino acid with a small uncharged side chain prior to the N-terminus of the mature protein. There is some sequence homology between the signal sequences for closely related zeins [82]. Thus, it appears that the features of this protein segment that have been conserved in evolution are the nature of the amino acids and perhaps the tertiary structure of the signal sequence rather than the actual amino acid sequence.

By analogy with animal systems, the function of signal sequences in plant storage proteins is to facilitate the translocation of the storage protein into the lumen of the endoplasmic reticulum (ER) as the first step in intracellular transport. Protein folding and disulfide bond formation are considered to occur within the lumen of the endoplasmic reticulum, and may be assisted by molecular chaperones and by the enzyme protein disulfide isomerase respectively [83]. The precise mechanism of intracellular transport of storage proteins from their site of synthesis to their site of deposition are still largely unknown but a two-way hypothesis has been proposed by Shewry [45].

3.8.2 Fish Proteins

The proteins in fish muscle tissue can be divided into three groups:

- *Structural proteins* (actin, myosin, tropormyosin, and actomyosin) which constitute 70%–80% of the total protein content (compared with 40% in mammals). These proteins are soluble in neutral salt solutions of fairly high ionic strength (0.5 M).
- *Sarcoplasmic proteins* (myoalbumin, globulin, and enzymes) which are soluble in neutral salt solutions of low ionic strength (<0.15 M). This fraction constitutes 25%–30% of the protein.
- *Connective tissue proteins* (collagen), which constitute approximately 3% of the protein in teleostei and about 10% in elasmobranchii (compared with 17% in mammals).

Structural proteins make up the contractile apparatus responsible for muscle movement. The amino-acid composition is roughly similar to corresponding proteins in mammalian muscle, although the physical properties can differ slightly.

When the proteins are denatured under controlled conditions, their properties may be utilized for technological purposes. A good example is the production of surimi-based products in which the gel-forming ability of the myofibrillar proteins is used. After salt and stabilizers are added to a washed, minced preparation of muscle proteins and after a controlled heating and cooling procedure, the proteins form a very strong gel.

The majority of sarcoplasmic proteins are enzymes participating in cell metabolism, such as the anaerobic energy conversion from glycogen to ATP. If the organelles within the muscle cells are broken, this protein fraction may also contain the metabolic enzymes localized inside the endoplasmatic reticulum, mitochondria, and lysosomes.

The fact that the composition of the sarcoplasmic protein fraction changes when the organelles are broken was suggested as a method for differentiating fresh from frozen fish, under the assumption that the organelles were intact until freezing. However, it was later stated that these methods should be used with great caution as some of the enzymes are liberated from the organelles during iced storage of fish as well.

The proteins in the sarcoplasmic fraction are excellently suited to distinguishing fish species as each species has a characteristic band pattern when separated by the isoelectric focusing method.

The chemical and physical properties of collagen proteins are different in tissues such as skin, swim bladder, and the myocommata muscle. In general, collagen fibrils form a delicate network structure with varying complexity in the different connective tissues in a pattern similar to that found in mammals. However, the collagen in fish is much more thermolabile and contains fewer, but more labile, cross-links than collagen from the warm-blooded vertebrates.

Different fish species contain varying amounts of collagen in body tissues. This has led to a theory that the distribution of collagen may reflect the swimming behavior of the species. Furthermore, the varying amounts and varying types of collagen in different fishes may also have an influence on the textural properties of fish muscle.

Fish proteins contain all the essential amino acids and, like milk, eggs, and mammalian meat proteins, have a very high biological value (Table 3.11) [84].

Cereal grains are usually low in lysine and/or sulfur-containing amino acids (methionine and cysteine), whereas fish protein is an excellent source of these amino acids. A supplement of fish can therefore significantly raise the biological value in cereal-based diets.

In addition to the fish proteins already mentioned there is a renewed interest in specific protein fractions that can be recovered from by-products, particularly in the viscera. One such example is the basic protein or protamines found in the milt of the male fish, which can contain as much as 65% arginine. The best sources are salmonids and herring, whereas ground fish such as cod are not found to contain protamines. The extreme basic character of protamines makes them interesting for several reasons. As they adhere to most other proteins less basic, they can enhance the functional properties of other food proteins (if all the lipids present in the milt are removed from the protein preparation to avoid off-flavor development in the finished products). But the most promising feature of basic proteins is their ability to prevent growth of microorganisms.

3.8.2.1 Non-Protein Nitrogen (NPN)-Containing Compounds in Foods

The N-containing extractives are defined as the water-soluble, low molecular weight, nitrogen-containing compounds of non-protein nature. This NPN-fraction (non-protein nitrogen) constitutes from 9% to 18% of the total nitrogen in the teleosts.

TABLE 3.11
Essential Amino Acids from Animal Protein

Essential Amino Acids (Percentage) in Various Proteins

Amino Acid	Fish	Milk	Beef	Eggs
Lysine	8.8	8.1	9.3	6.8
Tryptophan	1.0	1.6	1.1	1.9
Histidine	2.0	2.6	3.8	2.2
Phenylalanine	3.9	5.3	4.5	5.4
Leucine	8.4	10.2	8.2	8.4
Isoleucine	6.0	7.2	5.2	7.1
Threonine	4.6	4.4	4.2	5.5
Methionine-cystine	4.0	4.3	2.9	3.3
Valine	6.0	7.6	5.0	8.1

The major components in this fraction are volatile bases such as ammonia and trimethylamine oxide (TMAO), creatine, free amino acids, nucleotides, and purine bases, and in the case of cartilaginous fish, urea.

Quantitatively, the main component of the NPN-fraction is creatine. In resting fish, most of the creatine is phosphorylated and supplies energy for muscular contraction.

TMAO represents a characteristic and important element of the NPN-fraction in marine species. It is found in all marine fish species in quantities from 1% to 5% of the muscle tissue (dry weight) but is virtually absent from freshwater species and from terrestrial organisms. One exception was found in a study of Nile perch and tilapia from Lake Victoria, where as much as 150–200 mg TMAO/100 g of fresh fish was found.

The amount of TMAO in the muscle tissue depends on the species, season, fishing ground, etc. In general, the highest amount is found in elasmobranchs and squid (75–250 mg N/100 g); cod have somewhat less (60–120 mg N/100 g) while flatfish and pelagic fish have the least. Pelagic fish (sardines, tuna, and mackerel) have their highest concentration of TMAO in the dark muscle while demersal, white-fleshed fish have a much higher content in the white muscle.

The NPN-fraction also contains a fair amount of free amino acids. These constitute 630 mg/100 g light muscle in mackerel (*Scomber scombrus*), 350–420 mg/100 g in herring (*Clupea harengus*) and 310–370 mg/100 g in capelin (*Mallotus villosus*). The relative importance of the different amino acids varies among species. Taurine, alanine, glycine, and imidazole-containing amino acids seem to dominate in most fish. Of the imidazole-containing amino acids, histidine has attracted much attention because it can be decarboxylated microbiologically to histamine. Active, dark-fleshed species such as tuna and mackerel contain a high content of histidine.

3.8.3 MILK PROTEIN FRACTIONS

The nitrogen content of milk is distributed among caseins (76%), whey proteins (18%), and NPN (6%). This does not include the minor proteins and glycoproteins that are associated with the fat globule membranes (FGM). This nitrogen distribution can be determined by the *Rowland fractionation* [85] method; given in Table 3.12.

1. Precipitation at pH 4.6—separates caseins from whey nitrogen.
2. Precipitation with sodium acetate and acetic acid (pH 5)—separates total proteins from whey NPN.

TABLE 3.12
Concentration of Proteins in Milk

	Grams/Liter	% of Total Protein
Total protein	33	100
Total caseins	26	79.5
Alpha s1	10	30.6
Alpha s2	2.6	8.0
Beta	9.3	28.4
Kappa	3.3	10.1
Total whey proteins	6.3	19.3
Alpha lactalbumin	1.2	3.7
Beta lactoglobulin	3.2	9.8
BSA	0.4	1.2
Immunoglobulins	0.7	2.1
Proteose peptone	0.8	2.4

3.8.3.1   Milk Enzymes

Enzymes are a group of proteins that have the ability to catalyze chemical reactions and the speed of such reactions. The action of enzymes is very specific. Milk contains both *indigenous* and *exogenous* enzymes. Exogenous enzymes mainly consist of heat-stable enzymes produced by psychrotrophic bacteria: lipases, and proteinases. There are many indigenous enzymes that have been isolated from milk. The most significant group is the hydrolases:

- Lipoprotein lipase
- Plasmin
- Alkaline phosphatase

Lipoprotein lipase (LPL): A lipase enzyme splits fats into glycerol and free fatty acids. This enzyme is found mainly in the plasma in association with casein micelles. The milk-fat is protected from its action by the FGM. If the FGM has been damaged, or if certain cofactors (blood serum lipoproteins) are present, the LPL is able to attack the lipoproteins of the FGM. Lipolysis may be caused in this way.

Plasmin: Plasmin is a proteolytic enzyme; it splits proteins. Plasmin attacks both β-casein and α_{s2}-casein. It is very heat stable and responsible for the development of bitterness in pasteurized milk and UHT processed milk. It may also play a role in the ripening and flavor development of certain cheeses, such as Swiss-cheese.

Alkaline phosphatase: Phosphatase enzymes are able to split specific phosporic acid esters into phosphoric acid and the related alcohols. Unlike most milk enzymes, it has a pH and temperature optima differing from physiological values; pH of 9.8. The enzyme is destroyed by minimum pasteurization temperatures and therefore, a phosphatase test can be done to ensure proper pasteurization.

3.8.3.2   Cheese Proteins

Amino acids: Cheese contains at least 20 amino acids. In the cheese made from pasteurized milk as the ripening age increases the concentration of the amino acids also increases. The most commonly identified amino acids in cheeses are as follows: Asp, Glu,Asn, Thr,Val, Phe, Lys, Gly, Ala, Met, Leu, His, Tyr, Pro, Ile, Ser, Gln, Tau (Taurine), GABA (Gama Amino butyric acis), Arg, Cys, Trp, Orn (Ornithine) [86].

"Proteose" is a water-soluble compound that is produced during digestion by the hydrolytic breakdown of proteins during cheese ripening.

"Peptones" are small polypeptides that are intermediate products in the hydrolysis of proteins. The term is often used for any partial hydrolysate of proteins, e.g., bacteriological peptone, which is used as a medium for the growth of microorganisms. Peptones are water-soluble protein derivatives obtained by the partial hydrolysis of a protein by an acid or enzyme during digestion.

3.8.3.3   Bioactive Peptides in Fermented Milk Products

Bioactive peptides have been defined as specific protein fragments that have a positive impact on body functions or conditions and may influence health. Upon oral administration, bioactive peptides, may affect the major body systems—namely, the cardiovascular, digestive, immune, and nervous systems. The beneficial health effects may be classified as antimicrobial, antioxidative, antithrombotic, antihypertensive, antimicrobial, and immunomodulatory [87,88].

The activity of these biofunctional peptides is based on their inherent amino acid composition and sequence. The size of active sequences may vary from 2 to 20 amino acid residues, and many peptides are known to have multifunctional properties [89], e.g., peptides from the sequence 60–70 of β-casein show immunostimulatory, opioid, and angiotensin I converting enzyme (acetyl choline esterase [ACE]) -inhibitory activities. This sequence has been defined as a strategic zone [90,91]. The sequence is protected from proteolysis because of its high hydrophobicity and the presence of proline residues. Other examples of the multi-functionality of milk-derived peptides include the

α_{s1}-casein fraction 194–199 showing immunomodulatory and ACE-inhibitory activity, the opioid peptides α- and β-lactorphin also exhibiting ACE-inhibitory activity and the calcium-binding phosphopeptides (CPPs), which possess immunomodulatory properties [87].

3.8.3.4   Source of Bioactive Peptides

Milk is a rich source of protein. Casein and whey proteins are the two main protein groups in milk; caseins comprise about 80% of the total protein content in bovine milk and are divided into α-, β-, and κ-caseins. Whey protein is composed of β-lactoglobulin, α-lactalbumin, immunoglobulins (IgGs), glycomacropeptides, bovine serum albumin, and minor proteins such as lactoperoxidase, lysozyme, and lactoferrin. Each of the subfractions found in casein or whey has its own unique biological properties. Milk proteins can be degraded into numerous peptide fragments by enzymatic proteolysis and serve as sources of bioactive peptides.

3.8.3.5   Production of Bioactive Peptides

Bioactive peptides are inactive within the sequence of the parent protein and can be released in three ways: (a) enzymatic hydrolysis by digestive enzymes, (b) food processing, and (c) proteolysis by enzymes derived from microorganisms or plants.

3.8.3.6   Occurrence of Bioactive Peptides in Fermented Dairy Products

Bioactive peptides can be generated during milk fermentation by the proteolytic activity of starter cultures. As a result, peptides with various bioactivities can be found in the end-products, such as various cheeses and fermented milks. These traditional dairy products may under certain conditions have specific health effects when ingested as part of the daily diet. A list of bioactive peptides found in dairy products is given in Table 3.13.

3.8.4   MUSCLE PROTEINS

We normally recognize three basic types of vertebrate muscle:

The *voluntary skeletal muscle* is that which is under conscious control. Each fiber is an enormous, multi-nucleate cell, formed by fusing hundreds of *myoblasts* end-to-end. They show a striated pattern, reflecting the regular arrangement of sarcomeres within each cell.

The *cardiac muscle* is similar to the skeletal muscle but is not under conscious control. These mono-nucleate cells are much smaller, but still show a striated pattern.

The *smooth muscle* is closer to non-muscle cells. No regular striations are visible and the contractions are much slower. Smooth muscle is found in the blood vessels, gut, skin, eye pupils, urinary, and reproductive tracts.

3.8.4.1   Skeletal Muscle Fiber Types

Voluntary muscles contain a variety of *fiber types* which are specialized for particular tasks. Most muscles contain a mixture of fiber types although one type may predominate. The pattern of gene expression within each voluntary muscle cell is governed by the firing pattern of its single motor neurone. Motor neurones branch within their target muscle and thereby control several muscle fibers, called a *motor unit*. The high precision eye muscles have only a few fibers in each motor unit, but the muscles in the back have thousands. All the cells in a motor unit contract in unison and they all belong to the same fiber type [101] (see Table 3.14).

3.9   AMINO ACID PROFILES OF FOOD PROTEINS

Table 3.15 shows representative amino acid profiles of some common foods and dietary protein supplements. The percentages are averages of several commercial products. Casein and whey are milk

TABLE 3.13
Bioactive Peptides Identified from Milk Products

Product	Origin	Biofunctional Role	Reference
Cheddar cheese	α_{s1}- and β-casein fragments	Several phosphopeptides with a range of properties including the ability to bind and solubilize minerals	[92]
Italian cheeses: Mozzarella, Crescenza, Italico, Gorgonzola	β-CN f (58–72)	ACE inhibitory	[93]
Yoghurt-type products	α_{s1}-, β-, and κ-CN fragments	ACE inhibitory	[94]
Gouda cheese	α_{s1}-CN f (1–9), β-CN f (60–68)	ACE inhibitory	[95]
Festivo cheese	α_{s1}-CN f (1–9), f (1–7), f (1–6)	ACE inhibitory	[96]
Emmental cheese	α_{s1}- and β–casein fragments	Immunostimulatory, mineral binding and solubilizing, antimicrobial	[97]
Manchego cheese	Ovine α_{s1}-, α_{s2}-, and β-casein fragments	ACE inhibitory	[98]
Sour milk	β-CN f (74–76, f (84–86), κ-CN f (108–111)	ACE inhibitory/ antihypertensive	[99]
Dahi	Ser-Lys-Val-Tyr-Pro	ACE inhibitory	[100]

Note: The major antimicrobial proteins of milk are lysozyme, lactoferrin (Lf), lactoperoxidase (LP), and the immunoglobulins.

proteins. Casein is the protein that precipitates from milk when curdled with rennet; it is the basis for making cheese. Whey is the watery part of milk that remains after the casein is separated [84].

The amino acid analyses of food products report *cystine* instead of *cysteine*. Cystine is an amino acid that is formed from the oxidation of two molecules of cysteine.

$$\text{HOOC-CH(NH}_2)\text{CH}_2\text{-S-S-CH}_2\text{CH(NH}_2)\text{COOH}$$

Cystine

Egg white protein is considered to have one of the best amino acid profiles for human nutrition. Plant proteins generally have lower content of some essential amino acids such as lysine and methionine. Soy protein is one of the best plant proteins, but nevertheless, the most prominent difference in it is the proportion of the essential sulfur-containing amino acid methionine. Egg white protein has approximately three times more methionine than is found in soy protein. The yeast information is for "brewer's yeast" (*Saccharomyces cervisiae*).

In the animal kingdom, peptides and proteins regulate metabolism and provide structural support. The cells and the organs of the human body are controlled by peptide hormones (see Chapter 7 for more details and Table 3.4). Insufficient protein in the diet may prevent the body from producing adequate levels of peptide hormones and structural proteins to sustain normal bodily functions. Individual amino acids serve as neurotransmitters and modulators of various physiological processes, while proteins catalyze most chemical reactions in the body, regulate gene expression, regulate the immune system, form the major constituents of muscle, and are the main structural elements of cells. Deficiency of good

TABLE 3.14
Skeletal Muscle Proteins

Component	M. Weight (kDa)	Function
Actin	40	Major component of the thin filaments. Actin is an important cytoskeletal protein in non-muscle cells. Six human isoforms are known
α-Actinin	102	Cross-links actin at the Z disks
Cap-Z		Caps the plus ends of actin filaments at the Z-disk
Caldesmon	150	Thin filament regulation in smooth muscle
Desmin	53	Intermediate filament protein. The muscle form links together adjacent myofibrils at the Z-disk
Dystrophin	426	Anchors some actin filaments to the sarcolemma (resembles spectrin in red cells). Defective or absent in muscular dystrophies. Utrophin is similar
Myomesins	190	Bind to titin
	165	Cross link adjacent thick filaments into a hexagonal array in the middle of each sarcomere
Myosin	200	Adult fast skeletal muscle heavy chains
	200	Adult skeletal muscle heavy chains
	200	Embryonic skeletal muscle heavy chains
	200	Fetal skeletal muscle heavy chains
	200	Cardiac muscle alpha (atrium) heavy chains
	200	Cardiac muscle beta (ventricle+type 1 skeletal) heavy chains
	200	Perinatal skeletal muscle heavy chains
	200	Smooth muscle heavy chains
	200	Extra-ocular muscle heavy chains
	20	Fast skeletal alkali light chains
	20	Cardiac/skeletal/smooth DTNB light chains
	20	Ventricular/slow skeletal alkali light chains
	20	Cardiac atrial/fetal skeletal alkali light chains
	20	Fetal skeletal DTNB light chains
Nebulin	700	Actin binding protein, defines the length of the thin filaments
Titin	2700	Elastic element, links thick filaments to Z-lines. Titin is the largest protein in the human genome
Tropomyosin	2×35	Thin filament component, seven repeats, each with a major and a minor actin binding site
Tropomodulin		Caps the minus ends of thin filaments
Troponin-C	18	Cardiac and slow twitch skeletal muscle
		Fast twitch skeletal muscle
		Binds calcium and troponin-T
Troponin-I	22	Slow twitch skeletal
		Fast twitch skeletal
		Cardiac muscle
		Binds to actin and troponin-T. Inhibits the actin–myosin interaction unless calcium is bound to troponin-C
Troponin-T	37	Slow skeletal
		Cardiac
		Fast skeletal
		Binds to tropomyosin, troponin-I, and troponin-C
Vinculin	130	Binds to α-actinin

TABLE 3.15
Percentage (%) by Weight of Amino Acid in a Food Protein

	Protein							
Amino Acid	Egg White	Tuna	Beef	Chicken	Whey	Casein	Soy	Yeast
Alanine	6.6	6.0	6.1	5.5	5.2	2.9	4.2	8.3
Arginine	5.6	6.0	6.5	6.0	2.5	3.7	7.5	6.5
Aspartic acid	8.9	10.2	9.1	8.9	10.9	6.6	11.5	9.8
Cystine	2.5	1.1	1.3	1.3	2.2	0.3	1.3	1.4
Glutamic acid	13.5	14.9	15.0	15.0	16.8	21.5	19.0	13.5
Glycine	3.6	4.8	6.1	4.9	2.2	2.1	4.1	4.8
Histidine[a]	2.2	2.9	3.2	3.1	2.0	3.0	2.6	2.6
Isoleucine[a]	6.0	4.6	4.5	5.3	6.0	5.1	4.8	5.0
Leucine[a]	8.5	8.1	8.0	7.5	9.5	9.0	8.1	7.1
Lysine[a]	6.2	9.2	8.4	8.5	8.8	3.8	6.2	6.9
Methionine[a]	3.6	3.0	2.6	2.8	1.9	2.7	1.3	1.5
Phenylalanine[a]	6.0	3.9	3.9	4.0	2.3	5.1	5.2	4.7
Praline	3.8	3.5	4.8	4.1	6.6	10.7	5.1	4.0
Serine	7.3	4.0	3.9	3.4	5.4	5.6	5.2	5.1
Threonine[a]	4.4	4.4	4.0	4.2	6.9	4.3	3.8	5.8
Tryptophan[a]	1.4	1.1	0.7	1.2	2.2	1.3	1.3	1.6
Tyrosine	2.7	3.4	3.2	3.4	2.7	5.6	3.8	5.0
Valine[a]	7.0	5.2	5.0	5.0	6.0	6.6	5.0	6.2

[a] Essential amino acids.

quality protein in the diet may contribute to seemingly unrelated symptoms such as sexual dysfunction, blood pressure problems, fatigue, obesity, diabetes, frequent infections, digestive problems, and bone mass loss leading to osteoporosis. Severe restriction of dietary protein causes kwashiorkor which is a form of malnutrition characterized by loss of muscle mass, growth failure, and decreased immunity.

Allergies are generally caused by the effect of foreign proteins on the human body. Proteins that are ingested are broken down into smaller peptides and amino acids by digestive enzymes called "proteases." Allergies to foods may be caused by the inability of the body to digest specific proteins. Cooking denatures (inactivates) dietary proteins and facilitates their digestion. Allergies or poisoning may also be caused by exposure to proteins that bypass the digestive system by inhalation, absorption through mucous tissues, or injection by bites or stings. Spider and snake venoms contain proteins that have a variety of neurotoxic, proteolytic, and hemolytic effects.

Many structures of the body are formed from protein. *Hair* and *nails* are made of *keratins* which are long protein chains containing a high percentage (15%–17%) of the amino acid cysteine. Keratins are also components of animal claws, horns, feathers, scales, and hooves. *Collagen* is the most common protein in the body and comprises approximately 20%–30% of all body proteins. It is found in tendons, ligaments, and many tissues that serve structural or mechanical functions. Collagen consists of amino acid sequences that coil into a triple helical structure to form very strong fibers. Glycine and proline account for about 50% of the amino acids in collagen. *Gelatin* is produced by boiling collagen for a long time until it becomes water soluble and gummy. *Tooth enamel* and *bones* consist of a protein matrix (mostly collagen) with dispersed crystals of minerals such as apatite, which is a phosphate of calcium. By weight, bone tissue is 70% mineral, 8% water, and 22% protein. *Muscle* tissue consists of approximately 65% *actin* and *myosin*, which are the contractile proteins that enable muscle movement. *Casein* is a nutritive phosphorus-containing protein present in milk. It makes up approximately 80% of the protein in milk and contains all the common amino acids.

3.10 RECENT DEVELOPMENTS IN PROTEIN NUTRITIONAL QUALITY EVALUATION

3.10.1 PROTEIN FUNCTIONS IN HUMAN NUTRITION

Proteins act as *enzymes*, *hormones*, and *antibodies*. They maintain fluid balance and acid and base balance. They also transport substances such as oxygen, *vitamins*, and *minerals* to target cells throughout the body. Structural proteins, such as collagen and keratin, are responsible for the formation of bones, teeth, hair, and the outer layer of skin, and they help maintain the structure of blood vessels and other tissues. In contrast, motor proteins use *energy* and convert it into some form of mechanical work (e.g., dividing cells and contracting muscle).

Enzymes are proteins that facilitate chemical reactions without being changed in the process. The inactive form of an enzyme is called a proenzyme. Hormones (chemical messengers) are proteins that travel to one or more specific target tissues or organs, and many have important regulatory functions. *Insulin*, for example, plays a key role in regulating the amount of *glucose* in the blood. The body manufactures antibodies (giant protein molecules), which combat invading antigens. Antigens are usually foreign substances such as *bacteria* and *viruses* that have entered the body and could potentially be harmful. Immunoproteins, also called immunoglobulins or antibodies, defend the body from possible attack by these invaders by binding to the antigens and inactivating them.

Proteins help to maintain the body's fluid and electrolyte balance. This means that proteins ensure that the proper types and amounts of fluid and minerals are present in each of the body's three fluid compartments. These fluid compartments are *intracellular* (contained within cells), *extracellular* (existing outside the cell), and *intravascular* (in the blood). Without this balance, the body cannot function properly.

Proteins also help to maintain balance between acids and bases within body fluids. The lower a fluid's pH, the more acidic it is. Conversely, the higher the pH, the less acidic the fluid is. The body works hard to keep the pH of the blood near 7.4 (neutral). Proteins also act as carriers, transporting many important substances in the bloodstream for delivery throughout the body. For example, a *lipoprotein* transports fat and cholesterol in the blood.

Proteins are vital to basic cellular and body functions, including cellular regeneration and repair, tissue maintenance and regulation, *hormone* and enzyme production, fluid balance, and the provision of energy.

3.10.1.1 Cellular Repair and Tissue Provisioning

Protein is an essential component for every type of cell in the body, including muscles, bones, organs, tendons, and ligaments. Protein is also needed in the formation of enzymes, *antibodies*, hormones, blood-clotting factors, and blood-transport proteins. The body is constantly undergoing renewal and repair of tissues. The amount of protein needed to build new tissue or maintain structure and function depends on the rate of renewal or the stage of growth and *development*. For example, the intestinal tract is renewed every couple of days, whereas blood cells have a life span of 60–120 days. Furthermore, an infant will utilize as much as one-third of the dietary protein for the purpose of building new connective and muscle tissues.

3.10.1.2 Hormone and Enzyme Production

Amino acids are the basic components of hormones that are essential chemical signaling messengers of the body. Hormones are secreted into the bloodstream by endocrine glands, such as the thyroid gland, adrenal glands, pancreas, and other ductless glands, and regulate bodily functions and processes. For example, the hormone *insulin*, secreted by the pancreas, works to lower the blood glucose level after meals. Insulin is made up of 48 amino acids.

Enzymes, which play an essential *kinetic* role in *biological* reactions, are composed of large protein *molecules*. Enzymes facilitate the rate of reactions by acting as *catalysts* and lowering the activation energy barrier between the reactants and the products of the reactions. All chemical reactions that occur during the digestion of food and the *metabolic* processes in tissues require enzymes. Therefore, enzymes are vital to the overall function of the body, and thereby indicate the fundamental and significant role of proteins.

3.10.1.3   Fluid Balance

The presence of blood protein molecules, such as *albumins* and *globulins*, are critical factors in maintaining the proper fluid balance between cells and extracellular space. Proteins are present in the capillary beds, which are one-cell-thick vessels that connect the arterial and venous beds, and they cannot flow outside the capillary beds into the tissue because of their large size. Blood fluid is pulled into the capillary beds from the tissue through the mechanics of oncotic pressure, in which the pressure exerted by the protein molecules counteracts the *blood pressure*. Therefore, blood proteins are essential in maintaining and regulating fluid balance between the blood and tissue. The lack of blood proteins results in clinical *edema*, or tissue swelling, because there is insufficient pressure to pull fluid back into the blood from the tissues. The condition of edema is serious and can lead to many medical problems.

3.10.1.4   Energy Provision

Protein is not a significant source of energy for the body when there are sufficient amounts of *carbohydrates* and fats available, nor is protein a storable energy, as in the case of fats and carbohydrates. However, if insufficient amounts of carbohydrates and fats are ingested, protein is used for the energy needs of the body. The use of protein for energy is limited by the ability of the liver to convert amino acids to glucose and glycogen [102].

The use of proteins as a source of energy is not necessarily economical for the body, because tissue maintenance, growth, and repair are compromised to meet energy needs. If taken in excess, protein can be converted into body fat. Protein yields as much usable energy as carbohydrates—4 kcal/g (kilocalories per gram). Although not the main source of usable energy, protein provides the essential amino acids that are needed for adenine, the nitrogenous base of ATP, as well as other nitrogenous substances, such as creatine phosphate (nitrogen is an essential element for important compounds in the body) [84].

3.10.1.5   Protein Requirement and Nutrition

The recommended protein intake for an average adult is generally based on body size: 0.8 g/kg of body weight is the generally recommended daily intake. The recommended daily allowances of protein do not vary in times of strenuous activities or exercise, or with progressing age. However, there is a wide range of protein intake which people can consume according to their period of development. For example, the recommended allowance for an infant up to 6 months of age, who is undergoing a period of rapid tissue growth, is 2.2 g/kg. For children ages 7 through 10, the recommended daily allowance is around 36 total grams, depending on body weight. Pregnant women need to consume an additional 30 g of protein above the average adult intake for the nourishment of the developing fetus [103].

3.10.1.6   Sources of Protein

Good sources of protein include high-quality protein foods, such as meat, poultry, fish, milk, egg, and cheese, as well as prevalent low-quality protein foods, such as fresh vegetables and fruits, except *legumes* (e.g., navy beans, pinto beans, chick peas, soybeans, and split peas), which are high in protein. Cereals are also a good source of proteins [104]. Protein content of the selected foods are given in Table 3.16 [84].

TABLE 3.16
Protein Content of Representative Foods in Human Diet

Food	Protein (g)
Milk, 244 g (8 oz)	8.0
Cheddar cheese, 84 g (3 oz)	21.3
Egg, 50 g (1 large)	6.1
Apple, 212 g (1, 3 1/4 in. diameter)	0.4
Banana, 74 g (1, 8 3/4 in. long)	1.2
Potato, cooked, 136 g (1 potato)	2.5
Bread, white, slice, 25 g	2.1
Fish, cod, poached, 100 g (3 1/2 oz)	20.9
Oyster, 100 g (3 1/2 oz)	13.5
Beef, pot roast, 85 g (3 oz)	22.0
Liver, pan fried, 85 g (3 oz)	23.0
Pork chop, bone in, 87 g (3.1 oz)	23.9
Ham, boiled, 2 pieces, 114 g	20.0
Peanut butter, 16 g (1 tablespoon)	4.6
Pecans, 28 g (1 oz)	2.2
Snap beans, 125 g (1 cup)	2.4
Carrots, sliced, 78 g (1/2 cup)	0.8

Source: U.S. Department of Agriculture, *Composition of Foods*, USDA Handbooks 8(1–20), U.S. Government Printing Office, Washington, DC, 1986. www.nal.usda.gov/fnic/foodcomp

3.10.1.7 Protein–Calorie Malnutrition

The nitrogen balance index (NBI) is used to evaluate the amount of protein used by the body in comparison with the amount of protein supplied from daily food intake [105]. The body is in the state of nitrogen (or protein) equilibrium when the intake and usage of protein is equal. The body has a *positive nitrogen balance* when the intake of protein is greater than that expended by the body. In this case, the body can build and develop new tissue. Since the body does not store protein, the overconsumption of protein can result in the excess amount to be converted into fat and stored as *adipose tissue*. The body has a *negative nitrogen balance* when the intake of protein is less than that expended by the body. In this case, protein intake is less than required, and the body cannot maintain or build new tissues.

A *negative nitrogen balance* represents a state of protein deficiency, in which the body is breaking down tissues faster than they are being replaced. The ingestion of insufficient amounts of protein, or food with poor protein quality, can result in serious medical conditions in which an individual's overall health is compromised. The *immune system* is severely affected; the amount of blood *plasma* decreases, leading to medical conditions such as *anemia* or edema; and the body becomes vulnerable to *infectious diseases* and other serious conditions. Protein malnutrition in infants is called *kwashiorkor*, and it poses a major health problem in developing countries, such as Africa, Central and South America, and certain parts of Asia. An infant with kwashiorkor suffers from poor muscle and tissue development, loss of appetite, mottled skin, patchy hair, diarrhea, edema, and, eventually, death (similar symptoms are present in adults with protein deficiency). Treatment or prevention of this condition lies in adequate consumption of protein-rich foods [106].

Proteins are not alike. They vary according to their origin (animal, vegetable), amino acid composition (particularly their relative content of essential amino acids), digestibility, texture, etc. Good quality proteins are those that are readily digestible and contain the essential amino acids in quantities that correspond to human requirements.

Humans require certain minimal quantities of essential amino acids from a biologically available source as part of a larger protein/nitrogen intake. The required amounts of these amino acids vary with age, physiological condition, and state of health. It is therefore important to be able to discriminate with both accuracy and precision the relative efficiency with which individual protein sources can meet human biological needs. This efficiency also has direct implications for the commercial value of the protein product [101,107].

Clinical human studies that measure growth and/or other metabolic indicators provide the most accurate assessment of protein quality. For reasons of both cost and ethics, such techniques cannot be used. Consequently, assay techniques designed to measure the effectiveness of a protein in promoting animal growth have been utilized. Since 1919, the protein efficiency ratio (PER) method, which measures the ability of a protein to support growth in young, rapidly growing rats, has been used in many countries because it was believed to be the best predictor of clinical tests. However, after decades of use, it is now recognized that PER overestimates the value of some animal proteins for human growth while it underestimates the value of some vegetable proteins for that purpose. The rapid growth of rats (which increases the need for essential amino acids) in comparison to human growth rates is the reason for this discrepancy [108].

For some time the use of an amino acid score has been advocated as an alternative to the PER. Although clearly the quality of some proteins can be assessed directly by using amino acid score values, but that of others cannot be assessed because of poor digestibility and/or bioavailability. Consequently, both amino acid composition and digestibility measurements are considered necessary to predict accurately the protein quality of foods for human diets [109] (Table 3.17).

On the other hand, the methods currently used for measuring protein quality of foods were established when information was not extensively available on human amino acid requirements. Therefore, while results were not grossly in error, they did not accurately reflect human requirements. Since most of these methods use a rat assay, they in large part measure the amino acid

TABLE 3.17
Essential Amino Acid Requirements and Casein Content for Children and Rats

Essential Amino Acid	Children (2–5 years) [110]	Laboratory Rat[a] (mg/g protein) [111]	Casein [108]
Arginine	—	50	37
Histidine	19	25	32
Isoleucine	28	42	54
Leucine	66	62	95
Lysine	58	58	85
Methionine and cystine	25	50[b] [112]	35
Phenylalanine and tyrosine	63	66	111
Threonine	34	42	42
Tryptophan	11	12.5	14
Valine	35	50	63

[a] Based on a protein requirement of 12% plus an ideal protein (100% true digestibility and 100% biological value).

[b] A lower rat requirement of 40 mg/g protein for methionine and cystine has also been reported.

requirements of the rat rather than the human. This is particularly misleading, since the rat appears to have a much higher requirement for sulfur amino acids than does the human (Table 3.16). In addition to the higher requirement for sulfur amino acids, the rat also has a higher requirement for histidine, isoleucine, threonine, and valine.

3.10.1.8   Need for International Standardized Procedure

The Codex Committee [113] on Vegetable Proteins (CCVP), while elaborating general guidelines for the utilization of vegetable protein products in foods, felt the need for a suitable indicator to express protein quality. It pointed out at its first session that PER might not be the most suitable means for protein quality evaluation. In the successive two sessions, the committee considered the suitability of other indicators such as the relative net protein ratio (RNPR) (a rat assay procedure) and the amino acid composition data (amino acid scores) corrected for crude protein digestibility/amino acid availability, but no decision was taken. At its fourth session, the committee noted improvements made in amino acid analysis and amino acid requirement pattern and discussed initial data from ongoing comparative studies organized by the United States Department of Agriculture (USDA) involving amino acid availability, nitrogen digestibility, and protein nutritional assessment based on amino acid composition data. The committee concluded that an amino acid scoring procedure, corrected for true digestibility of protein and/or bioavailability of limiting amino acids, is the preferred approach for assessing protein quality of vegetable protein products and other food products. References [102,114] and recent improvements in amino acid methodology enabled the FAO to endorse the use of the suggested pattern of amino acid requirements of a 2–5 year old child as the reference for calculating amino acid scores [110,115]. The committee agreed that amino acid scoring (based on the amount of the single most limiting amino acid) corrected for true digestibility of protein (as determined by the rat balance method) is the most suitable routine method for assessing the protein quality of most vegetable protein products and other food products [113]. Because the methodology used to measure protein quality had broad implications beyond its purview, the CCVP recognized the need for the wider scientific community to address issues such as human requirements for essential amino acids, amino acid evaluation methodology, protein digestibility, and amino acid availability.

3.10.1.9   Joint FAO/WHO Expert Consultation on Protein Quality Evaluation

The consultation was convened for the task of

- Reviewing present knowledge of protein quality evaluation
- Discussing various techniques used in evaluating protein quality
- Specifically evaluating the method recommended by the CCVP, i.e., amino acid score corrected for digestibility [116]

The consultation reviewed in particular the scientific basis for the adoption of the protein digestibility-corrected amino acid score method. It recognized that the most serious problem with the use of a rat growth assay in predicting protein quality in food is that rats have a higher requirement than humans for some amino acids. The PER is the official method for assessing protein quality of foods in Canada and the United States, but it has been severely criticized for not meeting the criteria for a valid routine test [117]. A major criticism of the PER assay is that it does not properly credit protein used for maintenance purposes. A protein source may not support growth and may have a PER near zero yet may still be adequate for maintenance purposes. Because of the error introduced by not making allowance for maintenance, PER values are not proportional to protein quality, i.e., a PER of 2.0 cannot be assumed to be twice as good as a PER of 1.0. The lack of proportionality of protein quality makes the PER method unsuitable for the calculation of utilizable protein, as in protein rating (protein in a reasonable daily intake, mass×PER).

The nutritive value of a protein depends on its capacity to provide nitrogen and amino acids in adequate amounts to meet the requirements of an organism. Thus, in theory the most logical approach for evaluating protein quality is to compare amino acid content (taking bioavailability into account) of a food with human amino acid requirements. A number of comparisons have been made using reference patterns such as those derived from egg and milk protein. The first major change in procedure was substitution of a provisional pattern of amino acid requirements for the egg protein standard.

A hypothetical reference protein derived from the pattern of human amino acid requirements was proposed as a standard for comparison. Shortcomings have been recognized and progress has been made in accurately evaluating human amino acid requirements. Equally critical for success is the ability to obtain precise measurements of amino acid content in the protein sources. Finally, to improve on accuracy of scoring procedures, chemically determined amino acid contents may have to be corrected for digestibility or biological availability.

3.10.1.9.1   Conclusions and Recommendations

Methodology for determining the amino acid composition of proteins. The consultation concluded that modern amino acid analysis can provide data with repeatability within a laboratory of about 5% and reproducibility between laboratories of about 10%. It recommended that this variability be considered acceptable for the purposes of calculating the amino acid score. To achieve such results requires careful attention to many aspects of the protocols, including replicating the complete analytical procedure [118].

The consultation also made the following recommendations:

1. Further studies should be undertaken to standardize the hydrolytic and oxidation procedures and improve accuracy of the procedures to further reduce inter-laboratory variation.
2. Amino acid data should be reported as mg amino acid per g N or be converted to mg amino acid per g protein by use of the factor 6.25. No other food-specific protein factor should be used.
3. FAO should update their publication *Amino acid content of foods and biological data on proteins* [115] and commission new analyses of foods where there are insufficient and unreliable data.
4. Reliable national tables of amino acid composition of products that have been clearly defined in terms of composition and processing should be developed.

3.10.1.9.1.1   Amino Acid Scoring Pattern   The consultation evaluated the existing evidence and arguments about the use of amino acid scoring patterns to evaluate protein quality and concluded that at present there is no adequate basis for the use of different scoring patterns for different age groups with the exception of infants. Therefore, it decided to make the following recommendations:

1. The amino acid composition of human milk should be the basis of the scoring pattern to evaluate protein quality in foods for infants less than 1 year of age.
2. The amino acid scoring pattern proposed by FAO/WHO/UNU [110] for children of preschool age should be used to evaluate dietary protein quality for all age groups except infants.
3. The recommendations made here for the two amino acid scoring patterns to be used for infants and for all other ages must be deemed as temporary until the results of further research either confirm their adequacy or demand a revision.
4. Further research must be carried out to confirm the currently accepted values of requirements of infants and preschool-aged children that are the basis for the scoring patterns recommended by this consultation.
5. Further research must be carried out to define the indispensable amino acids (IAA) requirements of school-aged or adolescent children and of adults.

6. Given the urgency of these research needs and the magnitude of the task required, it is recommended that an FAO/WHO coordinated international research program be established immediately to assist in the determination of human amino acid needs.

3.10.1.9.1.2 Digestibility of Proteins   The consultation discussed in detail the various methods used for determining the digestibility of proteins and made the following recommendations:

1. Studies should be undertaken to compare protein digestibility values of humans and rats from identical food products.
2. Extensive evaluation of existing *in vitro* and *in vivo* methods in foods indicates that the rat balance method is the most suitable practical method for predicting protein digestibility by humans. Therefore, when human balance studies cannot be used, the standardized rat fecal-balance method of [119] or [109] is recommended.
3. Since the true digestibility (the proportion of food nitrogen that is absorbed) is of crude protein is a reasonable approximation of the true digestibility of most amino acids (as determined by the rat balance method), it is recommended that amino acid scores be corrected only for the true digestibility of protein.

$$\frac{A}{I} = \frac{I - (F - F \cdot k)}{I}$$

where
   A is the absorbed nitrogen
   F is the fecal nitrogen
   $F \cdot k$ is the metabolic nitrogen
   I is the nitrogen intake

4. For new or novel products or processes, digestibility values must be determined. However, established digestibility values of well-defined foods may be taken from a published data base for use in the amino acid scoring procedure. A database should be established for all raw and processed products.
5. Further research is encouraged to perfect and evaluate the most promising *in vitro* procedures for estimating protein digestibility, such as those of [107,119].

Based on the above conclusions, the consultation agreed that the protein digestibility-corrected amino acid score method was the most suitable approach for routine evaluation of protein quality for humans and recommended that it be adopted as an official method at the international level.

The report of the consultation contains details of the recommended methodology for the evaluation of protein quality and a practical guide on how to apply this methodology for individual foods as well as food mixtures.

3.10.2   PROTEIN DIGESTIBILITY CORRECTED AMINO ACID SCORE

The Protein Digestibility Corrected Amino Acid Score (PDCAAS) is superior to other methods for evaluating the protein quality of food proteins for humans because it measures the quality of a protein based on the amino acid requirements of a 2–5 year old child (the most demanding age group), adjusted for digestibility [120]. It has replaced the Protein Evaluation Ratio (PER).

Adoption of PDCAAS allows evaluation of food protein quality based on the needs of humans. Isolated soy protein such as Solae™, with a PDCAAS of 1.0, is a complete protein and has the same score as milk protein and egg white.

PDCAAS is based on a food protein's amino acid content, its true digestibility, and its ability to supply indispensable amino acids in amounts adequate to meet the amino acid requirements of a 2–5 year old child—the age group used as the standard.

The highest PDCAAS value that any protein can achieve is 1.0. This score means that after digestion of the food protein, it provides per unit of protein, 100% or more of the indispensable amino acids required by the 2–5 year old child. A score above 1.0 is rounded down to 1.0. Any amino acids in excess of those required to build and repair tissue would not be used for protein synthesis, but would be catabolized and eliminated from the body or stored as fat [121].

The following steps are necessary to calculate the PDCAAS of a food protein.

1. The food protein must be analyzed for its proximate composition, its nitrogen content.
2. Protein content is calculated by multiplying the nitrogen content by 6.25.
3. The food protein is analyzed to determine its indispensable amino acid content.
4. The uncorrected amino acid score is calculated by dividing the milligrams of a particular indispensable amino acid in 1 g of the test protein by the milligrams of the indispensable amino acids in 1 g of the reference protein which is the amino acid requirement pattern for the 2–5 year old child.
5. The digestibility of food protein needs to be determined. The classical procedure for determining digestibility is a rat balance method which has been recommended as the most suitable method for humans by the FAO/WHO Expert Consultation.
6. The PDCAAS is calculated by multiplying the lowest uncorrected amino acid score by the food protein's digestibility.

For example, isolated soy protein has 26 mg of histidine per gram of protein. The digestibility factor of 97% means that out of 26 mg of histidine reaching the intestinal tract, 25.2 mg are absorbed. A 2–5 year old child requires 19 mg of histidine per gram of protein, giving isolated soy protein a PDCAAS of 1.3 for this amino acid. This means that isolated soy protein provides 130% of the histidine required by the reference pattern, which reflects the requirements of a 2–5 year old child.

The PDCAAS for a food protein is equal to the lowest score for a single indispensable amino acid (or amino acid pair). In this case, methionine and cystine have a PDCAAS of 1.0, which becomes the score for the entire protein. Had all of the individual amino acid scores exceeded 1.0, the PDCAAS would still have been rounded down to 1.0, the highest PDCAAS possible.

The PDCAAS is currently used for labeling protein on food products for adults and for children over 1 year of age.

The FDA gave two reasons for adopting the PDCAAS over PER.

1. PDCAAS is based on human amino acid requirements, which makes it more appropriate for humans than a method based on the amino acid needs of animals.
2. The Food and Agricultural Organization/World Health Organization (FAO/WHO) had previously recommended PDCAAS for regulatory purposes. FAO/WHO is a recognized international organization, which is experienced in establishing these types of standards.

Isolated soy protein has a PDCAAS equal to the protein in milk and in egg white.

As an example, an adult who needs 50 g of protein per day could satisfy all of his or her protein needs by consuming 50 g of a certain brand of isolated soy protein such as Solae. Whole wheat has a PDCAAS of 0.40. This means that it would take about 125 g of protein from whole wheat to supply the amounts of all the indispensable amino acids provided by 50 g of this isolated soy protein.

A PDCAAS value of 1 is the highest and 0 the lowest as Table 3.18 demonstrates the ratings of commons foods.

3.10.3 Biological Value of Protein

The *biological value* (BV) of protein is a measure of the proportion of absorbed protein from a food which becomes incorporated into the proteins of the organism's body. It summarizes how readily the broken down protein can be used in protein synthesis in the cells of the organism. Proteins are the major source of nitrogen food, unlike carbohydrates and fats [122]. This method assumes protein is the only source of nitrogen and measures the proportion of this nitrogen absorbed by the body which is then excreted. The remainder must have been incorporated into the proteins of the organism's body. A ratio of nitrogen incorporated into the body over nitrogen absorbed gives a measure of protein 'usability' - the BV.

Unlike some measures of protein usability, biological value does not take into account how readily the protein can be digested and absorbed (largely by the small intestine). This is reflected in the experimental methods used to determine BV [123].

BV confusingly uses two similar scales:

TABLE 3.18
Protein Digestibility Corrected Amino Acid Scores (PDCASS) of Selected Proteins

Whey	1.0
Egg white	1.0
Casein	1.0
Milk	1.0
Soy protein isolate	1.00
Beef	0.92
Soybean	0.91
Kidney beans	0.68
Rye	0.68
Whole wheat	0.54
Lentils	0.52
Peanuts	0.52
Zeitan	0.25

1. The true percentage utilization (usually shown with a percent symbol).
2. The percentage utilization relative to a readily utilizable protein source, often egg (usually shown as unitless).

These two values will be similar but not identical.

The BV of a food varies greatly and depends on a wide variety of factors. In particular the BV value of a food varies depending on its preparation and the recent diet of the organism. This makes reliable determination of BV difficult and of limited use—fasting prior to testing is universally required in order to make the values reliable.

BV is commonly used in nutrition science in many mammalian organisms, and is a relevant measure in humans. It is a popular guideline for protein choice in bodybuilding [124].

For accurate determination of BV

1. The test organism must only consume the protein or mixture of proteins of interest (the test diet).
2. The test diet must contain no non-protein sources of nitrogen.
3. The test diet must be of suitable content and quantity to avoid use of the protein primarily as an energy source.

These conditions mean the tests are typically carried out over the course of over 1 week with strict diet control. Fasting prior to testing helps produce consistency between the subjects (it removes recent diet as a variable).

There are two scales on which BV is measured; percentage utilization and relative utilization. By convention percentage BV has a percent sign (%) suffix and relative BV has no unit.

Biological value is determined based on the following formula

$$BV = (N_r/N_a)100$$

where
N_a is the nitrogen absorbed in proteins on the test diet
N_r is the nitrogen incorporated into the body on the test diet

However direct measurement of N_r is essentially impossible. It will typically be measured indirectly from nitrogen excretion in urine. Fecal excretion of nitrogen must also be taken into account—this protein is not absorbed by the body and so not included in the calculation of BV.

$$BV = \left[\left(N_i - N_{e(f)} - N_{e(u)} - N_b \right) / N_i - N_{e(f)} \right] 100$$

where

N_i is the nitrogen intake in proteins on the test diet
$N_{e(f)}$ is the nitrogen excreted in feces whilst on the test diet
$N_{e(u)}$ is the nitrogen excreted in urine whilst on the test diet
N_b is the nitrogen excreted on a protein free diet
Note:
$N_r = N_i - N_{e(f)} - N_{e(u)} - N_b$
$N_a = N_i - N_{e(f)}$

This can take any value of 100 or less, including negative. A BV of 100% indicates complete utilization of a dietary protein, i.e., 100% of the protein ingested and absorbed is incorporated into proteins into the body. Negative values are possible if excretion of nitrogen exceeds intake in proteins. All non-nitrogen containing diets have negative BV. The value of 100% is an absolute maximum, no more than 100% of the protein ingested can be utilized (in the equation above $N_{e(u)}$, $N_{e(f)}$, and N_b cannot go negative, setting 100% as the maximum BV.

Due to experimental limitations, BV is often measured *relative* to an easily utilizable protein. Normally egg protein is assumed to be the most readily utilizable protein and given a BV of 100. For example: Two tests of BV are carried out on the same person; one with the test protein source and one with a reference protein (egg protein).

$$\text{relative BV} = \left(BV_{(test)} / BV_{(egg)} \right) 100$$

where

$BV_{(test)}$ is the percentage BV of the test diet for that individual
$BV_{(egg)}$ is the percentage BV of the reference (egg) diet for that individual

This is not restricted to values of less than 100. The percentage BV of egg protein is only 93.7% which allows other proteins with true percentage BV between 93.7% and 100% to take a relative BV of over 100. For example, whey protein takes a relative BV of 104, while its percentage BV is under 100%.

The principal advantage of measuring BV relative to another protein diet is accuracy; it helps account for some of the metabolic variability between individuals. In a simplistic sense the egg diet is testing the maximum efficiency of the individual to take up protein, the BV is then provided as a percentage taking this as the maximum.

TABLE 3.19
Biological Values of Select Food Proteins

Isolated whey: 100
Whey protein concentrate: 104
Whole bean: 96
Whole soybean: 96
Human milk: 95
Chicken egg: 94
Soybean milk: 91
Cow milk: 90
Casein: 77
Cheese: 84
Rice: 83
Defatted soy flour: 81
Fish: 76
Beef: 74.3
Immature bean: 65
Full-fat soy flour: 64
Soybean curd (*tofu*): 64
Wheat gluten: 64
Whole wheat: 64
White flour: 41

Source: Johnston, T.K., Nutritional implications of vegetarian diets, in *Modern Nutrition in Health and Disease*, 9th edn, Shills, M.E. et al., Eds., Williams & Wilkins, Baltimore, MD, 1999.

Three major properties of a protein source affect its BV:

- Amino acid composition, and the limiting amino acid, which is usually lysine
- Preparation (cooking)
- Vitamin and mineral content

Amino acid composition is the principal effect. All proteins are made up of combinations of the 21 biological amino acids. Some of these can be synthesized or converted in the body, whereas others cannot and must be ingested in the diet. These are known as essential amino acids (EAAs), of which there are nine in humans. The number of EAAs varies according to species (see below).

EAAs missing from the diet prevent the synthesis of proteins that require them. If a protein source is missing critical EAAs, then its biological value will be low as the missing EAAs form a bottleneck in protein synthesis. For example, if a hypothetical muscle protein requires phenylalanine (an essential amino acid), and then this must be provided in the diet for the muscle protein to be produced. If the current protein source in the diet has no phenylalanine in it, the muscle protein cannot be produced, giving a low usability and BV of the protein source.

In a related way if amino acids are missing from the protein source which are particularly slow or energy consuming to synthesize this can result in a low BV.

Common foodstuffs and their biological values: Note: this scale uses 100 as 100% of the nitrogen incorporated, as given in Table 3.19.

Common foodstuffs and their values: Note: These values use "whole egg" as a value of 100, so foodstuffs that provide even more nitrogen than whole eggs, can have a value of more than 100. 100 does not mean that 100% of the nitrogen in the food is incorporated into the body, and not excreted, as in other charts [124].

TABLE 3.20
Net Protein Utilization of Some Food Proteins

Proteins	NPU Value
Whole egg	95.0
Dried milk powder	75.0
Beef muscle	72.0
Gelatin	5.0
Wheat germ	67.0
Soya flour	56.0
Wheat gluten	38.0
Rice gluten	35.0
Corn	55.0

3.10.4 Net Protein Utilization

The *net protein utilization*, or NPU, is the ratio of amino acid converted to proteins to the ratio of amino acids supplied. This figure is somewhat affected by the salvage of essential amino acids within the body, but is profoundly affected by the level of limiting amino acids within a food [125].

Experimentally, this value can be determined by determining dietary protein intake and then measuring nitrogen excretion. One formula for NPU is

$$NPU = \left(\left[0.16 \times \left(24\ h\ \text{protein intake in grams} \right) \right] - \left[\left(24\ h\ \text{urinary urea nitrogen} \right) + 2 \right] \right.$$

$$\left. - \left[0.1 \times \left(\text{ideal body weight in kilograms} \right) \right] \right) \Big/ \left[0.16 \times \left(24\ h\ \text{protein intake in grams} \right) \right]$$

As a value, NPU can range from 100 to 0, with a value of 100 indicating 100% utilization of dietary nitrogen as protein and a value of 0 an indication that none of the nitrogen supplied was converted to protein. Values of some foods, such as, eggs or milk, rate as close to 100 on an NPU chart [126]. Selected NPU values of the proteins are given in Table 3.20. Net protein utilization requirements does not change in young or elderly [127].

REFERENCES

1. Venter, J.C. et al. The sequence of the human genome. *Science* 291: 1304–1351, February 16, 2001.
2. Reiland, S. et al. Large-scale *Arabidopsis* phosphoproteome profiling reveals novel chloroplast kinase substrates and phosphorylation networks. *Plant Physiol.* 150(2): 889–903, 2009.
3. Crick, F.H.C. The structure of the hereditary material. *Scientific Am.* 191: 54–61, 1954.
4. Krasilnikov A.S., Yang X.J., Pan T., and Mondragon A. Crystal structure of the specificity domain of ribonuclease P. *Nature* 421: 760–764, 2003.
5. Shewry, P.R., Napier, J.A., and Tatham, A.S. Seed storage proteins: Structures and biosynthesis. *Plant Cell* 7(7): 945–956, [PubMed] July 1995.
6. Robert, B. Bio production of therapeutic proteins in the 21st century and the role of plants and plant cells as production platforms. *Ann. N.Y. Acad. Sci.* 1102: 121–134, 2007.
7. Stryer, L. *Biochemistry*, 2nd edn, W. H. Freeman & Company, New York, 1981.
8. Nelson, D.L. and Cox, M.M. *Lehninger Principles of Biochemistry*, 3rd edn, Worth Publishers, New York. ISBN 1-57259-153-6, 2000.
9. IUPAC/IUBMB. http://www.iupac.org/ http://www.chem.qmul.ac.uk/iubmb/ (2008)
10. Kyte, J. and Doolittle, R.F. A simple method for displaying the hydropathic character of a protein. *J. Mol. Biol.* 157(1): 105–132, PMID 7108955, 1982.
11. Derewenda, U., Derewenda, Z., Dodson, G.G., Hubbard, R.E., and Korber, F. Molecular structure of insulin: The insulin monomer and its assembly. *Br. Med. Bull.* 45: 4–18, 1989.
12. Geng, L. and Ju, G. [The discovery of pituitary adenylate cyclase activating polypeptide (PACAP) and its research progress]. Sheng li ke xue jin zhan [*Prog. Physiol.*] 28(1): 29–34, PMID 10921074, 2000.
13. Colmers W.F. and El Bahn, B. Neuropeptide Y and epilepsy. *Epilepsy Curr./Am. Epilepsy Soc.* 2(3): 53–58, PMID 15309085, 2003.
14. Sandström O. and El-Salhy, M. Ontogeny and the effect of aging on pancreatic polypeptide and peptide YY. *Peptides* 23(2): 263–267, PMID 11825641, 2002.
15. Pasieka, J.L. and Hershfield, N. Pancreatic polypeptide hyperplasia causing watery diarrhea syndrome: A case report. *Can. J. Surg.* 42(1): 55–58, 1999 (review).
16. Jeppesen, P.B. Clinical significance of GLP-2 in short-bowel syndrome. *J. Nutr.* 133(11): 3721–3724, PMID 14608103, 2004.
17. Schneider, H.G. and Lam, Q.T. Procalcitonin for the clinical laboratory: A review. *Pathology* 39(4): 383–390, PMID 17676478, 2007.
18. Chakrabartty, A., Ananthanarayan, V.S., and Choy, L.H., Structure-function relationships of a winter flounder antifreeze polypeptide. *J. Biol. Chem.* 264:11307–11312, 1989.
19. Pauling, L. and Wilson, E.B. Jr. *Introduction to Quantum Mechanics, with Applications to Chemistry*, Dover, The McGraw-Hill Companies, New York, 1985.
20. Stryer, L. *Biochemistry*, 3rd edn, W.H. Freeman, New York, 1988.
21. McKee, T. and McKee, J.R. *Biochemistry: An Introduction*, WCB/McGraw-Hill, Boston, MA, 1999.
22. Zubay, G.L. and Atkinson, D.E. *Biochemistry*, 2nd edn, Macmillan Pub. Co., Collier Macmillan, New York/London, 1988.
23. Voet, D. and Voet, J.G. *Biochemistry*, 3rd edn, John Wiley & Sons, Hoboken, NJ, 2004.
24. Mathews, C.K. *Biochemistry*, 4th edn, Benjamin Cummings, San Francisco, CA, 2008.
25. Berg, J. *Biochemistry and Lecture Notebook and MBB222 Course Outline*. W. H. Freeman & Company, Holtzbrinck Publishers, New York, 2004.
26. Swanson, T.A. *Biochemistry and Molecular Biology*, 4th edn, Lippincott Williams & Wilkins, Philadelphia, PA, 2007.
27. Pelley, J.W. *Elsevier's Integrated Biochemistry*, Mosby, Philadelphia, PA, 2007, 215pp.
28. Boyer, R. *Biochemistry Laboratory: Modern Theories and Techniques*, Benjamin Cummings, San Francisco, CA, 2006.
29. Kuchel, P.W. *Schaum's Outline of Theory and Problems of Biochemistry*, 2nd edn, McGraw-Hill, New York, 1998, 559pp.
30. Berg, J.M., Tymoczko, J.L., and Stryer, L. *Biochemistry*, 5th edn, W.H. Freeman, New York, 2002.
31. Huang, K.S., Bayley, H., Liao, M.J., London, E., and Khorana, H.G. Refolding of an integral membrane protein. Denaturation, renaturation, and reconstitution of intact bacteriorhodopsin and two proteolytic fragments. *Biol. Chem.* 256(8): 3802–3809, 04, 1981.
32. Lennarz, W.J. *The Biochemistry of Glycoproteins and Proteoglycans*, Plenum Press, New York, 1980.
33. Hamilton, S. et al. Production of complex human glycoproteins in yeast. *Science* 301: 1244, 2003.

34. Madan Babu, M. and Sankaran, K. DOLOP—Database of bacterial lipoproteins. *Bioinformatics* 18: 641–643, 2002.
35. O'Keefe, J.H. Jr, Cordain, L., Harris, L.H., Moe, R.M., and Vogel, R. Optimal low-density lipoprotein is 50 to 70 mg/dl: Lower is better and physiologically normal. *J. Am. Coll. Cardiol.* 43(11): 2142–2146, PMID 15172426, 2004.
36. Wong, L.H. et al. Centromere RNA is a key component for the assembly of nucleoproteins at the nucleolus and centromere. *Genome Res.* 17: 1146–1160, 2007.
37. Leslie, M. Isolating SRP. *J. Cell Biol.* 171(1): 13, 2005.
38. Hirose-Kumagai, A., Oda-Tamai, S., and Akamatsu, N. The interaction between nucleoproteins and thyroid response element (TRE) during regenerating rat liver. *Biochem. Mol. Biol. Int.* 35(4): 881–888, April 1995.
39. Barton, K.A., Thompson, J.F., Madison, J.T., Rosenthal, R., Jarvis, N.P., and Beachy, R.N. The biosynthesis and processing of high molecular weight precursors of soybean glycinin subunits. *Biol. Chem.* 257(11): 6089–6095, June 1982.
40. Harding, H.P., Zhang, Y., and Ron, D. Protein translation and folding are coupled by an endoplasmic-reticulum-resident kinase. *Nature* 397(6716): 271–274, January 21, 1999.
41. Shewry, P.R., Tatham, A.S., Forde, J., Kreis, M., and Miflin, B.J. The classification and nomenclature of wheat gluten proteins: a reassessment. *J. Cer. Sci.* 4: 97–106, 1986.
42. Shoup, F.K., Pomeranz, Y., and Deyoe, C.W. Amino acid composition of wheat varieties and flours varying widely in bread-making potentialities. *J. Food Sci.* 31: 94–101, 1966.
43. Shewry, P.R., Halford, N.G., Field, J.M., and Tatham, A.S. The structure and functionality of wheat gluten proteins. *Proceedings of the 38th Australian Cereal Chemistry Conference*, Sydney, NSW, Murray, L. ed., Royal Australian Chemical Institute, Melbourne, VIC, Australia, 1989.
44. Schofield, J.D., Bottomley, R.C., Timms, M.F., and Booth, M.R.,The effect of heat on wheat gluten and the involvement of sulphydryl-disulphide interchange reaction. *J. Cereal Sci.* 1: 24, 1983.
45. Shewry, P.R. Biological and evolutionary aspects of cereal seed storage proteins. In: Shewry, P.R. and Stobart, K. (Eds.), *Seed Storage Compounds: Biosynthesis, Interactions, and Manipulations*, Clarendon Press, Oxford, U.K., 1993.
46. Smith, J.S.C. and Lester, R.N. Biochemical systematics and evolution of Zea, Tripsacum, and related genera. *Econ. Bot.* 34: 201, 1980.
47. Spencer, D. and Higgins, T.J.V. Seed maturation and deposition of storage proteins. In: Smith, H. and Grierson, D. (Eds.), *The Molecular Biology of Plant Development*, Blackwell, Oxford, U.K., 1982.
48. Stevens, D.J. Reaction of wheat proteins with sulphite. III. Measurement of labile and reactive disulphide bonds in gliadin and in the protein of aleurone cells. *J. Sci. Food Agric.* 24: 279–283, 1973.
49. Woychick, J.H., Boundy, J.A., and Dimler, R.J., Starch gel electrophoresis of wheat gluten proteins with concentrated urea. *Arch. Biochem. Biophys.* 94: 477–482, 1961.
50. Tatham, A.S., Shewry, P.R., and Belton, P.S. Structural studies of cereal prolamines, including wheat gluten. In: Pomeranz, Y. (Ed.), *Advances in Cereal Science and Technology*, American Association of Cereal Chemists, St Paul, MN, 1990.
51. Taylor N.W. and Cluskey, J.E. Wheat gluten and its glutenin component: Viscosity, diffusion and sedimentation studies. *Arch. Biochem. Biophys.* 97: 399–405, 1962.
52. Taylor, W. and Thorton, J. Recognition of super-secondary structure in proteins. *J. Mol. Biol.* 173: 487–514, 1984.
53. Thornton, J.M. Disulphide bridges in globular proteins. *J. Mol. Biol.* 151: 261–287, 1981.
54. Lew, E.J.-L., Kuzmicky D.D., and Kasarda, D.D. Characterization of low molecular weight glutenin subunits by reversed-phase high-performance liquid chromatography, sodium dodecyl sulfate-polyacrylamide gel electrophoresis, and N-terminal amino acid sequencing, *Cereal Chem.*, 69(5): 508–515, 1992.
55. Kasarda D.D., Autran, J.-C., Lew, E.J.-L., Nimmo C.C., and Shewry P.R. N-terminal amino acid sequences of ω-gliadins and ω-secalins. Implications for the evolution of prolamin genes, *Biochim. Biophys. Acta* 747(1–2): 138–150, 1983.
56. Tatham, A.S., Masson, P., and Popineau, Y. Conformational studies of peptides derived from the enzymic hydrolysis of a gamma-type gliadin. *J. Cereal Sci.* 11: 1–13, 1990.
57. Scheets, K., Rafalski, J.A., Hedgcoth, C., and Soll, D.G. Heptapeptide repeat structure of a wheat gamma-gliadin. *Plant Sci. Lett.* 37: 221–225, 1985.
58. Thomson, N.H., Miles, M.J., Tatham, A.S., and Shewry, P.R., Molecular images of cereal prolamines by STM. *Ultramicroscopy* 42–44: 1204–1213, 1992.

59. Madeka, H. and Kokini, J.L. Effect of glass transition and crosslinking on rheological properties of zein: Development of a preliminary state diagram. *Cereal Chem.* 73(4): 433–438, 1996.

60. Halford, N.G. et al. Analysis of HMW glutenin subunits encoded by chromosome 1A of bread wheat (*Triticum aestivum* L.) indicates quantitative effects on grain quality. *TAG Theor. Appl. Genet.* 83(3): 373–378, January 1992.

61. Payne P.I. Genetics of wheat storage proteins and the effect of allelic variation on bread-making quality. *Annu. Rev. Plant Physiol.* 38: 141–153, June 1987.

62. Ewart, J.A.D. Glutenin structure. *J. Sci. Food Agric.* 30: 482–492, 1979.

63. Ewart, J. A. D. A modified hypothesis for the structure and rheology of glutelins. *J. Sci. Food Agric.* 23: 687–699, 1972.

64. Kasarda, D.D., Bernardin, J.E., and Nimmo, C.C. In: Pomeranz, Y. (Ed.), *Advances in Cereal Science and Technology*, American Association of Cereal Chemists, St Paul, MN, 1976.

65. Khan, K. and Bushuk, W. Studies of Glutenin. XII. Comparison by sodium dodecyl sulfate-polyacrylamide Gel Electrophoresis of unreduced and reduced Glutenin from various isolation and purification procedures. *Cereal Chem.* 56: 63–68, 1979.

66. Schofield, J.D., Bottomley, R.C., Timms, M.F., and Booth, M.R. The effect of heat on wheat gluten and the involvement of sulphydryl-disulphide interchange reaction. *J. Cereal Sci.* 1: 241–253, 1983.

67. Krull, L.H., Wall, J.S., Zobel, H., and Dimler, R.J. Synthetic polypeptides containing side-chain amide groups. Water soluble Polymers. *Biochemistry* 5: 1521–1527, 1966.

68. Bollecker, S., Viroben, G., Popineau, Y., and Gueguen, J. Acid deamidation and enzymic modification at pH 10 of wheat gliadins: Influence on their functional properties. *Sci. Aliments* 10: 343–356, 1990.

69. Popineau, Y. and Thebaudin, J.Y. Functional properties of enzymatically hydrolyzed glutens. In: Bushuk, W. and Tkachuk, R. (Eds.), *Gluten Protein*, American Association of Cereal Chemists, St Paul, MN, 1990.

70. Hamada, J.S. and Marshall, W.E. Preparation and functional properties of enzymatically deamidated soy proteins. *J. Food Sci.* 54: 598–601, 1989.

71. Finley, J.W. Deamidated gluten: A potential fortifier for fruit juices. *J. Food Sci.* 40: 1283–1285, 1975.

72. Wright, H.T. and Robinson, A.B. Cryptic amidase active sites catalyze deamidation in proteins. In: Kaplan, N.O. and Robinson, A.B. (Eds.), *From Cyclotrons to Cytochromes*, Academic Press, New York, 1982.

73. Metwalli, A.A.M. and van Boekel, M.A.J.S. On the kinetics of heat-induced deamidation and breakdown of caseinate. *Food Chem.* 61(1–2): 53–61, January 1998.

74. Robinson, A.B. and Rudd, C. Deamidation of glutaminyl and asparaginyl residues in peptides and proteins. *Curr. Top. Cell. Reg.* 8: 248, 1974.

75. Esen, A., Bietz, J.A., Paulis, J.W., and Wall, J.S. Tandem repeats in the N-terminal sequence of a proline-rich protein from corn endosperm. *Nature* 296: 678–679, April 15, 1982.

76. Bietz, A. Cereal prolamin evolution and homology revealed by sequence analysis. *Biochem. Genet.* 20(11–12), December 1982.

77. Wrigley, C.W. and Shepherd, K.W. Electrofocusing of grain proteins from wheat genotypes. *Ann. N. Y. Acad. Sci.* 209: 154–162, 1973.

78. Brown, J.W.S., Law, C.N., Worland, A.J., and Flavell, R.B. Genetic variation in wheat endosperm proteins. An analysis by two-dimensional electrophoresis using intervarietal chromosomal substitution lines. *Theor. Appl. Genet.* 59: 361–371, 1981.

79. Payne, P.I., Holt, L.M., and Law, C.N. Structural and genetic studies on the high-molecular-weight subunits of wheat glutenins. Part I. Allelic variation in subunits amongst varieties of wheat (*Triticum aestivum*). *Theor. Appl. Genet.* 60: 229–236, 1981.

80. Sears, E.R. Wheat cytogenetics. *Ann. Rev. Genet.* 3(3): 451–468, 1969.

81. Von Heijne, G. Patterns of amino acids near signal-sequence cleavage sites. *Eur. J. Biochem.* 133: 17–21, 1983.

82. Spena, A., Viotti, A., and Pirrotta, V. A homologous repetitive block structure underlies the heterogeneity of heavy and light chain zein genes. *EMBO J.* 1: 1589–1594, 1982.

83. Gething, M.J. and Sambrook, J. Protein folding in the cell. *Nature* 355: 33–45, 1992.

84. U.S. Department of Agriculture. *Composition of Foods*, USDA Handbooks 8(1–20), U.S. Government Printing Office, Washington, DC, 1986. www.nal.usda.gov/fnic/foodcomp

85. Rowland, S.J. The determination of the nitrogen distribution in milk. *J. Dairy Res.* 9: 42, 1938.

86. Garcia-Palmer F.J., Serra N., Palou A., and Gianotti, M. Free amino acids as indices of Mahon cheese ripening. *J. Dairy Sci.* 80(9): 1908–1917, 1997

87. FitzGerald, R.J. and Meisel, H. Milk protein derived inhibitors of angiotensin-I-converting enzyme. *Br. J. Nutr.* 84: S33–S37, 2000.

88. Korhonen, H. and Pihlanto, A. Food-derived bioactive peptides—opportunities for designing future foods. *Curr. Pharma. Des.* 9: 1297–1308, 2001.
89. Meisel, H. and FitzGerald, R.J. Biofunctional peptides from milk proteins. Mineral binding and cytomodulatory effects. *Curr. Pharmaceut. Des.* 9: 1289–1295, 2003.
90. Migliore-Samour, D., Floch, F., and Jollés, P. Biologically active casein peptides implicated in immunomodulation. *J. Dairy Res.* 56: 357–362, 1989.
91. Meisel, H. Overview on milk protein-derived peptides. *Int. Dairy J.* 8: 363–373, 1998.
92. Singh, T.K., Fox. P.F., and Healy, A. Isolation and identification of further peptides in the diafiltration retentate of the water-soluble fraction of Cheddar cheese. *J. Dairy Res.* 64: 433–443, 1997.
93. Smacchi, E. and Gobbetti, M. Peptides from several Italian cheeses inhibitory to proteolytic enzymes of lactic acid bacteria, *Pseudomonas fluorescens* ATCC 948 and to the angiotensin I-converting enzyme. *Enzyme Microb. Technol.* 22: 687–694, 1998.
94. Nakamura, Y., Yamamoto, N., Sakai, K., and Takano. T. Antihypertensive effect of sour milk and peptides isolated from it that are inhibitors to angiotensin I-converting enzyme. *J. Dairy Sci.* 78: 1253–1257, 1995.
95. Saito, T., Nakamura, T., Kitazawa, H., Kawai, Y., and Itoh. T. Isolation and structural analysis of antihypertensive peptides that exist naturally in Gouda cheese. *J. Dairy Sci.* 83: 1434–1440, 2000.
96. Ryhänen, E.L. and Pihlanto-Leppälä, A., and Pahkala, E. A new type of ripened; low-fat cheese with bioactive properties. *Int. Dairy J.* 11: 441–447, 2001.
97. Gobetti, M., Minervini, F., and Rizzello C.G. Angiotensin I-converting-enzyme-inhibitory and antimicrobial bioactive peptides. *Int. J. Dairy Technol.* 57: 173–188, 2004.
98. Gomez-Ruiz, J.A., Ramos M., and Recio, I. Angiotensin-converting enzyme-inhibitory peptides in Manchego cheeses manufactured with different starter cultures. *Int. Dairy J.* 12: 697–706, 2002.
99. Ashar, M.N. and Chand, R. Fermented milk containing ACE-inhibitory peptides reduces blood pressure in middle aged hypertensive subjects. *Milchwissenschaft* 59: 363–366, 2004.
100. Phelan, M., Aherne, A., FitzGerald, R.J., and O'Brien, N.M. Casein-derived bioactive peptides: Biological effects, industrial uses, safety aspects and regulatory status, *Int. Dairy J.* 19(10): 551–642, 2009.
101. Illingworth, J.A. *Muscle Structure and Function*, School of Biochemistry and Molecular Biology, Leeds, U.K., http://www.bmb.leeds.ac.uk/illingworth/muscle/index.htm, 2008.
102. Bodwell, C.E., Carpenter, K.J., and McDonough, F.E. A collaborative study of methods of protein evaluation: introductory paper. *Plant Foods Hum. Nutr.* 39: 3–11, 1989.
103. Wardlaw, G.M. and Kesse, M. *Perspectives in Nutrition*, 5th edn, McGraw-Hill, Boston, MA, 2002.
104. USDA National Nutrient Database for Standard Reference, Release 20, United States Department of Agriculture. Last modified on September 26, 2007.
105. FAO, *Maize in Human Nutrition*, *FAO Food and Nutrition Series*, No. 25, ISBN 92-5-103013-8, FAO, Rome (Italy), 1992.
106. Coss-Bu, J. Resting energy expenditure and nitrogen balance in critically ill pediatric patients on mechanical ventilation. *Nutrition*, 14(9): 649–652, 1998.
107. Satterlee, L.D., Marshall, H.F., and Tennyson, J.M. Measuring protein quality. *J. Am. Oil Chem. Soc.* 56: 103–109, 1979.
108. Steinke, F.H., Prescher, E.E., and Hopkins, D.T. Nutritional evaluation (PER) of isolated soybean protein and combinations of food proteins. *J. Food Sci.* 45: 323–327, 1980.
109. McDonough, F.E., Sarwar, G., Steinke, F.H., Slump, P., Garcia, S., and Boisen, S. In vitro assay for protein digestibility: Interlaboratory study. *J. Assoc. Off. Anal. Chem.* 73: 622–625, 1990.
110. FAO/WHO/UNU. Energy and protein requirements. Report of a Joint FAO/WHO/UNU Expert Consultation, WHO Tech. Rep. Ser. No. 724, WHO, Geneva, Switzerland, 1985.
111. National Research Council. *Nutrient Requirements for Laboratory Animals*, No. 10, National Academy of Sciences, Washington, D.C., 1978.
112. Sarwar, G., Peace, R.W., and Botting, H.G. Corrected relative net protein ratio (CRNPR) method based on differences in rat and human requirements for sulfur amino acids. *J. Assoc. Off. Anal. Chem.* 68: 689–693, 1985.
113. Codex Alimentarius Commission, Document ALINORM 89/30, FAO, Rome, Italy, 1989.
114. McDonough, F.E. et al. In vivo rat assay for true protein digestibility: Collaborative study. *J. Assoc. Off. Anal. Chem.* 73: 801–805, 1990.
115. FAO. Amino acid content of foods and biological data on proteins. Nutr. Div. Pub. 24. FAO, Rome, Italy, 1970.
116. Pedersen, B. and Eggum, B.O. Prediction of protein digestibility by an in vitro enzymatic pH-stat procedure. *Z. Tierphysiol. Tierernähr. Futtermlttelkd.* 49: 265–277, 1983.

117. Sarwar, G. and McDonough, F.E. Evaluation of protein digestibility-corrected amino acid score method for assessing protein quality of foods. *J. Assoc. Off. Anal. Chem.* 73: 347–356, 1990.
118. Boutrif, E. Food Quality and Consumer Protection Group, Food Policy and Nutrition Division, FAO, Rome: Recent Developments in Protein Quality Evaluation. *Food, Nutrition and Agriculture*, Issue 2/3, 1991.
119. Eggum, B.O. *A Study of Certain Factors Influencing Protein Utilization in Rats and Pigs*. Publ. 406, National Institute of Animal Science, Copenhagen, Denmark, 1973.
120. Millward, D.J., Layman, K.D., Tomé D., and Schaafsma, G. Protein quality assessment: Impact of expanding understanding of protein and amino acid needs of optimal health. *Am. J. Clin. Nutr.* 87(5): 1576S–1581S, 2008.
121. Schaafsma, G. Center of Expertise Nutrition, DMV International-Campina Melkunie, 6700 AA, Wageningen, the Netherlands The Protein Digestibility-Corrected Amino Acid Score. *J. Nutr.* 130(7): 1865S–1867S, 2000.
122. Berdanier, C.D. *CRC Desk Reference for Nutrition*, 2nd edn, CRC Press, Boca Raton, FL, 2005.
123. Srikantia, S.G. University of Mysore, Mysore: The Use of Biological Value of a Protein in Evaluating Its Quality for Human Requirements, Joint FAO/WHO/UNU Expert Consultation on Energy and Protein Requirements EPR 81 29, Rome, August 1981.
124. Johnston, T.K. Nutritional implications of vegetarian diets. In: Shills, M.E. et al. (Eds.), *Modern Nutrition in Health and Disease*, 9th edn, Williams & Wilkins, Baltimore, MD, 1999.
125. Mohammed H. Rahman, Iqbal Hossain, and Moslehuddin, Nutritional evaluation of sweet lupin (*Lupinus angustifolius*): Net protein utilization (NPU) nitrogen balance and fractionation studies. *Br. J. Nutr.* 77(03): 443–457, March 1997.
126. Miller, D.S. and Bender, A.E. The determination of the net utilization of proteins by a shortened method. *Br. J. Nutr.* 9: 382–388, 1955.
127. Campbell, W.W., Johnson, C.A., McCabe, G.P., and Carnell, N.S. Dietary protein requirements of younger and older adults. *Am. J. Clin. Nutr.* 88(5): 1322–1329, 2008.

4 Enzymes Applied in Food Technology

Dimitris G. Arapoglou, Athanasios E. Labropoulos,
and Theodoros H. Varzakas

CONTENTS

4.1 INTRODUCTION TO ENZYMES

Enzymes are special kinds of proteins that consist of long chains of amino acids bound together by peptide bonds. They are found in all living cells, performing specific functions, i.e., controlling the metabolic processes, converting nutrients into energy, breaking down food materials, etc. For example, the enzymes pepsin, trypsin, and peptidases break down proteins into amino acids, the lipases split fats into glycerol and fatty acids, and the amylases break down starch into simple sugars.

Enzymes are some of the proteins that are made by cells from amino acids. These are linked together by strong, covalent bonds, which are referred to as "peptide bonds." Thus each enzyme (as a protein) has a different amino acid sequence called "primary structure" (Figure 4.1). Some of the amino acid (side chain) residues (cysteine) form disulfide bridges that are not as strong as the peptide bond but can help to make the enzyme, for example, heat stable. Sometimes, amino acids connect together (wear manner) with a hydrogen atom through interactions called "hydrogen bonding." These create various shapes or conformations or structures, which are referred as (1) "secondary structure," if the hydrogen bonding exists between nitrogen and oxygen atoms in the peptide chain; (2) "tertiary structure" if there are interactions between the amino acid side groups; and (3)

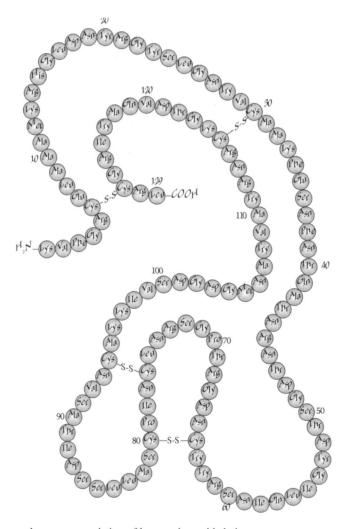

FIGURE 4.1 Enzyme lysozyme consisting of long amino acid chains.

"quaternary structure" if the polypeptides aggregate together into complexes. Some enzymes are made by cells in an inactive form called proenzymes or zymogens, which protect the cells from unregulated activities.

Lysozyme, depicted in Figure 4.1, is among the most extensively and thoroughly studied enzymes. It is produced in both plant and animal tissues. It cleaves polysaccharide chains found in the cell walls of certain bacteria by hydrolyzing the glycosidic bonds between neighboring hexosyl residues.[1]

Egg white lysozyme has a molecular weight of about 14,600 and consists of a single polypeptide chain of 129 amino acids. The polypeptide contains three, short helical regions, and a segment arranged in the form of a β-pleated sheet. All the polar residues are located on the enzyme's surface. Hydrogen bonds occur in the polypeptide chain.[2] Lysozyme is a hydrophilic, positively charged globular protein with four disulfide bonds.[3] The enzyme is oval in shape with a deep cleft across its midline (containing the active site) that divides the molecule into two parts. This cleft consists of hydrophobic groups that form van der Waals forces with the substrate. In the catalytic site (active site), two acidic amino acids, aspartic acid (Asp 52) and glutamic acid (Glu 35), exist, which are involved in catalysis. These two acids are attached to secondary structure components in their two-separated domains.[4]

Biotechnology has managed nowadays to isolate genes from cells by introducing mutation, making copies, and reinserting them back to the cells. These pieces of DNA are mixed and reformed in different ways, making up the so-called recombinant DNA. This leads to a new protein (enzyme), which is specific for a particular action. Thus, microorganisms can create gene coding for animal and plant enzymes. However, bacteria are less adapted to these modifications compared to fungi and yeasts (e.g., bacteria make incorrect disulfide bridges and will not glycosylate proteins). Today, there are many patents being used to make recombinant enzymes (e.g., certain lipases) by the use of genetic engineering.

Enzymes are biocatalysts and involve in the speed up of slow biochemical processes. In general, enzymes are released again after a reaction ceases and can continue in another reaction. Practically, this process cannot go forever, since more catalysts have limited stabilities and, slowly, they become inactive. In the food industry, enzymes are often used once and then they are discarded. In comparison to inorganic catalysts, i.e., acids, bases, metals, and metal oxides, enzymes have very specific functions. Enzyme actions are limited to specific bonds in their reactions with various compounds. An enzyme molecule usually binds to the substrate(s) and a specialized part(s) of it to catalyze the substrate into a product. For each type of reaction in a cell, there is a different enzyme. The specific actions of enzymes in industrial processes usually obtain high production yields with a minimum level of by-products.

Environmental conditions (temperature, pressure, and acidity) play an important role in enzyme functionality. For example, some enzymes function better at temperatures ranging between 30°C and 70°C and at nearly the neutral pH, while others function optimally at other specific conditions. Lately, specific enzymes have been developed that can work at extreme conditions, i.e., resistant to high heat, high pressure, and corrosion, for specific applications. Currently, enzyme immobilization and other new technologies have succeeded in obtaining enzyme processes suited for potential energy savings, making investments on specific equipment, and also with a wide range of applications in the food industry. This can be due to their efficiency, specific actions, the mild conditions in which they work, and their high biodegradability.

Enzyme technology is the application of free enzymes and whole-cell biocatalysts in the production of foods and services. A more narrow definition limits enzyme technology to the technological concepts that allow the use of enzymes in competitive large-scale bioprocesses. Enzyme technology is an interdisciplinary field, recognized by the Organization for Economic Cooperation and Development (OECD) as an important component of sustainable industrial development.

The choice of getting (buying) an enzyme is dependent on the following factors[5,6]:

- Cost benefit (adding value or reducing production cost)
- Availability, consistency, and quality support (reputation of suppliers)
- Activity (specific substrate alteration by pH, ions, temperature, and inhibitors)
- Ability to modify a reaction's quality (quality measurement and understanding of an enzyme can control its activity more precisely)

4.1.1 ENZYME HISTORY

2000 BC. The Egyptians and Sumerians developed various and specific fermentations for use in brewing, bread making, and cheese making.

800 BC. Calves' stomachs and the enzyme chymosin were used for cheese making.

1878. The components of yeast cells, which cause fermentation, were identified and the term "enzyme" was coined. The word itself means "in yeast" and is derived from the Greek *en* meaning "in" and *zyme* meaning "yeast" or "leaven."

1926. Enzymes were first known to be proteins.

1980s. Enzyme preparations were developed to improve the digestibility and nutrient availability of certain animal feeds.

1982. The first food application of a product of gene technology, i.e., α-amylase, took place.

1988. Recombinant chymosin was approved and introduced in Switzerland, marking an early approval of a product of gene technology for food application.

1990s. Two food processing aids obtained using gene technology were applied, i.e., an enzyme for use in cheese making (United States), and a yeast for use in baking (United Kingdom).

2000s. Advances with enzymes (e.g., immobilized enzymes) are taking place with many applications to food industry (flavors, aromas, etc.)

4.1.2 ENZYME CHARACTERISTICS

In general, living systems control their activities through the aid of enzymes, which are protein molecules, i.e., biological catalysts with the following three main characteristics.[7] First, the main function of an enzyme is to increase the rate of a reaction. Most cellular reactions occur about a million times faster than they would in the absence of an enzyme. Second, most enzymes act specifically with only one reactant called "substrate" in order to produce a product (Figure 4.2). The third characteristic is that enzymes are regulated from a state of low activity to high activity and vice versa. The individuality of a living cell is mostly due to the unique set of some 3000 enzymes programmed genetically to control the activities of a living system. If even one enzyme is missing or defective, the results can be disastrous. Most of the information and knowledge we have today about enzymes has been made possible through the isolation technologies of enzymes from cells that are made to work in a test tube environment. In addition, extensive work has also been done by x-ray diffraction techniques to elucidate the three-dimensional structure of some enzymes.[8] The ribbon and backbone

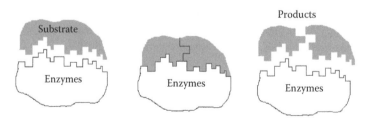

FIGURE 4.2   An enzyme functionality in its ribbon and backbone forms.

form of the enzyme carboxypeptidase (shown in Figure 4.2) hydrolyzes a peptide at the carboxyl or the C terminal end of the chain.

Some of the characteristics of enzymes are the following:

1. High activity
2. Selectivity
3. Regiospecificity
4. Stereo-specificity
5. Controllability
6. Environmentally friendly

High activity. Enzymes can increase the rate of a reaction millions of times by lowering the activation energy of the reaction like conventional chemical catalysts. The maximum rate of conversion of a substrate to a product by a molecule of an enzyme is known as the "turnover number" (K_{cat}). For example, the K_{cat} for catalase, which catalyzes the conversion of hydrogen peroxide to O_2 and H_2O, is approximately 600,000 molecules per second per molecule of enzyme.

Selectivity. The catalytic active site of an enzyme, which is the result of folding of the polypeptide chain, fits only to a small (specific) number of substrates for the conversion to products. Since an enzyme acts only on one compound of a complex mixture, it can control a reaction independently of many others occurring in the complex. Minor changes, like increasing the temperature, in the structure of the substrate may cause the enzyme to be unable to recognize the substrate and convert it to corresponding products.

Regiospecificity. An enzyme can detect differences in the spatial arrangement of atoms in a compound. For example, monoxygenase enzymes oxidize methyl cyclohexane, but different enzymes are found to oxidize either the methyl substitute or the methylene ring group. This has potential application in the production of synthetic fragrances and flavors.

Stereospecificity. Enzymes are highly stereospecific in the choice of substrate or in the formed product. Their selectivity for the substrate usually extends to discernment between different forms (chiral) of the substrate molecule (D or L forms). For example, bacterial glucose oxidase will use only D-glucose.

Controllability. The catalysis of reactions by enzymes can be controlled by the amount of substrate and by other factors (temperature, pH, etc.). An enzyme is active (switched on) in the presence of a correct amount of the substrate and inactive (turned off) when the substrate is not available. For example, the response of certain bacteria to the presence of lactose induces the formation of β-galactosidase (lactase), which hydrolyses lactose to glucose and galactose used as food by organisms.

Environmental factors. Reactions catalyzed by enzymes (biocatalysts with high specificity) are less environmentally polluting due to the lower by-product produced. The synthesized products can be classified as "natural" when substrates are naturally derived, and they are valuable to the food industry as they are not chemically manufactured.

4.1.3 ENZYME CONFIGURATION

The activity of an enzyme depends on a specific protein molecule but other molecules and ions affect the enzyme structure, facilitating their activity as catalysts. In this case, the enzyme consisted of a protein, and a combination of one or more parts are called "cofactors." When an extra molecule is covalently attached to the enzyme, it is referred as the "prosthetic group." An example of this group is the flavine adenine dinucleotide (FAD), which is a part of glucose oxidase, that acts to generate H_2O_2.

The polypeptide or protein part of the enzyme is called an apoenzyme (Figure 4.3) and may be inactive in its original

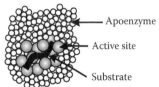

FIGURE 4.3 An enzyme and its associated parts.

synthesized structure. The inactive form of the apoenzyme is known as a proenzyme or zymogen. The proenzyme may contain several extra amino acids in the protein that are removed, and allows the final, specific tertiary structure to be formed before it is activated as an apoenzyme.

A cofactor is usually a nonprotein substance, which may be organic, and is called a coenzyme. A coenzyme is often derived from a vitamin. Another type of cofactor is an inorganic metal ion called a metal ion activator. The inorganic metal ions may be bonded through coordinate covalent bonds. Examples are (1) calcium (Ca^{2+}) ions bound to α-amylase, which degrades starch; (2) iron (Fe^{2+}) ions bound to catalase, which degrades hydrogen peroxide to O^{2-} and H_2O; and (3) nicotinamide adenine dinucleotide (NAD^+) associated with dehydrogenase, which is involved in the citric acid (Krebs) cycle.

The type of association between the cofactor and the apoenzymes varies but both come together only during the course of a reaction. The bonds sometimes are loose and come together only during the course of a reaction, while at other times, they are firmly bound together by covalent bonds. The role of a cofactor is to activate the protein by changing its geometric shape, or by actually participating in the overall reaction. The whole enzyme contains a specific geometric shape called the active site where the reaction takes place. The molecule acted upon is called the substrate.

4.1.4 SOURCE OF ENZYMES

Commercial sources of enzymes are any living organism, i.e., animals, plants, and microbes. These naturally occurring enzyme sources are quite readily available for the commercial productivity of sufficient quantities for food applications and/or other industrial uses. Table 4.1 shows some important representative industrial enzymes from animal, plant, and microbial sources. Of the number of these industrial enzymes, the majority (more than half come from fungi and yeast and a third from bacteria) come from microorganisms and the reminder, a minority, from animals (8%) and plants (4%). A number of enzymes have found use, also, in chemical analyses and medical diagnosis. However, enzymes from microbes are preferred than from animals and plants because (1) they are cheaper to produce; (2) they are more predictable and controllable; (3) they are reliable supplies of raw materials of constant composition; and (4) plant and animal tissues contain more harmful materials (phenolics, inhibitors, etc.)

In general, the majority of microbial enzymes come from a limited number of genera, of which *Aspergillus*, *Bacillus*, and other *Saccharomyces* predominate. Examples of such strains used in the food industry are for the production of high-fructose corn syrup (HFCS) using pullulanase and glucose isomerase in starch hydrolyses.[9] Both producers and users of industrial enzymes might wish to share the common goal of economy, effectiveness, and safety. The development of commercial enzyme production requires high skills in (1) developing new/improved enzymes, (2) fermentation technology, and (3) regulatory aspects. However, by isolating microbial strains that produce the desired enzyme and optimizing the conditions for growth, commercial quantities can be obtained. This technique is well known for more than 3000 years and is called fermentation. Today, such fermentation processes are carried out in contained vessels where once fermentation is completed, the microorganisms are destroyed and the enzymes are isolated and further processed for commercial use.

Today, microbial enzymes are being employed, which have a high specific activity per unit dry weight and they can be grown in large volumes using fermentation techniques.[10] Furthermore, microbial enzymes are not subject to seasonal fluctuations, like plant and animal enzymes. Microbial enzymes possess more useful properties, for example, they have a broad range of specificities and some can tolerate extremes of pH, while others are heat stable. Catalase, originating from beef, is a microbial enzyme possessing certain advantages and more specifically the pH range (for higher than 60% activity) for a microbial enzyme is 4.5–8, whereas that from a fungal source has a respective range of 2–10, and the temperature range (for higher than 60% activity) of a microbial enzyme is 10°C–45°C, whereas that of a fungal source has a respective range of 10°C–55°C. Industrial genetic techniques enable the

---

TABLE 4.1

Some Representative Industrial Enzymes with Their Sources and Application

Enzyme	EC Number	Source	Product Application
a. Animal enzymes			
Rennet	3.4.23.4	Abomasums Fourth stomach of ruminant animals	Cheese
Catalase	1.11.1.6	Liver	Food
Lipase	3.1.1.3	Pancreas	Food
b. Plant enzymes			
α-Amylase	3.2.1.1	Malted barley	Brewing
Lipoxygenase	1.13.11.12	Soybeans	Food
Papain	3.4.22.2	Papaya	Meat
c. Bacterial enzymes			
β-Amylase	3.2.1.1	*Bacillus*	Starch
Asparaginase	3.5.1.1	*E. coli*	Health
Glucose isomerase	5.3.1.5	*Bacillus*	Fructose syrup
Pullulanase	3.2.1.41	*Klebsiella*	Starch
d. Fungal enzymes			
α-Amylase	3.2.1.1	*Aspergillus*	Baking
Lactase	3.2.1.23	*Aspergillus*	Dairy
Pectinase	3.2.1.15	*Aspergillus*	Drinks
e. Yeast enzymes			
Invertase	3.2.1.26	*Saccharomyces*	Confectionary
Lactase	3.2.1.23	*Kluyveromyces*	Dairy
Lipase	3.1.1.3	*Candida*	Food

Sources: Data from NC-IUBMB, Enzyme nomenclature: Recommendations of the Nomenclature Committee of the International Union of Biochemistry and Molecular Biology on the nomenclature and classification of enzyme-catalyzed reactions by Nomenclature Committee of the International Union of Biochemistry and Molecular Biology, available online at www.chem.qmul.ac.uk/iubmb/ enzyme/; NC-IUBMB, Symbolism and terminology in enzyme kinetics Nomenclature Committee of the International Union of Biochemistry, available online at www.chem.qmul.ac.uk/iubmb/kinetics/ prepared by G.P. Moss department of chemistry, Queen Mary University of London, U.K., 1981.

---

large production of plant and animal enzymes with all the benefits of microbial enzymes.[11] Microbial enzymes are intracellularly (soluble and membrane bound) or extracellularly derived. Extracellular enzymes are much more stable and, thus, they have better industrial applications.

4.1.5 CLASSIFICATION OF ENZYMES

Enzyme classification is primarily based on the recommendations of the Nomenclature Committee of the International Union of Biochemistry and Molecular Biology (IUBMB)[12], and it describes each type of characterized enzyme for which an EC (Enzyme Commission) number has been provided. EC classes define enzyme function based on the reaction, which is catalyzed by the enzyme. The classification scheme is hierarchical, with four levels. There are six broad categories of function at the top of this hierarchy and about 3500 specific reaction types at the bottom. EC classes are expressed as a string of four numbers separated by periods. EC class strings with fewer than four numbers refer to an internal node in the tree, implicitly including all of the subclasses and leaves below it. The numbers specify a path down the hierarchy, with the leftmost number identifying the highest level.

4.1.5.1 Oxidoreductases

These enzymes catalyze oxidation–reduction reactions in which hydrogen or oxygen atoms or electrons are transferred between molecules. This large class includes dehydrogenases (hydride transfer), oxidases (electron transfer to molecular oxygen), oxygenases (oxygen transfer from molecular oxygen), and peroxidases (electron transfer to peroxide). An example is shown in Figure 4.4, with the glucose oxidase (EC 1.1.3.4; systematic name, β-D-glucose-oxygen 1-oxidoreductase) reaction and its products.

4.1.5.2 Transferases

These enzymes catalyze the transfer of an atom or group of atoms (e.g., acyl-, alkyl-, and glycosyl-), from one molecular and/or functional groups to another, but excluding such transfers as are classified in the other groups (e.g., oxidoreductases and hydrolases). An example is shown in Figure 4.5, with the aspartate aminotransferase (EC 2.6.1.1; systematic name, L-aspartate: 2-oxoglutarate aminotransferase; also called glutamic-oxaloacetic transaminase or simply GOT) reaction and its product.

4.1.5.3 Hydrolases

These involve hydrolytic reactions and their reversal (degradation of H_2O to OH^- and H^+ products). This is presently the most commonly encountered class of enzymes within the field of enzyme technology and includes the esterases, glycosidases, lipases, and proteases. An example of the chymosin (EC 3.4.23.4; no systematic name declared; also called rennin) reaction and its product are shown in Figure 4.6.

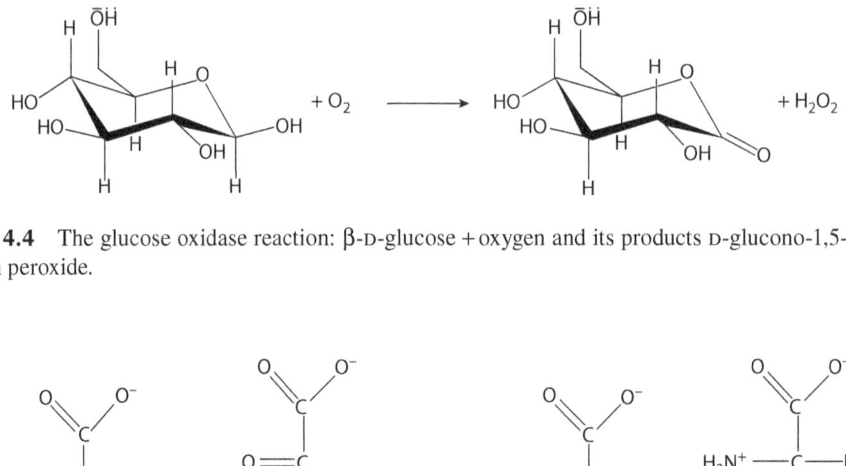

FIGURE 4.4 The glucose oxidase reaction: β-D-glucose + oxygen and its products D-glucono-1,5-lactone + hydrogen peroxide.

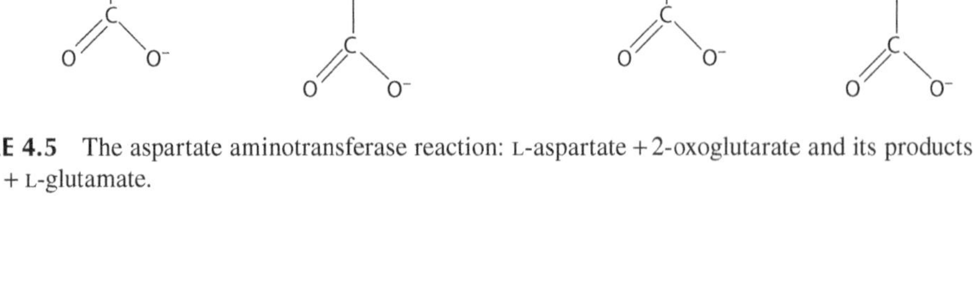

FIGURE 4.5 The aspartate aminotransferase reaction: L-aspartate + 2-oxoglutarate and its products oxaloacetate + L-glutamate.

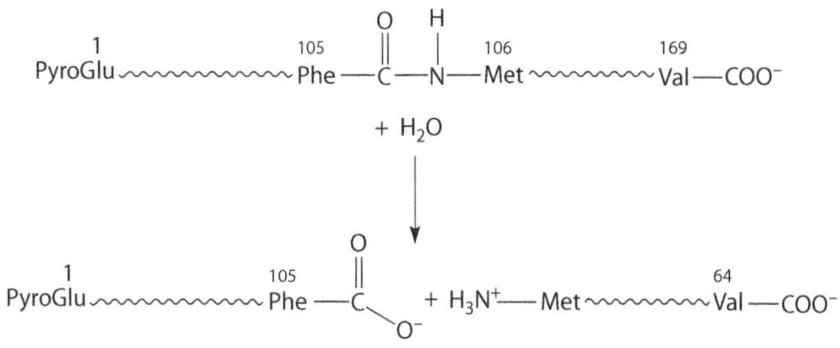

FIGURE 4.6 The chymosin reaction: κ-casein + water and its products *para*-κ-casein + caseino macropeptide.

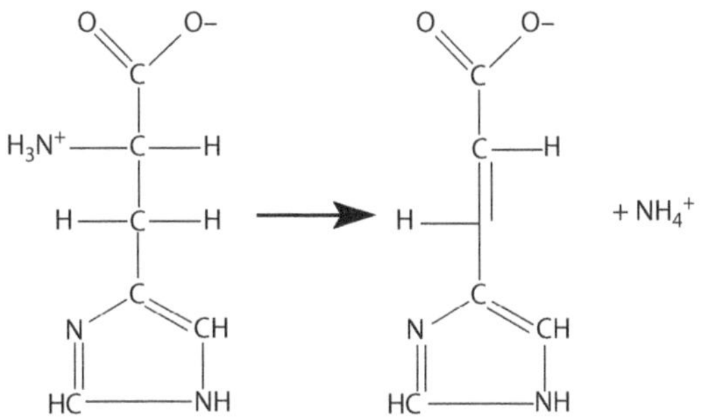

FIGURE 4.7 The histidine ammonia-lyase reaction: L-histidine and its products urocanate + ammonia.

4.1.5.4 Lyases

Lyases involve elimination reactions in which a group of atoms is removed from the substrate. These catalytic reactions require the addition of groups to a double bond or the formation of a double bond (e.g., C=C, C=N, C=O). This includes the aldolases, decarboxylases, dehydratases, and some pectinases but does not include hydrolases. An example is the histidine ammonia-lyase (EC 4.3.1.3; systematic name, L-histidine ammonia-lyase, which is also called histidase) reaction with its products, shown in Figure 4.7.

4.1.5.5 Isomerases

Isomerases catalyze molecular isomerizations and include the epimerases, racemases, and intramolecular transferases. An example is shown in Figure 4.8, with the xylose isomerase (EC 5.3.1.5; systematic name, D-xylose ketol-isomerase; commonly called glucose isomerase) transformation of α-D-glucopyranose to α-D-fructofuranose.

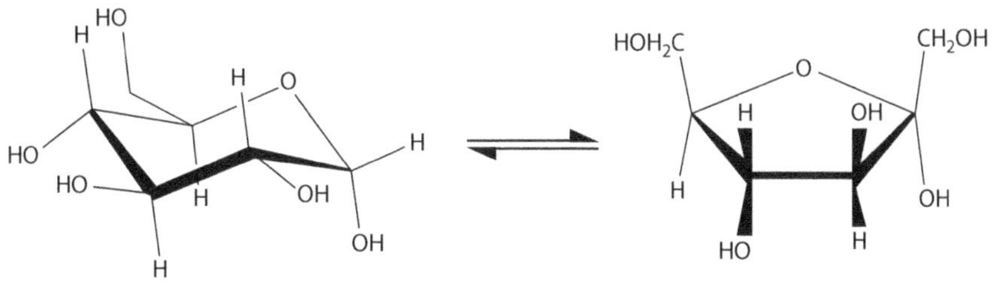

FIGURE 4.8 The xylose isomerase reaction: α-D-glucopyranose and its product α-D-fructofuranose.

4.1.5.6 Ligases

Ligases catalyze the condensation of two molecules together with the cleavage of ATP or another pyrophosphate bond. They, also known as synthetases, form a relatively small group of enzymes, which involve the formation of a covalent bond joining two molecules together, coupled with the hydrolysis of a nucleoside triphosphate. An example is the glutathione synthase (EC 6.3.2.3; systematic name, α-L-glutamyl-L-cysteine:glycine ligase (ADP-forming), which, also, called glutathione synthetase) reaction, shown in Figure 4.9 with its products.

Each enzyme category (e.g., EC 1) is divided into subclasses (e.g., EC 1.2) and its subclass is subdivided into sub-subclasses (e.g., EC 1.2.2). The last (fourth) number of the IUBMB nomenclature refers to the particular enzyme (e.g., EC 1.2.2.3). For example, the enzyme EC 1.10.3.2 (laccase) represents an *oxidoreductase* enzyme (EC 1) that acts on diphenols and related substances as donors (EC 1.10), with oxygen as the acceptor (EC 1.10.3). Table 4.2 shows the six EC enzyme classes with their functionalities.

In industry, the two major groups of enzyme categories, which are being used, are (1) the oxidoreductases, for example, catalase and glucose oxidase and (2) the hydrolases, for example, amylases, proteases, cellulase, invertase, etc. Transferases, lyases, and ligases are not used extensively, while some isomerases have been extremely useful in the industries (production of syrups). In industry, there are the following three main ways in which enzymes can be used as (1) soluble (fungal α-amylase—baking); (2) immobilized (bacterial glucose isomerase—syrups, e.g., HFCS); and (3) nonproliferating whole cells (*Pseudomonas putida*—adipic acid). Soluble and immobilized enzymes are intracellular or extracellular, while nonproliferating whole cells are nondividing cells. Examples of these enzymes are given bellow. The soluble enzymes are not used in continuous processes but in batch-type reactors because they are sacrificed after the reaction is completed and are not economically justified for recovery. These enzymes, usually, diffuse in the substrate and need to be rinsed or inactivated to control the reaction.[13]

The immobilized enzymes are those enzymes that retain their catalytic activity, while their movement is physically restricted or localized in a defined space.[14] For example, an immobilized enzyme, within a matrix, should be able to diffuse the substrate and product in and out of the matrix, freely.[15] In other words, an immobilized enzyme is retained in the reactor while the desired product freely moves out, allowing a continual production of the desired product. Researchers, enzymologists, and biotechnologists are very much interested in immobilized enzymes and more specifically soluble enzymes as heterogeneous catalysts. Nonproliferating whole cells are those that are metabolically active but are starved of an essential nutrient for growth. They perform multi-sequence reactions.[16,17]

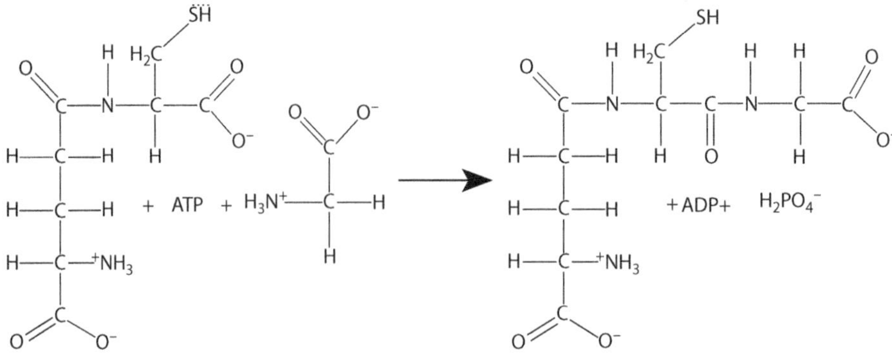

FIGURE 4.9 The glutathione synthase reaction: ATP + γ-L-glutamyl-L-cysteine + glycine and its products ADP + phosphate + glutathione.

TABLE 4.2
The EC Enzyme Classes with Their Activities

EC 1	Oxidoreductases
EC 1.1	Acting on CH-OH groups of donors
EC 1.2	Acting on aldehyde or oxo groups of donors
EC 1.3	Acting on CH-CH groups of donors
EC 1.4	Acting on CH-NH_2 groups of donors
EC 1.5	Acting on CH-NH groups of donors
EC 1.6	Acting on NADH or NADPH
EC 1.7	Acting on other nitrogenous compounds as donors
EC 1.8	Acting on sulfur groups of donors
EC 1.9	Acting on hemi groups of donors
EC 1.10	Acting on diphenols and related substances as donors
EC 1.11	Acting on a peroxide as acceptors
EC 1.12	Acting on hydrogen as donors
EC 1.13	Acting on single donors, incorporation of molecular oxygen (oxygenases)
EC 1.14	Acting on paired donors, incorporation or reduction of molecular oxygen
EC 1.15	Acting on superoxide radicals as acceptors
EC 1.16	Oxidizing metal ions
EC 1.17	Acting on CH or CH_2 groups
EC 1.18	Acting on iron-sulfur proteins as donors
EC 1.19	Acting on reduction flavodoxin as donors
EC 1.20	Acting on phosphorus or arsenic in donors
EC 1.21	Acting on X-H and Y-H to form X-Y bonds
EC 1.97	Other oxidoreductases

EC 2	Transferases
EC 2.1	Transferring one-carbon groups
EC 2.2	Transferring aldehyde or ketonic groups
EC 2.3	Acyltransferases
EC 2.4	Glycosyltransferases
EC 2.5	Transferring alkyl or aryl groups, other than methyl groups
EC 2.6	Transferring nitrogenous groups
EC 2.7	Transferring phosphorus-containing groups
EC 2.8	Transferring sulfur-containing groups
EC 2.9	Transferring selenium-containing groups

EC 3	Hydrolases
EC 3.1	Acting on ester bonds
EC 3.2	Glycosidases
EC 3.3	Acting on ether bonds
EC 3.4	Acting on peptide bonds (peptidases)
EC 3.5	Acting on carbon bonds, other than peptide bonds
EC 3.6	Acting on acid anhydrides
EC 3.7	Acting on carbon-carbon bonds
EC 3.8	Acting on halide bonds
EC 3.9	Acting on phosphorus-nitrogen bonds
EC 3.10	Acting on sulfur-nitrogen bonds
EC 3.11	Acting on carbon-sulfur bonds
EC 3.12	Acting on sulfur-sulfur bonds
EC 3.13	Acting on carbon-sulfur bonds

(continued)

TABLE 4.2 (continued)
The EC Enzyme Classes with Their Activities

EC 4	Lyases
EC 4.1	Carbon-carbon lyases
EC 4.2	Carbon-oxygen lyases
EC 4.3	Carbon-nitrogen lyases
EC 4.4	Carbon-sulfur lyases
EC 4.5	Carbon-halide lyases
EC 4.6	Phosphorus-oxygen lyases
EC 4.99	Other lyases
EC 5	**Isomerases**
EC 5.1	Recamases and epimerases
EC 5.2	*cis–trans*-Isomerases
EC 5.3	Intramolecular isomerases
EC 5.4	Intramolecular transferases (mutases)
EC 5.5	Intramolecular lyases
EC 5.99	Other isomerases
EC 6	**Ligases**
EC 6.1	Forming carbon-oxygen bonds
EC 6.2	Forming carbon-sulfur bonds
EC 6.3	Forming carbon-nitrogen bonds
EC 6.4	Forming carbon-carbon bonds
EC 6.5	Forming phosphoric ester bonds
EC 6.6	Forming nitrogen-metal bonds

Sources: NC-IUBMB, Enzyme nomenclature: Recommendations of the Nomenclature Committee of the International Union of Biochemistry and Molecular Biology on the nomenclature and classification of enzyme-catalyzed reactions, available online at www.chem.qmul.ac.uk/iubmb/enzyme/

4.1.6 ENZYME UNITS

The amount of enzyme present or used in a process is difficult to determine in absolute terms (i.e., grams), as its purity is often low and a proportion may be in an inactive or partially active state.[18] Other interesting parameters are the activity of the enzyme preparation and the activities of any contaminating enzymes. These activities are usually measured in terms of the activity unit "U," which is defined as the amount that will catalyze the transformation of $1\,\mu mol$ of the substrate per minute under standard conditions. Typically, this represents 10^{-6} to 10^{-11} kg for pure enzymes and 10^{-4} to 10^{-7} kg for industrial enzyme preparations.

Another unit of recommended enzyme activity is the "katal (kat)," which is defined as the amount that will catalyze the transformation of one mole of substance per second ($1\,kat = 60,000,000\,U$). This is an impracticable unit with very limited acceptance.

The activity holds major interest since it measures the content of the enzyme when the enzyme is to be used in a process. However, enzymes are usually marketed in terms of activity rather than weight. The specific activity (e.g., $U\,kg^{-1}$) is a parameter of interest as an index of purity but otherwise is of less importance. There is a problem with the definitions of activity (the rather vague notion of "standard conditions"). These are meant to refer to optimal conditions, especially with regard to pH, ionic strength, temperature, substrate concentration, and the presence and concentration of cofactors and coenzymes. However, these optimal conditions vary between laboratories and between suppliers. They also depend on the specific application in which the enzyme is to be used. Additionally, preparations of the same notional specific activity may differ with respect to stability and be capable of very different total catalytic productivity. This is the total substrate converted to the product during the lifetime of the catalyst, under specified conditions. These conditions for maximum initial activity are not necessarily those for maximum stability. Therefore, biotechnologists and processors should be careful when considering, these factors when the most efficient catalyst for a particular purpose is to be chosen.

4.2 ENZYME KINETICS

Enzymes are known as protein catalysts that, like all other catalysts, speed up the rate of a chemical reaction without being used up in the process. A *catalyst* is a substance that increases the rate of a reaction without modifying the overall standard Gibbs energy change in the reaction (Figure 4.10). An uncatalyzed reaction requires higher activation energy than does a catalyzed reaction.[19] This definition is equivalent to the statement that the catalyst does not appear in the stoichiometric expression of the complete reaction. Catalysts are said to exert a *catalytic action*, and a reaction in which a catalyst is involved is called a *catalyzed reaction*.

Kinetic equations are commonly expressed in terms of the *amount-of-substance concentrations* of the chemical species involved. The amount-of-substance concentration is the amount of substance (for which the SI unit is the mole, symbol mol) divided by the volume. As it is the only kind of concentration commonly used in biochemistry it is usually abbreviated to *concentration* and this shorter form will be used in the remainder of this document without further discussion. The unit almost invariably used for concentration is mol dm^{-3}, which is alternatively written as mol L^{-1}, mol l^{-1}, or simply KMK (molar).[20,21]

Added substances sometimes increase or decrease the rate of an enzyme-catalyzed reaction without interacting with the enzyme itself. These substances may interact with substrates or with modifiers or effectors that are already present in the system. Such substances are referred to as *activators* or *inhibitors*, but not as enzyme activators, enzyme inhibitors, modifiers, or effectors.[22]

An *inhibitor* is a substance that diminishes the rate of a chemical reaction and the process is called *inhibition*. In enzyme-catalyzed reactions an inhibitor frequently acts by binding to the enzyme, in which case it may be called an enzyme inhibitor. An *activator* is a substance, other

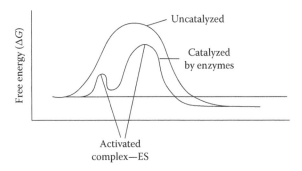

FIGURE 4.10 Higher activation energy is required at the uncatalyzed reaction compared to the catalyzed reaction.

than the catalyst or one of the substrates, that increases the rate of a catalyzed reaction. An activator of an enzyme-catalyzed reaction may be called an *enzyme activator* if it acts by binding to the enzyme.

A typical overall enzyme-catalyzed reaction involving a single substrate and a single product may be written as[23]

Enzyme + substrate ⇌ (enzyme − substrate complex) → enzyme + product

$$E + S \underset{k_{-1}}{\overset{k_{+1}}{\rightleftharpoons}} ES \xrightarrow{k_{+2}} P + E \qquad (4.1)$$

where k_{+1}, k_{-1}, and k_{+2} are the respective rate constants, typically having values of 10^5–10^8 M^{-1} s^{-1}, 1–10^4 s^{-1}, and 1–10^5 s^{-1} respectively.

The rate at which an enzyme works is influenced by the following:[24]

1. The concentration of substrate molecules (the more of them available, the quicker the enzyme molecules collide and bind with them). The concentration of substrate is designated [S] and is expressed in the unit of "molarity."
2. The temperature. As the temperature rises, molecular motion—and hence collisions between enzyme and substrate—speed up. This cannot be infinite since enzymes are proteins that have an upper limit beyond which the enzyme becomes denatured and ineffective.
3. The presence of inhibitors. They are classified in the following two categories:
   a. *Competitive inhibitors* are molecules that bind to the same site as the substrate preventing the substrate from binding—but are not changed by the enzyme.
   b. *Noncompetitive inhibitors* are molecules that bind to some other site on the enzyme reducing its catalytic power.
4. pH. The conformation of a protein is influenced by pH and as enzyme activity is crucially dependent on its conformation, its activity is likewise affected.

The study of the rate at which an enzyme works is called enzyme kinetics.[25] Following that, enzyme kinetics will be examined as a function of the concentration of the substrate available to the enzyme.

First let us set up a series of tubes containing graded concentrations of substrate, [S].[26] At time zero, we add a fixed amount of the enzyme preparation. Over the next few minutes, we measure the concentration of the product formed. If the product absorbs light, we can easily do this in a spectrophotometer. Early in the run, when the amount of substrate is in substantial excess to the amount of enzyme, the rate we observe is the initial velocity of V_i. By plotting V_i as a function of [S], the following are observed (Figure 4.11).

- At low values of [S], the initial velocity, V_i, rises almost linearly with increasing [S].
- But as [S] increases, the gains in V_i level off (forming a rectangular hyperbola).
- The asymptote represents the maximum velocity of the reaction, designated V_{max}.
- The substrate concentration that produces a V_i that is one-half of V_{max} is designated the Michaelis–Menten constant, K_m.

K_m is an inverse measure of the affinity or strength of binding between the enzyme and its substrate. The lower the K_m, the greater the affinity (so the lower the concentration of substrate needed to achieve a given rate).

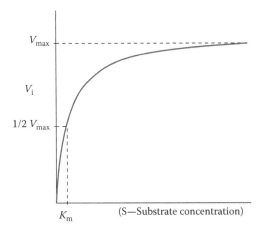

FIGURE 4.11 Initial velocity, V_i (moles/time), as function of substrate concentration, [S] (molar).

The Michaelis–Menten equation (Equation 4.2) is simply derived from enzyme kinetics equations, by substituting K_m for

$$\frac{K_{-1} + K_{+2}}{K_{+1}} \tag{4.2}$$

K_m is known as the **Michaelis constant** with a value typically in the range 10^{-1}–10^{-5} M. When $K_{+2} \ll K_{-1}$, K_m equals the dissociation constant (K_{-1}/K_{+1}) of the enzyme–substrate complex. The velocity at the reaction can be expressed by the following equations:

$$V = K_{+2}[ES] = \frac{K_{+2}[E]_0[S]}{[S] + K_m} \tag{4.3}$$

or, more simply

$$V = \frac{V_{max}[S]}{[S] + K_m} \tag{4.4}$$

where V_{max} is the maximum rate of reaction, which occurs when the enzyme is completely saturated with substrate (i.e., when [S] is very much greater than K_m, V_{max} equals $K_{+2}[E]_0$, as the maximum value [ES] can have is $[E]_0$ when $[E]_0$ is less than $[S]_0$). The above equation may be rearranged to show the dependence of the rate of reaction on the ratio of [S] to K_m as follows:

$$V = \frac{V_{max}}{1 + \dfrac{K_m}{[S]}} \tag{4.5}$$

The rectangular hyperbolic nature of the relationship, having asymptotes at $v = V_{max}$ and $[S] = K_m$, is shown in the equation:

$$(V_{max} - v)(K_m + [S]) = V_{max} K_m \tag{4.6}$$

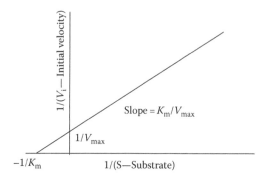

FIGURE 4.12 A double-reciprocal or Lineweaver–Burk relationship of $1/V_i$ versus $1/[S]$.

Plotting the reciprocals (Figure 4.12) of the "same data points" yields a "double-reciprocal" or Lineweaver–Burk plot. This provides a more precise way to determine V_{max} and K_m.

- V_{max} is determined by the point where the line crosses the $1/V_i = 0$ axis (so the [S] is infinite).
- The magnitude represented by the data points in this plot "decrease" from the lower left to upper right.
- K_m equals V_{max} times the slope of line, which is easily determined from the intercept on the x-axis.

4.2.1 INHIBITION

The inhibition is divided into the following two categories:[27]

1. Competitive, when the substrate and inhibitor compete for binding to the same active site
2. Noncompetitive, when the inhibitor binds somewhere else on the enzyme molecule reducing its efficiency

The distinction can be determined by plotting enzyme activity with and without the inhibitor present as shown in Figure 4.13.

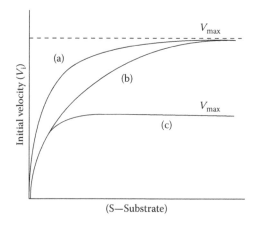

FIGURE 4.13 Diagrams show the distinction between competitive (b)/noncompetitive (c) inhibition and the curve with no inhibition (a).

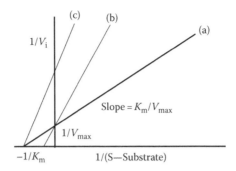

FIGURE 4.14 Results due to competitive (b) noncompetitive/(c) inhibition by the Lineweaver–Burk plot.

4.2.1.1 Competitive Inhibition

In the presence of a competitive inhibitor, it takes a higher substrate concentration to achieve the same velocities that were reached in its absence. So while V_{max} can still be reached if sufficient substrate is available, one-half V_{max} requires a higher [S] than before and thus K_m is larger (Figure 4.13).

4.2.1.2 Noncompetitive Inhibition

With noncompetitive inhibition, enzyme molecules that have been bound by the inhibitor are not counted so the following (Figure 4.14) are displayed by the Lineweaver–Burk plot.

1. Enzyme rate (velocity) is reduced for all values of [S], including
2. V_{max} and one-half V_{max}, but
3. K_m remains unchanged because the active site of those enzyme molecules that have not been inhibited is unchanged.

4.3 ENZYME APPLICATIONS

Centuries ago, man has used naturally occurring microorganisms (bacteria, yeasts, and moulds) and the enzymes they produce to make foods such as bread, cheese, beer, wine, and others. For example in bread-making the enzyme "amylase" is used to break down flour into sugars, which are transformed by yeast into alcohol and carbon dioxide which makes the bread rise.

Today, enzymes are used for an increasing range of applications, i.e., bakery, cheese making, starch processing and production of fruit juices and various other products.[28] They involve in improvements of texture, appearance, nutritional value, and can produce desirable flavors and aromas. Food enzymes come from animal and plant origin, a starch-digesting enzyme, amylase, can be obtained from germinating barley seeds.[29] But most come from a number of beneficial microorganisms.

Enzymes in food and other production have the following advantages:

- They are alternatives to traditional chemical-based technology, and can replace synthetic chemicals in many processes.
- They allow real advances in the environmental performance of production processes, through lower energy consumption and biodegradability.
- They are more specific in their action than other synthetic chemicals and therefore have fewer side reactions and waste byproducts.
- Processes, which use enzymes, give higher quality products and reduces the pollution.
- They allow processes to be performed, which would otherwise be impossible. For example the production of clear apple juice concentrate relies on the use of the enzyme "pectinase."

TABLE 4.3
Marketed Enzymes and Uses in Food Production

Principal Enzyme Activity	Applications
α-Acetolactate decarboxylase	Brewing
α-Amylase	Baking, brewing, distilling, starch
Amyloglucosidase	Alcohol production, baking, starch
Catalase	Mayonnaise, dairy
Cellulase	Brewing
Chymosin	Cheese
β-Glucanase	Brewing, juice
α-Glucanotransferase	Starch
Glucose isomerase	Starch
Glucose oxidase	Baking, egg mayonnaise, juice
Hemicellulase	Baking
Inulinases	Inulin
Lipase	Fats, oils
Lactase	Lactose-free milk product
Maltogenic amylase	Baking, starch, brewing
Microbial rennet	Dairy
Pectinase	Wine and fruit juice
Papain	Meat
Phytase	Starch
Protease	Baking, brewing, diary, distilling, fish, meat, starch, vegetable
Pullulanase	Brewing, starch
Rennet	Meat
Xylanase	Baking, starch

Source: EUFIC, European Food Information Council, Application of food biotechnology: Enzymes, available online at www.eufic.org/it/tech/tech01e.htm, 2005.

The enzymes are produced from microorganisms with a variety of application in the food industries (Table 4.3). The main types of microorganisms that produce enzymes include species of *Bacillus*, *Aspergillus*, *Streptomyces*, and *Kluyveromyces*. They are grown by fermentation in large vats or fermenters with capacities of up to 150,000 L; affecting fermentation factors, such as temperature, nutrients, and air supply, are adjusted to suit their optimal development. As in other food processes, strict rules of hygiene should be followed in fermentation. After the process is completed, the fermented material contains a product, which includes enzymes, nutrients, and microbes. Passing it through a series of filters to remove impurities and extracting the enzyme purify this material.

Each week, every European eats an average of 11 kg of food. Enzymes, the naturally occurring proteins which allow all the biochemical processes of life to occur, are found in all food raw materials. When purified and used in food preparation, some of these enzymes offer benefits such as improved flavor, texture, and digestibility.[30]

4.3.1 Advances in Enzymatic Technology

Since the early years of the 1980 decade, enzyme producing companies, have been using genetic engineering techniques to improve production efficiency and quality, and to develop new products. There are certain advantages for both industry and consumers, with major improvements in enzyme production giving better products and processes. Table 4.3 shows some of the enzymes produced for the food industry with their specialized applications.

However, this progress has been slowed due to the fact that the debate on some other, more controversial applications of engineering and/or biotechnology in food production–such as genetic engineering in animals–is continuing throughout Europe.

Currently, modern biotechnology has been used to give a number of advances in enzymatic production technology such as the following[8]:

- Improved productivity and cost-effectiveness in existing processes by producing enzymes more efficiently. For example the amount of raw materials, energy, and water needed to make a product can be reduced approximately by one-half, by changing from a traditional strain of microbe to a genetically modified one.
- Enzyme processors can tailor their enzymes more precisely to customer demands for products with specific properties.
- Production industries can supply enzymes which otherwise could not be produced in large enough quantities, giving the consumer access to a wider variety of products. An example is the amylase-based product, which makes bread stay fresh for longer.

4.3.2 ALLERGIC EFFECTS

To date, there have been no reports of consumer allergies to enzyme residues in food consumption. The levels of enzyme residues appearing in foods are so low that they are highly unlikely ever to cause allergies. Like all proteins, enzymes can cause allergic effects when people have been sensitized through exposure to large quantities. For this reason, enzyme companies take a variety of protective measures and some enzymes are produced as liquids, granules, capsules, or as immobilized preparations so that exposure is limited to sensitized people (workers and consumers).[32]

4.4 SPECIFIC APPLICATION IN THE FOOD INDUSTRY

4.4.1 ENZYMES IN THE FOOD INDUSTRY

The properties of enzymes make them ideal tools for the manipulation of biological material and, therefore, their use is well established within the food industry.[29] The very old processes of brewing and cheese making rely on enzymatic activity at various stages of manufacture. The addition of enzymes from other sources (exogenous enzymes) dates from the beginning of the century when enzymes were used in the leather and brewing industry. Surprisingly, only a small number of enzymes are used commercially. This is due to several factors including instability of the enzyme during processing, unsuitable reaction conditions or the big cost involved in obtaining large amounts of a sufficiently pure enzyme. The application of new biotechnological techniques helps to overcome these problems.

It is evident that new processes are increasingly important since consumers demand more improved and novel products and there is pressure to use and upgrade cheaper raw materials. Today industrial enzymes are used in the production of diverse products such as cheese, sausages, baked goods, juices, and egg white substitutes. They are used to reduce viscosity, improve extractions, carry out bioconversions, enhance separations, develop functionality, and create or intensify flavor.[33] Enzymes can also be very important tools for protein modification since they provide fast reaction rates, mild conditions and high specificity which lead to economic, health, and safety benefits.[34]

Enzymes could also be used to modify insoluble biopolymers and this is very important to the food industry. Examples include pectinase use in fruit juice and nectar production, food (e.g., fish, meat, and plant) protein hydrolyses, the production of yeast hydrolysates, the hydrolysis and recovery of proteins from by-product sludges, the recovery of meat scraps from bones and carcasses, solids' waste treatment, low temperature starch gelatinization and hydrolysis, and cellulose hydrolysis.[35] Some of the main objectives of these processes are to increase process yield or product quality; to modify functionality of proteins; to produce higher value products (e.g., peptides and flavors); to increase material digestibility or availability. The common perspective of these processes is that they all involve the action of hydrolases on insoluble substrates.

The problems of choosing one of the several commercial enzymes for a specific food system application are (1) wide ranges in enzyme activities assayed by different methods, (2) batch to batch variability, (3) different pH and temperature profiles depending on the source, (4) activators or inhibitors necessary for the reaction, and (5) compatibility of combining two or more enzymes.[6]

Enzymes have also been used to increase oil and protein release. Reports on rapeseed, sunflower, coconut, olive, and cottonseed have shown a high oil extraction yield using enzymes in the aqueous medium. Enzymes can be used to assist in the extraction of oils.[36] It has been shown that a mixture of cell wall–degrading enzymes such as hemicellulases, cellulases, and pectinases can materially assist in oil extraction, particularly under mild conditions, by liquefying the structural cell wall components of the oilseed.

Enzyme infusion is a way to create changes in foods or food ingredients without a lot of processing while simultaneously discouraging consumer concerns already existing over genetically engineered foods, although consumers might not want enzymes either.[37] It is difficult, though, to characterize and monitor enzyme movement within and between cells. Enzyme infusion should be used in many cases to increase productivity, reduce waste, energy, and water consumption. However, much improvement is necessary as well as new infusion methods and development of new commercial enzymes for the potential of this technique to be realized.[38]

4.4.2 Examples of Enzymes Used in the Food Industry

4.4.2.1 Cellulases

Cellulases can be produced by submerged or solid state fermentation. The latter is very advantageous since it gives higher enzymatic production and protein rate and employs natural cellulosic wastes as a substrate in contrast to the necessity of using pure cellulose in submerged fermentation. Cellulases play a crucial role in the degradation of compounds, leading to the production of ethanol or other useful compounds such as glucose and high fructose syrup.[39] Other applications are in the livestock industry (food additives for pig production), textile industry (cotton and linen processing), detergent industry, and agricultural waste treatment. Currently, they account for 40% of the total cost of biomass conversion so that an increase of their catalytic efficiency would result in significant savings for the industry. This increase could be achieved through protein engineering. They are also used in the extraction of green-tea components, modification of food tissues, removal of soybean seed coat, improvement of cattle feed quality, recovery of juice as well as other products from plant tissues, and they could be a component of digestive aid.[40] Cellulases also have been used to improve the palatability of low-quality vegetables, increase the flavor of mushrooms, promote the extraction of natural products, and alter the texture of foods.[41]

4.4.2.2 Xylanases

Some uses of xylanases are presented in the following. Microbial hemicellulases can reduce the levels of barley β-glucans, which are compounds that could be a nuisance during beer brewing since they may clog pumps and filters in breweries.[42,43] They also can be used to improve the properties of dough used in the production of baked goods. Fungal pentosanase can hydrolyze wheat hemicellulose, leading to a coarser bread crumb. Treatment of bread dough with purified fungal xylanase decreases dough strength yielding loaves with an open crumb structure. Xylanase preparations can also degrade wheat gums and fibers during the processing of wheat starch and gluten. Hemicellulases may be used along with pectinases and cellulases for the maceration of fruit and vegetable materials and for the clarification of juices and wines. They can also aid the extraction of coffee, marine and freshwater animals, plant oils, and starch, and are responsible for the modification of the textural and staling properties of baked products. Their presence may improve the production efficiency of agricultural silage and the nutritive properties of grain feed. Other applications involve extraction of pigments, flavors, and spices.[44] Hemicellulases (endo-β-xylanases) can be used in the bleaching of Kraft pulp to reduce consumption of chlorine chemicals in the bleaching process, and lower the adsorbable organic halogen (AOX) of the effluents as well as increase the brightness of the pulp that is very important in the development of totally chlorine-free bleaching sequences. Xylanases act mainly on the relocated, reprecipitated xylan on the surface of the pulp fibers.[45] Current applications involve (1) enzymatic hydrolysis of agro-wastes for further bioconversion into alcohol fuels, (2) enzymatic treatment of animal feeds to release free pentose sugars, (3) production of dissolving pulps, yielding cellulose for rayon production, and (4) bio-bleaching of wood pulps for paper production.[46]

4.4.2.3 Cell Wall–Degrading Enzyme

Cell wall–degrading enzymes may be used to extract vegetable oil in an aqueous process by liquefying the structural cell wall components of the oilseed. This concept has been investigated in rapeseed oil extraction and has been commercialized in connection with olive oil processing. Enzymes have already been used to increase oil yield by breaking down the olive pulp. They actually break down the pectin in olives naturally. A commercial new enzyme, cytolase O, increases yields by 5% and improves extra virgin olive oil quality.[47] Cytolase O breaks down the pulp, and natural antioxidants are released which resist rancidity and prolong olive oil's shelf life.

4.4.3 FOOD APPLICATION AND BENEFITS

4.4.3.1 Sugar Syrups

During the nineteenth century, boiling starch with strong acids like sulfuric acid were found to produce sugar syrups. This harsh process became a predominant method to develop a number of starch syrups. However, by the middle of the twentieth century, enzymes were rapidly supplementing the use of strong acids in the manufacture of sugar syrups.[48]

The sucrose industry is a comparatively minor user of enzymes but provides few historically significant and instructive examples of enzyme technology. The hydrolysis "inversion" of sucrose, completely or partially, to glucose and fructose provides sweet syrups that are more stable (i.e., less likely to crystallize) than pure sucrose syrups.[49,50] The most familiar "golden syrup" produced by acid hydrolysis of one of the less pure streams from the cane sugar refinery but other types of syrup are produced using yeast (*Saccharomyces cerevisiae*) invertase.

Other problems involving dextran and raffinose require the development of new industrial enzymes. A dextran is produced by the action of dextransucrase (EC 2.4.1.5) from *Leuconostoc mesenteroides* on sucrose and found as slime on damaged cane and beet tissues, especially when processing is delayed in hot and humid climates.[51,52] Raffinose, which consists of sucrose with α-galactose attached through its C-4 atom to the 1 position on the fructose residue, is produced at low temperatures in sugar beet. Both dextran and raffinose have the sucrose molecule as part of their structure and both inhibit sucrose crystal growth. This produces plate-like or needle-like crystals that are not readily harvested by equipment designed for the approximately cubic crystals otherwise obtained. Dextran can produce extreme viscosity in process streams, and can even bring a plant to a stop. Extreme dextran problems are frequently solved by the use of fungal dextranases produced from the *Penicillium* species.

The use of enzymes provides many advantages, including higher quality products, energy efficiency, and a safer working environment. Processing equipment also lasts longer since the milder conditions reduce corrosion.

4.4.3.2 Corn Syrups

The development of the various types of corn syrups, maltodextrins, and HFCS from corn starch sources could be called one of the greatest achievements in the sugar industry. Corn starch can be hydrolyzed into glucose relatively easily, but it was not until the 1970s that it became a commercially major product bringing about changes in the food industry. The starch is processed and refined from the kernels of corn by using a series of processes such as: steeping (swelling the kernel), separation, and grinding processes to separate the starch from the other parts of the kernel that is used for animal feed. The starch is hydrolyzed using acid, acid–enzyme, or enzyme–enzyme catalyzed processes. The first enzyme is generally a thermally stable α-amylase, which produces about 10%–20% glucose. Further treatment with the enzyme glucoamylase yields 93%–96% glucose.

One of the triumphs of enzyme technology so far has been the development of glucose isomerase, which in turn led to the commercialization of high fructose corn syrups. Now it is known that several types of bacteria can produce such glucose isomerases. They are resistant to thermal denaturation and will act at very high substrate concentrations, with the additional benefit of substantially stabilizing the enzymes at higher operational temperatures. The vast majority of glucose isomerases are retained within the cells that produce them but need not be separated and purified before use. All glucose isomerases are used in immobilized forms. Although different immobilization methods have been used for enzymes from different organisms, the principles of use are very similar. The corn syrup is then converted to fructose in a batch process to make 42% fructose syrup. Although this syrup can be made chemically with sodium hydroxide, the extremely high alkalinity limits the yield since large amounts of byproducts are formed. Because of these limitations, the use of enzymes with greater specificity and mild use-conditions emerged as the production method of choice. Lately, the production of HFCS is a major product in the syrup industry, which converts large quantities of corn (maize) and other botanical starches to this and other useful sweeteners.[53–55] These sweeteners are used in soft drinks, candies, baking, jams, and jellies and many other foods.

The environmental benefits are (1) reduced use of strong acids and bases, (2) reduced energy consumption (less greenhouse gas), (3) less corrosive waste, and (4) safer production environment for workers. The consumer benefits are (1) sweetener availability and stable prices (i.e., due to the ability to source from starch as an alternative to sugar cane and sugar beets), (2) consistent, and (3) higher quality syrups.

4.4.3.3 Dairy

4.4.3.3.1 Cheeses

An enzyme mixture from the stomach of calves and other ruminant mammals "Rennet" is a critical element in cheese making.[56,57] The enzyme Rennet has been the principal ingredient facilitating the separation of the curd (cheese) from the whey for thousands of years. Many things have been learned about the functional attributes of the enzyme rennet and other cheese-making enzymes since they were first used in the food production. A purified form of the major enzyme rennet—chymosin—is produced microbially from genetically modified microorganisms modified to contain the gene for calf chymosin. Today, it is commercially available without the need for sacrificing young calves and other animals. This kind of chymosin is the same as that isolated directly from calves.[58]

The environmental benefits are (1) cheese makers are no longer dependent upon enzymes recovered from slaughtered calves and lambs for production of rennet needed for most cheese making processes, and (2) based on current demands for chymosin, commercial needs for rennet could not be met from animal sources. The consumer benefits are (1) plentiful, consistently high quality enzyme chymosin is available at low prices that help assure availability of excellent cheeses at a reasonable cost and (2) people who follow kosher and vegetarian eating practices can consume cheese since the enzyme is from a microbe and not a calf.

4.4.3.3.2 Flavors

One of the major expenses in cheese making is the cost of storing cheeses as they age and develop their distinctive flavors. Cheddar takes six months to a year to mature, while Parmesan takes a full year. During that time off-flavors and bitterness, the most common Cheddar defects, may develop. Cheese making begins when processors add starter cultures to warm milk. The cultures contain strains of bacteria that produce enzymes that break down the proteins in milk, giving the cheese flavor and helping it ripen more quickly. In addition to the starter culture, cheese makers sometimes use a culture of the bacterium *Lactobacillus helveticus* to reduce bitterness and enhance flavor.[59] However, if the genes that produced the key enzyme were part of the bacteria in the starter culture, cheese makers would reduce costs by not having to use additional cultures.

The varied selections of cheeses enjoyed today are due in part to the action of enzymes "lipases." The lipases contribute to the distinctive flavor development during the ripening stage of cheese production.[60,61] Lipases are a class of enzymes that can act on the butterfat of cheese in order to produce flavors that are characteristic of different types of cheeses. Specific lipases are responsible for the flavors we enjoy in cheeses ranging from the piquant flavor (typical of Romano and provolone cheeses) to the distinct flavors of blue[62] and Roquefort cheeses. The environmental benefits are less discarded cheeses and less wasted production. The consumer benefits are the wide variety of flavors and the high quality of a variety of cheeses.

4.4.3.3.3 Lactose-Free Dairy Products

A significant part of the adult population is unable to consume natural dairy products as they cause gastrointestinal (GI) upset in the form of bloating, gas, or diarrhea, or a combination of GI symptoms.

The enzyme "lactase" that occurs naturally in the intestinal tract of children and many adults, is either absent or not present in sufficient quantity in lactose intolerant adults.[63,64] Lactase converts the milk sugar found in dairy products, (i.e., as milk, ice cream, yogurt, and cheese), to two readily digestible sugars, glucose and galactose. Without the enzyme lactase, the lactose in dairy foods digest in the intestines, producing undesirable side effects.

People who could not consume dairy products in their diet can enjoy these nutritious foods due to the commercial availability of the digestive enzyme, lactase. Today, many products present in the dairy case of supermarkets are labeled "lactose-free" as the result of pretreatment of the milk (final product) with the enzyme, lactase. Moreover, the enzyme lactase is available at retail stores for use in treating lactose containing dairy products in homes.

The environmental benefits are (1) elimination of dairy product wastes and (2) better use of products. The benefits for the consumer are (1) a large number of adults (app. 20%–30% in the United States) are lactose intolerant, (2) many individuals (lactose intolerant) can now enjoy the nutritional benefits and sensory pleasure of dairy products without gastrointestinal side effects, and (3) a large number of people can select lactose-free or low-lactose dairy products or can add commercially available lactase to dairy products in their homes.

4.4.3.4 Baking Products

Some of the earliest applications of industrial enzymes have been in the baking industry.[65–67] One of the more commonly used baking enzymes is fungal amylase. Fungal amylases have been used for over 50 years to modify flours. Bakers have used enzymes to produce more uniform dough and products. This has expanded to the modern use of maltogenic-amylases in the baking industry to extend shelf-life of various types of breads. Today enzymes are used to replace bromates, chemicals that are becoming increasingly undesirable to some consumer groups. Certain enzymes, which are naturally occurring protein molecules manufactured by all living organisms, can similarly and safely be used to enhance and control nonenzymatic browning, increase loaf volume, improve crumb structure, improve texture and provide gluten modification.

There are countless strains of enzymes used in the baking industry. The major enzyme classifications include xylanase, which affects water absorption and gluten formation by working on the non-starch polysaccharides; amylase, which extends the shelf lives of products; protease, which reduces mix times and improves pan flow and gas retention on high-speed production lines; and lipase, which provides an emulsifying effect, creating a fine texture and crumb.

4.4.3.4.1 *Bromate Replacers*

Bread production is often reliant upon oxidative compounds that can help in forming the right consistency of the dough. It is well known that chemical oxidants such as bromates, azodicarbonamide, and ascorbic acid have been widely used to strengthen gluten when making bread.[68] However, potassium bromate has also been used for this purpose for many years, as it was the first inorganic compound to be used for improving flour quality. Over the years, bromate has been used to bake bread of a consistently high quality with a high consumer acceptance. Recent studies, however, have questioned the use of bromate in bread and its use has been abandoned in many countries around the world.

Enzymes such as glucose oxidase have been used to replace the unique effect of bromate.[69] This way, enzymes can help the bakery production to make breads that are up to the quality standards of consumers' demand.[70,71] The environmental benefits are the less chemical usage, and the elimination of environmental chemical contamination. The consumer benefits are the elimination of bromate without sacrificing the quality of the bread.

4.4.3.4.2 *Softer Bread Products*

Consumers enjoy soft bread. To ensure high-quality bread, enzymes are often used to modify the starch that in turn keeps the bread softer for a longer period of time.

The staling of white bread is considered to be related to a change in the starch. Over time, the moisture in the starch becomes unbound when starch granules revert from a soluble to an insoluble form. When the starch can no longer "hold" water, it loses its flexibility and the bread becomes hard and brittle. This results in a subsequent reduction in taste appeal of the bread and it is termed "stale." By choosing the right enzyme, the starch can be modified during baking to retard staling. The bread stays soft and flavorful for a longer time: 3–6 days. The environmental benefits are less waste, and better use of raw material and the consumer enjoys better quality bread.

4.4.3.5 Alcoholic Drinks

4.4.3.5.1 *Low Calorie Beer*

Recently, "light" beers, of lower calorific content, have become more popular. These require a higher degree of saccharification at lower starch concentrations to reduce the alcohol and total solids contents of the beer.[72]

Calorie-conscious consumers can enjoy reduced calorie beer due to the use of special enzymes in the brewing process. Major ingredients used in the production of beer include, barley, rice, and other grains. The grains are essential components in the conversion of carbohydrates to alcohol during yeast fermentation.[73] First, simple carbohydrates are converted to alcohol followed by conversion of carbohydrates of increasing complexity, until the desired alcohol content is achieved. The remaining carbohydrate remains as a component of the finished product. By using enzymes (glucoamylase and/or fungal a-amylase during the fermentation) to transform the complex carbohydrates to simpler sugars, the desired alcohol content can be achieved with a smaller amount of added grain. This results in a beer with fewer carbohydrate calories and ultimately, a lower calorie beer. The environmental benefits are lower agricultural demand for grains used in brewing and the consumer benefits are the good tasting products and the lower calorie beers.

4.4.3.5.2 Clear Fruit Juice

Enzymes play a very important role in fruit juice technology. In fact, pectolytic and amylolytic enzyme preparations have been used for more than 60 years primarily in apple and berry juice production.

One of the major problems in the preparation of fruit juices and wine is cloudiness due primarily to the presence of pectins.[74] Pectin imparts a cloudy appearance to the juice and results in an appearance and mouth-feel that many consumers do not find appealing. These consist primarily of α-1,4-anhydrogalacturonic acid polymers, with varying degrees of methyl esterification. They are associated with other plant polymers and, after homogenization, with the cell debris. The cloudiness that they cause is difficult to remove except by enzymic hydrolysis.[75] Such treatment also has the additional benefits of reducing the solution viscosity, increasing the volume of juice produced (e.g., the yield of juice from white grapes can be raised by 15%), subtle but generally beneficial changes in the flavor and, in the case of wine-making, shorter fermentation times. Insoluble plant material is easily removed by filtration, or settling and decantation, once the stabilizing effect of the pectins on the colloidal haze has been removed.

Commercial pectolytic enzyme preparations are produced from *Aspergillus niger* and consist of a synergistic mixture of enzymes: polygalacturonase (EC 3.2.1.15), responsible for the random hydrolysis of 1,4-α-D-galactosiduronic linkages; pectinesterase (EC 3.2.1.11), which releases methanol from the pectyl methyl esters, a necessary stage before the polygalacturonase can act fully (the increase in the methanol content of such treated juice is generally less than the natural concentrations and poses no health risk); pectin lyase (EC 4.2.2.10), which cleaves the pectin, by an elimination reaction releasing oligosaccharides with non-reducing terminal 4-deoxymethyl-α-D-galact-4-enuronosyl residues, without the necessity of pectin methyl esterase action; and hemicellulase (a mixture of hydrolytic enzymes including: xylan endo-1,3-α-xylosidase, EC 3.2.1.32; xylan 1,4-α-xylosidase, EC 3.2.1.37; and α-L-arabinofuranosidase, EC 3.2.1.55), strictly not a pectinase but its adventitious presence is encouraged in order to reduce hemicellulose levels. The optimal activity of these enzymes is at a pH between 4 and 5 and generally below 50°C.[76–78]

The environmental benefits are (1) reducing waste and (2) controlling costs. The use of enzymes in juice processing helps assure that the maximum amount of juice is removed from the fruit. The consumer benefits are (1) aesthetically pleasing, (2) clear, (3) sediment-free juices, and (4) many varieties of fruit are widely available at retail stores.

4.4.3.6 Other Food Products

4.4.3.6.1 Meat Tenderizing

Some cuts of meat are more tender than others. Meat is mostly protein, indeed a rather complex set of proteins with defined structure(s). The major meat proteins responsible for tenderness are the myofibrillar proteins and the connective tissue proteins. Protease enzymes are used to modify these proteins. In fact, proteases like papain [79–81] and bromelain [82] have been used to tenderize tougher cuts of meat for many years. This can be a difficult process to control since there is a fine line between tender and mushy meat. To improve this process, more specific proteases have also been introduced to make the tenderizing process more robust.[83]

The environmental benefits are (1) less waste and (2) better use of raw materials. The consumer benefits are (1) the ability to tenderize tougher cuts of meat, (2) making less expensive cuts, and (3) more attractive menu items.

4.4.3.6.2 Confectionery

Confectionery (soft candy and other treats) made with sugar, i.e., soft-centered candies such as chocolate covered cherries often have a short shelf-life because the sugar, sucrose, contained in the product begins to crystallize soon after the confectionery is produced. A similar change occurs in soft cookies and other specialty bakery items. An enzyme, invertase, converts the sucrose to two simple sugars, glucose and fructose, and thus prevents the formation of sugar crystals that otherwise would severely shorten the shelf-life of the product or make some products virtually unavailable at reasonable prices.[35,72]

The environmental benefits are (1) enzymes replace hydrochloric acid in the manufacturing process, (2) eliminate the need for harsh chemical processing, (3) reduce the risk to the environment, and (4) provide a safer workplace due to the elimination of the use of a strong acid. The consumer benefits are (1) confections and specialty baked goods with excellent mouthfeel and taste characteristics are readily available at reasonable cost due to the use of the enzyme invertase; (2) soft centers of fine chocolates remain smooth and creamy; and (3) some candies stay chewy and soft cookies are available on the grocer's shelves, rivaling homemade versions because of this special food enzyme.

REFERENCES

1. Madhusudan and Vijajan, M., Additional binding sites in lysozyme. X-ray analysis of lysozyme complexes with bromophenol red and bromophenol blue, *Protein Eng.*, 5(5), 399, 1992.
2. Stryer, L., Catalytic strategies, in *Biochemistry*, 4th edn., Part II: *Proteins: Conformation, Dynamics and Function*, W.H. Freeman & Co., New York, 1995, pp. 207–216, Chapter 9.
3. Xu, S. and Damodaran, S., Kinetics of adsorption of proteins at the air-water interface from a binary mixture, *Langmuir*, 10, 472, 1994.
4. Poole, P.L., The role of hydration in lysozyme structure and activity: Relevance in protein engineering and design, *J. Food Eng.*, 22, 349, 1994.
5. Wells, J.A. and Estell, D.A., Subtilisin-an enzyme designed to be engineered, *Trends Biochem. Sci.*, 13, 291, 1988.
6. West, S., The enzyme maze: How to overcome problems to choosing the right enzyme for your process, *Food Technol.*, 42, 98, 1988.
7. Ophardt, C., Virtual chembook, available online at www.elmhurst.edu/~chm/vchembook/, 2003.
8. Van Beilen, J.B. and Li, Z., Enzyme technology: An overview, *Curr. Opin. Biotechnol.*, 13, 338, 2002.
9. Reilly, P.J., Enzymic degradation of starch, in *Starch Conversion Technology*, Van Beynum, G.M.A. and Reels, A. (eds.), Marcel Dekker Inc., New York, 1984, pp. 101–142.
10. Baret, J.L., Large-scale production and application of immobilised lactase, *Methods Enzymol.*, 136, 411, 1987.
11. Peberdy, J.F., Genetic engineering in relation to enzymes, in *Biotechnology*, Vol. 7a, *Enzyme Technology*, Kennedy, J.F. (ed.), VCH Verlagsgesellschaft mbH, Weinheim, Germany, 1987, pp. 325–344.
12. NC-IUBMB, Enzyme nomenclature: Recommendations of the Nomenclature Committee of the International Union of Biochemistry and Molecular Biology on the nomenclature and classification of enzyme-catalyzed reactions, Nomenclature Committee of the International Union of Biochemistry and Molecular Biology, available online at www.chem.qmul.ac.uk/iubmb/enzyme/
13. Luisi, P.L. and Laane, C., Solubilization of enzymes in apolar solvents via reverse micelles, *Trends Biotechnol.*, 4, 153, 1986.
14. Kricka, L.J. and Thorpe, G.H.G., Immobilised enzymes in analysis, *Trends Biotechnol.*, 4, 253, 1986.
15. Daniels, M.J., Industrial operation of immobilised enzymes, *Methods Enzymol.*, 136, 356, 1987.
16. Lowe, C.R., Immobilised coenzymes, in *Topics in Enzyme and Fermentation Biotechnology*, vol. 5, Wiseman, A. (ed.), Ellis Horwood Ltd., Chichester, U.K., 1981, pp. 13–146.
17. Manson, M.A. and Mosbach, K., Immobilised active coenzymes, *Methods Enzymol.*, 136, 3, 1987.
18. Enzyme Technical Association., Web site www.enzymetechnicalassoc.org/index.html
19. Koshland, D.E., Jr., The comparison of non-enzymic and enzymic reaction velocities, *J. Theor. Biol.*, 2, 75, 1962.
20. International Union of Pure and Applied Chemistry, Symbolism and terminology in chemical kinetics, *Pure Appl. Chem.*, 53, 753, 1981.
21. NC-IUBMB, Symbolism and terminology in enzyme kinetics Nomenclature Committee of the International Union of Biochemistry, available online at www.chem.qmul.ac.uk/iubmb/kinetics/, prepared by G.P. Moss Department of Chemistry, Queen Mary University of London, London, U.K., 1981.
22. Cornish-Bowden, A. and Endrenyi, L., Robust regression of enzyme kinetic data, *Biochem. J.*, 234, 21, 1986.
23. Fersht, A., *Enzyme Structure and Mechanism*, 2nd edn., W.H. Freeman & Co., New York, 1985.

24. Cornish-Bowden, A., *Principles of Enzyme Kinetics*, Butterworth & Co., London, U.K., 1976.

25. Henderson, P.J.F., Statistical analysis of enzyme kinetic data, *Techniques in the Life Sciences, Biochemistry* vol. B1/11, *Techniques in Protein and Enzyme Biochemistry*—Part II, Elsevier/North-Holland Biomedical Press, Amsterdam, the Netherlands, 1978, pp. B113/1–41.

26. Kimball's, J.W., Kimball's biology pages, available online at http://users.rcn.com/jkimball.ma.ultranet/BiologyPages/E/EnzymeKinetics.html, 2005.

27. Cornish-Bowden, A., A simple graphical method for determining the inhibition constants of mixed, uncompetitive and non-competitive inhibitors, *Biochem. J.*, 137, 143, 1974.

28. Pariza, E.W. and Johnson, E.A., Evaluating the safety of microbial enzyme preparations used in food processing: Update for a new century, *Regul. Toxicol. Pharmacol.*, 33, 173, 2001.

29. Tucker, G.A. and Woods, L.F.J., *Enzymes in Food Processing*, Blackie & Son Ltd., Glasgow and London, U.K., 1991.

30. Peppler, H.J. and Reed, G., Enzymes in food and feed processing, in *Biotechnology*, Vol. 7a: *Enzyme Technology*, Kennedy, J.F. (ed.), VCH Verlagsgesellschaft mbH, Weinheim, Germany, 1987, pp. 547–603.

31. EUFIC, European Food Information Council, Application of food biotechnology: Enzymes, available online at www.eufic.org/it/tech/tech01e.htm, 2005.

32. Fontaine, J.F. and Pauli G., Allergic cross-reactions: From theory to clinical practice, *Revue Française d'Allergologie et d'Immunologie Clinique*, 46, 484, 2006.

33. Klacik, M.A., Enzymes in food processing, *Chem. Eng. Prog.*, 84, 25, 1989.

34. Enzyme Technical Association, *Enzyme. A Primer on Use and Benefits Today and Tomorrow*, Enzyme Technical Association, Washington, DC, June 2001, p. 34.

35. Chaplin, M.F. and Bucke, C., *Enzyme Technology*, Cambridge University Press, Cambridge, U.K., 1990, ISBN 0 52134429 8, available online at www.lsbu.ac.uk/biology/enztech/index.html

36. Christensen, F.M., Enzyme technology versus engineering technology in the food industry, *Biotechnol. Appl. Biochem.*, 11, 249, 1989.

37. McArdle, R.N. and Culver, C.A., Enzyme Infusion: A developing technology, *Food Technol.*, 94, 85, 1994.

38. Varzakas, T.H., Leach, G.C., Israilides, C.J., and Arapoglou, D., Theoretical and experimental approaches towards the determination of solute effective diffusivities in foods, *Enzyme Microbiol. Technol.*, 37, 29, 2005.

39. Klyosov, A.A., Trends in biochemistry and enzymology of cellulose degradation, *Biochem. J.*, 29(47), 10577, 1990.

40. Lonsane, B.K. and Ghildyal, N.P., Exoenzymes, in *Solid Substrate Cultivation*, Doelle, H.W., Mitchell, D.A., and Rolz, C.E. (eds.), Elsevier Science Publishers, Essex, U.K., 1991.

41. Bigelis, R., Carbohydrases, in *Enzymes in Food Processing*, 3rd edn., Nagodawithana, T. and Reed, G. (eds.), Academic Press Inc., San Diego, CA, 1993, pp. 144–197.

42. Varzakas, T., Uptake of endoglucanase and endoxylanase from *A niger* by soybean seeds, *Deutsche Lebensmittel-Rundschau*, 97(12), 465, 2001.

43. Varzakas T.H., Arapoglou D., and Israilides C.J., Infusion of an endoglucanase and an endoxylanase from *Aspergillus niger* in soybean, *Food Sci. Technol. LWT*, 38, 239, 2005.

44. Hazlewood, G.P. and Gilbert, H.J., Molecular biology of hemicellulases, in *Hemicellulose and Hemicellulases*, Chapter 5, Coughlan, M.P. and Hazlewood, G.P. (eds.), Portland Press Ltd., London, U.K., 1993, pp. 103–126.

45. Viikari, L., Kantelinen, A., Sundquist, J., and Linko, M., Xylanases in bleaching. From an idea to the industry, *FEMS Microbiol. Rev.*, 13, 335, 1994.

46. Vandamme, E.J. and Soetaert, W., Biotechnical modification of carbohydrates, *FEMS Microbiol. Rev.*, 16, 163, 1995.

47. Anon., New oil extraction enzyme provides greater olive oil yields and improved oil quality, *Food Marketing and Technology International*, December 1993, Key No. 36720, 1993

48. Fullbrook, P.D., The enzymic production of glucose syrups, in *Glucose Syrups: Science and Technology*, Dziedzic S.Z. and Kearsley, M.W. (eds.), Elsevier Applied Science, London, U.K., 1984, pp. 65–115.

49. Bothast, R.J., Nichols, N.N., and Dien, B.S., Fermentations with new recombinant organisms, *Biotechnol. Prog.*, 15, 867, 1999.

50. Furegon, L., Peruffo, A.D.B., and Curioni, A., Immobilization of rice limit dextrinase on γ-alumina beads and its possible use in starch processing, *Proc. Biochem.*, 32, 113, 1997.

51. Mariana, S., Alírio, R., and José A. Teixeira production of dextran and fructose from carob pod extract and cheese whey by *Leuconostoc mesenteroides* NRRL B512(f), *Biochem. Eng. J.*, 25, 1, 2005.

52. Vediashkina, T.A., Revin, V.V., and Gogotov, I.N., Optimizing the conditions of dextran synthesis by the bacterium *Leuconostoc mesenteroides* grown in a molasses-containing medium, *Prikladnaia Biokhimiia i Mikrobiologiia*, 41, 409, 2005.

53. Bandlish, R.K., Michael Hess, J., Epting, K.L., Vieille, C., and Kelly, R.M., Glucose-to-fructose conversion at high temperatures with xylose (glucose) isomerases from *Streptomyces murinus* and two hyperthermophilic Thermotoga species, *Biotechnol. Bioeng.*, 80, 185, 2002.

54. Kaneko, T., Takahashi, S., and Saito, K., Characterization of acid-stable glucose isomerase from *Streptomyces* sp., and development of single-step processes for high-fructose corn sweetener (HFCS) production, *Biosci. Biotechnol. Biochem.*, 64, 940, 2000.

55. Visuri, K. and Klibanov, A.M., Enzymic production of high fructose corn syrup (HFCS) containing 55% fructose in aqueous ethanol, *Biotechnol. Bioeng.*, 30, 917, 1987.

56. Pirisi, A., Pinna, G., Addis, M., Piredda, G., Mauriello, R., De Pascale, S., Caira, S., Mamone, G., Ferranti, P., Addeo, F., and Chianese, L., Relationship between the enzymatic composition of lamb rennet paste and proteolytic, lipolytic pattern and texture of PDO Fiore Sardo ovine cheese, *Int. Dairy J.*, 17, 143, 2007.

57. Prados, F., Pino, A., Rincón, F., Vioque, M., and Fernández-Salguero, J., Influence of the frozen storage on some characteristics of ripened Manchego-type cheese manufactured with a powdered vegetable coagulant and rennet, *Food Chem.*, 95, 677, 2006.

58. Seker, S., Beyenal, H., Ayhan, F., and Tanyolac, A., Production of microbial rennin from *Mucor miehei* in a continuously fed fermenter, *Enzyme Microbiol. Technol.*, 23, 469, 1998.

59. Kenny, O., FitzGerald, R.J., O'Cuinn, G., Beresford, T., and Jordan, K., Autolysis of selected *Lactobacillus helveticus* adjunct strains during Cheddar cheese ripening, *Int. Dairy J.*, 16, 797, 2006.

60. Hasan, F., Shah A.A., and Hameed A., Industrial applications of microbial lipases, *Enzyme Microbiol. Technol.*, 39, 235, 2006.

61. Kilcawley, K.N., Wilkinson, M.G., and Fox, P.F., Enzyme-modified cheese, *Int. Dairy J.*, 8, 1, 1998.

62. Tomasini, A., Bustillo G., and Lebeault, J.M., Production of blue cheese flavour concentrates from different substrates supplemented with lipolyzed cream, *Int. Dairy J.*, 5, 247, 1995.

63. Rosado, J.L. and Mimiaga, C., Lactose-free formula products available in Mexico. Their significance and applications, *Revista De Investigacion Clinica; Organo Del Hospital De Enfermedades De La Nutricion*, 48, 67, 1996.

64. Nipat, S., Yothi, T., Orapin, T., and Wandee, V., Randomized, double-blind clinical trial of a lactose-free and a lactose-containing formula in dietary management of acute childhood diarrhea, *J. Med. Assoc. Thai. Chotmaihet Thangphaet*, 87, 641, 2004.

65. Elms, J., Robinson, E., Mason, H., Iqbal, S., Garrod, A., and Evans, G.S., Enzyme exposure in the British baking industry, *Ann. Occup. Hyg.*, 50, 379, 2006.

66. Collins, T., Hoyoux, A., Dutron, A., Georis, J., Genot, B., Dauvrin, T., Arnaut, F., Gerday, C., and Feller, G., Use of glycoside hydrolase family 8 xylanases in baking, *J. Cereal Sci.*, 43, 79, 2006.

67. Ozge, K.S., Gülüm, S., and Serpil, S., Usage of enzymes in a novel baking process, *Die Nahrung*, 48, 156, 2004.

68. Kurokawa, Y., Maekawa, A., Takahashi, M., and Hayashi, Y., Toxicity and carcinogenicity of potassium bromate-a new renal carcinogen, *Environ. Health Perspect.*, 87, 309, 1990.

69. Okolie, N.P. and Osarenren, E.J., Toxic bromate residues in Nigerian bread, *Bull. Environ. Contam. Toxicol.*, 70, 443, 2003.

70. Bonet, A., Rosell, C.M., Caballero, P.A., Gómez, M., Pérez-Munuera, I., and Lluch, M.A., Glucose oxidase effect on dough rheology and bread quality: A study from macroscopic to molecular level, *Food Chem.*, 99, 408, 2006.

71. Rasiah, I.A., Sutton, K.H., Low, F.L., Lin, H.M., and Gerrard, J.A., Crosslinking of wheat dough proteins by glucose oxidase and the resulting effects on bread and croissants, *Food Chem.*, 89, 325, 2005.

72. Godfrey, T. and West, S., *Industrial Enzymology*, Macmillan Press Ltd., London, U.K., 1996.

73. James, J.A. and Lee, B.H., Glucoamylases: Microbial sources, industrial applications and molecular biology—A review, *J. Food Biochem.*, 21, 1, 1997.

74. Kashyap, D.R., Vohra, P.K., Chopra, S., and Tewari, R., Applications of pectinases in the commercial sector: A review. *Bioresour. Technol.*, 77, 215, 2001.

75. Busto, M.D., García-Tramontín, K.E., Ortega, N., and Perez-Mateos, M., Preparation and properties of an immobilized pectinlyase for the treatment of fruit juices, *Bioresour. Technol.*, 97, 1477, 2006.

76. Spagna, G., Barbagallo, R.N., Greco, E., Manenti, I., and Pifferi, P.G., A mixture of purified glycosidases from *Aspergillus niger* for oenological application immobilised by inclusion in chitosan gels, *Enzyme Microbiol. Technol.*, 30, 80, 2002.
77. Spagna, G., Pifferi, P.G., and Gilioli, E., Immobilization of a pectinlyase from *Aspergillus niger* for application in food technology, *Enzyme Microbiol. Technol.*, 17, 729, 1995.
78. Le Traon-Masson, M.P. and Pellerin, P., Purification and characterization of two beta-D-glucosidases from an *Aspergillus niger* enzyme preparation: Affinity and specificity toward glucosylated compounds characteristic of the processing of fruits, *Enzyme Microbiol. Technol.*, 22, 374, 1998.
79. Kaul, P., Sathish, H.A., and Prakash, V., Effect of metal ions on structure and activity of papain from *Carica papaya*, *Die Nahrung*, 46, 2, 2002.
80. Sinha, R.C. and Bahadur, K., Comparative study of the proteolytic activities of papain and trypsin against casein, egg albumin and meat extract, *Enzymologia*, 19, 319, 1958.
81. Pawar, V.D., Surve, V.D., and Machewad, G.M., Tenderization of chevon by papain and trypsin treatments, *J. Food Sci. Technol.*, 40, 296, 2003.
82. Kim, H.J. and Taub, I.A., Specific degradation of myosin in meat by bromelain, *Food Chem.*, 40, 337, 1991.
83. Qihe, C., Guoqing, H., Yingchun, J., and Hui, N., Effects of elastase from a *Bacillus* strain on the tenderization of beef meat, *Food Chem.*, 98, 624, 2006.

5 Lipids, Fats, and Oils

Ioannis S. Arvanitoyannis, Theodoros H. Varzakas,
Sotirios Kiokias, and Athanasios E. Labropoulos

CONTENTS

5.1   INTRODUCTION

Lipids consist of a broad group of compounds that are generally soluble in organic solvents but only sparingly soluble in water. They are major components of adipose tissue, and together with proteins and carbohydrates, they constitute the principal structural components of all living cells. Glycerol esters of fatty acids, which make up to 99% of the lipids of plant and animal origin, have been traditionally called fats and oils.[1] The difference between oils and fats is that fats are solids at room temperatures.[2]

TABLE 5.1
Lipid Classification

Classification	Categories	Characteristics
Acyl residue		
	Simple lipids (not saponifiable)	
	Free fatty acids	—
	Isoprenoid lipids	Steroids, carotenoids, monoterpenes
	Tocopherols	—
	Acyl lipids (saponifiable)	
	Mono-, di-, triacylglycerols	Fatty acids, glycerol
	Phospholipids (phosphatides)	Fatty acids, glycerol or sphingosine, phosphoric acid, organic base
	Glycolipids	Fatty acids, glycerol or sphingosine, mono-, di-, or oligosaccharide
	Diol lipids	Fatty acids, ethane, propane, or butane diol
	Waxes	Fatty acids, fatty alcohol
	Sterol esters	Fatty acids, sterol
Neutral-polar	Glycerophospholipids	Fatty acids ($>C_{12}$)
	Glyceroglycolipids	Mono-, di-, or triacylglycerols
	Sphingophospholipids	Sterols, sterol esters
	Sphingoglycolipids	Carotenoids, waxes, tocopherols

The majority of lipids are derivatives of fatty acids. In these so-called acyl lipids, the fatty acids are present as esters, and in some minor lipid groups in amide forms (Table 5.1). The acyl residue influences strongly the hydrophobicity and the reactivity of the acyl lipids. Some lipids act as building blocks in the formation of biological membranes, which surround cells and subcellular particles. Primarily, triacylglycerols are deposited in some animal tissues and organs of some plants. Lipid content in such storage tissues can rise to 15%–20% or higher and so serve as a commercial source for isolation of triacylglycerols.[3]

5.1.1 FATTY ACIDS

Fatty acid is a carboxylic acid often with a long, unbranched aliphatic chain, which is either saturated or unsaturated. Carboxylic acids as short as butyric acid (four carbon atoms) are considered to be fatty acids, whereas fatty acids derived from natural fats and oils may be assumed to have at least eight carbon atoms, e.g., caprylic acid (octanoic acid). Fatty acids are aliphatic monocarboxylic acids derived from or contained in an esterified form in an animal or vegetable fat, oil, or wax. Natural fatty acids commonly have a chain of 4–28 carbons (usually unbranched and even numbered), which may be saturated or unsaturated. By extension, the term is sometimes used to embrace all acyclic aliphatic carboxylic acids.[4]

5.1.2 SATURATED FATTY ACIDS

Saturated fatty acids do not contain any double bonds or other functional groups along the chain. The term "saturated" refers to hydrogen, in that all carbons (apart from the carboxylic acid [–COOH] group) contain as many hydrogens as possible. Saturated fatty acids form straight chains and, as a result, can be packed together very tightly, allowing living organisms to store chemical energy very densely. The fatty tissues of animals contain large amounts of long-chain saturated fatty acids.[5]

Fatty acids have an "-oic" suffix to the name of the acid but the suffix is usually "-ic." The shortest descriptions of fatty acids include only the number of carbon atoms and double bonds in them (e.g., C18:0 or 18:0). C18:0 means that the carbon chain of the fatty acid consists of 18 carbon atoms, and there are no (zero) double bonds in it, whereas C18:1 describes an 18-carbon chain with one double bond in it. Each double bond can be in either a cis- or trans- conformation and in a different position with respect to the ends of the fatty acid; therefore, not all C18:1s, for example, are identical. If there are one or more double bonds in the fatty acid, it is no longer considered saturated, but rather mono- or polyunsaturated.[6] The characteristics of saturated fatty acids are given in Table 5.2.

5.1.3 UNSATURATED FATTY ACIDS

Unsaturated fatty acids are of similar form, except that one or more allyl functional groups exist along the chain, with each alkene substituting a single-bonded "–CH$_2$–CH$_2$–" part of the chain with a double-bonded "–CH=CH–" portion. The two next carbon atoms in the chain that are bound to either side of the double bond can occur in a cis or trans configuration.[7]

A cis configuration means that adjacent carbon atoms are on the same side of the double bond. The rigidity of the double bond freezes its conformation and, in the case of the cis isomer, causes the chain to bend and restricts the conformational freedom of the fatty acid. The more double bonds the chain has in the cis configuration, the less flexibility it has. When a chain has many cis bonds, it becomes quite curved in its most accessible conformations.[8] For example, oleic acid has one double bond, and linoleic acid with two double bonds has a more pronounced bend. α-Linolenic acid, with three double bonds, favors a hooked shape. The effect of this is that in restricted environments, such as when fatty acids are part of a phospholipid in a lipid bilayer, or triglycerides in lipid droplets, cis bonds limit the ability of fatty acids to be closely packed, and therefore could affect the melting temperature of the membrane or of the fat.[9]

A trans configuration, by contrast, means that the next two carbon atoms are bound to opposite sides of the double bond. As a result, they do not cause the chain to bend much, and their shape is similar to straight saturated fatty acids.[10] In most naturally occurring unsaturated fatty acids, each double bond has $3n$ carbon atoms after it, for some n, and all are cis bonds. Most fatty acids in the trans configuration (trans fats) are not found in nature and are the result of human processing. The differences in geometry between the various types of unsaturated fatty acids, as well as between saturated and unsaturated fatty acids, play an important role in biological processes, and in the construction of biological structures (such as cell membranes).[11] The characteristics of unsaturated fatty acids are given in Table 5.3.

5.1.4 ACYLGLYCEROLS

Neutral fats are mono-, di-, and tri-esters of glycerol with fatty acids, and are termed monoacylglycerols, diacylglycerols, and triacylglycerols, respectively. They are designated as neutral lipids. Edible oils or fats consist nearly completely of triacylglycerols.[3] Although glycerol by itself is a completely symmetrical molecule, the central carbon atom acquires chirality (asymmetry) if one of the primary hydroxyl groups (on carbons 1–3) is esterified, or if the two primary hydroxyls are esterified to different acids.[1]

5.2 MAJOR OILS AND FATS

All oils and fats, with their high carbon and hydrogen content, can be traced back to organic sources. Oils and fats are also produced by plants, animals, and other organisms through organic processes and these oils are remarkable in their diversity.[12] Oils are fats that are liquids at room temperature. Solid fats are fats that are solids at room temperature, like butter and shortening. Solid fats come from many animal foods and can be made from vegetable oils through a process called hydrogenation.[13]

TABLE 5.2
Saturated Fatty Acids

IUPAC Name	Common Name	Abbreviation	Chemical Structure	Structure	Melting Point (°C)
Butanoic acid	Butyric acid	4:0	$CH_3(CH_2)_2COOH$		−7.9
Pentanoic acid	Valeric acid	5:0	$CH_3(CH_2)_3COOH$		−34.5
Hexanoic acid	Caproic acid	6:0	$CH_3(CH_2)_4COOH$		−3.9
Heptanoic acid	Enanthic acid	7:0	$CH_3(CH_2)_5COOH$		−7.5
Octanoic acid	Caprylic acid	8:0	$CH_3(CH_2)_6COOH$		16.3
Nonanoic acid	Pelargonic acid	9:0	$CH_3(CH_2)_7COOH$		12.4
Decanoic acid	Capric acid	10:0	$CH_3(CH_2)_8COOH$		31.3
Dodecanoic acid	Lauric acid	12:0	$CH_3(CH_2)_{10}COOH$		44
Tetradecanoic acid	Myristic acid	14:0	$CH_3(CH_2)_{12}COOH$		54.4
Hexadecanoic acid	Palmitic acid	16:0	$CH_3(CH_2)_{14}COOH$		62.9

(continued)

TABLE 5.2 (continued)
Saturated Fatty Acids

IUPAC Name	Common Name	Abbreviation	Chemical Structure	Structure	Melting Point (°C)
Heptadecanoic acid	Margaric acid	17:0	$CH_3(CH_2)_{15}COOH$		61.3
Octadecanoic acid	Stearic acid	18:0	$CH_3(CH_2)_{16}COOH$		69.6
Eicosanoic acid	Arachidic acid	20:0	$CH_3(CH_2)_{18}COOH$		75.4
Docosanoic acid	Behenic acid	22:0	$CH_3(CH_2)_{20}COOH$		80.0
Tetracosanoic acid	Lignoceric acid	24:0	$CH_3(CH_2)_{22}COOH$		84.2
Hexacosanoic acid	Cerotic acid	26:0	$CH_3(CH_2)_{24}COOH$		87.7

TABLE 5.3
Unsaturated Fatty Acids

Common Name	Abbreviation	Chemical Structure	Structure	Family (ω)	Family (Δ)	Melting Point (°C)
Myristoleic acid	14:1	$CH_3(CH_2)_3CH=CH(CH_2)_7COOH$		ω-5	cis-Δ^9	—
Palmitoleic acid	16:1	$CH_3(CH_2)_5CH=CH(CH_2)_7COOH$		ω-7	cis-Δ^9	0.1
Oleic acid	18:1	$CH_3(CH_2)_7CH=CH(CH_2)_7COOH$		ω-9	cis-Δ^9	13.4
Linoleic acid	18:2	$CH_3(CH_2)_4CH=CHCH_2CH=CH(CH_2)_7COOH$		ω-6	cis, cis-Δ^9, Δ^{12}	−5
α-Linolenic acid	18:3	$CH_3CH_2CH=CHCH_2CH=CHCH_2$ $CH=CH(CH_2)_7COOH$		ω-3	cis, cis, cis-Δ^9, Δ^{12}, Δ^{15}	−11
γ-Linolenic acid	18:3	$(CH=CHCH_2)_3CH_2CH_2CH_2COOH$		ω-6	cis, cis, cis-Δ^6, Δ^9, Δ^{12}	—
Arachidonic acid	20:4	$CH_3(CH_2)_4CH=CHCH_2CH=CHCH_2$ $CH=CHCH_2CH=CH(CH_2)_3COOH$		ω-6	cis, cis, cis, cis-Δ^5, Δ^8, Δ^{11}, Δ^{14}	−49.5
Eicosapentaenoic acid (EPA)	20:5	$CH_3CH_2CH=CHCH_2CH=CH$ $CH_2CH=CHCH_2CH=CHCH_2$ $CH=CH(CH_2)_3COOH$		ω-3	cis, cis, cis, cis, cis-Δ^5, Δ^8, Δ^{11}, Δ^{14}, Δ^{17}	—

(continued)

TABLE 5.3 (continued)
Unsaturated Fatty Acids

Common Name	Abbreviation	Chemical Structure	Structure	Family (ω)	Family (Δ)	Melting Point (°C)
Erucic acid	22:1	$CH_3(CH_2)_7CH=CH(CH_2)_{11}COOH$		ω-9	cis-Δ^{13}	34.7
Docosahexaenoic acid (DHA)	22:6	$CH_3CH_2CH=CHCH_2CH=CHCH_2$ $CH=CHCH_2CH=CHCH_2CH=CH$ $CH_2CH=CH(CH_2)_2COOH$		ω-3	cis, cis, cis, cis, cis, cis-Δ^5, Δ^8, Δ^{11}, Δ^{14}, Δ^{17}	—
Nervonic acid	24:1	$(CH_2)_{12}COOH$		ω-9	cis-Δ^{15}	42.5

5.2.1 Oils and Fats of Vegetable Origin

The vegetable oils may be subdivided into three categories: (1) by-products, where the crop is grown for another purpose other than seed oil, e.g., cotton (fabric), (2) three crops, which are generally slow to mature but then produce crops regularly for many years (olive, palm, and coconut), (3) crops, which have to be replanted each year to produce an annual harvest and where decisions about cultivation are made each sowing season by a large number of individual farmers (rape, sunflower, sesame, etc.).[14] Typical fatty acid compositions of vegetable oils and fats are summarized in Table 5.4.

5.2.1.1 Olive Oil

Over 750 million olive trees are cultivated worldwide, about 95% of those in the Mediterranean region. Most of the global production comes from Southern Europe, North Africa, and the Middle East. Of the European production, 93% comes from Spain, Italy, Turkey, and Greece. Spain's production alone accounts for 40%–45% of the world production, which was 2.6 million metric tons in 2002.[15] In olive oil–producing countries, the local production is generally considered the finest.

The olive oil extraction is carried out with technological industrial processes (continuous or discontinuous), even though the quality and the quantity of the obtained oil are still to be optimized.[16] The most traditional way of making olive oil is by grinding olives. Green olives produce bitter oil and overly ripened olives produce rancid oil, so care is taken to make sure the olives are perfectly ripened. First, the olives are ground into an olive paste using large mills. The olive paste generally stays under the mills for 30–40 min. The oil collected during this part of the process is called virgin oil. After grinding, the olive paste is spread on fiber disks, which are stacked on top of each other, and then placed into the press. Pressure is then applied onto the disk to further separate the oil from the paste. This second step produces a lower grade of oil.[17] The production of olive oil is shown in Figure 5.1.

The oil is characterized by a high level of oleic acid with Codex ranges of 8%–20% for palmitic acid, 55%–83% for oleic acid and 4%–21% for linoleic acid.[18] Extra virgin olive oil has a perfect flavor and odor with a maximum acidity of 1% (as oleic acid). Fine virgin oil also has a perfect flavor and odor with a maximum acidity of 2%. Semi-fine or ordinary virgin oil has good flavor and odor and a maximum acidity of 3.3% with a 10% margin of tolerance. Virgin olive oil with an off-flavor or off-odor and acidity >3.3% is designated lampante. Refined olive oil, obtained from virgin olive oil by refining methods which do not affect fatty acid or glycerol ester composition, should have acidity <0.5%.[19]

5.2.1.2 Corn Oil

Corn oil is oil extracted from the germ of corn (maize). Its main use is in cooking, where its high smoke point makes it a valuable frying oil. It is also a key ingredient in some margarines. Corn oil has a milder taste and is less expensive than most other types of vegetable oils.[20]

A major vegetable oil, with a production of around 2 million tonnes per annum from maize (*Zea mays*), is obtained by wet milling, particularly in the United States. The major acids are palmitic (9%–17%), oleic (20%–42%), and linoleic (39%–63%), and the major triacylglycerols.[21] Refined corn oil is 99% triglyceride, with proportions of approximately 59% polyunsaturated fatty acid, 24% monounsaturated fatty acid, and 13% saturated fatty acid.[22]

5.2.1.3 Soybean Oil

The soybean or soya bean (*Glycine max*) is a species of legume native to East Asia. In processing soybeans for oil extraction, selection of high quality, sound, clean, dehulled, yellow soybeans is very important. To produce soybean oil, the soybeans are cracked, adjusted for moisture content, rolled into flakes, and solvent-extracted with commercial hexane. The oil is then refined, blended for different applications, and sometimes hydrogenated.[23] In the past, hydrogenation was used to reduce

TABLE 5.4
Typical Fatty Acid Composition of the Main Vegetable Oils and Fats

Kind of Vegetable Oil and Fats	Saturated (g/100 g)	Monounsaturated (g/100 g)	Polyunsaturated (g/100 g)	Fatty Acids 16:1 (wt%)	18:0 (wt%)	18:1 (wt%)	18:2 (wt%)	18:3 (wt%)
Olive oil	14.0	69.7	11.2	10	2	78	7	1
Corn oil	12.7	24.7	57.8	13	3	31	52	1
Soybean oil	14.5	23.2	56.5	11	4	23	53	8
Sunflower oil	11.9	20.2	63.0	6	5	20	60	—
Cottonseed oil	25.5	21.3	48.1	23	2	17	56	—
Wheat germ oil	18.8	15.9	60.7	16	2	14	60	5
Rapeseed/canola oil	5.3	64.3	24.8	3	1	16	14	10
Palm oil	45.3	41.6	8.3	41	4	31	12	—
Safflower oil	10.2	12.6	72.1	7	3	14	75	—
Coconut oil	85.2	6.6	1.7	81	4	5	1	—
Cocoa butter	24.2	21.7	48.9	26	34	20	43	—
Sesame oil	16.4	40.1	42.5	10	4	46	46	—

Sources: Gunstone, F.D., *The Chemistry of Oils and Fats—Sources, Composition, Properties and Uses*, CRC Press, Boca Raton, FL, p. 7, 2004; FSA (Food Standards Agency), Fats and oils. *McCance and Widdowson's the Composition of Foods*, 5th edn, Royal Society of Chemistry, Cambridge, U.K., 1991; Altar, T., More than you wanted to know about fats/oils, Sundance Natural Foods, 1995. Available at: http://www.efn.org/~sundance/fats_and_oils.html (accessed 20/2/2008).

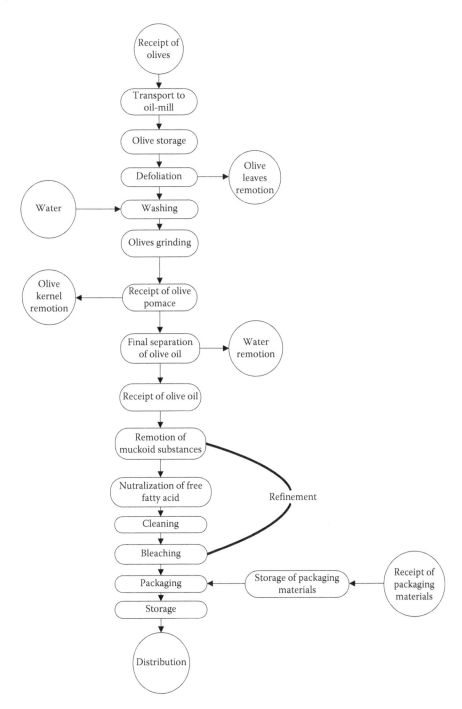

FIGURE 5.1 Flow diagram of olive oil production.

the unsaturation in linolenic acid, but this produced an unnatural trans fatty acid, having a trans fat configuration, whereas in nature the configuration is cis.[24]

The seed oil contains palmitic (about 11%, range 7%–14%), oleic (about 20%, range 19%–30%), linoleic (about 53%, range 44%–62%), and linolenic acids (about 7%, range 4%–11%). It also contains the saturated fatty acids: 4% stearic acid and 10% palmitic acid.[25]

After suitable modification it can be used as a solvent, a lubricant, and as biodiesel. Valuable by-products recovered during refining include lecithin, tocopherols, and phytosterols. Attempts to modify the fatty acid composition by seed breeding or genetic modification are directed to reducing the level of saturated acid or linolenic acid, or increasing the content of stearic acid.[26]

In the 2002–2003 growing season, 30.6 million metric tons of soybean oil were produced worldwide, constituting about half of the worldwide edible vegetable oil production, and thirty percent of all fats and oils produced, including animal fats and oils derived from tropical plants.[15] Soybean oil is produced in larger amounts than any other traded oil (about 23 million tonnes a year) and is grown particularly in the United States, followed by Brazil, Argentina, and China.[25]

5.2.1.4 Sunflower Oil

Sunflower oil is the nonvolatile oil expressed from sunflower (*Helianthus annuus*) seeds. A major vegetable oil (about 9 million tonnes per annum), it is extracted from the seed of *Helianthus annuus* grown mainly in the USSR, Argentina, Western and Eastern Europe, China, and the United States.[26] Sunflower oil is commonly used in food as frying oil, and in cosmetic formulations as an emollient.

Sunflower oil contains predominantly linoleic acid in triglyceride form. The British Pharmacopoeia lists the following profile: palmitic acid (4%–9%), stearic acid (1%–7%), oleic acid (14%–40%), linoleic acid (48%–74%).[27] There are several types of sunflower oils produced, such as high linoleic, high oleic, and mid oleic. High linoleic sunflower oil typically has at least 69% linoleic acid. High oleic sunflower oil has at least 82% oleic acid. Variation in fatty acid profile is strongly influenced by both genetics and climate. Sunflower oil also contains lecithin, tocopherols, carotenoids, and waxes. Sunflower oil's properties are typical of vegetable triglyceride oil.[18] High-oleic varieties (*Sunola* or *Highsun*, *NuSun*) with about 85% and 60% oleic acid have been developed and find use as sources of oleic acid in enzymatically modified products.[28]

5.2.1.5 Cottonseed Oil

Cottonseed oil is a vegetable oil extracted from the seeds of the cotton plant after the cotton lint has been removed. It must be refined to remove gossypol, a naturally occurring toxin that protects the cotton plant from insect damage.[29] In its natural unhydrogenated state, cottonseed oil, like all vegetable oils, has no cholesterol. It also contains no trans fatty acids. Further, these polyunsaturated fats can potentially go rancid during the extraction process.[30]

A major vegetable oil (4 million tonnes per annum) obtained as a by-product in the production of cotton and grown mainly in China, the United States, the USSR, India, and Pakistan. It ranks fifth among vegetable oils. Cottonseed oil is rich in palmitic acid (22%–26%), oleic acid (15%–20%), and linoleic acid (49%–58%), and contains a 10% mixture of arachidic acid, behenic acid, and lignoceric acid. It also contains about 1% sterulic and malvalic acids in the crude oil. The latter are identified by the Halphen test. The cyclopropene acids are undesirable components, but they are largely removed during refining, particularly deodorization, and also during hydrogenation. They are not considered to present any health hazard in cottonseed oil.[31]

5.2.1.6 Wheat Germ Oil

Wheat germ oil is extracted from the germ of the wheat kernel, which makes up only 2.5% by weight of the kernel. Wheat germ oil is particularly high in octacosanol—a 28 carbon long-chain saturated primary alcohol found in a number of different vegetable waxes. Octacosanol has been studied as an exercise-enhancing and a physical performance–enhancing agent. As a cooking oil, wheat germ oil is strongly flavored, expensive, and easily perishable.[32]

Oil from wheat germ (the embryo of the seed of *Triticum aestivum*) is rich in linoleic acid (ca. 60%), also contains α-linolenic acid (ca. 5%), 16% palmitic acid, and 14% oleic acid. The oil is rich in tocopherols and shows high vitamin E activity.[18]

5.2.1.7 Rapeseed or Canola Oil

Canola or rapeseed (*Brassica napus* or *B. campestris*) is a bright yellow-flowering member of the Brassicaceae (also known as the mustard) family. It is cultivated for the production of animal feed, vegetable oil for human consumption, and biodiesel. Worldwide, canola was the third leading source of vegetable oil in 2000, after soy and palm oils. Canola is also the world's second leading source of protein meal.[33]

Typically, this oil was rich in erucic acid, which is still available from high-erucic rapeseed oil (HEAR) or from crambe oil. Erucic acid is mildly toxic to humans in large doses but is used as a food additive in smaller doses. The variety low in erucic acid (<5% or <2%) and also in glucosinolates (LEAR, double zero) is now more important. The oil typically contains palmitic (4%), stearic (2%), oleic (56%), linoleic (26%), and linolenic acids (10%).[34] Rapeseed lends itself to genetic manipulation and rapeseed oil containing a lower level of linolenic acid or higher levels of lauric, stearic, or oleic acid or new acids, such as δ-linolenic, ricinoleic, or vernolic acids, are being developed for commercial exploitation.[18]

Canada and the United States produce between 7 and 10 million metric tons of canola seed per year.[34] Annual Canadian exports total 3–4 million metric tons of the seed, 700,000 metric tons of canola oil, and 1 million metric tons of canola meal. The United States is the net consumer of canola oil. The major customers of canola seed are Japan, Mexico, China, and Pakistan, while the bulk of canola oil and meal goes to the United States, with smaller amounts shipped to Taiwan, Mexico, China, and Europe. The world production of rapeseed oil in 2002–2003 was about 14 million metric tons.[15]

5.2.1.8 Palm and Palm Kernel Oil

Palm oil is a form of edible vegetable oil obtained from the fruit of the oil palm tree (*Elaeis guineensis*). The palm fruit is the source of both palm oil (extracted from palm fruit) and palm kernel oil (extracted from the fruit seeds). Palm oil itself is reddish because it contains a high amount of β-carotene. It is used as a cooking oil, to make margarine, and is a component of many processed foods.[35] Boiling it for a few minutes destroys the carotenoids, and the oil becomes colorless. Palm oil is one of the few vegetable oils relatively high in saturated fats, and thus semisolid at room temperature.[36]

Palm oil contains almost equal proportions of saturated (palmitic 48%, stearic 4%, and myristic 1%) and unsaturated acids (oleic 37% and linoleic 10%). The oil can be fractionated to give palm stearin, palm olein, and palm mid fraction.[37] It is used mainly for food purposes but has some nonfood uses. Valuable by-products obtained from palm oil are carotene, tocopherols and tocotrienols (vitamin E), and palm-fatty acid distillate (PFAD). Palm kernel oil is lauric oil, similar in composition to coconut oil (lauric acid 50% and myristic acid 16%) and contains palmitic acid (8%), capric acid (3%), caprilic acid (3%), stearic acid (2.5%), oleic acid (15%), and linoleic acid (2.5%).[38,39]

Palm oil and palm kernel oil are composed of fatty acids, esterified with glycerol just like any ordinary fat. Both are high in saturated fatty acids, about 50% and 80%, respectively. The oil palm gives its name to the 16-carbon saturated fatty acid, palmitic acid, found in palm oil; monounsaturated oleic acid is also a constituent of palm oil while palm kernel oil contains mainly lauric acid.[40]

Palm oil products are made using milling and refining processes, first, using fractionation, then crystallization and separation processes to obtain a solid stearin and a liquid olein.[41] By melting and degumming, impurities can be removed and then the oil filtered and bleached. Next, physical refining removes odors and coloration, to produce refined bleached deodorized palm oil (RBDPO), and free pure fatty acids, used as an important raw material in the manufacture of soaps, washing powder, and other hygiene and personal care products.[42]

5.2.1.9   Safflower Oil

Safflower (*Carthamus tinctorius*) is a highly branched, herbaceous, thistle-like annual, usually with many long, sharp spines on the leaves. Safflower is grown particularly in India. Traditionally, the crop was grown for its seeds, used for coloring and flavoring foods, and making red (carthamin) and yellow dyes.[43]

Safflower oil is flavorless and colorless, and is nutritional. It is used mainly as a cooking oil and for the production of margarine. It may also be taken as a nutritional supplement.[44] There are two types of safflowers that produce different kinds of oil, one high in monounsaturated fatty acids (oleic acid) and the other high in polyunsaturated fatty acids (linoleic acid).[45] Normally, it is a linoleic-rich oil (about 75% linoleic acid) with LLL (47%), LLO (19%), and LLS (18%) as the major triacylglycerols. Safflower oil rich in oleic acid (about 74%) has also been developed (Saffola).[46]

5.2.1.10   Coconut Oil

Coconut oil, also known as coconut butter, is a tropical oil with many applications. It is extracted from copra, which means dried coconut and is a product of the coconut palm (*Cocos nucifera*). Coconut oil (about 3.1 million tonnes per annum) comes mainly from Indonesia and the Philippines. Coconut oil constitutes seven percent of the total export income of the Philippines, the world's largest exporter of the product. Coconut oil was developed as a commercial product by merchants in the South Seas and South Asia in the 1860s.[47]

Coconut oil is a fat consisting of about 90% saturated fat. The oil contains predominantly medium chain triglycerides, with roughly 92% saturated fatty acids, 6% monounsaturated fatty acids, and 2% polyunsaturated fatty acids.[48] It is particularly rich in lauric acid (47%), myristic acid (8%), and caprylic acid (8%), although it contains seven different saturated fatty acids in total. The oil finds extensive use in the food industry and also, usually after conversion to the alcohol (dodecanol), in the detergent, cosmetic, and pharmaceutical industries. The only other commercially available lauric oil is palm kernel oil but there also exists laurate-canola and cuphea species.[49]

Among the most stable of all oils, coconut oil is slow to oxidize, and thus resistant to rancidity, lasting up to 2 years due to its high saturated fat content. It is best stored in solid form, below 24.5°C in order to extend shelf life.[50] However, unlike most oils, coconut oil will not be damaged by warmer temperatures. Virgin coconut oil is derived from fresh coconuts (rather than dried).[51] Most oils marketed as "virgin" are produced by one of three ways: (1) quick drying of fresh coconut meat which is then used to press out the oil, (2) wet-milling (coconut milk), with this method the oil is extracted from fresh coconut meat without drying first. Coconut milk is expressed first by pressing. The oil is then further separated from the water. The methods which can be used to separate the oil from the water include boiling, fermentation, refrigeration, enzymes, and mechanical centrifuge[52] and (3) wet-milling (direct micro-expelling), in this process, the oil is extracted from fresh coconut meat after the adjustment of the water content, then the pressing of the coconut flesh results in the direct extraction of the free-flowing oil.[47]

5.2.1.11   Cocoa Butter

The cocoa bean (*Theobroma cacao*) is the source of two important ingredients of chocolate: cocoa powder and a solid fat called cocoa butter.[53] To evaluate the oxidative behavior of cocoa butter, the autoxidation of refined and unrefined butter samples is accelerated (oxidized at day light at room temperature and at 90°C). The quantity of certain aldehydes formed during the oxidation of cocoa butter is examined by gas chromatography. The oxidation stability of butter is evaluated over a 12 week period.[54]

The usefulness of cocoa butter for this purpose is related to its fatty acid and triacylglycerol compositions. The major triacylglycerols are symmetrical disaturated oleic glycerol esters of the type SOS and include POP (18%–23%), POSt (36%–41%), and StOSt (23%–31%).[55,56] Cocoa butter commands a good price and cheaper alternatives have been developed. The annual production of cocoa beans is about 2.7 million tonnes with 45%–48% of cocoa butter.[55]

5.2.1.12 Sesame Oil

Sesame oil (also known as gingelly oil or til oil) is an organic oil of the plant *Sesamum indicum* grown mainly in India and China but also in Myanmar (Burma), Sudan, and Mexico.[57] The sesame seeds are protected by a capsule, which does not burst open until the seeds are completely ripe. The ripening time tends to vary. For this reason, the farmers cut the plants by hand and place them together in an upright position to carry out ripening for a few days. The seeds are only shaken out onto a cloth after all the capsules have opened.[58]

The annual production is about 0.7 million tonnes. The seed has 40%–60% oil with almost equal levels of oleic acid (range 35%–54%, average 46%), linoleic acid (range 39%–59%, average 46%), palmitic acid (7%–12%), palmitoleic acid (trace to 0.5%), stearic acid (3.5%–6%), linolenic acid (trace to 1%), and eicosenoic acid (trace to 1%). The oil contains sesamin (0.5%–1.1%) and sesamolin (0.3%–0.6%) and has high oxidative stability due to the presence of natural antioxidants.[57]

5.2.2 OILS AND FATS OF ANIMAL ORIGIN

Animal fats are rendered tissue fats that can be obtained from a variety of animals. Examples of edible animal fats are butter, lard (pig fat), tallow, ghee, and fish oil. They are obtained from fats in the milk, meat, and under the skin of the animal.[61] Typical fatty acid composition of some animal fats and oils are summarized in Table 5.5.

5.2.2.1 Butter

Butter, a water-in-oil emulsion, comprises of >80% milk fat, but also contains water in the form of tiny droplets, perhaps some nonmilk fat, with or without salt (sweet butter); texture is a result of working/kneading during processing at appropriate temperatures, to establish a fat crystalline network that results in desired smoothness (compare butter with melted and recrystallized butter).[62] It is used as a spread, a cooking fat, or a baking ingredient.[63]

The most common form of butter is made from cows' milk, but it can also be made from the milk of other mammals, including sheep, goats, buffalo, and yaks. Salt, flavorings, or preservatives are sometimes added to butter. When refrigerated, butter remains a solid, but softens to a spreadable consistency at room temperature, and melts to a thin liquid consistency at 32°C–35°C. Butter generally has a pale yellow color, but varies from deep yellow to nearly white.[64]

TABLE 5.5
Typical Fatty Acid Composition of the Main Animal Oils and Fats

Kind of Animal Oils and Fats	Fatty Acids							
	Saturated (g/100 g)	Monounsaturated (g/100 g)	Polyunsaturated (g/100 g)	14:0 (wt%)	16:0 (wt%)	18:0 (wt%)	18:1 (wt%)	18:2 (wt%)
Butter	54	19.8	2.6	12	26	11	28	2
Lard	40.8	43.8	9.6	2	26	11	44	11
Tallow	50	42	4	26	26	14	47	3
Fish oil	26	25	35	9	17	10	10	22

Sources: Gunstone, F.D., *The Chemistry of Oils and Fats—Sources, Composition, Properties and Uses*, CRC Press, Boca Raton, FL, p. 7, 2004; Wikipedia, Fat, 2007, Available at: http://en.wikipedia.org/wiki/Fat (accessed 20/2/2008).

Production of butter is about 5.8 million tonnes a year on a fat basis. As a water-in-oil emulsion it contains 80%–82% milk fat and 18%–20% of an aqueous phase. It is produced throughout the world (6–7 million tonnes a year). Butterfat is very complicated in its fatty acid and triacylglycerol composition. In addition to the usual C_{16} and C_{18} acids, it contains short-chain and medium-chain acids (C_4–C_{14}), a range of trans monoene acids, mainly 18:1, and oxygenated and branched-chain acids. The trans acids represent 4%–8% of the total acids.[65] Butter contains some cholesterol (0.2%–0.4%). Spreads with lower levels of fat are also available. Butter that spreads directly from the refrigerator is made by removing some of its higher melting glycerol esters or by blending with a vegetable oil.[66] The flow diagram of butter manufacture is shown in Figure 5.2.

5.2.2.2 Lard

Lard refers to pig fat in both its rendered and unrendered forms. Lard was commonly used as a cooking fat or shortening, or as a spread similar to butter. Its use in contemporary cuisine has diminished because of health concerns posed by its saturated fat content and its often negative image. The culinary qualities of lard vary somewhat depending on the part of the pig from which the fat is taken and how the lard is processed.[67]

Lard can be obtained from any part of the pig as long as there is a high concentration of fatty tissue. The highest grade of lard, known as leaf lard, is obtained from the "flare" fat deposit surrounding the kidneys and inside the loin. The next highest grade of lard is obtained from fatback, the hard fat between the back skin and flesh of the pig.[68] The lowest grade (for purposes of rendering into lard) is obtained from the soft caul fat surrounding digestive organs, such as small intestines, though caul fat is often used directly as a wrapping for roasting lean meats or in the manufacture of pates.[69]

Industrially produced lard is rendered from a mixture of high and low quality fat sources from throughout the pig. It is typically hydrogenated, and often treated with bleaching and deodorizing agents, emulsifiers, and antioxidants.[68] Such treatments make the lard shelf stable.[70] The available quantities of lard are about 6 million tonnes a year. The fat contains palmitic (20%–32%), stearic (5%–24%), oleic (35%–62%), and linoleic (3%–16%) acids as major components. It is unusual in that 70% of the palmitic acid is in the sn-2 position. It also contains cholesterol (0.37%–0.42%).[34]

5.2.2.3 Tallow

Animal edible tallow is normally obtained from beef but also from sheep and goats, processed from suet. Unlike suet, tallow can be stored for extended periods without the need for refrigeration to prevent decomposition, provided it is kept in an airtight container to prevent oxidation. It is used in animal feed, to make soap, for cooking, as bird feed, and was used for making candles. It can be used as a raw material for the production of biodiesel and other oleochemicals.[71]

The annual production of tallow is about 7 million tonnes. Tallow contains mainly saturated acids (50%): palmitic acid (26%), stearic acid (14%), and myristic acid (3%). It also contains mono-unsaturated acids (42%), especially oleic acid (47%) and palmitoleic acid (3%), and polyunsaturated acids (4%), especially linoleic acid (3%) and linolenic acid (1%).[72] Also present are odd-chain, branched-chain, and trans fatty acids, and cholesterol (0.08%–0.14% in beef tallow and 0.23%–0.31% in mutton tallow).[73]

5.2.2.4 Ghee

Ghee is a solid fat-based product made in India from cow- or buffalo-ripened milk. It is less perishable than butter and, therefore, more suitable for a tropical climate.[74] Milk is curdled. The curd is then manually churned until it precipitates butter and leaves behind some whey. The butter is then heated on a low flame until a layer of white froth covers the surface. This state indicates the end of the process, and the liquid obtained on filtering the suspension is pure ghee.[75] Ghee is made by simmering unsalted butter in a large pot until all water has boiled off and protein

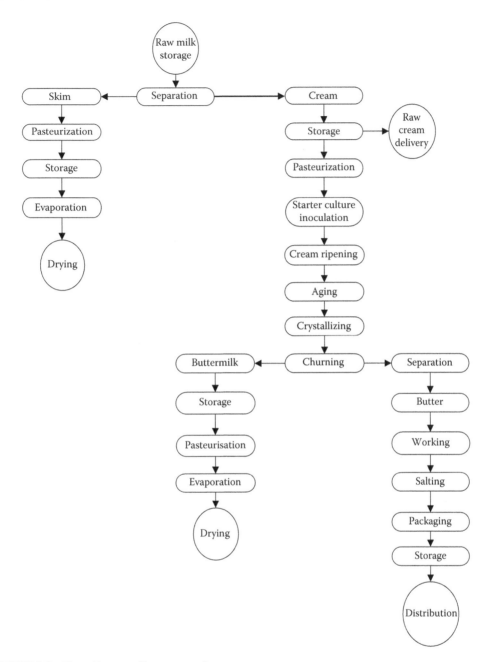

FIGURE 5.2 Flow diagram of butter manufacture.

has settled to the bottom. The cooked and clarified butter is then spooned off to avoid disturbing the milk solids on the bottom of the pan. Unlike butter, ghee can be stored for extended periods without refrigeration, provided it is kept in an airtight container to prevent oxidation and remains moisture-free.[74]

5.2.2.5 Fish Oil

Fish oil is the lipid extracted from the body, muscle, liver, or other organ of fish. The major producing countries are Japan, Chile, Peru, Denmark, and Norway and the main fish sources are herring (*Clupea harengus*), menhaden (*Brevoortia tyrannus*), capelin (*Mallotus villosus*), anchovy (*Engraulis encrasicolus*), sardine (*Sardina pilchardus*), tuna (*Opuntia tuna*), and cod (*Gadus morhua* liver).[76]

Fish oil is recommended for a healthy diet because it contains the ω-3 fatty acids eicosapentaenoic acid (EPA) and docosahexaenoic acid (DHA), precursors to eicosanoids that reduce inflammation throughout the body.[77] However, fish do not actually produce ω-3 fatty acids, but instead accumulate them from either consuming microalgae that produce these fatty acids, as is the case with prey fish like herring and sardines, or, as is the case with fatty predatory fish, by eating prey fish that have accumulated ω-3 fatty acids from microalgae. Such fatty predatory fish like mackerel, lake trout, albacore tuna, and salmon may be high in ω-3 fatty acids, but due to their position at the top of the food chain, these species can accumulate toxic substances (mercury, dioxin, PCBs, and chlorates).[78]

Fish oils contain a wide range of fatty acids from C_{14} to C_{26} in chain length with 0–6 double bonds. The major acids include saturated (14:0, 16:0, and 18:0), monounsaturated (16:1, 18:1, 20:1, and 22:1) and *n*–3 polyene members (18:4, 20:5, 22:5, and 22:6). Fish oils are easily oxidized and are commonly used in fat spreads only after partial hydrogenation.[79] However, they are the most readily available sources of *n*–3 polyene acids, especially, EPA and DHA, and with appropriate refining procedures and antioxidant addition these acids can be conserved and made available for use in food. The long-chain polyene acids are valuable dietary materials and there is a growing demand for high quality oil rich in EPA and DHA.[80]

In 2005, fish oil production declined in all the main producing countries with the exception of Iceland. The production estimate was about 570,000 tonnes in the five main exporting countries (Peru, Denmark, Chile, Iceland, and Norway) in 2005, a 12% decline from the 650,000 tonnes produced in 2004. Peru continues to be the main fish oil producer in the world, with about one fourth of the total fish oil production. Though Peruvian catches of fish were destined for reduction in 2005 it was more or less in line with the 2004 result, however, fish oil production declined from 350,000 tonnes to 290,000 tonnes, due to the lower fat content of the fish. In the recent summer months, the fat content was as low as 2% as compared to 4% in 2004.[81]

5.3 PHYSICAL PARAMETERS

5.3.1 Crystallization, Melting Point, and Polymorphism

Long-chain compounds frequently exist in more than one crystalline form. This property of polymorphism is of both scientific and technical interest and consequently these compounds have more than one melting point.[82]

5.3.1.1 Crystallization

Crystallization, is a nonchemical process used by the vegetable oil industry to obtain triacylglycerides (TAGS) fractions with the given phase change properties. Overall, four different events are involved during crystallization, namely supercooling, nucleation, crystal growth, and crystal ripening. Blends of palm stearin in sesame oil were used as a model system. The results showed that the involvement of molecular local-order and sporadic nucleation, depend on the cooling rate used and the extent of supercooling during vegetable oil crystallization. Additionally, during crystal growth the involvement of secondary crystallization is definite.[83]

A crystal nucleus is the smallest crystal that can exist in a solution and is dependent on the concentration and temperature. Spontaneous nucleation rarely occurs in fats. Instead heterogeneous nucleation occurs on solid particles or on the walls of the container. Once the crystals are formed, fragments that drop-off may either redissolve or act as nuclei for further crystal formation.[84]

5.3.1.2 Melting Point

The melting point of a crystalline solid is the temperature range at which it changes its state from solid to liquid. Although the phrase would suggest a specific temperature, most crystalline compounds actually melt over a range of a few degrees or less. At the melting point, the solid and liquid phases exist in equilibrium at a total pressure of 1 atm.[85] The melting points of the acids with an even number of carbon atoms in the molecule and their methyl esters plotted against a chain-length fall on the smooth curves lying above similar curves for the odd acids and their methyl esters. Odd acids melt lower than even acids with one less carbon atom. The two curves for saturated acids converge at 120°C–125°C.[86]

To reduce the melting point of a tallow-rapeseed oil mixture, the triglyceride composition of the mixture was altered by enzymatic interesterification in a solvent-free system. The interesterification and hydrolysis were followed by melting point profiles and by free fatty acid determinations. The degree of hydrolysis was linearly related to the initial water content of the reaction mixture. The rate of the interesterification reaction was influenced by the amount of enzyme but not much by temperature, between 50°C and 70°C. The melting point reduction achieved by interesterification depended on the mass fractions of the substrates: the lower the mass fraction of tallow, the larger the reduction of the melting point.[87]

5.3.1.3 Polymorphism

Polymorphism is the phenomenon where a compound can precipitate to form numerous crystal structures. The different crystalline structures each have different physical properties, which can change the use of the chemical. The physical properties that may differ from one polymorphism to another include: solubility, density, melting point, and even color. One of the variables that affect the crystallization process is the solvent that is used in the precipitation. The solvent may cause less stable polymorphisms to form instead of those that are more stable.[88] Each polymorphic form, sometimes termed polymorphic modification, is characterized by specific properties, such as x-ray spacings, specific volume, and melting point.[89] Transformation of one polymorphic form into another can take place in the solid state without melting. Two crystalline forms are said to be "monotropic" if one is stable and the other metastable throughout their existence and regardless of the temperature change. The temperature, at which the relative stability changes, is known as the transition point.[90]

The polymorphism of rapeseed oils with high and low erucic acid content was investigated using differential scanning calorimetry and x-ray diffraction. Both oils were hydrogenated to various iodine values. The fatty acid pattern showed that erucic acid is slowly saturated. The melting curves were followed by DSC and Pulsed NMR. For low iodine value the low erucic acid rapeseed oil exhibits a second melting peak owing to the appearance of new triglycerides with different properties. Samples of hydrogenated rapeseed oils were aged at 20°C and 29°C.[91]

5.3.2 Density, Viscosity, and Refractive Index

5.3.2.1 Density

Density is defined as mass per unit volume, i.e., metric ton per barrel. The oil industry in different parts of the world uses different units of measurements. For example, in Europe, the metric ton (a mass unit) is generally accepted as the unit of measurement.[92] Density may not seem an exciting physical property to many technologists, but it is very important in the trading of oils since shipments are sold on a weight basis but measured on a volume basis. These two values are related by density, so it is important to have correct and agreed values for this unit. Density is not the same for all oils as it depends on the fatty acid composition and minor components, as well as the temperature.[93]

5.3.2.2 Viscosity

The combination of molecular size and autocohesive character produces a property in all fluids known as "viscosity." It can be defined either as a resistance to flow or as a resistance to the movement of something through that fluid. Both of these definitions represent the resistance of the molecules of the fluid to separate from each other or "sheer."[94] Viscosity is temperature dependant. By heating the olive oil, it becomes more and more water-like in its consistency. Alternatively, viscosity increases as temperature decreases, and oils become more solid-like in character.[95]

The viscosity of a vegetable oil depends on its chemical composition (iodine value and saponification value) and the temperature of measurement. Equations have been derived which permit the calculation of viscosity from the knowledge of the other three parameters.[96]

5.3.2.3 Refractive Index

Refractive index is the ratio of the velocity of light (of specific wavelength) in air to the velocity in the substance of interest. Refractive index may also be defined as the sine of the angle of incidence divided by the sine of the angle of refraction, as light passes from air into the substance. Refractive index is a fundamental property used in conjunction with other properties to characterize hydrocarbons and their mixtures.[97]

5.4 CHEMICAL PARAMETERS

5.4.1 Oxidation

Oxidation of unsaturated fatty acids is the main reaction responsible for the degradation of lipids. Indeed, the oxidation level of oil and fat is an important quality criterion for the food industry. Oxidation of oils not only produces rancid flavors but can also decrease the nutritional quality and safety by the formation of oxidation products, which may play a role in the development of diseases.[98]

Many methods have been developed to access the extent of oxidative deterioration, which are related to the measurement of the concentration of primary or secondary oxidation products or of both. The most commonly used are peroxide value (PV) that measures volumetrically the concentration of hydroperoxides, anisidine value (AV), spectrophotometric measurement in the UV region and gas chromatographic (GC) analysis for volatile compounds.[99] Vibrational spectroscopy, because of its high content in molecular structure information, has also been considered to be useful for the fast measurement of lipid oxidation. In contrast to the time consuming chromatographic methods, modern techniques of IR and Raman spectrometry are rapid and do not require any sample preparation steps prior to analysis. These techniques have been used to monitor oil oxidation under moderate and accelerated conditions and the major band changes have been interpreted.[100]

Raman spectroscopy is poorly explored in its application to edible oils. It has been basically applied to the characterization and authentication, and to the quantitative analysis of total unsaturation, cis/trans isomers, and free fatty acid content.[101,102]

5.4.1.1 Autoxidation

Autoxidation is the process in foods and bulk lipids, which leads to rancidity. Rancidity is the spoiled off-flavor obtained by subjective organoleptic appraisal of food.[103] Autoxidation is the oxidative deterioration of unsaturated fatty acids via an autocatalytic process consisting of a free radical mechanism. This indicates that the intermediates are radicals (odd electron species) and that the reaction involves an initiation step and a propagation sequence, which continues until the operation of one or more termination steps. Autoxidation of lipid molecules is briefly described by reactions 1–3.[104]

1. Initiation $RH \Rightarrow R^{\bullet} + H^{\bullet}$
   or $X^{\bullet} + RH \Rightarrow R^{\bullet} + XH$
2. Propagation $R^{\bullet} + O_2 \Rightarrow ROO^{\bullet}$
   $ROO^{\bullet} + RH \Rightarrow ROOH + R^{\bullet}$
3. Termination $R^{\bullet} + R^{\bullet} \Rightarrow R\text{--}R$
   $R^{\bullet} + ROO^{\bullet} \Rightarrow ROOR$
   $ROO^{\bullet} + ROO^{\bullet} \Rightarrow ROOR + O_2$

In the initiation step, hydrogen is abstracted from an olefinic acid molecule (RH) to form alkyl radicals ($R^{\bullet}$), usually in the presence of a catalyst, such as metal ions, light, heat, or irradiation, at a relatively slow rate. The duration of the initiation stage varies for different lipids and depends on the degree of unsaturation and on the presence of natural antioxidants.[105]

In the propagation sequence, given an adequate supply of oxygen, the reaction between alkyl radicals and molecular oxygen is very fast and peroxyl radicals are formed ($ROO^{\bullet}$). These react with another fatty acid molecule producing hydroperoxides (ROOH) and new free radicals that contribute to the chain by reacting with another oxygen molecule. Hydroperoxide molecules can decompose in the presence of metals to produce alkoxyl radicals ($RO^{\bullet}$), which cleave into a complex mixture of aldehydes and other products, i.e., secondary oxidation products.[106] The mutual annihilation of free radicals is known as the termination stage, when the free radicals $R^{\bullet}$ and $ROO^{\bullet}$ interact to form stable, non-radical products. The rate of oxidation of fatty acids increases with their degree of unsaturation. The relative rate of autoxidation of oleate, linoleate, and linolenate is in the order of 1:40 to 50:100 on the basis of oxygen uptake and 1:12:25 on the basis of peroxide formation.[107]

5.4.1.2   Azo-Initiated Oxidation

An intrinsic problem in the determination of rate constants in lipid oxidation is the uncertainty about the rate of initiation Ri, according to the reaction:

$$RH + X^{\bullet} \Rightarrow R^{\bullet} + XH$$

One possible way of overcoming this problem is to introduce into the reaction mixture a compound that decomposes at a constant rate to free radicals ($X^{\bullet}$) capable of extracting a hydrogen atom from the PUFAs (RH) and consequently initiating the autoxidation process. The compounds most frequently used for this are the so-called azo-initiators (X–N=N=X), which thermally decompose to highly reactive carbon-centered radicals.[108]

Therefore, azo-initiators are useful for *in vitro* studies of lipid peroxidation generating free radicals according to the following reaction:

$$X - N = N - X \Rightarrow 2X^{\bullet} + N_2$$

The water-soluble azo-initiator AAPH[2,2Azo-bis(2-amidinopropane)dihydrochloride] can be used to produce radicals in the aqueous phase, whereas the lipid-soluble AMVN [2,2′-azo-bis-(2,4-dimethylvaleronitrile)] can be used to produce radicals in the lipid phase.[109] AAPH decomposes with a first-order rate constant of $K_d = 6.6 \times 10^{-5}$/min at 37°C, and the flux of the free radicals is proportional to the AAPH concentration.[110]

5.4.1.3   Photosensitized Oxidation

Photooxidation involves the direct reaction of light-activated, singlet oxygen ($^1O_2$) with unsaturated fatty acids and the subsequent formation of hydroperoxides. In the most stable triplet state (two unpaired electrons in a magnetic field), oxygen is not very reactive with unsaturated compounds. Photosensitized oxidation involves reaction between a double bond and highly reactive singlet oxygen (paired electrons and no magnetic moment) produced from an ordinary triplet oxygen by light in the presence of a sensitizer, such as chlorophyll, erythrosine, or methylene blue.[111]

$$1.\ ^1\text{sens} + h\nu \Rightarrow {}^1\text{sens}^* \Rightarrow {}^3\text{sens}^* + {}^3O_2$$

$$2.\ ^3\text{sens}^* + {}^3O_2 \Rightarrow {}^1\text{sens} + {}^1O_2$$

The singlet oxygen oxidation differs from autoxidation in several important respects: (1) it is an ene and not a radical chain reaction, (2) it gives products which are similar in type but not identical in structure to those obtained by autoxidation, and (3) it is a quicker reaction and its rate is related to the number of double bonds rather than the number of doubly activated allylic groups.[105]

5.4.1.4 Metal Catalyzed Oxidation

Many natural oils contain metals such as cobalt iron, magnesium, and copper, which possess two or more valence states with a suitable oxidation–reduction potential and can serve as excellent pro-oxidants in lipid oxidation reactions.[112] Contamination of oils with specific metals (copper, iron, etc.) can also occur during the refining procedure.

Metals can initiate fatty acid oxidation by a reaction with oxygen. The anion thus produced can either lose an electron to give a singlet oxygen or react with a proton to form a peroxyl radical, which is a good chain initiator.[113]

$$M^{n+} + O_2 \Rightarrow M^{(n+1)+} + O_2^{\cdot-} \Big\langle {}^{^1O_2}_{OH^{\cdot}}$$

Many oxygenated complexes of transition metals have now been isolated and used as catalysts for the oxidation of olefins and recent evidence supports the initiation of autoxidation through the formation of a metal hydroperoxide catalyst complex.

Once a small amount of hydroperoxides is formed, the transition metals can promote the decomposition of the preformed hydroperoxides due to their unpaired electrons in the 3d and 4d orbitals. A metal, capable of existing in two valence states typically acts as:

$$M^{n+} + ROOH \Rightarrow RO^{\cdot} + OH^- + M^{(n+1)+} \text{—the reduced metal ion is oxidized}$$

$$M^{(n+1)+} + ROOH \Rightarrow ROO^{\cdot} + H^+ + M^{n+} \text{—the oxidized metal is reduced}$$

$$2ROOH \Rightarrow RO^* + ROO^{\cdot} + H_2O \text{—net reaction}$$

In a system containing multivalent metal ions, such as Cu^+ and Cu^{2+} or Fe^{2+} and Fe^{3+}, the hydroperoxides decompose readily with the formation of both $RO^{\cdot}$ and $ROO^{\cdot}$ as the metal ions undergo oxidation–reduction.[114]

5.4.1.5 Enzyme Catalyzed Oxidation

The basic chemistry of enzyme catalyzed oxidation of food lipids such as in cereal products, or in many fruits, and vegetables is the same as for autoxidation, but the enzyme lipoxygenase (LPX) is very specific for the substrate and for the method of oxidation.[115] Lipoxygenases are globulins with molecular weights ranging from $0.6–1 \times 10^5$ Da, containing one iron atom per molecule at the active site.

LPX type 1, from many natural sources e.g. potato, tomato, and soybean, prefers polyunsaturated fatty acids that have as their best substrate linoleic acid, which is oxidized to 9 and 13 hydroperoxides. These hydroperoxides suffer fragmentation to give short-chain compounds (hexanal, 9-oxononanoic acid, 2-nonenal), some of which have marked and characteristic odors. LPX type 2 (present in gooseberry, soybean, and legumes), catalyzes the oxidation of acylglycerols, whereas cooxidation of other plant components e.g., carotenoids may also occur.[116] In that case hydroperoxides may suffer enzyme-catalyzed reactions to give a mixture of products as shown in the following scheme:

$$RH + O_2 \xrightarrow{\text{Lipoxygenase}} ROOH \xrightarrow[\text{reactions}]{\text{Enzymatic}} ROH, RCHO, ROOR, RCO_2H$$

5.4.1.6   Decomposition of Hydroperoxides

A large body of scientific evidence suggests that the loss of food palatability as a result of lipid oxidation is due to the production of short chain compounds from the decomposition of the hydroperoxides. The volatile compounds produced from the oxidation of edible oils are influenced by the composition of the hydroperoxides and the positions of oxidative cleavage of double bonds in the fatty acids.[117]

A variety of compounds such as hydrocarbons, alcohols, furans, aldehydes, ketones, and acid compounds are formed as secondary oxidation products and are responsible for the undesirable flavors and odors associated with rancid fat.[118] The off-flavor properties of these compounds depend on the structure, concentrations, threshold values, and food systems. Aliphatic aldehydes are the most important volatile breakdown products because they are major contributors to unpleasant odors and flavors in food products.

5.4.1.7   Physical Aspects—Lipid Oxidation in Food Emulsions

An emulsion is a dispersion of droplets of one liquid in another liquid with which it is not miscible. A system that consists of oil droplets dispersed in an aqueous phase is called an oil-in-water (or O/W) emulsion. A system that consists of water droplets dispersed in an oil phase is called a water-in-oil (or W/O) emulsion. The material that makes up the droplets in an emulsion is referred to as the dispersed or internal phase; whereas the material that makes up the surrounding liquid is called the continuous or external phase.[119]

Emulsions are thermodynamically unstable systems because of the positive energy required to increase the surface area between the oil and water phases.[120] It is thermodynamically favored for the emulsions to reduce that surface area by separating into a system that consists of a layer of oil (lower density) on top of a layer of water (higher density), unless they are effectively stabilized. Generally, the stability of food emulsions is complex because it covers a large number of phenomena, including flocculation, coalescence, creaming, and final phase separation.[121]

A material that lowers the interfacial energy between the two immiscible phases of an emulsion system (e.g., oil and water) and thereby facilitates the dispersion of one phase in another is called an emulsifier. There are different types of emulsifiers used in various industrial applications, such as monoglycerides, polyglycerol esters, lecithin etc. The size of droplets in an emulsion influences many of their sensory and bulk physicochemical properties including rheology, appearance, mouthfeel, and stability.[122] It is therefore important for food manufacturers to carefully control the size of the droplets in a food product and to have analytical techniques to measure the droplet size. Typically, the droplets in a food emulsion are somewhere in the range of 0.1 to 50 μm. The size of the droplets produced in an emulsion depends on many different factors, including the emulsifier type and concentration, homogenization conditions, and physicochemical properties of bulk liquids.[123]

To create an emulsion either industrially or in the laboratory, it is necessary to use the appropriate homogenization procedure.[124] If an application does not require that the droplets in an emulsion be particularly small, it is usually the easiest to use a high-speed blender. When the involving ingredients are limited in availability or are expensive, an ultrasonic piezoelectric transducer can be ideally used, whereas for monodispersed emulsions a membrane homogenizer is preferred.

Because many foods are emulsified materials (e.g., milk, mayonnaise, coffee creamers, salad dressings, butter, and baby foods), an understanding of the mechanisms of lipid oxidation in emulsions is crucial for the formulation, production, and storage of food products.[125] Oil-in-water emulsions consist of three different components: water (the continuous phase), oil (the dispersed phase), and surface-active agent (the interface).

In such a system, the rate of oxidation is influenced by the emulsion composition (relative concentrations of substrate and emulsifier) and especially, by the partition of the emulsifier between the interface and the water phases.[126] Other factors influencing lipid oxidation in emulsions are particle size of the oil droplets, the ratio of oxidizable to non-oxidizable compounds in the emulsion droplets, and the packing properties of the surface-active molecules.[127] In addition, the amount and composition of the oil phase in an emulsion are important factors that influence oxidative stability, formation of volatiles, and partition of the decomposition products, between the oil and water phase.

Antioxidant behavior is more complex when evaluated in emulsions than in oils because more variables influence lipid oxidation including emulsifier, pH, and buffer systems.[128] Hence, limited research has been done so far to systematically evaluate antioxidant activity with respect to the interaction between these system dependent variables. Natural antioxidants have been particularly difficult to evaluate in oils and emulsions because of the complex interfacial phenomena affecting the partition of the antioxidants in the different phases.[129]

5.4.2 ANTIOXIDANT ACTIVITY OF CAROTENOIDS

5.4.2.1 Carotenoids as Radical Scavengers—Mechanism of Antioxidant Action

A large body of scientific evidence suggests that carotenoids scavenge and deactivate free radicals both in vitro and in vivo.[130] According to Bast et al.,[131] their antioxidant action is determined by: (1) electron transfer reactions and the stability of the antioxidant free radical, (2) the interplay with other antioxidants, and (3) their action with active oxygen. Moreover, the antioxidant activity of carotenoids is characterized by literature data for (a) their relative rate of oxidation by a range of free radicals, or (b) their capacity to inhibit lipid peroxidation in multilamellar liposomes.[132] According to Mordi,[133] the antioxidant activity of carotenoids is a direct consequence of the chemistry of their long polyene chain: a highly reactive, electron-rich system of conjugated double bonds susceptible to attack by electrophilic reagents, and forming stabilized radicals. Therefore, this structural feature is mainly responsible for the chemical reactivity of carotenoids toward oxidizing agents and free radicals, and consequently, for any antioxidant role.

β-Carotene has received considerable attention in recent times as a putative chain breaking antioxidant, although it does not have the characteristic structural features associated with conventional primary antioxidants, but its ability to interact with free radicals including peroxyl radicals is well documented.[134] Burton and Ingold[135] were the first researchers that investigated the mechanisms by which β-carotene acts as a chain breaking antioxidant. In fact, the extensive system of conjugated double bonds makes carotenoids very susceptible to radical addition, a procedure which eventually leads to the free radical form of the carotenoid molecule.

The carbon-centered radical is resonance stabilized to such an extent, that when the oxygen pressure is lowered the equilibrium of reaction 2 shifts sufficiently to the left, to effectively lower the concentration of the peroxyl radicals and hence reduces the amount of autoxidation in the system.[136] Furthermore, the β-carotene radical adduct can also undergo termination by a reaction with another peroxyl radical (reaction 3).

Reactions:

1. β-CAR + ROO˙ → ROO—β—CAR˙

2. ROO—β—CAR˙ + O_2 ↔ ROO—β—CAR—OO˙

3. ROO—β—CAR˙ + ROO˙ → inactive products

According to Edge et al.,[137] the reactivity of β-carotene toward peroxyl radicals and the stability of the resulting radical give the molecule its antioxidant capability. The first feature, results in its competing with other lipids for the peroxyl radicals, whereas the second characteristic leads to the formation of the stable carbon-centered β-carotene radical especially at low oxygen partial pressure. In fact, in living organisms the partial pressure of oxygen, in the capillaries of the active muscle is only about 20 Torr, while in the tissue it must be considerably lower.[135] In addition, the effect of partial oxygen pressure on β-carotene antioxidant capacity was demonstrated by Kasaikina et al.[138]

5.4.2.2 Reactivity of Carotenoids Toward Free Radicals—Effect of Structure on Scavenging Activity

Woodall et al.[139] suggested that the different reactivities of the carotenoids against free radicals can be partly attributed to the differences in electron distribution along the polyene chain of different chromophores that would alter the susceptibility of free radical addition to the conjugated double bond system. However, other factors must be also considered. Stearic hindrance, hydrogen abstraction from the allylic position to the polyene chain (C-4 of β-carotene and its derivatives, end of lycopene), would reduce radical scavenging activity. In addition, the stability of the polyene radical is important in determining the rate of the loss of carotenoids and hence it affects their antioxidant activity.

Britton[140] observed that all carotenoids react rapidly with the oxidizing agents and free radicals, though the reactivity depends on the length of the polyene chromophore and, to some extent, on the nature, of the end groups. Calculations have shown that electron density in the carotenoid polyene chain is greater toward the end of the chromophore. These electron rich sites are likely to be preferred for reactions with free radical species. As a consequence, β-carotene and its 3,3′diol (zeaxanthin), which have the same chromophore, have very similar electron density profiles, and therefore scavenging activities.[141] In addition it is proposed that the substituents that do not contribute to the chromophore can also have a significant influence on the rate and course of the oxidation reactions.[142] Ketocarotenoids such as, canthaxanthin and astaxanthin, were generally found to react more slowly with free radicals than β-carotene. The differences in reactivity are attributed to the presence of the hydroxyl or keto-groups in the allylic C-4 and C-4′ position, preventing hydrogen abstraction from these positions to give a resonance stabilized neutral radical. The rate of reaction between the carotenoids and phenoxyl radical was found to increase with the increase in the number of conjugated double bonds, and to decrease in the presence of hydroxyl, and especially keto-groups.[143] Mortensen et al.[144] proposed that the substitution of the hydrogens with the carbonyl groups at the 4 position increases the overall peroxyl radical trapping ability due to the electron withdrawing character of the carbonyl oxygen atoms, which substantially reduces the unpaired electron density of the carbon-centered radical and reduces its reactivity with oxygen. Terao[145] studied the antioxidant activity of β-carotene and related carotenoids on the free radical oxidation of methyl linoleate in solution. Canthaxanthin, astaxanthin, and other carotenoids containing the oxo-groups in the 4 position of the β-ionone ring, were more effective antioxidants than β-carotene under these conditions. In addition, Mortensen et al.[146] carried out a pulse radiolysis study, to determine the mechanisms and relative rates of nitrogen dioxide ($NO_2^{\cdot}$), thiyl ($RS^{\cdot}$), and sulfonyl ($RSO_2^{\cdot}$) radical scavenging by different carotenoids.

Haila et al.[147] examined the effect of solvent polarity on the antioxidant effect of different carotenoids. It was found that in acetone in air, all carotenoids except β-carotene reduced the number of spin adducts formed by about 20%, whereas in the less polar solvent toluene most carotenoids lowered the number of spin adducts to a lesser extent than in acetone.

5.4.2.3 Oxidative Degradation of Carotenoids by Free Radicals

To understand the mechanism of antioxidant activity of carotenoids it is important to analyze the oxidation products that are formed during their action as antioxidants.[148] The most important part of the molecule involved in those reactions is the polyene chain. This is a highly reactive electron rich system, susceptible to attack by electrophilic reagents such as peroxyl radicals, and thus responsible for the instability of the carotenoids toward oxidation.

Recent work in the area has concentrated on the reactions of carotenoids with peroxyl radicals, generated mainly by the thermal decomposition of azo-initiators that lead to a variety of products.[149] Most of these products seem to be apocarotenals or apocarotenons of various chain lengths produced by cleavage of a double bond in the polyene chain, such as β-apo-12′-carotenal, β-apo-14′-carotenal, β-apo-10-carotenal, and β-apo-13-carotenone. Kennedy and Liebler[134] reported that 5,6-epoxy-β,β-carotene and 15,15′-epoxy-β,β-carotene and several unidentified polar products were formed by the peroxyl radical oxidation of β-carotene by the peroxyl radicals.

Recent research has shown that the antioxidant activity of carotenoids may shift to prooxidant activity depending mainly on the biological environment in which they act.[150] The pro-oxidant potency is affected by several factors including oxygen tension, and interaction with other carotenoids etc. Burton and Ingold[135] found that β-carotene at concentrations higher than 5 mM acted as a prooxidant at high oxygen pressure (>20% O_2). They explained the change from antioxidant to prooxidant action as being due to the formation of an adduct between peroxy radicals and the carotenoid to form an alkyl radical, which was delocalized and hence stable at low oxygen pressure, but further addition with oxygen could yield a carotenoid-peroxyl radical, which was no longer stabilized by the conjugation.

Kennedy and Liebler[134] suggested that the loss in antioxidant efficiency of β-carotene at high pO_2 may be due to β-carotene autoxidation rather than to formation of a β-carotene-oxygenated radical. The increased autoxidation would consume β-carotene without scavenging the peroxyl radicals and would attenuate β-carotene's antioxidant efficiency. Also, Niki and co-workers[151] found that β-carotene acts more efficiently as an antioxidant in the absence of oxygen (inhibiting 70% lipid oxidation) and this effect decreases rapidly to about 50% inhibition, on increasing the partial pressure of oxygen to 8 mmHg.

Farombi and Burton,[152] examined the effects of several carotenoids on autoxidized triglycerides and concluded that potential prooxidant effects could occur at high concentrations. Moreover, Henry et al.,[153] suggested that at concentrations >500 ppm both β-carotene and lycopene acted as prooxidants by decreasing the induction time significantly, during the heat catalyzed oxidation of safflower seed oil. Several studies have shown that the presence of primary antioxidants and especially tocopherols stabilize carotenoids so they exhibit synergistic antioxidant character, instead of the prooxidant action that carotenoids would present individually.

5.4.2.4 Oxygen Quenching Activity of Carotenoids on Oil-Photooxidation

As already discussed, singlet molecular oxygen ($^1O_2$) participates in the photooxidation of vegetables oils and oil containing foods, and various chromophoric impurities such as chlorophylls are believed to act as photosensitizers, thereby generating this state of oxygen by transfer of the excitation energy.[154] The discovery that carotenoids deactivate singlet molecular oxygen was an important advancement in understanding their technological and biological effects. The mechanism by which they protect lipid systems from damage due to photooxidation appears to depend largely on physical quenching and to a much lesser extent on chemical reaction.[155] According to Bradley and Min,[156] the addition of various carotenoids to foods containing unsaturated oils improves their shelf life.

The mechanism by which carotenoids, and especially β-carotene, which has been widely studied, act as oxygen quenchers can be summarized as follows: In the presence of β-carotene, the singlet oxygen will preferentially transfer exchange energy to produce the triplet state carotene, while oxygen comes back to its ground energy state and therefore is inactivated (reaction 1). The transfer of energy from the singlet oxygen to the carotenoid takes place through an exchange electron transfer mechanism.[157] The triplet state β-carotene releases energy in the form of heat, and the carotenoid is returned to its normal energy state (reaction 2). In this way carotenoids act so effectively, that one carotenoid molecule is able to quench up to 1000 molecules of singlet oxygen.[158]

Reactions:

$$1. \ ^1O_2 + \beta\text{-carotene} \rightarrow \ ^3\beta\text{-carotene}^* + \ ^3O_2$$

$$2. \ ^3\beta\text{-carotene}^* \rightarrow \beta\text{-carotene} + heat$$

Lee and Min[159] conducted a study to assess the effects of lutein, zeaxanthin, lycopene, isozeaxanthin, and astaxanthin on the chlorophyll sensitized photooxidation of soybean oil. They found that as the carotenoid concentration increased, the oil PVs decreased, whereas, the antioxidant effect increased with an increased number of double bonds.

Chen et al.[160] found that capsanthin which contains 11 conjugated double bonds, a conjugated keto group, and a cyclopentane ring, had higher antioxidant activity than β-carotene, which also contains 11 double bonds but neither of the other functional groups. Nielsen et al.[161] also found a higher antioxidant effect for canthaxanthin (and carbonyl-containing carotenoids in general) toward photodegradation (transient absorption spectroscopy study) than β-carotene and other carotenoids without the constituent groups.

5.4.3 Natural Antioxidants Tested

5.4.3.1 Tocopherols and Tocotrienols

Tocopherols and tocotrienols comprise the group of eight chromanol homologs that possess vitamin E activity in the diet. They are natural monophenolic compounds with varying antioxidant activities.[162] The α-, β-, γ-, and δ-tocopherols are characterized by a saturated side chain consisting of three isoprenoid units, whereas their corresponding tocotrienols have double bonds at the 3′, 7′, and 11′ position of the isoprenoid side chain (Figure 5.3).

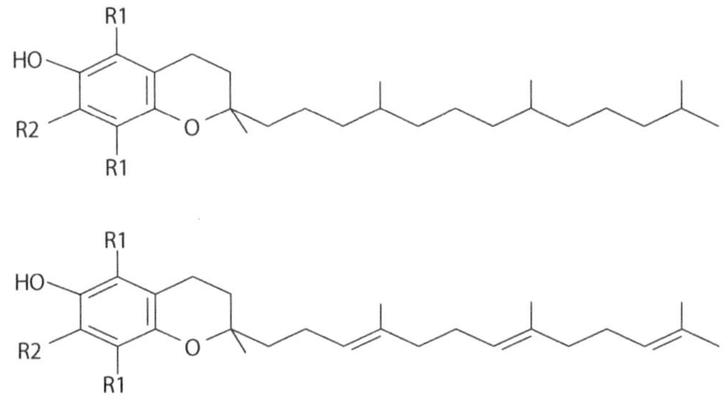

FIGURE 5.3 Chemical structures of E vitamins.

TABLE 5.6
Approximate Contents of Tocopherols Found in Vegetable Oils (mg/kg)

Oil	Tocopherols (mg/kg)			
	α	β	γ	δ
Olive	1–240	Trace	—	—
Corn	60–260	—	400–900	1–50
Soybean	30–120	0–20	250–930	50–450
Sunflower	350–700	20–40	10–50	1–10
Cottonseed	40–560	—	270–410	—
Wheat germ	560–1200	660–810	260	270
Palm	180–260	Trace	320	70
Coconut	5–10	—	5	5

Source: Modified from Schuler, P., Natural antioxidants exploited commercially, *Food Antioxidants*, Hudson, F.B., ed., Elsevier Applied Sciences, London, U.K., 1990.

With regard to vitamin E activity α-tocopherol is the most potent homologue. Vegetable foods (cereal products, oil-seeds, peas, beans, and carrots) contain various tocopherols and tocotrienols in important quantities.[163] In vegetable oils, tocopherol content depends very much on the growing conditions of the plant from which the oil is extracted as well as on the processing and storage of the oil and losses of 30%–40% may occur during oil refining. The range of tocopherol contents in vegetable oils is given in Table 5.6.

Tocopherols are absorbed from the small intestine and secreted into the lymph in the chylomicrons produced in the intestinal wall. The bile produced by the liver, emulsifies the tocopherols incorporating them into micelles, along with other fat-soluble compounds and thereby facilitating their absorption.[164]

Tocopherols behave as chain breaking antioxidants by competing with the substrate (RH) for the chain peroxyl radicals (ROO*) that are normally present in the highest concentration in the system.[165] Tocopherols donate hydrogen to a peroxyl radical resulting in an α-tocopherol semiquinone radical (reaction 1). This may further donate hydrogen to produce methyl tocopherol quinone or react with another α-tocopheryl semiquinone radical (reactions 2 and 3) to produce an α-tocopherol dimer which also possesses antioxidant activity.[166]

Reactions:

$$1. \ ROO^\bullet + \alpha\text{-toc} \ \Rightarrow \ ROOH + \alpha\text{-toc-semiq}^\bullet$$

$$2. \ \alpha\text{-toc-semiq}^\bullet + ROO^\bullet \ \Rightarrow \ ROOH \ \text{and methyl-tocqin}$$

$$3. \ \alpha\text{-toc-semiq}^\bullet + \alpha\text{-toc-semiq}^\bullet \ \Rightarrow \ \alpha\text{-toc–dimer}$$

The antioxidant activities of tocopherols have been investigated in various test systems, including vegetable oils, animal fats, emulsions, PUFAs etc.[167] It has been proposed that the relative antioxidant activity of different tocopherols depends on temperature, lipid composition, physical state (bulk phase, emulsion), and tocopherol concentration.[168] On the basis of hydroperoxide formation, α-tocopherol was reported to have antioxidant activity at low concentrations, but this changed to prooxidant at high concentrations. γ-tocopherol retained antioxidant activity at higher concentrations than α-tocopherol, though its effectiveness was not increased with concentration in soybean oil.[169] In aqueous linoleic acid micelles solution, α-tocopherol was a more effective antioxidant than γ-tocopherol.[170] In addition α-tocopherol is considered as the major antioxidant of olive oil with an activity dependent on both concentration and temperature.[171] Generally, it has been found that under common test conditions in oils, the antioxidant activity decreases from δ-tocopherol to α-tocopherol.[163] In food manufacturing practice, it is recommended to keep the amount of total α-tocopherol (natural or added) at levels between 50 and 500 mg/kg, depending on the kind of food product, as at sufficiently higher levels of addition the effect may become prooxidant.

5.4.3.2 Ascorbic Acid and Ascorbyl Palmitate

L-Ascorbic acid, or vitamin C, is very widespread in nature, and it is gaining importance as a versatile natural food additive due both to its vitamin activity and its ability to improve the quality and extend the shelf life of many food products. In addition, vitamin C is recognized as an antioxidant nutrient with multi-functional effects depending on the conditions of the food system.[172]

Ascorbic acid occurs in all the tissues of living organisms where it is responsible for the normal functioning of important metabolic processes. Vegetable foodstuffs (oranges, blackcurrants, parsley, green peppers, etc.) are the richest sources of vitamin C, but animal products contain relatively little ascorbic acid. L-Ascorbic acid is a six carbon weak acid with a pK_a of 4.2, which is reversibly oxidized due to its enediol structure with the loss of an electron to form the free radical semihydroascorbic acid.[173]

Compared with other radical species, this radical is relatively stable and its further oxidation results in dehydroascorbic acid (DHASc) that probably exists in vivo in multiple forms[174] and can be reduced back to ascorbic acid by the same intermediate radical (Figure 5.4). In food systems, ascorbic acid is a secondary antioxidant that can scavenge oxygen, act synergistically with chelators, and regenerate primary antioxidants.[175] Depending on conditions, ascorbic acid can act by several mechanisms: (a) hydrogen donation to regenerate the stable antioxidant radical, (b) metal inactivation to reduce the initiation of the metals, (c) hydroperoxide reduction to produce stable alcohols by non-radical processes, and (d) oxygen scavenging.[172]

In aqueous systems containing metals, ascorbic acid may also act as a prooxidant by reducing the metals that become active catalysts of oxidation in their lower valences. However, in the absence of added metals, ascorbic acid is an effective antioxidant at high concentrations.[176] The action of ascorbic acid in lipid autoxidation is dependent on concentration, the presence of metal ions, and other antioxidants.[177] It has been shown that ascorbates can protect plasma and LDL lipids from peroxidative damage, and may inhibit the binding of copper ions to LDL.[178] Mixtures

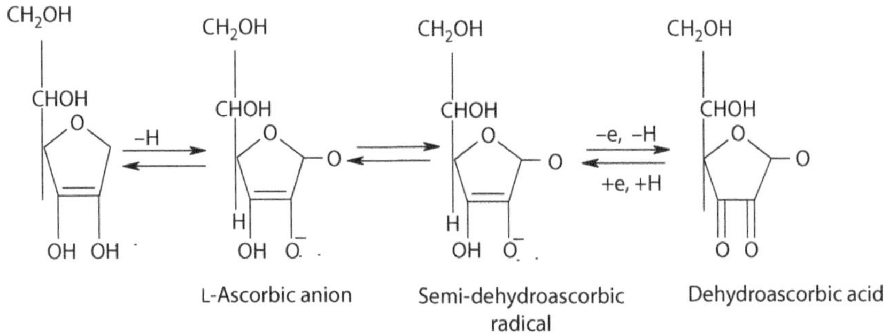

FIGURE 5.4 Oxidation reactions of vitamin C.

of tocopherols and ascorbic acid exhibit a synergistic effect. Vitamin C donates a hydrogen ion to tocopherols, so that tocopheroxyl radicals (TOC·), are reduced back to tocopherols (TOC) according to the equation:

$$TOC^· + Asc \Rightarrow TOC + DHAsc.$$

By this synergistic mechanism, tocopherols and ascorbic acid can mutually reinforce one another by regenerating the oxidized form of the other. Radical exchange reactions among lipid radicals, tocopherols, and ascorbic acid are the basis of numerous approaches for stabilizing oil and foods with their mixtures.[179] It is however important to note that vitamin C is not soluble in the lipid phase that is most susceptible to oxidation. This was the reason why L-ascorbic esters were developed, e.g., ascorbyl palmitate that has a lipid solubility superior to that of ascorbic acid.[162] Mixtures of ascorbyl palmitate with tocopherols are well known for their synergistic activity.

5.4.3.3 Olive-Oil Phenolics

The stability of olive oil to autoxidation depends on several factors including fatty acid composition content, natural antioxidants, free acidity, and PV.[180] Virgin olive is unrefined oil that contains phenolic compounds that are usually removed from other edible oils in the refining stages. Together with α-tocopherol, the fraction of polar phenols contributes significantly to olive oil stability.[181] The polar fraction of olive oil is a complex one and contains alcohols (4-hydroxyphenylethanol or tyrosol, 3,4-dihydroxyphenylethanol or hydroxytyrosol), vanillic, caffeic, and other phenolic acids, as well as a number of not fully identified aglycons of secoiridoid glycosides.[182] Phenolic antioxidants inhibit autoxidation of lipids (RH) by trapping intermediate peroxyl radicals in two ways:

$$i. ROO^· + ArOH \Rightarrow ROOH + ArO^·$$

$$ii. ROO^· + ArO^· \Rightarrow ROO–ArO$$

First, the peroxyl radical abstracts a hydrogen atom from the phenolic antioxidant to yield a hydroperoxide and aroxyl radical that subsequently undergoes radical coupling to give peroxide products.[183] The rate of oxidation of a lipid inhibited by a phenolic antioxidant requires consideration of the following reactions too:

$$iii. ArO^· + ROOH \Rightarrow ROO^· + ArOH$$

$$iv. 2ArO^· \Rightarrow Non\text{-}radical\ products$$

$$v. ArO^· + RH \Rightarrow ArOH + R^·$$

In diphenols with antioxidant activity (hydroxytyrosol, oleuropein, and caffeic acid), the rates of reactions (ii) and (iv) greatly exceed the rates of reactions (iii) and (v). However, for monopherols including tyrosol, reactions (iii) and (v) are relatively rapid and consequently the molecule lacks effective antioxidant activity.[184]

The antioxidant activity of phenolic compounds is related to their structure, in particular the number of hydroxy-substituents in the aromatic ring and the nature of the substituents in the *para* or *ortho* position. In particular the addition of a hydroxyl or methoxy group in the ortho position increases the antioxidant activity due to a strong electron delocalizing effect and this helps to explain why ferulic acid is more active than *p*-coumaric acid.[185] Chimi et al.[183] found that during the autoxidation of linoleic acid dispersed in an aqueous media, the order of activity was: tyrosol < caffeic acid < oleuropein < hydroxytyrosol.

5.5 LEGISLATION FOR OILS AND FATS

5.5.1 EU Legislation for Oils and Fats

This Directive 76/621/EEC (entry into force 28/7/1976) shall apply: (a) to oils, fats, and mixtures thereof which are intended as such for human consumption, (b) to compound foodstuffs to which oils, fats, or mixtures thereof have been added and the overall fat content of which exceeds 5%. Member States may, however, also apply the provisions of this Directive to these foodstuffs when their fat content is equal to or less than 5%. The sampling procedures and methods of analysis necessary to establish the level of erucic acid of the products referred to in this Directive shall be determined in accordance with the procedure laid down in this Directive. The Commission shall examine as soon as possible the grounds given by the Member State concerned and consult the Member States within the Standing Committee on Foodstuffs, and shall then deliver its opinion forthwith and take the appropriate measures.[186]

According to Directive 96/3/EC (entry into force 17/2/1996) the bulk transport in sea-going vessels of liquid oils or fats which are to be processed, and which are intended for or likely to be used for human consumption, is permitted in tanks that are not exclusively reserved for the transport of foodstuffs, subject to the following conditions: (I) that, where the oil or fat is transported in a stainless steel tank, or tank lined with epoxy resin or technical equivalent, the immediately previous cargo transported in the tank shall have been a foodstuff, or a cargo from the list of acceptable previous cargoes set out in the Annex, (II) that, where the oil or fat is transported in a tank of materials other than those in point (I), the three previous cargoes transported in the tanks shall have been foodstuffs, or from the list of acceptable previous cargoes set out in the Annex. The captain of the sea-going vessel transporting, in tanks, bulk liquid oils and fats intended for or likely to be used for human consumption shall keep accurate documentary evidence relating to the three previous cargoes carried in the tanks concerned, and the effectiveness of the cleaning process applied between these cargoes.[187]

Following Regulation (EEC) No. 136/66 (entry into force 1/10/1966) the Council shall fix a single production target price, a single market target price, a single intervention price, and a single threshold price for olive oil for the Community. Olive oil bought in by intervention agencies shall not be sold by them on the Community market on terms which might impede price formation at the level of the market target price. When olive oil is exported to third countries: (1) the difference between prices within the Community and prices on the world market may be covered by a refund where the former are higher than the latter, and (2) a levy equal at most to the difference between prices on the world market and prices within the Community may be charged where the former are higher than the latter. Virgin olive oil produced by mechanical processes and free from any admixtures of other types of oil or of olive oil extracted in a different manner. Virgin olive oil is classified as follows: (a) Extra: olive oil of absolutely perfect flavor, with a free fatty acid content expressed as oleic acid of not more than 1 g per 100 g, (b) Fine: olive oil with the same characteristics as "Extra" but with a free fatty acid content expressed as oleic acid of not more than 1.75 g per 100 g, (c) Ordinary: olive oil of good flavor with a free fatty acid content expressed as oleic acid of not more than 3.73 g per 100 g, (d) Lampante: off-flavor olive oil or olive oil with a free fatty acid content expressed as oleic acid of more than 3.73 g per 100 g.[188]

Regulation (EEC) No. 2568/91 (entry into force 8/9/1991) makes clear that the characteristics of the oils shall be determined in accordance with the methods of analysis set out below: (a) for the determination of the free fatty acids, expressed as the percentage of oleic acid, (b) for the determination of the peroxide index, (c) for the determination of aliphatic alcohols, (d) for the determination of the sterol content, (e) for the determination of erythrodiol and uvaol, (f) for the determination of the saturated fatty acids in position 2 of the triglyceride, (g) for the determination of the trilinolein content, (h) for spectrophotometric analysis, (i) for the determination of the fatty acid composition, (j) for the determination of the volatile halogenated solvents, (k) for the evaluation of the organoleptic characteristics of virgin olive oil, and (l) for proof that refining has taken place.[189]

Regulation (EC) No. 2991/94 (entry into force 1/1/1996) laid down standards for: milk fats, fats, and fats composed of plant and/or animal products with a fat content of at least 10% but less than 90% by weight, intended for human consumption. The fat content excluding salt must be at least two-thirds of the dry matter. The Regulation applies also to products which remain solid at a temperature of 20°C, and which are suitable for use as spreads. The products may not be supplied or transferred without processing to the ultimate consumer either directly or through mass caterers, unless they meet the requirements set out. It has no application to the designation of products the exact nature of which is clear from traditional usage and/or when the designations are clearly used to describe a characteristic quality of the product and to concentrated products (butter, margarine, blends) with a fat content of 90% or more. The following information must be indicated in the labeling and presentation of the products: the sales description, the total percentage fat content by weight at the time of production for products, the vegetable, milk, or other animal fat content in decreasing order of weighted importance as a percentage by total weight at the time of production for compound fats, and the percentage salt content must be indicated in a particularly legible manner in the list of ingredients.[190]

According to Regulation (EC) No. 2815/98 (entry into force 31/10/2001) the designation of origin shall relate to a geographical area and may mention only: (a) a geographical area whose name has been registered as a protected designation of origin or protected geographical indication and/ or, (b) for the purposes of this Regulation: a Member State, the European Community, a third country. The designation of origin, where this indicates the European Community or a Member State shall correspond to the geographical area in which the "extra virgin olive oil" or "virgin olive oil" was obtained. However, in the case of blends of "extra virgin olive oils" or "virgin olive oils" in which more than 75% originates in the same Member State or in the Community, the main origin may be designated provided that it is followed by the indication "selection of (extra) virgin olive oils more than 75% of which was obtained in… (designation of origin)." An extra virgin or virgin olive oil shall be deemed to have been obtained in a geographical area for the purposes of this paragraph only if that oil has been extracted from olives in a mill located within that area.[191]

Regulation (EC) No. 1019/2002 (entry into force 1/11/2002) laid down specific standards for retail-stage marketing of the olive oils and olive-pomace oils. Oils shall be presented to the final consumer in packaging of a maximum capacity of 5 L. Such packaging shall be fitted with an opening system that can no longer be sealed after the first time it is opened and shall be labeled. However, in the case of oils intended for consumption in mass caterers, the Member States may set a maximum capacity exceeding 5 L. Certain categories of oil are: (a) extra virgin olive oil, (b) virgin olive oil, (c) olive oil composed of refined olive oils and virgin olive oils, and (d) olive-pomace oil. Only extra virgin and virgin olive oil may bear a designation of origin on the labeling.[192]

Some representative points and comments (repeals, modifications, and amendments) of the directive/regulations for oils and fats are given in Table 5.7.

5.5.2 U.S. Legislation Related to Oil

An international agreement on olive oil and table olives (1986) aims for the modernization of olive cultivation, olive oil extraction, and table olive processing: (a) to encourage research and development to elaborate techniques that could: (i) modernize olive husbandry and the olive-products industry through technical and scientific planning, (ii) improve the quality of the products obtained

TABLE 5.7
Directive and Regulations (Main Points and Comments) with Regard to Oils and Fats

Directive/Regulation	Title	Main Points	Comments
Directive 76/621/EEC (entry into force 28/7/1976)	The fixing of the maximum level of erucic acid in oils and fats intended as such for human consumption and in foodstuffs containing added oils and fats	• The aim of this Directive is to protect consumers from the possible negative effects of erucic acid in oils and fats. • The level of erucic acid contained in the products concerned must not exceed 5% of the total level of fatty acids in the fat component.	There is an amendment for this Directive which is Regulation (EC) No. 807/2003 (entry into force 5/6/2003).
Directive 96/3/EC (entry into force 17/2/1996)	Hygiene of foodstuffs with regard to the transport of bulk liquid oils and fats by sea	• Definition of equivalent conditions to ensure the protection of public health and the safety and wholesomeness of the foodstuffs concerned. • The bulk transport in sea-going vessels of liquid oils or fats is permitted in tanks that are not exclusively reserved for the transport of foodstuffs. • Register of the three previous cargoes carried in the tanks concerned, and the effectiveness of the cleaning process applied between these cargoes.	
Regulation (EEC) No. 136/66 (entry into force 1/10/1966)	Establishment of a common organization of the market in oils and fats	• An export refund may be paid on exports of olive oil and rapeseed oil to cover the difference between the world and Community market prices. • Marketing standards, in particular covering labeling and quality grading, may be laid down for olive oil. • Olive oil used for humanitarian aid purposes is purchased on the Community market or comes from intervention stocks.	There are 26 Regulations which amend the Regulation (EEC) No. 136/66. The last amendment is Regulation (EC) No. 2826/2000 (entry into force 1/1/2001).
Regulation (EEC) No. 2568/91 (entry into force 8/9/1991)	The characteristics of olive oil and olive-residue oil and on the relevant methods of analysis	• The characteristics of the oils shall be determined in accordance with the methods of analysis. • The content of free fatty acids is expressed as acidity calculated conventionally. • All the equipment used shall be free from reducing or oxidizing substances.	

(continued)

TABLE 5.7 (continued)
Directive and Regulations (Main Points and Comments) with Regard to Oils and Fats

Directive/Regulation	Title	Main Points	Comments
Regulation (EC) No. 2991/94 (entry into force 1/1/1996)	Marketing standards for olive oil	• Application to milk fats and fats composed of plant and/or animal products. • Not applicable to concentrated products. • The products are supplied to the consumer only after processing. • Products imported into the Community must comply with the Directive.	
Regulation (EC) No. 2815/98 (entry into force 31/10/2001)	Marketing standards for olive oil	• The designation of origin shall relate to a geographical area. • The "extra virgin olive oil" and "virgin olive oil" shall be packaged in an establishment approved for that purpose. • Designation of origin checks in the packaging plants.	
Regulation (EC) No. 1019/2002 (entry into force 1/11/2002)	Standards for spreadable fats	• Specific standards for retail-stage marketing of the olive oils and olive-pomace oils. • Oils shall be presented to the final consumer in packaging of a maximum capacity of 5L. • Specific way of labeling.	The Regulation (EC) No. 1964/2002 (entry into force 1/7/2002) makes clear that the products which were legally produced, were labeled and got into circulation before 1/1/2003 will be marketed until the consumption of the stocks.

TABLE 5.8
International Agreement for Olive Oil and Table Olives

Title	Year	Main Points	Comments
International agreement on olive oil and table olives	1986	• The objectives of this agreement were: international cooperation, the modernization of olive cultivation, olive oil extraction and table olive processing • Definitions (olive oil crop year, olive products, etc.)	There are two amendments for this agreement, the first one was in 1993 and the other one in 2000

therefrom, (iii) reduce the cost of production of the products obtained, particularly that olive oil, with a view to improving the position of that oil in the overall market for fluid edible vegetable oils, (iv) improve the situation of the olive products industry as regards the environment to abate any harmful effects, (b) to encourage the transfer of technology and training in the olive sector. "Olive oil" shall be restricted to oil obtained solely from the olive, to the exclusion of oil obtained by solvent or re-esterification processes and of any mixture with oils of other kinds. "Virgin olive oil" is oil which is obtained from the fruit of the olive tree solely by mechanical or other physical means under conditions, and particularly thermal conditions, that do not lead to the deterioration of the oil, and which has not undergone any treatment other than washing, decantation, centrifugation, and filtration, to the exclusion of oil obtained by solvent or re-esterification processes and of any mixture with oils of other kinds.[193,194] The main points of this agreement are given in Table 5.8.

5.5.3 CANADA LEGISLATION FOCUSED ON OIL

For Edible oil products Act (1990) "edible oil product" means a food substance of whatever origin, source, or composition that is manufactured for human consumption wholly or in part from a fat or oil other than that of milk. No person shall manufacture or sell an edible oil product, other than oleomargarine, manufactured by any process by which fat or oil other than that of milk has been added to or mixed or blended with a dairy product in such manner that the resultant edible oil product is an imitation of or resembles a dairy product. The Lieutenant Governor in Council may make regulations: (a) designating the edible oil products or classes of edible oil products to which this Act applies; (b) providing for the issue of licenses to manufacturers and wholesalers of any edible oil product and prescribing the form, terms, and conditions thereof and the fees to be paid therefore, and providing for the renewal, suspension, and cancellation thereof, (c) prescribing standards for the operation and maintenance of premises and facilities in which any edible oil product is manufactured, packed, or stored, (d) prescribing the standards of quality for and the composition of any edible oil product or class of edible oil product, (e) providing for the detention and confiscation of any edible oil product that does not comply with this Act and the regulations, (f) respecting the advertising of any edible oil product or class of edible oil product, (g) requiring and providing for the identification by labeling or otherwise of any edible oil product or class of edible oil product sold or offered for sale, (h) prescribing the powers and duties of inspectors and analysts, (i) prescribing the records to be kept by manufacturers and wholesalers of any edible oil product, (j) exempting any manufacturer, wholesaler, or retailer of any edible oil product from this Act and the regulations, and prescribing terms and conditions therefore, (k) respecting any matter necessary or advisable to carry out effectively the intent and purpose of this Act.[195,196] The main points of this Act are given in Table 5.9.

TABLE 5.9
Canadian Act for Oil

Title	Year	Main Points	Comments
Edible Oil Products Act	1990	• Definitions (edible oil product, etc.) • This act applies to every edible oil product and class of edible oil product designated in the regulations	This act was amended in 1994, 1999, and 2001

5.6 AUTHENTICITY OF OILS AND FATS

A variety of physical and chemical tests have been developed to determine the authenticity of fats and oils. Classical physical and chemical tests have been supplemented with, or supplanted by, newer chemical and instrumental techniques, including silver ion chromatography, stable carbon isotope ratio analysis, chemometrics, gas liquid chromatography with capillary or open-tubular columns, high-performance liquid chromatography, supercritical fluid chromatography, Fourier-transform and attenuated total reflectance, infrared spectrophotometry, mass spectrometry and nuclear magnetic resonance spectroscopy.[82,197,198]

5.6.1 Authenticity of Vegetable Oils and Fats

5.6.1.1 Olive Oil Authenticity

Olive oil is often illegally adulterated with other less expensive vegetable oils. Oils widely used for this purpose include olive pomace oil, corn oil, peanut oil, cottonseed oil, sunflower oil, soybean oil, and poppy seed oil.[199,200] Among the various chemical and physical methods employed toward the detection of the adulteration of olive oil by low-grade olive oils and seed oils are[200]: (a) Sterol analysis (presence of stigmasterol and β-sitosterol), (b) alkane analysis (C_{27}, C_{29}, and C_{31}), (c) wax and aliphatic alcohol analysis, (d) fatty acids/(with HPLC) trans fatty acid, and (e) Triacylglycerol.

Several chemical analyses and quality tests were applied to characterize its composition and to evaluate its quality and detect adulteration.[201] Interchange of research information and knowledge in the field of chemical analysis, quality, and packaging issues establish a worldwide acceptable product which will satisfy the consumer preferences and demands. A range of analytical methods for fraud detection, modified, and reappraised on a continuous basis to be a step ahead of those pursuing these illegal activities.[202]

Virgin olive oils were separated from the lower grade olive oils with a dedicated principal component analysis and isotopic data ($\delta^{13}C_{oil}$, $\delta^{13}C_{16:0}$, $\delta^{13}C_{18:1}$) and on the $\delta^{13}C_{16:0}$ vs. $\delta^{13}C_{18:1}$ covariations performed on olive oil samples. While the authenticity of 23 vegetables oils in Slovenia and Croatia was determined based on their fatty acids (olive, pumpkin, sunflower, maize, rape, soybean, and sesame oils) after having been separated by alkaline hydrolysis and derivatized to methyl esters for chemical characterization with the method just mentioned (GC/C/IRMS) prior to isotopic analysis.[203]

McKenzie et al.[204] developed a rapid and quantitative ^{13}C NMR spectroscopic method for the determination of the major fatty acids (oleic, linoleic, and saturated acids) contained as triacylglycerides in South African extra-virgin olive oils. With the judicious use of a shiftless NMR relaxation agent, Cr(acetylacetonate)3, it became possible to determine the principal fatty acids in 0.4 g of olive oil within 20 min of using five experimental ^{13}C NMR acquisition methods in an attempt to find the shortest possible means of quantitatively determining the major triacylglycerides without any loss in accuracy. The estimation of the distribution of oleic and linoleic acids in the naturally occurring triacylglycerides in olive oils, expressed as α/β ratios for the triacylglycerides, was also successfully carried out.

Fourier transform mid-infrared (FTIR), near-infrared (FTNIR), and Raman (FT-Raman) spectroscopy were used[205] for discrimination among 10 different edible oils and fats, and for comparing the performance of these spectroscopic methods in edible oil/fat studies. The FTIR apparatus was equipped with a deuterated triglycine sulfate (DTGS) detector, while the same spectrometer was also used for FT-NIR and FT-Raman measurements with additional accessories and detectors. The spectral features of edible oils and fats were studied and the unsaturation bond (C=C) in IR and Raman spectra was identified and used for the discriminant analysis. Linear discriminant analysis (LDA) and canonical variate analysis (CVA) were used for the discrimination and classification of different edible oils and fats based on spectral data. FTIR spectroscopy measurements in conjunction with CVA yielded about 98% classification accuracy of oils and fats followed by FT-Raman (94%) and FTNIR (93%) methods; however, the number of factors was much higher for the FT-Raman and FT-NIR methods.

One hundred and thirty-eight (138) oil samples were analyzed with visible (vis) and near-infrared (NIR) transflectance spectroscopy.[206] Forty-six of them were Greek pure extra virgin olive oils and the same oils adulterated with 1% (w/w) and 5% (w/w) sunflower oil. However, no significant difference was found between the spectrum of pure sunflower oil and that of olive oil, which can be detected by the naked eye. Olive and sunflower oils differ in their composition principally in their content of linoleic and oleic acids. Accordingly, typical figures for olive oil were quoted at 12.3% and 66.3%, respectively, while for sunflower oil the corresponding mean values of 66.2% and 25.1%, respectively.

Tay et al.[207] used the same method to detect the adulteration in olive oils. Single-bounce attenuated total reflectance measurements were taken on pure olive oil and olive oil samples adulterated with various concentrations of sunflower oil (20–100 mL vegetable oil/liter of olive oil). Twelve (12) principal components of the discriminant analysis and the PLS model were able to classify the samples as pure and adulterated olive oils based on their spectra and to verify the concentrations of the adulterant. Lai et al.[208] achieved the correct classification of 93% of extra virgin and refined oil samples in a calibration set, and 100% of samples in an independent validation set, despite these two types of oil being chemically and spectroscopically very similar. The determination of food authenticity and the detection of adulteration are problems of continuously increasing importance in the food industry. This is especially so for "added value" products, where the potential financial rewards for substitution with a cheaper ingredient, e.g., with a cheaper seed oil, are high. Fourier transform infrared spectroscopy combined with attenuated total reflectance and partial least squares regression could comprise two types of "contaminant" oil-refined olive and walnut.[209]

Twenty-four samples of four European virgin olive oil varieties Arbequina, Coratina, Koroneiki, and Picual, cultivated in Greece, Italy, and Spain-were analyzed for their chemical composition. Non-volatile compounds (31): fatty acids, sterols, alcohols, and methylsterols; and volatile ones (65): aldehydes, alcohols, furans, hydrocarbons, acids, ketones, and esters; were determined using GC and DHS-GC-MS methods, respectively.[210]

A simple method was developed for quantitative determination of polar compounds in fats and oils, using monostearin as an internal standard, with solid-phase extraction (silica cartridges) and subsequently separated by high-performance size-exclusion chromatography into triglyceride polymers, triglyceride dimers, oxidized triglyceride monomers, diglycerides, internal standard, and fatty acids. A pool of polar compounds was used to check linearity, (3.0%–5.7%) and accuracy (100%–103.3%) of the method, as well as the solid-phase extraction recovery (101%). Therefore, less than an hour was required to carry out the complete analytical procedure, including quantization of total level and distribution of polar compounds, and applied to oxidation studies, in order to follow-up the oxidative process during early and advanced stages, and to determine the action of antioxidants.[211]

A recent analytical procedure by Flores et al.[212] targeted the optimization of the interface performance in the online coupling of reversed phase liquid chromatography and gas chromatography (RPLC-GC) coupling by means of a horizontally positioned PTV (Programmed Temperature Vaporizer) injector. It improved the sensitivity achievable in the direct analysis of olive oil-mixtures with 5% or 12% of some virgin and refined hazelnut oils, respectively, based on the analysis of filbertone enantiomers within 30 min and without any kind of pretreatment.

The HPLC-determined triacylglycerol (TAG) profile of vegetable oils appeared to be an adequate method for quality control and a useful tool for authenticity issues. Sample preparation consisted in the dissolution of the oils in acetone and filtration (using a Kromasil 100 C_{18} column (at 25°C) and gradient elution with acetone and acetonitrile). Detection was accomplished with an evaporative light scattering detector (ELSD). The TAG peaks were identified using a categorical principal component analysis (CAT-PCA) for data simplification. The results depicted on a two dimensional plot explained 82.24% of the total variance taking into account the logarithms of Cronbach's α in relation to the homogeneous TAG (relative retention times to triolein) and their quantification was based on the internal normalization method.[213]

Total luminescence and synchronous scanning fluorescence spectroscopy techniques were used to characterize and differentiate edible oils, including soybean, sunflower, rapeseed, peanut, olive, grapeseed, linseed, and corn oils. Total luminescence spectra of all oils studied as n-hexane solutions exhibit an intense peak, which appears at 290 nm in excitation and 320 nm in emission, is attributed to tocopherols. Some of the oils exhibit a second long-wavelength peak, appearing at 405 nm in excitation and 670 nm in emission, belonging to pigments of the chlorophyll group. Bands attributed to tocopherols, chlorophylls, and unidentified fluorescent components were detected in the synchronous-scanning fluorescence spectra. Classification of oils based on their synchronous fluorescence spectra was performed using a non-parametrical k nearest neighbors method and linear discriminant analysis. Both methods provided very good discrimination of the oil classes with low classification error.[214]

Direct coupling of a headspace sampler to a mass spectrometer for the detection of adulterants in olive oil was the method that Marcos-Lorenzo et al.[215] used to evaluate samples of olive oils mixed with different proportions of sunflower oil and olive-pomace oil, respectively. Patterns of the volatile compounds in the original and mixed samples were generated with LDA chemometric technique, and the application of the linear discriminant analysis technique to the data was sufficient to differentiate the adulterated from the non-adulterated oils and to discriminate the type of adulteration. The results obtained revealed 100% success in classification and close to 100% in prediction, suggesting speed of analysis, low cost, and the simplicity of the measuring process. Table 5.10 summarizes all the quality control methods reported in the literature about the authentication of olive oil providing information regarding the country, target compounds, adulterant oil and adulteration detection threshold, and multivariate analysis methods.

Virgin olive oil is a complex aggregate of different substances requiring thorough and meticulous methodology steps in order for most of them to be determined. This powerful identity offered a special position among the edible oils and certainly a special high price. Thus, the customers' demands become more intense and the corresponding legislation is clear. The application of "protected designation of origin" (PDO) requires numerous investigations regarding determination of quality and genuineness, since *Olea europaea* L. has a considerably great number of cultivars often influenced by environmental conditions and locations and cases of homonymy and synonymy among them. However, pedoclimatic aspects together with olive ripeness, harvest of olives, and the olive extraction system determine the chemical composition and sensory descriptors outlining each cultivar.[216]

An on-line LC–GC method was used to assess the n-alkane composition of forty (40) olive oil samples obtained from three different cultivars from a restricted grove zone in Croatia. Olive samples were handpicked at three different ripening stages during four consecutive years and trivial differences were found according to the "period of harvesting". LDA verified the efficiency of the n-alkane composition as a means of variety differentiation, despite the influence of the production

TABLE 5.10
Country, Target Compounds, Adulterant oils, Adulteration Detection Thresholds, and Multivariate Analysis Methods for Olive Oil Authentication

Country	Extraction Method	Target Compound	Adulterant Oil	Adulteration Detection Threshold	QC Detection Method	Multivariate Analysis
Spain, Italy, Greece, France, Switzerland, Morocco, United States, Tunisia[203]	—	FAMEs	Maize, groundnut, sunflower, hazelnut, walnut, rape, and vegetable oils	Very sensitive (<0.05% wt)	GC-MS CSIA GC-C-IRMS	PCA
Slovenia, Croatia[203]		FA/FAMEs	—	Reliable classification	GC-C-IRMS	PCA
South Africa[204]	40% CDCl$_3$	Oleic, linoleic, and saturated acids	Fake or mixed with sunflower	—	HR-FT-^{13}C-NMR	—
___[205]	No treatment	IR and Raman spectral fingerprint	Eight vegetable oils	Classification	FT-IR	LDA
Greece[206]	No treatment	Spectral vibrations and absorbance peaks	Sunflower	1%	Vis NIR	HCA, SIMCA, PLS
___[208]	No treatment	Spectral vibrations and peaks	Seven vegetable oils	—	FT-IR	PLS, DA, PCA
___[209]	No treatment	Spectral vibration and peaks	Refined OO walnut oil	SEP 0.92%	FT-IR	PLS
Spain, Italy, Greece[210]	Esterification—TLC DHS	Chemical compounds, sensory, and consumers' attributes	Twelve virgin olive oils	Classification	GC-FID GC-MS-FID	Stweart–Newman–Keuls test
___[211]	SPE	Polar compounds distribution	Olive, refined sunflower, soybean, and 17 used frying oils	Classification	HPSEC	—
___[212]	No pre-treatment/SPE	R- and S-filbertone enantiomers	Hazelnut	5%, 7%	On-line HPLC-GC/ MDGC	—

(continued)

TABLE 5.10 (continued)

Country, Target Compounds, Adulterant oils, Adulteration Detection Thresholds, and Multivariate Analysis Methods for Olive Oil Authentication

Country	Extraction Method	Target Compound	Adulterant Oil	Adulteration Detection Threshold	QC Detection Method	Multivariate Analysis
Portugal[213]	Acetone/CH_3CN dilution	TAG profile	Sunflower, corn, peanut, soybean, hazelnut, walnut, and sesame	100% classification	HPLC-ELSD	DA, ANOVA, PCA, PWC
Russia[214]	1% hexane solution	EEMatrices	Vegetable oils	100% correct classification	TLS SFS	LDA, Knn
Spain[246]	NT	Fluorescence spectrum (EEMatrices)	Olive oils/Pomace oil	5% good classification	EEFS	PCA, U-PCA, PARAFAC, LDA, N-PLS
Italy[248]	No treatment	^{13}C resonances of oleyl, linoleyl, and linolenyl acids	Soya oil	5.9%–20.2%	^{13}C-NMR-DEPT	—

year and year-variety interaction. As the number of groups (varieties) considered was three, two discriminant functions were calculated by the program as linear combinations of the chemical descriptors.[217]

Fifteen (15) samples of two cultivars and five different areas of Spain were examined by Guadaramma et al.[218] with an electronic nose apparatus, apart from the organoleptic characterization of olive oil, for their cultivar and variety identity. The instrument worked using an array of electrodeposited conducting polymer-based sensors. The 16 selected sensors showed a stability of at least 4–6 months; a reversibility of <10 min and good reproducibility. Such an array was able to distinguish among not only olive oils of different qualities (extra virgin, virgin, ordinary, and lampante) with 90% confidence ellipses which were perfectly separated from each other.

Downey et al.[219] classified extra virgin olive oils from the eastern Mediterranean on the basis of their geographic origin with visible and near-infrared reflectance spectra. Classification strategies included partial least-squares regression, factorial discriminant analysis, and k-nearest neighbors analysis. Discriminant models were developed and evaluated using spectral data in the visible (400–750 nm), near-infrared (1100–2498 nm), and combined (400–2498 nm) wavelength ranges. Data pretreatments were applied and good results were obtained using factorial discriminant analysis on raw spectral data over the combined wavelength range; thus, a correct classification rate of 93.9% was achieved on a prediction sample set.

Olive oil fatty acids stand for very important components because some of olive oil's most beneficial properties (i.e., heart attack prevention) are attributed to them. Furthermore, it was repeatedly demonstrated that fatty acids profile varies considerably between olive oil and seed oil thus making this determination very crucial. Triacylglycerols (TGs) comprise a major part of naturally occurring fats and oils and the analysis of intact TGs was successfully conducted with the application of capillary gas-liquid chromatography, high-performance chromatography in normal and reversed phase mode, thin-layer chromatography, and supercritical fluid chromatography, with emphasis on detection systems widely used in laboratories, i.e., FID, UV, RI, and ELSD.[220]

Muik et al.[221] reported two direct, reagent-free methods to estimate the FFA content in olive oil and olives using Fourier transform Raman spectrometry and multivariate analysis in combination with partial least squares (PLS) regression. Oils were directly investigated in a simple flow cell. Although the results are less accurate compared to the official method, the proposed methods are well-suited for quality control in process monitoring. Due to its simplicity, the method for olive oil analysis allows the on-line measurement of the produced oil. The method for olive oil analysis is much faster than the official one, because sample preparation is reduced to a minimum (milling). Ninety percent of the oil samples and 80% of the olives were correctly classified, demonstrating that the proposed procedures can be used for screening of good quality olives before processing, as well as, for the on-line control of the produced oil.

The separation of fatty acids (as fatty acid methyl esters, FAMEs, or free fatty acids, FFAs) with supercritical fluid chromatography (SFC) is reviewed by Seronans and Ibanez.[222] Different analytical approaches about both FAMEs and FFAs separation are presented, among these approaches, the tuning of the mobile and the stationary phase are reviewed for open tubular, packed, and packed capillary SFC. The approach of tuning the polarity of the stationary phase as a way of increasing the range of polar compounds analyzed with SFC using pure CO_2 was widely discussed in this review for compounds such as FFAs avoiding the drawbacks associated with the use of modifiers in SFC. The applications of the analysis of FAMEs and FFAs in different foods are also reviewed.

As extensively reviewed by Meier-Augestein[223], the Compound-Specific Isotope Analysis (CSIA) of fatty acids, a relatively young analytical method, has increasingly gained considerable ground as the opted method in areas where an accurate and precise knowledge of isotopic composition at natural abundance level is required. The CSIA of fatty acids at natural abundance level provides information on biogenetic and geographic origin of lipids and oils that is invaluable and almost indispensable nowadays for authenticity control and fraud detection in food analysis. In combination with naturally enriched or stable isotope labeled precursors, CSIA of fatty acids has also gained increasing importance in biochemical, medical, and geochemical applications as it offers a reliable and risk-free alternative to the use of radioactive tracers.

Ranalli et al.[224] applied HRGC for the determination of the contents of triacylglycerols, diacylglycerols, and fatty acid composition to three kinds of olive fruit oils (pulp, seed, and whole fruit) coming from seven major Italian olive varieties. The three fruit oil kinds contained identical fatty acid species but of different concentrations. Pulp and whole fruit oils were richer in individual monounsaturated fatty acids (eicosenoic acid excepted), as well as in total monounsaturated fatty acids (MUFA) essentially due to higher contents of oleic acid (the major fatty acid component found in the three fruit oil kinds). The two fruit oil kinds in question were also richer in total saturated fatty acids (SFA), essentially due to higher contents of palmitic acid (the major saturated fatty acid component found in the three fruit oil kinds), even though their content of stearic acid (another important saturated fatty acid) was lower. Seed oil was richer in total polyunsaturated fatty acids (PUFA) (even though its linolenic acid content was lower), because of higher contents of linoleic acid, the major fatty acid component of the PUFA fraction). The major fatty acids of olive oil in parallel with the extraction/determination methods and their amounts (% w/w as oleic acid) are presented in Table 5.11.

Phenolic compounds stand for one of the most important classes of compounds since they react with free radicals thus reducing considerably the occurrence of cancer and heart attack. The identification of phenolic compounds' species is a very important "marker" toward olive oil authentication. They include hydrophilic phenols, the most abundant natural antioxidants of virgin olive oil (VOO), tocopherols, and carotenes. In this prevalent class are embodied phenolic alcohols, phenolic acids, flavonoids, lignans, and secoiridoids. Secoiridoids, including aglycon derivatives of oleuropein, demethyloleuropein, and ligstroside, are the most abundant phenolic antioxidants of VOO. Their amounts are greatly influenced by agronomic and technological parameters such as olive cultivar, the ripening stage of fruit, and the malaxation conditions and the extraction- systems' methodologies.[225, 226]

Twenty-one (21) different phenols and polyphenols were determined[227] with capillary zone electrophoresis (CZE-DAD-UV (200 nm); tyrosol (10 mg/mL); 2,3-dihydroxyphenylethanol (20 mg/mL); oleuropein glycoside (20 mg/mL); hydroxytyrosol (10 mg/mL); dihydrocaffeic acid (20 mg/mL); cinnamic acid (20 mg/mL); 4-hydroxy-phenylacetic acid (20 mg/mL); gentisic acid (20 mg/mL); taxifolin (15 mg/mL); syringic acid (20 mg/mL); ferulic acid (20 mg/mL); luteolin (100 mg/mL); o-coumaric acid (20 mg/mL); p-coumaric acid (20 mg/mL); quercetin (50 mg/mL); vanillic acid (20 mg/mL); 4-hydroxybenzoic acid (20 mg/mL); caffeic acid (20 mg/mL); 3,4-dihydroxyphenylacetic acid (20 mg/mL); gallic acid (50 mg/mL); protocatechuic acid (20 mg/mL).

Brenes et al.[228] employed two HPLC variants; one equipped with the UV detection system and proposed the other with electrochemical (EC) detection. An effectiveness test was simultaneously performed on the three most applicable extraction methods, SPE, LLE, and DMF (proposed) based on the treatment of the extracted oil with 2NHCl followed by the analysis of phenols in the aqueous phase. The conclusion was that 15%–40% of phenols remained unextracted when the liquid/liquid extraction method was applied with 80% methanol. Solid phase extraction (C_{18} cartridge) succeeded in retaining most phenols in the cartridge, but the recovery yield from the sorbent material was low. However, the new extraction method, based on the use of N,N-dimethylformamide (DMF) as an extraction solvent, achieved a complete extraction of phenols from oils.

Solid phase extraction was used in the analysis of phenol compounds of ten Italian virgin olive oil samples.[229] In doped refined olive oil samples a comparison between liquid/liquid and SPE extraction evidenced higher recovery when the C_{18} sorbent phase was employed, whereas, in the case of total suppression of the residual silanolic group (C_{18}EC), only contradictory data was obtained. The same procedures were applied to the genuine virgin olive oil samples and the main observation with the standards was that the higher values recorded from the C_{18} sorbent phase compared to the C_{18}EC and L/L extraction procedures. Satisfactory results were also obtained in the detection of ligstroside aglycon but for the oleuropein aglycon the quantitative is not completely reliable for the overlap of some

TABLE 5.11
Country, Extraction Method, and Amounts of Olive Oil Fatty Acids

Country	Extraction Method	Amount (% w/w as Oleic Acid)	Species				Method
—[211]	TLC/SPE	1	—				HP-SEC
Spain[221]	Without pre-treatment	0.20–6.14	—				FT-Raman
Italy[224]	Converted to FAMEs	ΣSFA 16.79 ± 1.30	Palmitic (16:0)				HR-GC-FID
		ΣMUFA 74.63 ± 6.38	Palmitoleic (16:1)				^{13}C-NMR
		ΣPUFA 8.58 ± 0.76	Heptadecanoic (17:0, 17:1)				
			Stearic (18:0)				
			Oleic (18:1)				
			Linoleic (18:2)				
			Linolenic (18:3)				
			Arachidonic (20:0)				
			Eicosenoic (20:1)				
			Behenic (22:0)				
Spain, Italy, Morocco[251]	AEDA	—	FA (% wt)	I	SP	M	HRGC-MS
			14:0	nd	nd	nd	
			16:0	10.7	8.6	10.9	
			16:1	0.5	0.5	0.8	
			18:0	2.4	4.4	2.5	
			18:1	79.0	81.4	70.4	
			18:2	6.6	4.3	14.3	
			18:3	0.4	0.5	1.0	
			20:0	0.5	0.4	0.2	
			20:1	0.3	0.2	0.2	
Italy[255]	DMC/titanium silicate powder FAMEs	—	Palmitic (%) 11.6 ± 0.5				On-line Py-GC-MS
			Stearic (%) 5.3 ± 0.2				
			Oleic (%) 78.5 ± 0.5				
			Linoleic (%) 3.6 ± 0.5				
			Linolenic (%) 1.0 ± 0.6				

Fatty Acids

unknown no-phenol compounds. Thus, for the quantification procedure, it is important to analyze the sample with GC-MS previously in order to ascertain whether other no-phenol compounds were present. An overview of the analysis of phenolic compounds is presented in Table 5.12.

The unique and delicate flavor of olive oil is attributed to a series of volatile components like aldehydes, alcohols, esters, hydrocarbons, ketones, furans, and other compounds.[230] Most of them were quantitated and identified by gas chromatography-mass spectrometry in high-quality olive oils indicating the close relation to its sensory quality. Hexanal, trans-2-hexenal, 1-hexanol, and 3-methylbutan-1-ol are the major volatile compounds of olive oil. Since volatile flavor compounds are formed in the olive fruit through an enzymatic process they are affected by the olive cultivar, origin, the maturity stage of the fruit, storage conditions of the fruit, and olive fruit processing, as well as its taste and aroma.[231]

Vichi et al.[232] used SPME technique for the qualitative and semi-quantitative analysis of virgin olive oil volatile compounds and compared the behavior of four fiber coatings, choosing a divinyl-benzene–carboxen–polydimethylsiloxane fiber coating to be the most suitable for this analysis. The SPME method, coupled to GC with MS and flame ionization detection, was applied to virgin olive oil samples and more than 100 compounds were isolated and characterized. The main volatile compounds present in the oil samples were determined quantitatively and the presence of some of these compounds in virgin olive oil has not previously been reported. This is the case for some hydrocarbons such as 2- and 3-methylpentane, 1-acetylcyclohexene, 1-methyl-3-(hydroxyethyl)propadiene and (E)-4,8-dimethyl-1,3,7-nonatriene, which gave chromatographic peaks of considerable area and were detected in all the samples analyzed.

Gaja et al.[233] determined volatile compounds in different edible oils (i.e., olive oil, almond oil, hazelnut oil, peanut oil, and walnut oil) using simultaneous distillation-solvent extraction followed by gas chromatographic–mass spectrometric analysis. An alternative approach which allows the direct injection of the oil sample (i.e., without any kind of pretreatment) and involves the use of on-line coupled reversed phase liquid chromatography-gas chromatography (RP-LC-GC) was also considered. The advantages of this approach is in avoiding off-line pre-separation for the analysis of volatile compounds associated with the lipophilic nature of most volatile compounds and, enhancing the reliability of the analysis.

Stella et al.[234] has made a selection of an array of conducting polymer sensors and tested it with extra-virgin olive oil samples as a first step toward the development of an electronic nose dedicated to the detection of olive oil aroma. Different sensors produced by both electrochemical and chemical techniques were initially exposed to a set of pure substances present in the headspace of extra-virgin olive oil and are meaningful for the evaluation of its overall organoleptic characteristics. Four sensors showing the best sensitivity to these standard substances were chosen to carry out further experiments on the samples of commercial olive oil. Two different experimental set-ups and protocols for olive oil sampling were tested and compared, providing evidence on the best procedure needed to handle this foodstuff and on the possibility of using a dedicated sensing system for practical purposes in the olive oil industry. Based on this, three different extra-virgin Italian types of olive oil can be easily distinguished with an array of four sensors and it is also possible to detect the changes in the aromatic content of the headspace after handling of the samples. Different samples of the same oil show reproducible responses.

The color of olive oil is one of the most important quality characteristics and a factor of acceptability among consumers. In addition, much attention is also paid to the deep green color in oil, like in virgin olive oils. The drupe of olive trees which contains the green pigment of the plants is responsible for their color and consequently of the olive oil. The characteristic green color is due to the chlorophyllic pigments, chlorophylls, pheophytins, and pyropheophytins.[235,236] Since natural chlorophyllic pigments are totally absent in the refined olive oil and, frequently, the color of this oil could be obtained by synthetic chlorophyll pigments as copper chlorophyll derivatives, the analysis of pigment compounds is of great importance. Moreover, it could be used as a powerful tool for olive oil authenticity and quality control.[237,238]

TABLE 5.12
Country, Extraction, and Detection Methods of Phenolic Compounds

Country	Extraction Method	Phenolic Compounds	Detection Method	Method
Italy[227]	LLE (olive oil) SLE (olive oil)	Tyrosol, hydroxytyrosol, vanillic acid, deacetoxy-oleuropein aglycon, pinoresinol, and acetoxypinoresinol	UV (280/200 nm)	HPLC-MSD CZE-HPLC
Italy[227]	LLE Hexane 60% aqueous MeOH MeOH/H$_2$O (1:1 v/v)	Tyrosol, 2,3-dihydroxyphenylethanol, oleuropein glycoside, hydroxytyrosol, dihydrocaffeic acid, cinnamic acid, 4-hydroxy-phenylacetic acid, gentisic acid, taxifolin, syringic acid, ferulic acid, luteolin, o-coumaric acid, p-coumaric acid, quercetin, vanillic acid, 4-hydroxy-benzoic acid, caffeic acid, 3,4-dihydroxyphenylacetic acid, gallic acid, and protocatechuic acid	UV–vis (200 nm) ESI-MSD	CZE-DAD HPLC
Spain[228]	DMF extraction LL (80% MeOH) SPE (C$_{18}$ cartridge)	Hydroxytyrosol, tyrosol, vanillic acid, vanillin, 4-acetoxyethyl-1,2-dihydroxybenzene, p-coumaric acid, ferulic acid, dialdehydic form of elenolic acid linked to hydroxytyrosol, dialdehydic form of elenolic acid linked to tyrosol, 1-acetoxypinoresinol, pinoresinol, oleuropein aglycon, luteolin, ligstroside aglycon, and apigenin	EC detector UV (280 nm)	RP-HPLC
Spain[256]	LLE Hexane 60% aqueous MeOH MeOH/H$_2$O (1:1 v/v)	Protocatechuic acid, dopac, 4-hydroxybenzoic acid, gallic acid, caffeic acid, vanillic acid, p-coumaric acid, o-coumaric acid, ferulic acid, gentisic acid, 4-hydroxy-phenylacetic acid, sinapinic acid, trans-cinnamic acid, and (+)-taxifolin	UV (210 nm) UV–vis (210 and 275 nm simultaneously)	Co-electrosmotic-CZE CZE-DAD

The detection and quantification of tocopherols, carotenoids, and chlorophylls in vegetable oil were effectively used for authentication purposes. The presence of tocopherols, carotenoids, and chlorophylls influence the oxidative stability of vegetable oils and their potential health benefits. Puspitasari-Nienaber et al.[239] demonstrated the application of a rapid and reliable analysis method of direct injection of C-30 RP-NPLC with electrochemical detection for the simultaneous analysis of the above mentioned substances. Aliquots of vegetable oils were dissolved in appropriate solvents and injected directly without saponification, thus preventing sample loss or component degradation. Thus the effective separation of tocopherols, carotenoids, and chlorophylls was achieved.

Buldini et al.[240] developed a procedure for the ion chromatographic determination of total chlorine, phosphorus, and sulfur, and of iron, copper, nickel, zinc, cobalt, lead, and cadmium in edible vegetable oils and fats after the complete removal of the organic matrix by saponification followed by oxidative UV photolysis. The method was simple and required fewer reagents compared with other sample pre-treatment procedures. Saponification lasted for half an hour since the addition of ethanol and potassium hydroxide.

Jimenez et al.[241] also used the ICP-MS method for the determination of Al, Ba, Bi, Cd, Co, Cu, Mn, Ni, Pb, Sn, and V. The main differences that they initiated focused on the on-line formation of olive oil-in-water emulsions, the considerable time-gain, and the automatic sample preparation process. Among the various experimental parameters studied and optimized for the development of this method were: emulsifier concentration at the mixing point, emulsifier concentration in the carrier solutions in the valves, injected sample emulsifier volumes, emulsion formation flow rate, design of the FIA manifold used (emulsion formation, reactor length, and size of the different connections), and the radiofrequency power in the plasma.

Anthemidis et al.[242] made a significant improvement by adding a simple on-line emulsion magnetic-stirring microchamber system for the continuous production of emulsion with Triton X-100 and subsequent multi-element analysis by inductively coupled plasma atomic emission spectrometry (ICP-AES). In all the investigated conditions, the resulting emulsions remained stable for at least 30 min, thereby confirming that the stirring chamber was very effective for on-line emulsification. The on-line oil emulsion preparation procedure was simpler, more effective, considerably less time consuming, and less labor intensive. The optimum concentration of oil in emulsion for maximum sensitivity was 50% (v/v) thus increasing the ICP-AES's sensitivity. The performance of the system was demonstrated for the additional determination of Ag, Al, B, Ba, Bi, Ca, Cd, Co, Cr, Cu, Fe, Ga, In, Mg, Mn, Ni, Pb, Tl, and Zn in olive, sunflower, and corn oil with good agreement between calibration curves for oil emulsion and the aqueous ones for most of the investigated spectral lines.

The concentrations of seventeen (17) elements (As, Ba, Ce, Co, Cr, Cs, Eu, Fe, Hg, K, Na, Rb, Sb, Sc, Se, Sr, and Zn) were determined by Iskander[243] using the Instrumental Neutron Activation Analysis in almond, sunflower, peanut, sesame, linseed, soy, corn, and olive oils, as well as in three margarine brands. The concentrations of As, Ba, Ce, Cs, Eu, Hg, Rb, Se, and Sr were found to be under the detection limits of the applied experimental conditions. Chromium was detected only in one margarine sample (171 μg/g); Sb only in corn oil (18 ng/g) and Sc only in linseed oil (19 ng/g). Cobalt, Fe, K, Na, and Zn were detected in all oil and margarine samples that were investigated. The concentration ranges for Co, Fe, K, Na, and Zn in oils were: 0.016–0.053, 4.45–19.1, 5.93–47.2, 2.44–12.9, and 0.48–1.54 μg/g, respectively, and always within the ranges reported in the literature for edible oils and fats.

Galeano-Diaz et al.[244] applied the same method (Ad-SSWV-complex of Cu-DCDT) for copper determination in several olive oils that promoted the appearance of a peak at −0.570 V. The extraction process of Cu from olive oil was carried out with HCl 1 M as the most suitable concentration for this determination. Any possible interference from other ions present in the olive oil was focused on iron since it is the most abundant; therefore, samples of 10 ng mL^{-1} of Cu including different amounts of iron were prepared in ratios (w/w) of Cu:Fe that varied from 1:1 to 1:100. The voltammetric signal obtained for Cu was compared with the corresponding sample's ion signal, without iron; showing that there was considerable affect by the iron ions if the difference was 5% at least and no interference existed in the concentration range assayed. Application of the method to five commercial olive oils samples revealed that Cu amounts were very similar toward those obtained with AAS analysis.[245]

Guimet et al.[246] used two potential multiway methods for the discrimination between virgin olive oils and pure olive oils; the unfold principal component analysis (U-PCA) and parallel factor analysis (PARAFAC), for the exploratory analysis of these two types of oils. Both methods were applied to the excitation-emission fluorescence matrices (EEM) of olive oils and followed the comparison of the results with the ones obtained with multivariate principal component analysis (PCA) based on a fluorescence spectrum recorded at only one excitation wavelength.

Fourier transform (FT) Raman spectrometry in combination with partial least squares (PLS) regression was used for direct, reagent-free determination of free fatty acid (FFA) content in olive oils and olives. Oils were directly investigated by means of a simple flow cell. Both external and internal (leave-one-out) validation were used to assess the predictive ability of the PLS calibration models for FFA content (in terms of oleic acid) in oil and olives in the range 0.20%–6.14% and 0.15%–3.79%, respectively. The final PLS regression model for determining the FFA content in olive oil was built using five latent variables in the region 1200–1800 cm^{-1}. Therefore, the root mean square error of calibration (RMSEC) and the correlation coefficient between actual and predicted value for the calibration set (r) were calculated as 0.27% and 0.985%, respectively.[221]

Total luminescence and synchronous scanning fluorescence spectroscopy techniques were used in conjunction with non-parametrical k nearest neighbors method and linear discriminant analysis in order to differentiate several edible oils, including olive oil. The k nearest neighbors (kNN) method was applied using the entire spectra as the input; while the linear discrimination method employed selected excitation/emission wave length pairs as the input. Both methods provided very good discrimination between the oil classes with low classification error. The peaks that appeared at 290 nm in excitation and 320 nm in emission, attributed to tocopherols, while some of the oils exhibited a second long-wavelength peak, appearing at 405 nm in excitation and 670 nm in emission, belonging to pigments of the chlorophyll group. The results obtained demonstrated the capability of the fluorescence techniques toward characterization and differentiation of vegetable oils.[214]

A rapid head-space analysis instrument for the analysis of the volatile fractions of 105 extra virgin olive oils coming from five different Mediterranean areas was put forward by Cerrato-Oliveros and his co-workers.[247] The rough information collected by this system was unraveled and interpreted with well-known multivariate techniques of display (principal component analysis), feature selection (stepwise linear discriminant analysis), and classification (linear discriminant analysis). 93.4% of the samples were correctly classified and 90.5% correctly predicted by the cross-validation procedure, whilst 80.0% of an external test set, aiming at full validation of the classification rule, were correctly assigned.

Vlahov et al.[248] applied the NMR methodology to measure the high-field (500 MHz) ^{13}C spectra at natural abundance in olive oil samples using the distortionless enhancement by polarization transfer (DEPT) pulse sequence in order to improve the signal-to-noise ratio of the spectra, optimized the acquisition, processing, and integration parameters for the validation of the quantitative measurements of the intensities of ^{13}C resonances of the whole olive oil spectrum. The object of the PCA is that of reducing the number of original variables ($N = 49$) to a small number of indices (called the principal components) that are linear combinations of the original variables. The principal component analysis, applied to the ^{13}C intensity data, correlated the oil samples (fatty acid composition of the triglyceride fraction) from the same geographical area provided the composition of the olive oil was monovarietal.

Delicious taste and aroma are synonyms to fresh and good quality of almost any foodstuff and, undoubtedly, affect the consumer's final opinion and choice. Sensory characteristics of olive oil, a major component in the Mediterranean diet, represent the sum of the sensory parameters that integrate its complex and specific qualitative profile including various intrinsic and extrinsic factors such as cultivar (genetic), weather conditions, stage of ripeness, methods and/or system of harvest and storage (Table 5.13). Regulations and standards related to olive oil sensory evaluation include

TABLE 5.13

Sensory Panel Composition, Attributes Used, and Results of Sensory Evaluation of Various Cultivars of Olive Oils

Cultivar	Panel Number and Training	Range	Attributes	Results
Arbequina, Coratina, Koroneiki, Picual[210]	10 SP/full trained	1–5	Olive fruity (green), strength of olive, salty taste, apple, strength of olive, olives taste, other ripe fruits, banana skins, green leaf taste, green tomato, grass taste, bitter, sweet, green banana (not ripe), pungent, hay/composite, dried green herbs, perfumery, minced pepper, rancid, red chilli, pepper, olive fruity (ripe), grassy, cream/butter, olive fruity (ripe and green), almond, coconut, throat-catching, caramel, thickness, grotty, bitter, pungent, velvet like, pungent odor intensity, sticky, sea breeze on the beach, slightly burned/toasted, allowable, prickling, ash tray, rough, glue with ethylacetate, twig, refinery, pine/harshy, bitter, lemon, astringent, orange, soft fruits, fruity, candies (fruit), cooling/evaporating wild flowers in spring time, fermenting fruit, cocoa butter/white chocolate, farm, putty/linseed oil after taste, ripe black olives, oil for salads (soybean oil), used frying oil, green olives, tallow, trany, cut green grassy, cod liver oil, dry wood, artichoke, nuts, dusty, medicine, dry, yeast, earthy, sharp/etching, bitter, taste intensity, pungent/ sharp throat, and rough	Slightly lighter in taste and scented, slightly sweet and palatable, mild olive flavor, pleasant, no bitter after taste, no harshness, mellow vegetable taste, quite fruity, fresher taste, etc.
	10 I-1/trained	1–5		Tastes like grass, a little harsh, unpleasant bitter taste, a bit fatty, yuk, sweetly perfumed, etc.
	14 GR/trained	1–5		Smells green and fresh, smells strong, pleasant fresh smell, pleasant odor, oily green smell, nice sweet, etc.
	11 I-2/full trained	1–5		Fruity smell, pungent smell, slightly lemon, smell of olives, strong smell but not quite so offensive, fresh smell, etc.
	9 UK/trained	100 mm		
	8 ND/trained	130 mm		

Cultivar	Panel	Scale	Attributes	Description
VVO[224] (new cv-I-77)	12/trained	1–5	Flavor Aroma Odor Taste Aftertaste mouth feel	Good quality and stability of its aroma and flavor, medium bitterness index and good overall acceptability
Various Italian cvs[226]	8/fully trained	1–9	Various	In accordance to the DHS-HR-GC
Corratina, Ogliarola, Maiatica[250]	34 (15 females and 19 males, 25–50 years old)	1–7	Apple, cut grass, almond, artichoke, green olive, tomato, pungency, astringency, and bitterness	—
	10/trained	1–9	Pungent, fruity, green leaf-like, fatty, apple like, black currant like, black olive like	
VOO (SP) (I) (M)[251]	9/experienced	0–3		Intense smell of black currants. Apple-like and green notes close to eating black olives
Coratina, Koroneiki, Picual[257]	12/trained	0–5	Cut green lawn, green leaf or twig, green olives, wild flowers, green, banana or its skin, green tomato, almond, bitter almond, apple, walnut husk, artichoke, green hay, butter/cream, bitter, pungent, sweet	High intensities of bitter almond and pungent, high intensities of green leaf or twig and green olives, high intensities of butter/cream and sweet, while they are significantly less pungent

tests by panels—comprising trained assessors of different number and nationalities—that score certain attributes in a given scale (structured vs. unstructured). A large number of volatile aromatic compounds, as well as, non-volatile ones strongly affect the human gustative and olfactory system and build up the final classification of olive oil. These include several groups of organic substances like aldehydes, alcohols, anthocyanins, flavonoids, phenolic acids, etc. However, it is a slow and expensive method that is hardly afforded by small retailers and sellers; therefore, rapid and economic modern techniques earn space in this field.[249]

Caporale et al.[250] explored the appropriateness of several sensory descriptors in evaluating the typicality of certain extra virgin olive oils. They analyzed the impact of information about the origin of the product on the sensory profile perception, and how the effect of sensory expectations can influence liking and "typicality" responses for the experimental oils obtained from a defined cultivar. The panel was constituted with consumers familiar with several typical extra virgin olive oils produced in Lucania. Sensory attributes were as follows: apple, cut grass, almond, artichoke, green olive, tomato, pungency, astringency, and bitterness. The appropriateness response was rated on a 7-point scale ranging from 1 for "not at all appropriate" to 7 for "extremely appropriate." Results revealed that bitterness and pungency proved to be the most appropriate sensory descriptors of certain typical olive oils.

Reiners et al.[251] employed a sensory analysis to check whether the results of the instrumental analyses were in agreement with their flavor profiles. The sensory panel used, consisted of nine experienced assessors, seven males and two females, aged 25–35 years. The sample was sniffed by the panelists (nasal evaluation) and then rinsed into the mouth (retronasal evaluation). The intensities of the odor characteristics of the olive oils scored as above on a category scale from 0 to 3.0. After an outlier test, the results were expressed as means (standard deviations). Apple-like and green notes were characteristic for the Italian oil, whereas the Spanish oil smelled intensely like black currants and the flavor of the Moroccan oil came close to eating black olives. Olive oils of several types and cultivars were sensory evaluated for main attributes by various composition panels giving important results.

The presence of nucleic acids in extra virgin olive oil was verified in order to determine the cultivar of the origin of the olives used for the production.[252] DNA extraction methods based on cetyltrimethyl-ammonium bromide (CTAB) were employed for extra virgin olive oils of four Italian cultivars (Casaliva, Moraiolo, Leccino, and Taggiasca) and their leaves; plus four commercial extra virgin olive oils. The extracted DNA solution was prepared of chloroplast and nuclear origin DNA; therefore, it was possible to amplify the cloned cultivar RAPD and AFLP fragments homologous to the nuclear DNA of other species. The DNA, amplifiable with a PCR technique and electrophoretic analysis, showed that it is, in most of the samples, in the 2,000–10,000 bp size range, even if in some of the samples, partially degraded DNA in the 100–1000 bp range is also found.

DNA technology is based on the presence of DNA in all olive oil samples and even in refined oil, in which the quantity greatly depends on the oil processing technology and oil conservation conditions. Therefore, among other supports tried for DNA retaining different techniques were checked (silica extraction, hydroxyapatite, magnetic beads, and spun column) for DNA preparation from variable amounts of oil. At this stage, it was usable for amplification through PCR technology and especially, with the magnetic beads. The final method used magnetic beads. DNA was released from the beads in a buffer and did not contain compounds inhibiting PCR amplification using several different SSR primer sets. The SSRs markers were obtained by amplifying short (130, up to 250 bp) and average sized fragments.[253]

Pafundo et al.[254] traced DNA with PCR markers such as the amplified fragment length polymorphisms (AFLPs) on four monovarietal olive oils obtained from cultivars from France (Salonenque and Tanche) and Spain (Arbequina and Hojiblanca). Fluorescent optimized AFLPs for the characterization of olive oil DNA, to obtain highly reproducible, high quality fingerprints by testing the following parameters: the concentrations of dNTPs and labeled primer, the kind of *Taq* DNA polymerase and thermal cycler, and the quantity of the DNA employed. It was found that the correspondence of fingerprinting comparison of oils and plants was close to 70% and the DNA extraction from olive oil was the limiting step for the reliability of the AFLP profiles, due to the complex matrix analyzed.

5.6.1.2 Maize Oil Authenticity

Maize oil is considered to be a premium vegetable oil and there is a temptation for unscrupulous producers to adulterate maize oil with cheaper oils. Determining the authenticity of maize oil with traditional methods was problematic because its fatty acid composition overlaps with that of several other vegetable oils. In addition, the concentration of sterols in maize oils is very large in comparison with that of other vegetable oils, so that the sterol composition of any blend will comprise predominantly of those from maize oil. It is possible to form blends of oils whose characteristics according to traditional analysis are very similar to those of pure maize oil. Stable carbon isotope ratio analysis can be used to overcome these difficulties. Maize is a C_4 plant whereas all other commercial oil-bearing plants use the C_3 biosynthetic pathway. Addition of, or substitution with, oil from virtually any other source, to maize oil, would therefore give rise to a $\delta^{13}C$ value that was more negative than authentic maize oil.[258]

Woodbury et al.[259] constructed theoretical mixing curves showing the variation in $\delta^{13}C$ values of individual fatty acids produced by mixing maize with rapeseed oil. The prepared oil blends were saponified and methylated using boron trifluoridemethanol complex. The derivatization method used ensured that the carbon incorporated into the fatty acids on methylation came only from one source, namely the methanol of the BF_3–MeOH complex. Good agreement was found between the experimental $\delta^{13}C$ values and the theoretical mixing curves. A blind trial was then conducted using oil blends containing maize oil and an adulterant C_3 vegetable oil in an attempt to determine the concentration of the adulterant oil. The experimentally predicted values obtained for low concentrations of adulterant oil (<10% w/w) were within 1% of the actual value. At higher concentrations, 15%–20% adulterant oil, the values obtained were less accurate because groundnut oil was the adulterant in the blind samples rather than rapeseed which was used for the calibration.

5.6.1.3 Rapeseed Oil Authenticity

Seven parameters of physicochemical properties, such as acid number, color, density, refractive index, moisture and volatility, saponification value and PV, were measured for quality and adulterated soybean, as well as quality and rancid rapeseed oils. The chemometric methods were then applied for qualitative and quantitative discrimination and prediction of the oils by methods such as exploratory principal component analysis (PCA), partial least squares (PLS), radial basis function-artificial neural networks (RBF-ANN), and multi-criteria decision making methods (MCDM), PROMETHEE and GAIA.[260]

In general, the soybean and rapeseed oils were discriminated by PCA, and the two spoilt oils behaved differently with the rancid rapeseed samples exhibiting more object scatter on the PC-scores plot than the adulterated soybean oil. For the PLS and RBF-ANN prediction methods, suitable training models were devised, which were able to predict satisfactorily the category of the four different oil samples in the verification set. Rank ordering with the use of MCDM models indicated that the oil types can be discriminated on the PROMETHEE II scale. For the first time, it was demonstrated how ranking of oil objects with the use of PROMETHEE and GAIA could be utilized as a versatile indicator of quality performance of products on the basis of a standard selected by the stakeholder.

5.6.1.4 Sesame Oil Authenticity

Triacylglycerol (TG) composition of authentic and adulterated sesame oils with perilla was studied by using reversed phase liquid chromatography. Triacylglycerols were separated according to their equivalent carbon number. 1,2-Dilinoleoyl-3-oleoyl-*rac*-glycerol (LLO) was the most predominant in sesame oils, and its concentration was 7.690%–14.097% (w/w). The other abundant components were 1,2-dioleoyl-3-linoleoyl-*rac*-glycerol (OOL) and trioleoylglycerol (OOO), but trilinoleoylglycerol (LLL) and 1,2-dioleoyl-3-stearoyl-*rac*-glycerol (OOS) were minor. The remarkable differences between sesame and perilla oils are the amounts of LLO and trilinolenoylg-lycerol (LnLnLn). LnLnLn/LLO ratio for authentic sesame oils were less than 0.029. LnLnLn/LLO ratio looks promising as a detection index for adulterated sesame oils mixed with more than 5% perilla oil. Authentic and adulterated sesame oils with perilla were differentiated by applying the principal component analysis and discriminant analysis to liquid chromatographic data sets.[261] The minimum detectable limit for adulterated sesame oils was estimated to about 5% in mixing ratio. The chemometric technique provides a valuable tool to detect adulteration of oil.

5.6.1.5 Mustard Oil Authenticity

The determination of mustard oil in other edible oils is based on the detection and estimation of allyl-isothiocyanate, a volatile constituent present in mustard oil but not in other edible oils. The Association of Official Analytical Chemists (AOAC) method consists of distilling the sample, and precipitating the allylisothiocyanate as a black precipitate and dark color with silver nitrate. The intensity of the dark color and the amount of black precipitate formed are directly related to the amount of mustard oil present.[262] The detection sensitivity is about 0.05% of mustard oil in other edible oils. Erucic acid is characteristic of mustard and rape, and hence the estimation of erucic acid number by selective oxidation to dihydroxybehenic acid with $KMnO_4$ can be used as an index of the purity of rapeseed and mustard oils.[263]

Linseed oil in mustard oil can be evaluated quantitatively by reacting with bromine in chloroform and then treating with alcohol and ether. A calibration curve prepared by plotting the percentage of precipitate (v/v) against percentage of linseed oil in mustard oil is almost linear and can be used as a standard.[264]

5.6.1.6 Cocoa Butter Authenticity

Increases in the price of cocoa butter have led to a search for alternative fats for use in manufacturing chocolate. Biino and Carlisi[265] suggested determining illipe butter in cocoa butter blends by measuring the sigmasterol/campesterol ratio and the POS/POP and SOS/POP triglyceride ratios.

Derbesy and Richert[266] reported that HPLC of the unsaponifiables (sterol and triterpene alcohol fraction) revealed less than 1% of shea butter in cocoa butter. With measurement at 280 nm, one specific peak area was linearly related to the percentage of shea butter. Homberg and Bielefeld[267] suggested that sterols, methyl sterols, and triterpene alcohols be determined to identify cocoa butter adulteration.

On-line liquid chromatography–gas chromatography (LC–GC) has been applied to the analysis of steryl esters in cocoa butter. Separation of the steryl esters was achieved after on-line transfer to capillary GC. HPLC removes the large amount of triglycerides and pre-separates the components of interest, thus avoiding time-consuming sample preparation prior to GC analysis. The identities of the compounds were confirmed by GC–MS investigation of the collected HPLC fraction and by comparison of the mass spectra (chemical ionization using ammonia as ionization gas) to those of the synthesized reference compounds. Using cholesteryl laurate as an internal standard, steryl esters were quantified in commercial cocoa butter samples, the detection limit being 3 mg/kg and the quantification limit 10 mg/kg, respectively. Only slight differences in percentage distributions of steryl esters depending on the geographical origin of the material were observed. The patterns were shown to remain unchanged after deodorization. The method described might be a valuable tool for authenticity assessment of cocoa butter.[268]

Geeraert and DeSchepper[269] pointed out that triglyceride separations obtained by capillary GLC on polar columns are based predominantly on molecular weight differences (carbon number determinations), while reversed-phase HPLC fractionations (equivalent carbon number separations) are based on the degree of unsaturation of the triglycerides. They also noted that the optimization of HPLC conditions could refine the separations obtained, and proposed an optimized reversed-phase

HPLC analysis of cocoa butter and chocolate products (UV detection) to determine 5% or less of CBE in cocoa butter, based on monitoring the POP:POS:SOS ratio of the test sample.

Jeffrey[270] described a high speed HPLC procedure for high resolution separations of triglycerides in vegetable fats and cocoa butter equivalents. The HPLC system consisted of a ternary liquid chromatograph, flame ionization detector, and 100×4.6 mm internal diameter column packed with $3\,\mu m$ spherical silica, loaded with 10% silver nitrate. Trisaturated and major isomeric unsaturated triglycerides were eluted within 12 min. A change of solvent composition allowed elution of triglyceride with up to nine double bonds within the same time frame.

5.6.1.7   Palm, Palm Kernel, and Coconut Oils Authenticity

To identify palm oil fractions in palm oil the iodine value was plotted against the slip (melting) point. Carbon number analysis, an excellent technique for analyzing lauric oil, has distinguished between coconut and palm kernel oils, which have similar fatty acid compositions. The composition of triglycerides with carbon numbers 32 to 42 is normalized to 100 to give "K" values, and K34+K40 is plotted against K36+K38 to distinguish clearly between the two oils.[271] Sassano and Jeffrey[272] used capillary column GLC to characterize the triglycerides of palm oil and palm oil fractions.

Comparison of the sterols and sterol esters of coconut and palm kernel oils have showed sufficient differences to form a basis for distinguishing between the two oils.[273] Sterols were isolated as the digitonides and analyzed using packed-column GLC. Sterol esters, separated by preparative TLC, were analyzed by temperature-programmed capillary GLC (50% phenyl: 50% methylpolysiloxane stationary phase) and reversed-phase HPLC. Palm kernel oil displayed two major peaks, apparently due to campesteryl myristate and unresolved sitosteryl myristate/avenasteryl palminate, which were present at lower levels in the coconut oil. In addition, variations in the concentrations of other components were observed.

The authenticity of single seed vegetable oils that utilize the C_3 photosynthetic pathway was investigated using gas chromatography-combustion-stable isotope ratio mass spectrometry (GC-C-SIRMS). Samples of authentic groundnut, palm, rapeseed, and sunflower oils were derivatized to form fatty acid methyl esters (FAMEs) and their carbon isotope ratios ($^{13}C/^{12}C$) determined. In-house reference materials (IHRMs) and internal standards were used routinely to monitor the extraction procedure and SIRMS measurement. These materials demonstrated the consistent performance of the technique. The $\delta^{13}C\%$ data for the authentic vegetable oil fatty acids fell into the narrow range of -27.6% to -32.1%. However, the values within the oil varieties considered were significantly different.[274] The data from sunflower oils were such that they could be separated from the other varieties by canonical discriminant analysis. The determination of fatty acid carbon isotope ratios may therefore provide an additional indication of the varietal authenticity of oils which use the C_3 photosynthetic pathway.

5.6.2   Authenticity of Animal Oils and Fats

Animal fats, in contrast to vegetable fats and oils, contain significant amounts of fatty acids having odd-numbered carbon chains and branched chains and a high concentration (up to 1000 mg/kg) of cholesterol.[275,276] In addition, animal fats are almost free of tocopherols, whereas vegetable oils contain up to 1000 mg/kg or more of tocopherols. Animal fats often contain appreciable levels of saturated fatty acids in the 2-position of the triglycerides in contrast to vegetable oils that have very small amounts of saturated acids in the 2-position. Fish oils contain a wide range of fatty acids, including odd-numbered carbon chain fatty acids and mono-, di-, and polyunsaturated fatty acids with more than 18 carbon atoms. Accordingly, animal fats and fish oils can be identified in vegetable oils by the determination of fatty acid, triglyceride, and sterol composition and fatty acids at the triglyceride 2-position.

A method for species identification from pork and lard samples using polymerase chain reaction (PCR) analysis of a conserved region in the mitochondrial (mt) cytochrome b (cyt b) gene has been developed. Genomic DNA of pork and lard were extracted using Qiagen DNeasy Tissue Kits and subjected to PCR amplification targeting the mt cyt b gene. The genomic DNA from lard was found to be of good quality and produced clear PCR products on the amplification of the mt cyt b gene of approximately 360 base pairs. To distinguish between species, the amplified PCR products were cut with the restriction enzyme BsaJI resulting in porcine-specific restriction fragment length polymorphisms (RFLP). The cyt b PCR-RFLP species identification assay yielded excellent results for identification of the pig species. It is a potentially reliable technique for detection of pig meat and fat from other animals for Halal authentication.[277]

The triacylglycerol (TAG) compositions by carbon number during ripening of two Protected Designation of Origin (PDO) cheeses were analyzed using short capillary column gas chromatography. Lipolysis levels were high in the Cabrales (blue cheese produced from cows' milk or from blends of cows' with goats' milk) and Majorero goats' milk cheeses at the end of ripening, with free fatty acid (FFA) levels of around 24,000 ppm and significant changes in the TAG composition. The level of lipolysis in an industrial blue cheese made from ewes' milk was low, with an FFA value of around 6000 ppm and no significant changes in the TAG composition during ripening. The TAG values recorded for each cheese sample were substituted into the multiple regression equations that have been proposed for use in detecting foreign fats in milk fat. The values thus obtained were within the established ranges in early ripening. In the cheeses with high lipolysis levels during ripening, some of the values obtained fell outside the established ranges.[278] These equations can be potentially useful for detecting foreign fats in these cheeses, when employed early in the ripening period. Furthermore, it is important to take into account that before coming to a conclusion about cheese authenticity, several individual samples should be analyzed.

5.6.2.1 Butter Authenticity

Ascertainment of traditional butter authenticity implies the development of an analytical method for detecting illegal addition of cream from rennet-whey cream (RWC) to milk cream.[279] The reference HPLC method adopted for detecting the presence of rennet-whey solids in skim milk powder (EC Reg. No. 213/2001) is based on the determination of non-glycosylated caseinomacropeptide A (CMP A, i.e., k-CN$_A$ f 106–169). In this paper, the same method, coupled to ESI-MS, has been applied to the water phase of butter in order to detect CMP A deriving from usage of RWC for butter manufacturing. The reliability of this approach has been evaluated by studying the effect of both natural creaming and cream ripening in originating CMP A or peptides with CMP-like chromatographic behavior. Results demonstrated that peptides other than CMP A, and interfering in the HPLC profile, can form in cream after prolonged ripening with commercial starters of lactic acid bacteria. Only LC/MS can unequivocally show the presence of non-glycosylated CMP A and hence the usage of RWC in the manufacturing of traditional butter. In this regard, monitoring of multi-charged ions at m/z 1697.5 and 2263.2 was adopted for the recognition of monophosphorylated CMP A.

5.6.2.2 Lard Authenticity

Detection of suet in lard can be achieved by the Borner value, fatty acid composition and "S" ratio.[280] The incorporation of palmitic acid, oleic acid, and linoleic acid into the 2-position of the triglycerides of beef and pork fat is shown to be highly correlated to the corresponding acid contents of the triglycerides. On the basis of various regression equations obtained for pork and beef fats, the percent adulteration of pork fat with beef tallow may be determined. The method involves GC determination of the triglyceride fatty acids, and analysis of the fatty acid contents in the monoglycerides obtained after the lipase treatment of the fat. Results indicate that mixing beef fat with 10% porcine fat may be accurately estimated.[281] The myristic: palmitic acid ratio is considered to be a better indicator than the Borner index. A value of the ratio of <6 indicates that the sample is pure. At ratios >6, values for myristoleic acid, pentadecanoic, and isopentadecanoic may be used in the calculation of "S" ratio, when values greater than 10 indicate a genuine lard.[282]

Adulteration of beef fat by pork lard has received attention recently. The detection can be made at not less than the 5% level by the determination of trans fatty acids.[283] Fatty acid composition, triglyceride composition and physical characteristics are the usual approaches used to detect these cases. For instance, comparative GC studies on the fatty acid composition of lard and goose dripping revealed differences in the content of stearic and oleic acids.[284] The widely different fatty acid composition between beef and lard, notably the high percentage of saturated 18:0 is useful in distinguishing the two fats and their blends.

5.7 FUNCTIONAL AND HEALTH PROPERTIES OF FATS AND OILS IN FOODS

5.7.1 FUNCTIONAL ISSUES

The fats and oils found in foods are very important for the functionality and health issues of human diet. Fats and oils are incorporated into processed foods for a variety of reasons. They act, usually, as lubricants, enhancers of eating properties, and nutrition providers and are important ingredients required in bakery products to provide the following functional properties:

- Aerating or creaming
- Creating structure or building up body
- Emulsifying fats/oils with water
- Acting as moisture barriers
- Acting as a heat conductor during frying

The functional ability of fats and oils depends on the following factors:

- Sources of oils and fats (e.g., soy lean, tallow, etc.)
- Molecular make up
- Solids
- Conditions in chilling and tempering
- Shipping and storage conditions
- Quality and quantity of emulsifiers used

In some shortenings, the base fats or oils contribute all the functionality. For example, a shortening will cream in cookie dough or make tender biscuits only if the base fat has the proper ratio of solids to liquids. However, the crystal form of the shortening is of great importance.

Examination of some visual depictions can make the terms of solids, crystals, structure, lubricity, etc. more understandable.

A good solid plastic shortening is the one with 80% oil held between crystals of solid fat. This shortening could be icing, all purpose, or multipurpose shortening. It is interesting to note that the solids in a fat (shortening) provide structure and creaming, and the liquid oil provides lubricity. When air is incorporated into butter or icing during the mixing process, it is trapped between conglomerates of crystals. However, crystals of proper size and configuration have a so called "Beta Prime" crystal form.

Heating and melting fats results in some oil release which creates softer shortenings (5% solids) with reduced structural (creaming) properties, but with increased lubricating properties in foods.

Relatively large clusters of crystals (25% solids) resulted in products similar to the ones made from lard. Crystals of this type of form are not suggested in cakes, icings, and cookies because they do not incorporate air during mixing. Such "loose" shortenings are favorable for producing tender and flaky pie crusts.

TABLE 5.14
Consistency Range (in Units) of Old- and
New-Generation Margarines at a 5°
C-Increment Temperature Range

Temperature (°C)	Old Margarine	New Margarine
5	218	119
10	154	91
15	90	63
20	47	46
25	23	27

Shortenings with fine crystals having the so called "Beta crystal" structure tend to entrap the oil tightly. These shortenings provide reduced levels lubricating and creaming properties since they are dry (hard) and crumbly (brittle).

However, the role of solid content in fat and oils is very important in special applications. A liquid oil, therefore is a product with virtually very low or no solids at room temperatures. Usually, oils are used at certain bakery foods which require lubrication, but never in icings which require structure (Table 5.14).

Products having oil contents of 95%–100%—usually called fluid "Cake" shortenings (Table 5.14)—consist of liquid oil with added emulsifiers. The emulsifiers in these fluid "Cake" shortenings provide all the favorable creaming properties and contribute to proper 'moist and melt in mouth' eating properties of cakes and other products.

Cakes using fluid "Cake" shortenings usually receive their structure from other sources which are balancing the cake formulas adapting a different degree of toughness or tenderization.

Fluid "Bread" shortenings are products with solids (90%) added in the form of hard or fully hydrogenated fats (Figure 5.5). They are usually contributing to the structure of bread by strengthening the side walls of bread. The main function of these fat products in breads is to provide lubricity.

Another group of fats are the "all purpose" functional shortening or fats which contain 20%–25% solids at which there is sufficient oil for effective lubrication (mixing) and satisfying eating properties.

Shortenings or margarines have less liquid but higher solids for several reasons. First, to create a tightly knit crystal matrix exhibiting proper extrusion and spreading properties, while retaining the capability to act as a moisture barrier in the product and process developments; second,

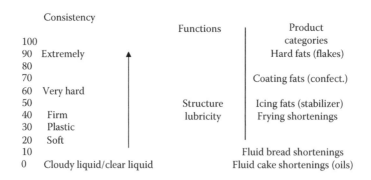

FIGURE 5.5   Consistency, functionality, and product categories of solid fat content at various percentages.

to be able to hold water when subjected to various food processes (e.g., extrusion of fat into doughs); and third, to contribute an adequate structure to (bakery) foods and improve the eating characteristics.

On the other hand, "icing" products are, also, fats with moderately high melting points and solid contents. They usually combine into other fats in which they act as stabilizers (Figure 5.5). But, their main purpose is to accelerate the setting (hardening) of the fats and to provide the following functional properties:

- Avoid stickiness on packaging
- Provide some gloss
- Raise the melting point.

In general, they are used to provide body/structure to the icings, but always at levels below 5% of the icing composition so as to avoid adverse effects on the eating characteristics.

Another category is the "coating fats" (Figure 5.5) that provide mainly a combination of structural and lubricating properties along with others (gloss or sheen). Their application to food products is as substitutes (e.g., for cocoa butter in chocolate) or pastel coatings applied to doughnuts and other related products for protection against drying out—acting as moisture barriers. They are usually made of specially hydrogenated soybean, cotton, and other oils, and act as moisture barriers.

Finally, the "hard" fats or "flakes" are a fat category (Figure 5.5) with the highest solid content. They are produced and sold mainly in flake forms for handling convenience and melting purposes. Because of their nature almost 100% solid crystals at ordinary temperatures are generally used to improve/enhance the structural and moisture barrier functionality of other fats and oils which may be too soft in their application to foods.

5.7.2 HEALTH ISSUES

Consumers, generally, consider saturated fats and cholesterol as emerging health issues. During the 1960s, partially hydrogenated oil began to replace animal fats, because they were able to contribute the same desirable characteristics in foods without providing high levels of saturated fat and cholesterol. Since then, many types of fats and oils from vegetables and seeds have been introduced to create healthier products. For example, in the 1980s, tropical fats lost favor because of their saturated fat content. Instead, olive oil became a nutritionist's favorite due to the various research studies that claimed a 8%–27% saturated and 55%–83% monounsaturated fatty acids content that drive the olive oil's cardiovascular benefits. But, cost and other attitudes kept it from meeting consumer's satisfaction, and the commercial gap for saturated vegetable and tropical fats and oils remained.[285]

Fat and oil manufacturers relied on tropical fat and oil replacement with soybean, corn, cotton, and canola oils that can mimic the characteristics of tropical oils (coconut and palm) and animal fats (lard and tallow). Some fats and oils are considered as healthier than others due to their higher level of polyunsaturated fatty acids relatively to their saturated fatty acids content. For example canola oil shows a healthier profile than soybean oil due to its overall fatty acid composition of 50% less saturated fat and higher monounsaturates. However, the marketing approach of one fat or oil over another stems from the trend toward products offering health claims, such as "heart-healthy", "may lower cholesterol", etc.

In general, fats and oils are both lipids and triglycerides that differ in fatty and other component matters.

Thus, saturated fatty acids that make up a triglyceride comprise of those with a full hydrogen atom to the long hydrocarbon chains (palmitic and stearic acids), while unsaturated are similar chains having double bonds and minus the hydrogen atoms. Further, some of these chains can be altered by adding hydrogen atoms through hydrogenation. Technically, a fat may contain some oil and vice versa, but saturated and highly hydrogenated fats are solid at room temperature, while unsaturated fats are liquid.

The health implication is that saturated fats lead to fatty deposits in human arteries with atherosclerosis effects and eventually can lead to heart attacks, while unsaturated fats do not have such effects. Thus, consumers should reduce or replace foods containing saturated fats, although there is evidence that some saturated fats have a neutral effect (e.g., olestra).

Unsaturated fatty acids are usually subjected to air oxidation reactions that create rancid taste effects. Therefore, hydrogenation of unsaturated fatty acids reduces these reactions and modifies their functionality. Nevertheless, how can one respond to the critics of hydrogenation who object to it from a health standpoint? Through the years nutrition biochemists have been concerned about the trans fatty acids formed during the hydrogenation process (hydrogen gas is bubbled into oils) that produces artificial trans fatty acids that accumulate in body tissues. These artificial fatty acids cause all kinds of diseases because they are foreign to the human immune system and unnatural.

In addition, hydrogenation produces, besides saturated fatty acids, changes in the crystallization and/or melting properties of fats, which do not behave like either an unsaturated or a saturated fat, but something in between. Thus, the trans fats may have a different (more or less adverse) reaction on health aspects than the saturated fats.

This creates a problem for their application in foods. The food industry is exploring ways to reduce or eliminate trans fats in the products, while maintaining the consumers' quality standards.[286]

The food industry, also, looks to modern technologies to find processes for healthier products using certain oil sources with different fatty acid compositions.

The Food and Drug Administration (FDA) of America has also, addressed the "trans issue" by setting labeling protocols for each manufactured food. The food industries are keen on introducing a portfolio of low/zero trans solutions in the food products by utilizing various alternative oil sources. These include soybean, canola, corn, cottonseed, palm, and sunflower oils for substituting the currently used liquid oils and stearic and poly unsaturated inter-esterified fats for shortening and margarine solutions. However, biotechnology has helped scientists to tinker with the fatty acid composition of oilseeds by utilizing conventional breeding and/or genetic engineering technologies to produce new and unique products that provide better health benefits and functional properties.

The food industry is concerned with utilizing an ingredient such as a no trans fatty acid oil that provides certain health and other functional issues that consumers desire without driving up the costs.

Finally, another class of fatty acids that are valuable for their nutritional and health attributes and less for their functional properties is the ω-3 long-chain poly unsaturated fatty acids (PUFAs) which are mainly found in fatty fish. Today, researchers suggest that ω-3s and ω-6s at certain ratios are beneficial for consumers (infants, children, and adults) preventing implications for cardiovascular and brain diseases.

The naturally occurring DHA and EPA that have been attributed to health benefits—surpassing even the patients' expectations—serve as ingredients for infant formulations in many countries. Potential applications for the above beneficiary ingredients are products such as yogurts, cheeses, and others.

Proper attention should be given to special fats for food formulations since they are heat sensitive and create oxidation problems. They have to be handled properly during food manufacture. The food industry, lately, has made significant advances in creating fats and oils to reduce and/or eliminate saturated and trans fatty acids or to advance nutritional profiles. Thus, consumers should realize that caloric intake and amount of trans fatty acids are the primary causes of obesity and cause cardiovascular diseases and other complications.

5.7.3 Process and Application Issues

The quality and specifications for use of a fat is based on the oil quality, and the relationship between the quantity and the type of crystals that are required for the particular end product. Thus, it is interesting to look into the developments of the hydrogenation of the base fats of shortenings and margarines.

A major breakthrough occurred when ordinary oils were hydrogenated to produce similar fat products to butters, (e.g., cocoa butter) by creating unique solid content and crystal behavior. This was one of the most important investigations in fat and oil technology several decades ago. The objective was to build significantly higher solids and more structure into the oil so that (1) excessively high melting point solids are not generated so that good (eating) characteristics during product applications are created, and (2) the plasticity of the fat products are improved over a broad level of temperatures (5°C–35°C).

These new technologies gave rise to some of the new generation fat products, which are in the market today. with some sophisticated improvements. Such fat products include the "puff pastry" shortenings and margarines, the "bakers" margarines and vegetable "pre-shortenings", which are interesting to see because of their specific properties and applications.

5.7.3.1 Shortenings "Puff Pastry" and Margarines

These products should take into consideration their functionality as applied to the consumers' acceptability (eating properties), because the high level melting point required (machinability) of the saturated fats is opposed to the good eating properties (edibility) required by the end products.

The new hydrogenation process technologies use formulations of pastry fat products characterized with good plasticity and structural characteristics at temperatures between 5°C–35°C without affecting the edible properties. Though lowering the melting point of fats can provide better mouthfeel acceptability it will create application problems by loosing solids, structure, and other properties at working temperatures.

A typical puff pastry fat may be considered as the one that possesses great structure, good moisture barrier, and eating properties with a solid fat index (SFI) constant (appr. 30% solids) at temperatures 10°C–85°C.

5.7.3.2 Baker's Margarines

These fat products have similar applications to those of the puff pastry, but a superior plastic range. A consistency demonstration of fats (margarines) produced by the old and the new generation technology is shown in Table 5.14.

In the table, for consistencies—the greater the number, the harder the product. So, the number 100 means a product excessively hard. It is noticeable that the change in the consistency of the new (92 units) over the old (195 units) generation fat products (margarines) is in the temperature range of 5°C–25°C.

5.7.3.3 Pie's Shortenings

These fat products (vegetable oil shortenings) are formulated in such a way that they have a required plasticity similar to lard at lower temperatures of 0°C–10°C. However, an ideal "pie" shortening exhibiting optimum plasticity at 5°C–25°C is a high level saturated (hard fat) product. Although these products have the required plasticity at lower temperatures, they do not offer an optimum structure and suitable edibility at room temperatures. The solids profile at a temperature range of 10°C–40°C is shown in Table 5.15.

TABLE 5.15
Solids' Profile of "Pie" Shortenings at a Temperature Range of 10°C–40°C

Temperature (°C)	Solids (%)
10	23.0
20	20.0
25	17.0
35	12.0
40	4.4

In conclusion, the new hydrogenation technologies have brought certain significant improvements on fats, for applications such as

1. Acceptable plasticity at broader temperature levels
2. Improved edibility (eating quality) with a reduced solid content in mouth-temperature
3. Enhanced structural properties at high temperatures (35°C)

These are some of the most interesting improvements that have made available a variety of choices of fats for various applications today.

REFERENCES

1. Nawar, W.W. Lipids (ed. O.R. Fennema), *Food Chemistry*, 3rd edition. Marcel Dekker, Inc., New York, p. 226. 1996.
2. http://en.wikibooks.org/wiki/Category:Fats_and_oils (accessed 13/2/2008).
3. Belitz, H.D., Grosh, W., and Schieberle, P. *Food Chemistry*, 3rd rev. edition. Springer-Verlag, Berlin, Germany, p. 157. 2004.
4. IUPAC. *The Gold Book*, 2nd edition. IUPAC Compendium of Chemical Terminology, International Union of Pure and Applied Chemistry. 1997.
5. AHA (American Heart Association). Trans fatty acids. 2008. Available at: http://www.americanheart.org/presenter.jhtml?identifier=3030450 (accessed 18/2/2008).
6. AHA (American Heart Association). The details: Fat and fatty acids. 2007. Available at: http://www.deliciousdecisions.org/ee/wbd_acids_def.html (accessed 16/2/2008).
7. Tedetti, M., Kawamura, K., Nakurawa, M., Joux, F., Charriere, B., and Sempere, R. Hydroxyl radical-induced photochemical formation of dicarboxylic acids from unsaturated fatty acid (oleic acid) in aqueous solution. *Journal of Photochemistry and Photobiology A: Chemistry*, **188**(1), 135–139. 2007.
8. Melechko, A.V. Fatty acid. 2007. Available at: http://human.freescience.org/htmx/fatty_acid.php (accessed 16/2/2008).
9. Harrison, K. Unsaturated fatty acid, linoleic acid, octadecadienoic acid. 2007. Available at: http://www.3dchem.com/moremolecules.asp?ID=382&othername=Unsaturated%20Fatty%20acid (accessed 18/2/2008).
10. Harrison, K. Docosahexaenoic acid, DHA, cervonic acid, unsaturated fatty acid. 2007. Available at: http://www.3dchem.com/molecules.asp?ID=379 (accessed 18/2/2008).
11. EU (European Union). Fatty acid. 2007. Available at: http://www.carnauba.eu/fatty_acid_en.html (accessed 16/2/2008).
12. Dupuy, N., Duponchel, L., Huvenne, J.P., Sombret, B., and Legrand, P. Classification of edible fats and oils by principal component analysis of Fourier transform infrared spectra. *Food Chemistry*, **57**(2), 245–251. 1996.
13. USDA (United States Department of Agriculture). What are "oils"? 2007. Available at: http://www.mypyramid.gov/pyramid/oils.html (accessed 19/2/2008).
14. Gunstone, F.D. and Hamilton, R.J. *Oleochemical Manufacture and Applications*. Sheffield Academic Press, Sheffield, U.K. 2001.
15. USDA Agricultural statistics. Statistics of oilseeds, fats and oils (Chapter III). 2005. Available at: http://www.usda.gov/nass/pubs/agr05/05_ch3.PDF (accessed 19/2/2008).
16. Chiacchierini, E., Mele, G., Restuccia, D., and Vinci, G. Impact evaluation of innovative and sustainable extraction technologies on olive oil quality. *Trends in Food Science and Technology*, **18**(6), 299–305. 2007.
17. Favati, F., Caporale, G., Monteleone, E., and Bertuccioli, M. Rapid extraction and determination of phenols in extra virgin olive oil. *Developments in Food Science*, **37**(1), 429–452. 1995.
18. Gunstone, F.D. and Herslof, B.G. *Glossary Lipid* 2. The Oily Press, P.J. Barnes and Associates, Bridgewater, U.K., p. 157. 2000.
19. Boskov, D. *Olive Oil: Chemistry and Technology*. AOCS Press, Champaign, IL, 1996.
20. Encyclopedia Britannica. Corn oil. 2008. Available at: http://www.britannica.com/eb/article-9026324/corn-oil (accessed 19/2/2008).

21. Strecker, L.R., Bieber, M.A., Maza, A., Grossberger, T., and Doskoczynski, W.J. Corn oil (ed. Y.H. Hui), *Bailey's Industrial Oil and Fat Products*, 5th edition, *Edible Oil and Fat Products: Oils and Oilseeds*. John Wiley & Sons, New York, Vol. 2, pp. 125–158. 1996.

22. Judd, J.T., Baer, D.J., Clevidence, B.A., Kris-Etherton, P., Muesing, R.A., and Iwane, M. Dietary cis and trans monounsaturates and saturated FA and plasma lipids and lipoproteins in men. *Lipids*, **37**(2), 123–131. 2002.

23. Wikipedia. Soybean—Soybean oil. 2007. Available at: http://en.wikipedia.org/wiki/Soya_oil (accessed 19/2/2008).

24. Tomas, A. Hydrogenation—Use in frying oils. 1998. SCI Lecture Papers Series. Available at: http://www.soci.org/SCI/publications/2001/pdf/pb92.pdf (accessed 19/2/2008).

25. Sipes, E.F. and Szuhaj, B.F. Soybean oil (ed. Y.H. Hui), *Bailey's Industrial Oil and Fat Products*, 5th edition, *Edible Oil and Fat Products: Oils and Oilseeds*. John Wiley & Sons, New York, Vol. 2, pp. 497–601. 1996.

26. Clark, J.P. Tocopherols and sterols from soybeans. *Lipid Technology*, **8**, 111–114. 1996.

27. British Pharmacopoeia Commission. *Ph. Eur. Monograph 1371*, British Pharmacopoeia. The Stationery Office, Norwich, U.K. 2005.

28. Davidson, H.F., Campbell, E.J., Bell, R.J., and Pritchard R.A. Sunflower oil (ed. Y.H. Hui), *Bailey's Industrial Oil and Fat Products*, 5th edition, *Edible Oil and Fat Products: Oils and Oilseeds*. John Wiley & Sons, New York, Vol. 2, pp. 603–689. 1996.

29. Encyclopedia Britannica. Cottonseed oil. 2007. Available at: http://www.britannica.com/eb/article-9026534/cottonseed (accessed 19/2/2008).

30. Wikipedia. Cottonseed oil. 2007. Available at: http://en.wikipedia.org/wiki/Cottonseed_oil (accessed 19/2/2008).

31. Jones, L.A. and King, C.C. Cottonseed oil (ed. Y.H. Hui), *Bailey's Industrial Oil and Fat Products*, 5th edition, *Edible Oil and Fat Products: Oils and Oilseeds*. John Wiley & Sons, New York, Vol. 2, pp. 159–240. 1996.

32. NutraSanusWheat Germ Oil (octacosanol) Information. 2008. Available at: http://www.nutrasanus.com/wheat-germ.html (accessed 19/2/2008).

33. Australian Government. What is "canola"? A problem with weeds—the canola story. Australian Biotechnology. 2007. Available at: http://www.biotechnologyonline.gov.au/foodag/weeds.cfm (accessed 19/2/2008).

34. Gunstone, F.D. *The Chemistry of Oils and Fats—Sources, Composition, Properties and Uses*. CRC Press, Boca Raton, FL, p. 7. 2004.

35. deMan, L. and deMan, J.M. Functionality of palm oil in margarines and shortenings. *Lipid Technology*, **6**, 5. 1994.

36. Catharina, Y.W. Ang, Keshun, L., and Huang, Y.W. *Asian Foods: Science and Technology*. Technomic, Lancaster, PA. 1999.

37. Basiron, Y. Palm oil (ed. Y.H. Hui), *Bailey's Industrial Oil and Fat Products*, 5th edition, *Edible Oil and Fat Products: Oils and Oilseeds*. John Wiley & Sons, New York, Vol. 2, pp. 271–375. 1996.

38. Basiron, Y. Palm oil beyond 2000. *INFORM*, **11**, 30–33. 2000.

39. Norulaini, N.A.N., Zaidul, I.S. Md., Anuar, O., and Omar, A.K.M. Supercritical enhancement for separation of lauric acid and oleic acid in palm kernel oil (PKO). *Separation and Purification Technology*, **35**(1), 55–60. 2004.

40. Wikipedia. Palm oil. 2007. Available at: http://en.wikipedia.org/wiki/Palm_oil#_note-5 (accessed 19/2/2008).

41. Norulaini, N.A.N., Ahmad, A., Omar, F.M., Banana, A.A.S., Zaidul, I.S.M., and Kadir, M.O.A. Sterilization and extraction of palm oil from screw pressed palm fruit fiber using supercritical carbon dioxide. *Separation and Purification Technology*, **60**(3), 272–277. 2008.

42. Kritchevsky, D., Tepper, S.A., Kuksis, A., Wright, S., and Czarnecki, S.K. Cholesterol vehicle in experimental atherosclerosis. 22. Refined, bleached, deodorized (RBD) palm oil, randomized palm oil and red palm oil. *Nutrition Research*, **20**(6), 887–892. 2000.

43. Zohary, D. and Hopf, M. *Domestication of plants in the Old World*, 3rd edition. Oxford University Press, Oxford, U.K., p. 211. 2000.

44. Zahran, A.M., Omran, M.F., Mansour, S.Z., and Ibrahim, N.K. Effectiveness of *Carthamus tinctorius* L. in the restitution of lipid composition in irradiated rats. *Egyptian Journal of Radiation Science and Applications*, **20**(1), 75–94. 2007.

45. Smith, J.R. *Safflower*. AOCS Press, Champaign, IL, 1996.

46. Smith, J. Safflower (ed. Y.H. Hui), *Bailey's Industrial Oil and Fat Products*, 5th edition, *Edible Oil and Fat Products: Oils and Oilseeds*. John Wiley & Sons, New York, Vol. 2, pp. 411–455. 1996.

47. Wikipedia. Coconut oil. 2007. Available at: http://en.wikipedia.org/wiki/Coconut_oil#_note-1 (accessed 20/2/2008).

48. USDA. Nutrient Database for Standard Reference, Release 16. 2003. Available at: http://www.nal.usda.gov/fnic/foodcomp/Data/SR16-1/reports/s16-1f04.pdf (accessed 20/2/2008).

49. Canapi, E.C., Augustin, Y.T.V., Moro, E.A., Pedrosa, E., and Bendano, M.L.J. Coconut oil (ed. Y.H. Hui), *Bailey's Industrial Oil and Fat Products*, 5th edition, *Edible Oil and Fat Products: Oils and Oilseeds*. John Wiley & Sons, New York, Vol. 2, pp. 97–124. 1996.

50. Villarino, B.J., Dy, L.M., and Lizada, M.C.C. Descriptive sensory evaluation of virgin coconut oil and refined, bleached and deodorized coconut oil. *LWT—Food Science and Technology*, **40**(2), 193–199. 2007.

51. Jayadas, N.H. and Prabhakaran Nair, K. Coconut oil as base oil for industrial lubricants—Evaluation and modification of thermal, oxidative and low temperature properties. *Tribology International*, **39**(9), 873–878. 2006.

52. Guarte, R.C., Mühlbauer, W., and Kellert, M. Drying characteristics of copra and quality of copra and coconut oil. *Postharvest Biology and Technology*, **9**(3), 361–372. 1996.

53. Shukla, V.K.S. Cocoa butter properties and quality. *Lipid Technology*, **7**, 54–57. 1995.

54. Hashim, L., Hudiyono, S., and Chaveron, H. Volatile compounds of oxidized cocoa butter. *Food Research International*, **30**(3–4), 163–169. 1997.

55. Padley, F.B. *Lipid Technologies and Applications* (eds. F.D. Gunstone and F.B. Padley). Marcel Dekker, New York, pp. 391–432. 1997.

56. Lawler, P.J. and Dimick, P.S. *Food Lipids: Chemistry, Nutrition and Biotechnology* (eds. C.C. Akoh and D.B. Min). Marcel Dekker, New York, pp. 229–250. 1998.

57. Deshpande, S.S., Deshpande, U.S., and Salunkhe, D.K. Sesame oil (ed. Y.H. Hui), *Bailey's Industrial Oil and Fat Products*, 5th edition, *Edible Oil and Fat Products: Oils and Oilseeds*. John Wiley & Sons, New York, Vol. 2, pp. 457–495. 1996.

58. Oplinger, E.S., Putnam, D.H., Kaminski, A.R., Hanson, C.V., Oelke, E.A., Schulte, E.E., and Doll, J.D. *Sesame, Alternative Field Crops Manual*. University of Wisconsin-Madison and University of Minnesota, St. Paul, MN, 1997.

59. FSA (Food Standards Agency). Fats and oils. *McCance and Widdowson's the Composition of Foods*, 5th edition. Royal Society of Chemistry, Cambridge, U.K. 1991.

60. Altar, T. More than you wanted to know about fats/oils. Sundance Natural Foods. 1995. Available at: http://www.efn.org/~sundance/fats_and_oils.html (accessed 20/2/2008).

61. Wikipedia. Fat. 2007. Available at: http://en.wikipedia.org/wiki/Fat (accessed 20/2/2008).

62. University of Guelph. Butter manufacture. 2005. Available at: http://www.foodsci.uoguelph.ca/dairyedu/butter.html (accessed 20/2/2008).

63. de Greyt, W. and Huyghebaert, A. Food and non-food applications of milk fat. *Lipid Technology*, **5**, 138–140. 1993.

64. Sweeney, R. The history of butter. 2007. Available at: http://www.associatedcontent.com/article/307688/the_history_of_butter_you_dont_know.html (accessed 20/2/2008).

65. Hettinga, D. Butter (ed. Y.H. Hui), *Bailey's Industrial Oil and Fat Products*, 5th edition, *Edible Oil and Fat Products: Specialty Oils and Oil Products*. John Wiley & Sons, New York, Vol. 3, pp. 1–63. 1996.

66. Flack, E. *Lipid Technologies and Applications* (eds. F.D. Gunstone and F.B. Padley). Marcel Dekker, New York, pp. 305–327. 1997.

67. Wikipedia. Lard. 2007. Available at: http://en.wikipedia.org/wiki/Lard (accessed 20/2/2008).

68. Ockerman, H.W. and Basu, L. Edible rendering—Rendered products for human use (ed. D.L. Meeker), *Essential Rendering*. National Renders Association, Arlington, VA, pp. 95–110. 2006.

69. Davidson, A. Lard. *The Penguin Companion to Food*. Penguin Books, New York, pp. 530–531. 2002.

70. Matz, S.A. Lard. *Bakery Technology and Engineering*. Springer, New York, p. 81. 1991.

71. Judd, B. Biodiesel from tallow. Energy efficiency and conservation authority. 2002. Available at: http://www.eeca.govt.nz/eeca-library/renewable-energy/biofuels/report/biodiesel-from-tallow-report-02.pdf (accessed 21/2/2008).

72. USDA. National Nutrient Database for Standard Reference, Release 20. 2007. Available at: http://www.nal.usda.gov/fnic/foodcomp/cgi-bin/list_nut_edit.pl (accessed 21/2/2008).

73. Love, J.A. Tallow (ed. Y.H. Hui), *Bailey's Industrial Oil and Fat Products*, *Edible Oil and Fat Products: Chemistry, Properties and Health Effects Edible Oil and Fat Products*. John Wiley & Sons, New York, Vol. 1, pp. 1–18. 1996.

74. Achaya, K.T. *Lipid Technologies and Applications* (eds. F.D. Gunstone and F.B. Padley). Marcel Dekker, New York, pp. 369–390. 1997.

75. Kumara, M.V., Sambaiaha, K., and Lokesh, B.R. Hypocholesterolemic effect of anhydrous milk fat ghee is mediated by increasing the secretion of biliary lipid. *The Journal of Nutritional Biochemistry*, **11**(2), 69–75. 2000.

76. Pigott, G.M. Fish oil (ed. Y.H. Hui), *Bailey's Industrial Oil and Fat Products*, 5th edition, *Edible Oil and Fat Products: Specialty Oils and Oil Products*. John Wiley & Sons, New York, Vol. 3, pp. 225–254. 1996.

77. Moffat, C. Fish oil triglycerides: A wealth of variations. *Lipid Technology*, **7**, 125–129. 1995.

78. EPA (Environmental Protection Agency). Fish consumption advisories. 2007. Available at: http://www.epa.gov/mercury/advisories.htm (accessed 21/2/2009).

79. Mourente, G., Dick, J.R., Bell, J.G., and Tocher, D.R. Effect of partial substitution of dietary fish oil by vegetable oils on desaturation and β-oxidation of $[1-^{14}C]18{:}3n{-}3$ (LNA) and $[1-^{14}C]20{:}5n{-}3$ (EPA) in hepatocytes and enterocytes of European sea bass (*Dicentrarchus labrax* L.). *Aquaculture*, **248**(1–4), 173–186. 2005.

80. Madsen, S. Development of a reduced-fat spread enriched with long-chain *n*–3 fatty acids. *Lipid Technology*, **10**, 129–132. 1998.

81. van der Roest, J., Kleter, G., Marvin, H.J.P., de Vos, B., Hurkens, R.R.C.M., Schelvis-Smit, A.A.M., and Booij, K. Options for pro-actively identifying emerging risk in the fish production chain. RIKILT Report 2007.006, Institute of Food Safety, Wageningen, the Netherlands, 2007.

82. Larsson, K. *Lipids: Molecular Organisation, Physical Functions and Technical Applications*. The Oily Press, Dundee, Scotland. 1994.

83. Toro-Vazquez, J.F., Dibildox-Alvarado, E., Charo-Alonso, M.A., Gomez-Aldapa, C., and Herrera-Coronado, V. The rheology of vegetable oils during crystallisation. *Food Engineering: Rheology and Texture*, IFT Annual Meeting, New Orleans, LA. 2001.

84. Lawler, P.J. and Dimick, P.S. Crystallisation and polymorphism in fats (eds. C.C. Akoh and D.B. Min). *Food Lipids: Chemistry, Nutrition and Biotechnology*, 2nd edition. Marcel Dekker, New York, pp. 275–300. 2002.

85. Brown, R.J.C. and Brown, R.F.C. Melting point and molecular symmetry. *Journal of Chemical Education*, **77**(6), 724. 2000.

86. Coupland, J.N. and McClements, D.J. Physical properties of liquid edible oils. *Journal of the American Oil Chemists' Society*, **74**, 1559–1564. 1997.

87. Forssell, P., Kervinen, R., Lappi, M., Linko, P., Suortti, T., and Poutanen, K. Effect of enzymatic inter-esterification on the melting point of tallow-rapeseed oil (LEAR) mixture. *Journal of the American Oil Chemists' Society*, **69**(2), 126–129. 1992.

88. Felfe, K. and Gwozdz, K. Polymorphisms of crystalline solids. ENVE, 101. 2009. Available at: http://www.iit.edu/~felfkri/report.htm2001 (accessed 22/2/2009).

89. D'Souza, V., de Man, J.M., and de Man, L. Short spacings and polymorphic forms of natural and commercial solid fats: a review. *Journal of the American Oil Chemists' Society*, **67**, 835–843. 1990.

90. Hagemann, J.W. Thermal behavior of acylglycerides (eds. N. Garti and K. Sato). *Crystallization and Polymorphism of Fats and Fatty Acids*. Marcel Dekker, New York, pp. 9–95. 1988.

91. Cossement, M., Michaux, M., Lognay, G., Gibon, V., and Deroanne, C. Effect of erucic acid on the polymorphism of hydrogenated rapeseed oil. *Fat Science Technology*, **92**(6), 229–231. 2006.

92. OECD (Organisation for Economic Co-operation and Development). Densities of oil products—Special paper 9. *Energy Statistics Working Group Meeting*. IEA, Paris, France, p. 2. 2004.

93. Pantzaris, T.P. The density of oils in the liquid state (with special reference to palm oil), 1984. Record number MY8505603, http://www.lib.upm.edu.my.

94. Ronneboog, R.M.J. Oil viscosity. 2008. Available at: http://www.scienceiq.com/ShowFact.cfm?ID=466 (accessed 22/2/2009).

95. Noureddini, H., Toeh, B.C., and Clements, L.D. Viscosities of oils and fatty acids. *Journal of the American Oil Chemists' Society*, **69**, 1189–1191. 1992.

96. Fisher, C.H. Correlating viscosity with temperature and other properties. *Journal of the American Oil Chemists' Society*, **75**, 1229–1232. 1998.

97. PERC (Petroleum Research Center). Refractive index of oils. 2000. Available at: http://www.perc.utah.edu/Capabilities/OilChar/RI/index.htm (accessed 22/2/2009).

98. Kanazawa, A., Sawa, T., Akaike, T., and Maeda, H. Dietary lipid peroxidation products and DNA damage in colon carcinogenesis. *European Journal of Lipid Science and Technology*, **104**, 439–447. 2002.

99. Frankel, E.N. *Lipid Oxidation*. The Oily Press Ltd., Glasgow, U.K., pp. 79–98. 1998.
100. van de Voort, F.R., Ismail, A.A., Sedman, J., and Emo, G. Monitoring the oxidation of edible oils by Fourier transform infrared spectroscopy. *Journal of the American Oil Chemists' Society*, **71**(3), 243–253. 1994.
101. Marigheto, N.A., Kemsley, E.K., Defernez, M., and Wilson, R.H. A comparison of mid-infrared and Raman spectroscopies for the authentication of edible oils. *Journal of the American Oil Chemists' Society*, **75**(8), 987–992. 1998.
102. Muik, B., Lendl, B., Malina-Diaz, A., and Ayora-Canada, M.J. Direct monitoring of lipid oxidation in edible oils by Fourier transform Raman spectroscopy. *Chemistry and Physics of Lipids*, **134**, 173–182. 2005.
103. Hamilton, R.J. *Rancidity in Foods* (eds. J.C. Allen and R.J. Hamilton), 3rd edition. Chapman & Hall, London, U.K., pp. 1–22. 1994.
104. Chan, H.W. *Autoxidation of Unsaturated Lipids*. Academic Press, London, U.K. 1987.
105. Gunstone, F.D. *Fatty Acid and Lipid Chemistry*. Chapman & Hall, Glasgow, U.K. 1996.
106. Allen, J.C. and Hamilton, R.J. *Rancidity in Foods*, 3rd edition. Chapman & Hall, London, U.K. 1994.
107. Hsieh, R.J. and Kinsella, E.J. Oxidation of PUFAs. Mechanisms products and inhibition with emphasis on fish oil. *Advances in Food and Nutrition Research*, **33**, 233–241. 1985.
108. Liegeois, C., Lernseau, G., and Collins, S. Measuring antioxidant efficiency of wort, malt, and hops against the 2,2,-azobis (2-qmidinopropane) dihydrochloride-induced oxidation of an aqueous dispersion of linoleic acid. *Journal of Agriculture and Food Chemistry*, **48**, 1129–1134. 2000.
109. Esterbauer, H. The chemistry of oxidation of lipoproteins (eds. C. Rice-Evans and K.R. Bruckdorfer). *Oxidative Stress, Lipoproteins, and Cardiovascular Dysfunction*. Portland Press, London, U.K. 1980.
110. Niki, E. Antioxidants in relation to lipid peroxidation. *Chemistry and Physics of Lipids*, **44**, 227–253. 1987.
111. David, W.R. and Dorr, A.L. *Food Lipid Chemistry, Nutrition and Biotechnology* (eds. C.C. Akoh and B.D. Min). Marcel Dekker, New York, 1997.
112. Frankel, E.N. Volatile lipid oxidation products. *Progress in Lipid Research*, **22**, 1–33. 1982.
113. Mistry, B.S. and Min, B.D. Oxidised flavor compounds in edible oils. Developments in food science (ed. G. Charalambous), *Off flavors in Foods and Beverages*. Elsevier, Amsterdam, the Netherlands. 1992.
114. Chen, J.H. and Schanus, E.G. The inhibitory effect of water on the Co^+ and Cu^{2+} catalyzed decomposition of methyl linoleate hydroperoxides. *Lipids*, **27**, 234–239. 1992.
115. Yamamoto. S. Enzymatic lipid peroxidation, reactions of mammalians lipoxygenases. *Free Radical Biology and Medicine*, **10**, 149–159. 1991.
116. Canfield, M.L. and Valenzuela, J.G. Cooxidations: Significance to carotenoid action in vivo. *Annals of the New York Academy of Sciences*, **691**, 191–199. 1993.
117. Frankel, E.N. Recent advances in lipid oxidation. *Journal of the Science of Food and Agriculture*, **54**, 495–511. 1991.
118. Min, B.D. and Bradley, D.G. Fat and oil flavors. *Encyclopedia of Food Science and Technology*. Wiley, New York. 1992.
119. Akoh, C.C. and Min, B.D. *Food Lipid Chemistry, Nutrition and Biotechnology*. Marcel Dekker, New York. 1997.
120. Dickinson, E. *Introduction to Food Colloids*. Oxford University Press, Oxford, U.K. 1997.
121. Friberg, E. and Larsson K. *Food Emulsions*, 3rd edition. Marcel Dekker, New York. 1996.
122. Walstra, P. Formation of emulsions. *Encyclopedia of Emulsion Technology*. Marcel Dekker, New York. 1996.
123. McClements, J.D. Lipid based emulsions and emulsifiers (eds. C.C. Akoh and B.D. Min), *Food Lipid Chemistry, Nutrition and Biotechnology*. Marcel Dekker, New York. 1997.
124. Pandolfe, W.D. Effect of premix conditions, concentration and oil level on the formation of o/w emulsions by homogenization. *Journal of Dispersion Science and Technology*, **16**, 633–668. 1995.
125. Poriginelbi, I., Nawar, W.W., and Chinchoti, P. Oxidation of linoleic acid in emulsions. Effects of substrate, emulsifiers, and sugar concentration. *Journal of the American Oil Chemists' Society*, **76**, 131–136. 1999.
126. Coupland, I.N. and McClements, J.D. Lipid oxidation in food emulsions. *Trends in Food Science and Technology*, **7**, 83–91. 1996.
127. Labuza, T.P. Kinetics of lipid oxidation in foods. *Critical Reviews in Food Technology*, **2**, 355–405. 1971.
128. Barkley, L.R.C. and Vingst, M.R. Membrane peroxidation: Inhibiting effects of water soluble antioxidants on phospholipids of different charge types. *Free Radical Biology and Medicine*, **16**, 779–788. 1994.

129. Frankel, E.N., Huang, W., Kanner, J., and German, B. Interfacial Phenomena in the evaluation of antioxidants: bulk oils vs emulsions. *Journal of Agriculture and Food Chemistry*, **42**, 1054–1059. 1994.

130. Tessa. J., Land, J., Shcalch, W., Truscott, G., and Tinkler, H.J. Interactions between carotenoids and the CCl_3O_2 radical. *American Chemistry Society*, **117**, 8322–8326. 1995.

131. Bast, A., Haanen, G.R., and VandenBerg, H. Antioxidant effects of carotenoids. *International Journal for Vitamin and Nutrition Research*, **68**, 399–403. 1998.

132. Soffers, A.M.F., Boerma, M.G., Laanre, C., and Rietjens, I.M. Antioxidant activities of carotenoids: Quantitative relationships between theoretical calculations and experimental. *Free Radical Research*, **30**, 233–240. 1999.

133. Mordi, R.C. Carotenoids-function and degradation. *Chemistry and Industry*, **110**, 79–83. 1993.

134. Kennedy, T.A. and Liebler, D.C. Peroxyl radical scavenging by β-carotene in lipids bilayers. *Journal of Biological Chemistry*, **267**, 4658–4662. 1992.

135. Burton, W.G. and Ingold, K.U. Beta carotene: An unusual type of lipid antioxidant. *Science*, **224**, 569–573. 1984.

136. Burton, W.G. Antioxidant action of carotenoids. *British Journal of Nutrition*, **119**, 109–111. 1988.

137. Edge, R., Truscoot, T.G., and McGarvey, D.J. The carotenoids as antioxidants—A review. *Journal of Photochemisrty and Photobiology*, **41**, 89–200. 1997.

138. Kasaikina, O.T., Kartaseva, Z.S., Lobanova, T.V., and Sirota, T.V. Effect of environmental factors on the β-carotene reactivity toward oxygen and free radicals. *Biologiche Membrany*, **15**, 168–176. 1998.

139. Woodall, A.A., Britton, G., Jackson, M.J., and Weesie, S.W.M. Oxidation of carotenes by free radicals-relationship between structure and reactivity. *Biochemica et Biophysica Acta Subjects*, **1336**, 33–42. 1997.

140. Britton, G. Structure and properties of carotenes in relation to function. *FASEB Journal*, **15**, 1551–1558. 1995.

141. Woodall, A.A., Britton, G., and Jackson, M.J. Dietary supplementation with carotenoids: Effects on α-tocopherol levels and susceptibility of tissues to oxidative stress. *British Journal of Nutrition*, **76**, 307–317. 1996.

142. Halliwell, B., Gutteridge, J.M., and Gross, C.E. Free radical antioxidants and human disease. Where are we now? *Journal of Laboratory and Clinical Medicine*, **119**, 598–620. 1992.

143. Mortensen, A. and Skibsted, L.H. Relative stability of carotenoid radical cations and homologues tocopheroxyl radicals. *FEBS Letters*, **417**, 261–266. 1997.

144. Mortensen, A., Skibsted, L.H, Sampson, J., and Everett, A.S. Comparative mechanisms and rates of free radical scavenging by carotenoid antioxidants. *FEBS Letters*, **418**, 91–97. 1997.

145. Terao, J. Antioxidant activity of β-carotene-related carotenoids in solution. *Lipids*, **24**, 659–663. 1989.

146. Mortensen, A. and Skibsted, L.H. Reactivity of β-carotene towards peroxyl radicals studied by later flash state photolysis. *FEBS Letters*, **426**, 392–396. 1998.

147. Haila, K.M., Nielsen, B.R., Heinonen, M.I., and Skibsted, L.H. Carotenoid reaction with free radicals in acetone and toluene at different oxygen partial pressures. *Zeitschrift fur Lebensmitteln*, **204**, 81–87. 1997.

148. Yamauchi, R., Miyake, N., and Kato, K. Products formed by peroxyl radical mediated initiated oxidation of canthaxanthin, in benzene and in methyl linoleate. *Journal of Agriculture and Food Chemistry*, **41**, 708–713. 1993.

149. Palozza, P. and Krinsky, N.I. β-Carotene and α-tocopherol are synergistic antioxidants. *Archives of Biochemistry and Biophysics*, **297**, 184–187. 1992.

150. Palozza, P. Pro-oxidant actions of carotenoids in biological systems. *Nutrition Reviews*, **56**, 257–265. 1998.

151. Niki, E., Noguchi, N., Tsuchihashi, H., and Goton, N. Interactions among vitamin E, vitamin C, and β-carotene. *American Journal of Clinical Nutrition*, **62**, 1322s–1326s. 1995.

152. Farombi, E.O. and Burton, G. Antioxidant activity of palm oil carotens in organic solution-effect of structural and chemical reactivity. *Food Chemistry*, **64**(3), 315–321. 1999.

153. Henry, L.K., Gatignani, G.L., and Scwharz, S. The influence of carotenoids and tocopherols on the stability of safflower seed oil during heat-catalyzed oxidation. *Journal of the American Oil Chemists' Society*, **75**, 1399–1402. 1998.

154. Terao, J. Inhibitory effects of carotenoids on singlet oxygen-initiated photooxidation of methyl-linolate and soybean oil. *Journal of Food Processing Preservation*, **4**, 79–93. 1990.

155. Dimascio, P., Sunduist, A., and Sies, H. Assay of lycopene and other carotenoids as singlet oxygen quenchers. *Methods in Enzymology*, **213**, 40. 1992.

156. Bradley, D.G. and Min, B.D. Singlet oxygen oxidation in foods. *Critical Reviews in Food Science and Nutrition*, **31**, 216–218. 1992.
157. Reisch, W.D., Dorris, A., and Eitenmiller, R. Antioxidants (eds. C.C. Akoh and B.D. Min), *Food Lipid Chemistry, Nutrition and Biotechnology*. Marcel Dekker, New York. 1997.
158. Halliwell, B. and Gutteridge, J. *Free radicals in Biology and Medicine*, 2nd edition. Clarendon Press, Oxford, U.K. 1995.
159. Lee, H.S. and Min, B.D. Effects, quenching mechanisms, and kinetics of carotenoids in chlorophyll-sensitized photo-oxidation of soyabean. *Journal of Agriculture and Food Chemistry*, **38**, 1630–1634. 1990.
160. Chen, J.H., Lee, T.C., and Ho, C.T. Antioxidant effect and kinetics study of capsanthin on the chlorophyll-sensitized photooxidation of soybean oil and selected flavor compounds. *ACS Symposium Series*, **660**, 188–198. 1997.
161. Nielsen, B.R., Mortensen, A., Jorgenesen, K., and Skibsted, L.H. Singlet versus triplet reactivity in photodegradation of C_{40} carotenoids. *Journal of Agriculture and Food Chemistry*, **44**, 2106–2113. 1997.
162. Eitenmiller, R. and Laden, W.O. Vitamins (eds. J.J. Jern and W.C. Ikins), *Analyzing Food for Nutrition Labeling and Hazardous Contaminants*. Marcel Dekker, New York. 1995.
163. Schuler, P. Natural antioxidants exploited commercially (ed. F.B. Hudson), *Food Antioxidants*. Elsevier Applied Sciences, London, U.K. 1990.
164. Behrens, W.A. and Madere, R. Interrelationship and competition of α-, and γ-tocopherols at the level of intestinal absorption, plasma transport and liver uptake. *Nutrition Research*, **45**, 891–897. 1983.
165. Frankel, E.N., Huang, W.S., Aeschabach, R., and Prior E. Antioxidant activity of a rosemary extract and its constituents, carnosic acid, carnosol, and rosemarin acid, in bulk oil and oil-in-water emulsion. *Journal of Agriculture and Food Chemistry*, **44**, 131–35. 1996.
166. Kamal-Eldin, A. and Appelwist, L.A. The chemistry and antioxidant properties of tocopherols and tocotrienols. *Lipids*, **31**, 671–675. 1996.
167. Gottstein, T. and Gross, W. Model study of different antioxidant properties of α- and γ- tocopherols in fats. *Food Science and Technology*, **92**, 139–144. 1990.
168. Huang, W., Frankel, E.N., and German, B. Interfacial phenomena in the evaluation of antioxidants: Bulk oils vs emulsions. *Journal of Agriculture and Food Chemistry*, **42**, 1054–1059. 1994.
169. Jung, M.Y. and Min, D.B. Effects of α-, γ-, and δ-tocopherols on oxidative stability of soybean oils. *Journal of food Science*, **55**, 1464–1465. 1992.
170. Pryor, W.A., Cornicall, J.A., Dorall, L.I., and Tait, B. A rapid screening test to determine the antioxidant activity of natural and synthetic antioxidants. *Journal of Organic Chemistry*, **58**, 3521–3522. 1993.
171. Marinova, E.M. and Yanislieva, N.V. Effect of temperature on antioxidant action of inhibitors in lipid autoxidation. *Journal of the Science of Food and Agriculture*, **60**, 313–318. 1992.
172. Frankel, E.N. Antioxidants in lipid foods and their impact on food quality. *Food Chemistry*, **57**, 51–56. 1996.
173. Johnson, E.L. Food Technology of the antioxidant nutrients. *Critical Reviews in Food Science and Nutrition*, **35**, 149–153. 1995.
174. Buettner, G.R. and Moseley, P.L. EPR spin trapping of free radicals produced by bleomycin and ascorbate. *Free Radical Research Communications*, **19**, 589–593. 1993.
175. Madhavi, L.D., Singhal, S.R., and Kulkavni, R.P. Technological aspects of foods antioxidants (ed. L.D. Madhavi), *Food Antioxidants: Technological Toxicological and Health Perspectives*. Marcel Dekker, New York. 1996.
176. Porter, L.W. Recent trends in food application of antioxidants (eds. Simic et al.), *Autoxidation in Food and Biological Systems*. Plenum Press, New York. 1980.
177. Ueda, S., Hayashi, T., and Namiki, M. Effect of ascorbic acid on lipid autoxidation in a model food system. *Agricultural and Biological Chemistry*, **50**, 1–7. 1986.
178. Stait, E.S. and Leake, S.D. Ascorbic acid can either increase or decrease LDL modification. *FEBS Letters*, **34**, 263–267. 1994.
179. Loliger, J. Natural antioxidants. *Lipid Technology*, **3**, 58–61. 1993.
180. Tsimidou, M. Phenolic compounds and stability of virgin oil. Part-1. *Food Chemistry*, **45**, 141–144. 1992.
181. Aparicio, R. and Morales, M.T. Relationship between volatile compounds and sensory attributes of olive oil. *Journal of the American Oil Chemists' Society*, **73**, 1253–1264. 1996.
182. Blekas, G. and Bosku, D. Antioxidative activity of 3,4, dihydrophenyl acetic acid and α-tocopherol in the triglyceride matrix of olive oil. Effect of acidity. *Grasas y aceites*, **49**, 34–37. 1998.
183. Chimi, H., Gillard, J., Gillard, P., and Pahman, M. Peroxyl and hydroxyl radical scavenging activity of some natural antioxidants. *Journal of the American Oil Chemists' Society*, **68**, 307–312. 1991.

184. Cuvelier, M.E., Richarld, H., and Berset, C. Comparison of the antioxidant action of some acid-phenols-structure/activity relationships. *Bioscience Biotechnology and Biochemistry*, **56**, 324–335. 1994.

185. Chen, J.H., Lee, T.C., and Ho, C.T. Antioxidant activities of caffeic acid and its related hydrocynamic acid compounds. *Journal of Agriculture and Food Chemistry*, **45**, 2374–2378. 1997.

186. Directive 76/621/EEC. 2008. Available at: http://eur-lex.europa.eu/LexUriServ/LexUriServ.do?uri=CELEX:31976L0621:EN:HTML (accessed 25/2/2008).

187. Directive 96/3/EC. 2008. Available at: http://europa.eu.int/smartapi/cgi/sga_doc?smartapi!celexplus!prod!DocNumber&1g=en&type_doc=Directive&an_doc=96&nu_doc=3 (accessed 25/2/2008).

188. Regulation No. 136/66/EEC. 2009. Available at: http://europa.eu.int/smartapi/cgi/sga_doc?smartapi!celexplus!prod!DocNumber&1g=en&type_doc=Regulation&an_doc=66&nu_doc=136 (accessed 25/2/2009).

189. Regulation (EEC) No 2568/91. 2009. Available at: http://europa.eu.int/smartapi/cgi/sga_doc?smartapi!celexplus!prod!DocNumber&1g=en&type_doc=Regulation&an_doc=91&nu_doc=2568 (accessed 25/2/2009).

190. Regulation (EC) No.2991/94. 2009. Available at: http://europa.eu.int/smartapi/cgi/sga_doc?smartapi!celexplus!prod!DocNumber&1g=en&type_doc=Regulation&an_doc=94&nu_doc=2991 (accessed 25/2/2009).

191. Regulation (EC) No.2815/98. 2009. Available at: http://europa.eu.int/smartapi/cgi/sga_doc?smartapi!celexplus!prod!DocNumber&1g=en&type_doc=Regulation&an_doc=98&nu_doc=2815 (accessed 25/2/2009).

192. Regulation (EC) No.1019/2002. 2009. Available at: http://europa.eu.int/smartapi/cgi/sga_doc?smartapi!celexplus!prod!DocNumber&1g=en&type_doc=Regulation&an_doc=2002&nu_doc=1019 (accessed 25/2/2009).

193. International agreement on olive oil and table olives. 2009. Available at: http://r0.unctad.org/commodities/agreements/oliveen.pdf (accessed 25/2/2009).

194. International agreement on olive oil and table olives. 2009. Available at: http://www.amazon.com/gp/product/9211122791/103–4320927–0003049?v=glance&n=283155 (accessed 25/2/2009).

195. Edible oil products Act. 2009. Available at: http://www.e-laws.gov.on.ca/DBLaws/RepealedStatutes/English/90e01_e.htm (accessed 25/2/2009).

196. Edible oil products Act. 2009. Available at: http://www.canlii.org/on/laws/sta/e-1/20041201/whole.html (accessed 25/2/2009).

197. Christie, W.W. *Advances in Lipid Methodology—Two*. Oily Press, Ayr, Scotland. 1993.

198. Lawson, H.W. *Food Oils and Fats: Technology, Utilization and Nutrition*, 2nd edition. Chapman & Hall, New York, 1994.

199. Aparicio, R., Calvente, J.J., and Morales, M.T. Sensory authentication of European extra-virgin olive oil varieties by mathematical procedures. *Journal of the Science of Food and Agriculture*, **72**, 435–447. 1996.

200. Kiritsakis, A. and Christie, W.W. Analysis of edible oils (eds. J. Harwood and R. Aparicio), *Handbook of Olive Oil: Analysis and Properties*. Aspen Publishers, Gaithersburg, MD, pp. 129–158. 2000.

201. Kiritsakis, A., Kanavouras, A., and Kiritsakis, K. Chemical analysis, quality control and packaging issues of olive oil. *European Journal of Lipid Science and Technology*, **104**, 628–638. 2002.

202. Ulberth, F. and Buchgraber, M. Authenticity of fats and oils. *European Journal of Lipid Science and Technology*, **102**, 687–694. 2000.

203. Spangenberg, J.E. Authentication of vegetable oils by bulk and molecular carbon isotope analyzes with emphasis on olive oil and pumpkin seed oil. *Journal of Agriculture and Food Chemistry*, **49**, 1534–1540. 2001.

204. McKenzie, J.M. and Koch, K.R. Rapid analysis of major components and potential authentication of South African olive oils by quantitative ^{13}C nuclear magnetic resonance spectroscopy. *South African Journal of Science*, **100**, 349–354. 2004.

205. Yang, H., Irudayaraj, J., and Paradkar, M.M. Discriminant analysis of edible oils and fats by FTIR, FT-NIR and FT-Raman spectroscopy. *Food Chemistry*, **93**, 25–32. 2005.

206. Downey, G., McIntyre, P., and Davis, A.N. Detecting and quantifying sunflower oil adulteration in extra virgin olive oils from the eastern Mediterranean by visible and near infrared spectroscopy. *Journal of Agriculture and Food Chemistry*, **50**, 5520–5525. 2002.

207. Tay, A., Singh, R.K., Krishnan, S.S., and Gore, J.P. Authentication of olive oil adulterated with vegetable oils using Fourier transform infrared spectroscopy. *Zeitschrift fur Lebensmittel Untersuchung und Forschung*, **35**, 99–103. 2002.

208. Lai, Y.W., Kemsley, E.K., and Wilson, R.H. Potential of FTIR spectroscopy for the authentication of vegetable oils. *Journal of Agriculture and Food Chemistry*, **42**, 1154–1159. 1994.

209. Lai, Y.W., Kemsley, E.K., and Wilson, R.H. Quantitative analysis of potential adulterants of extra virgin olive oil using infrared spectroscopy. *Food Chemistry*, **53**, 95–98. 1995.

210. Aparicio, R., Morales, M.T., and Alonso, M.V. Authentication of European virgin olive oils by their chemical compounds, sensory attributes, and consumers' attitudes. *Journal of Agriculture and Food Chemistry*, **45**, 1076–1083. 1997.

211. Marquez-Ruiz, G., Jorge, N., Martin-Polvillo, M., and Dobarganes, M.C. Rapid, quantitative determination of polar compounds in fats and oils by solid-phase extraction and size-exclusion chromatography using monostearin as internal standard. *Journal of Chromatography A*, **749**, 55–60. 1996.

212. Flores, G., Castillo, M.L.R., Herraiz, M., and Blanch, G.P. Study of the adulteration of olive oil with hazelnut oil by on-line coupled high performance liquid chromatographic and gas chromatographic analysis of filbertone. *Food Chemistry*, **97**(4), 742–749. 2006.

213. Cunha, S.C. and Oliveira, M.B.P.P. Discrimination of vegetable oils by triacylglycerols evaluation of profile using HPLC/ELSD. *Food Chemistry*, **95**(3), 518–524. 2005.

214. Sikorska, E., Gorecki, T., Khmelinskii, I.V., Sikorski, M., and Kozio, J. Classification of edible oils using synchronous scanning fluorescence spectroscopy. *Food Chemistry*, **89**, 217–225. 2005.

215. Marcos Lorenzo, I., Perez-Pavon, J.L., Fernandez-Laespada, M.E., Garcia-Pinto, C., and Moreno-Cordero, B. Detection of adulterants in olive oil by headspace–mass spectrometry. *Journal of Chromatography A*, **945**, 221–230. 2002.

216. Aparicio, A. and Luna, G. Characterisation of monovarietal virgin olive oils. *European Journal of Lipid Science and Technology*, **104**, 614–627. 2002.

217. Koprivnjak, O., Moret, S., Populin, T., Lagazio, C., and Conte, C.S. Variety differentiation of virgin olive oil based on n-alkane profile. *Food Chemistry*, **90**, 603–608. 2005.

218. Guadarrama, A., Rodriguez-Mendez, M.L.R., Sanz, C., Saja, J.A., and Ros, J.L. Electronic nose based on conducting polymers for the quality control of the olive oil aroma Discrimination of quality, variety of olive and geographic origin. *Analytica Chimica Acta*, **432**, 283–292. 2001.

219. Downey, G., McIntyre, P., and Davies, A.N. Geographic classification of extra virgin olive oils from the eastern Mediterranean by chemometric analysis of visible and near infrared spectroscopic data. *Applied Spectroscopy*, **57**, 158–163. 2003.

220. Buchgraber, M., Ulberth, F., Emons, H., and Anklam, E. Triacylglycerol profiling by using chromatographic techniques. *European Journal of Lipid Science and Technology*, **106**, 621–648. 2004.

221. Muik, B., Lendl, B., Molina-Diaz, A., and Ayora-Canada, M.J. Direct, reagent-free determination of free fatty acid content in olive oil and olives by Fourier transform Raman spectrometry. *Analytica Chimica Acta*, **487**, 211–220. 2003.

222. Senorans, F.J. and Ibanez, E. Analysis of fatty acids in foods by supercritical fluid chromatography. *Analytica Chimica Acta*, **465**, 131–144. 2002.

223. Meier-Augenstein, W. Stable isotope analysis of fatty acids by gas chromatography-isotope ratio mass spectrometry. *Analytica Chimica Acta*, **465**, 63–79. 2002.

224. Ranalli, A., Pollastri, L., Contento, S., Di Loreto, G., Iannucci, E., and Lucera, L. Acylglycerol and fatty acid components of pulp, seed, and whole olive fruit oils. Their use to characterize fruit variety by chemometrics. *Journal of Agriculture and Food Chemistry*, **50**, 3775–3779. 2002.

225. Servili, M. and Montedoro, G. Contribution of phenolic compounds to virgin olive oil quality. *European Journal of Lipid Science and Technology*, **104**, 602–613. 2002.

226. Angerosa, F., Mostallino, R., Basti, C., Vito, R., and Serraiocco, A. Virgin olive oil differentiation in relation to extraction methodologies. *Journal of the Science of Food and Agriculture*, **80**, 2190–2195. 2000.

227. Bonoli, M., Mantanucci, M., Toschi, T.G., and Lercker, G. Fast separation and determination of tyrosol, hydroxytyrosol and other phenolic compounds in extra-virgin olive oil by capillary zone electrophoresis with ultraviolet-diode array detection. *Journal of Chromatography A*, **1011**, 163–172. 2003.

228. Brenes, M., Garcia, A., Garcia, P., and Garrido, A. Rapid and complete extraction of phenols from olive oil and determination by means of a coulometric electrode array system. *Journal of Agriculture and Food Chemistry*, **48**, 5178–5183. 2000.

229. Liberatore, L., Procida, G., D'Alessandro, N., and Cichelli, A. Solid-phase extraction and gas chromatographic analysis of phenolic compounds in virgin olive oil. *Food Chemistry*, **73**, 119–124. 2001.

230. Kiritsakis, A.K. Flavor components of olive oil—A review. *Journal of the American Oil Chemists Society*, **75**(6), 673–681. 1998.

231. Kalua, C.M., Allen, M.S., Bedgood, D.R., Bishop, A.G., Prenzler, P.D., and Robards, K. Olive oil volatile compounds, flavor development and quality: A critical review. *Food Chemistry*, **100**(1), 273–286. 2007.

232. Vichi, S., Castellote, A.I., Pizzale, L., Conte, L.S., Buxaderas, S., and Lopez-Tamames, E. Analysis of virgin olive oil volatile compounds by headspace solid-phase microextraction coupled to gas chromatography with mass spectrometric and flame ionization detection. *Journal of Chromatography A*, **983**, 19–33. 2003.

233. Blanch, G.P., Gaja, M.M., Ruiz del Castillo, M.L., Herraiz, M., and Martinez Alvarez, R. Analysis of volatile compounds in edible oils using simultaneous distillation-solvent extraction and direct coupling of liquid chromatography with gas chromatography. *European Journal on Food Research and Technology*, **211**, 45–51. 2000.

234. Stella, R., Barisci, J.N., Serra, G., Wallace, G.G., and De Rossi, D. Characterisation of olive oil by an electronic nose based on conducting polymer sensors. *Sensors and Actuators: B Chemical*, **63**, 1–9. 2000.

235. Minguez-Mosquera, M.I., Gandul-Rojas, B., Garrido-Fernandez, J., and Gallardo-Guerrero, L. Pigments present in olive oil. *Journal of the American Oil Chemists Society*, **67**, 192–196. 1990.

236. Minguez-Mosquera, M.I., Rejano, L., Gandul, B., Sanchez, A.H., and Carrido, J. Colour-pigment correlation in virgin olive oil. *Journal of the American Oil Chemists Society*, **68**, 332–336. 1991.

237. Gandul-Rojas, B. and Minguez-Mosquera, M.I. Chlorophyll and carotenoid composition in virgin olive oil from various Spanish varieties. *Journal of the Science of Food and Agriculture*, **72**, 31–39. 1996.

238. Gandul-Rojas, B., Roca-Lopez Cepero, M., and Minguez-Mosquera, M.I. Use of chlorophyll and carotenoid pigment composition to determine authenticity of virgin olive oil. *Journal of American Oil Chemists Society*, **77**(8), 853–858. 2000.

239. Puspitasari-Nienaber, N.L., Ferruzzi, M.G., and Schwarz, S.J. Simultaneous detection of tocopherols, carotenoids, and chlorophylls in vegetable oils by direct injection C-30 RPHPLC with coulometric electrochemical array detection. *Journal of the American Oil Chemists Society*, **79**(7), 633–640. 2002.

240. Buldini, P.L., Ferri, D., and Lal Sharma, J. Determination of some inorganic species in edible vegetable oils and fats by ion chromatography. *Journal of Chromatography A*, **789**(1–2), 549–555. 1997.

241. Jimenez, M.S., Velarte, R., and Castillo, J.R. On-line emulsions of olive oil samples and ICP-MS multi-elemental determination. *Journal of Analytical Atomic Spectrometry*, **18**(9), 1154–1162. 2003.

242. Anthemidis, A., Arvanitidis, V., and Stratis, J.A. On-line emulsion formation and multi-element analysis of edible oils by inductively coupled plasma atomic emission spectrometry. *Analytica Chimica Acta*, **537**(1–2), 271–278. 2005.

243. Iskander, F.Y. Determination of 17 elements in edible oils and margarine by instrumental neutron-activation analysis. *Journal of the American Oil Chemists Society*, **70**(8), 803–805. 1993.

244. Galeano Diaz, T., Guiberteau, A., Lopez Soto, M.D., and Ortiz, J.M. Determination of copper with 5,5-dimethylcyclohexane- 1,2,3-trione 1,2-dioxime 3-thiosemicarbazone in olive oils by adsorptive stripping square wave voltammetry. *Food Chemistry*, **96**(1), 156–162. 2006.

245. De Leonardis, A., Macciolo, V., and De Felice, M. Copper and Iron determination in edible vegetable oils by graphite furnace atomic absorption spectrometry after extraction with diluted nitric acid. *International Journal of Food Science and Technology*, **35**(4), 371–375. 2000.

246. Guimet, F., Boque, R., and Ferre, J. Application of unfold principal component analysis and parallel factor analysis to the exploratory analysis of olive oils by means of excitation-emission matrix fluorescence spectroscopy. *Analytica Chimica Acta*, **515**, 75–85. 2004.

247. Cerrato Oliveros, C., Boggia, R., Casale, M., Armanino, C., and Forina, M. Optimisation of a new headspace mass spectrometry instrument. Discrimination of different geographical origin olive oils. *Journal of Chromatography A*, **1076**, 7–15. 2005.

248. Vlahov, G., Shaw, A.D., and Kell, D.B. Quantitative ^{13}C NMR method using the DEPT pulse sequence for the determination of the geographical origin (DOP) of olive oils. *Magnetic Resonance in Chemistry*, **39**, 689–695. 2001.

249. Arvanitoyannis, I.S. and Vlachos, A. Implementation of physicochemical and sensory analysis in conjunction with multivariate analysis towards assessing olive oil authentication/adulteration. *Critical Reviews in Food Science and Nutrition*, **47**, 441–498. 2007.

250. Caporale, G., Policastro, S., Carlucci, A., and Montelone, E. Consumer expectations for sensory properties in virgin olive oils. *Food Quality and Preference*, **17**, 116–125. 2006.

251. Reiners, J. and Grosch, W. Odorants of virgin olive oils with different flavor profiles. *Journal of Agricultural and Food Chemistry*, **46**, 2754–2763. 1998.

252. Busconi, M., Foroni, C., Corradi, M., Bongiorni, C., Cattapan, F., and Fogher, C. DNA extraction from olive oil and its use in the identification of the production cultivar. *Food Chemistry*, **83**, 127–134. 2003.

253. Breton, C., Claux, D., Metton, I., Skorski, G., and Berville, A. Comparative study of methods for DNA preparation from olive oil samples to identify cultivar SSR alleles in commercial oil samples: Possible forensic applications. *Journal of Agriculture and Food Chemistry*, **52**, 531–537. 2004.

254. Pafundo, S., Agrimont, C., and Marmiroli, N. Traceability of plant contribution in olive oil by amplified fragment length polymorphisms. *Journal of Agriculture and Food Chemistry*, **53**, 6995–7002. 2005.

255. Fabbri, D., Baravelli, V., Chiavari, G., and Prati, S. Profiling fatty acids in vegetable oils by reactive pyrolysis-gas chromatography with dimethyl carbonate and titanium silicate. *Journal of Chromatography A*, **1083**, 52–57. 2005.

256. Carrasco-Pancorbo, A., Segura-Carretero, A., and Fernandez-Gutierrez, A. Co-electroosmotic capillary electrophoresis determination of phenolic acids in commercial olive oil. *Journal of Separation Science*, **28**(9–10), 925–934. 2005.

257. Stefanoudaki, E., Kotsifaki, F., and Koutsaftaki, A. Sensory and chemical profiles of three European olive varieties (*Olea europea* L.): An approach for the characterization and authentication of the extracted oils. *Journal of the Science of Food and Agriculture*, **80**, 381–389. 2000.

258. Kelly, S.D. and Rhodes, C. Emerging techniques in vegetable oil analysis using stable isotope ratio mass spectrometry. *Grasas y Aceites*, **53**(1), 34–44. 2002.

259. Woodbury, S.E., Evershed, R.P., Rossell, J.B., Griffith, R.E., and Farnell, P. Detection of vegetable oil adulteration using gas chromatography/isotope ratio mass spectrometry. *Analytical Chemistry*, **67**, 2685–2690. 1995.

260. Zhang, G., Ni, Y., Churchill, J., and Kokot, S. Authentication of vegetable oils on the basis of their physico-chemical properties with the aid of chemometrics. *Talanta*, **70**(2), 293–300. 2006.

261. Lee, D.S., Lee, E.S., Kim, H.J., Kim, S.O., and Kim, K. Reversed phase liquid chromatographic determination of triacylglycerol composition in sesame oils and the chemometric detection of adulteration. *Analytica Chimica Acta*, **429**(2), 321–330. 2001.

262. Mitra, S.N., Roy, B.R., and Sengupta, P.N. Detection of mustard oil in other edible oils. *Current Science*, **27**(6), 221. 1958.

263. Benn, M.H. A new mustard oil glucoside synthesis: The synthesis of glucotropaeolin. *Canadian Journal of Chemistry*, **41**(11), 2836–2838. 1963.

264. Ministry of Health and Family Welfare Government of India. Manual of methods of analysis of foods—Oils and fats. 2009. Available at: http://mohfw.nic.in/Methods%20of%20Analysis%20-2005.%20oils%20and%20fats%20-%20Sep%2005.pdf (accessed 5/3/2009).

265. Biino, L. and Carlisi, E. Research on illipe fat and cocoa butter. *Rivista Italiana Sostanze Grasse*, **48**(4), 170–176. 1971.

266. Derbesy, M. and Richert, M.T. Detection of shea butter in cocoa butter. *Oleagineux*, **34**(8/9), 405–409. 1979.

267. Homberg, E. and Bielefeld, B. Main components of 4-methyl-sterol and triterpene fraction of twelve vegetable fats and their influence on sterol analysis. *Fat Science and Technology*, **92**(12), 478–480. 1990.

268. Kamm, W., Dionisi, F., Fay, L.B., Hischenhuber, C., Schmarr, H.G., and Engel, K.H. Analysis of steryl esters in cocoa butter by on-line liquid chromatography–gas chromatography. *Journal of Chromatography A*, **918**(2), 341–349. 2001.

269. Geeraert, E. and DeSchepper, D. Structure elucidation of triglycerides by chromatographic techniques. Part 2: RP HPLC of triglycerides. *Journal of High Resolution Chromatography and Chromatography Communications*, **6**(3), 123–132. 1983.

270. Jeffrey, B.S.J. Silver complexation liquid chromatography for fast high-resolution separations of triacylglycerols. *Journal of the American Oil Chemists Society*, **68**(5), 289–293. 1991.

271. Rossell, J.B. Purity criteria in edible oils and fats. *Fat Science and Technology*, **93**(4), 526–531. 1991.

272. Sassano, G.J. and Jeffrey, B.S.J. Gas chromatography of triacylglycerols in palm oil fractions with medium-polarity wide-bore columns. *Journal of the American Oil Chemists Society*, **70**(11), 1111–1114. 1993.

273. Gordon, M.H. and Griffith, R.E. A comparison of the steryl esters of coconut and palm kernel oils. *Fat Science and Technology*, **94**(6), 218–221. 1992.

274. Kelly, S., Parker, I., Sharman, M., Dennis, J., and Goodall, I. Assessing the authenticity of single seed vegetable oils using fatty acid stable carbon isotope ratios ($^{13}C/^{12}C$). *Food Chemistry*, **59**(2), 181–186. 1997.

275. Enser, M. Animal carcass fats and fish oils (eds. J.B. Rosssell and J.L.R. Pritchard), *Analysis of Oilseeds, Fats and Fatty Foods*. Elsevier Applied Science Publishers, London, U.K., pp. 329–394. 1991.

276. Krishnamurthy, M.N. Updated methods for detection of adulterants contaminants in edible oils and fats: A critical evaluation. *Journal of Food Science and Technology (India)*, **30**(4), 231–238. 1993.

277. Aida, A.A., Che Man, Y.B., Wong, C.M.V.L., Raha, A.R., and Son, R. Analysis of raw meats and fats of pigs using polymerase chain reaction for Halal authentication. *Meat Science*, **69**(1), 47–52. 2005.

278. Fontecha, J., Mayo, I., Toledano, G., and Juarez, M. Use of changes in triacylglycerols during ripening of cheeses with high lipolysis levels for detection of milk fat authenticity. *International Dairy Journal*, **16**(12), 1498–1504. 2006.

279. De Noni, I. and Resmini, P. Identification of rennet-whey solids in "traditional butter" by means of HPLC/ESI-MS of non-glycosylated caseinomacropeptide. *Food Chemistry*, **93**(1), 65–72. 2005.
280. Wolff, J.P. Development of the enzymatic determination for very long polyunsaturated chain cholesterol esters. *Revue Francaise des Corps Gras*, **10**, 187–195. 1963.
281. Verbeke, R. and Brabander, H.F. Proceedings of the European meeting of meat research workers. *Food Science and Technology*, **13**, 767–772. 1979.
282. Singhal, R.S., Kulkarni, P.R., and Rege, D.V. *Handbook of Indices of Food Quality and Authenticity.* Woodhead Publishing Limited, Cambridge, U.K., p. 345. 1997.
283. International Standard. *Animal and Vegetable Oils: Determination of Composition of the Sterol Fraction—Method Using Gas Chromatography*, p. 7. 1991.
284. Stull, J.W. and Brown, W.H. Fatty acid composition of milk. II. Some differences in common dairy breeds. *Journal of Dairy Science*, **47**, 1412. 1964.
285. Bauerlain, R.J. Fats and oils. Proceedings of the 61st annual meeting, ASBE, Chicago, IL, 1985.
286. Spizziri, J. Health plans for fats and oils. *Food Product Design*, 2004, http://www.foodproductdesign.com/articles/2004.

6 Nucleic Acid Biochemistry: Food Applications

Işıl A. Kurnaz and Çağatay Ceylan

CONTENTS

6.1 BASIC GENETIC TECHNIQUES

6.1.1 NUCLEIC ACID DETECTION AND ANALYSIS

Nucleic acids are composed of nucleotides, the building blocks, and essentially contain a five-carbon sugar, a phosphate group, and nitrogen-containing bases. There are two major groups of nucleic acids, DNA, or deoxyribonucleic acid; and RNA, or ribonucleic acid. DNA serves as the genetic material, in other words, genes are made up of DNA. The DNA molecule has four major building blocks, adenine (A), thymine (T), cytosine (C), and guanine (G), attached to a deoxyribose sugar. RNA is also composed of nucleotide building blocks, with one letter difference: it uses A, C, and G, but uracil (U) instead of thymine, all attached to a ribose sugar. Both DNA and RNA are polar compounds due to the many negative charges they contain in the phosphate groups of the sugar–phosphate backbone of each DNA and RNA strand [1].

The isolation of DNA, which is suitable for digestion with restriction enzymes (which will be discussed in Section 6.1.7), is an essential requirement for both genetic engineering techniques as well as for the analysis of genetic makeup of industrial products. The isolation of DNA from various organisms requires specific protocols, and there are various companies that provide extraction kits. One vital element to remember (if one wishes to study with pure DNA preparations) is the contamination of the sample with other nucleic acids, namely RNA, as well as other macromolecules such as protein and polysaccharides. In order to avoid this, most extraction protocols include RNase for the removal of RNA, and proteinase K enzyme to remove protein contaminants [2].

Once the DNA is obtained, however, there are various methods with which one analyzes the extracted nucleic acid. The most common method for visualizing and analyzing DNA is gel electrophoresis. A gel is a colloid in a solid form. The term "gel electrophoresis" refers to the migration of a charged molecule across a span of gel when placed in an electrical field. A molecule's properties determine how fast a molecule migrates through this electrical field, in a gelatinous environment. Since DNA molecules carry an overall negative charge due to the phosphate groups in their backbone, the molecule moves toward the positive electrode when placed into an electrical field in a rate inversely proportional to size: the smaller the DNA molecule, the faster it will move, thereby allowing for a size-based separation of the molecules. However, one must note that this property is very similar in RNA molecules (although secondary structures in RNA molecules or in single-stranded DNA molecules would affect migration in an electrical field); therefore, the DNA preparation must not contain RNA molecules for a correct analysis of the sample [2].

6.1.2 Polymerase Chain Reaction

Polymerase chain reaction, or PCR, is a commonly used technique that serves to amplify a specific DNA region. It involves *in vitro* amplification of new DNA strands from a template DNA such as genomic DNA, cDNA, or plasmid DNA, where in about 30 PCR reaction "cycles" or rounds, one generates 2^{30} new DNA "amplicons" from each template molecule as shown in Figure 6.1 [1–3].

Each PCR cycle consists of three main stages: (1) denaturation at around 95°C, (2) annealing of primers to the template (usually at around 45°C–60°C), and (3) elongation or polymerization at around 72°C by the enzyme *Taq* DNA polymerase.

Genetically modified organisms (GMOs) have been introduced to the food market based on the advances in genetic engineering technology; however, the requirement to inform the consumer about the origin of these food items brings about the question of how to analyze and screen them. PCR-based methods lie at the center of high-throughput GMO identification and

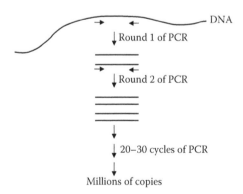

FIGURE 6.1 Polymerase chain reaction (PCR). The region of interest is amplified from a template DNA using specific primers (short horizontal arrows). After 20–30 rounds or cycles of PCR, the targeted DNA region is amplified 220–230 times.

quantification processes [4]. Real-time PCR methods are a recent addition to these techniques, where the polymerization reaction is monitored in real time, as the reaction progresses. A variety of systems have been devised by different companies in these past few years, including Taqman probes or Molecular beacons [2]. Essentially, real-time or quantitative PCR systems are based on the detection and quantitation of the amount of PCR products through a signal from a fluorescent molecule. One can measure the amount of fluorescence emitted at each cycle, thereby monitoring the progress of the reaction real-time at each cycle.

6.1.3 Gene Sequencing

In order to understand the function of a gene and improve its function through genetic modifications, the first place to start is the sequence of that gene. Researchers have devised methodologies to understand the base-by-base sequence of nucleotides in a given DNA molecule. The common techniques include the Maxam–Gilbert sequencing (not discussed in this chapter) and the Sanger method of sequencing that is mentioned below.

The basic principle of the Sanger method of DNA sequencing is essentially similar to the PCR technique. The DNA is denatured, and the DNA template (green strand in the above picture) is annealed to the sequencing primer (short red strand). The primer is extended using DNA polymerase and deoxynucleotides (in the 5′ to 3′ direction), until terminating nucleotides are "randomly" incorporated, and the polymerization reaction is stopped. The extended DNA polymers are then run through a sequencing gel, and individual fragments are visualized, and the DNA sequence is read [1–3].

In this method, the termination nucleotides used to stop the polymerization reaction are "dideoxynucleotides" (ddGTP, ddATP, ddCTP, and ddTTP). Since these nucleotides lack both oxygens in the ribose sugar, the sugar–phosphate bond cannot be formed, and the polymerization reaction terminates. In order to exploit this property to randomly terminate the sequencing reaction, the initial polymerization reaction is briefly carried out in the presence of a radioactively labeled deoxynucleotide (e.g., $^{35}SdCTP$) and deoxynucleotides (e.g., dATP, dTTP, and dGTP), and afterward the reaction is divided into four tubes, each containing only one type of dideoxynucleotide (ddATP tube, ddTTP tube, etc.). The reaction products in the four tubes are loaded onto different wells of the sequencing gel, exposed to an x-ray film, and visualized through autoradiography, and read as in Figure 6.2.

In "automated sequencing," the deoxynucleotides are conjugated to a special dye that fluoresces upon excitation with laser. Each of the four nucleotides have their own unique color, which means the reaction can be carried out in a single tube, as opposed to four tubes in manual sequencing. As the sequencing reaction passes through the gel, each fluorescence intensity is measured by a single peak, which is then interpreted by the computer and the sequence is thus generated. For convincing results, and to avoid any ambiguity, usually both the DNA strands are sequenced and analyzed by the computer [2].

6.1.4 Genome Projects

With the recent developments in molecular techniques, genomics has gained a substantial interest. Genomics is mainly concerned with the identification and analysis of the total genomic information in a given organism, and genome sequencing of commercially valuable crops such as rice has been completed [5] (The International Rice Genome Sequencing Project, 2007) while others such as corn, potato, tomato, and soybean are still ongoing [6].

The Institute for Genomic Research (TIGR; http://www.tigr.org/plantProjects.shtml) plant genome projects not only focus on the genome sequencing of commercially important crops such as *Arabidopsis*, maize, rice, and plant but also the sequencing of genomes of common plant pathogens.

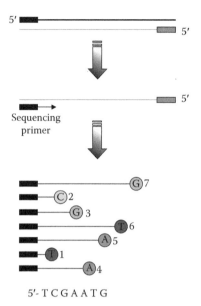

FIGURE 6.2 Sequencing reaction. The region of interest is denatured to generate a single-stranded template (light gray line; second line from top), and a sequencing primer (black bar) is used for the PCR-like elongation from this template. The elongation is randomly terminated as indicated by the letters at the end of each amplified molecule. The sequence is "read" as indicated below the figure.

It should also be noted at this stage—in relation to Section 6.1.7.2—that natural breeding of plants to generate a new plant variety versus transgenic plant production that is patentable are two separate events that need to be distinguished. Genome sequencing projects, and plant functional genomics which follows from it, would also prove useful in the identification of GMOs by comparison with the annotated genomes.

6.1.5 EXPRESSION PROFILING: DNA MICROARRAYS

While the genomic information is the same among all cells of the same organism, the set of genes that are expressed in different tissues can be vastly different. This differential expression is important for the adaptive response of cells and organisms under different conditions. Therefore, it is necessary to control the expression of these genes in the right cell type and at the right time. This elaborate control, called gene regulation, is achieved through certain DNA elements preceding each gene, called promoters, and illustrated in Figure 6.3 [1,3].

Microarray is a method commonly employed to study the differential regulation of genes, although this is by no means the only application. The standard method used for gene expression profiling is as follows: mRNA from different cells or tissues is isolated in order to compare for any alterations in transcriptional control. These mRNA samples are then reverse transcribed into complementary DNA (cDNA) molecules and, subsequently, labeled by different fluorescent dye molecules (commonly red and green fluorescent dyes for different mRNA samples). These cDNA samples are then mixed, and hybridized onto a microarray, either custom printed or commercially obtained. Upon incubation, the microarray slides are scanned for red and green fluorescence, either sequentially or simultaneously, depending on the company, and the fluorescence intensities are compared and analyzed for similarities or differences in the original mRNA amounts as shown in Figure 6.4 [2].

The microarray analysis can, for instance, be used to study the differences in gene expression in plants exposed to environmental stress, such as drought, cold, heat, or pathogens, and the information thus obtained could be used to design effective preventive measures to such environmental

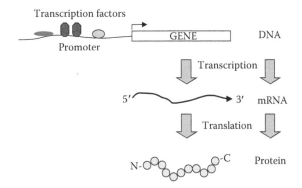

FIGURE 6.3 An overview of gene expression. A gene is typically regulated by an upstream promoter through the action of transcription factors, resulting in the synthesis of an mRNA from the DNA template in a process known as transcription. This mRNA will later on get translated into a protein through the action of ribosomes.

stresses to commercially important crops. The recent increase in rice prices due to water shortage is likely to have a tremendous effect on a large portion of the world population, and any preventive measure to increase the tolerance of rice to such environmental stress, or genetic engineering of rice so as to minimize water requirement without compromising on the nutritious quality, would be extremely beneficial to the world at large.

However, one must also note that microarray technology is not purely utilized to study gene expression profiling; it is also possible to use this technique to identify single nucleotide polymorphisms (SNPs), and also to identify GMOs through genome comparisons.

6.1.6 Protein Detection and Analysis

Proteins are perhaps the most diverse of all biological molecules, composed of building blocks called amino acids. There are 20 different amino acids that possess many ionizable groups, and, therefore, at any pH, they exist in solution as electrically charged species. Furthermore, in any given protein, these amino acids may be present in any combination. Due to this diversity, proteins serve many different functions in the cell, such as structural components of the cell, cellular recognition or cell-to-cell signaling, or as enzymes, carrying out biological reactions [3].

In order to study the function of a protein with the ultimate aim of understanding and improving its function, one must first isolate the protein in question, either as crude lysates from an appropriate source organism or as highly purified preparations through centrifugation, chromatography, and fractionation. Once the protein in question is obtained, the next step would be to analyze it using gel electrophoresis, Western blots or other means, depending on the ultimate objective.

Unlike DNA which possesses an overall negative charge due to the uniform distribution of negatively charged phosphate groups, the amino acid diversity among different proteins results in a difference in the overall charge of each protein, and depending on this charge the proteins may migrate either toward the cathode or the anode when placed in an electrical field. Also complicating the picture is the secondary and tertiary interactions in protein molecules that result in a three-dimensional shape of the protein, which affects the migration of the molecule through this gelatinous medium [2].

Therefore, unlike DNA gel electrophoresis, two kinds of protein gel electrophoresis are possible: native gel electrophoresis, where the net charge and shape of the protein contribute to the migration properties of the molecules, and denaturing gel electrophoresis, where all the secondary interactions have been disrupted and an overall negative charge has been introduced to the molecule, thereby allowing for a size-based separation. The latter is commonly known as an SDS-PAGE (sodium

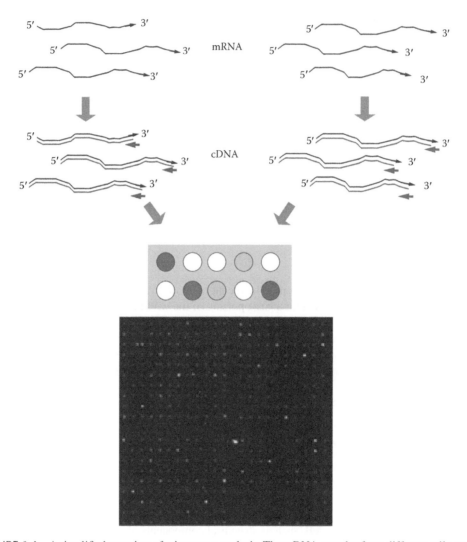

FIGURE 6.4 A simplified overview of microarray analysis. The mRNA samples from different cells are isolated and purified, and converted into cDNA through the reverse transcription reaction. The cDNAs are labeled with different fluorescent dyes, mixed and hybridized to the microarray slide, which contains unique DNA sequences representative of different genes on each spot. If the cDNA contains a complementary sequence to that on the spot, then the fluorescence intensity can be obtained through scanning. If the same cDNA appears on both red- and green-labeled samples, then the merged fluorescence images appear as yellow, indicating there is no difference in the expression level of that gene in these two cell types. (Courtesy of Caglayan and Kurnaz, unpublished data.)

dodecyl sulfate polyacrylamide gel electrophoresis). Protein samples are loaded onto an SDS-PAGE gel, "run" in an electrical field, and visualized through the use of the Coomassie blue dye.

Western blotting is another technique used to analyze proteins. The term "blotting" is used to describe the process of transferring molecules separated on a gel to a membrane surface for subsequent identification. Western blotting uses the antigen–antibody interaction specificity to detect proteins immobilized on membrane surfaces through the use of specific antibodies. The presence of the protein is indirectly determined by visualizing the specific antibodies on the membrane surface, since the antibodies used are conjugated to either enzymes (such as horseradish peroxidase) or chemiluminescent dyes [2].

6.1.7 Recombinant DNA Technology

Genetic engineering, alternatively called recombinant DNA technology, is the name given to all the techniques used in the laboratory-based manipulation of genes. The tools needed for genetic engineering or gene cloning have been identified since the past 20 years, the main components being the restriction enzymes and plasmid vectors [2].

Restriction enzymes are bacterial enzymes that recognize special DNA sequences and cut, or digest, the DNA to smaller fragments, as shown in Figure 6.5. They are present in the bacteria to actually prevent bacteriophages (or bacterial viruses) from infecting the bacteria, hence the name "restriction" [1–3].

Vectors are special DNA sequences used to "carry" the gene to be cloned. The most commonly used ones are derived from bacterial plasmids, or small circular DNA molecules replicating within the bacteria, independent of chromosomal replication [1–3].

Vectors are used to insert foreign DNA into the organism of study, be it bacteria, yeast, insect, or mammalian cells, where this newly engineered "recombinant DNA" can duplicate every time the host cell divides, as illustrated in Figure 6.6. If bacteria are used as host cells, millions and billions of copies can be generated in a short period. Since bacteria divide by binary fission and produce identical progeny, all the bacteria thus produced will contain exactly the same recombinant DNA molecule, hence these bacteria are referred to as clones or colonies, and this procedure is known as gene cloning.

Such cloning procedures can be complemented with transgenic plant or animal production techniques in order to improve the amount or quality of food matter, for example, metabolic engineering of the caffeine synthesis pathway in tea, which is currently being undertaken worldwide, by overexpressing a cloned enzyme is in order to either decrease or eliminate the caffeine amount [7]. Baker's yeast, *Saccharomyces cerevisiae*, for instance, has been engineered to express a cloned and improved version of the cyclodextrin glucanotransferase gene for enhancement of the baking process [8]. In other plant biotechnology studies, researchers from India have cloned and characterized most of the genes encoding legumin storage proteins in the chickpea [9].

6.1.7.1 Animal Biotechnology

Transgenic animals for commercially important protein production have been under scrutiny for a very long time. One of the first transgenic animals reported was sheep, in 1982, for the production of human growth hormone [10]. Today, the types of animals that can be genetically modified are essentially endless, from cattle to pigs, and from chicken to fish, and the traits that researchers and

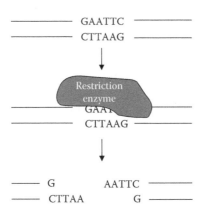

FIGURE 6.5 Restriction enzymes act at specific sequences (called restriction sites) and "digest" or cleave the double-stranded DNA into two separate fragments at the site of digestion.

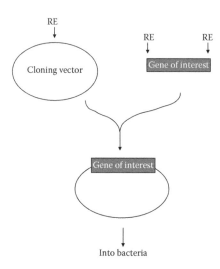

FIGURE 6.6 General cloning strategy. The gene of interest and the cloning vector are both digested with the same restriction enzyme(s) and ligated. The construct thus produced is then transferred into bacteria and amplified (the "cloning" step).

companies are trying to improve include the quality of the meat, the rate of growth, resistance to diseases, and production of a pharmaceutical, for instance insulin in milk.

The procedure for generating transgenic animals is in theory very simple—the egg and sperm are allowed to fertilize, and upon blastocyst formation, the inner cell mass (ICM) of the embryo (containing embryonic stem cells) is isolated. These cells are then transfected or microinjected with foreign DNA containing the gene(s) of interest, and the recombinant cells are afterward selected [11]. These selected, recombinant ES cells are then transferred back into the blastocyst, and thereafter the *in vitro*-manipulated embryo is transplanted into the surrogate mother. The babies born to the mother are thereafter analyzed for carrying the transgene, and those that have the transgene are selected and maintained [2].

Researchers today exploit this technology for a variety of purposes: many studies are ongoing for the production of recombinant human proteins in the mammary glands of transgenic livestock mammals, and the first marketing approval of one such recombinant protein was given in 2006 [12,13]. In other studies, transgenic pigs have been constructed for the production of the human complement regulatory protein, CD59, toward biomedical applications [14].

6.1.7.2 Plant Biotechnology

With the recent advances in recombinant DNA technology, it has long become possible, and even routine, to transform plants with foreign genes in order to improve yield and quality, or to increase herbicide- or stress resistance in agricultural crops, such as potato, rice, or wheat. Plant viruses, for instance, cause significant damage to commercially important crops, and to combat this researchers have developed virus-resistant transgenic plants. Similarly, certain plants may have an enzyme for the detoxification of a herbicide, which can be cloned in another commercially important crop to confer herbicide resistance. Alternatively, transgenic plants can be used as "bioreactors" to produce economically important molecules.

A naturally occurring phenomenon is the transfer of a stable Ti (tumor inducing) plasmid from the bacterium *A. tumafaciens* into the plant cell, resulting in an oncogenic transformation in the plant. Although the exact mechanism of this DNA transfer is still under study, derivatives of this plasmid have been largely used to introduce foreign DNA into plants. *Agrobacterium* carrying this recombinant plasmid can be directly applied onto a wound site, or *in vitro* with protoplasts or leaf discs. The leaf discs that are infected with the recombinant *Agrobacterium* are then propagated in a selection medium (such as one that contains kanamycin as a selection marker), and are analyzed for gene expression [2].

Such transgenic plants have been of extreme interest for biotechnology companies since the 1980s due to their economic impact: in theory, it is possible to produce genetically engineered plants with stress tolerance or improved nutritional value, or even with the production of high-value proteins such as certain pharmaceuticals. Indeed, perhaps the most famous of all these recombinant plants is the FLAVR-SAVR tomato developed by Calgene, Inc., Davis, California (reviewed by the U.S. Food and Drug Administration, Center for Food Safety and Applied Nutrition, 1994). This tomato was genetically engineered in 1990—by genetically engineering an antisense gene to the polygalactouronase enzyme under kanamycin resistance—to remain on the vine longer and ripen to full flavor before its harvest so that it survives its trip to the market without getting crushed. The general strategy until then was to harvest the tomatoes while they were green, and to induce the ripening by treating the tomatoes with ethylene gas on the shelf.

In more recent studies, tobacco plants have been genetically modified so as to induce osmotic stress tolerance through the overexpression of a gene called *osmotin* [15] and transgenic papaya have been generated, in Hawaii, in order to confer resistance to the papaya ringspot virus [16]. In yet other studies, *Arabidopsis thaliana* has been genetically engineered to express human insulin, which could be recovered from the seeds for pharmaceutical applications [17].

However, although the techniques for generating genetically engineered agricultural crops are largely established, one of the major concerns in this area has been the unintended spreading of these transgenic crops into the food supply, and seed dispersal [18]. In fact, one major company in the United States, ProdiGene Inc., was fined by the U.S. government for not taking effective measures to prevent its transgenic crop (producing a pharmaceutical) from entering the food supply [19].

6.2 APPLICATIONS IN FOOD SCIENCE AND TECHNOLOGY

6.2.1 LATERAL GENE TRANSFER

Since food is one of the materials human beings most frequently touch and internalize, the idea of material transfer, including the transfer of nucleic acids even in the form of intact genes, called lateral gene transfer, was put forward. The initial hypotheses regarding this were advocated as the endosymbiont hypotheses in the early 1970s. They stressed the idea that endocytosis created a brand new way for early eukaryotes to evolve by the acquired ability to consume solid food. The same idea was improved by saying that unicellular eukaryotes evolved through their ability to intake bacteria as food and that this ability presented them with an added benefit in terms of metabolic abilities [20].

In 1997, Schubbert et al. [21] published their experimental study on the subject. In their elegantly chosen model system, M13mp18 DNA was used as the test model with no homology to mouse DNA. As opposed to control animals, a significant portion of the foreign DNA was detected in sections of the small intestine and blood of the test animals. The foreign DNA fragments were traced in peripheral leukocytes by PCR and localized using fluorescent *in situ* hybridization in the columnar epithelial cells. One out of thousand white blood cells was found to contain the foreign DNA using fluorescent *in situ* hybridization. A 1299 bp foreign genome was re-cloned in a vector genome. Another clone contained mouse DNA, bacterial DNA, and virus DNA. Two more clones were found to contain viral genes of considerable length.

The idea of lateral gene transfer from prokaryotes to eukaryotes may not be limited only to food grade microorganisms. All living organisms share the same open system, and many eukaryotes have developed life styles shared with prokaryotes and viruses.

6.2.2 NUTRIGENOMICS

The term "nutrigenomics" refers to the effect of dietary traits, functional foods, or supplements on the expression of genes in the genome. Possible changes in the expression profiles in turn affect the proteome and, finally, the health of human and animal populations. It is becoming possible to change the dietary habits according to the health conditions of persons by imposing the uptake of specific micronutrients found in foods. The uptake of specific micronutrients is believed to establish genomic stability by nourishing DNA replication and repair, since these micronutrients provide the cofactor requirements of these fundamental metabolic-genetic activities of survival. These micronutrients (vitamins and minerals) are generally required as cofactors for enzymes. The critical aim of nutrigenomics is to find the optimum nutritional requirements to prevent DNA damage and aberrant gene expression for genetic subgroups and achieve the ability to continue to replace senescent cells in the body with fresh cells bearing normal genotypes and gene expression patterns [22]. Chapter 18 discusses nutrigenomics in greater detail.

6.2.3 GENOMICS IN FOOD SCIENCE

Genomics is the scientific field interested in the sequence and structure of the genetic background of organisms. Functional genomics deals with the functional impact of the genetic information confined in genomics. Since organisms are in a dynamic relationship with their environment, food constitutes one of the ways through which organisms, and hence their genomes and their interpretations, interact with their environments. One of the current uses of functional genomics in food science is guiding the approaches to prevent food spoilage stemming from several food pathogens. This will help process scientists to understand the survival behavior of microorganisms under stress conditions during food processing in complex food structures. This behavior should be difficult to predict, since foods are quite often heterogeneous in structure. The data from DNA microarray experiments have been used to predict the sporulation behavior of *Bacillus subtilis* and *Bacillus sporothermodurans* under different environmental and process conditions. The total RNA of the target microorganisms were extracted and used for transcriptome analysis to evaluate the effect of heat treatment. This approach, along with the use of artificial intelligence methods, makes it possible to apply the same approach to the total food chain. In addition, it allows the design of novel quality control systems and the development of integrated food processing systems, microbiological detection systems, novel antimicrobial materials, and novel diagnostic systems [23,24].

These scientific developments will turn the food industry from a rather low-technology field to a high-tech/high value one via the use of functional foods. Functional foods will be particularly important for the rapidly growing elderly population [25].

6.2.4 NUTRITION AND CANCER

Cancer is a group of diseases, all involving an uncontrolled cell proliferation, that sometimes metastasizes. Many cancer types are thought to be formed from the interplay between genetic susceptibility and environmental factors. One of the environmental factors affecting the molecular reactions resulting in cancer formation is nutritional background. Nutrition is the process of taking in and utilizing necessary chemicals as foods in the body. Since cell growth in general is a continuous process requiring nutrition as the main driving force for the maintenance of the organisms' basic life processes, cancer has always been associated with nutrition directly or indirectly.

The role of nutrition in the formation or prevention of cancer has always been speculated by the society. To be able to prove or disprove this speculation requires scientific research. Actually, this is a hot scientific topic investigated by several scientific groups and laboratories vigorously. In one of the reviews dealing with such studies, Leppert et al. (2006) summarized the data gathered till now on bladder cancer, since it is a commonly used model to test similar arguments. In their review, they argue that bladder cancer was responsive to efforts to prevent or delay it, as supported by pre-clinical and clinical data [26]. However, they also indicate that the chemo-preventive effects of many natural products, such as vitamins and herbal components, lack conclusive evidence. In many respects, smoking was found to be the number one risk factor in bladder cancer formation [27] and that the risk can be exacerbated by smoking [28], although the exact biochemical mechanism of cancer formation is still unknown. Similar to the cancerous chemicals in cigarette smoke, the exposure to various chemicals and toxins, such as arsenic, found in water was found to trigger cancer in certain parts of the world [29].

Foods are a mixture of several different chemical species. Research will focus more on the microelements of foods rather than on global and generalized aspects, as the level of experimental research increases. Within this context, increasing the use of food additives places them in the center of attention of the public, especially in terms of health concerns.

As the largest study ever undertaken to study the relationships between dietary habits and disease states including cancer, The European Prospective Investigation into Cancer and Nutrition found a reverse correlation between fluid intake and bladder cancer frequency [30]. As expected, higher caloric intake was correlated with higher incidence of bladder cancer in men older than 60 [31]. Similar results were obtained with fruit and vegetable, green tea polyphenol, and vitamin supplement consumption, despite the fact that the results were sometimes unexpected for different cancer cases [32–34].

Although for some cancers preventive data often advise reduction of alcohol, red meat, and animal fat intake, and increasing the consumption of vegetables, fruits, and fibrous materials, the scientific basis for these recommendations appears to be sparse for breast cancer [35]. Similarly, nitrate contamination found in drinking water was positively correlated with the number of the several cancer incidents in some population-based studies but some analogous studies failed to produce similar results [36,37]. Although population-based statistical studies are valuable to correlate the occurrence of an incident in a population, in scientific terms, the hypotheses should be rigorously verified at the level of molecular studies.

6.2.5 EDIBLE VACCINES

Use of recombinant DNA technology in health and medicine opened new avenues in the application of agro-biotechnology via the development of new generation vaccines. This new technology allowed introducing to the plant genome the genes that code the proteins of interest for their use as antigens, because of the inconveniences in their classical use. Therefore, the term "edible vaccines" applies to the use of the edible parts of the plants genetically modified for the purpose of producing antigens of a pathogen against which the consumer is intended to be protected [38]. Actually, the use of plants as bioreactors to express foreign protein antigens induced by plant transgenic vectors is an example of the utilization of plants for the production of several biomolecules, including antibodies and other bioactive molecules [39].

Due to the nature of edible vaccine materials (foods), the only delivery method of these antigens is the oral route. Most of the commercial vaccines available are either attenuated or inactivated microorganisms, which are delivered via injection as a more direct way of presenting antigenic determinant regions (epitopes) to the immune system. Oral intake of antigens, therefore, presents a difficulty in the quality of the antigens presented due to the harsh conditions these molecules experience in the gastrointestinal system of the animal or human host. Another disadvantage of the method is the irregular heterogeneous distribution of the expressed antigen molecules among different tissues of vegetables. However, the method can be used advantageously for the delivery of vaccines in an inexpensive and common way. One other advantage of the method is that it does not require expensive refrigeration cycles after production and the ease of production and distribution, which is already available for many commonly produced and consumed vegetables, such as wheat and maize [40]. They also provide a more convenient way of inoculation when compared to traditional injection methods. The trials for these vaccine candidates are in progress [41,42]. For this purpose, till now many different plant systems have been used, including fruits (banana and papaya), crops (maize, alfalfa, wheat, and rice), and vegetables. In addition to the aforementioned transgenic approaches, a plant virus-based system was also used to propagate specifically single-stranded RNA viruses with positive polarity. The system offers a rapid production and is being used by a company on an industrial scale [43,44].

6.2.6 DNA Comet Assay

The DNA comet assay is one of the analytical detection methods adopted as a European standard by the European Committee for Standardization. It is a rapid and simple test not requiring expensive equipment. The method is used in food analyses to detect whether DNA has been treated by ionizing radiation. The fragmentation occurring after the radiation treatment can be studied by microgel electrophoresis of single cells or nuclei. DNA fragments or relaxed loops of DNA will migrate and form a tail in the direction of the anode, giving the appearance of a comet. Irradiated DNA contains larger comets than non-irradiated DNA and appears homogeneous [45]. The method has been tested successfully for several DNA-containing foods, such as meat [46], poultry [47], and fresh fruits [48].

6.2.7 Use of Nucleic Acid Markers for Food Applications

The origin and nature of the food material is of public concern for health and cultural reasons. Various DNA-based molecular techniques have been studied so far to accomplish the difficult task of determining the nature of foods. As a biological macromolecule, DNA offers more possibilities when compared to proteins in terms of testing advantages. First, DNA is more heat stable than many proteins; therefore, it can endure many heat transfer-based operations. Second, DNA contains more data than proteins for many authentication studies. Additionally, DNA is found in many foods and can be analyzed through several well-developed methods [49].

One of the initial methods used for food analysis was DNA hybridization. The method was based on the principle that labeled DNA from an organism should hybridize to another DNA molecule from the same source. The use of nylon membranes was necessary to provide a solid attachment environment for the DNA molecules. This method was further improved with the use of satellite sequences to distinguish meat products from closely related animals, such as sheep and goat, even if they were heat processed [50].

The cumbersome hybridization procedures led to the use of PCR-based identification methods. Short, synthetic oligonucleotide primers allowed the amplification of the desired genomic regions of the target species. The amplification products were then analyzed using many different methods.

Other methods that did not require sequence information but produced species-specific electrophoretic gel patterns were also developed. Single-strand conformational polymorphism analysis is one such method, which was used to discriminate fish and meat products. In this method, the same region of the DNA is amplified, double-stranded amplification products are denatured and single-stranded products are allowed to form their sequence-dependent secondary structures, and their corresponding migration patterns on the gel are observed [51].

Random amplified polymorphic DNA analysis is one method used for the same purpose but based on the utilization of short arbitrary (about 10 bases) primers yielding a number of electrophoretic species-specific amplification products for given reaction conditions. The method was applied in species determination in foods, especially in meat products [52].

6.2.8 DNA Barcoding for Food Applications

Since it is very important to know which food material is consumed in terms of regulatory issues as well as consumer awareness, a standard way of species definition at the molecular level is sought.

As recombinant DNA technology progressed at a very high speed, new cloning and high-throughput sequencing technologies, such as emulsion PCR, revealed complex genomic sequence data from tiny amounts of material in a very short time [53].

DNA barcoding, which is a diagnostic technique for species and cultivar identification, is performed using a short fragment of DNA. Although its use in taxonomy and identification has been controversial, it is mostly devised for biodiversity studies [54]. The method, using the mitochondrial *coxI* gene, is well established for animals [55]. However, studies are in progress to further strengthen the method for animals. In such a work, Hajibabaei et al. (2007) suggested the *cytochrome b* gene in addition to the well-established T *coxI* gene [56]. They also investigated the possible use of array-based DNA chips for the same purpose.

Several plant genomic regions, including portions of the plastid genome and intron regions, have been proposed to be candidates, but none has proved sufficient enough to encode for the rich variety of plant species on earth for the purpose of DNA barcoding of plant species [57–59]. Lahaye et al. (2008) identified a portion of the plastid *matK* gene as a universal DNA barcode for biodiversity studies, by studying more than 1600 plant species with the help of several bioinformatics tools [54]. Therefore, more work is required to ensure a complete coverage of plants and plant-originated foods.

6.2.9 Nucleic Acid Detection of Food-Borne Pathogens

Food-borne diseases are mainly caused by either bacteria or viruses. The detection of these pathogens is difficult for two reasons. First, food has a heterogeneous structure that presents compositional difficulties in the purification of the agent's components. Second, the presence of indigenous microorganisms can hinder the detection of pathogenic microorganisms of lower magnitude. Therefore, specific enrichment steps should be carried out before final steps are taken. In addition to traditional microbiological methods, the invention of PCR has revolutionized the detection methods. PCR methodology has been widely used for the detection of several bacterial and viral pathogens from foods. One other molecular method employed for the detection of pathogenic bacteria in food is colony hybridization. In this method, the colony is hydrolyzed by alkali or detergent treatment and the hybridization of single-stranded DNA with a labeled DNA probe is allowed. One of the widely used targets of such probes is variable ribosomal RNA sequences [60].

Real-time PCR has been the technique of choice, providing the advantages of accurate and fast measurements along with the quantification of the target molecules. It is commonly used for both food authentication and pathogen detection purposes. The method monitors the fluorescence emitted from a fluorescent reporter during each cycle of a proceeding reaction. Several real-time PCR fluorescent reporters have been used in biotechnology in different forms of primers/probes. The method is also used for quantification purposes, such as measuring mRNA expression levels, DNA copy number, transgene copy number, allelic discrimination, and microbial load/titer in a sample [61].

One of the applications of this method is in food authentication. Lopez-Calleja et al. (2007) used the real-time PCR method to quantitatively detect goat milk present in sheep milk based on the amplification of the 12S ribosomal RNA gene, in order to prevent this type of food adulteration [62]. Similarly, Chisholm et al. (2005) used the method to detect horse and donkey meat in commercial products [63]. They used primers designed for the mitochondrial *cytochrome b* gene. This method, which enabled the detection of 1 pg donkey and 25 pg horse DNA templates, worked satisfactorily with model food samples.

Real-time PCR has also been successfully used for the purpose of food-borne pathogen detection. The method provided a sensitive and also very rapid detection of *Campylobacter jejuni* in chicken rinses in 90 min of analysis time, when compared to the traditional 5–7 days of plating to selective media, including enrichment and sub-culturing steps [64]. O'Grady et al. (2008) used the method as a novel way of detecting *Listeria monocytogenes* with high specificity among other

Listeria and bacterial species in food samples [65]. They used a 162 bp fragment of the *ssr A* gene as a nucleic acid diagnostic target. A new, rapid real-time PCR method for the detection of *L. monocytogenes* was found to have the potential to be used as an alternative to the standard method for food quality assurance [66]. The technique has been used for the detection of other common bacterial food-borne bacterial pathogens such as *Brucella* species [67] and *Coxiella burnetii* [68]. The real-time PCR technique is also used for the detection of food-borne viruses, such as the hepatitis A virus [69], hepatitis E virus [70], noroviruses [71,72], rotaviruses [73], and bacteriophages [74].

REFERENCES

1. Klug, W.S., Cummings, M.R., and Spencer, C., *Concepts of Genetics*, 8th Edn., Pearson Education, Upper Saddle River, NJ, 2005.
2. Twyman, R.M. and Primrose, S.B., *Principles of Gene Manipulation and Genomics*, 7th Edn., Wiley Blackwell, Oxford, U.K., 2006.
3. Alberts, B., ed., *Molecular Biology of the Cell*, 4th Edn., Garland Publishers, New York, 2002.
4. Elenis, D.S. et al., Advances in molecular techniques for the detection and quantification of genetically modifies organisms. *Anal. Bioanal. Chem.* 392, 347, 2008.
5. International Rice Genome Sequencing Project, The map-based sequence of the rice genome, *Nature*, 436, 793, 2007.
6. Whitelaw, C.A. et al., Enrichment of gene-coding sequences in maize by genome filtration, *Science*, 302, 2118, 2003.
7. Yadav, S.K. and Ahuja, P.S., Towards generating caffeine-free tea by metabolic engineering, *Plant Foods Hum. Nutr.*, 62,185, 2007.
8. Shim, J.H. et al., Improved bread-baking process using *Saccharomyces cerevisiae* displayed with engineered cyclodextrin glucanotransferase, *J. Agric. Food Chem.*, 55, 4735, 2007.
9. Mandaokar, A.D. and Koundal, K.R., Cloning and characterization of legumin storage protein genes of chickpea (*Cicer arietinum* L.), *J. Plant Biochem. Biotechnol.*, 9, 7, 2000.
10. Rexroad, C.E. et al., Production of transgenic sheep with growth-regulating genes, *Mol. Reprod. Dev.*, 1, 164, 1989.
11. Hammer, R.E. et al., Production of transgenic rabbits, sheep, and pigs by microinjection, *Nature*, 315, 680, 1985.
12. Salamone, D. et al., High level expression of bioactive recombinant human growth hormone in the milk of a cloned transgenic cow, *J. Biotechnol.*, 124, 469, 2006.
13. Boesze, Z., Baranyi, M., and Whitelaw, C.B., Producing recombinant human proteins in the milk of livestock species, *Adv. Exp. Med. Biol.*, 606, 357, 2008.
14. Deppenmeier, S. et al., Health status of transgenic pigs expressing the human complement regulatory protein CD59, *Xenotransplantation*, 13, 345, 2006.
15. Barthakur, S., Babu, V., and Bansal, K.C., Over-expression of osmotin induces proline accumulation and confers osmotic stress tolerance in transgenic tobacco, *J. Plant Biochem. Biotechnol.*, 10, 31, 2000.
16. Gonsalves, D., Resistance to papaya ringspot virus, *Annu. Rev. Phytopathol.*, 36, 415, 1998.
17. Nykiforuk, C.L. et al., Transgenic expression and recovery of biologically active recombinant human insulin from *Arabidopsis thaliana* seeds, *Plant Biotechnol.*, 4, 77, 2006.
18. Lin, C. et al., A built-in strategy for containment of transgenic plants: Creation of selectively terminable transgenic rice, *PLoS One*, 3(3), e1818, 2008.
19. Fox, J.L., Puzzling industry response to ProdiGene fiasco, *Nat. Biotechnol.*, 21, 3, 2003.
20. Doolittle, W.F., You are what you eat: A gene transfer rachet could account for bacterial genes in eukaryotic nuclear genomes, *Trends Genet.*, 14, 307, 1998.
21. Schubbert, R. et al., Foreign (M13) ingested by in mice reaches peripheral leukocytes, spleen, and liver via the intestinal wall mucosa and can be covalently linked to the mouse DNA, *Proc. Natl. Acad. Sci. U.S.A.*, 94, 961, 1997.
22. Fenech, M., Genome health nutrigenomics and nutrigenetics—Diagnosis and nutritional treatment of genome damage on an individual basis, *Food Chem. Toxicol.*, 46, 1365, 2008.
23. Oomes, S.J.C.M. and Brul, S., The effect of metal ions commonly present in food on gene expression of sporulating *Bacillus subtilis* cells in relation to spore wet heat resistance, *Innov. Food Sci. Emerg. Technol.*, 5, 307, 2004.

24. Brul, S. et al., The impact of functional genomics on microbiological food quality and safety, *Int. J. Food Microbiol.*, 112, 195, 2006.
25. Verrips, C.T., Warmoeskerken, M.M.C.G., and Post, J.A., General introduction to the importance of genetics in food biotechnology and nutrition, *Curr. Opin. Biotechnol.*, 12, 483, 2001.
26. Leppert, J.T. et al., Prevention of bladder cancer: A review, *Eur. Urol.*, 49, 226, 2006.
27. Wynder, E.L. and Goldsmith, R., The epidemiology of bladder cancer: A second look, *Cancer*, 40, 1246, 1977.
28. Evans, C.P. et al., Bladder cancer: Management and future directions, *Eur. Urol. Suppl.*, 6, 365, 2007.
29. Chiou, H.Y. et al., Incidence of transitional cell carcinoma and arsenic in drinking water: a follow-up study of 8,102 residents in an arseniasis-endemic area in northeastern Taiwan, *Am. J. Epidemiol.*, 153, 411, 2001.
30. Michaud, D.S. et al., Fluid intake and the risk of bladder cancer in men, *New Engl. J. Med.*, 340, 1390, 1999.
31. Vena, J.E. et al., Coffee, cigarette smoking, and bladder cancer in western New York, *Ann. Epidemiol.*, 3, 586, 1993.
32. Yalçin, O. et al., The levels of glutathione peroxidase, vitamin A, E, C and lipid peroxidation in patients with transitional cell carcinoma of the bladder, *BJU Int.*, 93, 863, 2004.
33. Steinmaus, C.M., Nunez, S., and Smith, A.H., Diet and bladder cancer: A meta-analysis of six dietary variables, *Am. J. Epidemiol.*, 151, 693, 2000.
34. Sun, C.L. et al., Green tea, black tea and breast cancer risk: A meta-analysis of epidemiological studies, *Carcinogenesis*, 27, 1310, 2006.
35. Hanf, V. and Gonder, U., Nutrition and primary prevention of breast cancer: Foods, nutrients and breast cancer risk, *Eur. J. Obstet. Gyn. R. B.*, 123, 139, 2005.
36. Weyer, P. et al., Municipal drinking water nitrate level and cancer risk in older women: The Iowa Women's Health Study, *Epidemiology*, 12, 327, 2001.
37. Gulis, G., Czompolyova, M., and Cerhan, J.R., An ecologic study nitrate in municipal drinking water and cancer incidence in Trnava district, Slovakia, *Environ. Res. Section A*, 88, 182, 2002.
38. Mercenier, A., Wiedermann, U., and Breiteneder, H., Edible genetically modified microorganisms and plants for improved health, *Curr. Opin. Biotechnol.*, 12, 510, 2001.
39. Wieland, H.W. et al., Plant expression of chicken secretory antibodies derived from combinatorial libraries, *J. Biotechnol.*, 122, 382, 2006.
40. Cebadera, E. and Camara, M., Plants as biofactories: Edible vaccines production, in green biotechnology, *J. Biotechnol.*, 131S, 43, 2007.
41. Ko, K. and Koprowski, H., Plant biopharming of monoclonal antibodies, *Virus Res.*, 111, 93, 2005.
42. Ma, J.K.C. et al., Antibody processing and engineering in plants, and new strategies for vaccine production. *Vaccine*, 23, 1814, 2005.
43. Gleba, Y., Klimyuk, V., and Marillonnet, S., Magnifection—A new platform for expressing recombinant vaccines in plants, *Vaccine*, 23, 2042, 2005.
44. Mei, H. et al., Research advances on transgenic plant vaccines, *Acta Genet. Sin.*, 33, 285, 2006.
45. Delincee, H., Khan, A.A., and Cerda, H., Some limitations of the comet assay to detect the treatment of seeds with ionising radiation, *Eur. Food Res. Technol.*, 216, 343, 2003.
46. Marin-Huachaca, N. et al., Use of the DNA Comet Assay to detect beef meat treated by ionizing radiation, *Meat Sci.*, 71, 446, 2005.
47. Villavicencio, A.L.C.H. et al., Identification of irradiated refrigerated poultry with the DNA comet assay, *Radiat. Phys. Chem.*, 71, 187, 2004.
48. Marin-Huachaca, N.S. et al., Detection of irradiated fresh fruits treated by e-beam or gamma rays, *Radiat. Phys. Chem.*, 63, 419, 2002.
49. Lockley, A.K. and Bardsley, R.G., DNA-based methods for food authentication, *Trends Food Sci. Technol.*, 11, 67, 2000.
50. Chikuni, K. et al., Species identification of cooked meats by DNA hybridization assay, *Meat Sci.*, 27, 119, 1990.
51. Rebhein, H. et al., Fish species identification in canned tuna by PCR-SSCP: Validation by a collaborative study and investigation of intra-species variability of the DNA-patterns, *Food Chem.*, 64, 263, 1999.
52. Martinez, I. and Yman, I.M., Species identification in meat products by RAPD analysis, *Food Res. Int.*, 31, 459, 1999.
53. Williams, R. et al., Amplification of complex gene libraries by emulsion PCR, *Nat. Methods*, 3, 545, 2006.

54. Lahaye, R. et al., DNA barcoding the floras of biodiversity hotspots, *Proc. Natl. Acad. Sci. U.S.A.*, 105, 2923, 2008.
55. Hebert, P.D.N. et al., Biological identifications through DNA barcodes, *P. R. Soc. B.*, 270, 313, 2003.
56. Hajibabaei, M. et al., Design and applicability of DNA arrays and DNA barcodes in biodiversity monitoring, *BMC Biol.*, 5:24, 2007.
57. Kress, W.J. et al., Use of DNA barcodes to identify flowering plants, *Proc. Natl. Acad. Sci. U.S.A.*, 102, 8369, 2005.
58. Kress, W.J. and Erickson, D.L., A two-locus global DNA barcode for land plants: The coding *rbcL* gene complements the non-coding *trnH-psbA* spacer region, *PloS One*, 2(6), e508, 2007.
59. Taberlet, P. et al., Power and limitations of the chloroplast *trnL* (UAA) intron for plant DNA barcoding, *Nucleic Acids Res.*, 35, e14, 2007.
60. Olsen, J.E., DNA-based methods for detection of food-borne bacterial pathogens, *Food Res. Int.*, 33, 257, 2000.
61. Ginzinger, D.G., Gene quantification using real-time quantitative PCR: An emerging technology hits the mainstream, *Exp. Hematol.*, 30, 503, 2002.
62. Lopez-Calleja, I. et al., Quantitative detection of goats' milk in sheep's milk by real-time PCR, *Food Control*, 18, 1466, 2007.
63. Chisholm, J. et al., The detection of horse and donkey using real-time PCR, *Meat Sci.*, 70, 727, 2005.
64. Debretsion, A. et al., Real-time PCR assay for rapid detection and quantification of *Campylobacter jejuni* on chicken rinses from poultry processing plant, *Mol. Cell. Probe*, 21, 177, 2007.
65. O'Grady, J.O. et al., Rapid real-time PCR detection of *Listeria monocytogenes* in enriched food samples based on the *ssrA* gene, a novel diagnostic target, *Food Microbiol.*, 25, 75, 2008.
66. O'Grady, J. et al., Rapid detection of *Listeria monocytogenes* in food using culture enrichment combined with real-time PCR, *Food Microbiol.*, 26, 4, 2009.
67. Bounaadja, L. et al., Real-time PCR for identification of *Brucella* spp.: A comparative study of *IS711*, *bcsp31* and *per* target genes, *Vet. Microbiol.*, 2009, doi:10.1016/j.vetmic.2008.12.023.
68. Howe, G.B. et al., Real-time PCR for the early detection and quantification of *Coxiella burnetii* as an alternative to the murine bioassay, *Mol. Cell. Probes*, 2009, doi:10.1016/j.mcp.2009.01.004.
69. Dubois, E. et al., Detection and quantification by real-time RT-PCR of hepatitis A virus from inoculated tap waters, salad vegetables, and soft fruits: Characterization of the method performances, *Int. J. Food Microbiol.*, 117, 141, 2007.
70. Gyarmati, P. et al., Universal detection of Hepatitis E virus by two real-time PCR assays: TaqMan® and Primer-Probe Energy Transfer, *J. Virol. Methods*, 146, 226, 2007.
71. Baert, L., Uyttendaele, M., and Debevere, J., Evaluation of viral extraction methods on a broad range of Ready-To-Eat foods with conventional and real-time RT-PCR for Norovirus GII detection, *Int. J. Food Microbiol.*, 123, 101, 2008.
72. Gentry, J. et al., A rapid and efficient method for quantification of genogroups I and II norovirus from oysters and application in other complex environmental samples, *J. Virol. Methods*, 156, 59, 2009.
73. Jothikumar, N., Kang, G., and Hill, V.R., Broadly reactive TaqMan assay for real-time RT-PCR detection of rotavirus in clinical and environmental samples, *J. Virol. Methods*, 155, 126, 2009.
74. Wolf, S. et al., Detection and characterization of F+ RNA bacteriophages in water and shellfish: Application of a multiplex real-time reverse transcription PCR, *J. Virol. Methods*, 149, 123, 2008.

7 Hormones: Regulation of Human Metabolism

Ayhan Karakoç

CONTENTS

7.1 HORMONES

7.1.1 DEFINITION OF HORMONE

A hormone is a chemical messenger through which messages are transmitted between different cells or tissues in a multicellular organism. Hormones are synthesized and secreted by specialized cells known as endocrine glands from which they are transported to their target tissues through the blood stream. This is the classical definition of the endocrine effects of hormones. The location of classical endocrine glands is shown in Figure 7.1.

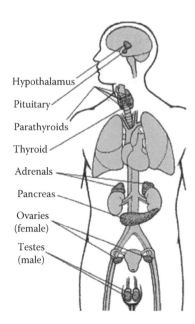

FIGURE 7.1   The location of classical endocrine glands in the human body (www.drstandley.com).

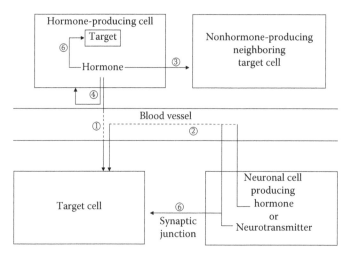

FIGURE 7.2   Types of hormone action. See text for the details of pathways.

However, hormones may act on the neighboring nonhormone-producing cells. This type of inter-action between the hormone-producing cells and the target tissues is termed the paracrine effect. Some hormones are released into the intercellular compartment where they act on receptors found on the same cells secreting them; this action is called the autocrine effect. Hormones can also act within the cell without being released, an effect known as intracrine effect (Figure 7.2).

Classical hormones like the thyroid hormones are released from the endocrine gland (thyroid gland) and transported through the blood stream to target tissues found in various parts of the body (Figure 7.2, pathway 1). The hypothalamic-releasing hormones are synthesized by hypothalamic neurons and carried in the blood stream to the target cells in the pituitary (Figure 7.2, pathway 2). Also, dopamine, which is a neurotransmitter and transported to the pituitary via the hypophy-seal-portal system, acts as a hormone in the pituitary gland (Figure 7.2, pathway 2). The action of sex steroids in the gonads is also an example of the paracrine effect (Figure 7.2, pathway 3). For example, testosterone is secreted into the blood stream, but also acts locally in the gonads to regulate spermatogenesis. The autocrine effects of hormones may be important in promoting cancer cell growth (Figure 7.2, pathway 4). The intracrine effect is also very important to the action of hormonal activity as in the regulation of insulin release; insulin regulates its own secretion by an intracrine effect (Figure 7.2, pathway 5). Finally, some hormonally active neurotransmitters (e.g., norepinephrine) are released into synaptic junctions by the nerve cells that are in contact with the target cells (Figure 7.2, pathway 6).

7.1.2 FEEDBACK CONTROL OF HORMONES

Hormones are necessary for maintaining the consistency of the internal environment (homeostasis), a requirement for the integrity of the organism as a whole. All endogenous or exogenous stimuli that affect homeostasis generate a hormonal response in an attempt to protect the organism. For example, in the starved state, hormones act to reduce the basal metabolic rate. The secretion of several hormones is affected by the change in the exogenous environment as in the course of diurnal rhythm of hormone secretion (changes in the amount of secreted hormones according to the day and night cycle). Calcium homeostasis is very important for many functions of the organism. The organism tries to reset the homeostatic state toward normal by hormonal action in response to a change in the calcium homeostasis.

Homeostasis is maintained by a negative feedback mechanism. An external or internal signal informs the central nervous system about the alteration in the exogenous or endogenous environment. This information is transmitted from the central nervous system through the hypothalamus to the pituitary and then to the peripheral endocrine glands. The hormonal product of the peripheral endocrine gland or physiologic actions induced by the hormone can produce feedback inhibition of the stimulus at any level (Figure 7.3).

Disappearance of the negative feedback control of hormones leads to uncontrolled hormone secretion. For example, ovarian failure—as it happens in menopause—is caused by the luteinizing hormone and follicle-stimulating hormone levels' increase due to the lack of negative feedback effects of the gonadal hormones.

Under certain circumstances, a positive feedback control of hormones exists. Estradiol has a negative feedback control on luteinizing hormone secretion. However, estradiol at high concentrations has a positive feedback action at the level of the hypothalamus and pituitary. High estradiol levels lead to mid-cycle luteinizing hormone surges by increasing gonadotropin-releasing hormone secretion from the hypothalamus and the sensitivity of the pituitary to gonadotropin-releasing hormone secretion [1]. This luteinizing hormone surge causes ovulation.

7.1.3 TYPES OF HORMONES

Hormones belong to the major classes of biologic molecules. Hormones can be glycoproteins, polypeptides or peptide derivatives, amino acid analogs, or lipids. Some examples of hormones from each class of biologic molecules are given in Table 7.1.

7.1.4 ACTIONS OF HORMONES

Hormones, in general, regulate the activity of the target tissues and affect all types of body processes. Hormones change the cellular metabolism in their target tissues. They can regulate the rate of enzymatic reactions by the phosphorylation and dephosphorylation of proteins. Hormones facilitate the movement of several ions between body compartments by altering membrane permeability. Hormones can also activate genes to influence gene expression and protein synthesis.

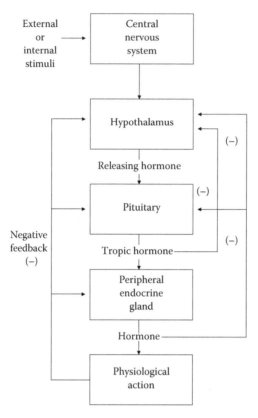

FIGURE 7.3   The negative feedback inhibition of hormonal action at each stimulatory level.

TABLE 7.1
Examples of Hormones with Different Molecular Structures

Structure	Hormone
Glycoprotein	Thyroid-stimulating hormone
	Luteinizing hormone
	Follicle-stimulating hormone
Polypeptide or peptide derivative	Growth hormone
	Adrenocorticotropic hormone
	Thyroid-releasing hormone
Amino acid analog	Thyroid hormones
	Epinephrine
	Norepinephrine
Steroid	Cortisol
	Aldosterone
	Estrogens
	Testosterone
	Progesterone
Fatty acid	Prostaglandins
	Retinoic acid

Hormones are responsible for several functions of the body:

1. Anterior pituitary hormones are tropic hormones that regulate the secretion of peripheral hormones. These tropic hormones also stimulate the growth of the peripheral endocrine glands. The thyroid-stimulating hormone has growth stimulatory effects on the thyroid gland. The adrenocorticotropic hormone results in bilateral hypertrophy of adrenal glands in Cushing's disease (a pituitary adenoma which secretes adrenocorticotropic hormone autonomously).
2. Hormones play a crucial role in extrauterine growth and development. Growth hormone, growth factors (like fibroblast growth factor and transforming growth factors), thyroid hormones, sex steroids, glucocorticoids, and insulin control extrauterine growth and development of the organism [2].
3. Reproductive hormones, follicle-stimulating hormone, and luteinizing hormone from the pituitary, chorionic gonadotropin from the placenta, and sex steroids (androgens, estrogens, progesterone) from the gonads are all crucial for pregnancy as well as sexual differentiation and development [2]. They are also responsible for the morphological changes and secondary sex character differences between males and females and the sex-related behavioral characteristics.
4. The heart, as an endocrine organ itself, secretes the atrial natriuretic peptide, which has several important influences on the cardiovascular system. In addition to atrial natriuretic peptide, catecholamines, thyroid hormones, mineralocorticoids, sex steroids, and angiotensin II are other hormones also known to exert their effects on the cardiovascular functions and blood pressure. Hormones regulate the contraction and dilatation of the vascular bed. They may also affect the contraction of other smooth muscles other than the vascular smooth muscle. Oxytocin stimulates contraction of the myoepithelium in the mammary gland that is necessary for milk ejection.
5. Many hormones influence membrane permeability, thereby regulating ion transport and water metabolism across the membrane. The antidiuretic hormone increases reabsorption of water in the kidney.
6. Hormones affect cancer cell proliferation and exocrine secretions.

The actions of hormones on intermediary metabolism are discussed in detail below.

Hormones released from the endocrine glands and those secreted by the organs other than the classical endocrine glands as well as their main functions are listed in Table 7.2. It should be borne in mind that numerous hormones and hormone-like substances might not have been mentioned in Table 7.2.

7.1.5 PRODUCTION, TRANSPORT, AND METABOLISM OF HORMONES

Hormonal response is regulated by hormone concentration. Hormones are produced in variable amounts according to the needs of the organism. Quantitatively, hormones are secreted in nanograms by the hypothalamus, in microgram amounts by the pituitary, and up to milligram amounts by the peripheral endocrine glands daily. However, they may be present in trace amounts in the plasma because of the very large distribution space. In general, hormones and their actions have short half-lives, and hormonal activity can be initiated or terminated by altering hormone concentration.

A variety of internal (e.g., blood glucose level for insulin and glucagon) or external (e.g., stress for catecholamines) stimuli trigger hormone synthesis and secretion. Hormones have a basal secretion rate. This basal secretion is necessary for the maintenance of receptors in the target cell and to keep the tissue primed for hormones.

Production of hormones composed of proteins does not require any special production pathway; growth hormone and prolactin are produced like the other protein molecules in the body.

TABLE 7.2

Hormones Released by Classical Endocrine Glands and by Nonclassical Hormone-Producing Tissues

	Hormone	Function
Endocrine Gland (Classical)		
Hypothalamus	Gonadotropin-releasing hormone	Stimulation of follicle-stimulating hormone and luteinizing hormone
	Growth hormone-releasing hormone	Stimulation of growth hormone
	Somatostatin (GH_IH)	Inhibition of growth hormone
	Thyrotropin-releasing hormone	Stimulation of thyroid-stimulating hormone
	Corticotropin-releasing hormone	Stimulation of adrenocorticotropic hormone
	Dopamine	Tonic inhibition of prolactin secretion
Anterior pituitary (adenohypophysis)	Follicle-stimulating hormone	Follicular growth, estradiol production, spermatogenesis
	Luteinizing hormone	Ovulation, estradiol and progesterone production, testosterone production
	Growth hormone	Mainly growth stimulation
	Thyroid-stimulating hormone	Thyroid hormone production
	Adrenocorticotropic hormone	Glucocorticoid production
	Prolactin	Milk production
Posterior pituitary (neurohypophysis)	Antidiuretic hormone	Water reabsorption in the kidney
	Oxytocin	Regulation of parturition and milk ejection
Parathyroid gland	Parathyroid hormone	Regulation of calcium and phosphorus metabolism
Thyroid gland	Triiodothyronine and tetraiodothyronine	Mainly control of metabolic rate
Adrenal cortex	Cortisol	Effects on carbohydrate metabolism
	Aldosterone	Sodium retention
Adrenal medulla	Epinephrine and norepinephrine	Reactions to all types of stresses
Gonads	Estrogens, androgens, and progestins	Sexual development, pregnancy, growth, bone metabolism, sexual behavior
Pancreas	Insulin and glucagon	Regulation of glucose metabolism
Nonclassical Endocrine Tissues		
Adipose tissue	Leptin	Appetite and weight control
Gastrointestinal tract	Gastrin, secretin, GIP, and CCK	Gastric acidity and digestion
Kidney	Erythropoietin	Erythropoiesis
Heart	Atrial natriuretic peptide	Effects on cardiovascular function
In many tissues/cells	Eicosanoids	Control on smooth muscle contraction
	Growth factors	Cell growth and differentiation

Some peptide hormones like insulin, glucagon, and adrenocorticotropic hormone are synthesized as larger molecules called pro-hormones, which are later converted to their active hormones by proteolytic enzymes [3].

Some specialized enzymes synthesize amino acid analog hormones. Thyroid hormones are produced by iodination and coupling of tyrosine residues in the thyroglobulin, and catecholamines are produced from phenylalanine through a special enzymatic pathway [3].

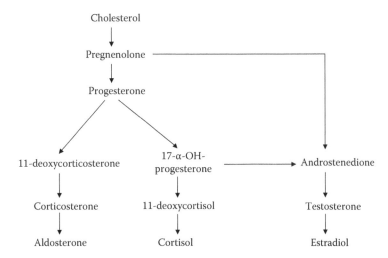

FIGURE 7.4 Main pathways for the synthesis of steroid hormones. Presence or absence of different enzymatic reactions responsible for the steroid hormone synthesis determines the type of end-product at that organ.

Steroid hormone production occurs on the smooth endoplasmic reticulum and in the mitochondria of the hormone-producing cells. Cholesterol is the precursor of all steroid hormones. The adrenal cortex produces the glucocorticoids, mineralocorticoids, and androgens, whereas the testes synthesize testosterone and the ovaries produce mainly estrogen and progesterone. The relative concentration of the various enzymes determines the final steroid hormone products [4]. Figure 7.4 shows the main pathways for the synthesis of steroid hormones.

The active hormone, 1,25-dihydroxycholecalciferol, is also synthesized from cholesterol through three steps in the skin, liver, and kidneys, respectively. The eicosanoid hormones are locally produced within cell membranes from 20-carbon fatty acids, such as arachidonic acids derived from membrane lipids.

In general, the endocrine glands secrete active hormones, while some are converted to active forms in the peripheral tissues. For example, testosterone is converted to dihydrotestosterone in its target tissues.

Protein and peptide hormones are generally synthesized several days before their release and are stored in granules within the hormone-producing cells. Steroids on the other hand are not stored but synthesized and secreted immediately when a stimulus for their release occurs.

Most peptide hormones are not bound to carrier proteins in the circulation. Exceptions to this are the growth hormone, the antidiuretic hormone, oxytocin, and insulin-like growth factor (IGF)-I and IGF-II. In contrast, steroid and thyroid hormones are bound to plasma proteins. For example, cortisol-binding globulin binds cortisol; sex hormone-binding globulin binds testosterone and estradiol, while thyroxin-binding globulin binds thyroxin. In general, the unbound (free) form of hormone is the active form. Transport proteins may regulate hormone distribution to the various tissues, may form a depot for continuous slow release, or may decrease the rate of clearance of the hormones from the plasma.

The clearance of released hormone from the circulation is critical for the regulation of hormone concentration. The rate of clearance may vary from a few minutes for polypeptide hormones, to a few hours for steroid and glycoprotein hormones, to days for thyroid hormones. The peptide hormones are cleared from the circulation mostly by proteolytic mechanisms in lysosomes after their uptake by cells through binding to cell-surface receptors and nonreceptor hormone-binding sites. Steroid hormones are bound to carrier proteins in the blood as mentioned above. Binding to these carrier proteins is necessary since the steroid hormones are lipophilic. Only 5%–10% of the hormone is present in the unbound form. Steroid hormones are degraded in the liver and in the kidney and are excreted by the kidney in urine or by the liver in bile salts.

7.1.6 Mechanism of Hormone Action

The manner in which hormones are delivered to the target cells and the presence of specific receptors on the target cells determines the selectivity of hormonal activity. Special delivery systems, like the hypophyseal-portal system which links the hypothalamus to the pituitary, ensure the delivery of sufficient amounts of hormones to the target tissues. Beyond the delivery system, the concentration of specific receptors on the target tissue remains the primary determinant of the sensitivity of the tissue to a hormone [2].

Hormones bind specifically to hormone receptors with high affinity. Binding of a hormone to its receptor initiates a hormonal response. There are two general types of receptors: cell-surface receptors and intracellular receptors. While protein and peptide hormones generally interact with cell-surface receptors, steroid and thyroid hormones act on intracellular receptors.

7.1.6.1 Cell-Surface Receptors

Cell-surface receptors have ligand recognition domains on the outer surface of the cell membrane, one or more membrane-spanning domains, and a ligand-regulated intracytoplasmic effector domain [3]. Cell-surface receptors transmit the outer signal into the cell.

Cell-surface receptors use secondary messengers to generate a cell response. Cyclic AMP, cyclic GMP, and Ca^{2+} ions act as secondary messengers. Secondary messengers activate protein kinase A and protein kinase C pathways, both of which can phosphorylate and activate intracellular proteins, leading to mediation of cellular response (Figure 7.5).

7.1.6.2 Intracellular Receptors

Steroid and thyroid hormones act via intracellular receptors. Receptors may be found in the cytoplasm or in the nucleus. Glucocorticoid, mineralocorticoid, and androgen receptors are located in the cytoplasm, while estrogen, progesterone, thyroid hormone, 1,25-dihydroxycholecalciferol, and retinoic acid receptors are found in the nucleus.

Intracellular receptors have ligand-binding (hormone-binding) domains and DNA-binding domains. The DNA-binding domain is the region where the hormone–receptor complex binds to

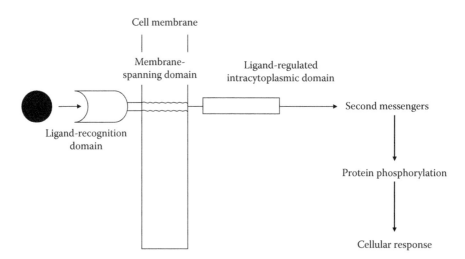

FIGURE 7.5 Structure and function of cell-surface receptors.

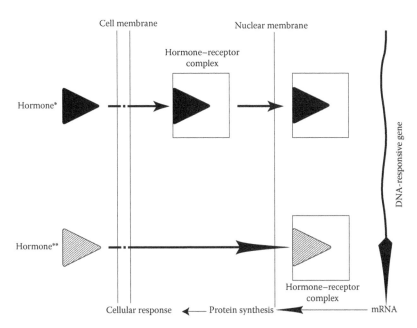

FIGURE 7.6 Structure and function of intracellular receptors. *Glucocorticoid, mineralocorticoid and androgen receptors are located in the cytoplasm. **Estrogen, progesterone, thyroid hormone, 1,25-dihydroxy-cholecalciferol and retinoic acid receptors are located in the nucleus.

hormone-responsive genes to stimulate transcription. Hormone-responsive genes contain specific hormone-responsive elements located at the binding sites. Activation of these hormone-responsive genes by the hormone–receptor complex leads to transcription of specific genes, which encode various proteins. As a result, synthesized proteins generate biological responses (Figure 7.6).

7.1.7 LIGAND CHARACTERISTICS ON HORMONE ACTION

Ligands may have agonistic, antagonistic, or partial agonistic–antagonistic effects on hormone receptors. An agonist binds to its receptor and forms a cellular response as expected. Some synthetic hormone analogs may have more potent activity than the natural hormone. An antagonist binds to a receptor but fails to produce any response. The antagonist usually competes with the agonist for binding to the receptor, thereby occupying the space available for the agonist on the receptors. Some antagonists, however, do not competitively prevent cellular responses this way. They may do so by other mechanisms. Partial agonists or antagonists on the other hand bind to receptors and form a response that is less than that of a full agonist [3]. A partial agonist blocks the binding of an agonist so it acts like a partial antagonist. For example, plant estrogens (phytoestrogens) are partial agonists of estrogen receptors [3]. Some hormone analogs may have agonistic effects in a tissue, but may have antagonistic effects in another tissue and they are called as mixed agonists–antagonists [3].

These properties of ligands are very important due to their exposure to various exogenous hormones and substances that act like hormones.

7.2 HORMONAL REGULATION OF HUMAN METABOLISM

7.2.1 GENERAL CHARACTERISTICS OF HUMAN METABOLISM

The term metabolism refers to all the chemical reactions that produce energy and build up complex molecules required for normal functioning by using this energy in an organism. These processes provide growth, structural stability, and response to environmental changes in organisms. A striking feature of metabolism is the similarity of the basic metabolic pathways between even the most complex and simplest species of living organisms. For example, the series of chemical steps in a pathway such as the citric acid cycle is universal among living cells in the unicellular bacteria as in multicellular organisms [5].

Energy metabolism is primarily concerned with heat production in an organism, while intermediary metabolism is related to the whole chemical reactions in a complex multicellular organism. Intermediary metabolism has not been thought of as a separate event from the energy metabolism at the cellular level. Moreover, the two mechanisms always act in a continuous manner in concert. In general, the term metabolism is interpreted to mean intermediary metabolism.

In a differentiated organism, each tissue must be provided with fuels that it can utilize for its own energy needs to perform its function. For example, muscles need to generate adenosine triphosphate (ATP) for their mechanical work of contraction, and the liver needs ATP for the synthesis of plasma proteins and fatty acids, gluconeogenesis, or for the production of urea for the excretion of nitrogenous compounds.

All metabolic reactions fall into one of two general categories: catabolic and anabolic reactions. The synthesis of larger molecules from smaller ones and the building of tissues are known as anabolic reactions. The process of breaking down tissue—larger complex molecules—into more simpler and smaller forms on the other hand is termed catabolism. Growth or weight gain occurs when anabolism exceeds catabolism. On the other hand, if catabolism occurs more rapidly than anabolism, weight loss occurs, as in periods of starvation and disease.

Both anabolic and catabolic processes include a vast number of different chemical reactions, albeit a number of common features. Most of the metabolic processes occur mainly in the cytoplasm, but can occur inside intracellular organelles, such as the mitochondria. Anabolic and catabolic reactions involve the action of enzymes and the utilization of energy. The metabolism of the whole body is controlled in an integrated fashion by the action of hormones and/or the nervous system.

Cells capture and store the energy released in catabolic reactions through the use of chemical compounds known as energy carriers. One of the most important energy carriers is adenosine triphosphate. This nucleotide is used to transfer chemical energy between different chemical reactions. Cells generally do not store large amounts of ATP, although the human body can generate enough energy to synthesize about its own weight in ATP every day. On the contrary, it is continuously regenerated in variable amounts according to the needs of the cells. ATP acts as a bridge between catabolism and anabolism, with catabolic reactions generating ATP and anabolic reactions consuming it. It also serves as a carrier of phosphate groups in phosphorylation reactions.

Nicotinamide adenine dinucleotide (NAD) is an important coenzyme that acts as a hydrogen acceptor. Many types of dehydrogenases remove electrons from their substrates and reduce NAD^+ into NADH. This reduced form of the coenzyme then serves as a substrate for any of the reductases in the cell that need to reduce their substrates. Nicotinamide adenine dinucleotide exists in two related forms in the cell: NADH and NADPH. The NAD^+/NADH is used in catabolic reactions, while $NADP^+$/NADPH is used in anabolic reactions.

7.2.1.1 Anabolic Processes

Anabolic processes use substrates in the diet to synthesize complex molecules such as cell membranes, store these nutrients for later use when needed, and synthesize hormones and proteins which are secreted from cells.

Anabolism involves three basic stages. The production of precursors, such as amino acids, monosaccharides, and nucleotides, is the first step. The second step involves their transformation into reactive forms using energy from ATP. Complex molecules such as proteins, polysaccharides, lipids, and nucleic acids are then synthesized from these precursors in the final stage.

7.2.1.2 Catabolic Processes

Catabolism is the set of metabolic processes involved in the breakdown of energy-containing components of the diet to provide energy for cells. These catabolic processes require the presence of appropriate enzymes.

Catabolic reactions in humans can be distinguished into three main stages. In the first, large organic molecules, such as proteins, polysaccharides, and lipids, are broken down into their smaller components outside cells. Next, these smaller molecules are converted mainly to acetyl coenzyme A in cells accompanied by some energy release. At the last stage, the acetyl group on the coenzyme A is then oxidized to water and carbon dioxide, releasing energy that is stored by reducing the coenzyme NAD^+ into NADH.

The major pathway of catabolism of monosaccharides, such as glucose and fructose, is glycolysis. Monosaccharides are converted into pyruvate with the generation of ATP [6]. Pyruvate is an intermediate in several metabolic pathways, but the majority of it is converted to acetyl coenzyme A, which enters the citric acid cycle. Although more ATP is generated in the citric acid cycle, the most important product is NADH, which is derived from NAD^+ as the acetyl coenzyme A is oxidized. This oxidation releases carbon dioxide as a waste product. An alternative route for glucose catabolism is the pentose phosphate pathway, in which pentose sugars such as ribose is produced.

The generation of glucose from compounds like pyruvate, lactate, glycerol, and amino acids is called gluconeogenesis. Although body fat represents a huge store of energy, in humans the fatty acids cannot be converted to glucose through gluconeogenesis because humans cannot convert acetyl coenzyme A into pyruvate. After long-term starvation, humans produce ketone bodies from fatty acids to replace glucose in tissues such as the brain that cannot metabolize fatty acids. Fats are catabolized by hydrolysis to free fatty acids and glycerol. The glycerol enters glycolysis and the fatty acids are broken down by beta oxidation to release acetyl coenzyme A which enters the citric acid cycle.

Amino acids are either used to synthesize proteins and other biomolecules or oxidized to urea and carbon dioxide as a source of energy. The oxidation pathway starts with the removal of the amino group by a transaminase. The amino group then enters the urea cycle, leaving a deaminated carbon skeleton in the form of a keto acid. Several of these keto acids are intermediates in the citric acid cycle [7]. The gluconeogenic amino acids can also be converted into glucose through gluconeogenesis.

Another pathway where catabolism supplies energy is by oxidative phosphorylation. In oxidative phosphorylation, the electrons removed from the food molecules in pathways, such as the citric acid cycle, are transferred to oxygen and the energy released is used to make ATP. This is done by a series of proteins in the membranes of the mitochondria called the electron transport chain. These proteins use the energy released from the reduced electron carriers NADH and $FADH_2$ (reduced flavin adenine dinucleotide) to pump protons across the membrane [8]. Pumping protons out of the mitochondria creates a proton concentration gradient across the membrane generating an electrochemical gradient. This force drives the protons back into the mitochondria. The flow of protons turns adenosine diphosphate into ATP. Figure 7.7 shows the catabolic processes in the body.

7.2.2 Control of Metabolism

For the body to function efficiently there has to be an effective means of controlling and integrating the metabolic processes occurring in all the cells, tissues, and organs. This integration and control is mainly achieved by circulating hormones, with their release being regulated in turn partly by the nervous system and partly by direct effects of substances in the blood on the endocrine glands.

The brain utilizes only glucose to meet its energy requirements under normal conditions. However, the brain can adapt during fasting, to use ketone bodies instead of glucose as a major fuel [9].

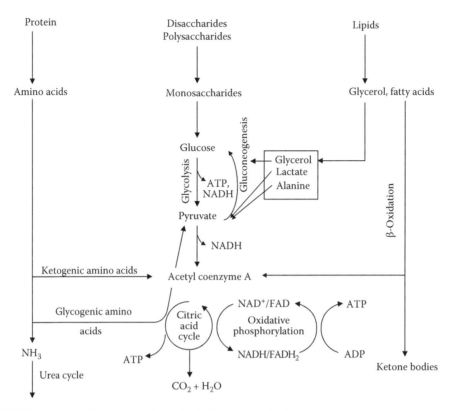

FIGURE 7.7 The catabolic pathways in the body. See the text for details.

Muscles can utilize glucose, fatty acids, or ketone bodies. Fatty acids represent the main source of energy for muscles in the resting state while glucose serves as the primary energy source in exercising muscles. In the early stages of exercise, glucose is supplied from the muscles' own glycogen stores. During exercise, the rate of glycolysis exceeds the rate of the citric acid cycle; as a result lactate accumulates and is released. Another metabolic product is alanine, produced via transamination from pyruvate. Both lactate and alanine are transported to the liver and converted to glucose through gluconeogenesis. Muscle proteins are another source of energy but the breakdown of muscle proteins to meet energy demand is an undesirable event, so protein breakdown occurs in order to survive [9].

Adipose tissue is the major energy store in humans. Synthesis and breakdown of triglycerides are constant events in adipose tissue. Fatty acid and glycerol mobilization is stimulated by activation of hormone-sensitive lipase [9].

The liver is the first organ that all digested molecules pass through. It takes up these molecules for metabolic processes. The liver is the major site for fatty acid synthesis. Glucose is produced from hepatic glycogen stores and from gluconeogenesis. The liver uses lactate from muscle, glycerol from adipose tissue, and the amino acids that are not included in protein synthesis for gluconeogenesis. Liver also synthesizes ketone bodies from fatty acids.

7.2.2.1 Hormones in the Regulation of Metabolism

Hormones involved in metabolism are responsible for the regulation of fuel storage in excess of current requirements as well as their mobilization upon increased demand [10]. Blood glucose levels are maintained within narrow limits despite wide fluctuations in glucose intake for proper functioning of the body, particularly of the nervous system. Insulin is the sole regulator of anabolic reactions whereas glucagon, epinephrine, norepinephrine, cortisol, and growth hormone are the hormonal regulators of catabolic reactions.

Insulin is a polypeptide hormone synthesized and secreted by the β cells of the islets of Langerhans in the pancreas. In addition to a variety of signals like the hormones related to digestion such as secretin, gastrin, glucagon-like peptide and many others, and amino acids like arginine, the major regulator signal for insulin release is the blood glucose levels.

The primary targets for insulin are the skeletal and cardiac muscles, adipose tissue, and liver. Insulin has anabolic effects on carbohydrate, lipid, and protein metabolism in these tissues. Insulin induces anabolic reactions while inhibiting catabolic reactions. It also has important effects on growth.

Insulin secretion stimulates glucose uptake by insulin-sensitive tissues which is the rate-limiting step in glucose utilization and storage. Insulin increases glycolysis, glycogenesis, and lipogenesis while inhibiting gluconeogenesis, glycogenolysis, and lipolysis. Insulin increases the activity of various enzymes like glycogen synthetase in glycogen production and inactivates some enzymes like glycogen phosphorylase. Insulin also regulates the expression of various genes that encode liver enzymes. It down regulates the transcription of gluconeogenic enzymes like fructose-1,6-bisphosphatase and glucose-6-phosphatase, and also increases transcription of glycolytic enzymes such as pyruvate kinase and lipogenic enzymes such as fatty acid synthase [10].

Insulin increases glucose uptake, and also promotes fatty acid uptake by enhancing lipoprotein lipase activity in adipocytes. Insulin inhibits lipolysis through the inhibition of hormone-sensitive lipase in adipose tissue [10].

Insulin has anabolic effects on protein metabolism by stimulating amino acid uptake and protein synthesis and by inhibiting protein breakdown [10].

Glucagon is synthesized and secreted by the α cells of the islets of Langerhans in the pancreas. Glucagon acts in the opposite direction to insulin and its secretion is regulated by glucose level and insulin. Hypoglycemia stimulates and hyperglycemia inhibits its secretion. Glucagon stimulates glycogenolysis and gluconeogenesis in the liver. It also stimulates the production of ketone bodies.

Epinephrine is released by the adrenal medulla and norepinephrine is secreted by nerve terminals adjacent to target cells. They are released in response to acute and chronic stresses such as hypoglycemia and pain. Epinephrine and norepinephrine have catabolic actions in order to supply energy in stress conditions and they also produce glucose to prevent hypoglycemia. Catecholamines suppress insulin secretion. They increase hepatic glycogenolysis, gluconeogenesis, and also gluconeogenic precursors such as lactate, alanine, and glycerol. Catecholamines also stimulate mobilization of fatty acids and glycerol from adipose tissue by activating hormone-sensitive lipase.

Hypocortisolemia is a life-threatening condition and cortisol is essential for survival under stress. Cortisol enhances metabolic reactions rather than initiating them.

Cortisol stimulates glycogenesis by increasing glycogen synthetase and inhibiting glycogen phosphorylase. It also stimulates gluconeogenic enzyme transcription mainly glucose-6-phosphatase and phosphoenolpyruvate carboxykinase, so it increases hepatic glucose production. Cortisol inhibits the uptake and utilization of glucose in muscles and the adipose tissue. It induces lipolysis and increases the concentrations of fatty acids and glycerol in blood. Its major catabolic effect is to facilitate the breakdown of proteins in muscle and connective tissue. This results in an increase in the level of amino acids in the blood which are used for gluconeogenesis [10].

Growth hormone has important effects on catabolism and glucose homeostasis. IGF-I is a growth factor induced by the growth hormone and mediates most of the growth-promoting effects of the hormone.

In general, growth hormone decreases glucose utilization and increases lipolysis resulting in an increase in blood glucose levels. Growth hormone also is an anabolic hormone as it stimulates protein synthesis and muscle mass. It causes positive nitrogen balance in the body. Growth hormone increases fatty acids and glycerol levels by lipolysis via the activation of hormone-sensitive lipase. As a result of lipolysis, the increased glycerol level provides substrates for gluconeogenesis and stimulates glycogenesis in the liver. Glucose catabolism in the liver and glucose uptake by peripheral tissues are decreased by the growth hormone.

Growth hormone stimulates amino acid transport into the muscle. The ultimate metabolic actions of growth hormone in normal conditions are for the synthesis of proteins in the body by regulating metabolic processes mentioned just above.

The thyroid hormone increases basal metabolic rate (resting energy expenditure). It has both direct actions and indirect effects on the metabolism by modifying the activities of other hormones such as catecholamines. The thyroid hormone also promotes growth. The thyroid hormone's action on metabolism is dose dependent.

In general, the thyroid hormone increases glucose level by glycogenolysis and gluconeogenesis, stimulates lipolysis especially in hyperthyroidism. In normal conditions, the thyroid hormone has stimulatory effects on the synthesis of proteins but also leads to catabolic reactions in muscle under hypo- and hyperthyroid states [10].

7.3   ANIMAL HORMONES AND HORMONES USED IN FARM ANIMALS

This section focuses not only on the similarities and differences between hormones in humans and other nonprimate mammals but also mainly on hormones used in animal production and on their possible health impacts.

Hormones in humans, mammals, and other vertebrates are nearly identical in chemical structure and function. Although steroid hormones are completely identical among mammals, the structure of peptide hormones frequently differs among the animal species. This difference in structure can limit the ability of the peptide hormones from one species to another. The structure of the bovine growth hormone for example is different from that of the humans. As a result binding to the receptor does not occur; thus making the bovine growth hormone biologically inactive in humans. But it is not the case for every peptide hormone, for example, porcine and beef insulins differ from human insulin by one and three amino acids, respectively. These insulins were used for many years in humans with great success.

Steroid hormones are not extensively digested, can be absorbed intact, and can be effective when taken orally but peptide hormones like insulin and the growth hormone are degraded and have no effect when taken orally.

7.3.1   HORMONES USED IN FARM ANIMAL PRODUCTION

Hormone-dependent sex difference in the growth rate is a well-known phenomenon. Naturally occurring sex steroids estradiol, progesterone, and testosterone have growth-promoting effects in man and animals.

The growth rate and feed conversion efficiency are higher in intact males than in castrated animals. A number of different approaches may be taken to improve the conversion of animal feed into meat; one of which is the application of hormones. The hormonal approach includes administration of anabolic sex steroids to either support the animal's steroid production rate or to replace steroids lost through castration. The growth hormone is another hormone used in animal production.

Diethylstilbestrol and hexoestrol were administered to cattle increasingly from the mid-1950s either as feed additives or implants. This approach resulted in a 10%–15% increase in daily weight gain, improvement in feed conversion efficiency, and lean/fat mass ratio. As a result of hormone application, the energy required per unit weight of meat produced has reduced [11].

In general, anabolic hormones stimulate increased incorporation of amino acids into protein in the muscles and mobilize fat stores resulting in an increased growth rate and leaner carcasses [12].

Anabolic hormones will be effective if nutrition of the animal is adequate. Inadequate food intake leads to a stress condition in the animal which stimulates glucocorticoid secretion. Glucocorticoid secretion in turn decreases protein synthesis and increases amino acid catabolism [12].

7.3.1.1 Natural Steroids

The natural steroids used for anabolic purposes in farm animals are estradiol-17β or its benzoic and propionic acid esters, progesterone, and testosterone.

Estradiol-17β, a potent anabolic agent in ruminants, is administered as an ear implant. It is usually combined with testosterone or progesterone for the main purpose of decreased release rate of estradiol thereby prolonging the duration of the effectiveness of the implant. Estradiol-17β in farm animals with low endogenous estrogen production such as veal calves, lambs, heifers, and steers increases growth by 5%–15% [12].

Estradiol has both direct and indirect effects. Estrogen receptors were found in bovine skeletal muscle [13] and the linear relation between the receptor concentration and growth response was documented [14]. In addition, estradiol may act by the stimulation of growth hormone secretion [15] and IGF-1 production [16].

Androgens are classical anabolic agents in humans, but they are less effective in farm animals like cattle due to their lower androgen receptor concentrations [14]. Testosterone is not used on its own as an anabolic agent in farm animals because of the inadequate delivery systems available for achieving effective concentrations. It is used in conjunction with estradiol. Its major role may be to slow down the release rate of estradiol.

Although some findings indicated anabolic actions of progesterone, there is no unambiguous data suggesting progesterone is anabolic in farm animals [12]. Its major use is to slow the release of estradiol from implants.

7.3.1.2 Synthetic Steroids

Synthetic steroids are trenbolone acetate—containing androgenic and melengestrol acetate—that has progestogenic effects.

Trenbolone acetate (TBA) is currently the only synthetic androgen approved for use for growth promotion in cattle, and it is also used to a lesser extent in sheep but not in pigs. The TBA is very efficient because of its multiple hormonal activities. The TBA strongly binds to androgen, progestin, and glucocorticoid receptors [17]. It has less androgenic but greater anabolic activity than testosterone. It has also antiglucocorticoid property, so both anabolic activity as an androgen and anticatabolic activity as an antiglucocorticoid make it a strong growth promoter [12]. The TBA has significant anabolic effects on its own in female cattle and sheep, but in castrated males, maximal response is achieved when used in conjunction with estrogens.

Melengestrol acetate is an orally active synthetic progestogen which increases the growth rate and feed efficiency in heifers. It is not effective in pregnant or spayed heifers or in steers. Its mode of action is to suppress ovulation presumably by suppressing luteinizing hormone pulse frequency; however, large follicles develop, which can increase concentrations of estradiol and growth hormone, and hence growth [18,19].

7.3.1.3 Synthetic Nonsteroidal Estrogens

Stilbene estrogens (either diethylstilbestrol [DES] or hexoestrol) have been banned in most countries as anabolic agents because of residue and food safety concerns.

The discovery of a naturally occurring estrogen—zearalenone (produced by the fungi *Fusarium* spp.)—led to the development of the synthetic analog zeranol. Among synthetic estrogens, only zeranol is used for cattle fattening in certain countries. Zeranol is estrogenic and has a weak affinity for the uterine estradiol receptor. It increases nitrogen retention, growth rate, and feed conversion. However, lower responses are seen in heifers. In addition to zeranol, its metabolites β-zearalanol (taleranol) and zearalanone also contribute to the total hormonal activity.

7.3.1.4 Growth Hormone

Both bovine and porcine growth hormones are used for meat production. When administered to cattle, the growth hormone increases the growth rate (5%–10%), feed conversion efficiency, and the carcass lean mass to fat ratio. The gender has little effect on response in cattle. The growth hormone improves growth and feed efficiency in sheep but not in poultry, and has dramatic growth-promoting effects in pigs. It induces growth by reducing glucose utilization by the adipose tissue, by decreasing fatty acid synthesis, by decreasing hepatic amino acid degradation, by supplying more glucose and amino acids to the muscles, by increasing IGF-I synthesis, and by direct effect on the muscles [12]. The effects of the growth hormone are largely additional to those obtained from steroid implants. The growth hormone has been approved for commercial use in some countries to increase milk production.

The hormone preparations used as growth promoters are listed in Table 7.3.

7.3.2 Impact of Anabolic Agents on Human Health

All foodstuffs of animal origin contains natural sex steroids at varying concentrations depending on the kind of tissue, species, gender, age, and physiological stage of the animal. Estradiol-17β is the most potent estrogen in animals and metabolized to estrone or estradiol-17α that are

TABLE 7.3
Hormonally Active Substances Used in Animal Production

Substances	Dose Levels	Form	Main Use—Animals	Trade Name
Estrogens Alone				
DES	10–20 mg/day	Feed additive	Steers, heifers	
DES	30–60 mg/day	Implant	Steers	
DES		Oil solution	Veal calves	
Hexoestrol	12–60 mg	Implant	Steers, sheep, calves, poultry	
Zeranol	12–36 mg	Implant	Steers, sheep	Ralgro
Gestagens Alone				
Melengestrol acetate	0.25–0.50 mg/day	Heifers		
Androgens Alone				
TBA	300 mg	Implant	Heifers, culled cows	Finaplix
Combined Preparations				
DES + testosterone	25 mg; 120 mg	Implant	Calves	Rapigain
DES + methyl-testosterone		Feed additive	Swine	Maxymin
Hexoestrol + TBA	30–45 mg; 300 mg	Implant	Steers	
Zeranol + TBA	36 mg; 300 mg	Implant	Steers	
Estradiol-17β + TBA	20 mg; 140 mg	Implant	Bulls, steers, calves, sheep	Revalor
Estradiol-17β benzoate + testosterone propionate	20 mg; 200 mg	Implant	Heifers, calves	Synovex H Implix BF
Estradiol-17β benzoate + progesterone	20 mg; 200 mg	Implant	Steers	Synovex S Implix BM

Source: Velle, W., FAO Animal Production and Health Paper: 31, The use of hormones in animal production, Department of Physiology, Veterinary College of Norway, Oslo, Norway, 1981.

Note: DES, diethylstilbestrol; TBA, trenbolone acetate.

less orally active than estradiol-17β. The metabolites of estradiol-17β-estrone and estradiol-17α contribute to the total estrogenic activity in animals [20]. The lowest estrogen concentrations were reported in steers (<3 pg/g) in some studies [20]. But some authors concluded that estradiol-17β and estrone concentrations in steers were in the same range in heifers and cycling cows (5–15 pg/g) [21].

With the rising use of anabolic agents in animal production opposition to their use also has increased. The DES was forbidden in 1979 as an anabolic agent in all farm animals in the USA. The use of anabolic hormones in meat production has officially been prohibited in the EU since 1989.

Estradiol, progesterone, and testosterone are natural hormones occurring in both humans and animals in identical molecular forms. Therefore, the naturally occurring sex hormones taken up by the ingestion of the animal tissue have the same biological activity as the endogenously produced hormones in humans. A person can be exposed to sex hormones not only by ingestion of meat, but also by dairy products and eggs.

The estradiol-17β concentrations in treated animals in their muscle, fat, liver, and kidney are 3.7-, 6.4-, 4.2-, and 6.2-fold, respectively, compared with untreated animals [20]. Estrone concentrations are also 1.2, 1.9, 1.4, and 1.6 times greater in muscle, fat, liver, and kidney, respectively [20]. The concentrations of estradiol-17α are 10–100-fold higher than estradiol-17β and estrone and the concentration of estradiol-17α cannot be ignored although this is not the case in most of the residue studies [20]. According to JECFA (The Joint Food and Agricultural Organization/World Health Organization [FAO/WHO] expert committee on food additives), the residue levels of estradiol, estrone, progesterone, and testosterone in treated cattle were twofold or higher than untreated calves, heifers, and steers [22].

Whether the levels of residues of sex steroids and their metabolites in edible tissues from treated animals are significantly higher than the control animals or not, the average amount of residue intake is the important point that must be considered to comprehend the impact of these hormones on human health. Also, whether the levels of the residues of the sex steroids and their metabolites have a significant impact on the consumer's health or not depends on the exogenous doses supplied in relation to the physiological levels of these hormones in the consumer [23].

In the U.S. Food and Drug Administration's guideline for toxicology testing, the conclusion about the use of natural sex steroids in meat production is that, there will be no physiological effect in individuals chronically ingesting animal tissues that contain an increase of the endogenous steroid equal to 1% or less of the amount in micrograms produced by daily synthesis in the segment of the population with the lowest daily production [24]. The lowest daily production for estradiol and progesterone is seen in prepubertal boys with that for testosterone established in prepubertal girls. In the case of estradiol, the estimated production rate was 6.5 μg/day in prepubertal boys [24], and the maximum acceptable daily intake was therefore estimated to be 65 ng/day [25]. However, the maximum acceptable daily intake remains questionable. Daily production rates are calculated according to the following equation:

Daily production rate (μg/day) = plasma concentration (μg/ml) × metabolic clearance rate (ml/day) [25].

The concentrations of plasma estradiol in prepubertal children were highly overestimated in earlier studies using radioimmunoassay. Klein et al. found that levels of estradiol in prepubertal boys may be more than 100 times lower with ultrasensitive and highly specific assays than previously reported [26]. Besides, the metabolic clearance rate used for the calculation of daily production rate of estradiol was based on values obtained from adults. As a result, the maximum acceptable daily intake may be significantly lower than the previously reported value. The reported residue levels still remain uncertain, with only limited data on the levels of metabolites of the anabolic steroids, and children are extremely sensitive to very low levels of sex steroids. The average concentration of estradiol-17β in meat from treated cattle was estimated to be 20 ng/500 g [20]. In the light of results from the study of Klein et al. [26], the maximum acceptable daily intake for estradiol was reduced to 400 pg/day [23]. This level corresponds to ingestion of treated meat as low as 10 g/day [20]. Due to these reasons, continuous exposure to low dose exogenous sex steroids may cause diseases like precocious puberty, spermatic abnormalities, infertility, and hormone-related cancers like breast and prostate cancers. Gynecomastia and elevated serum estradiol levels in boys and girls attending a school in Milan were reported. Although it was never confirmed, poultry and beef from the school cafeteria were the suspected source of this exogenous estrogen [27,28].

7.4 PLANT HORMONES

Hormones in plants are called phytohormones. This group includes auxin, cytokinin, the gibberellins, abscisic acid, ethylene, the brassinosteroids, and jasmonic acid. With the notable exception of the steroidal hormones of the brassinosteroids, plant hormones bear little resemblance to their animal counterparts [29]. Rather, they are relatively simple, small molecules such as ethylene and indole-3-acetic acid, the primary auxin in the majority of plant species [29]. Plant hormones are chemicals that affect flowering; aging; root growth; distortion and killing of leaves, stems, and other parts; prevention or promotion of stem elongation; color enhancement of fruit; prevention of foliation and/or leaf fall; and many other conditions. Very small concentrations of these substances produce major growth changes. Phytohormones exert their effects via specific receptor sites in target cells, similar to the mechanism found in animals.

Hormones are produced naturally by plants, while plant growth regulators are applied to plants by humans. Plant growth regulators may be synthetic compounds (e.g., Cycocel) that mimic naturally occurring plant hormones, or they may be natural hormones that are extracted from plant tissue (e.g., indole-3-acetic acid).

These substances are usually applied at concentrations measurable in parts per million (ppm) and in some cases parts per billion (ppb). These growth-regulating substances most often are applied as a spray to foliage or as a liquid drench to soil around a plant's base. Generally, their effects are short lived, and they may need to be reapplied in order to achieve the desired effect.

There are five groups of plant-growth-regulating compounds: auxin, gibberellin, cytokinin, ethylene, and abscisic acid. For the most part, each group contains both naturally occurring hormones and synthetic substances.

Auxins are responsible for bending toward a light source (phototropism), downward root growth in response to gravity (geotropism), promotion of apical dominance (the tendency of an apical bud to produce hormones that suppress growth of the buds below it on the stem), flower formation, fruit set and growth, formation of adventitious roots.

Auxin is the active ingredient in most rooting compounds in which cuttings are dipped during vegetative propagation.

Gibberellins stimulate cell division and elongation, break seed dormancy, and speed up germination. If the seeds of some species are difficult to germinate; you can soak them in a gibberellin solution to get them started.

Unlike other hormones, cytokinins are found in both plants and animals. They stimulate cell division and often are included in the sterile media used for growing plants from tissue culture. If the growth-regulating compounds in a medium is high in cytokinins but low in auxin, the tissue culture explant (small plant part) will produce numerous shoots. On the other hand, if the mixture has a high ratio of auxin to cytokinin, the explant will produce more roots. Cytokinins also are used to delay aging and death (senescence).

Ethylene is unique in that it is found only in the gaseous form. It induces ripening, causes leaves to droop (epinasty) and drop (abscission), and promotes senescence. Plants often increase ethylene production in response to stress, and ethylene often is found in high concentrations within cells at the end of a plant's life. The increased ethylene in leaf tissue in the fall is part of the reason leaves fall off trees. Ethylene also is used to ripen fruits (e.g., green bananas).

Abscisic acid is a general plant-growth inhibitor. It induces dormancy and prevents seeds from germinating; causes abscission of leaves, fruits, and flowers; and causes stomata to close. High concentrations of abscisic acid in guard cells during periods of stress induced by drought probably have a role in stomatal closure.

2,4-Dichlorophenoxyacetic acid (2,4-D) is a commonly used herbicide. Herbicides are chemicals used to control weed growth. 2,4-D belongs to the group of related synthetic herbicides called chlorophenoxy herbicides. The chemical structure of 2,4-D resembles indoleacetic acid, a naturally occurring hormone produced by plants to regulate their own growth. This resemblance allows 2,4-D to artificially regulate plant growth.

In mammals, 2,4-D disrupts energy production [30], and causes cellular mutations which can lead to cancer. Numerous epidemiological studies have linked 2,4-D to non-Hodgkin's lymphoma (NHL) among farmers [31,32].

As an endocrine disrupter, 2,4-D causes suppression of thyroid hormone levels, increases thyroid gland mass, and decreases weight of the ovaries and testes [33]. 2,4-D causes slight decreases in testosterone release and significant increases in estrogen release from testicular cells [34]. Male farm sprayers exposed to 2,4-D had lower sperm counts and more spermatic abnormalities compared to men who were not exposed to this chemical [35].

ACKNOWLEDGMENTS

I would like to thank to Professor Dr. Hatice Pasaoglu and Professor Dr. Nuri Cakir for their critical reviews.

REFERENCES

1. Karakoç, A. and Arslan, M., Reproductive hormones in females and hormone replacement therapy, in *Phytoestrogens in Functional Foods*, Yildiz, F., Ed., Taylor & Francis Group, Boca Raton, FL, pp. 169–208, 2006.
2. Frohman, L.A. and Felig, P., Introduction to the endocrine system, in *Endocrinology & Metabolism*, Felig, P. and Frohman, L.A., Eds., 4th edn., McGraw-Hill Companies, Inc., Columbus, OH, pp. 3–17, 2001.
3. Baxter, J.D., Ribeiro, R.C.J., and Webb, P., Introduction to endocrinology, in *Basic & Clinical Endocrinology*, Greenspan, F.S. and Gardner, D.G., Eds., 7th edn., McGraw-Hill Companies, Inc., Columbus, OH, pp. 1–37, 2004.
4. Gordon, N.G., Biosynthesis, secretion and metabolism of hormones, in *Endocrinology & Metabolism*, Felig, P. and Frohman, L.A., Eds., 4th edn., McGraw-Hill Companies, Inc., Columbus, OH, pp. 29–48, 2001.
5. Smith, E. and Morowitz, H., Universality in intermediary metabolism, *Proc Natl Acad Sci U S A*, 101, 13168–13173, 2004.
6. Bouché, C., Serdy, S., Kahn, C., and Goldfine, A., The cellular fate of glucose and its relevance in type 2 diabetes, *Endocr Rev*, 25, 807–830, 2004.
7. Brosnan, J., Glutamate, at the interface between amino acid and carbohydrate metabolism, *J Nutr*, 130 (4S Suppl), 988S–990S, 2000.
8. Schultz, B. and Chan, S., Structures and proton-pumping strategies of mitochondrial respiratory enzymes, *Annu Rev Biophys Biomol Struct*, 30, 23–65, 2001.
9. Mathews, C.K. and van Holde, K.E., Integration and control of metabolic processes, in *Biochemistry*, The Benjamin/Cummings Publishing Company, Inc., Redwood City, CA, pp. 779–812, 1990.
10. Greenway, S.C., Hormones in human metabolism and disease, in *Functional Metabolism: Regulation and Adaptation*, Storey, K.B., Ed., John Wiley & Sons, Inc., Hoboken, NJ, pp. 271–294, 2005.
11. Velle, W., FAO Animal Production and Health Paper: 31. The use of hormones in animal production, Department of Physiology, Veterinary College of Norway, Oslo, Norway, 1981.
12. Meyer, H.H.D., Biochemistry and physiology of anabolic hormones used for improvement of meat production, *APMIS*, 109, 1–8, 2001.

13. Meyer, H.H.D. and Rapp, M., Estrogen receptor in bovine skeletal muscle, *J Anim Sci*, 60, 94–300, 1985.

14. Sauerwein, H. and Meyer, H.H.D., Androgen and estrogen receptors in bovine skeletal muscle: Relation to steroid induced allometric muscle growth, *J Anim Sci*, 67, 206–212, 1989.

15. Davis, S.L., Ohlson, D.L., Klindt, J., and Anfinson, N.S., Episodic growth hormone secretory patterns in sheep: Relationship to gonadal steroids, *Am J Physiol*, 233, E519–523, 1977.

16. Sauerwein, H., Meyer, H.H.D., and Schams, D., Divergent effects of oestrogens on the somatotropic axis in male and female calves, *J Reprod Develop*, 38, 271–278, 1992.

17. Bauer, E.R.S., Daxenberger, A., Petri, T., Sauerwein, H., and Meyer, H.H.D., Characterisation of the affinity of different anabolics and synthetic hormones to the human androgen receptor, human sex hormone binding globulin and to the bovine gestagen receptor, *APMIS*, 108, 838–846, 2000.

18. Henricks, D.M., Brandt, R.T. Jr., Titgemeyer, E.C., and Milton, C.T., Serum concentrations of trenbolone 17β and estradiol-17β and performance of heifers treated with trenbolone acetate, melenges trol acetate, or estradiol-17β, *J Anim Sci*, 75, 2627–2633, 1997.

19. Hageleit, M., Daxenberger, A., Kraetzl, W.D., Ketler, A., and Meyer, H.H.D., Dose-dependent effects of melengestrol acetate (MGA) on plasma levels of estradiol, progesterone and luteinizing hormone in cycling heifers and influences on oestrogen residues in edible tissues, *APMIS*, 108, 847–854 (8), 2000.

20. Daxenberger, A., Ibarreta, D., and Meyer, H.H.D., Possible health impact of animal oestrogens in food, *Human Reprod Update*, 7, 340–355, 2001.

21. Henricks, D.M., Gray, S.L., and Hoover, J.L.B., Residue levels of endogenous estrogens in beef tissues, *J Anim Sci*, 57, 247–255, 1983.

22. The joint FAO/WHO Expert Committee on Food Additives. Residues of some veterinary drugs in animals and foods. Food and Agriculture Organization of the United Nations FAO Food and Nutrition paper 41. 1988.

23. Andersson, A. and Skakkebaek, N.E., Exposure to exogenous estrogens in food: Possible impact on human development and health, *Eur J Endocrinol*, 140, 477–485, 1999.

24. U.S. Food and Drug Administration. Guideline 3, part 2: Guideline for toxicological testing. www.fda.gov 1–5, 1999 (accessed March 22, 2008).

25. Aksglaede, L., Juul, A., Leffers, H., Skakkebaek, N.E., and Andersson, A., The sensitivity of the child to sex steroids: Possible impact of exogenous estrogens, *Human Reprod Update*, 12, 341–349, 2006.

26. Klein, K.O., Baron, J., Colli, M.J., McDonnell, D.P., and Cutler, G.B. Jr., Estrogen levels in childhood determined by an ultrasensitive recombinant cell bioassay, *J Clin Invest*, 94, 2475–2480, 1994.

27. Scaglioni, S., Di Pietro, C., Bigatello, A., and Chiumello, G., Breast enlargement at an Italian school, *Lancet*, 1, 551–552, 1978.

28. Fara, G.M., Del Corvo, G., Bernuzzi, S., Bigatello, A., Di Pietro, C., Scaglioni, S., and Chiumello, G., Epidemic of breast enlargement in an Italian school, *Lancet*, 2, 295–297, 1979.

29. Gray, W.M., Hormonal regulation of plant growth and development, *PLoS Biol*, 2, 9, e311, 2004.

30. Zychlinkski, L. and Zolnierowicz, S., Comparison of uncoupling activities of chlorophenoxy herbicides in rat liver mitochondria, *Toxicol Lett*, 52, 25–34, 1990.

31. Zahm, S.H., Mortality study of pesticide applicators and other employees of a lawn care service company, *J Occup Environ Med*, 39, 1055–1067, 1997.

32. Fontana, A., Picoco, C., Masala, G., Prastaro, C., and Vineis, P., Incidence rates of lymphomas and environmental measurements of phenoxy herbicides: Ecological analysis and case-control study, *Arch Environ Health*, 53, 384–387, 1998.

33. Charles, J.M., Cunny, H.C., Wilson, R.D., and Bus, J.S., Comparative subchronic studies on 2,4-D, amine, and ester in rats, *Fundam Appl Toxicol*, 33, 161–165, 1996.

34. Liu, R.C., Hahn, C., and Hurt, M.E., The direct effect of hepatic peroxisome proliferators on rat leydig cell function in vitro, *Fundam Appl Toxicol*, 30, 102–108, 1996.

35. Lerda, D. and Rizzi, R., Study of reproductive function in persons occupationally exposed to 2,4-D, *Mutat Res*, 262, 47–50, 1991.

8 Physiologically Bioactive Compounds of Functional Foods, Herbs, and Dietary Supplements

Giovanni Dinelli, Ilaria Marotti, Sara Bosi, Diana Di Gioia, Bruno Biavati, and Pietro Catizone

CONTENTS

8.1   FUNCTIONAL FOODS

There is widespread recognition that diet plays an important role in the incidence of many diseases. Whereas basic nutrients, including vitamins and minerals, are important for growth and development, the focus of functional foods is to provide health benefits beyond those provided by basic nutrients. Although the mechanisms are not completely clear, when eaten on a regular basis as part of a varied diet, functional foods may lower the risk of developing diseases such as cancer or heart disease.

8.1.1   Physiologically Active Food Components

Several thousands of physiologically active compounds have been identified in functional foods. Each functional food has a different mixture of these active components, which usually are responsible for giving the food its aroma, flavor, and color. The concentration in food of these components may vary depending on the plant or animal variety, maturity, and growth location. Moreover, environmental conditions, such as storage, sunlight, processing, and cooking, may affect the chemical nature, bioactivity, and bioavailability of the many compounds in foods.

Several categories of physiologically active components from plant and animal sources, known as phytochemicals and zoochemicals, respectively, have been described. Some of these are discussed in detail in the following sections.

8.1.1.1   Phytochemicals

Overwhelming evidence from epidemiological, *in vivo*, *in vitro*, and clinical trial data indicates that a plant-based diet can reduce the risk of chronic disease, particularly cancer. In 1992, a review of 200 epidemiological studies showed that cancer risk in people consuming diets high in fruits and vegetables was only one-half than in those consuming few of these foods.[1] It is now clear that there are components in a plant-based diet other than traditional nutrients that can reduce cancer risk. Steinmetz and Potter[2] identified more than a dozen classes of these biologically active plant chemicals, now known as "phytochemicals."

8.1.1.1.1   Terpenes

Terpenes are a large class of compounds made up of single or multiple hydrocarbon units. They include four groups, namely the carotenoids (Figure 8.1), limonoids, saponins, and chromanols (Figure 8.2).

8.1.1.1.1.1 Carotenoids   Carotenoids are a class of natural fat-soluble pigments found principally in plants, algae, and photosynthetic bacteria. They are responsible for many of the red, orange, and yellow hues of plant leaves, fruits, and flowers,[3] as well as the colors of some birds, insects, fish, and crustaceans. Some familiar examples of carotenoid coloration are the oranges of carrots and citrus fruits, the reds of peppers and tomatoes, and the pinks of flamingoes and salmon.[4] From the

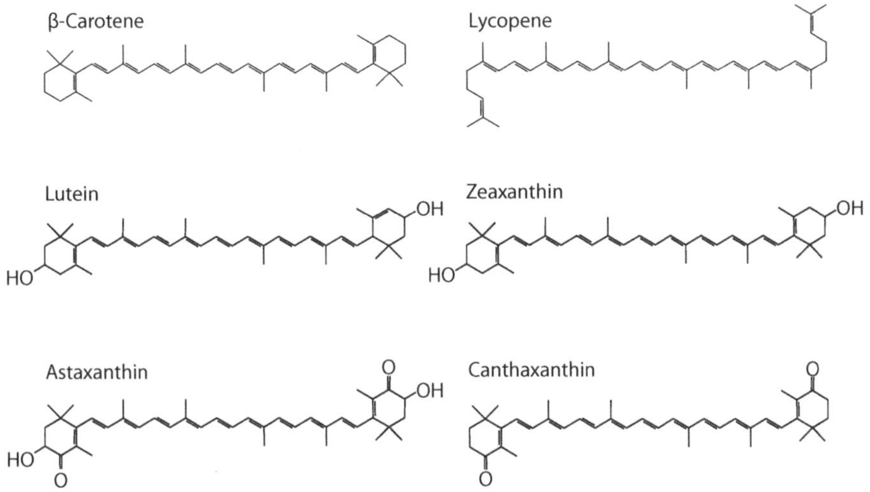

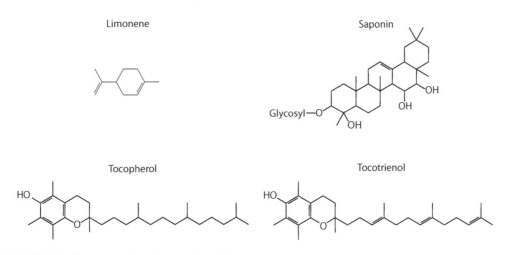

β-Carotene
Lycopene

Lutein
Zeaxanthin

Astaxanthin
Canthaxanthin

FIGURE 8.1 Chemical structures of major carotenoids.

Limonene
Saponin
Glycosyl—O

Tocopherol
Tocotrienol

FIGURE 8.2 Some examples of noncarotenoid terpenes.

plant products commonly consumed by humans, more than 600 different carotenoids (the number includes cis–trans isomeric forms) have been isolated to date.

Chemically, carotenoids are classified in two main groups: carotenes and xanthophylls. Carotenes refer to the carotenoids that contain only carbon and hydrogen (β-carotene and lycopene). Xanthophylls refer to compounds that contain, in addition, a hydroxyl group (lutein, zeaxanthin, and β-cryptoxanthin), a keto group (canthaxanthin), or both (astaxanthin). The structure of a carotenoid ultimately determines the potential biological functions that pigment may have. The distinctive pattern of alternating single and double bonds in the polyene backbone of carotenoids is what allows them to absorb excess energy from other molecules, while the nature of the specific end groups on carotenoids may influence their polarity. The former may account for the antioxidant properties of biological carotenoids, while the latter may explain the differences in the ways that individual carotenoids interact with biological membranes.[5]

In human beings, carotenoids can serve several important functions. The most widely studied and well-understood nutritional role for carotenoids is their provitamin A activity. Vitamin A, which has many vital systemic functions in humans, can be produced within the body from certain carotenoids, notably β-carotene.[6] Dietary β-carotene is obtained from a number of fruits and vegetables, such as carrots, spinach, peaches, apricots, and sweet potatoes.[7] Other provitamin A carotenoids include α-carotene (found in carrots, pumpkins, and red and yellow peppers) and β-cryptoxanthin (from oranges, tangerines, peaches, nectarines, and papayas).

Carotenoids play also an important potential role in human health by acting as biological antioxidants, protecting cells and tissues from the damaging effects of free radicals and singlet oxygen. Lycopene, the hydrocarbon carotenoid that gives tomatoes their red color, is particularly effective at quenching the destructive potential of singlet oxygen.[8] It is most known for its association with a decreased risk of developing prostate cancer in men. Lutein and zeaxanthin (Figure 8.1), xanthophylls found in corn and in leafy greens such as kale and spinach, are believed to function as protective antioxidants in the macular region of the human retina, helping preventing age-related macular degeneration (AMD).[9] Astaxanthin, found mostly in red yeasts and red algae, is now fed to salmon, trout, crabs, krill, and shrimp in "fish farms" to provide the red and pink color of their natural red-algae-eating wild brethren. This most powerful of the carotenoid antioxidants has been shown to enhance secondary immune response in humans, and help reduce symptoms of *Helicobacter pylori* infections and rheumatoid arthritis.[10] Other health benefits of carotenoids that may be related to their antioxidative potential include enhancement of immune system function,[11] protection from sunburn,[12] and inhibition of the development of certain types of cancers.[13]

8.1.1.1.1.2 Limonoids   Limonoids form an important class of monoterpenes naturally found in the peels of citrus fruits. They appear to be specifically directed to the protection of lung tissue. In one study, a standardized extract of D-limonene, α-pinene, and eucalyptol was effective in clearing congestive mucus from the lungs of patients with chronic obstructive pulmonary disease.[14] Limonoids and perillyl alcohol, monoterpenes found in mandarin oranges, appear to have specific cancer and cardioprotective effects.[15] In animal studies, results suggest that the chemotherapeutic activity of these monoterpenes can be attributed to the induction of both Phase I and Phase II detoxification enzymes in the liver.[16] These enzymes are part of the body's protection against harmful substances.

8.1.1.1.1.3 Saponins   Saponins are found primarily in legumes, with the greatest concentration occurring in soybeans. Recent experimental investigations suggest that saponins have cholesterol-lowering, anticancer, and immunostimulatory properties. Anticancer properties of saponins appear to be the result of antioxidant effects, immune modulation, and regulation of cell proliferation.[17] Animals have reduced high cholesterol levels when fed with either soy protein, daidzein (a soy isoflavone), or soy germ.[18]

8.1.1.1.1.4 Chromanols   The most familiar are the tocotrienols and the tocopherols (also known as vitamin E) (see also Section 8.2.1.2.3). They have a chromanol ring, with a hydroxyl group which can donate a hydrogen atom to reduce free radicals and a hydrophobic side chain which allows for the penetration into biological membranes. These two naturally occur in palm oils and whole grain germ and/or bran, yet research has shown that the biologic functions of tocopherols and tocotrienols are unrelated. Tocotrienols appear to inhibit breast cancer cell growth, whereas tocopherols have been mostly studied for their cardiovascular health effects.[19]

8.1.1.1.2   Polyphenols

Thousands of molecules having a polyphenol structure (i.e., several hydroxyl groups on aromatic rings) have been identified in higher plants, and several hundred are found in edible plants. These compounds may be classified into different groups as a function of the number of phenol rings that they contain and of the structural elements that bind these rings to one another.[20] Distinctions are thus made among phenolic acids (hydroxybenzoic acids and hydroxycinnamic acids), flavonoids, stilbenes, and lignans (Figure 8.3).

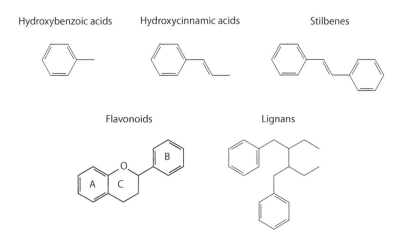

FIGURE 8.3 Chemical structures of major classes of polyphenols.

8.1.1.1.2.1 Phenolic Acids Phenolic acids can be distinguished into two main classes: derivatives of benzoic acid and derivatives of cinnamic acid. The hydroxybenzoic acid content of edible plants is generally very low, with the exception of certain red fruits, black radish, and onions, which can have concentrations of several tens of milligrams per kilogram fresh weight.[21] Tea leaves are an important source of gallic acid: they may contain up to 4.5 g/kg fresh weight.[22] Additionally, hydroxybenzoic acids are components of complex structures such as hydrolyzable tannins (gallotannins in mangoes and ellagitannins in red fruit such as strawberries, raspberries, and blackberries).[23] Because these hydroxybenzoic acids, both free and esterified, are found in only a few plants eaten by humans, they have not been extensively studied and are not currently considered to be of great nutritional interest.

The occurrence of hydroxycinnamic acids in human food is more common than hydroxybenzoic acids and consists mainly of *p*-coumaric, caffeic, and ferulic acids. These acids are rarely found in the free form, except in processed food that has undergone freezing, sterilization, or fermentation.[20] The types of fruit having the highest concentrations (blueberries, kiwis, plums, cherries, apples) contain 0.5–2 g hydroxycinnamic acids/kg fresh weight.[24] *p*-Coumaric acid can be found in a wide variety of edible plants such as peanuts, tomatoes, carrots, and garlic. It has antioxidant properties and is believed to lower the risk of stomach cancer by reducing the formation of carcinogenic nitrosamines.[25,26]

Caffeic acid frequently occurs in fruits, grains, and vegetables as simple esters with quinic acid (forming chlorogenic acid) or saccharides, and are also found in traditional Chinese herbs.[27] Chlorogenic acid is found in particularly high concentrations in coffee: green coffee beans typically contain 6%–7% of this component (range: 4%–10%) and a cup of instant coffee (200 mL) contains 50–150 mg of chlorogenic acid.[28] This compound, long known as an antioxidant, also slows the release of glucose into the bloodstream after a meal.[29] Ferulic acid is the most abundant phenolic acid found in cereal grains. The main food source of ferulic acid is wheat bran (5 g/kg) and it may represent up to 90% of total polyphenols.[30,31] As ferulic acid is found predominantly in the outer parts of the grain, the ferulic acid content of different wheat flours is directly related to levels of sieving.[32] Rice and oat flours contain approximately the same quantity of phenolic acids as wheat flour (63 mg/kg), although the content in maize flour is about three times as high.[21]

8.1.1.1.2.2 Flavonoids Flavonoids are polyphenolic compounds sharing a common structure consisting of two aromatic rings (A and B) that are bound together by three carbon atoms that form an oxygenated heterocycle (ring C) (Figure 8.3). They may be divided, according to the oxidation level of the C ring, into 14 subclasses; the most common being flavonols, flavones, isoflavonoids

Subclass	Basic Structure	Example	Sources	Quantities (mg/100 g or mL)
Flavones		Apigenin	Celery stalks	13–15
			Celery hearts	15–19
Flavanones		Naringenin	Citrus fuits	18–47
			Tomato	2–4
Flavonols		Quercetin	Cranberry	14–27
			Onions	28–49
			Apple	2–7
			Red wine	1–6
Anthocyanidins		Cyanidin	Cranberry	46–172
			Red raspberry	23–59
			Red wine	0.1–10
Flavanols		Epicatechin (monomer)	Apple	5–15
			Chocolate	20–400
			Red wine	2–45
			Black tea	7–35
		Proanthocyanidin (dimer)	Chocolate	50–1180
			Apple	49–104
			Red wine	10–57
Isoflavones		Glycitein	Uncooked soybean	22–45
			Soy proteins	50–200

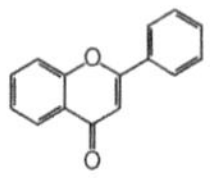

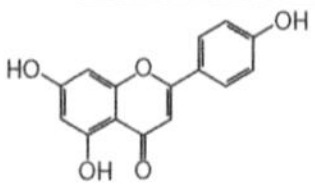

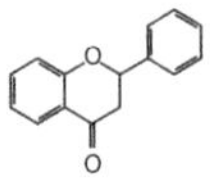

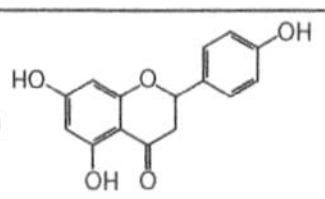

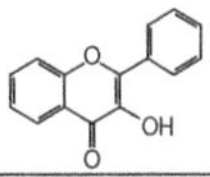

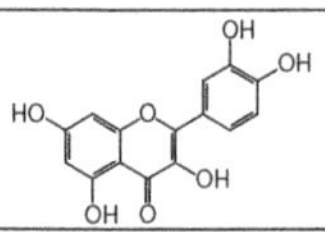

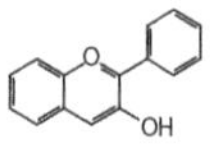

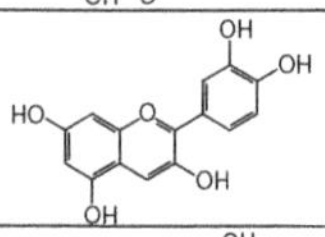

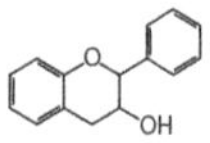

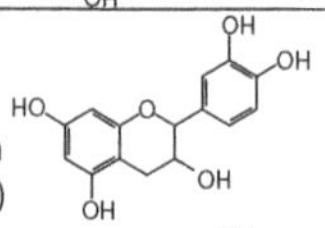

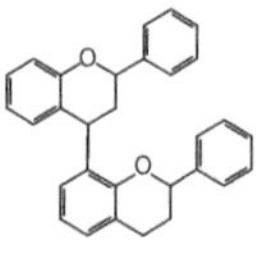

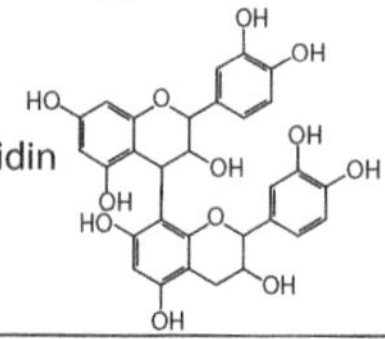

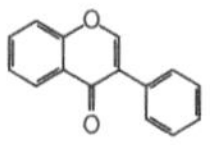

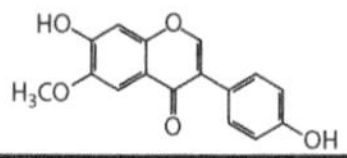

FIGURE 8.4 Basic structures and examples of the main subclasses of dietary flavonoids.

(isoflavones, coumestans), flavanones, anthocyanidins, and flavanols (catechins and proanthocyanidins) (Figure 8.4).[33]

Flavonols are the most ubiquitous flavonoids in foods, and the main representatives are quercetin and kaempferol. The richest sources are onions (up to 1.2 g/kg fresh weight), curly kale, leeks, broccoli, and blueberries.[20] Recently the presence of kaempferol has been reported in the outer part of seeds as well as in seedlings of some Italian common bean (*Phaseolus vulgaris* L.) landraces.[34,35] These compounds are present in glycosylated forms. The associated sugar moiety is very often glucose or rhamnose, but other sugars may also be involved (e.g., galactose, arabinose, xylose, glucuronic acid). Fruit often contains between 5 and 10 different flavonol glycosides.[24] These flavonols accumulate in the outer and aerial tissues (skin and leaves) because their biosynthesis is stimulated by light. Marked differences in concentration exist between pieces of fruit on the same tree and even between different sides of a single piece of fruit, depending on exposure to sunlight.[36] Similarly, in leafy vegetables such as lettuce and cabbage, the glycoside concentration is 10 times higher in the green outer leaves than in the inner light-colored leaves.[37] This phenomenon also accounts for the higher flavonol content of cherry tomatoes than of standard tomatoes, because they have different proportions of skin to whole fruit.

Flavones are much less common and were identified in sweet red pepper (luteolin) and celery (apigenin).[38] Cereals such as millet and wheat contain *C*-glycosides of flavones.[39–41]

Citrus fruits are the main food source of flavanones. The main aglycones are naringenin in grapefruit, hesperetin in oranges, and eriodictyol in lemons. Flavanones are generally glycosylated by a disaccharide at position 7: either a neohesperidose, which imparts a bitter taste (such as to naringin in grapefruit), or a rutinose, which is flavorless. Orange juice contains between 200 and 600 mg hesperidin/L and 15–85 mg narirutin/L, and a single glass of orange juice may contain between 40 and 140 mg flavanone glycosides.[42] Because the solid parts of citrus fruit, particularly the albedo (the white spongy portion) and the membranes separating the segments, have a very high flavanone content, the whole fruit may contain up to five times as much as a glass of orange juice.

Isoflavonoids are a large and very distinctive subclass of the flavonoids. These compounds differ structurally from other classes of the flavonoids in having the phenyl ring (B-ring) attached at the 3- rather than at 2-position of the heterocyclic ring. In addition, the isoflavonoids differ by their greater structural variation and the greater frequency of isoprenoid substitution.[43] Isoflavones constitute the largest group of natural isoflavonoids and are the most investigated for their structural similarities to estrogens. Although they are not steroids, they have hydroxyl groups in positions 7 and 4 in a configuration analogous to that of the hydroxyls in the estradiol molecule. This confers them pseudohormonal properties, including the ability to bind to estrogen receptors, and they are consequently classified as phytoestrogens. The most interesting compounds with regard to estrogenicity are genistein, daidzein, glycitein, biochanin A and formononetin (Figure 8.5). Genistein is one of the most active principles with high binding affinity for the estrogen receptor.[44] The methoxy derivative, biochanin A, does not bind to the estrogenic receptor but is estrogenic *in vivo*.[45] Daidzein (4′,7-dihydroxyisoflavone) has a higher binding affinity for the estrogen receptor than its methoxy derivative, formononetin, but both are weak estrogens *in vivo*.[105] Glycitein (4′,5,7-trihydroxyisoflavone) has the highest estrogenicity and can be more easily absorbed than daidzein and genistein.[46]

Isoflavones are found almost exclusively in leguminous plants. Legumes, particularly soybean (*Glycine max* L.) and its processed products, are the richest sources of isoflavones; mainly genistein, daidzein, and glycitein, in the human diet.[47] Recently soy isoflavones have been reported also in 2–3

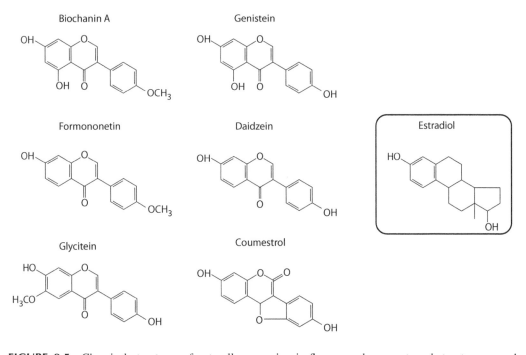

FIGURE 8.5 Chemical structures of naturally occurring isoflavone and coumestan phytoestrogens and endogenous estradiol.

day old seedlings of Italian common bean landraces, even if at concentrations four- to sixfold lower than those detected in soybean sprouts.[35]

Another group of isoflavonoids are coumestans. Coumestrol (3,9-dihydroxy-6H-benzofuro[3,2-c] [1]benzopyran-6-one) (Figure 8.5), the most potent of coumestans, has higher binding affinity for the estrogen receptor than genistein.[48] This is consistent with the receptor binding model that appears to depend on a phenolic group in the 4′ position of isoflavones and the 12′ position of coumestans. The main dietary source of coumestrol, is legumes; however low levels have been reported in brussel sprouts and spinach.[49–51] Clover and soybean sprouts are reported to have the highest concentration, 28 and 7 mg/100 g dry weight, respectively; mature soybeans only have 0.12 mg/100 g dry weight.[49–51]

Flavanols exist in both the monomer form (catechins) and the polymer form (proanthocyanidins). In contrast to other classes of flavonoids, flavanols are not glycosylated in foods. Catechins are found in many types of fruit, especially in apricots (250 mg/kg fresh weight). They are also present in red wine (up to 300 mg/L) and chocolate.[52,53] However, tea is by far the richest source: young shoots contain 200–340 mg of catechin, gallocatechin, and their galloylated derivatives/g of dry leaves.[54] An infusion of green tea contains 1 g/L catechins.[55] In black tea, their content is reduced to about half this value due to their oxidation into more complex polyphenols during fermentation.[56] Proanthocyanidins, which are also known as condensed tannins, are dimers, oligomers, and polymers of catechins that are bound together by links between C4 and C8 (or C6). Through the formation of complexes with salivary proteins, condensed tannins are responsible for the astringent character of fruit (grapes, peaches, kakis, apples, pears, and berries) and beverages (wine, cider, tea, and beer) and for the bitterness of chocolate.[57] It is difficult to estimate the proanthocyanidin content of foods because proanthocyanidins have a wide range of structures and molecular weights. The only available data concern dimers and trimers, which are as abundant as the catechins themselves.[58]

Anthocyanins are water-soluble pigments that give raspberries, blueberries, strawberries, cherries, grapes, radishes, and red cabbages their deep red, blue, and purple color pigments.[59] They exist in different chemical forms, both colored and uncolored, according to pH. Although they are highly unstable in the aglycone form (anthocyanidins), while they are in plants, they are not subjected to degradation factors such as light, pH, and oxidation conditions. Degradation is prevented by glycosylation, generally with glucose at position 3, and esterification with various organic acids (citric and malic acids) and phenolic acids. In addition, anthocyanins are stabilized by the formation of complexes with other flavonoids (copigmentation).[20] In the human diet, anthocyanins are found in red wine (average content of 26 mg/L),[52] certain varieties of cereals, and certain leafy and root vegetables (aubergines, cabbage, beans, onions, and radishes), but they are most abundant in fruits such as strawberries (0.15 mg/g fresh fruit) and cherries (0.45 mg/g fresh fruit).[60] Cyanidin is the most common anthocyanidin in foods. These values increase as the fruit ripens. Anthocyanins are found mainly in the skin except for certain types of red fruit in which they also occur in the flesh (cherries and strawberries). Wine contains 200–350 mg anthocyanins/L, and these anthocyanins are transformed into various complex structures as the wine ages.[61,62]

Consumption of flavonoid-rich foods is associated with a lower incidence of heart disease, ischemic stroke, cancer, and other chronic diseases.[63–66] For example, 7 of 12 epidemiological studies evaluating the risk of coronary heart disease reported protective effects of dietary flavonoids.[67] Additional studies also found inverse associations between flavonoid intake and the risk of stroke,[68,69] and lung and colorectal cancers.[68,70,71] Because these chronic diseases are associated with increased oxidative stress and flavonoids are strong antioxidants *in vitro*, it has been suggested that dietary flavonoids exert health benefits through antioxidant mechanisms.[72–74] However, a recent study reported that many of the biological effects of flavonoids appear to be related to their ability to modulate cell signaling pathways, rather than their antioxidant activity.[75] Unlike in the controlled conditions of a test tube, flavonoids are poorly absorbed by the human body (less than 5%), and most of what is absorbed is quickly metabolized and excreted. The huge increase in antioxidant capacity of blood seen after the consumption of flavonoid-rich foods is not caused directly by the

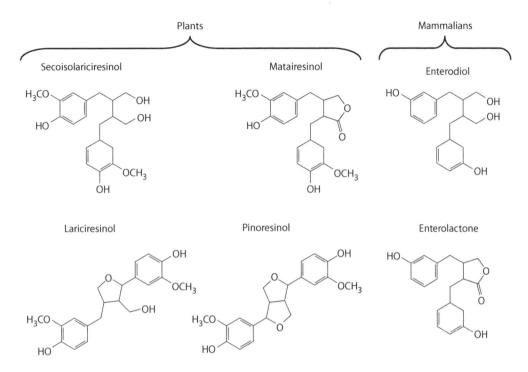

FIGURE 8.6 Structures of plant and mammalian lignans.

flavonoids themselves, but most likely is due to the fact that the body recognizes flavonoids as foreign compounds and through different mechanisms, they could play a role in preventing cancer or heart disease.

8.1.1.1.2.3 Lignans Lignans are polyphenolic compounds derived from the combination of two phenylpropanoid C6–C3 units (Figure 8.6).[33] They may occur glycosidically bound to various sugar residues, esterified or as structural subunits of biooligomers.[76–78] Flaxseed (*Linum usitatissimum* L.) is known as the richest dietary source of lignans, with glycosides of secoisolariciresinol and matairesinol as the major compounds (370 and 1 mg/100 g, respectively). Also lignan concentrations in sesame seeds (29 mg/100 g, mainly pinoresinol and lariciresinol) were reported to be relatively high.[79] Significant amounts of secoisolariciresinol (21 mg/100 g of dry weight) were found in pumpkin seeds. Other cereals (triticale and wheat), leguminous plants (lentils, soybeans), fruits (pears, prunes), and certain vegetables (garlic, asparagus, and carrots) also contain traces of these same lignans, but concentrations in flaxseed are about 1000 times as high as concentrations in these other food sources.[80] When ingested, secoisolariciresinol and matairesinol are metabolized by bacteria in the gastrointestinal tract and converted into the mammalian lignans enterodiol (END) and enterolactone (ENL), respectively (Figure 8.7).[81] After the conversion, END is oxidized to ENL.[82] END and ENL are hormone-like compounds that have the ability to bind to estrogen receptors with low affinity and with weak estrogen activity.[80] Recently, the plant lignans pinoresinol, lariciresinol, syringaresinol, 7-hydroxymatairesinol, and arctigenin were identified as further precursors of mammalian lignans. The first three are predominantly present in cereals, particularly in whole-grain rye products.[83] More recently, an investigation of the lignan profile of some Italian soft wheat (*Triticum aestivum* L.) cultivars highlighted the high content and unique composition in lignans (arctigenin, hinokinin, and syringaresinol) of old cultivars with respect to modern ones.[84] Lignans possess several biological activities, such as antioxidant and (anti)estrogenic properties, and thus reduce the risk of certain hormone-related cancers as well as cardiovascular diseases (CVD).[80,85–87]

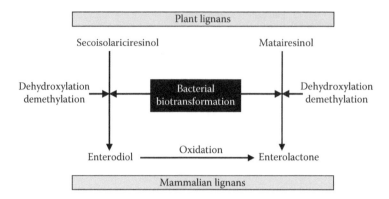

FIGURE 8.7 Bacterial conversion of secoisolariciresinol and matairesinol (plant lignans) in mammalian lignans.

8.1.1.1.2.4 Stilbenes Stilbenes are mainly constituents of the heartwood of the genera *Pinus* (*Pinaceae*), *Eucalyptus* (*Myrtaceae*), and *Maclura* (*Moraceae*). Although stilbene aglycones are common in heartwood, plant tissues may contain stilbene glycosides.[33] One of these, resveratrol (3,4′,5-trihydroxystilbene) (Figure 8.8), is found largely in the skins of red grapes and its amount in red wine range between 0.3 and 7 mg/L.[88] Resveratrol came to scientific attention few years ago as a possible explanation for the "French paradox," which is the low incidence of heart disease among the French, who eat a relatively high-fat diet.[89] Subsequently, reports on the potential for resveratrol to inhibit the development of cancer and extend lifespan in cell culture and animal models have continued to generate scientific interest.[90,91] However, more recent studies demonstrate that the attenuation of coronary heart disease risk in wine drinkers is probably due as much to their lifestyle consumption of high amounts of polyphenol-containing fruits and vegetables as to wine.[39]

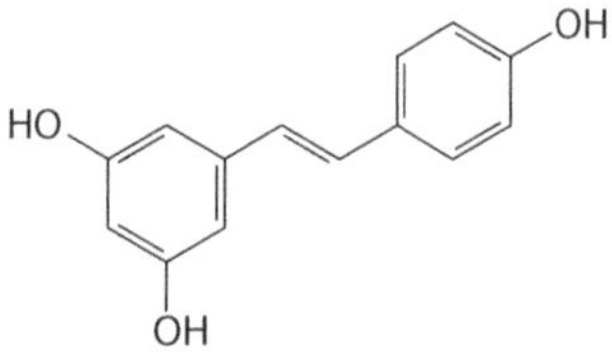

FIGURE 8.8 Chemical structure of resveratrol.

8.1.1.1.3 Phytosterols and Phytostanols

Phytosterols are triterpenes similar in structure (the four-ring steroid nucleus, the 3β-hydroxyl group, and often a 5,6-double bond) and function (stabilization of the phospholipid bilayers in cell membranes) to cholesterol. Whereas the cholesterol side chain is comprised of 8 carbon atoms, most phytosterol side chains contain 9 or 10 carbon atoms.[92] In plants, more than 200 different types of phytosterols have been reported, the most abundant being β-sitosterol, campesterol, and stigmasterol. Ergosterol is the principal sterol of yeast and is found in corn, cotton seed, peanut, and linseed oils.[93] Structures of the most common phytosterols found in food are included in Figure 8.9. In nature, sterols can be found as free sterols or as four types of conjugates in which the 3β-hydroxyl group is esterified to a fatty acid or a hydroxycinnamic acid, or glycosylated with a hexose (usually glucose) or a 6-fatty acyl hexose. Glycosides are the most common form found in cereals.[94] Phytostanols, a fully saturated subgroup of phytosterols, are intrinsic constituents of cereals (corn, wheat, rye, and rice), fruits, and vegetables, but their concentrations are generally lower than those of unsaturated plant sterols.[95,96]

The most important natural sources of plant sterols in human diets are oils and margarines, although they are also found in a range of seeds, legumes, vegetables, and unrefined vegetable oils.[95,97–100] Cereal products are a significant source of plant sterols, their contents, expressed on a fresh weight basis, being higher than in vegetables.

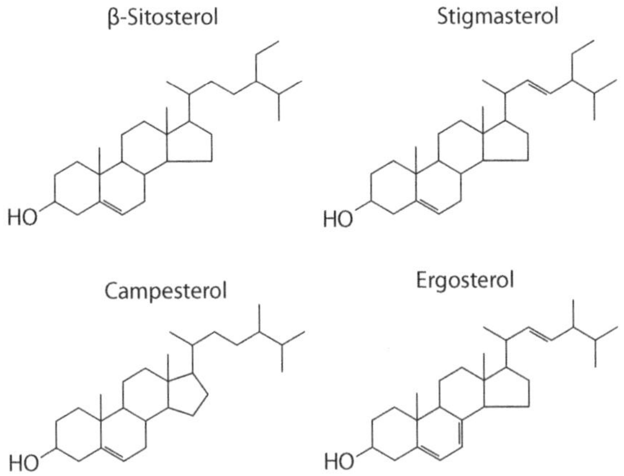

FIGURE 8.9 Chemical structures of some representative phytosterols.

Phytosterols have been the object of increasing interest given their cholesterol-lowering properties. In addition, they possess anti-inflammatory,[101–103] antiatherogenicity,[104] anticancer, and antioxidative activities.[105,106]

Their capacity to decrease serum low-density lipoprotein (LDL) cholesterol levels and, thus, in protecting against CVD, has led to the development of functional foods enriched with plant sterols. At present, several functional products, such as spreadable fats, yoghurts, and milk, with free phytosterols or phytosteryl fatty acid esters or phytostanyl fatty acid esters added at high levels, are available in the market, especially in several European countries.[96]

8.1.1.1.4 Organosulfur Compounds

Naturally occurring sulfur-containing compounds are found especially in the cruciferous vegetables, such as broccoli, brussels sprouts, cabbage, kale, and turnips (*Brassica* spp.), and the onion (*Allium* spp.) and mustard (*Sinapis* spp.) families. The sulfur compounds in these groups are slightly different and, consequently, each has specific health benefits.

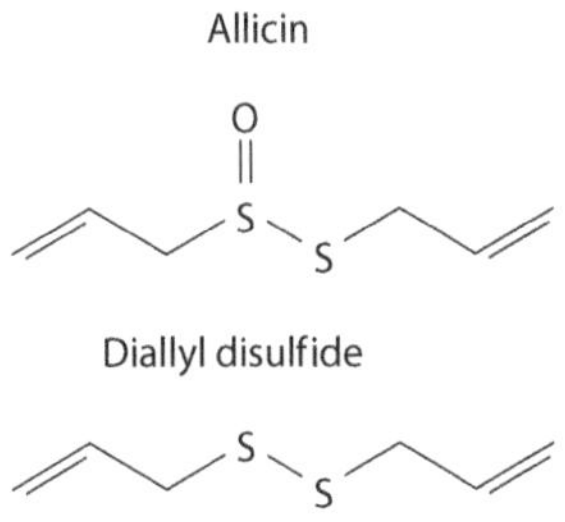

Thiosulfonates are most notably found in onions and garlic as well as in chives, leeks, and shallots. When the plants are cut or smashed, sulfur compounds release biotransformation products, including allicin, ajoene, allylic sulfides (Figure 8.10), vinyl dithin, and D-allyl mercaptocysteine. Some of these are considered to have antiatherosclerotic and anticancer activities, especially for those of the gastrointestinal tract. This latter effect may be due to the garlic's ability to inhibit the activity of *H. pylori*, the bacterium that causes stomach ulcers.[107,108] Other effects are antibacterial, antiviral, and antifungal.[109] Allicin, allyl sulfides, and allyl mercaptocysteine are also strong antioxidants. Specific allylic sulfides block the activity of toxins produced by bacteria and viruses.[110] Garlic and onions, like their cruciferous relatives, can also selectively alter liver detoxification enzyme systems to reduce toxic by-products.[111] Finally, garlic powder has been shown in numerous studies to lower cholesterol, often by as much as 10%.[112]

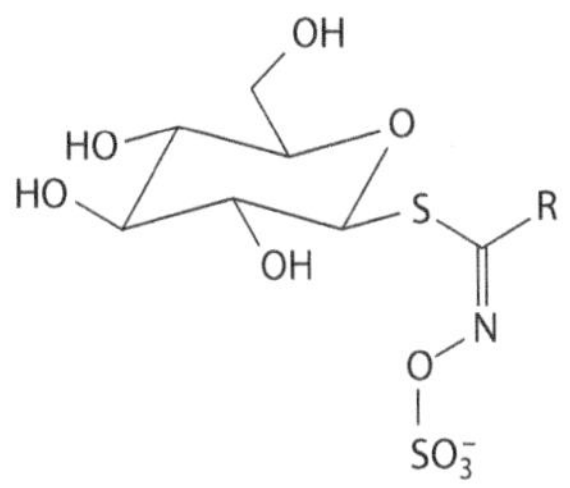

Glucosinolates (Figure 8.11) are a class of about 100 naturally occurring thioglucosides that are characteristic of the *Cruciferae* and

FIGURE 8.10 Chemical structures of allicin and diallyl disulfide, the two major organosulfur compounds in garlic.

FIGURE 8.11 Basic structure of glucosinolates.

related families. At present, the diets of people in many parts of the world include considerable amounts of cruciferous crops and plants. These range from the consumption of processed radish and wasabi in the Far East to that of cabbage and traditional root vegetables in Europe and North America. Other crops, such as rapeseed, kale, and turnip may also contribute indirectly to the human food chain since they are extensively used as animal feedstuffs. Because epidemiological studies provide some evidence that diets rich in cruciferous vegetables are associated with lower risk of several types of cancer, scientists are interested in the potential cancer-preventive activities of compounds derived from glucosinolates.[113] Glucosinolates are converted into several biotransformation products in the human body, particularly indole-3-carbinol (I3C), thiosulfonates, and isothiocyanates. Among these, indole-3-carbinol, derived from the enzymatic hydrolysis of the indole glucosinolate, glucobrassicin,[114] is particularly protective against hormone-induced cancers, such as breast cancers. Analogously, sulforaphane, a potent isothiocyanate, not only inhibits tumors but also can be helpful in maintaining stomach health (it has antibacterial activity against the ulcer-causing *H. pylori*).

8.1.1.1.5 Omega-3 Fatty Acids—ALA

Omega-3 fatty acids are a class of polyunsaturated fatty acids (PUFAs). The term omega-3 ("n-3," "ω-3") signifies that the first double bond exists as the third carbon–carbon bond from the terminal methyl end (ω) of the carbon chain (Figure 8.12). Flaxseed and its oil contain a mixture of PUFAs. They are particularly rich in α-linolenic acid (ALA, C18:3), an essential omega-3 fatty acid, and linoleic acid, an essential omega-6 fatty acid. α-Linolenic acid constitutes 57% of the total fatty acids in flaxseed, making it a very rich source of omega-3. By comparison, α-linolenic acid constitutes only 11% of the total fatty acids in canola oil. Other sources of ALA are soybeans, soybean oil, pumpkin seeds, pumpkin seed oil, walnuts, and walnut oil. Bioactive properties and health-promoting effects on humans are described in Section 8.1.1.2.1.

8.1.1.2 Zoochemicals

Although the vast number of naturally occurring health-enhancing substances are of plant origin, there are a number of physiologically active components in animal products that deserve attention for their potential role in optimal health.

8.1.1.2.1 Omega-3 Fatty Acids—DHA and EPA

Although present in some plant sources (see Section 8.1.1.5), omega-3 fatty acids are predominantly found in fatty fish such as salmon, tuna, mackerel, sardines, and herring.[115] The high content of omega-3 fatty acids in marine lipids is suggested to be a consequence of cold-temperature adaptation in which omega-3 PUFAs remain liquid and oppose any tendency to crystallize.[116] Another potential source of omega-3 fatty acids is the New Zealand green-lipped mussel (*Perna canaliculus*), used for centuries by the Maories to promote good health. The major PUFAs derived from fish oils are eicosapentaenoic acid (EPA, C20:5) and docosahexaenoic acid (DHA, C22:6).

It has been suggested that the Western-type diet is currently deficient in omega-3 fatty acids, which is reflected in the current estimated omega-6 (linoleic acid, arachidonic acid) to omega-3 dietary ratio of about 20–25:1, compared to the 1:1 ratio on which humans evolved.[117] This has

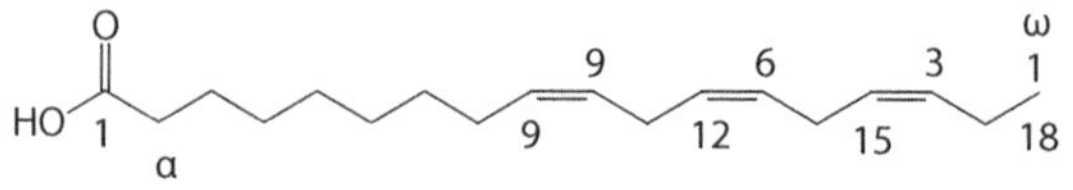

FIGURE 8.12 Chemical structure of α-linolenic acid. Although chemists count from the carbonyl carbon (or alpha carbon), physiologists count from the omega carbon. Note that from the omega end, the first double bond appears as the third carbon–carbon bond, hence the name "omega-3."

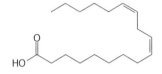

FIGURE 8.13 Chemical structure of linoleic acid.

prompted researchers to examine the role of omega-3 fatty acids in the so-called diseases of civilization, particularly cancer and CVD, and more recently, in early human development. That n-3 fatty acids may play an important role in CVD was first brought to light in the 1970s when Bang and Dyerberg[118] reported that Eskimo populations, consuming large amounts of traditional marine mammals and fish, had little mortality from coronary heart disease. Recently, extensive research indicates that omega-3 fatty acids reduce inflammation and help prevent certain chronic illnesses such as arthritis.[119–121]

Among animal-derived foods, eggs are excellent dietary sources of several essential and nonessential components, including n-3 PUFAs. In particular, n-3 PUFA-enriched eggs can be produced by modifying hens' diets.[122] Each one of these modified eggs contains about 350 mg of n-3 PUFA, relatively to the standard eggs that contain about 60 mg, and three of these enriched eggs provide approximately the same amount of n-3 PUFA as one meal of fish.[123]

8.1.1.2.2 Conjugated Linoleic Acid

Conjugated linoleic acid (CLA) was first identified as a potent antimutagenic agent in grilled beef.[124] CLA is a mixture of structurally similar forms of linoleic acid (*cis*-9, *trans*-11, octadecadienoic acid) (Figure 8.13) and occurs particularly in large quantities in dairy products and foods derived from ruminant animals.[125] Interestingly, CLA increases in foods that are cooked and/or otherwise processed. This is significant in view of the fact that many mutagens and carcinogens have been identified in cooked meats.[123]

The inhibition of mammary carcinogenesis in animals is the most extensively documented physiological effect of CLA,[126] and there is also emerging evidence that CLA may decrease body fat in humans,[127] and increase bone density in animal models.[128]

8.1.1.2.3 Probiotics and Prebiotics

Probiotics are defined as "live microbial feed supplements which beneficially affect the host animal by improving its intestinal microbial balance."[129] Of the beneficial microorganisms inhabiting the human gastrointestinal tract, lactic acid bacteria (e.g., *Bifidobacterium* and *Lactobacillus*) have attracted the most attention.[130] The use of probiotic cultures, particularly *Lactobacillus acidophilus* and *Bifidobacterium*, underwent a boom in Europe in the late 1980s and early 1990s, especially in France. More recently, newer and more complex strains have been introduced. LC1 is a range of fermented dairy products from Chambourcy (Nestlè) containing a new strain of *L. acidophilus*, which is claimed to reinforce the body's natural defense mechanisms.[131] Gaio yoghurt from the Danish company MD Foods contains Causido, a Caucasian lactic acid culture, which is claimed to have cholesterol-lowering properties. Mona (the Netherlands) produces the Vifit range of dairy products containing *L. caseii GG*, for which beneficial effects have been reported such as increased resistance of gut flora against invading microorganisms and prevention/effective treatment of diarrhea. Danone's Actimel is a yoghurt drink containing *L. caseii Actimel*, which is claimed to "balance the intestinal flora and support the body's natural powers of resistance."[132]

Although a variety of health benefits have been attributed to probiotics, their anticarcinogenic, hypocholesterolemic, and antagonistic actions against enteric pathogens and other intestinal organisms have received much interest.[133] Although a number of human clinical studies have assessed the cholesterol-lowering effects of fermented milk products,[130] results are equivocal. Study outcomes have been complicated by inadequate sample sizes, failure to control nutrient intake and energy expenditure, and variations in baseline blood lipids. More evidence supports the role of probiotics in cancer risk reduction, particularly colon cancer.[133] This observation may be due to the fact that lactic acid cultures can alter the activity of fecal enzymes (e.g., β-glucuronidase, azoreductase, and nitroreductase) that are thought to play a role in the development of colon cancer. Relatively less attention has been focused on the consumption of fermented milk products and breast cancer risk, although an inverse relationship has been observed in some studies.[134,135]

In addition to probiotics, there is a growing interest in fermentable carbohydrates that feed the good microflora of the gut. These prebiotics, defined by Gibson and Roberfroid[136] as "nondigestible food ingredients that beneficially affect the host by selectively stimulating the growth and/or activity of one or a limited number of bacteria in the colon and thus improves host health," may include starches, dietary fibers, other nonabsorbable sugars, sugar alcohols, and oligosaccharides.[137] Of these, oligosaccharides have received the most attention, and numerous health benefits have been attributed to them.[138] Oligosaccharides consist of short-chain polysaccharides composed of 3 and 10 simple sugars linked together. They are found naturally in many fruits and vegetables (including banana, garlic, onions, milk, honey, and artichokes). A commercial company manufactures Raftiline products that consist of six different types of inulin extracted from chicory roots. From inulin, Orafti produces Raftilose, which consists of oligofructose liquids and powders. Both inulin and oligofructose stimulate the development of bifidobacteria in the human colon.[139] Borculo Whey Products (the Netherlands) has developed a prebiotic based on lactose called Elix'or. This contains galacto-oligosaccharides, which act as growth promoters for several beneficial intestinal bacteria such as *Bifidobacterium* and *Lactobacilli*. In contrast, hardly any putrefactive microorganisms are able to utilize galacto-oligosaccharides.[132] The prebiotic concept has been further extended to encompass the concept of synbiotics, a mixture of pro- and prebiotics that beneficially affects the host by improving the survival and implantation of live microbial dietary supplements, by selectively stimulating the growth of one or a limited number of health-promoting bacteria, and thus improving host welfare.[136] Products that contain both pre- and probiotics are still rare in Europe.[131] One of the first was "Fyos," a fermented milk drink launched by Nutricia in 1994 in Belgium. It contained *L. caseii* and inulin.[132]

8.2 DIETARY SUPPLEMENTS

Dietary supplements are products containing a broad category of nutrients and other bioactive substances that contribute significantly to total dietary intakes. Their formulations vary from single components to combinations of many different vitamins and elements. They are typically taken by mouth as a pill, capsule, tablet, liquid, or powder.

Dietary supplements, particularly those containing nutrients (vitamins and minerals), provide a means of consuming specific nutrients that otherwise might be low or lacking in reducing diets. Nutrients are essential for human health and adverse effects can result from intakes that are too low as well as too high. The balance between deficiency and toxicity is often expressed as in Figure 8.14. Within the acceptable range of intake, neither deficiency nor toxicity would occur in normally susceptible people. Estimated average requirement (EAR) is the average daily nutrient intake level that meets the needs of 50% of the "healthy" individuals in a particular age and gender group. It is based on a given criteria of adequacy which will vary depending on the specified nutrient. Therefore, estimation of requirement starts by stating the criteria that will be used to define adequacy and then establishing the necessary corrections for physiological and dietary factors. Once a mean requirement value is obtained from a group of subjects, the nutrient intake is adjusted for interindividual variability to arrive at a recommendation. The recommended dietary allowance (RDA) is the average daily dietary intake level that is sufficient to meet the nutrient requirement of nearly all (97%–98%) healthy individuals in a specific life stage and gender group. Upper limits (ULs) of nutrient intake have been set for some micronutrients and are defined as the maximum intake from food, water, and supplements that is unlikely to pose the risk of adverse health effects from excess, in almost all apparently healthy individuals in an age- and sex-specific population group. The UL is not intended to be a recommended level of intake, and there is no established benefit for healthy individuals if they consume a nutrient in amounts above the RDA.[140]

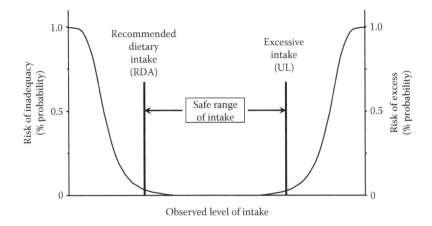

FIGURE 8.14 The concept of safe range of intakes. The safe range of intakes is associated with a very low probability/risk for the individual of either inadequacy or excess of the nutrient. All intakes in the range are equally beneficial and safe.

Dietary supplements may be grouped into three major categories related to dietary function or origin: (1) substances with established nutritional function, such as vitamins and minerals, (2) other substances with a wide variety of origins and physiologic roles (comprising ergogenic supplements), and (3) botanical products and their concentrates and extracts.[141]

8.2.1 VITAMINS

The term "vitamin" is derived from the words vital and amine, because vitamins are required for life and were originally thought to be amines. Although not all vitamins are amines, they are organic compounds required by humans in small amounts from the diet. An organic compound is considered a vitamin if a lack of that compound in the diet results in overt symptoms of deficiency.[142] To date, 13 essential vitamins have been identified. Each of these is soluble in either water or fat. The fat-soluble vitamins are vitamins A, D, E, and K (Figure 8.15). The water-soluble vitamins (Figure 8.16) are vitamin C and the B-complex vitamins, which include vitamins B_1 (thiamin), B_2 (riboflavin), B_3 (niacin), B_5 (pantothenic acid), B_6 (pyridoxine), B_{12} (cyanocobalamin), biotin, and folic acid.[143] Fat-soluble vitamins are capable of being stored in the body and, therefore, can accumulate. They are not easily destroyed by heat during cooking or processing or through exposure to air. In contrast, water-soluble vitamins usually do not accumulate in the body because they are stored only in small quantities. However, this property allows deficiencies of water-soluble vitamins to occur faster than deficiencies of fat-soluble vitamins. Excessive dosages of water-soluble vitamins generally are excreted in the urine but can lead to toxicity if the dosages are high enough.[144] A list of fat- and water-soluble vitamins, along with deficiency diseases, RDA, and UL, are given in Table 8.1.

The reason that the list of vitamins seems to skip directly from E to the rarely mentioned K is that the vitamins corresponding to the "letters" F–J were either reclassified over time, discarded as false leads, or were renamed because of their relationship to vitamin B, which became a "complex" of vitamins. Table 8.2 lists chemicals that had previously been classified as vitamins, as well as the earlier names of vitamins that later became part of the B-complex.

Vitamins are essential organic compounds that perform numerous and diverse metabolic functions, often serving as enzymatic cofactors. With some exceptions, vitamins or their precursors must be obtained from food or supplements. The main exceptions are vitamin D, a hormone-like

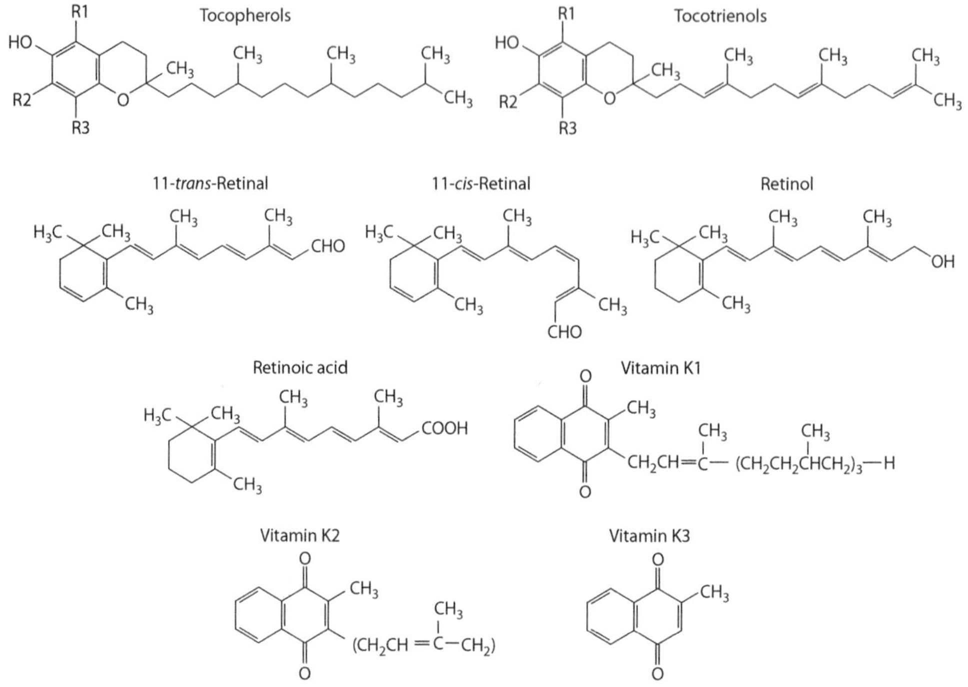

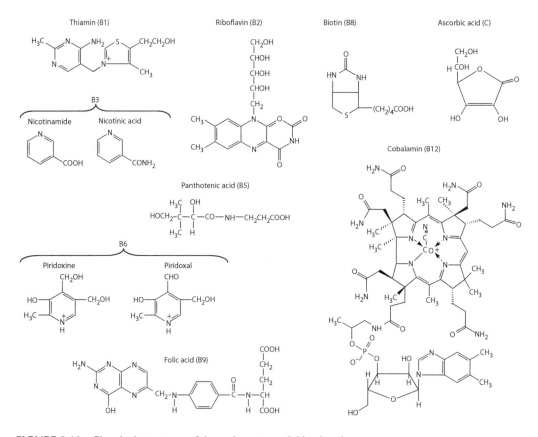

FIGURE 8.15 Chemical structures of the major fat-soluble vitamins.

FIGURE 8.16 Chemical structures of the major water-soluble vitamins.

TABLE 8.1

Chemical Name, Solubility, Deficiency Disease, Daily Recommended Dietary Allowance (RDA), and Daily Upper Intake Level (UL) of Vitamins

Vitamins	Chemical Name	Solubility	Deficiency Disease	RDA (mg)	UL Day (mg)
A	Retinoids (retinol, retinal, retinoic acid, 3-dehydroretinol and its derivatives)	Fat	Night-blindness, keratomalacia	0.9	3
B_1	Thiamin	Water	Beriberi	1.2	n.d.
B_2	Riboflavin	Water	Ariboflavinosis	1.3	n.d.
B_3	Niacin	Water	Pellagra	16	35
B_5	Pantothenic acid	Water	Paresthesia	5	n.d.
B_6	Pyridoxine	Water	Anemia	1.3–1.7	100
B_7	Biotin	Water	n.a.	0.03	n.d.
B_9	Folic acid	Water	Deficiency during pregnancy is associated with birth defects	0.4	1
B_{12}	Cyanocobalamin	Water	Megaloblastic anemia	0.0025	n.d.
C	Ascorbic acid	Water	Scurvy	90	2000
D_2–D_4	Lumisterol, ergocalciferol, cholecalciferol, dihydrotachysterol, 7-dehydrocholesterol	Fat	Rickets	0.005–0.01	0.05
E	Tocopherol, tocotrienol	Fat	Deficiency is very rare, mild hemolytic anemia in newborn infants	15	1000
K	Naphthoquinone	Fat	Bleeding diathesis	0.12	n.d.

Source: Adapted from different sources.

n.a. = not available, n.d. = not determined.

vitamin that can be synthesized in the skin upon exposure to sunlight and other forms of ultraviolet radiation, and therefore is not absolutely required in the diet, and vitamin K, which can be synthesized by intestinal microflora. Some vitamins have multiple chemical forms or isomers that vary in biological activity. For example, four forms of tocopherol occur in food, with α-tocopherol having the greatest vitamin E activity. Numerous carotenoids occur in nature and in food, but only a few have vitamin A activity. Some carotenoids, such as lycopene, do not possess vitamin A activity but, nonetheless, have important health benefits.

The popular literature has emphasized that natural vitamins are superior to synthetic ones, even though the active ingredient is often chemically identical in both forms. However, added folic acid

TABLE 8.2
Previous and Current Nomenclature of Vitamins

Previous Vitamin Name	Chemical Name	Current Vitamin Name	Reason for Name Change
B_4	Adenine	n.a.	No longer classified as a vitamin
B_8	Adenylic acid	n.a.	No longer classified as a vitamin
F	Essential fatty acids	n.a.	Needed in large quantities, does not fit definition of vitamin
G	Riboflavin	B_2	Reclassified as B-complex
H, I	Biotin	B_7	Reclassified as B-complex
J	Catechol, flavin	n.a.	No longer classified as a vitamin
L_1	Orthoaminobenzoic acid, anthranilic acid	n.a.	No longer classified as a vitamin
L_2	Adenyl thiomethylpentose	n.a.	No longer classified as a vitamin
M	Folic acid	B_9	Reclassified as B-complex
P	Flavonoids	n.a.	No longer classified as a vitamin
PP	Niacin	B_3	Reclassified as B-complex
R, B_{10}	Pteroylmonoglutamic acid	n.a.	No longer classified as a vitamin
S, B_{11}	Pteroylheptaglutamic acid	n.a.	No longer classified as a vitamin
U	Allantoine	n.a.	No longer classified as a vitamin

Source: Adapted from different sources.
n.a. = not available.

in dietary supplements or fortified foods is better absorbed than naturally occurring folates in foods. Pregnant women administered synthetic folic acid (pteroylglutamic acid) had higher serum folate levels than women given the natural conjugated form of the vitamin.[145]

The mechanisms by which vitamins prevent illnesses are not well understood, and the amounts needed to lower risks for certain disease conditions may be higher than the current recommended levels for preventing nutritional deficiencies. For example, the Institute of Medicine recommends that to prevent neural tube birth defects, women of child-bearing age should consume 400 μg of folic acid per day (but not more than 1000 μg/day) from fortified foods and/or dietary supplements in addition to folates obtained from a varied diet.[146] Ascorbic acid intakes of 80–200 mg daily (8–20 times the amounts needed to prevent scurvy) may be necessary to enhance certain physiological functions and minimize specific disease risks.[147]

8.2.1.1 Water-Soluble Vitamins

8.2.1.1.1 Thiamin (Vitamin B₁)

Description. Thiamin is a water-soluble B-complex vitamin, previously known as vitamin B_1 or aneurine.[148] Isolated and characterized in the 1930s, thiamin was one of the first organic compounds to be recognized as a vitamin.[149] Thiamin occurs in the human body as free thiamin and its phosphorylated forms: thiamin monophosphate (TMP), thiamin triphosphate (TTP), and thiamin pyrophosphate (TPP).

Sources. A varied diet should provide most individuals with adequate thiamin to prevent deficiency. Whole grain cereals, legumes (e.g., beans and lentils), nuts, lean pork, and yeast are rich sources of thiamin.[148] Because most of the thiamin is lost during the production of white flour and polished (milled) rice, white rice and foods made from white flour (e.g., bread and pasta) are fortified with thiamin.

Supplements. Thiamin is available in nutritional supplements and for fortification as thiamin hydrochloride and thiamin nitrate.[150]

8.2.1.1.2 Riboflavin (Vitamin B₂)

Description. Riboflavin is a water-soluble B-complex vitamin, also known as vitamin B_2. In the body, riboflavin functions in the mitochondrial electron transport system as the coenzymes, flavin adenine dinucleotide and flavin mononucleotide.[151]

Sources. Riboflavin is found in a variety of foods, including dairy products, meat, vegetables, and cereals. Therefore, riboflavin deficiency is uncommon in Western countries. Riboflavin is easily destroyed by exposure to light. Up to 50% of the riboflavin in milk contained in a clear glass bottle can be destroyed after 2 h of exposure to bright sunlight.[152]

Supplements. The most common forms of riboflavin available in supplements are riboflavin and riboflavin-5'-monophosphate. Riboflavin is most commonly found in multivitamin and vitamin B-complex preparations.[150]

8.2.1.1.3 Niacin (Vitamin B₃)

Description. Niacin, also known as vitamin B_3, refers to nicotinic acid and nicotinamide, which are both used by the body to form the coenzymes, nicotinamide adenine dinucleotide (NAD) and nicotinamide adenine dinucleotide phosphate (NADP). Neither form is related to the nicotine found in tobacco, although their names are similar.[153]

Sources. Niacin and substances that are convertible to niacin are found naturally in meat (especially red meat), poultry, fish, legumes, and yeast. In addition to preformed niacin, some L-tryptophan found in the proteins of these foods is metabolized to niacin. Niacin is also present in cereal grains, such as corn and wheat. However, consumption of corn-rich diets has resulted in niacin deficiency in certain populations. The reason for this is that niacin exists in cereal grains in bound forms, such as the glycoside niacytin, which exhibit little or no nutritional availability. Interestingly, niacin deficiency is not common in Mexico and Central America even though the diets of those in these countries are based on corn. Alkaline treatment, such as soaking corn in a lime solution—the process used by the populations of Mexico and Central America in the production of corn tortillas—yields release of bound niacin and increased availability of the vitamin.

Supplements. Niacin supplements are available as nicotinamide or nicotinic acid. Nicotinamide is the form of niacin typically used in nutritional supplements and in food fortification. Nicotinic acid is available over the counter and with a prescription as a cholesterol-lowering agent.[150]

The nomenclature for nicotinic acid formulations can be confusing. Nicotinic acid is available over the counter in an "immediate-release" (crystalline), and "slow-release" or "timed-release form." A shorter acting, timed-release preparation referred to as "intermediate release" or "extended release" nicotinic acid is available by prescription.[154] Due to the potential for side effects, medical supervision is recommended for the use of nicotinic acid as a cholesterol-lowering agent.

8.2.1.1.4 Pantothenic Acid (Vitamin B₅)

Description. Also known as vitamin B_5, pantothenic acid is essential to all forms of life.[155] Pantothenic acid is found throughout living cells in the form of coenzyme A (CoA), a vital coenzyme in numerous chemical reactions.[156] The term "pantothenic acid" is derived from the Greek word *pantos*, meaning everywhere.

Sources. Pantothenic acid is widely distributed in plant and animal food sources, where it occurs in both bound and free forms. Rich sources of the vitamin include organ meats (liver, kidney), egg yolk, avocados, cashew nuts and peanuts, brown rice, soya, lentils, broccoli, and milk. Pantothenic acid is synthesized by intestinal microflora, and this may also contribute to the body's pantothenic acid requirements.

Supplements. Supplements commonly contain pantothenol, a more stable alcohol derivative, which is rapidly converted by humans to pantothenic acid. Calcium and sodium D-pantothenate, the calcium and sodium salts of pantothenic acid, are also available as supplements.[155]

8.2.1.1.5 Pyridoxine, Pyridoxal, and Pyridoxamine (Vitamin B_6)

Description. There are six forms of vitamin B_6: pyridoxal (PL), pyridoxine (PN), pyridoxamine (PM), and their phosphate derivatives: pyridoxal 5′-phosphate (PLP), pyridoxine 5′-phosphate (PNP), and pyridoxamine 5′-phosphate (PMP). PLP is the active coenzyme form and has the most importance in human metabolism.[157]

Sources. Foods rich in vitamin B_6 include white meat (poultry and fish), bananas, liver, whole-grain breads and cereals, soybeans, and vegetables. Certain plant foods contain a unique form of vitamin B_6 called pyridoxine glucoside. This form of vitamin B_6 appears to be only about half as bioavailable as vitamin B_6 from other food sources or supplements. In most cases, including foods in the diet that are rich in vitamin B_6, it should supply enough to prevent deficiency. However, those who follow a very restricted vegetarian diet might need to increase their vitamin B_6 intake by eating food, fortified with vitamin B_6, or by taking a supplement.

Supplements. Vitamin B_6 is available in nutritional supplements principally in the form of pyridoxine hydrochloride. Pyridoxal-5′-phosphate is also available as a nutritional supplement. Pyridoxine hydrochloride is available in multivitamin and multivitamin/multimineral products as well as products that, in addition to vitamins and minerals, contain other nutritional substances. Single-ingredient pyridoxine products are also available.[150]

8.2.1.1.6 Biotin (Vitamin B_8 or Vitamin H)

Description. Also known as vitamin B_8 or H, this vitamin is of great importance for the biochemistry of the human organism. Biotin is the cofactor for a small group of enzymes that catalyze carboxylation, decarboxylation, and transcarboxylation reactions in carbohydrate and fatty acid metabolism. Deficiency of this vitamin is rare in humans, but can be induced in special circumstances: in individuals with inborn errors of biotin metabolism, in individuals taking certain medications, and in some women during pregnancy.[158,159]

Sources. A balanced diet usually contains enough biotin as it is found in many foods (for example bread, egg yolks, fish, legumes, meat, dairy products, and nuts). In the intestines, bacteria produce a small amount of biotin, which may be absorbed and contribute to daily needs.[160]

Supplements. Biotin is available in multivitamin and multivitamin/multimineral products as well as in single ingredient products. The major benefit of biotin as a dietary supplement is in strengthening hair and nails. Some skin disorders, such as "cradle cap," improve with biotin supplements. Biotin has also been used to combat premature graying of hair, though it is likely to be useful only for those with a low biotin level. Biotin has been used for people in weight-loss programs to help them metabolize fat more efficiently.

8.2.1.1.7 Folic Acid (Vitamin B_9)

Description. The term "folic acid" is used to denote pteroylmonoglutamic acid. The term "folate" refers to the naturally occurring folates in foods, which are pteroylpolyglutamic acids with two to eight glutamic acid groups attached to the primary structure.[161] The folate participates in several key biological processes, including the synthesis of DNA, RNA, and proteins. It is necessary for DNA replication and repair, the maintenance of the integrity of the genome, and is involved in the regulation of gene expression. A deficiency of folic acid impairs DNA synthesis and cell division: the common clinical manifestation of severe folic acid deficiency is megaloblastic anemia.[162] In particular, folic acid is very important for all women before conception and very early in pregnancy in order to decrease the risk of having babies with neural tube birth defects.[163]

Sources. Green leafy vegetables (foliage) are rich sources of folate and provide the basis for its name (from Latin, *folium*-leaf). Citrus fruit juices, legumes, and fortified cereals are also excellent sources of folate.[164]

Supplements. The principal form of supplementary folate is folic acid. It is available in single ingredient and combination products, such as B-complex vitamins and multivitamins.[150]

8.2.1.1.8   Cobalamin (Vitamin B₁₂)

Description. Vitamin B_{12} is the largest and most complex of all the vitamins. It is unique among vitamins in that it contains a metal ion, cobalt. For this reason "cobalamin" is the term used to refer to compounds having B_{12} activity. Methylcobalamin and 5-deoxyadenosyl cobalamin are the forms of vitamin B_{12} used in the human body.[153] The form of cobalamin used in most supplements, cyanocobalamin, is readily converted to 5-deoxyadenosyl and methylcobalamin.

Sources. Only bacteria can synthesize vitamin B_{12}. Vitamin B_{12} is present in animal products such as meat, poultry, fish (including shellfish), and to a lesser extent milk, but it is not generally present in plant products or yeast.[153] Fresh pasteurized milk contains $0.9\,\mu g$ per cup and is an important source of vitamin B_{12} for some vegetarians.[165] Those vegetarians who eat no animal products need supplemental vitamin B_{12} to meet their requirements.

Supplements. Cyanocobalamin (readily converted in the body to the bioavailable forms 5-deoxyadenosyl and methylcobalamin) is the principal form of vitamin B_{12} used in supplements but methylcobalamin is also available. Cyanocobalamin is available by prescription in an injectable form and as a nasal gel for the treatment of pernicious anemia. Over the counter preparations containing cyanocobalamin include multivitamin, vitamin B-complex, and vitamin B_{12} supplements.[150]

8.2.1.1.9   Ascorbic Acid (Vitamin C)

Description. The name "ascorbic acid" is derived from "a-" and "scorbuticus" (scurvy) as a shortage of this molecule may lead to scurvy. In humans, vitamin C is a highly effective antioxidant, acting to lessen oxidative stress, a substrate for ascorbate peroxidase, as well as an enzyme cofactor for the biosynthesis of many important biochemicals. Most mammalian species synthesize L-ascorbic acid from D-glucose in the liver.[166] The enzyme that catalyzes the final step, L-gulono-1,4-lactone oxidase, is missing or mutated in humans, other primates, guinea pig, and certain fruit bats. Therefore, vitamin C must be introduced through the diet.

Sources. The best food sources of vitamin C are citrus fruits, berries, melons, tomatoes, potatoes, green peppers, and leafy green vegetables. Vitamin C is sensitive to air, heat, and water, so it can easily be destroyed by prolonged storage, overcooking, and processing of foods. It is also present in some cuts of meat, especially liver.

Supplements. Vitamin C is the most widely taken dietary supplement. It is available in many forms including caplets, tablets, and capsules, drink mix packets, in multivitamin formulations, in multiple antioxidant formulations, as a chemically pure crystalline powder, timed release versions, and also including bioflavonoids, such as quercetin, hesperidin, and rutin. The use of vitamin C supplements with added bioflavonoids and, often, flavors and sweeteners, can be problematic at gram dosages, since those additives are not so well studied as vitamin C.[167] In supplements, vitamin C most often comes in the form of various mineral ascorbates, as they are easier to absorb, more easily tolerated, and provide a source of several dietary minerals.

8.2.1.2 Fat-Soluble Vitamins

8.2.1.2.1 Vitamin A

Description. Vitamin A is a generic term for a large number of related compounds. Retinol (an alcohol) and retinal (an aldehyde) are often referred to as preformed vitamin A. Retinal can be converted by the body to retinoic acid, the form of vitamin A known to affect gene transcription. Retinol, retinal, retinoic acid, and related compounds are known as retinoids. β-Carotene and other carotenoids that can be converted by the body into retinol are referred to as provitamin A carotenoids. Vitamin A has many functions in the body. In addition to helping the eyes adjust to light changes, vitamin A plays an important role in bone growth, tooth development, reproduction, cell division, and gene expression. Also, the skin, eyes, and mucous membranes of the mouth, nose, throat, and lungs depend on vitamin A to remain moist.

Sources. Free retinol is not generally found in foods. Retinyl palmitate, a precursor and storage form of retinol, is found in foods from animals. Plants contain carotenoids, some of which are precursors for vitamin A (e.g., α-carotene and β-carotene). Yellow and orange vegetables contain significant quantities of carotenoids. Green vegetables also contain carotenoids, though the pigment is masked by the green pigment of chlorophyll.[168] Vitamin A is measured in retinol equivalents (REs), which allows the different forms of vitamin A to be compared. One RE equals 1 μg of retinol or 6 μg of β-carotene. Vitamin A is also measured in international unit (IU) with 1 μg RE equivalent to 3.33 IU.

Supplements. The principal forms of preformed vitamin A (retinol) in supplements are retinyl palmitate and retinyl acetate. β-Carotene is also a common source of vitamin A in supplements, and many supplements provide a combination of retinol and β-carotene.[150]

8.2.1.2.2 Vitamin D

Description. There are two forms of the vitamin. Vitamin D_2 (ergocalciferol) is derived from ergosterol in the diet, whereas vitamin D_3 (cholecalciferol) is derived from cholesterol via 7-dehydrocholesterol photochemical reactions using ultraviolet B (UV-B) radiation from sunlight. It is the only vitamin the body manufactures naturally and is technically considered a hormone. However, there are conditions where the synthesis of vitamin D_3 in the skin is not sufficient to meet physiological requirements. Humans who are not exposed to sufficient sunlight due to the reasons of geography and shelter or clothing require dietary intake of vitamin D. Vitamin D plays a critical role in the body's use of calcium and phosphorous. It increases the amount of calcium absorbed from the small intestine and helps form and maintain bones. Children especially need adequate amounts of vitamin D to develop strong bones and healthy teeth.

Sources. The primary food sources of vitamin D are milk and other dairy products fortified with vitamin D. Vitamin D is also found in oily fish (e.g., herring, salmon, and sardines) as well as in cod liver oil. In addition to the vitamin D provided by food, the body obtains vitamin D through the skin, which makes vitamin D in response to sunlight.

Supplements. Supplemental vitamin D is available as vitamin D_2 (ergocalciferol) or vitamin D_3 (cholecalciferol). Usually, vitamin D is present in a multivitamin, multimineral preparation. Typical dosage is 200–400 IU (5–10 μg) daily. Most vitamin D supplements available without a prescription contain cholecalciferol (vitamin D_3). Multivitamin supplements for children generally provide 200 IU (5 μg) and multivitamin supplements for adults generally provide 400 IU (10 μg) of vitamin D. Single ingredient vitamin D supplements may provide 400–1000 IU of vitamin D, but 400 IU is the most commonly available dose. A number of calcium supplements may also provide vitamin D.

8.2.1.2.3 Vitamin E

Description. The term "vitamin E" describes a family of eight antioxidants, four tocopherols (α-, β-, λ-, and δ-) and four tocotrienols (also α-, β-, λ-, and δ-). α-Tocopherol is the only form of vitamin

E that is actively maintained in the human body and is, therefore, the form of vitamin E found in the largest quantities in the blood and tissue.[169] The main function of vitamin E is to maintain the integrity of the body's intracellular membrane by protecting its physical stability and providing a defense line against tissue damage caused by oxidation.

Sources. Vitamin E is found in plants, animals, and in some green, brown, and blue/green algae. The richest sources of the vitamin are found in unrefined edible vegetable oil, including wheat germ, safflower, sunflower, cottonseed, canola, and olive oils. In these oils, approximately 50% of the tocopherol content is in the form of α-tocopherol. Soybean and corn oils contain about ten times as much γ-tocopherol as they do α-tocopherol. Palm, rice bran, and coconut oils are rich sources of tocotrienols. α-Tocopherol is the major form of vitamin E in animal products and is found mainly in the fatty portion of the meat. Other foods containing vitamin E include unrefined cereal grains, fruits, nuts, and vegetables.[170]

Dietary vitamin E requirements frequently are expressed as milligram α-tocopherol equivalents, but food and supplement labels typically use IU to express vitamin E activity, and 1.0 IU is equivalent to 1.0 mg of all-racemic-α-tocopherol acetate. The relative activity of all racemic-α-tocopherol is set at 74% that of RRR-α-tocopherol, considered to have the highest bioavailability, and thus as the standard against which all the others must be compared.[171,172]

Supplements. The vitamin E used in supplements and food additives generally is esterified with acetate, nicotinate, or succinate to prevent oxidation and prolong its shelf life. These esters are hydrolyzed easily by the gut to yield the bioactive form of the vitamin.

8.2.1.2.4 Vitamin K

Description. The "K" is derived from the German word *koagulation*. Coagulation refers to blood clotting, because vitamin K is essential for the functioning of several proteins involved in blood clotting.[153]

Sources. There are two naturally occurring forms of vitamin K. Plants synthesize phylloquinone, also known as vitamin K_1. Bacteria synthesize a range of vitamin K forms, using repeating 5-carbon units in the side chain of the molecule. These forms of vitamin K are designated menaquinone-n (MK-n), where n stands for the number of 5-carbon units. MK-n is collectively referred to as vitamin K_2.[173] K_3 is synthetic menadione. When administered, vitamin K_3 is alkylated to one of the vitamin K_2 forms of menaquinone. Green leafy vegetables and some vegetable oils (soybean, cottonseed, canola, and olive) are major contributors of dietary vitamin K.

Supplements. There is no typical dosage for vitamin K. Some multivitamin preparations contain vitamin K as vitamin K_1 or vitamin K_2. In Japan, vitamin K, usually in the form of vitamin K_2, is used for the management of osteoporosis. The fermented soybean product natto is rich in menaquinone-7 or vitamin K_2. The bacteria that is used in the preparation of natto, *Bacillus natto*, is also used in Japan as a dietary supplement source of vitamin K_2.[174]

8.2.2 Minerals

Human body uses minerals to activate the enzymes, hormones, and other molecules that participate in the function and maintenance of life processes. Minerals are absolutely essential, as they cannot be synthesized by the body, but must be accumulated from the trace amounts present in soil or mineral rich waters.[175]

The number of different minerals necessary to maintain good health is still unknown to science. While dietary deficiencies of only a few minerals such as iron or iodine can be linked directly to disease, some authorities claim that as many as 60 elements are necessary for optimum longevity and quality of life in humans. What has become more widely recognized is that a number of minerals are required by the body, and deficiencies in these minerals may not produce obvious symptoms but can still result in poor health or a shortened life expectancy. Even some minerals once considered solely "toxic" have now been identified as important in supporting longevity and quality of life. Selenium was long viewed as a toxic compound, but several international studies have demonstrated that dietary selenium markedly reduces the incidences of and death from cancer. Some forms of chromium are also considered toxic, but trivalent chromium is a mineral associated with fat metabolism and has been linked to the regulation of blood glucose levels in the body.

Mineral supplements are sold as organic or inorganic salts of metals synthesized by the chemical industry. Because minerals are generally recovered by mining ores or from byproducts of chemical processes, their use as nutritional supplements may not result in a mineral form that can be taken up by the body—one that is readily "bioavailable." Some metals such as selenium and chromium are also offered as complexes with yeast to give a more organic nature to the products.[176]

There are two groups of minerals, major minerals and trace minerals. Major minerals (also known as macrominerals, macroelements, or bulk minerals) are needed in the diet in amounts of 100 mg or more each day. They include calcium, magnesium, sodium, potassium, phosphorus, and chlorine. Macrominerals are present in virtually all cells of the body, maintaining general homeostasis, and are required for normal functioning. Acute imbalances of these minerals can be potentially fatal, although nutrition is rarely the cause of these cases. Trace minerals (also known as microminerals) are micronutrients that are chemical elements. They include iron, chromium, copper, iodine, manganese, selenium, zinc, and molybdenum. They are dietary minerals needed by the human body in very small quantities (generally less than 100 mg/day) (Table 8.3).

There are several other minerals that may be essential for humans, but research has not established their importance, including tin, nickel, silicon, and vanadium. There are also minerals found in the body that are regarded as contaminants including lead, mercury, arsenic, aluminum, silver, cadmium, barium, strontium, and others.

8.2.2.1 Major Minerals

8.2.2.1.1 Calcium (Ca)

Description. Calcium is the most common mineral in the human body. About 99% of the calcium in the body is found in bones and teeth, while the other 1% is found in the blood and soft tissue. Calcium levels in the blood and fluid surrounding the cells (extracellular fluid) must be maintained within a very narrow concentration range for normal physiological functioning. The physiological functions of calcium are so vital to survival that the body will demineralize bone to maintain normal blood calcium levels when calcium intake is inadequate. Thus, adequate dietary calcium is a critical factor in maintaining a healthy skeleton.[177]

Sources. Dairy products represent rich and absorbable sources of calcium, but certain vegetables and grains also provide calcium. However, the bioavailability of that calcium must be taken into consideration. While the calcium rich plants in the kale family (broccoli, bok choy, cabbage, mustard, and turnip greens) contain calcium that is as bioavailable as that in milk, some food components have been found to inhibit the absorption of calcium. Oxalic acid, also known as oxalate, is the most potent inhibitor of calcium absorption, and is found in high concentrations in spinach and rhubarb and somewhat lower concentrations in sweet potato and dried beans.[142] Phytic acid is a less potent inhibitor of calcium absorption than oxalate. Yeast possesses an enzyme (phytase) which breaks down phytic acid in grains during fermentation, lowering the phytic acid content of breads and other fermented foods. Only concentrated sources of phytate such as wheat bran or dried beans substantially reduce calcium absorption.[177]

Supplements. The two main forms of calcium found in supplements are carbonate and citrate. Calcium carbonate is the most common because it is inexpensive and convenient. The absorption

TABLE 8.3

Function, Deficiency Disease, Main Sources, and Daily Recommended Dietary Allowance (RDA) of Minerals

Mineral	Function	Deficiency Disease	Main Sources	RDA (mg)
Calcium	Cellular biochemistry signaling	Rickets, osteoporosis	Milk, butter, cheese, sardines, green leafy vegetables, citrus fruit	800
Chromium	Metabolism of sugar	Adult onset diabetes	Yeast, black pepper, liver, wholemeal bread, beer	0.12
Copper	Component of important enzymes	Anemia, Menkes syndrome	Green vegetables, fish, oysters, liver	1.2
Fluorine	Component of tooth enamel	Tooth decay, osteoporosis	Seafood, tea	n.a.
Iodine	Involved in formation of thyroid hormones	Goiter, cretinism	Seafood, table salt	0.15
Iron	An essential component of many proteins	Anemia	Liver, kidney, green leafy vegetables, egg yolk, fruit, potatoes	14
Magnesium	An essential component of many enzymes	Irregular heartbeat, muscle weakness, insomnia	Green leafy vegetable, nuts, grain	300
Manganese	An essential cofactor of many enzymes	Skeletal deformities, growth impairment	Legumes, cereals, green leafy vegetables, tea	n.a.
Molybdenum	Involved in chemical pathways	Irritability, irregular heart beat	Legumes, cereals, liver, kidney	n.a.
Phosphorus	Component of DNA, cellular energy	Muscular weakness, bone pain, appetite loss	Meat, poultry, fish, eggs, beans, milk	800
Potassium	Abundant monovalent inorganic cation in cells	Irregular heartbeat, muscle weakness, fatigue	Vegetables, meat, oranges, bananas, bran	3500
Selenium	Antioxidant	No known deficiency symptoms	Seafood, cereals, meat, egg yolk, garlic	0.055
Sodium	Diverse roles including signal transduction	Impaired acid–base balance	Table salt	1500
Zinc	Component of many proteins including insulin	Impaired wound healing, appetite loss, impaired sexual development	Meat, grain, legumes, oysters, milk	15

Source: Adapted from different sources.
n.a. = not available.

of calcium citrate is similar to calcium carbonate. For instance, a calcium carbonate supplement contains 40% calcium while a calcium citrate supplement only contains 21% calcium. However, you have to take more pills of calcium citrate to get the same amount of calcium as you would get from a calcium carbonate pill since citrate is a larger molecule than carbonate. One advantage of calcium citrate over calcium carbonate is better absorption in those individuals who have decreased stomach acid. Calcium citrate malate is a form of calcium used in the fortification of certain juices and is also well absorbed.[178] Other forms of calcium in supplements or fortified foods include calcium gluconate, lactate, and phosphate. Figure 8.17 compares the amount of calcium (elemental calcium) found in some different forms of calcium supplements.[179]

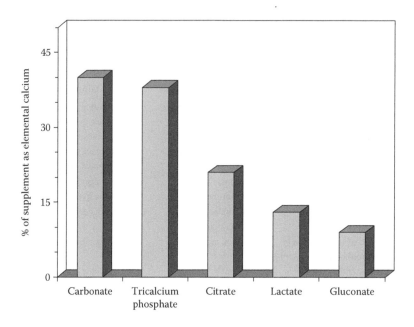

FIGURE 8.17 Comparison of calcium content of various supplements.

8.2.2.1.2 Magnesium (Mg)

Description. Magnesium plays important roles in the structure and the function of the human body. Over 60% of all the magnesium in the body is found in the skeleton, about 27% is found in muscle, while 6%–7% is found in other cells, and less than 1% is found outside of cells.[180]

Sources. Because magnesium is part of chlorophyll, the green pigment in plants, green leafy vegetables are rich in magnesium. Unrefined grains and nuts also have high magnesium content. Meats and milk have intermediate magnesium content, while refined foods generally have the lowest magnesium content. Water is a variable source of intake; harder water usually has a higher concentration of magnesium salts.[181]

Supplements. Oral magnesium supplements combine magnesium with another substance such as a salt. Examples of magnesium supplements include magnesium oxide, magnesium sulfate, and magnesium carbonate. Elemental magnesium refers to the amount of magnesium in each compound. Figure 8.18 compares the amount of elemental magnesium in different types of magnesium supplements.[182] The amount of elemental magnesium in a compound and its bioavailability influence the effectiveness of the magnesium supplement. Bioavailability refers to the amount of magnesium in food, medications, and supplements that is absorbed in the intestines and ultimately available for biological activity in our cells and tissues. Enteric coating of a magnesium compound can decrease bioavailability.[183] In a study that compared four forms of magnesium preparations, results suggested lower bioavailability of magnesium oxide, with significantly higher and equal absorption and bioavailability of magnesium chloride and magnesium lactate.[184] This supports the belief that both the magnesium content of a dietary supplement and its bioavailability contribute to its ability to replete deficient levels of magnesium.

8.2.2.1.3 Sodium (Na)

Description. Most of the sodium in the body (about 85%) is found in the fluids that surround the body's cells (such as blood and lymph fluid).

Sources. Almost all foods contain sodium naturally or as an ingredient, such as table salt (sodium chloride) or baking soda (sodium bicarbonate) added in processing or while cooking. Many

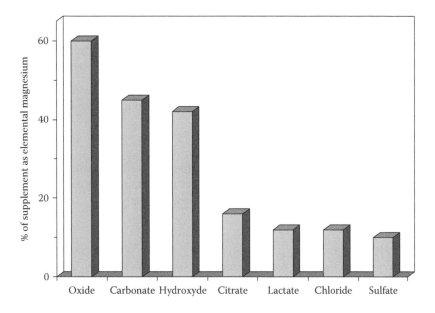

FIGURE 8.18 Comparison of magnesium content of various supplements.

medicines and other products also contain sodium, including laxatives, aspirin, mouthwash, and toothpaste.

Although scientists agree that a minimal amount of salt is required for survival, the health implications of excess salt intake represent an area of considerable controversy among scientists, clinicians, and public health experts.[185]

8.2.2.1.4 Potassium (K)

Description. Potassium is an essential mineral micronutrient in human nutrition; it is the major cation inside animal cells, and it is thus important in maintaining fluid and electrolyte balance in the body. Potassium is also important in allowing muscle contraction and the sending of all nerve impulses in animals.

Sources. The best dietary sources of potassium are fresh unprocessed foods, including meats, vegetables (especially potatoes), fruits (especially avocados and bananas), and citrus juices (such as orange juice). Most potassium needs can be met by eating a varied diet with adequate intake of milk, meats, cereals, vegetables, and fruits.

The 2004 guidelines of the Institute of Medicine specify an RDA of 4000 mg of potassium. However, it is thought that most Americans consume only half that amount per day.[186] Similarly, in the European Union, particularly in Germany and Italy, insufficient potassium intake is widespread.[187]

Supplements. Supplements of potassium in medicine are most widely used in conjunction with the most powerful classes of diuretics, which rid the body of sodium and water, but have the side effect of also causing potassium loss in urine. Potassium supplements are available as a number of different salts, including potassium chloride, citrate, gluconate, bicarbonate, aspartate, and orotate.[150]

8.2.2.1.5 Phosphorus (P)

Description. Phosphorus is an essential mineral that is required by every cell in the body for normal function.[188] The majority of the phosphorus in the body is found as phosphate (PO_4). Approximately 85% of the body's phosphorus is found in bone.[189]

Sources. Phosphorus is found in most foods because it is a critical component of all living organisms. Dairy products, meat, and fish are particularly rich sources of phosphorus. Phosphorus is also a component of many polyphosphate food additives, and is present in most soft drinks as phosphoric acid. The phosphorus in all plant seeds (beans, peas, cereals, and nuts) is present in a storage form of phosphate called phytic acid or phytate. Only about 50% of the phosphorus from phytate is available to humans because of the lack of enzymes (phytases) that liberate it from phytate.[190]

Supplements. Sodium phosphate and potassium phosphate salts are used for the treatment of hypophosphatemia, and their use requires medical supervision. Calcium phosphate salts are sometimes used as calcium supplements.[150]

8.2.2.1.6 Chloride (Cl)

Chloride is an anion, generally consumed as sodium chloride (NaCl) or table salt. There is a high correlation between the sodium and chloride contents of the diet, and only under unusual circumstances, levels of sodium and chloride vary in the diet independently.

8.2.2.2 Trace Minerals

8.2.2.2.1 Iron (Fe)

Description. Every cell in the body requires iron for a variety of functions. This versatile mineral is involved in oxygen transport (hemoglobin) and storage (myoglobin), is required by enzymes that produce energy for the cell, and it plays an important role in the function of the immune and central nervous systems.[191] Iron is required in relatively high doses to maintain proper nutrition. Of all the nutrients, the allowance for iron is the most difficult to obtain from dietary sources, and therefore is the most common single micronutrient deficiency in the world.[192]

Sources. The amount of iron in food (or supplements) that is absorbed and used by the body is influenced by the iron nutritional status of the individual and whether or not the iron is in the form of heme iron. Because it is absorbed by a different mechanism than nonheme iron, heme iron is more readily absorbed and its absorption is less affected by other dietary factors. Individuals who are anemic or iron deficient absorb a larger percentage of the iron they consume (especially nonheme iron) than individuals who are not anemic and have sufficient iron stores.[193,194]

Supplements. Supplemental iron is available in two forms: ferrous and ferric. Ferrous iron salts (ferrous fumarate, ferrous sulfate, and ferrous gluconate) are the best absorbed forms of iron supplements.[195] Elemental iron is the amount of iron in a supplement that is available for absorption. Figure 8.19 lists the percent elemental iron in these supplements.

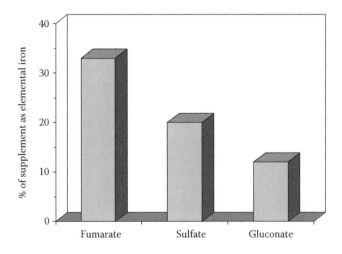

FIGURE 8.19   Comparison of iron content of various supplements.

8.2.2.2.2   Chromium (Cr)

Description. Trivalent chromium(III) is an essential trace metal and is required for the proper metabolism of sugar in humans.

Sources. The amount of chromium in foods is variable, and it has been measured accurately in relatively few foods. Presently, there is no large database for the chromium content of foods. Processed meats, whole grain products, ready-to-eat bran cereals, green beans, broccoli, and spices are relatively rich in chromium.

Supplements. Chromium(III) is available as a supplement in several forms: chromium chloride, chromium nicotinate, chromium picolinate, and high-chromium yeast. They are available as stand-alone supplements or in combination products. Chromium supplementation may improve glucose tolerance in people with Turner's syndrome, a disease linked with glucose intolerance. It is also important in insulin metabolism and may help in controlling blood sugar levels in diseases such as diabetes. Chromium picolinate is a widely available nutritional supplement. There is some evidence, including results from human studies, that it has a role in glucose homeostasis.[196] However, analogues of picolinic acid have been shown to induce profound alterations in the metabolism of serotonin, dopamine, and norepinephrine in brain. Thus, caution should be used with chromium picolinate supplements especially by individuals prone to behavioral disorders.[197]

8.2.2.2.3   Copper (Cu)

Description. Copper is an essential trace element for humans and animals. In the body, copper shifts between the cuprous (Cu^{1+}) and the cupric (Cu^{2+}) forms, though the majority of the body's copper is in the Cu^{2+} form. The ability of copper to easily accept and donate electrons explains its important role in oxidation-reduction reactions and the scavenging of free radicals.[198] Although Hippocrates is said to have prescribed copper compounds to treat diseases as early as 400 BC, scientists are still uncovering new information regarding the functions of copper in the human body.[199]

Sources. Copper is found in a wide variety of foods and is most plentiful in organ meats, shellfish, nuts, and seeds. Wheat bran cereals and whole grain products are also good sources of copper.[142]

Supplements. Copper supplements are available as cupric oxide, copper gluconate, copper sulfate, and copper amino acid chelates.[150]

8.2.2.2.4   Fluoride (F)

Description. Fluorine occurs naturally in the Earth's crust, water, and food as the negatively charged ion, fluoride (F^-). Fluoride is considered a trace element because only small amounts are present in the body (about 2.6 g in adults), and because the daily requirement for maintaining dental health is only a few milligrams a day. About 95% of the total body fluoride is found in bones and teeth.[200] Although its role in the prevention of dental caries (tooth decay) is well established, fluoride is not generally considered an essential mineral element because humans do not require it for growth or to sustain life.[201] However, if one considers the prevention of chronic disease (dental caries), an important criterion in determining essentiality, then fluoride might well be considered an essential trace element.[202]

Sources. The fluoride content of most foods is low (less than 0.05 mg/100 g). Rich sources of fluoride include tea, which concentrates fluoride in its leaves, and marine fish that are consumed with their bones (e.g., sardines). Foods made with mechanically separated (boned) chicken, such as canned meats, hot dogs, and infant foods also add fluoride to the diet.[203]

Supplements. Fluoride supplements are available only by prescription, and are intended for children living in areas with low water fluoride concentrations for the purpose of bringing their intake to approximately 1 mg/day.[204]

8.2.2.2.5 Iodine (I₂)

Description. Iodine, a nonmetallic trace element, is required by humans for the synthesis of thyroid hormones, triiodothyronine (T_3) and thyroxine (T_4). Iodine deficiency is an important health problem throughout much of the world. Most of the Earth's iodine is found in its oceans. In general, the older an exposed soil surface, the more likely the iodine has been leached away by erosion. Mountainous regions, such as the Himalayas, the Andes, and the Alps, and flooded river valleys, such as the Ganges, are among the most severely iodine-deficient areas in the world.[205]

Sources. The iodine content of most foods depends on the iodine content of the soil in which it was raised. Seafood is rich in iodine because marine animals can concentrate the iodine from seawater. Certain types of seaweed (e.g., *Undaria pinnatifida* or wakame) are also very rich in iodine. Processed foods may contain slightly higher levels of iodine due to the addition of iodized salt or food additives, such as calcium iodate and potassium iodate.[142]

Supplements. Potassium iodide is available as a nutritional supplement, typically in combination products, such as multivitamin/multimineral supplements. Iodine makes up approximately 77% of the total weight of potassium iodide.[150] Potassium iodide as well as potassium iodate may be used to iodize salt. Iodized vegetable oil is also used in some countries as an iodine source.[150,206]

8.2.2.2.6 Manganese (Mn)

Description. Manganese is a mineral element that is both nutritionally essential and potentially toxic. The derivation of its name from the Greek word for magic remains appropriate because scientists are still working to understand the diverse effects of manganese deficiency and manganese toxicity in living organisms.[207] Manganese plays an important role in a number of physiologic processes as a constituent of some enzymes and an activator of other enzymes.[201]

Sources. Rich sources of manganese include whole grains, nuts, leafy vegetables, and teas. Foods high in phytic acid, such as beans, seeds, nuts, whole grains, and soy products, or foods high in oxalic acid, such as cabbage, spinach, and sweet potatoes, may slightly inhibit manganese absorption. Although teas are rich sources of manganese, the tannins present in tea may moderately reduce the absorption of manganese.[208]

Supplements. Several forms of manganese are found in supplements, including manganese gluconate, manganese sulfate, manganese ascorbate, and amino acid chelates of manganese. Manganese is available as a stand-alone supplement or in combination products.[150] Relatively high levels of manganese ascorbate may be found in a bone/joint health product containing chondroitin sulfate and glucosamine hydrochloride.

8.2.2.2.7 Selenium (Se)

Description. Selenium is a comparatively rare, though widely distributed, element. Chemically, it is classified as a metalloid, with properties of both metals and nonmetals. Until the 1950s, Se was considered by most scientists to be a chemical curiosity, the compounds of which had offensive odors and were toxic. Subsequently, it was found that Se could prevent several diseases of farm animals.[209] Se was shown to have a number of important biological functions that depend on the activity of certain Se-containing proteins.[210] The first of these functional selenoproteins to be identified were the glutathione peroxidase (GSHPx) enzymes, which are part of the body's antioxidant defense system. Their role is to break down hydrogen peroxide and lipid hydroperoxides generated by free radicals, which can damage cell membranes and disrupt cellular functions. More than 30 other selenoproteins have been identified and their structures at least partly determined. Their variety indicates the wide range of biochemical pathways and physiological functions to which Se contributes.[209]

Sources. Since the soil is the primary source of Se in food, levels in plant foods and in the animals fed on them reflect the considerable variations that occur in soil Se throughout the world. Levels usually found in different foods are in the microgram range: ~10–500 µg/kg. Certain plants have the ability to accumulate Se from the soil and store it to high levels. Among these are Brazil nuts, which can contain more than 50 mg/kg of the element, 10^5 times the levels found in most other foods.[209]

A number of Se-fortified food products are available in certain countries. So far, apart from Se-enriched infant formulae, these appear to be mainly foods designed for special physiological purposes. Approval has been given, for example, by the Australia New Zealand Food Authority (ANZFA), which is responsible for food regulation in both countries, for the sale of Se-enriched sports foods. These include a variety of foods and drinks designed "to assist athletes achieve peak performance."[211] In Japan and some other Asian countries several different kinds of Se-enriched foods are marketed on the grounds of their health-enhancing properties. In China a Se-rich drink which, it is claimed, helps to prevent ageing and heart disease, uses Se-rich green tea as the source of the element.[212] Products based on garlic, which can contain relatively high levels of Se, as well as Se-rich nuts and other foods, are also available. An intriguing possibility is the development of products using Se-enhanced "designer" eggs from hens fed fortified layer rations.[213]

Supplements. Selenium supplements are available in several forms. Sodium selenite and sodium selenate are inorganic forms of selenium. Selenate is almost completely absorbed, but a significant amount is excreted in the urine before it can be incorporated into proteins. Selenite is only about 50% absorbed, but is better retained than selenate, once absorbed. Selenomethionine, an organic form of selenium that occurs naturally in foods, is about 90% absorbed. Selenomethionine and selenium-enriched yeast, which mainly supplies selenomethionine, are also available as supplements.[142]

8.2.2.2.8   Zinc (Zn)

Description. Zinc is an essential trace element for all forms of life and its importance is reflected by the numerous functions and activities over which it exerts a regulatory role.[214] It is believed to possess antioxidant properties, which protect against premature aging of the skin and muscles of the body. The significance of zinc in human nutrition and public health was recognized relatively recently. Clinical zinc deficiency in humans was first described in 1961, when the consumption of diets with low zinc bioavailability due to high phytic acid content was associated with "adolescent nutritional dwarfism" in the Middle East.[215] Since then, zinc insufficiency has been recognized by a number of experts as an important public health issue, especially in developing countries.[216]

Sources. Zinc is found in oysters, and to a far lesser degree in most animal proteins, beans, nuts, almonds, whole grains, pumpkin seeds, and sunflower seeds.[142]

Supplements. A number of zinc supplements are available, including zinc acetate, zinc gluconate, zinc picolinate, and zinc sulfate. Zinc picolinate has been promoted as a more absorbable form of zinc, but there is little data to support this idea in humans.[142]

8.2.2.2.9   Molybdenum

Description. Molybdenum is an essential trace element for virtually all life forms. It functions as a cofactor for a number of enzymes that catalyze important chemical transformations in the global carbon, nitrogen, and sulfur cycles.[217]

Sources. Legumes, such as beans, lentils, and peas, are the richest sources of molybdenum. Grain products and nuts are considered good sources, while animal products, fruits, and many vegetables are generally low in molybdenum.[218] Because the molybdenum content of plants depends on the soil molybdenum content and other environmental conditions, the molybdenum content of foods can vary considerably.[219]

Supplements. Molybdenum in nutritional supplements is generally in the form of sodium molybdate or ammonium molybdate.[150]

8.2.3 Ergogenic (Sport-Enhancing) Supplements

Substances such as creatine, caffeine, L-carnitine, aspartate, bee pollen, and specific amino acids have recently received scientific attention due to their possible influence on performance, fatigue, and recovery.[220]

8.2.3.1 Amino Acids

Amino acids are theorized to enhance performance in a variety of ways, such as an increasing secretion of anabolic hormones, modifying fuel use during exercise, prevent adverse effects of overtraining, and preventing mental fatigue.[221]

Creatine, or methylguanidine-acetic acid, is an amino acid that is synthesized from arginine and glycine in the liver, pancreas, and kidneys.[222] Creatine was first introduced as a potential ergogenic aid in 1993 as creatine monohydrate and is currently being used extensively by athletes.[223] Creatine provides instant additional energy to the muscles. Unlike the energy provided by carbohydrates or fats, which take some time to convert to useable energy, the energy for creatine can be converted almost instantaneously. Oral creatine supplementation is popular among athletes, but many studies have failed to show an ergogenic benefit. Furthermore, numerous side effects including renal abnormalities have been reported for creatine,[224] and there is growing concern about potential long-term adverse effects on various organ systems.[225]

Branched chain amino acids (BCAAs) are the essential amino acids, leucine, isoleucine, and valine. BCAAs are considered to be among the most beneficial and effective supplements in any sports nutrition program. They are needed for the maintenance of muscle tissue and appear to preserve muscle glycogen stores, and help prevent muscle protein breakdown during exercise. BCAA supplements may be used to prevent muscle loss at high altitudes, prolong endurance performance, and to provide the needed amino acids not provided by a vegetarian diet.

Potassium and magnesium aspartate are salts of aspartic acid, an amino acid. They have been used as ergogenics, possibly by enhancing fatty acid metabolism and sparing muscle glycogen utilization or by mitigating the accumulation of ammonia during exercise.[221]

8.2.3.2 Caffeine

Caffeine enhances the contractility of skeletal and cardiac muscle, and helps metabolize fat, thereby sparing muscle glycogen stores. It is also a central nervous system stimulant, which can aid in activities that require concentration. Many small studies using randomized, double-blind design have associated caffeine use with increased endurance times in athletes.[226,227]

8.2.3.3 L-Carnitine

Carnitine is a quaternary amine whose physiologically active form is β-hydroxy-γ-trimethylammonium butyrate. This is found in meats and dairy products and is synthesized in the human liver and kidneys from two essential amino acids: lysine and methionine. L-Carnitine is thought to be ergogenic in two ways. First, by increasing free fatty acid transport across mitochondrial membranes, carnitine may increase fatty acid oxidation and utilization for energy, thus sparing muscle glycogen. Second, by buffering pyruvate, and thus reducing muscle lactate accumulation associated with fatigue, carnitine may prolong exercise.[228]

8.2.3.4   Bee Pollen

Bee pollen contains trace amounts of minerals and vitamins and is very high in protein and carbo-hydrates. Bee pollen has been used to enhance energy, memory, and performance, although there is no scientific evidence that it does. Serious allergic reactions to bee pollen have been reported, including potentially life-threatening anaphylaxis.[229]

8.2.3.5   Whey Protein and Colostrum

Whey and colostrum are two forms of protein that are theorized to be ergogenic. Whey proteins are extracted from the liquid whey that is produced during the manufacture of cheese or casein, while colostrum is the first milk secreted by cows. Both are rich sources of protein, vitamins, and miner-als, but may contain various biologically active components, including growth factors. Research regarding the ergogenic effect of whey protein and colostrum supplementation is very limited and in general research findings are equivocal.[221]

8.2.4   OTHER DIETARY SUPPLEMENT CATEGORIES

8.2.4.1   Hormones

Hormones are not typically considered to be part of the diet, yet hormonal products represent an important segment of the dietary supplement market. Hormone supplements are promoted for main-taining or restoring physiological levels of hormones that decline with aging. Some preparations contain meat animal by-products such as ovaries, uteri, and testes. Plant-derived ingredients include phytoestrogens and other compounds that mimic mammalian hormones.

Dehydroepiandrosterone (DHEA), a naturally occurring precursor of endogenous testosterone synthesis, assumed national attention following its use in professional baseball. Unlike the paren-teral administration of testosterone, however, DHEA does not produce a notable elevation in serum testosterone levels or enhance skeletal muscle hypertrophy.[230] Several adverse effects were reported and the International Olympic Committee considers DHEA an illegal steroid, banning its use. A recent analysis of sixteen commercial DHEA products revealed that only half the products con-tained the actual amount of DHEA stated on the product label, with actual levels varying between 0% and 150% of the label content.[231]

Melatonin is an important hormone secreted by the pineal gland in the brain. Since its identifica-tion in 1958, studies have shown that melatonin actually regulates many of the other hormones in the body. These hormones control our circadian rhythm, the 24 h patterns that our bodies respond to everyday. The release of melatonin is stimulated by darkness and suppressed by light, so it helps control when we sleep and when we wake. Melatonin also controls the timing and release of female reproductive hormones, affecting menstrual cycles, menarche, and menopause. Overall levels of melatonin in the body also respond to the process of aging. Melatonin is available as tablets, cap-sules, and sublingual tablets to aid sleep, counter jet lag, and slow aging, but there are concerns about potential side effects.[232]

8.2.4.2   Cartilage

Cartilage is an elastic, translucent connective tissue in humans and animals. Most cartilage turns into bone as animals mature, but some remains in its original form in places such as the nose, ears, knees, and other joints. Two kinds of cartilage are used as nutritional supplements. Bovine cartilage (also called cow cartilage) comes from cattle, and shark cartilage is extracted from the heads and fins of sharks. Cartilage supplements started to be used to treat cancerous tumors when researchers observed that a protein in shark cartilage stopped the growth of new blood vessels (angiogenesis).[233,234] However, studies to prove that the process works in humans have not yet been published. Bovine cartilage has been shown to speed up wound healing and reduce inflammation.[235]

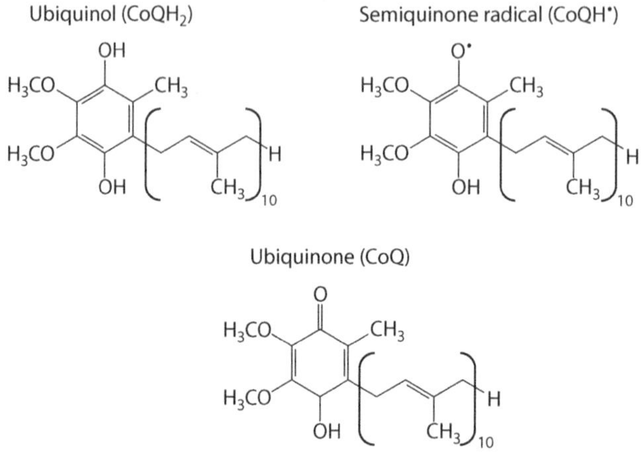

Ubiquinol (CoQH$_2$)   Semiquinone radical (CoQH•)

Ubiquinone (CoQ)

FIGURE 8.20 Coenzyme Q can exist in three oxidation states: the fully reduced ubiquinol form (CoQH$_2$), the radical semiquinone intermediate (CoQH•), and the fully oxidized ubiquinone form (CoQ).

8.2.4.3 Coenzyme Q$_{10}$

Coenzyme Q$_{10}$ is a member of the ubiquinone family of compounds. All animals, including humans, can synthesize ubiquinones; hence, coenzyme Q$_{10}$ cannot be considered a vitamin.[236] The name ubiquinone refers to the ubiquitous presence of these compounds in living organisms and their chemical structure, which contains a functional group known as a benzoquinone. Ubiquinones are fat-soluble molecules with anywhere from 1 to 12 isoprene (5-carbon) units. The ubiquinone found in humans, ubidecaquinone or coenzyme Q$_{10}$, has a "tail" of 10 isoprene units (a total of 50 carbon atoms) attached to its benzoquinone "head" (Figure 8.20).[237]

Rich sources of dietary coenzyme Q$_{10}$ include mainly meat, poultry, and fish. Other relatively rich sources include soybean and canola oils, and nuts. Fruits, vegetables, eggs, and dairy products are moderate sources of coenzyme Q$_{10}$. Coenzyme Q$_{10}$ is also available without a prescription as a dietary supplement.[142]

8.2.4.4 Choline

Although choline is not by strict definition a vitamin, it is an essential nutrient. Despite the fact that humans can synthesize it in small amounts, choline must be consumed in the diet to maintain health.[238] The majority of the body's choline is found in specialized fat molecules known as phospholipids, the most common of which is called phosphatidylcholine or lecithin.[239]

Very little information is available on the choline content of foods.[240] Most choline in foods is found in the form of phosphatidylcholine. Milk, eggs, liver, and peanuts are especially rich in choline. Phosphatidylcholine also known as lecithin contains about 13% choline by weight.

Choline salts, such as choline chloride and choline bitartrate are available as supplements. Phosphatidylcholine supplements also provide choline; however, they are only 13% choline by weight. Although the chemical term "lecithin" is synonymous with phosphatidylcholine, commercial lecithin preparations may contain anywhere from 20% to 90% phosphatidylcholine. Thus, lecithin supplements may contain even less than 13% choline.[150]

8.3 HERBAL SUPPLEMENTS

Herbs, also known as botanicals, are one of humanity's oldest health care tools, and the basis of many modern medicines. Primitive and ancient civilizations relied on herbs for healing, as do contemporary cultures throughout the world. In fact, the World Health Organization has estimated that 80% of the world's population continues to use traditional therapies, a major part of which are derived from plants.

Herbal supplements may be derived from conventional primary food sources (e.g., soy extracts containing isoflavones, tomato extracts rich in lycopene) or from secondary sources such as herbs and spices (e.g., garlic oil, rosemary extracts, and green tea extracts). Other botanical products may have no significant history of use as supplementary ingredients but may be derived from sources which have been used in herbal medicinal products in various regions of the world (e.g., *Ginkgo biloba*, Ginseng extract, St. John's wort). Finally, materials with no history of human use may be considered for use in botanical supplements, e.g., phytostanols derived as a by-product from wood and used in cholesterol-lowering products.[241]

Herbal products are often perceived by consumers as safe because they are "natural."[242] However, herbs may contain pharmacologically active chemical constituents, and, in essence, are drugs.[243] In this scenario, the safety of a botanical supplement depends on many factors such as its chemical makeup, how it works in the body, how it is prepared, and the dose used. It is important to recognize that there is little regulation and standardization of these products.[244,245] This leads to a situation in which there is a wide variability in the quantity, or even the complete absence, of the known or supposed active ingredient. Of 24 ginseng products assayed by a thin-layer chromatography spectrophotometric method, 8 (33%) did not contain any detectable panaxosides, which are considered to be the active components.[31] The total panaxosides' content in the remaining products ranged from 0.26–6.85 mg/250 mg sample.[246] In a study of 44 feverfew products, 14 (32%) did not contain the minimum of 0.2% parthenolide content that is proposed, albeit disputed, as the necessary primary active ingredient and concentration. Another 10 products (22%) did not contain any detectable levels of parthenolide.[247]

Moreover, a complete characterization of all the chemical constituents from a natural product is often unknown. Unlike pharmaceuticals, botanical products are complex mixtures in which the active ingredients may be unknown or only partially characterized. For several herbal products, the active ingredient or the quantity necessary for effectiveness has not been determined.[248,249] For example, some manufacturers standardize St. John's wort according to its hypericin content, which is only 1–10 identified active components in this herb.[250] Additionally, the chemical makeup of a natural product may vary depending on the part of the plant processed (stems, leaves, roots, etc.), seasonality, and growing conditions.[251]

Lack of government regulation (i.e., no requirement for companies to follow good manufacturing practices) and standardization coupled with poor quality control of the manufacturing process lead to a plausible potentiality for misidentification or cross contamination as well as to significant variations, in terms of active ingredient content and efficacy, of a particular preparation from one lot or brand to another.[252]

Although most botanical products are probably safe under most conditions, some are known to be toxic at high doses and others may have potentially adverse effects under some conditions. Many dangerous and lethal side effects have been reported from the use of herbal products.[253] These side effects may occur through several different mechanisms, including direct toxic effects of the herb, effects of contaminants, and interactions with drugs or other herbs.

Severe consequences following the consumption of certain herbal products have been reported. As an example a recent case study described a group of women in Brussels who developed rapid deterioration in their kidney function after taking herbal weight-loss products containing the herb *Aristolochia fangchi*.[254] A recent review of adverse events associated with the herb ephedra (*Ephedra sinica*) found that 31% of analyzed cases, including deaths and permanent disability, were definitely or probably related to the use of ephedra.[255]

Other concerns regarding the use of herbs include the possibility of herb–herb or herb–drug interactions and allergic reactions.[256] Any pharmacologically active agent has the potential to result in synergistic or antagonistic interaction when consumed with other pharmacologically active compounds. This is no less the case for medicinal herbs. For example, many herbal products such as garlic, ginger, ginseng, ginkgo, and feverfew possess antiplatelet properties that can be additive when used with drugs known to affect hemostasis as warfarin or heparin, albeit through different mechanisms.[257–260] Another example is Kava, used for its sedative and antianxiety properties, which can interact with barbiturates and benzodiazepines, prolonging sleep and even inducing coma.[261] However, the majority of suspected herb–drug interactions are identified through case reporting,[258] and it is therefore difficult to definitively determine if the herb–drug combination produced the observed side effect, or if patient characteristics or other factors were the causal factor. Few studies have examined herb–drug interactions, and therefore the risk of combining herbal products with prescription drugs is largely unknown.[262]

Adverse side effects have also been reported due to the contamination of herbal products with metals, unlabeled prescription drugs, microorganisms, or other substances. As an example a study examining the contents of more than 200 Asian patent medicines found that 25% contained high levels of heavy metals (such as lead, mercury, and arsenic) and another 7% contained undeclared pharmaceuticals.[263] Analogously, the popular herbal remedy PC-SPES, sold for "prostate health," has been found to contain various amounts of synthetic drugs, including the widely used anticoagulant drug warfarin.[264]

A variety of herbs and herbal extracts are available on the market: they contain different phytochemicals with biological activity that can provide therapeutic effects. Most herbal products are used for health maintenance or for benign, self-limited conditions such as lower blood cholesterol concentrations, prevent some types of cancer, and stimulate the immune system. Besides herbal products for medicinal uses there are many culinary herbs (i.e., basil, caraway, cilantro, coriander, cumin, dill, oregano, rosemary, sage, and thyme) that, if used generously to flavor food, provide a variety of active phytochemicals that promote health and protect against chronic diseases.[265] A list of most commonly purchased herbal supplements in the United States and Europe,[244] along with the used plant parts, common use, and putative active ingredients is given in Table 8.4.

8.3.1 ECHINACEA

Echinacea purpurea L. and other *Echinacea* spp. are herbs with a long history of use. In the early 1900s, *Echinacea* was the major plant-based antimicrobial medicine in use. With the development of sulfa drugs, the use of *Echinacea* declined rapidly. *Echinacea* appears to be useful in moderating the symptoms of common cold, flu, and sore throat. Its beneficial effect in the treatment of infections results from its ability to act as an anti-inflammatory agent and as an immunostimulant. It can promote the activity of lymphocytes, increase phagocytosis, and induce interferon production.[266] The most common types used in supplements are the roots and flowering leaves of *E. purpurea*, *E. pallida*, and *E. angustifolia*. The active ingredients have not been confirmed. Polysaccharides, caffeic acid derivatives, polyenes, and polyines are thought to be among the possible active ingredients.[267] Echinacea products include tablets, capsules, and liquids such as tinctures, extracts, and the expressed juice of the fresh flowering plant.

8.3.2 GINSENG

The most common ginseng found in supplements and added in food is *Panax ginseng* C.A. Meyer (Asian ginseng). Other types include the American ginseng (*Panax quinquefolius* L.) and the Siberian ginseng (*Eleuthrococcus senticosus*). Ginseng has been the subject of many health claims, including increased energy, improved physical performance (ergogenic effects), and memory enhancement. Ginseng has also been studied in relation to diabetes and cancer.[268] The active ingredients are thought to be saponins, also called ginsenosides (Figure 8.21). The amount of ginsenosides can vary with the age of the plant, method of preservation, and season of harvest.[269] Since different

TABLE 8.4

Most Commonly Used Herbal Supplements in United States and Europe

Botanical Name	Common Name	Parts Used	Proposed Use	Putative Active Ingredients
Echinacea purpurea L., *E. anguistifolia* DC, *E. pallida* Nutt.	Echinacea	Shoots, roots	Respiratory infections, immunostimulant	Chlorogenic acid, caffeic acid derivatives (echinacosides)
Panax ginseng C.A. Meyer	Ginseng	Roots	Fatigue and stress, physical performance	Ginsenosides, panaxans, sesquiterpenes
Serenoa repen (Bartran) small	Saw palmetto	Fruits	Benign prostatic hyperplasia, inflammations	Steroids (β-sitosterols), flavonoids
Ginkgo biloba L.	Ginkgo	Leaves	Dementia, cognitive impairment, mental fatigue	Terpene trilactones (ginkgolides), flavonol glycosides
Hypericum perforatum L.	St. John's wort	Shoots	Depression, antiviral agent	Hyperforin, hypericin, flavonol glycosides
Allium sativum L.	Garlic	Bulb, oil	Hypercholesterolemia, diabetes, hypertension, respiratory infections	Alliins, allicin, ajoens, oligosulfides
Hydrastis canadensis L.	Goldenseal	Rhizome, roots	Gastrointestinal disorders, respiratory infections	Hydrastine, berberine, canadine
Sylybum marianum L.	Milk thistle	Seeds	Liver disorders, lactation problems	Sylimarins, flavonoids (apigenin, chrysoeriol, quercetin, taxifolin)
Tanacetum parthenium L. Schultz-Bip	Feverfew	Green parts	Migraines, inflammation	Sesquiterpene lactones
Cimifuga racemosa Nutt.	Black cohosh	Roots	Premenstrual symptoms, dysmenorrhea, menopausal symptoms	Triterpene glycosides (cimifugaside, actein), isoflavones
Piper methysticum G. Forst	Kava-kava	Shoot	Sedative, euphoriant	Lactones (kavalactones)
Valeriana officinalis L.	Valerian	Roots	Sleep aid, anxiety	Valeric acids, valepotriates
Glycyrrhiza glabra L.	Licorice	Roots	Antiviral effects	Triterpenoid saponin (glycyrrhizin)
Glycine max L.	Soybean	Seeds, sprouts	Menopausal symptoms, heart health	Isoflavones

Source: Adapted from different sources.

forms of ginseng may contain varying amounts of ginsenosides, studying the effects of ginseng in humans can be difficult. As with most other herbs, ginseng is poorly regulated for quality, purity, and quantity of active substance. A survey of 50 ginseng products sold in stores found a large variation in ginsenoside concentrations and a complete absence of ginseng in some products.[270] Another study of 25 ginseng products also found significant variation in concentrations of active ingredients

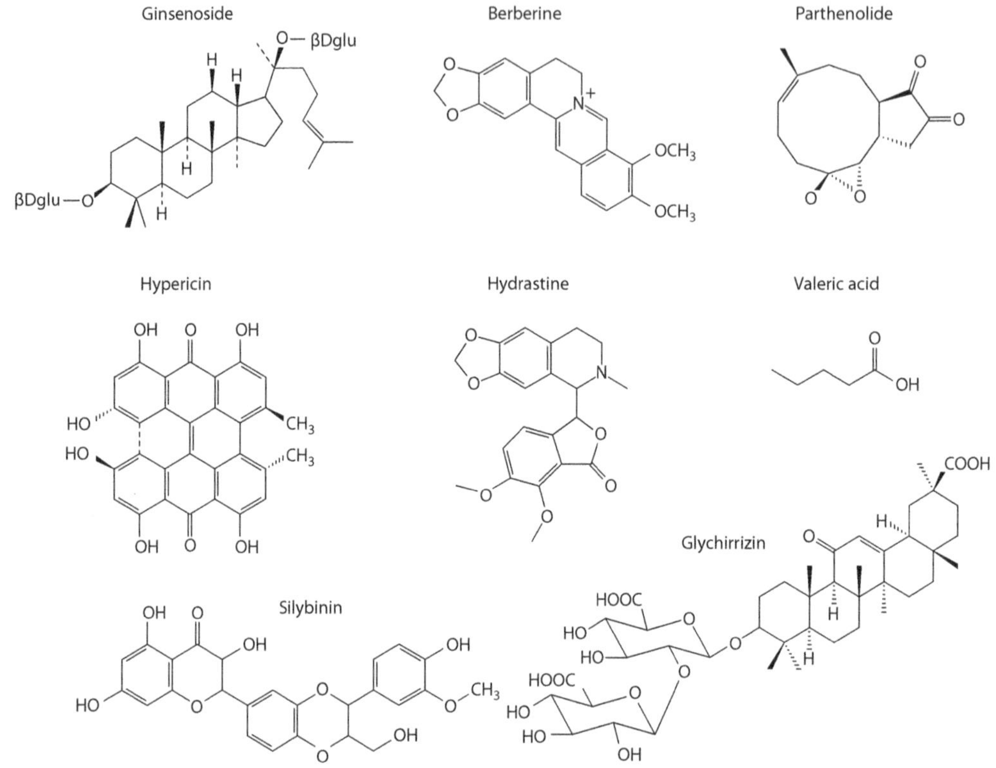

FIGURE 8.21 Chemical structures of bioactive compounds contained in some herbs.

among products.[271] Asian ginseng is available as whole root, powder, and in various forms, including "white and "red" ginseng. White ginseng is simply the dried root, while red ginseng is the root after a steaming-drying process that makes it rust-colored red. The product forms include tinctures, capsules, tablets, teas, and extracts.

8.3.3 SAW PALMETTO

Serenoa repens (Bartram) Small (saw palmetto) is an herb that is most commonly used to treat problems related to benign prostatic hyperplasia (BPH). The medicinal element of saw palmetto is taken from the partially dried ripe fruit of the American dwarf palm tree, which is indigenous to the coastal regions of the southern United States, from the Carolinas and Florida to California. BPH is a nearly universal result of the aging process in men. Saw palmetto is widely used in other countries; for example, it is used in 50% of treatments for BPH in Italy and in 90% of such treatments in Germany.[272] The active part of the plant is the sterols and free fatty acids found in the berry. It is unclear which components are the most active, and the mechanism of action is not fully understood. Some of the mechanisms proposed include anti-inflammatory activity,[273] blocked conversion of testosterone to dihydrotestosterone (DHT),[274,275] and prostate epithelial involution similar to effects noted with the use of finasteride (Proscar).[276] Supplements formulations (capsules, tablets, and tinctures) are made from the dried fruit, in whole or ground form.

8.3.4 GINKGO

Ginkgo products come from the leaves of *G. biloba* L. tree, the only surviving member of the Ginkgoaceae family. Most commercial leaf production is from plantations in South Carolina, France, and China. Ginkgo leaf extracts are thought to improve blood flow, especially to the brain and heart, by blocking the effects of platelet-activation factors (PAFs), and it may protect against oxidative cell damage from free radicals.[4] Claims of memory enhancement may make gingko particularly attractive for older adults.[277] The active constituents in Ginkgo are thought to be flavone glycosides and diterpenoids (ginkgolides). The flavonoids are thought to act as antioxidants, and the ginkgolides may act against PAF to reduce clotting time.[266,278] However, the potent antiplatelet effect of the ginkgolides could be responsible for bleeding complications that have been reported.[279] Ginkgo is one of the few herbs that use standardized products to achieve predictable results. Preparations are generally capsules and tablets containing highly concentrated leaf extract with known amounts of flavonol glycosides and ginkgolides.

8.3.5 St. John's Wort

St. John's wort (*Hypericum perforatum* L.) is a wild-growing herb with golden yellow flowers that has been used for centuries to treat mental disorders. Today, the herb is popular for the treatment of depression.[267] It is also used to treat anxiety, seasonal affective disorder, and sleep disorders. Although its active ingredients have not been confirmed, some researchers believe that they include hypericin (Figure 8.21) and hyperforin. Besides their antidepressive effects, hypericin and hyperforin have been reported to have potent antiretroviral activity against HIV.[267,280,281] Externally, St. John's wort oil is used for the treatment of wound, nerve pain, and first-degree burns. St. John's wort products are made from the dried herb and flowering tops, and include tea, capsules, tablets, tinctures, and extracts. Various interactions with prescription drugs such as antidepressant, coumarin-type anticoagulants, immunosuppressants, and certain HIV drugs have been reported.[282]

8.3.6 Garlic

Garlic (*Allium sativum* L.) has been used effectively as food and medicine for many centuries. Well-documented health benefits include reducing cholesterol and triglycerides in the blood (while increasing the high-density lipoprotein [HDL], so-called good cholesterol), reducing blood pressure, improving circulation, and helping to prevent bacterial and fungal infections, colds, and flu.[283–285] Garlic has also shown to have cancer-preventive potential.[286] Garlic's active ingredient is thought to be a sulfur-containing compound called alliin that when mechanically disrupted (e.g., chopped during food preparation) forms allicin (Figure 8.10), which gives garlic its characteristic odor. Garlic is available in many product forms, including fresh and dried garlic, capsules, odorless garlic tablets, and aged garlic extracts. Onions (*Allium cepa* L.) may also be considered natural anticlotting agents because they possess substances that have fibrinolytic activity and can suppress platelet aggregation.[284,287,288] A whole family of α-sulfinyl disulfides isolated from onions has been shown to strongly inhibit the arachidonic acid cascade in platelets.[288]

8.3.7 Goldenseal

Goldenseal (*Hydrastis canadensis* L.) is a valued medicinal plant that has become increasingly popular within the last decade.[289,290] Its rhizomes and roots contain the alkaloids hydrastine, berberine (Figure 8.21), canadine, and canadaline. Berberine produces antibiotic effects against bacteria (*Escherichia coli* and *H. pylori*), yeast (*Candida*), and protozoa,[291–293] while hydrastine acts as a uterine hemostatic and antiseptic. Canadine acts as a sedative and muscle relaxant.[294] Today, goldenseal is marketed as a tonic to aid digestion, soothe upset stomachs, and as an antibacterial and antiviral agent. Goldenseal supplements are available as tablets and capsules (containing the powdered root), liquid extracts, and glycerites (low-alcohol extracts). Goldenseal is most often combined with echinacea in preparations designed to strengthen the immune system.

8.3.8 Milk Thistle

Milk thistle (*Sylibum marianum* L.) extracts were used as early as the fourth century BC, became a favored medicine for hepatobiliary diseases in the sixteenth century, and have been used in Europe during this century.[295] The principal active components are flavonoligans, including silybin, silybinin (Figure 8.21), sylidianin, and silychrystin, known collectively as silymarin. Silybin is the component of the highest grade of biological activity. Silymarin can be found in all of the plant, but is concentrated in the fruits and seeds. Milk thistle is currently purported to have value as a liver protectant to lessen damage from potentially hepatotoxic drugs and for treating liver disorders including toxic liver damage caused by chemicals, *Amanita phalloides* mushroom poisoning, jaundice, chronic inflammatory liver disease, hepatic cirrhosis, and chronic hepatitis.[296]

8.3.9 Feverfew

Feverfew (*Tanacetum parthenium* (L.) Schultz-Bip) comes from the Latin word *febrifuge*, which means fever-reducing. Feverfew has been used for centuries in European folk medicine for reducing fever, for treating headaches, arthritis, and digestive problems. Modern work has supported the use of feverfew as a treatment for migraine headaches. By inhibiting the release of serotonin and prostaglandins, both of which are believed to aid the onset of migraines, feverfew limits the inflammation of blood vessels in the head.[297,298] The active ingredients in feverfew are sesquiterpene lactones, including parthenolide (Figure 8.21) and tanetin, both of which offer relief from migraine. Feverfew supplements are available fresh, freeze-dried, or dried and can be purchased in capsule, tablet, or liquid extract forms. Feverfew supplements should be standardized to contain at least 0.2% parthenolide.

8.3.10 Black Cohosh

More than two centuries ago, Native Americans discovered that the root of the black cohosh plant (*Cimicifuga racemosa* (L.) Nutt.) helped relieve menstrual cramps and symptoms of menopause such as hot flashes, irritability, mood swings, and sleep disturbances. Today, people use black cohosh for these same reasons. In fact, the herb has been widely used for more than 40 years in Europe and is approved in Germany for premenstrual discomfort, painful menstruation, and menopausal symptoms. A number of pharmacological studies also suggest a mild sedative and anti-inflammatory activity. Compounds in the root, a group of related triterpene glycosides, have been associated with its biological activity. The underground stems (rhizomes) and roots of black cohosh are commonly used fresh or dried to make strong teas (infusions), capsules, solid extracts used in pills, or liquid extracts (tinctures).

8.3.11 Licorice

The major component of licorice root (*Glycyrrhiza glabra* L.) is a sweet-tasting triterpenoid saponin (glycyrrhizin) (Figure 8.21) that along with its aglycone (glycyrrhetinic acid) have been reported to induce interferon activity and augment natural killer cell activity.[299] Hatano and colleagues reported that the chalcones in licorice possess antiviral activity against HIV.[300] Glycyrrhizin also has anti-inflammatory and antiallergic properties.[301]

8.3.12 Kava-Kava

Kava-kava, the extract of *Piper methysticum* G. Forst., has been used as a euphoriant and sedative for centuries in several South Pacific islands. Several lactones are the purported active ingredients, although their precise pharmacologic role remains undefined. Adverse reactions have been associated with heavy use of kava, but the effects of intermittent use are unknown.[302]

8.3.13 Valerian

The root of valerian (*Valeriana officinalis* L.) also has a long history of use as a sedative-hypnotic agent. Currently, it is widely used as a mild nerve sedative and sleep aid for insomnia, excitability, and exhaustion.[303] While various valepotriates, valeric acid (Figure 8.21), and other components have been suggested as possible active compounds for *Valeriana* species, no single compound or group of compounds has been proven responsible for its sedative effect.[304] Dried valerian root is available in whole, cut-and-sifted, and powdered form for teas, capsules, tablets, tinctures, extracts, and other preparations. Valerian may also be marketed in combination products, along with hops (*Humulus lupulus* L.), lemon balm (*Melissa officinalis* L.), St. John's wort, hawthorn (*Crataegus laevigata* Hawthorn), and other herbs.

8.3.14 Soybean

Soybean (*Glycine max* L.) has been a part of the human diet for almost 5000 years. Unlike most plant foods, soybean is high in protein and is considered equivalent to animal foods in terms of protein quality. Today, researchers are interested in both the nutritional value and the potential health benefits of soy. Several studies of populations in which soy foods are a major component of the diet (particularly Asian populations) suggest that soy may be beneficial for reducing menopausal symptoms, and the risk of heart disease and osteoporosis.[305] A possible relationship between dietary soy and the prevention of hormone-related cancers (such as breast, prostate, and endometrial cancer) has also been postulated.[306] Studies investigating the functional value of soy have focused on isoflavone content of soy foods. Isoflavones are phytoestrogens or plant compounds that weakly mimic the action of estrogen. Soy phytoestrogens are very complex; they bind to estrogen receptors and either enhance or diminish the effects of the hormone (see Section 8.1.1.1.2.2). Soy contains other nutrients in addition to protein and isoflavones, and some researchers believe that the healthful benefits of soy foods may be due to the combined action of these nutrients. Extracted phytoestrogens are marketed in numerous forms as dietary supplements, and generally they contain isoflavones derived from soybean. However, other common sources of phytoestrogens for supplement formulations are red clover (*Trifolium pratense* L.), alfalfa (*Medicago sativa* L.), and black cohosh.

REFERENCES

1. Block, G., Patterson, B., and Subar, A., Fruit, vegetables, and cancer prevention: A review of the epidemiological evidence, *Nutr. Cancer*, 18, 1, 1992.
2. Steinmetz, K.A. and Potter, J.D., Vegetables, fruit and cancer II. Mechanisms, *Cancer Causes Control*, 2, 427, 1991.
3. International Agency for Research on Cancer, *IARC Handbooks of Cancer Prevention: Carotenoids*, International Agency for Research on Cancer, Lyon, France, 1998.
4. Pfander, H., Carotenoids: An overview, *Meth. Enzymol.*, 213, 3, 1992.
5. Britton, G., Structure and properties of carotenoids in relation to function, *FASEB J.*, 9, 155, 1995.
6. Britton, G., Liaaen-Jensen, S., and Pfander, H., Carotenoids today and challenges for the future, in *Carotenoids: Isolation and Analysis*, vol. 1A, Britton, G., Liaaen-Jensen, S., and Pfander, H., Eds., Birkhäuser, Basel, Switzerland, 1995.
7. Perry, A., Rasmussen, H., and Johnson E.J., Xanthophyll (lutein, zeaxanthin) content in fruits, vegetables and corn and egg products, *J. Food Comp. Anal.*, 22, 9, 2009.
8. Di Mascio, P., Kaiser, S., and Sies, H., Lycopene as the most efficient biological carotenoid singlet oxygen quencher, *Arch. Biochem. Biophys.*, 274, 532, 1989.
9. Snodderly, D.M., Evidence for protection against age-related macular degeneration by carotenoids and antioxidant vitamins, *Am. J. Clin. Nutr.*, 62, 1448S, 1995.
10. Agarwal, S. and Rao, A.V., Carotenoids and chronic diseases, *Drug Metab. Drug Interact.*, 17, 189, 2000.
11. Bendich, A., Carotenoids and the immune response, *J. Nutr.*, 119, 112, 1989.

12. Mathews-Roth, M.M., Plasma concentration of carotenoids after large doses of beta-carotene, *Am. J. Clin. Nutr.*, 52, 500, 1990.
13. Nishino, H., Cancer prevention by carotenoids, *Mutat. Res.*, 402, 159, 1998.
14. Meister, R. et al., Efficacy and tolerability of myrtol standardized in long-term treatment of chronic bronchitis. A double-blind, placebo-controlled study. Study Group Investigators, *Arzneimittelforschung*, 49, 351, 1999.
15. Maltzman, T.H. et al., Effects of monoterpenes on in vivo DMBA-DNA adduct formation and on phase I hepatic metabolizing enzymes, *Carcinogenesis*, 12, 2081, 1991.
16. Elegbede, J. et al., Effects of anticarcinogenic monoterpenes on phase II hepatic metabolizing enzymes, *Carcinogenesis*, 14, 1221, 1993.
17. Rao, A.V., Anticarcinogenic properties of plant saponins, *Presented at Second International Symposium on the Role of Soy in Preventing and Treating Chronic Disease*, Brussels, Belgium, September 15–18, 1996.
18. Hendrick, S. et al., Are saponins and/or other soybean components responsible for hypocholesterolemic effects of soybean foods? *Presented at Third International Symposium on the Role of Soy in Preventing and Treating Chronic Disease*, Washington, DC, October 31–November 3, 1999.
19. Hayes, K.C. et al., Differences in the plasma transport and tissue concentrations of tocopherols and tocotrienols—Observations in humans and hamsters, *Proc. Soc. Exp. Biol. Med.*, 202, 353, 1993.
20. Manach, C. et al., Polyphenols: Food sources and bioavailability, *Am. J. Clin. Nutr.*, 79, 727, 2004.
21. Shahidi, F. and Naczk, M., *Food Phenolics, Sources, Chemistry, Effects, Applications*, Technomic Publishing Co. Inc., Lancaster, PA, 1995.
22. Tomas-Barberan, F.A. and Clifford, M.N., Dietary hydroxybenzoic acid derivatives and their possible role in health protection, *J. Sci. Food Agric.*, 80, 1024, 2000.
23. Clifford, M.N. and Scalbert, A., Ellagitannins—Occurrence in food, bioavailability and cancer prevention, *J. Food Sci. Agric.*, 80, 1118, 2000.
24. Macheix, J.-J., Fleuriet, A., and Billot, J., *Fruit Phenolics*, CRC Press, Boca Raton, FL, 1990.
25. Ferguson, L.R., Zhu, S., and Philip, H.J., Antioxidant and antigenotoxic effects of plant cell wall hydroxycinnamic acids in cultured HT-29 cells, *Mol. Nutr. Food Res.*, 49, 585, 2005.
26. Kikugawa, K. et al., Reaction of *p*-hydroxycinnamic acid derivatives with nitrite and its relevance to nitrosamine formation, *J. Agric. Food Chem.*, 31, 780, 1983.
27. Jiang, R.W. et al., Chemistry and biological activities of caffeic acid derivatives from *Salvia miltiorrhiza*, *Curr. Med. Chem.*, 12, 237, 2005.
28. Clifford, M.N., Chlorogenic acids and other cinnamates—Nature, occurrence and dietary burden, *J. Sci. Food. Agric.*, 79, 362, 1999.
29. Hemmerle, H. et al., Chlorogenic acid and synthetic chlorogenic acid derivatives: Novel inhibitors of hepatic glucose-6-phosphate translocase, *J. Med. Chem.*, 40, 137, 1997.
30. Kroon, P.A. et al., Release of covalently bound ferulic acid from fiber in the human colon, *J. Agric. Food Chem.*, 45, 661, 1997.
31. Lempereur, I., Rouau, X., and Abecassis, J., Genetic and agronomic variation in arabinoxylan and ferulic acid contents of durum wheat (*Triticum durum* L.) grain and its milling fractions, *J. Cereal Sci.*, 25, 103, 1997.
32. Hatcher, D.W. and Kruger, J.E., Simple phenolic acids in flours prepared from Canadian wheat: Relationship to ash content, color, and polyphenol oxidase activity, *Cereal Chem.*, 74, 337, 1997.
33. Dinelli, G. et al., Biosynthesis of polyphenol phytoestrogens in plants, in *Phytoestrogens in Functional Foods*, Yildiz, F., Ed., CRC Press, Boca Raton, FL, 2006, 19.
34. Dinelli, G. et al., Content of flavonols in Italian bean (*Phaseolus vulgaris* L.) ecotypes, *Food Chem.*, 99, 105, 2006.
35. Dinelli, G. et al., Compositional changes induced by UV-B radiation treatment of common bean and soybean seedlings monitored by capillary electrophoresis with diode array detection, *J. Sep. Sci.*, 30, 604, 2007.
36. Price, S.F. et al., Cluster sun exposure and quercetin in Pinot noir grapes and wine, *Am. J. Enol. Vitic.*, 46, 187, 1995.
37. Herrmann, K., Flavonols and flavones in food plants: A review, *J. Food Technol.*, 11, 433, 1976.
38. Hertog, M.G.L., Hollman, P.C.H., and Katan, M.B., Content of potentially anticarcinogenic flavonoids of 28 vegetables and 9 fruits commonly consumed in the Netherlands, *J. Agric. Food Chem.*, 40, 2379, 1992.
39. King, H.G.C., Phenolic compounds of commercial wheat germ, *J. Food Sci.*, 27, 446, 1962.

40. Feng, Y., McDonald, C.E., and Vick, B.A., C-glycosylflavones from hard red spring wheat bran, *Cereal Chem.*, 65, 452, 1988.

41. Sartelet, H. et al., Flavonoids extracted from Fonio millet (*Digitaria exilis*) reveal potent antithyroid properties, *Nutrition*, 12, 100, 1996.

42. Tomas-Barberan, F.A. and Clifford, M.N., Flavanones, chalcones and dihydrochalcones—Nature, occurrence and dietary burden, *J. Sci. Food Agric.*, 80, 1073, 2000.

43. Mazur, W. and Adlercreutz, H., Naturally occurring estrogens in food, *Pure Appl. Chem.*, 70, 1759, 1998.

44. Shutt, D.A. and Cox, R.I., Steroid and phyto-oestrogen binding to sheep uterine receptors *in vitro*, *J. Endocrinol.*, 52, 299, 1972.

45. Miksicek, R.J., Interaction of naturally occurring nonsteroidal estrogens with expressed recombinant human estrogen receptor, *J. Steroid Biochem. Mol. Biol.*, 49, 153, 1994.

46. Song, T.T., Hendrich. S., and Murphy, P.A., Estrogenic activity of glycitein, a soy isoflavone, *J. Agric. Food Chem.*, 47, 1607, 1999.

47. Kuhnle, G.G.C. et al., Variability of phytoestrogen content in foods from different sources, *Food Chem.*, 113, 1184, 2009.

48. Shemesh, M., Lindner, H.R., and Ayalon, N., Affinity of rabbit uterine oestradiol receptor for phytoestrogens and it's a competitive protein-binding radioassay for plasma coumestrol, *J. Reprod. Fert.*, 29, 1, 1972.

49. Knuckles, B.E., DeFremery, D., and Kohler, G.O., Coumestrol content of fractions obtained during wet processing of alfalfa, *J. Agric. Food Chem.*, 24, 1177, 1976.

50. Wang, G.S. et al., A simplified HPLC method for the determination of phytoestrogens in soybean and its processed products, *J. Agric. Food Chem.*, 38, 185, 1990.

51. Franke, A.A. et al., Quantification of phytoestrogens in legumes by HPLC, *J. Agric. Food Chem.* 42, 1905, 1994.

52. Frankel, E.N., Waterhouse, A.L., and Teissedre, P.L., Principal phenolic phytochemicals in selected California wines and their antioxidant activity in inhibiting oxidation of human low-density lipoproteins, *J. Agric. Food Chem.*, 43, 890, 1995.

53. Arts, I.C., Hollman, P.C., and Kromhout, D., Chocolate as a source of tea flavonoids, *Lancet*, 354, 488, 1999.

54. Hara, Y. et al., Special issue on tea, *Food Rev. Int.*, 11, 371, 1995.

55. Lee, M.-J. et al., Analysis of plasma and urinary tea polyphenols in human subjects, *Cancer Epidemiol. Biomark. Prev.*, 4, 393, 1995.

56. Ding, Z., Kuhr, S., and Engelhardt, U.H., Influence of catechins and theaflavins on the astringent taste of black tea brews, *Z. Lebensm. Unters. Forsch.*, 195, 108, 1992.

57. Santos-Buelga, C. and Scalbert, A., Proanthocyanidins and tannin-like compounds: Nature, occurrence, dietary intake and effects on nutrition and health, *J. Sci. Food Agric.*, 80, 1094, 2000.

58. DePascualTeresa, S., SantosBuelga, C., and RivasGonzalo, J.C., Quantitative analysis of flavan-3-ols in Spanish foodstuffs and beverages, *J. Agric. Food Chem.*, 48, 5331, 2000.

59. Mazza, G. and Miniati, E., *Anthocyanins in Fruits, Vegetables, and Grains*, CRC Press, Boca Raton, FL, 1993, 362.

60. Clifford, M.N., Anthocyanins in foods, *Presented at Symposium on Polyphenols and Anthocyanins as Food Colourants and Antioxidants*, Pfannhauser, W. and Strigl, A., Eds., Wien, Austria, 1996, 1.

61. Clifford, M.N., Anthocyanins—Nature, occurrence and dietary burden, *J. Food Sci. Agric.*, 80, 1063, 2000.

62. Es-Safi, N.E., Cheynier, V., and Moutounet, M., Interactions between cyanidin 3-*O*-glucoside and furfural derivatives and their impact on food color changes, *J. Agric. Food Chem.*, 50, 5586, 2002.

63. Mirmiran, P. et al., Fruit and vegetable consumption and risk factors for cardiovascular disease, *Metabolism*, 58, 460, 2009.

64. Joshipura, K.J. et al., Fruit and vegetable intake in relation to risk of ischemic stroke, *JAMA*, 282, 1233, 1999.

65. Riboli, E. and Norat, T., Epidemiologic evidence of the protective effect of fruit and vegetables on cancer risk, *Am. J. Clin. Nutr.*, 78, 559S, 2003.

66. Bosetti, C. et al., Flavonoids and breast cancer risk in Italy, *Cancer Epidemiol. Biomarkers Prev.*, 14, 805, 2005.

67. Arts, I.C. and Hollman, P.C., Polyphenols and disease risk in epidemiologic studies, *Am. J. Clin. Nutr.*, 81, 317S, 2005.

68. Knekt, P. et al., Flavonoid intake and risk of chronic diseases, *Am. J. Clin. Nutr.*, 76, 560, 2002.

69. Keli, S.O. et al., Dietary flavonoids, antioxidant vitamins, and incidence of stroke: The Zutphen study, *Arch. Intern. Med.,* 156, 637, 1996.

70. Hirvonen, T. et al., Flavonol and flavone intake and the risk of cancer in male smokers (Finland), *Cancer Causes Control,* 12, 789, 2001.

71. Arts, I.C. et al., Dietary catechins and cancer incidence among post-menopausal women: The Iowa Women's Health Study (United States), *Cancer Causes Control,* 13, 373, 2002.

72. Aviram, M. and Fuhrman, B., Wine flavonoids protect against LDL oxidation and atherosclerosis, *Ann. N. Y. Acad. Sci.,* 957, 146, 2002.

73. Rietveld, A. and Wiseman, S., Antioxidant effects of tea: Evidence from human clinical trials, *J. Nutr.,* 133, 3285S, 2003.

74. Serafini, M.J.A. et al., Inhibition of human LDL lipid peroxidation by phenol-rich beverages and their impact on plasma total antioxidant capacity in humans, *J. Nutr. Biochem.,* 11, 585, 2000.

75. Silvina-Lotito, B. and Frei, B., Consumption of flavonoid-rich foods and increased plasma antioxidant capacity in humans: Cause, consequence, or epiphenomenon? *Free Radic. Biol. Med.,* 41, 1727, 2006.

76. Kamal-Eldin, A. et al., An oligomer from flaxseed composed of secoisolariciresinoldiglucoside and 3-hydroxy-3-methyl glutaric acid residues, *Phytochemistry,* 58, 587, 2001.

77. Bambagiotti-Alberti, M. et al., Revealing the mammalian lignan precursor secoisolariciresinoldigluco-side in flax seed by ionspray mass spectrometry, *Rapid Commun. Mass Spectrom.,* 8, 595, 1994.

78. Coran, S.A., Giannellini, V., and Bambagiotti-Alberti, M., A novel monitoring approach for mammalian lignan precursors in flaxseed, *Pharm. Sci.,* 2, 529, 1996.

79. Milder, I.E.J. et al., Lignan contents of Dutch plant foods: A database including lariciresinol, pinoresinol, secoisolariciresinol and matairesinol, *Br. J. Nutr.,* 93, 393, 2005.

80. Adlercreutz, H. and Mazur, W., Phyto-oestrogens and Western diseases, *Ann. Med.,* 29, 95, 1997.

81. Mazur, W. et al., Isotope dilution gas chromatographic-mass spectrometric method for the determination of isoflavonoids, coumestrol, and lignans in food samples, *Anal. Biochem.,* 233, 169, 1996.

82. Borriello, S.P. et al., Production and metabolism of lignans by the human faecal flora, *J. Appl. Bacteriol.,* 58, 37, 1985.

83. Heinonen, S. et al., *In vitro* metabolism of plant lignans: New precursors of mammalian lignans entero-lactone and enterodiol, *J. Agric. Food Chem.,* 49, 3178, 2001.

84. Dinelli, G. et al., Lignan profile in seeds of modern and old Italian soft wheat (*Triticum aestivum* L.) cultivars as revealed by capillary electrophoresis-mass spectrometry analyses, *Electrophoresis,* 28, 4212, 2007.

85. Velentzis, L.S. et al., Do phytoestrogens reduce the risk of breast cancer and breast cancer recurrence? What clinicians need to know, *Eur. J. Cancer,* 44, 1799, 2008.

86. Raffaelli, B. et al., Enterolignans, *J. Chromatogr. B: Biomed. Sci. Appl.,* 777, 29, 2002.

87. Arts, I.C.W. and Hollman, P.C.H., Polyphenols and disease risk in epidemiological studies, *Am. J. Clin. Nutr.,* 81, 5317, 2005.

88. Vitrac, X. et al., Direct liquid chromatography analysis of resveratrol derivatives and flavanonols in wines with absorbance and fluorescence detection, *Anal. Chim. Acta,* 458, 103, 2002.

89. Criqui, M.H. and Ringel, B.L., Does diet or alcohol explain the French paradox? *Lancet,* 344, 1719, 1994.

90. Kundu, J.K. and Surh, Y.-J., Cancer chemopreventive and therapeutic potential of resveratrol: Mechanistic perspectives, *Cancer Lett.,* 269, 243, 2008.

91. Howitz, K.T. et al., Small molecule activators of sirtuins extend *Saccharomyces cerevisiae* lifespan, *Nature,* 425, 191, 2003.

92. Fernandes, P. and Cabral, J.M.S., Phytosterols: Applications and recovery methods, *Biores. Technol.,* 98, 2335, 2007.

93. Kritchevsky, D. and Chen, S.C., Phytosterols—Health benefits and potential concerns: A review, *Nutr. Res.,* 25, 413, 2005.

94. Moreau, R.A., Whitaker, B.D., and Hicks, K.B., Phytosterols, phytostanols, and their conjugates in foods: Structural diversity, quantitative analysis, and health-promoting uses, *Prog. Lipid Res.,* 41, 457, 2002.

95. Schwartz, H. et al., Tocopherol, tocotrienol and plant sterol contents of vegetable oils and industrial fats, *J. Food Comp. Anal.,* 21, 152, 2008.

96. Laakso, P., Analysis of sterols from various food matrices, *Eur. J. Lipid Sci. Technol.,* 107, 402, 2005.

97. Weihrauch, L. and Gardner, J.M., Sterol content of foods of plant origin, *J. Am. Diet. Assoc.,* 73, 39, 1978.

98. Normén, L. et al., Plant sterols in vegetables and fruits commonly consumed in Sweden, *Eur. J. Nutr.,* 38, 84, 1999.

99. Piironen, V., Toivo, J., and Lampi, A.M., Natural sources of dietary plant sterols, *J. Food Comp. Anal.*, 13, 619, 2000.
100. Phillips, K.M., Ruggio, D.M., and Ashraf-Khorassani, M., Phytosterol composition of nuts and seeds commonly consumed in the United States, *J. Agric. Food Chem.*, 53, 9436, 2005.
101. Gomez, M.A. et al., Study of the topical anti-inflammatory activity of *Achillea ageratum* on chronic and acute inflammation models, *Z. Naturforsch. [C]*, 54, 937, 1999.
102. Bouic, P.J.D., Sterols and sterolins: New drugs for the immune system? *Drug Discov. Today*, 7, 775, 2002.
103. Okoli, J. and Akah, P.A., Mechanisms of the anti–inflammatory activity of the leaf extracts of *Culcasia scandens* P. Beauv (Araceae), *Pharmacol. Biochem. Behavior*, 79, 473, 2004.
104. Nashed, B. et al., Antiatherogenic effects of dietary plant sterols are associated with inhibition of pro-inflammatory cytokine production in Apo E-KO mice, *J. Nutr.*, 135, 2438, 2005.
105. Ifere, G.O. et al., Differential effects of cholesterol and phytosterols on cell proliferation, apoptosis and expression of a prostate specific gene in prostate cancer cell lines, *Cancer Detect. Prev.*, 32, 319, 2009.
106. Berger, A., Jones, P.J.H., and Abumweis, S.S., Plant sterols: Factors affecting their efficacy and safety as functional food ingredients, *Lipids Health Dis.*, 3, 5, 2004.
107. Steinmetz, K.A. and Potter, J.D., Vegetables, fruit and cancer prevention: A review, *J. Am. Diet. Assoc.*, 96, 1027, 1996.
108. Dorant, E. et al., Garlic and its significance for the prevention of cancer in humans: A critical review, *Br. J. Cancer*, 67, 424, 1993.
109. Lash, L.J., Garlic dietary supplements: An assessment of product information provided by garlic manufacturers, *Minnesota Pharm.*, 53, 13, 1999.
110. Reuter, H.D. et al., Therapeutic effects and applications of garlic and its preparations, in *Garlic: The Science and Therapeutic Application of* Allium sativum *L and Related Species*, Koch, J.P. and Lawson, L.D., Eds., Williams & Wilkins, Baltimore, MD, 1996, 135.
111. Brady, J.F. et al., Inhibition of cytochrome P-450 2E1 by diallyl sulfide and its metabolites, *Chem. Res. Toxicol.*, 4, 642, 1991.
112. Silagy, C. and Neil, A., Garlic as a lipid lowering agent—A meta-analysis, *J. R. Coll. Phys. Lond.*, 28, 39, 1994.
113. Verhoeven, D.T. et al., A review of mechanisms underlying anticarcinogenicity by brassica vegetables, *Chem. Biol. Interact.*, 103, 79, 1997.
114. Kim, Y.S. and Milner, J.A., Targets for indole-3-carbinol in cancer prevention, *J. Nutr. Biochem.*, 16, 65, 2005.
115. Kris-Etherton, P.M. et al., Polyunsaturated fatty acids in the food chain in the United States, *Am. J. Clin. Nutr.*, 71, 179S, 2000.
116. Ackman, R.C., The year of fish oil, *Chem. Ind. Lond.*, 3, 139, 1988.
117. Simopoulos, A.P., Omega-3 fatty acids in health and disease and in growth and development, *Am. J. Clin. Nutr.*, 54, 438, 1991.
118. Bang, H.O. and Dyerberg, J., Plasma lipids and lipoproteins in Greenlandic west-coast Eskimos, *Acta Med. Scand.*, 192, 85, 1972.
119. Seki, H. et al., Omega-3 PUFA derived anti-inflammatory lipid mediator resolvin E1, *Prostaglandins Other Lipid Mediat.*, 2009, doi:10.1016/j.prostaglandins.2009.03.002
120. Cleland, L.G., James, M.J., and Proudman, S.M., Omega–6/omega–3 fatty acids and arthritis, in *Omega-6/Omega-3 Essential Fatty Acid Ratio: The Scientific Evidence*, Simopoulos, A.P. and Cleland, L.G., Eds., Krager, Basel, Switzerland, 2003, 152.
121. Okuyama, H., High n-6 to n-3 ratio of dietary fatty acids rather than serum cholesterol as a major risk factor for coronary heart disease, *Eur. J. Lipid Sci. Technol.*, 103, 418–422, 2001.
122. Lewis, N.M., Seburg, S., and Flanagan, N.L., Enriched eggs as a source of n-3 polyunsaturated fatty acids for humans, *Poult. Sci.*, 79, 971, 2000.
123. Prates, J.A.M. and Mateus, C.M.R.P., Functional foods from animal sources and their physiologically active components, *Revue Med. Vet.*, 153, 155, 2002.
124. Ha, Y.L., Grimm, N.K., and Pariza, M.W., Anticarcinogens from fried ground beef: Heat-altered derivatives of linoleic acid, *Carcinogenesis*, 8, 1881, 1987.
125. Parodi, P.W., *Advances in Conjugated Linoleic Acid Research*, vol. 1, AOCS Press, Champaign, IL, 1999.
126. Belury, M.A., Conjugated dienoic linoleate: A polyunsaturated fatty acid with unique chemo-protective properties, *Nutr. Rev.*, 53, 83, 1995.
127. Blankson, H. et al., Conjugated linoleic acid reduces body fat mass in overweight and obese humans, *J. Nutr.*, 130, 2943, 2000.

128. Wtkins, B.A., Yong, L., and Feifert, M.F., Bone metabolism and dietary conjugated linoleic acid, in *Advances in Conjugated Linoleic Acid Research*, vol. 1, Yurawecz, M.P. et al., Eds., AOCS Press, Champaign, IL, 1999, 253.

129. Fuller, R., History and development of probiotics, in *Probiotics*, Fuller, R., Ed., Chapman & Hall, New York, 1994, 1.

130. Sanders, M.E., Lactic acid bacteria as promoters of human health, in *Functional Foods—Designer Foods, Pharmafoods, Nutraceuticals*, Goldberg, I., Ed., Chapman & Hall, New York, 1994, 294.

131. Hilliam, M., Functional foods: The Western consumer viewpoint, *Nutr. Rev.*, 54, 189S, 1996.

132. Sheehy, P.J.A. and Morrissey, P.A., Nutritional aspects of food processing and ingredients, in *Functional Foods: Prospects and Perspectives*, Henry, C.J.K. and Heppel, N.J., Eds., Aspen Publishers, Gaithersburg, MD, 1998, Chap. 3.

133. Mital, B.K. and Garg, S.K., Anticarcinogenic, hypocholesterolemic, and antagonistic activities of *Lactobacillus acidophilus*, *Crit. Rev. Micro.*, 21, 175, 1995.

134. Talamini, R. et al., Social factors, diet and breast cancer in a northern Italian population, *Br. J. Cancer*, 49, 723, 1984.

135. van't Veer, P. et al., Consumption of fermented milk products and breast cancer: A case-control study in the Netherlands, *Cancer Res.*, 49, 4020, 1989.

136. Gibson, G.R. and Roberfroid, M.B., Dietary modulation of the human colonic microbiota: Introducing the concept of prebiotics, *J. Nutr.*, 125, 1401, 1995.

137. Gibson, G.R. et al., Fermentation of non-digestible oligosaccharides by human colonic bacteria, *Proc. Nutr. Soc.*, 55, 899, 1996.

138. Tomomatsu, H., Health effects of oligosaccharides, *Food Technol.*, 48, 61, 1994.

139. Coussement, P., Non-digestible oligosaccharides for functional foods, in *Proceedings of the First Vitafoods International Conference*, Blenford, D.E., Ed., FoodTech Europe, Copenhagen, Denmark, 1997, 15.

140. *Dietary Reference Intakes for Calcium, Phosphorous, Magnesium, Vitamin D, and Fluoride*, 2000, The National Academy Press, Washington, DC, http://www.nap.edu/openbook/0309063507/html/21.html (accessed March 12, 2009).

141. Hathcock, J., Dietary supplements: How they are used and regulated, *J. Nutr.*, 131, 1114S, 2001.

142. Higdon, J., *An Evidence-Based Approach to Vitamins and Minerals: Health Benefits and Intake Recommendations*, Thieme Medical Publishers, New York, 2003.

143. *The United States Pharmacopoeia Guide to Vitamins and Minerals*, Avon Books, New York, 1996, 41.

144. Machelle, M. and Seibel, M.D., The role of nutrition and nutritional supplements in women's health, *Fertil. Steril.*, 72, 579, 1999.

145. Neuhouser, M.L. et al., Absorption of dietary and supplemental folate in women with prior pregnancies with neural tube defects and controls, *J. Am. Coll. Nutr.*, 17, 625, 1998.

146. Food and Nutrition Board, Institute of Medicine, *Dietary Reference Intakes for Thiamin, Riboflavin, Niacin, Vitamin B6, Folate, Vitamin B12, Pantothenic Acid, Biotin, and Choline*, Standing Committee on the Scientific Evaluation of Dietary Reference Intakes, National Academy Press, Washington, DC, 1998, 425 pp.

147. Weber, P., Bendich, A., and Schalch, W., Vitamin C and human health—A review of recent data relevant to human requirements, *Int. J. Vitam. Nutr. Res.*, 66, 19, 1996.

148. Tanphaichitr, V., Thiamin, in *Nutrition in Health and Disease*, 9th edn, Shils, M, Ed., Williams & Wilkins, Baltimore, MD, 1999, 381.

149. Rindi, G., Thiamin, in *Present Knowledge in Nutrition*, 7th edn, Ziegler, E.E. and Filer, L.J., Eds., ILSI Press, Washington, DC, 1996, 160.

150. Hendler, S.S and Rorvik, D.R., Eds., *PDR for Nutritional Supplements*, Medical Economics Company Inc., Montvale, NJ, 2001.

151. Lukaski, H.C., Vitamin and mineral status: Effects on physical performance, *Nutrition*, 20, 632, 2004.

152. Bohles, H., Antioxidative vitamins in prematurely and maturely born infants, *Int. J. Vitam. Nutr. Res.*, 67, 321, 1997.

153. Brody, T., *Nutritional Biochemistry*, 2nd edn, Academic Press, San Diego, CA, 1999.

154. Knopp, RH., Evaluating niacin in its various forms, *Am. J. Cardiol.*, 86, 51L, 2000.

155. Plesofsky-Vig, N., Pantothenic acid, in *Nutrition in Health and Disease*, 9th edn, Shils, M., Ed., Williams & Wilkins, Baltimore, MD, 1999, 423.

156. Tahiliani, A.G. and Beinlich, C.J., Pantothenic acid in health and disease, *Vitam. Horm.*, 46, 165, 1991.

157. Leklem, J.E., Vitamin B-6, in *Handbook of Vitamins*, Machlin, L, Ed., Marcel Decker Inc., New York, 1991, 341.

158. McMahon, R.J., Biotin in metabolism and molecular biology, *Annu. Rev. Nutr.,* 22, 221, 2002.
159. Said, H.M., Biotin: The forgotten vitamin, *Am. J. Clin. Nutr.,* 75, 179, 2002.
160. Said, H.M. et al., Biotin uptake by human colonic epithelial NCM460 cells: A carrier-mediated process shared with pantothenic acid, *Am. J. Physiol.,* 275, 1365, 1998.
161. Herbert, V., Folic acid, in *Modern Nutrition in Health and Diseases,* 9th edn, Shils, M.E., Olson, J.A., and Ross, A.C., Eds., Williams & Wilkins, Philadelphia, PA, 1999, 433.
162. Herbert, V., Current concepts in therapy: Megloblastic anemia, *N. Engl. J. Med.,* 268, 201, 1963.
163. Shaw, G.M. et al., Periconceptional vitamin use, dietary folate, and the occurrence of neural tube defects, *Epidemiology,* 6, 219, 1995.
164. Food and Nutrition Board, Institute of Medicine, Folic Acid, in *Dietary Reference Intakes: Thiamin, Riboflavin, Niacin, Vitamin B-6, Vitamin B-12, Pantothenic Acid, Biotin, and Choline,* National Academy Press, Washington, DC, 1998, 193.
165. Food and Nutrition Board, Institute of Medicine, Vitamin B_{12}, in *Dietary Reference Intakes: Thiamin, Riboflavin, Niacin, Vitamin B-6, Vitamin B-12, Pantothenic Acid, Biotin, and Choline,* National Academy Press, Washington, DC, 1998, 306.
166. Padayatty, S. et al., Vitamin C as an antioxidant: Evaluation of its role in disease prevention, *J. Am. Coll. Nutr.,* 22, 18, 2003.
167. Johnston, C.S. and Luo, B., Comparison of the absorption and excretion of three commercially available sources of vitamin C, *J. Am. Diet Assoc.,* 94, 779, 1994.
168. Groff, J.L., *Advanced Nutrition and Human Metabolism,* 2nd edn, West Publishing, St. Paul, MN, 1995.
169. Traber, M.G., Utilization of vitamin E, *Biofactors,* 10, 115, 1999.
170. Sen, C.K., Khanna, S., and Roy, S., Tocotrienols: Vitamin E beyond tocopherols, *Life Sci.,* 78, 18, 2088, 2006.
171. Traber, M.G. and Packer, L., Vitamin E: Beyond antioxidant function, *Am. J. Clin. Nutr.,* 62, 1501S, 1995.
172. Traber, M.G. and Sies, H., Vitamin E in humans: Demand and delivery, *Ann. Rev. Nutr.,* 16, 321, 1996.
173. Shearer, M.J., Vitamin K, *Lancet,* 345, 229, 1995.
174. Sumi, H. et al., A novel fibrinolytic enzyme (nattokinase) in the vegetable cheese Natto; a typical and popular soybean food in the Japanese diet, *Experientia,* 43, 1110, 1987.
175. Lippard, S. J. and Berg, J. M., *Principles of Bioinorganic Chemistry,* University Science Books, Mill Valley, CA, 1994.
176. Wardlaw, G.M., *Perspectives in Nutrition,* 4th edn, WCB McGraw-Hill, Boston, MA, 1999.
177. Weaver, C.M. and Heaney, R.P., Calcium, in *Nutrition in Health and Disease,* 9th edn, Shils, M. et al., Eds., Williams & Wilkins, Baltimore, MD, 1999, 141.
178. Andon, M.B. et al., Calcium absorption from apple and orange juice fortified with calcium citrate malate (CCM), *J. Am. Coll. Nutr.,* 15, 313, 1996.
179. Levenson, D. and Bockman, R., A review of calcium preparations, *Nutr. Rev.,* 52, 221, 1994.
180. Shils, M.E., Magnesium, in *Handbook of Nutritionally Essential Minerals,* O'Dell, B.L. and Sunde, R.A., Eds., Marcel Dekker Inc., New York, 1997, 117.
181. Food and Nutrition Board, Institute of Medicine, Magnesium, in *Dietary Reference Intakes: Calcium, Phosphorus, Magnesium, Vitamin D, and Fluoride,* National Academy Press, Washington, DC, 1997, 190.
182. Klasco, R.K., Ed., *USP DI® Drug Information for the Healthcare Professional,* Thomson MICROMEDEX, Greenwood Village, CO, 2003.
183. Fine, K.D. et al., Intestinal absorption of magnesium from food and supplements, *J. Clin. Invest.,* 88, 296, 1991.
184. Firoz, M. and Graber, M., Bioavailability of US commercial magnesium preparation, *Magnes. Res.,* 14, 257, 2001.
185. Taubes, G., The (political) science of salt, *Science,* 281, 898, 1998.
186. Ascherio, A. et al., Intake of potassium, magnesium, calcium, and fiber and risk of stroke among U.S. men, *Circulation,* 98, 1198, 1998.
187. European Nutrition and Health Report, Energy and Nutrient Intake in the European Union, *Ann. Nutr. Metab.,* 48, 1, 2004.
188. Knochel, J.P., Phosphorus, in *Nutrition in Health and Disease,* 9th edn, Shils, M. et al., Eds., Williams & Wilkins, Baltimore, MD, 1999, 157.
189. Food and Nutrition Board, Institute of Medicine, Phosphorus, in *Dietary Reference Intakes: Calcium, Phosphorus, Magnesium, Vitamin D, and Fluoride,* National Academy Press, Washington, DC, 1997, 146.

190. National Research Council, Food and Nutrition Board, *Recommended Dietary Allowances*, 10th edn, National Academy Press, Washington, DC, 1989, 184.

191. Cook, J.D., Defining optimal body iron, *Proc. Nutr. Soc.*, 58, 489, 1999.

192. DeMayer, E.M. et. al., *A Guide for Health Administrators and Programme Managers*, WHO, Geneva, Switzerland, 1989.

193. Yip, R. and Dallman, P.R., Iron, in *Present Knowledge in Nutrition*, 7th edn, Ziegler, E.E. and Filer. L.J., Eds., ILSI Press, Washington, DC, 1996, 277.

194. Lynch, S.R., Interaction of iron with other nutrients, *Nutr. Rev.*, 55, 102, 1997.

195. Hoffman, R. et al., Disorders of iron metabolism: Iron deficiency and overload, in *Hematology: Basic Principles and Practice*, 3rd edn, Churchill Livingstone, Harcourt Brace & Co, New York, 2000, Chap. 26.

196. Fox, G.N. and Sabovic, Z., Chromium picolinate supplementation for diabetes mellitus, *J. Fam. Pract.*, 46, 83, 1998.

197. Reading, S.A., Chromium picolinate, *J. Fla. Med. Assoc.*, 83, 29, 1996.

198. Linder, M.C. and Hazegh-Azam, M., Copper biochemistry and molecular biology, *Am. J. Clin. Nutr.*, 63, 797S, 1996.

199. Turnlund, J.R., Copper, in *Nutrition in Health and Disease*, 9th edn, Shils, M. et al., Eds., Williams & Wilkins, Baltimore, MD, 1999, 241.

200. Cerklewski, F.L., Fluoride bioavailability—Nutritional and clinical aspects, *Nutr. Res.*, 17, 907, 1997.

201. Nielsen, F.H., Ultratrace minerals, in *Nutrition in Health and Disease*, 9th edn, Shils, M. et al., Eds., Williams & Wilkins, Baltimore, MD, 1999, 283.

202. Cerklewski, F.L., Fluoride—Essential or just beneficial, *Nutrition*, 14, 475, 1998.

203. Fein, N.J. and Cerklewski, F.L., Fluoride content of foods made with mechanically separated chicken, *J. Agric. Food Chem.*, 49, 4284, 2001.

204. Food and Nutrition Board, Institute of Medicine, Fluoride, in *Dietary Reference Intakes: Calcium, Phosphorus, Magnesium, Vitamin D, and Fluoride*, National Academy Press, Washington, DC, 1997, 288.

205. Hetzel, B.S. and Clugston, G.A., Iodine, in *Nutrition in Health and Disease*, 9th edn, Shils, M. et al., Eds., Williams & Wilkins, Baltimore, MD, 1999, 253.

206. Dunn, J.T., What's happening to our iodine? *J. Clin. Endocrinol. Metab.*, 83, 3398, 1998.

207. Keen, C.L. et al., Nutritional aspects of manganese from experimental studies, *Neurotoxicology*, 20, 213, 1999.

208. Kies, C., Bioavailability of manganese, in *Manganese in Health and Disease*, Klimis-Tavantzis, D.L., Ed., CRC Press Inc., Boca Raton, FL, 1994, 39.

209. Reilly, C., Selenium: A new entrant into the functional food arena, *Trends Food Sci. Technol.*, 9, 3, 114, 1998.

210. Arthur, R. and Beckett, G.J., Newer aspects of micronutrients in at risk groups: New metabolic roles for selenium, *Proc. Nutr. Soc.*, 53, 615, 1994.

211. Australia New Zealand Food Authority, Full Assessment Report P 92 Sports Foods, ANZFA, Canberra, Australia, 1997.

212. Reilly, C., *Selenium in Food and Health*, Blackie Academic and Professional, London, U.K., 1996.

213. Cantor, A.H., The role of selenium in poultry nutrition, in *Biotechnology in the Feed Industry, Proceedings of Alltech's 13th Annual Symposium*, Lyons, T.P. and Jacques, K.A., Eds., Nottingham University Press, Loughborough, U.K., 1997, 155.

214. Vallee, B.L. and Falchuk, K.H., The biochemical basis of zinc physiology, *Physiol. Rev.*, 73, 1993.

215. Prasad, A.S., Halsted, J.A., and Nadimi, M., Syndrome of iron deficiency anemia, hepatosplenomegaly, hypogonadism, dwarfism, and geophagia, *Am. J. Med.*, 31, 532, 1961.

216. Prasad, A.S., Zinc deficiency in humans: A neglected problem, *J. Am. Coll. Nutr.*, 17, 542, 1998.

217. Wuebbens, M.M. et al., Insights into molybdenum cofactor deficiency provided by the crystal structure of the molybdenum cofactor biosynthesis protein MoaC, *Structure Fold Des.*, 8, 709, 2000.

218. Food and Nutrition Board, Institute of Medicine, Molybdenum, in *Dietary Reference Intakes for Vitamin A, Vitamin K, Boron, Chromium, Copper, Iodine, Iron, Manganese, Molybdenum, Nickel, Silicon, Vanadium, and Zinc*, National Academy Press, Washington, DC, 2001, 420.

219. Mills, C.F., Davis, G.K., Molybdenum, in *Trace Elements in Human and Animal Nutrition*, 5th edn, Mertz, W., Ed., Academic Press, San Diego, CA, 1987.

220. Brouns, F., *Nutritional Needs of Athletes*, John Wiley & Sons Ltd., Chichester, U.K., 1993.

221. Williams, M., Dietary supplements and sports performance: Aminoacids, *Int. J. Sport Nutr.*, 2, 63, 2005.

222. Walker, J.B., Creatine: Biosynthesis, regulation and function, *Adv. Enzymol. Relat. Areas. Mol. Med.*, 50, 177, 1979.

223. Maughan, R.J., Creatine supplementation and exercise performance, *Int. J. Sport Nutr.*, 5, 94, 1995.
224. Pritchard, N.R. and Kaira, P.A., Renal dysfunction accompanying oral creatine supplements, *Lancet*, 351, 1252, 1998.
225. Juhn, M.S., Oral creatine supplementation. Separating fact from hype, *Phys. Sportsmed.*, 27, 47, 1999.
226. Graham, T.E. and Spriet, L.L., Performance and metabolic responses to a high caffeine dose during prolonged exercise, *J. Appl. Physiol.*, 71, 2292, 1991.
227. Kalmar, J.M. and Cafarelli, E., Effects of caffeine on neuromuscular function, *J. Appl. Physiol.*, 87, 801, 1999.
228. Vuchovich, M.D., Costill, D.L., and Fink, W.J., Carnitine supplementation: Effect on muscle carnitine and glycogen content during exercise, *Med. Sci. Sports. Exerc.*, 26, 1122, 1994.
229. Saris, W.H.M. et al., Functional food science and substrate metabolism. *Br. J. Nutr.*, 80, 47S, 1998.
230. King, D.S. et al., Effect of oral androstenedione on serum testosterone and adaptations to resistance training in young men, *J. Am. Med. Assoc.*, 281, 2020, 1999.
231. Parasrampuria, J., Schwartz, K., and Petesch, R., Quality control of dehydroepiandrosterone dietary supplement products, *J. Am. Med. Assoc.*, 280, 1565, 1998.
232. Grafias, S., Melatonin: A trusty travel companion?, *Phys. Sportsmed.*, 24, 19, 1996.
233. Horsman, M.R., Alsner, J., and Overgaard, J., The effect of shark cartilage extracts on the growth and metastatic spread of the SCCVII carcinoma, *Acta Oncol.*, 37, 441, 1998.
234. Miller, D.R. et al., Phase I/II trial of the safety and efficacy of shark cartilage in the treatment of advanced cancer, *J. Clin. Oncol.*, 16, 3649, 1998.
235. Prudden, J.F., The treatment of human cancer with agents prepared from bovine cartilage, *Biol. Response Mod.*, 4, 551, 1985.
236. Ernster, L. and Dallner, G., Biochemical, physiological and medical aspects of ubiquinone function, *Biochem. Biophys. Acta*, 1271, 195, 1995.
237. Crane, F.L., Biochemical functions of coenzyme Q10, *J. Am. Coll. Nutr.*, 20, 591, 2001.
238. Blusztajn, J.K., Choline, a vital amine, *Science*, 281, 794, 1998.
239. Zeisel, S.H., Choline and phosphatidylcholine, in *Nutrition in Health and Disease*, 9th edn, Shils, M. et al. Eds., Williams & Wilkins, Baltimore, MD, 1999, 513.
240. Zeisel, S.H., Choline: An essential nutrient for humans, *Nutrition*, 16, 669, 2000.
241. Walker, R., Criteria for risk assessment of botanical food supplements, *Toxicol. Lett.*, 49, 187, 2004.
242. Gesler, W.M., Therapeutic landscapes: Medical issues in light of the new cultural geography, *Soc. Sci. Med.*, 34, 735, 1992.
243. Winslow, L.C. and Kroll, D.J., Herbs as medicines, *Arch. Intern. Med.*, 158, 2192, 1998.
244. Foster, S. and Tyler, V.E., *Tyler's Honest Herbal: A Sensible Guide to the Use of Herbs and Related Remedies*, The Haworth Herbal Press, Inc., Binghamton, NY, 1999, 442 p.
245. Miller, L.G., Herbal medicinals. Selected clinical considerations focusing on known or potential drug–herb interactions, *Arch. Intern. Med.*, 158, 2200, 1998.
246. Liberti, L.E. and DerMarderosian, A.D., Evaluation of commercial ginseng products, *J. Pharmaceut. Sci.*, 67, 1487, 1978.
247. Heptinstall, S. et al., Extracts of feverfew may inhibit platelet behavior via neutralization of sulfhydryl groups, *J. Pharm. Pharmacol.*, 39, 459, 1987.
248. Awang, D.V.C., Feverfew products, *Can. Med. Assoc. J.*, 157, 510, 1997.
249. Blumenthal, M., Testing botanicals: A report on developing the scientific and clinical evidence to support the clinical use of heterogeneous botanical products, *HerbalGram*, 40, 43, 1997.
250. Newall, C., Anderson, L.A., and Phillipson, J.D., St. John's wort monograph, in *Herbal Medicines*, Newall, C., Anderson, L.A., and Phillipson, J.D., Eds., The Pharmaceutical Press, London, U.K., 1990, 250.
251. Chavez, M.L., Jordan, M.A., and Chavez, P.I., Evidence-based drug–herbal interactions, *Life Sci.*, 78, 2146, 2006.
252. Matthews, H.B., Lucier, G.W., and Fisher, K.D., Medicinal herbs in the United States: Research needs, *Environ. Health Perspect.*, 107, 10, 1999.
253. Ernst, E., Harmless herbs? A review of the recent literature, *Am. J. Med.*, 104, 170, 1998.
254. Nortier, J.L. et al., Urothelial carcinoma associated with the use of a Chinese herb (*Aristolochia fangchi*), *New Engl. J. Med.*, 342, 1686, 2000.
255. Haller, C.A. and Benowitz, N.L., Adverse cardiovascular and central nervous system events associated with dietary supplements containing ephedra alkaloids, *New Engl. J. Med.*, 343, 1833, 2000.
256. The Complete German Commission E Monographs, in *Therapeutic Guide to Herbal Medicines*, Blumenthal, M., Ed., Integrative Medicine Communications, Boston, MA, 1998.

257. Janetzky, K. and Morreale, A.P., Probable interactions between warfarin and ginseng, *Am. J. Health-Syst. Pharm.*, 54, 692, 1997.

258. Izzo, A.A. and Ernst, E., Interactions between herbal medicines and prescribed drugs: A systematic review, *Drugs,* 61, 2163, 2001.

259. Backon, J., Ginger: Inhibition of thromboxane synthetase and stimulation of prostacyclin: Relevance for medicine and psychiatry, *Med. Hypothesis*, 20, 271, 1986.

260. Rowin, J. and Lewis, S.L., Spontaneous bilateral subdural hematomas associated with chronic ginkgo biloba ingestion, *Neurology*, 46, 1775, 1996.

261. Almeida, J.C. and Grimsley, E.W., Coma from the health food store: Interaction between kava and alprazolam, *Ann. Intern. Med.*, 125, 940, 1996.

262. Bent, S.M.D. and Ko, R., Commonly used herbal medicines in the United States: A review, *Am. J. Med.*, 116, 478, 2004.

263. Ko, R., Adulterants in Asian patent medicines, *New Engl. J. Med.*, 339, 847, 1998.

264. Ko, R., Wilson, R.D., and Loscutoff, S., Pc-Spes, *Urology,* 61, 1292, 2003.

265. Craig, W.J., Health-promoting properties of common herbs, *Am. J. Clin. Nutr.*, 70, 491S, 1999.

266. Tyler, V., *Herbs of Choice. The Therapeutic Use of Phytomedicinals*, Haworth Press, New York, 1994.

267. Bruneton, J., *Pharmacognosy, Phytochemistry, Medicinal Plants*, 3rd edn, Intercept, Paris, France, 1999.

268. Zhang, Q.H. et al., Protective effects of total saponins from stem and leaf of *Panax ginseng* against cyclophosphamide-induced genotoxicity and apoptosis in mouse bone marrow cells and peripheral lymphocyte cells, *Food Chem. Toxicol.*, 46, 293, 2008.

269. Lieberman, H.R., The effects of ginseng, ephedrine, and caffeine on cognitive performance, mood and energy, *Nutr. Rev.*, 59, 91, 2001.

270. Cui, J. et al., What do commercial ginseng preparations contain? *Lancet*, 344, 134, 1994.

271. Harkey, M.R. et al., Variability in commercial ginseng products: An analysis of 25 preparations, *Am. J. Clin. Nutr.*, 73, 1101, 2001.

272. Scholtysek, C. et al., Characterizing components of the Saw Palmetto Berry Extract (SPBE) on prostate cancer cell growth and traction, *Biochemical and Biophysical Research Communications*, 379, 795, 2009.

273. Lowe, F.C. and Ku, J.C., Phytotherapy in treatment of benign prostatic hyperplasia: A critical review, *Urology*, 48, 12, 1996.

274. Briley, M., Carilla, E., and Roger, A., Inhibitory effect of Permixon on testosterone 5a-reductase activity of the rat ventral prostate, *Br. J. Pharmacol.*, 83, 401P, 1984.

275. Marks, L.S. et al., Tissue effects of saw palmetto and finasteride: Use of biopsy cores for *in situ* quantification of prostatic androgens, *Urology*, 57, 999, 2001.

276. Marks, L.S. et al., Effects of a saw palmetto herbal blend in men with symptomatic benign prostatic hyperplasia, *J. Urol.*, 163, 1451, 2000.

277. Blumenthal, M., The benefits of ginkgo in healthy people, *HerbalGram*, 67, 5, 2005.

278. Kleijnen, J. and Knipschild, P., *Gingko biloba*, Lancet, 340, 1136, 1992.

279. Rosenblatt, M. and Mindel, J., Spontaneous bilateral hyphema associated with ingestion of ginkgo biloba extract, *N. Engl. J. Med.,* 1108, 1997.

280. Meruelo, D., Lavie, G., and Lavie, D., Therapeutic agents with dramatic antiretroviral activity and little toxicity at effective doses: Aromatic polycyclic diones hypericin and pseudohypericin, *Proc. Natl. Acad. Sci. U. S. A.*, 85, 5230, 1988.

281. Degar, S. et al., Inactivation of the human immunodeficiency virus by hypericin: Evidence for photochemical alterations of p24 and a block in uncoating, *AIDS Res. Hum. Retroviruses,* 8, 1929, 1992.

282. Schulz, V., Safety of St. John's wort extract compared to synthetic antidepressants, *Phytomedicine*, 13, 199, 2006.

283. Warshafsky, S., Kramer, R.S., and Sivak, S.L., Effect of garlic on total serum cholesterol: A meta-analysis, *Ann. Intern. Med.*, 119, 599, 1993.

284. Kleijnen, J., Knipschild, P., and ter Riet, G.T., Garlic, onions and cardiovascular risk factors. A review of the evidence from human experiments with emphasis on commercially available preparations, *Br. J. Clin. Pharmacol.*, 28, 535, 1989.

285. Stevinson, C. et al., Garlic for treating hypercholesterolemia: A meta-analysis of randomized clinical trials, *Ann. Intern. Med.*, 133, 420, 2000.

286. Rassoul, F. et al., The influence of garlic (*Allium sativum*) extract on interleukin 1 alpha-induced expression of endothelial intercellular adhesion molecule-1 and vascular cell adhesion molecule-1, *Phytomedicine*, 13, 230, 2006.

287. Kendler, B.S., Garlic (*Allium sativum*) and onion (*Allium cepa*): A review of their relationship to cardio-vascular disease, *Prev. Med.*, 16, 670, 1987.

288. Kawakishi, S. and Morimitsu, Y., Sulfur chemistry of onions and inhibitory factors of the arachidonic acid cascade, in *Food Phytochemicals for Cancer Prevention I. Fruits and Vegetables*, Huang, M.T. et al., Eds., American Chemical Society, Washington, DC, 1994, 120.

289. Sinclair, A. and Catling, P. M., Cultivating the increasingly popular medicinal plant, Goldenseal: Review and update, *Am. J. Alternative Agric.*, 16, 131, 2001.

290. Sanders, S.M. and McGraw, J.B., Distribution, abundance, and population dynamics of Goldenseal (*Hydrastis canadensis* L.) in an Indiana Nature Preserve, USA, *Nat. Areas J.*, 22, 129, 2002.

291. Hwang, B.Y. et al., Antimicrobial constituents from goldenseal (the rhizomes of *Hydrastis canadensis*) against selected oral pathogens, *Planta Med.*, 69, 623, 2003.

292. Mahady, G.B. et al., *In vitro* susceptibility of *Helicobacter pylori* to isoquinoline alkaloids from *Sanguinaria canadensis* and *Hydrastis canadensis*, *Phytother. Res.*, 17, 217, 2003.

293. Inbaraj, J.J. et al., Photochemistry and photocytotoxicity of alkaloids from Goldenseal (*Hydrastis canadensis* L.). 2. Palmatine, hydrastine, canadine, and hydrastinine, *Chem. Res. Toxicol.*, 19, 739, 2006.

294. Purdue University, 2002, Purdue University Web site, Available at http://www.hort.purdue.edu/newcrop/med-aro/factsheets/GOLDENSEAL.html (accessed April 15, 2008).

295. Schuppan, D. et al., Herbal products for liver diseases: A therapeutic challenge for the new millennium, *Hepatology*, 30, 1099, 1999.

296. Saller, R., Meier, R., and Brignoli, R., The use of silymarin in the treatment of liver disease, *Drugs*, 61, 2035, 2001.

297. Johnson, E.S. et al., Efficacy of feverfew as prophylactic treatment of migraine, *BMJ*, 291, 569, 1985.

298. Pfaffenrath, V. et al., The efficacy and safety of *Tanacetum parthenium* in migraine prophylaxis—A double blind, multicentre, randomized placebo-controlled dose-response study, *Cephalagia*, 22, 523, 2002.

299. Abe, N., Ebina, T., and Ishida, N., Interferon induction by glycyrrhizin and glycyrrhetinic acid in mice, *Microbiol. Immunol.*, 26, 535, 1982.

300. Hatano, T. et al., Anti-human immunodeficiency virus phenolics from licorice, *Chem. Pharm. Bull.*, 36, 2286, 1988.

301. Shibata, S., Antitumor-promoting and anti-inflammatory activities of licorice principles and their modi-fied compounds, in *Food Phytochemicals for Cancer Prevention II. Teas, Spices and Herbs*, Huang, M.T. et al., Eds., American Chemical Society, Washington, DC, 1994, 308.

302. Mathews, J.D., Riley, M.D., and Fejo, L., Effects of the heavy usage of kava on physical health: Summary of a pilot survey in an aboriginal community, *Med. J. Aust.*, 148, 548, 1988.

303. Vorbach, E., Gortelmayer, R., and Runing, R., Treatment of insomnia: Efficacy and tolerance of a valer-ian extract, *Psychopharmakotherapie*, 3, 109, 1996.

304. Houghton, R.J., The biological activity of valerian and related plants, *J. Ethnopharmacol.*, 22, 121, 1988.

305. Kurzer, M.S., Phytoestrogen supplement use by women, *J. Nutr.*, 133, 1983S, 2003.

306. Barnes, S. et al., Soy isoflavonoids and cancer prevention: Underlying biochemical and pharmacological issues, in *Dietary Phytochemicals and Cancer Prevention*, Butrum, R., Ed., Plenum Press, New York, 1996, 87.

9 Flavor Compounds in Foods

Dilek Boyacioglu, Dilara Nilufer, and Esra Capanoglu

CONTENTS

9.1 INTRODUCTION

The flavor of food is the most important sensory attribute affecting the acceptance and preference of consumers. Standards organizations in many countries define flavor as a total impression of taste, odor, tactile, kinesthetic, temperature, and pain sensations perceived through tasting [1]. It is widely accepted that flavor includes the aromatics, such as olfactory perceptions caused by volatile substances; the tastes, such as gustatory perceptions (salty, sweet, sour, and bitter) caused by soluble substances [2]; and the chemical feeling factors that are perceived as astringency, spicy hot, cooling, and metallic flavor, stimulating the nerve ends in the membranes [3]. The volatile compounds have low molecular weights and may be lost during processing. However, some important components that are naturally present in foodstuffs may participate in flavor development during the processing as well.

Food products, whether fresh or processed, must have desirable flavors that are pleasant to the palate of the consumer. Flavorings are often added to foods, to create a totally new taste, to enhance or increase the perception of flavors already present, to replace unavailable flavors, to mask less desirable flavors that are naturally present in some processed foods, or, to supplement flavors already present and that had disappeared as a result of food processing [4].

Besides the sensory quality provided by flavor compounds, they also have antioxidant activity, antimicrobial activity, and health-promoting functions as well [5]. Antimicrobial activities of essential oils of various sources are the subjects of many studies and have been recognized for centuries [6–9]. Antioxidant activities have also been investigated and studied in many plants: wild plants *Pistacia lentiscus* L. (Anacardiaceae) and *Myrtus communis* L. (Myrtaceae) [10]; wild basil (*Ocimum gratissimum* L.) leaf [11], *P. lentiscus*, [12]; some herbs such as rosemary and sage [13], lavender [14], oregano and sage [15], capers [16], fruits such as eucalyptus [17], buntan (shaddock) (*Citrus grandis* Osbeck) [18], and sweet orange peels [19]. It is apparent that this topic is very large, and it certainly needs to be covered in a separate book.

The intent of this chapter is to summarize naturally occurring flavor compounds in aromatic foods, and the formation of flavor during processing of foods including the Maillard, lipid oxidation, microbial, and enzymatic reactions. A brief discussion on the impact of interactions between these components on food flavor and off-flavor formation is also included. At the end of chapter, the manufacturing of flavor compounds by the utilization of biotechnology and pyrolysis techniques is further explained.

9.2 NATURALLY OCCURRING FLAVOR COMPOUNDS

A natural flavor is a preparation obtained exclusively from natural sources (mostly vegetable, although certain meat derived flavors also qualify) that is processed to be fit for human consumption. Sometimes a natural flavor does not give the exact flavor impact desired, and the flavor chemist will "dress it up" by adding natural flavors from other sources. The result is called "with other natural flavors" (W.O.N.Fs) [20]. These flavors give a chance to the producer for obtaining a naturally stronger flavor with no need for artificial enhancers, and for labeling their products as "all natural" in flavor.

9.2.1 HERBS AND SPICES

Spices are aromatic vegetable substances used to provide flavor and aroma. A convenient classification for spices might be the following: (a) the tropical spices such as pepper and cinnamon, (b) herbs such as sage and rosemary, and (c) seed spices such as mustard and sesame [4].

Essential oils, also called volatile oils, are the distilled fraction of spices. They are sterile, soluble in oils, emulsifiable in other solutions, and stable during storage. A disadvantage of essential oils is that they do not contain the hydrophilic flavor components, fixed oils, or antioxidants found in oleoresins. They are primarily used to flavor and standardize oleoresins and seasoning blends [4].

Spice oleoresins are the solvent extracts of a spice containing both essential oils and other nonvolatile components [21]. Although oleoresins are fairly stable during storage, a weakness of oleoresins is that they are very viscous and thick, making them difficult to weigh, handle, and mix during processing [4].

The aroma of spice oil is dependent on its composition. The oxygenated compounds, like alcohols, esters, aldehydes, and ketones, have characteristic strong aromas, and these are very important

TABLE 9.1
Flavor and Flavor Character Impact Compounds of Some Spices and Herbs

Spice	Flavor	Chief Flavor Compounds
Anise	Distinctly licorice, sweet, warm	Anethole
Basil	Sweet aromatic, slightly pungent	1-Linalool, methyl chavicol
Bay	Delicate aromatic, spicy bitter	Cineole, terpineol, geraniol
Caraway	Aromatic, a bit sharp, sweet	Carvone, limonene
Cinnamon	Most fragrant, pungently sweet	Cinnamaldehyde, eugenol
Clove	Penetrating, distinctly	Eugenol, eugenylacetate
Cumin	Sweet, warm, slightly salty, mild caraway flavor	Cuminaldehyde
Garlic	Penetrating, distinctive	Allicin
Marjoram	Delicate aromatic, pleasantly bitter and sharp	Terpiene, terpineol, geraniol
Mint	Strong, tangy but cooling	Menthol and menthone
Mustard	Sharp, hot, pungent, tasting oil	Allyl isothiocyanate (black), p-hydroxy-benzylisothiocyanate (white)
Onion	Taste-tempting, distinctive	Volatile organic sulfur compounds
Saffron	Distinctive, pleasant, somewhat bitter concentrated	Volatile oil and crocin (color)
Turmeric	Slightly peppery, bitter, fresh, and aromatic	Turmerone and curcumin (color)
Vanilla	Characteristic, delicate, perfume, delightfully aromatic	Vanillin

Source: Lewis, Y.S., *Spices and Herbs for the Food Industry*, Food Trade Press, Orpington, U.K., 1984, 12.

factors contributing to the odor quality of the oils. The monoterpenes are weakly aromatic but are responsible for the spicy top notes in oils like pepper and ginger. Sesquiterpenes are still less aromatic but they make up the bulk of many spice oils. The hot taste in spices like pepper, ginger, and capsicum is due to piperine, gingerol, and capsaicin, respectively [22]. Besides the pungency and aroma factors, spices contain many other substances, like fats and oleoresins, that are necessary to give the natural flavor of the spices. Also some spices like paprika, turmeric, and saffron give attractive colors to foods. The flavor impact compounds of some spices and herbs are given in Table 9.1.

9.2.2 FRUITS AND VEGETABLES

The volatile compounds in vegetables and fruits arise from unsaturated fatty acids, amino acids, carbohydrates, and other generally universal components (Figure 9.1) [23]. The main difference in the flavor of fruits and vegetables is that fruits have higher concentrations of sugars. Fruits also generally contain higher concentrations of volatile oils than vegetables. The common components in fruit volatile oils are aliphatic esters, e.g., ethyl acetate, ethyl butyrate, ethyl hexaonate, butyl hexaonate, butyl acetate, and branched esters. Linalool and other terpenes are more common than in vegetables [24].

It is known that volatile compounds found in fruits are mainly derived from three biosynthetic pathways in many plants [25]: The formation of the hedonically important short-chain aldehydes and alcohols, such as *cis*-3-hexenol, takes place through the action of lipases, hydroperoxide lyases, and cleavage enzymes on lipid components, followed by the action of alcohol dehydrogenases [26].

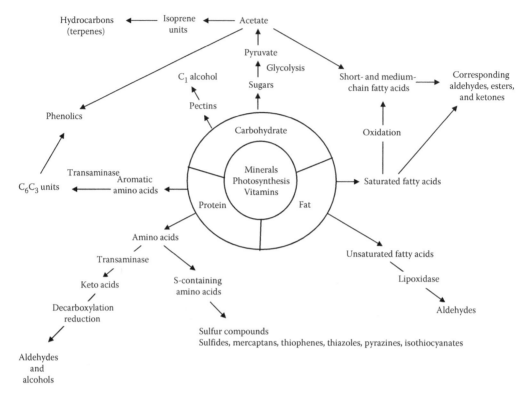

FIGURE 9.1 Formation of volatile aroma in fruits and vegetables. (From Eskin, N.A.M., *Biochemistry of Foods*, Second edition, Academic Press, Inc., San Diego, CA, 134, 1990. With permission.)

Other prominent flavor compounds such as eugenol, phenethyl alcohol, and guaiacol are derived from the shikimic acid pathway [27]. The isoprenoids such as β-ionone are probably produced by the degradation of longer terpenoids such as β-carotene and lycopene, respectively, whereas monoterpenes such as linalool are formed directly from geranyl diphosphate via the isoprenoid pathway [28].

In both fruits and vegetables, there are some aroma compounds present in the intact product; these are apparently produced in the normal metabolism of the plant by the enzymes. Many volatiles, on the other hand, are produced only when the raw vegetable or fruit is broken or blending [29]. When cells are ruptured, flavor precursors are released and exposed to the enzymes [5]. *Allium* species like garlic, onion, and green onion contain volatile sulfur compounds such as thiols, sulfides, disulfides, trisulfides, and thiosulfinates. When these plants are chopped or crushed, the allinase enzyme present in the tissue is activated and acts on alliin to produce allicin. Production of isothiocyanates in Brassicas like turnips, mustard, cabbage, broccoli, cauliflower, and Brussel sprouts occurs by enzymatic reactions on specific glycosides. Any process causing inactivation of this enzyme will decrease aroma production. In mushrooms, 1-octen-3-ol compound is responsible for the significant flavor and it is formed enzymatically (lipoxygenase, hydroperoxide lyase, and allene oxide synthases) from linoleic acid [5]. In some fruits like apples, grapes, and tomatoes, hexanal and hexenal compounds occur in the ruptured tissues. These compounds are derived from the action of 13-lipoxygenase. In some fruits, like grapes, papaya, nectarine, and tea leaves, monoterpenes can be found in two forms: free and glycosidic conjugates. Free forms involve compounds that under mild acid hydrolysis yield flavor volatiles. However, glycosides upon either acidic or enzymatic hydrolysis release aglycones, which produce the flavor [30].

The coupled oxidative reactions of carotenoids with linoleic acid and lipoxidase result in bleaching of the carotenoids. The oxidative breakdown products of carotenoids are somewhat related to the oxidative breakdown of unsaturated fatty acids. The tomato volatiles could result from the oxidative cleavage of lycopene, phytofluene, and phytoene. Terpenoids other than carotenoids, such as terpene and sesquiterpene hydrocarbons, could also undergo oxidative reactions. The typical aroma of tomatoes is attributed to aldehydes and ketones, short-chain alcohols and hydrocarbons, long-chain alcohols, and esters. Tomato volatiles can also be formed through amino acids such as leucine, alanine, and aspartic acids, by the action of specific amino transferases [31]. In addition, the available unsaturated fatty acids also seem to undergo oxidative breakdown during cooking through the decomposition of peroxides that are already present in the raw product [32].

Terpenes, the major compounds in citrus oils, contribute to the flavor of citrus fruits. (+)-Limonene, a monoterpene hydrocarbon possessing a lemon like odor, accounts for approximately 90% of most citrus oils. However, it is the oxygenated terpenes or terpenoids, representing only 5% of the oil, that provide the characteristic flavor of individual citrus species. Of these, neral and geraniol contribute to the distinctive flavor of lemons, and bicyclic sesquiterpene nootkatone contributes to the flavor of grape juice. These compounds undergo structural changes and hydration in the presence of air or dissolved oxygen [33]. Therefore, low-temperature vacuum evaporation processes yield citrus juice concentrates with superior quality.

Apple, pea, and quince aromas are formed by the mixture of a great number of volatile constituents. A range of esters gives an impression of ripeness in these fruits, and the lower molecular weight compounds an impression of natural freshness [34]. However, it is difficult to suggest the type of fruit depending on the presence of major constituents related to the characteristic aroma of fruits.

9.2.3 OTHER FLAVORED FOODS

9.2.3.1 Coffee

Coffee is a drink that is preferred for its stimulating effect and attractive flavor. The chemistry of coffee flavor and aroma has been extensively studied, and it is noted that coffee is highly complex in character involving 300 volatiles in unroasted and 850 in roasted coffee that are identified [35]. Coffee flavor is affected by botanical species, roasting, and brewing methods. Roasting is an important operation; its purpose is to develop flavor and to facilitate the grinding [36]. During roasting the chlorogenic acids are susceptible to the degradative changes and are converted into various alicyclic and phenolic acids. Of the generated acids, quinic acid emerges in the greatest quantity [37].

Caffeine is responsible for the bitterness of coffee brews. Other compounds associated with bitterness are the breakdown products of proteins, the various degradation products of chlorogenic acids, and caramelized sugars. Astringency is associated with dehydroxy groups in quinic acids. Also polysaccharides such as arabinogalactans and mannans may undergo condensation and complexing reactions with themselves or other components, such as proteins, during roasting. They may also be associated with some of the phenolic substances, and some melanoidin or caramelized substances may be bitter as well [37].

The volatiles of roasted coffee include the aliphatic compounds with different carbonyl compounds, and sulfur containing compounds, and alicyclic compounds containing ketones and aromatic benzenoid compounds. Among the character impact compounds, the (2-furyl)-methanethiol and kahwefuran (2-methyl-3-oxa-8-thiobicyclo-[3.3.0] octa-1,4-diene) are present in relatively high amounts. The burnt coffee-like character is mostly associated with 2-ethylfuran, *N*-ethyl formylpyrrole, thiobutyrolactone, and 2-acetyl-3-methylthiopen [35,37].

9.2.3.2 Tea

Tea is one of the most widely consumed beverages in the world. The essential oil is contained in both green leaves and manufactured tea, with tea leaf processing resulting in the formation of new aromatic substances responsible for the specific odor of the manufactured tea [38].

The main components of tea are essential oil (0.5%) which is formed during fermentation, caffeine (1.8%–5.0%), and tannins (13%–18%). Enzymatic formation of black tea aroma follows biosynthetic pathways. The main precursors are amino acids and carotenoids (β-carotene, lutein, neoxanthin, and violaxanthin). Fermentation significantly reduces carotenoids and forms ionone and terpenoid carbonyls as a result of primary oxidations. During firing, secondary epoxidation takes place and forms epoxyionone, dihydroactinidiole, and trimethyl substituted cyclohexanones [35].

Tea aroma is due to the volatile compounds of the essential oils that can be divided into primary (found in the fresh green leaf) and secondary (formed during tea leaf processing) compounds. The essential oil is composed of acids, phenols, bases, aldehydes, and a neutral essential oil fraction. Two pathways of aroma formation can be distinguished, enzymatic and thermal. Tea aroma begins to develop by the action of oxidative enzymes during withering of the tea. The process is enhanced by tissue maceration during rolling. In the macerated tissues of tea leaf, fermentation proceeds at a most rapid rate and the volatile components of essential oils are actively produced. Catechins have been found to play an important role in the process. They are oxidized by o-diphenol oxidase into o-quinones which, in their turn, are strong oxidizing agents for numerous compounds of fermented tea leaves. A proportion of the o-quinones is transformed into theaflavins, thearubigins, and theaflavic acids, which give black tea its characteristic color and, to a great extent, contribute to the taste of brewed tea. The interactions between o-quinones and amino acids, and carotenes and unsaturated fatty acids result in the formation of compounds contributing to the aroma of manufactured tea. The deamination of amino acids by simple polyphenols (pyrocatechin, pyrogallol) or by catechins and o-diphenol oxidase of the leaf results in the formation of corresponding aldehydes, NH_3, and CO_2. Aldehydes, like many other carbonyl compounds, are of great importance to tea aroma, since carbonyl compounds constitute up to 63% of the total amount of tea aroma constituents [38].

In addition to the enzymatic pathway of aroma formation, a thermal route also exists. At high temperatures, interactions of amino acids and sugars result in the formation of various aldehydes. After thermal treatment, the tea becomes more tasty and pleasant, and has a better aroma. An essential source of secondary volatiles, formed during tea leaf processing, is oxidative. o-Quinone resulting from the oxidation of catechins can oxidize, besides amino acids and carotenes, unsaturated fatty acids as well. Linoleic and linolenic acids can be converted into hexenal and $trans$-hex-2-enal, respectively, and in addition, small amounts of other volatile compounds, especially hexanoic acid and $trans$-hex-2-enoic acid, can be formed from the same acids, respectively. Also the monoterpene alcohols, linalool and geraniol, play an important role in the formation of the aroma of black tea [38].

9.2.3.3 Cocoa

Fresh beans of cocoa have the odor and taste of vinegar. The basic compounds present in cocoa mass are hexanal, pyrazines, butanal, butanoate, lactones, and aldehydes. Cocoa aroma depends on the harvesting, fermentation, drying, and roasting processes. Amino acids that are released during fermentation are the precursors of aroma compounds that are formed during roasting of the beans. The bitter taste is the result of theobromine and caffeine, and dioxopiperazines formed from the degradation of proteins during roasting [39]. During the drying and roasting processes, the development of sugar degradation products provide a characteristic flavor to the product [40].

9.2.3.4 Vanilla

Vanilla has a sweet, delicate, and creamy flavor. It enhances the flavor of other ingredients, although fresh vanilla has no taste [41]. Vanillin is the most abundant aromatic component of cured vanilla beans [35]. When picked, the green pods have no vanillin as it is bound to a sugar molecule. Vanilla is released after enzyme action during the curing (sweating, fermenting, and drying) process, resulting in aroma and taste. Other compounds found in vanilla are glucovanillin, vanillic acid, anisic acid, phenols, aldehydes, alcohols, lactones, acids, ether, and esters [41].

Natural vanilla is the second most valuable flavoring in the food industry and is derived from the fruits of the tropical orchid *Vanilla planifolia*. The mature, green vanilla beans have no characteristic aroma. The flavor develops during the postharvest processing of the beans (curing). Many aroma compounds of vanilla (vanillin, vanillic acid, *p*-hydroxybenzaldehyde, *p*-cresol, 2-phenylalcohol, anisaldehyde, guaiacol, phenyl-acetaldehyde, diacetyl, eugenol, and methyl-cinnamate) are formed during curing. Thermal processes, plant enzyme reactions, and microbial activities all take part in flavor generation. The major aroma precursors, glucosides (such as glucovanillin), lead to the formation of many compounds based on thermal processes and enzymatic and microbial reactions [42]. In recent years, the high price of natural vanillin extracted from vanilla pods has stimulated research for alternative means of natural vanillin production, such as utilizing ferulic acid for the bioconversion of vanillin using a recombinant strain of *Escherichia coli* [43,44].

9.2.3.5 Peppermint

Peppermint was discovered in the seventeenth century for its cooling effect and is used as a medicinal infusion. Nowadays, it is widely used in chewing gums, liqueurs, confectionery, oral toiletries, and medicines [35,41]. Peppermint flavor is obtained by the distillation of leaves and flowering tops of the freshly harvested herb [35]. Peppermint involves 0.5%–5% essential oil, which is pale yellow. It contains menthol, menthone, menthyl acetate, menthofurane, isomenthone, limonene, pulegone, and β-pinene. Menthol and menthyl acetate have a refreshing and cooling, pungent odor [41].

9.3 FLAVOR FORMATION DURING FOOD PROCESSING

Flavor can be produced by thermal reactions between naturally occurring compounds in foods, such as the creation of meat flavor by the thermal reactions of certain amino acids and sugars (the so-called Maillard reactions). These types of materials have been used by the industry for more than 100 years in savory applications [4]. Flavors generated during heating or processing by enzymatic reactions or by fermentation are generally considered to be "natural" flavors [45].

Other process ingredients such as hydrolyzed vegetable proteins (HVPs), autolyzed yeast extracts (AYE), and flavor enhancers may produce savory flavors. Hydrolyzed vegetable protein produced by chemical or enzymatic hydrolysis of vegetable proteins is a typical example of a process flavor. Autolysates are produced by allowing edible yeast (i.e., brewer's yeast) to rupture. After rupture, the normally present enzymes digest the cell's proteins, carbohydrates, and nucleic acids, producing flavor components [4]. Included in the category of savory flavors are the flavor enhancers. These include monosodium glutamate (MSG), disodium inosinate, and disodium guanylate. The distinctive taste of MSG has been named by a Japanese scientist, Ikeda, as "umami," which is derived from the Japanese word meaning "delicious" or "savory" [46]. The flavor enhancers are a separate category from flavors and require different labeling. Manufacturing of flavors using HVPs or AYE will be further discussed in this chapter.

The main reactions that lead to the formation of flavor can be listed as Maillard reactions, the Strecker degradation of amino acids, lipid oxidation and microbial and enzymatic reactions, and interactions between lipids, proteins, and carbohydrates.

9.3.1 MAILLARD REACTIONS

A great variety of chemical reactions take place during the processing of foods. The reactions between reducing sugars and α-amino acids, known as the Maillard reactions, are the most important ones [47]. The Maillard reactions and the Strecker degradation of α-amino acids are responsible for the formation of many heterocyclic compounds with distinctive aromas and low odor thresholds. Contrary to popular opinion, the Maillard reaction does not require high temperatures; thus sugars and amino acids even at refrigerated temperatures can show signs of nonenzymatic browning on storage. The reaction rate increases markedly with temperature, and the formation of volatile flavor compounds generally occurs at temperatures associated with cooking. Although the Maillard reaction does proceed in an aqueous solution, it occurs much more readily at low moisture levels [48].

The reaction products of the Maillard reaction, such as 1-amino-1-deoxy-2-ketose (Amadori product) or 2-amino-2-deoxyaldose (Heyns product), do not contribute to flavor directly but they are important precursors of flavor compounds [48]. These thermally unstable compounds undergo dehydration and deamination reactions to give numerous rearrangement and degradation products. The thermal degradation of such intermediates is responsible for the formation of volatile compounds that impart the characteristic burnt odor and flavor to various food products. For example, at temperatures above 100°C, enolization products (such as 1-amino-2,3-enediol and 3-deoxyosone) yield, upon further dehydration, furfural from a pentose and 5-hydroxy methylfurfural and 5-methylfurfural from a hexose [2].

In addition, peptides play significant roles in the flavor formation of meat, cocoa, and hydrolyzed vegetable proteins. The size of the peptide and the position of the peptide bonds play key roles in the mechanism of flavor formation. Thus, specific peptides may be responsible for the generation of characteristic aroma volatiles; the molecular structure of the peptide may also help to control how much of each compound is formed [49].

The Strecker degradation involves the oxidative deamination and decarboxylation of a α-amino acid in the presence of a dicarbonyl compound. The products formed from this reaction are an aldehyde containing one less carbon atom than the original amino acid and an α-aminoketone (Table 9.2). The Strecker degradation of methionine and cystein is a source of sulfur-containing intermediates (e.g., hydrogen sulfide and 2-methylthiopropanal = methional) [48].

The interactions between the products of the Maillard and Strecker degradation reactions may lead to the formation of many other flavor compounds. These include heterocyclic compounds such as pyrazines, oxazoles, thiophenes, and heterocyclic compounds with more than one sulfur atom. For example, the occurrence of oxazoles and oxazolines has been observed in food systems which have undergone moderate heating. Both classes of oxygen- and nitrogen-containing heterocycles possess potent sensory qualities at low concentrations [50]. Similarly, both the character and strength of polysulfide heterocyclics (such as thialdine) make them very important compounds in the overall flavor profile of roasted products where they occur [51].

There are also other interactions between a variety of volatile compounds formed during the reactions of aldehydes and amino acids and lipids available in food. Carbonyl compounds may

TABLE 9.2
Volatile Strecker Aldehydes from the Reaction of Amino Acids and α-Dicarbonyl Compounds

Amino Acid	Aldehyde
Glycine	Formaldehyde
Alanine	Acetaldehyde
Valine	Propanal
Leucine	3-Methylbutanal
Isoleucine	2-Methylbutanal
Serine	2-Hydroxyethanal
Threonine	2-Hydroxypropanal
Methionine	3-Methylthiopropanal

Source: Whitfield, F.B., *Crit. Rev. Food Sci. Nutr.*, 31, 1, 1992. With permission.

react with amino acids to form heterocyclic compounds such as pyridines, thiophenes, trithiolanes, tetrathianes, and pentathianes. In addition, the carbonyl compound itself can be converted to a number of new volatile components through degradation and condensation reactions. Such reactions could lead to the identification of many new potent flavor compounds [48].

Carbonyl compounds derived from lipid oxidation would also react with some sugar degradation products (e.g., furans) to form volatile compounds. However, compounds formed by the interaction of lipid and sugar degradation products could subsequently react with hydrogen sulfide or ammonia or with a combination of both of these amino acid degradation products to yield compounds with distinct aromas. Sugars react with aliphatic aldehydes and ketones to form acetals and ketals, and such carbonyl compounds could be derived from the oxidation of lipids. Interactions between products of the Maillard reaction and triglycerides or phospholipids may result in the formation of long-chain alkyl substituents. Maillard/lipid interaction products have also been identified in cooked products, such as beef, fried chicken, and French-fried potatoes [48].

9.3.2 Lipid Oxidation

Oxidation of unsaturated acyl chains of lipids is a major route to volatile compounds during cooking of fat-containing food of either animal or vegetable origin. The unsaturated fatty acids, readily susceptible to the attack by oxygen, form hydroperoxides which in themselves are odorless and tasteless. The compounds that influence the flavor of the product result from a further breakdown of these hydroperoxides, and, normally, include saturated and unsaturated aldehydes, alcohols, and ketones. The carbonyl compounds resulting from autoxidation impart specific flavors that are normally detrimental to food products (Table 9.3). It should be pointed out, however, that they may also contribute to the desirable characteristic flavor of foods [48].

Even though autoxidation plays a major role in the flavor stability of lipid-containing foods, the formation of breakdown products by nonoxidative mechanisms is also important. The formation of numerous compounds such as free volatile acids, lactones, methyl ketones, aldehydes may be generated without the presence of oxygen. Mechanisms involved here include hydrolysis of triglycerides, hydroxyacid glycerides, β-keto acid glycerides, and plasmalogens. The resulting end products are chemically similar to those resulting from oxidative reactions with lipids.

Deep-frying operations also affect the flavor of fried foods directly through absorption as a film on the surface. The frying process is a complicated thermal–chemical process that produces fried foods with desirable color, flavor, and texture. Food placed in hot fat is heated quickly to a point where water is vaporized, and the resulting steam causes a boiling action in the oil. This boiling action increases aeration in the oil, which results in an increased oxidation of the oil with the formation of the primary oxidation products—hydroperoxides. These hydroperoxides are extremely

TABLE 9.3
Some Volatile Carbonyl Compounds from
Autoxidized Unsaturated Fatty Acids

Oleic Acid	Linoleic Acid	Linolenic Acid
Octanal	Hexanal	2,4,7-Decatrienal
Nonanal	(Z) 2-Octenal	Hexenal
(E) 2-Undecanal	(E) 2-Octenal	
Decanal	(E,Z) 2,4-Decadienal	
Heptanal	(E,E) 2,4-Decadienal	
	(Z) 3-Hexenal	

Source: Whitfield, F.B., *Crit. Rev. Food Sci. Nutr.*, 31, 1, 1992.
        With permission.

unstable and decompose via fission, dehydration, and formation of free radicals to form a variety of chemical products, both volatile and nonvolatile decomposition products. In general, the oxidation products, such as alkanals, 2-alkenals, and 2,4-alkadienals, all with 7–11 carbons, are most likely to affect the sensory quality of the frying oil [52].

9.3.3 MICROBIAL AND ENZYMATIC REACTIONS

Fermentation is defined as the "slow decomposition process of organic substances induced by micro-organisms, or by complex nitrogenous substances (enzymes) of plant or animal origin." It can be described as a biochemical change that is brought about by the anaerobic or partially anaerobic oxidation of carbohydrates by either microorganisms or enzymes [53]. It is one of the oldest methods of processing and has many functions such as improving the taste of foods, enhancing the digestibility, preserving food from deterioration and degradation by organisms, and increasing nutritional value through the synthesis of essential amino acids and vitamins. Many fermented products contain lactic acid bacteria, but other bacteria, yeasts, and molds may be involved as well. In order to be able to produce products with consistent quality, starter cultures for the preparation of fermented food have been developed. Various fermented products originate from spontaneous fermentation of a raw material, such as milk, meat, or vegetables [54]. Bread wine, cheese, and yogurt are all familiar and popular fermented foods [55] but there are also many different products such as cultured milks, pickles, fermented soy products, meat products, etc. [56].

Industrially, fermentation is carried out using carefully selected microorganisms appropriate to the substrate under precisely specified conditions. There are various fermentation products including alcohol, glycerol, butyl alcohol, acetone, as well as lactic, acetic, citric, gluconic, and glutamic acids, many of which are also important in food flavorings [35].

The formation of flavors in a fermented product is a complex process involving many biochemical reactions. Three main pathways take place: the conversions of glucoside (glycolysis), lipids (lipolysis), and proteins (proteolysis) [54]. The enzymes involved in these pathways are predominantly derived from the starter cultures used in these fermentations. During glycolysis the main conversion obviously leads to the formation of lactate by lactic acid bacteria, but a fraction of pyruvate can be converted into various flavor compounds such as diacetyl, acetoine, 2,3-butanediol, ethanol, acetaldehyde, or acetic acid. Lipolysis results in the formation of free fatty acids, which can be precursors of different flavor compounds including methyl ketones, lactones, aldehydes, and alcohols. Lactic acid bacteria contribute relatively little to lipolysis but additional cultures (e.g., molds in the case of surface-ripened cheeses) often have high activities in fat conversion [57]. Moreover, proteolytic activity can lead to the formation of alcohols, sulfur compounds, amines, and some organic acids [58,59].

9.3.3.1 Dairy Products

Traditionally fermented dairy products have been used as beverages, meal components, and ingredients for many new products [60]. The formation of flavor in fermented dairy products is a result of reactions of milk components: lactose, fat, and casein. Particularly, the enzymatic degradation of proteins leads to the formation of key-flavor components that contribute to the sensory perception of the products [55]. Methyl ketones are responsible for the fruity, musty, and blue cheese flavors of cheese and other dairy products. Aromatic amino acids, branched-chain amino acids, and methionine are the most relevant substrates for cheese flavor development [55]. Volatile sulfur compounds derived from methionine, such as methanethiol, dimethylsulfide, and dimethyltrisulfide, are regarded as essential components in many cheese varieties [61]. Conversion of tryptophan or phenylalanine can also lead to benzaldehyde formation. This compound, which is found in various hard- and soft-type cheeses, contributes positively to the overall flavor [57,62]. The conversion of caseins is undoubtedly the most important biochemical pathway for flavor formation in several cheese types [62,63]. A good balance between proteolysis and peptidolysis prevents the formation of bitterness in cheese [64,65].

9.3.3.2   Meat Products

In the fermentation of meat, combinations of various bacterial species are used. In almost all cases, lactic acid bacteria are present in these starter cultures having *Lactobacillus sakei*, *Lb. curvatus*, *Lb. plantarum*, *Pediococcus pentasaceus*, and *P. acidilactici* as most important species in addition to *Micrococcus varians*, *Staphylococcus carnosus*, and *S. xylosus* species. These microorganisms contribute to the typical flavor of fermented meat products by the formation of esters and other aromatic compounds from amino acids. A number of fermented meat products also exhibit fermentation on the surface, helping flavor formation through specific lipolytic and proteolytic activities [55,66]. Also, amino acid catabolism by *Staphylococci* and *Lactobacilli* yields volatile products contributing to meat flavor [67]. Flavor compounds are generated during sausage fermentation by (a) the generation of flavor volatiles by lipolysis and hydrolysis of phospholipids, and which is followed by the oxidation of free fatty acids; (b) the production of organic acids by microorganisms; conversion of amino acids and peptides to flavor-active alcohols, aldehydes, and acids; and the modification of lipid oxidation products (e.g., esterification of acyl moieties or reduction of aldehydes); and (c) the addition of spices, smoking, or surface ripening with yeasts or molds depending on the product formula and maturation conditions [67].

9.3.3.3   Wine

During the manufacturing of another fermented product, namely wine, many microbial reactions that contribute to the product's quality take place during fermentation [54]. Quantitatively, volatiles formed during yeast fermentation represent the most important flavor components in wine. Ethanol is obviously an important volatile constituent of wine as it is produced from the catabolism of hexoses by *Saccharomyces cerevisiae* during fermentation. Numerous acetate esters and ethyl esters of fatty acids contribute characteristic fruity aromas to wines. Following the primary yeast fermentation, many wines undergo a secondary microbial, or malolactic fermentation. The main consequence of this fermentation is the conversion of malic acid to lactic acid that results in a decrease in wine acidity. Wines that have undergone malolactic fermentation also often have a characteristic buttery character due to the microbial formation of diacetyl, 2,3-butanedione [68]. The most abundant glycosidase in berries and grapes is β-glucosidase. During wine and juice production enzyme treatments also increase the concentration of volatile monoterpene flavor compounds [5].

9.3.3.4   Vegetables

The fermentation of vegetables, such as olives, cabbage, and pickles, belongs to very early practices. The production methods are still rather simple: basically, lactic acid bacteria such as *Lb. plantarum*, *Leuconostoc* spp., *Pediococcus*, and *Lactococcus* are used. The acid formation results in an extended shelf life, and also the flavor formation by the microorganisms adds to the overall quality of these products. For the fermentation of soy, which is a very common and traditional process in East Asia, various mold species are used, such as *Rhizopus oligosporus* and *R. oryzae*. These fermentations are often solid-state fermentations, for which the raw material is inoculated with microorganisms that grow at the surface of the raw material, making it essentially an aerobic fermentation process. For some of the products, there is an extensive proteolysis, leading to products like soy sauce, whereas in other fermentations (e.g., tempeh) only limited proteolysis occurs [54].

9.3.3.5   Olive

Flavor is closely connected with the composition of volatile compounds, and it is considered as a quality index for many fruit and vegetable products, playing an important role in consumers' acceptability [69]. Changes in olive aroma allow us to compare different cultivars and processing methods, as well as to follow the evolution of quality during processing and to check off-flavors occurring during fruit storage. Volatile compounds are not produced in significant amounts during fruit growth but arise during the climacteric stage of ripening [70]. Olive oil volatile compounds are produced by enzymatic pathways during the grinding process. Thus, in a good quality olive oil, the desired flavor is produced only by physical processes that activate endogenous enzymes and not by a fermentative process as in table olives. Flavor formation in table olives is a dynamic process that develops mainly during fermentation by indigenous lactic acid bacteria and yeasts, together with a variety of contaminating microorganisms, which produce volatile compounds from major fruit constituents through various biochemical pathways. Fermenting olives are typically very complex ecosystems with active enzyme systems from the ingredient material, affecting the metabolic activities of microorganisms [71]. Lactic acid bacteria directly influence the flavor of fruits, contributing to the development of sensory characteristics of fermented olives, but there are also less direct ways by which microorganisms affect flavor. Moreover alcohols, esters, aldehydes, and ketones, as well as acids, are known to be formed by microorganisms and affect the flavor of the product [71,72].

9.3.3.6 Bread

In general, the baking process influences the typical aroma of bread crust, while dough fermentation is fundamental for the development of crumb flavor. Compounds strongly affecting bread flavor are mainly organic acids, alcohols, esters, and carbonyls [73,74]. Bacterial proteolysis during sourdough fermentation was shown to contribute much more to the development of typical sourdough flavors of baked breads when compared to breads produced from chemically acidified or yeasted doughs [74]. The various flavor components formed in these products, such as aldehydes, alcohols, and pyrols, are partly the result of biochemical conversions by the starter cultures but also result from the baking process itself [1,75].

9.3.3.7 Tea

The original purpose of tea fermentation is to enhance the flavor of tea which is the most important attribute for grading [76]. Because "fermentation" in modern tea science is mainly defined as the oxidation of catechins, the contents of catechins or their oxidation products, theaflavins and thearubigins, may be the first consideration for setting quality standards [77]. The presence of high levels of simple (nongallated) flavan-3-ols and flavan-3-ol gallate esters are of critical importance for the production of black teas with high sensory evaluation, changing depending on the region of cultivation [77,78].

9.3.4 Interactions between Lipids, Proteins, and Carbohydrates

The quality of flavor in food is attributed to low concentrations of volatile compounds in its headspace. The headspace concentration of volatile flavors in foods is determined by several factors; vapor pressure of the flavor compound, its interaction with other components of the food, and temperature. Carbohydrates, fats, and proteins are all known to affect the vapor pressure of flavors. In addition to odor, the perceived flavor, i.e., taste, of foods is significantly affected by different rates and extent of flavor release (volatility and temperature) when food is chewed [79].

Proteins affect several sensory properties such as appearance, color, texture, and flavor. Although most proteins have little flavor, they influence perceived flavor because they may contain bound off-flavors; modify flavor by selective binding; produce off-flavors or act as precursors of flavors through the Strecker degradation of amino acids [79].

The binding ability of proteins to flavor compounds is related to the porous nature and denaturation of protein and to the size of the volatile compound. Aldehydes and ketones show considerable interaction with proteins. For example, the binding of saturated aldehydes and methyl ketones to whole native whey protein is of interest since it is then possible to alter the processing conditions for minimizing the presence of off-flavor compounds. Another milk protein, β-lactoglobulin, can bind to phenolic compounds such as p-nitrophenyl phosphate, p-nitrophenyl β-glucoronide, and p-nitrophenol, particularly, through the aromatic amino acids, tryptophan and phenylalanine [80]. α-Lactoglobulin can also readily bind the alkanone, 2-heptanone, 2-octanone, and 2-nonanon, possibly through hydrophobic interactions similar to the binding phenolic compounds [81].

Denaturation of proteins can enhance flavor absorption. This is related to the interactions of nonpolar flavor compounds with the hydrophobic regions of unfolded protein. The off-flavors (e.g., alkanones and alkanals) result from lipoxygenase initiated peroxidation of the unsaturated fatty acids in soybeans. The binding of these carbonyl compounds to main storage proteins (glycinin and conglycinin) can cause unacceptable beany flavors in products containing these proteins [82–84]. In aqueous systems the binding of aliphatic carbonyls to soy proteins is hydrophobic in nature and that the β-conglycinin component may be the protein fraction responsible for the off-flavor binding of soy proteins [83]. The hydrophobic binding of proteins to flavor molecules would result in a conformational change in the protein molecule [82].

Starch has been shown to form complexes with a wide variety of molecules. These chemicals involve many different functional groups, molecular sizes, and both polar and nonpolar molecules [85]. The helical structure of the amylose fraction has an important effect on flavor release from a food product. The polar hydroxyl groups are oriented to the outside of the coil, making it hydrophilic. The inside of the helix structure is hydrophobic due to the hydrogen and glycosides being oriented to the inside of the coil [79]. The hydrophobic flavor chemicals "nest" in the coil. The complexed flavor chemicals are generally not released during consumption except when the food is kept in the mouth for a relatively long period of time (more than 20 s), allowing the salivary amylose to breakdown the starch and release the flavor [80]. The amylopectin portion of starch shows little tendency to form complexes. However, it probably helps to stabilize the complex in starch solutions [85]. The thickened or viscous systems require a higher concentration of flavoring or sweetener to produce the desired sensory intensity than do aqueous or fluid systems. This may be partly due to the hindered diffusion of flavor molecules to the taste buds, physical coating of taste buds, and unavailability of flavor molecules as a result of complexing with starch or other polysaccharides [85].

Fats act as precursors to flavor development by interacting with proteins and other ingredients when heated. Off-flavors are not normally perceived in full-fat systems because most of them are fat-soluble. However, in the absence of fat, the vapor pressure of the aroma chemical in water is increased, resulting in a very intense perception of the off-flavor chemical. Flavor release is a critical factor governing smell and taste. The majority of flavor components are dissolved to some extent in the lipid phase of food—releasing the flavor slowly in the mouth and resulting in a pleasant aftertaste [75].

Fat and water are solvents for aroma chemicals, proteins, and carbohydrates. They can absorb, make complexes, and bind to aroma chemicals. A fat-soluble aroma chemical can be solubilized in a water and gum emulsion or in oil. However, the aromatic compound will be perceived more intense in the water-gum blend due to the higher vapor pressure in the headspace. On the other hand, aroma molecules surrounded by oil molecules will not volatilize (thus a reduction in the perceived intensity), resulting in a lower vapor pressure [75].

9.3.5 Formation of Off-Flavors

Off-flavors are widely defined as unpleasant odors or tastes resulting from the natural deterioration of a food. Off-flavors may arise in a food product through many routes. For example, the food may become tainted due to contamination from airborne, waterborne, or packaging related sources. Alternatively, microbial activity or chemical reactions in the food itself (e.g., lipid oxidation, nonenzymatic browning, or enzymatic reactions) may occur, resulting in off-flavors. Off-flavors entering a food product from airborne sources may occur in food having minimal packaging barriers. Waterborne off-flavors are most commonly due to the microbial growth [62].

Pesticides, disinfectants, and detergents comprise a group of chemicals that will occasionally contribute to off-flavors in foods. For example, chlorophenols may arise from a reaction between phenol and chlorine in the water supply or from the direct contamination of the food with chlorophenols, which are frequently used in algicides and fungicides [62].

Microbial activity can produce off-flavors in several ways; the production of undesirable primary metabolites, incidental conversion of food constituents of otherwise little flavor significance, or through residual enzyme activity after cell death [62].

Off-flavors generated by chemical or enzymatic reactions are particularly associated with the degradation of lipids, either by enzymatic or nonenzymatic reactions. A potent flavor compound, *trans*-4,5-epoxy-*E*-2-decenal, is responsible for off-flavors during the heat treatment of fats [86]. The catalysis of lipid oxidation by heme pigments also results in a state of rancid odor and flavor in the meat. The free radicals that are produced in oxidized lipids can oxidize and decompose heme pigments affecting the color of the product. The off-flavor of stored rice originates mainly from oxidative degradation of unsaturated fatty acids, especially linoleic acid [87]. The amount of hexanal, one of the main off-flavor agents has been found to be linearly proportional to the amount of oxidized linoleic acid [88]. Another off-flavor compound, *trans*-2-nonenal, is one of the main causes of beer staling. The generation of *trans*-2-nonenal in beer is a result of enzymatic or nonenzymatic oxidation of lipids and oxidized free fatty acids [89]. Lipolytic activity is a potential source of a variety of off-flavors in milk and also is known to impart characteristic flavors to cheese and other milk-containing products [90].

Lipoxygenase enzyme can oxidize polyunsaturated fatty acids into position-specific hydroperoxides that may be decomposed to produce various degradation products, such as aldehydes, alcohols, ketone, and hydrocarbons. These compounds influence the flavor quality of several foods rich in unsaturated fatty acids such as fish [90] and corn [91].

Proteolytic activities in foods may also result in off-flavor formation. The release of trichloroacetic acid–soluble free amino acids by proteolytic enzymes has been associated with off-flavor development [92]. On the other hand, some chemicals inherent in food products may be converted into off-flavor compounds during processing. For example, ferulic acid is a precursor of off-flavors in stored orange juice [93].

9.4 FLAVOR MANUFACTURING

9.4.1 BIOTECHNOLOGICAL PRODUCTION OF FLAVORS

Biotechnology is currently being applied to agriculture and food processing all over the world and has the potential to improve productivity and enhance quality [94]. Most recently, biotechnology has been used for the production of flavors. During the last few years, an increasing demand for natural flavors has stimulated a broad range of research activities in the development of flavors. Biotechnological processes play an important role in the production of flavors through biotransformation, *de novo* synthesis, traditional fermentations, and genetic engineering. Most of the flavors have been produced through microorganisms, plant tissue- and cell cultures, and using isolated enzymes [95]. However, the formation of volatile flavor compounds by fermentation on an industrial scale still needs to be developed. As a result, although more than hundreds of aroma compounds may be produced microbially, only a few of them are produced on an industrial scale. The reasons are economical problems, low transformation efficiency, and lack of information [5]. On the other hand, recent developments in transgenic research have produced data for the use of metabolic engineering of biosynthetic pathways to produce high-value secondary metabolites that can enhance the flavors of food products. Biotechnology is playing an increasingly important role in the breeding of food crops for enhanced flavor [96].

9.4.1.1 Flavor Biotransformation

Microorganisms and enzymes play very important roles in the production of flavor compounds in a wide variety of foodstuff. There are four main sources of enzymes in food: (a) enzymes inherent in the food; (b) enzymes arising from microbial contaminants; (c) enzymes of desirable microorganisms deliberately added to foods; and (d) enzymes that are intentionally added to foods [97].

Endogenous enzymes may contribute to the natural flavor of the food, but these are invariably destroyed during processing. Microbial contaminants may give rise to off-flavors in foods and the ultimate aim of preserving food is to reduce or eliminate entirely these undesirable microorganisms. Some microorganisms used in the processing of food produce flavor-active metabolites that belong to most classes of chemical compounds, including acids, alcohols, lactones, esters, aldehydes, ketones, etc. If these materials accumulate in the foodstuff, then they influence its final flavor. Interest in using the microbial approach for making flavors arises for several reasons. The microbial approach offers an opportunity to produce flavors of natural origin. The microbial approach often supplies complex flavor characteristics, which may be difficult to duplicate economically *via* compounding [97].

The production of certain aromatic flavor compounds by the use of whole microbial or alternative cell systems is the preferred route as they contain many enzymes that can act simultaneously or in a sequential manner. However, if the biotransformation is required, the use of isolated enzymes may be beneficial [98]. The use of isolated enzymes can preclude the formation of side-products and result in highly homogeneous products (Table 9.4) depending on the purity of the starting substrate [95]. For example, ethyl esters of short-chain fatty acids are a large group of flavor and fragrance compounds widely used in food, beverage, cosmetic, and pharmaceutical industries. Using lipase enzyme produced by *Rhizopus chinensis* as a substitute catalyst could be an efficient tool in mediating synthesis of flavor esters and catalyzing reactions more specifically with higher yield [99].

Many flavoring products such as protein-based flavors may be produced using a combination of enzymes to enable a high degree of hydrolysis. For example, vegetable proteins (sodium caseinate and soy isolates) and animal proteins (beef, gelatin, and fish) have been more extensively hydrolyzed using the combination of Flavorzyme™ (enzyme complex from *Aspergillus oryzae* containing endo and exo peptidase) and Alkalase™ (alkaline serine protease with endoprotease activity). The combination of proteases/peptidases as Flavorzyme with lipases can be utilized for the production of cheese flavor directly from milk. When gelatin is hydrolyzed to a degree of hydrolysis above 50%, the hydrolysates have a pronounced high degree of mouthfeel. Hydrolysis of milk proteins by Flavorzyme to a degree of hydrolysis above 70% leads to a cheese-like flavor [95].

With the introduction of commercial fungal proteolytic enzyme complexes, it is possible to produce protein hydrolysates with fewer undesired by-products, resulting in new opportunities and new products for savory ingredient manufacturers. Food ingredients like savory flavors and enhancers are important in the manufacture of soups, sauces, ready-to-eat meals, and snacks [99].

Enzymatically modified cheeses developed to accelerate the ripening and flavor building blocks can be produced by controlled proteolytic and/or lipolytic enzyme treatment of natural cheese. The most popular enzyme-modified cheeses include Cheddar, Swiss, Parmesan, Romano, Brick, and Blue cheeses [95].

TABLE 9.4
Flavor Production by Enzymatic Processes

Substrate	Enzyme	Flavoring Compound
FAD,[a] NAD,[b] O_2	Alcohol oxidase	Aromatic aldehydes
Amygdalin[c]	β-Glucosidase and mandelonitrile lyase	Benzaldehyde
Naringin and hesperidin[d]	Rhamnosidase	6-Deoxysugar rhamnose

Source: Hessing, M. and van der Lugt, J.P., *World Ing.*, 1, 14, 1995.

[a] Flavin adenine dinucleotide.
[b] Nicotinamide adenine dinucleotide.
[c] Cyanogenic glycoside.
[d] Bitter citrus-plant glycosides.

Fat-based flavors are produced by the modification of fats particularly in dairy products, like creams, butter fat, and cheese. Enzyme-modified creams are generally produced as a result of controlled lipase treatment of dairy cream. Similarly, lipases are also utilized for the manufacturing of enzyme-modified butterfat products from emulsified anhydrous butterfat.

Volatile aldehydes and alcohols are key compounds in the fresh and green sensorial notes of vegetables and fruits [59]. The characteristic aroma compounds responsible for green note include *trans*- and *cis*-2-hexenol, *trans*- and *cis*-3-hexenol (leaf alcohol), hexanol, hexanal, and *cis*-2-hexenal [100]. These compounds are biosynthetically produced using lipoxygenase pathway enzymes [5,101].

Due to the increasing use of bioflavors in food and a demand for new processes and natural products, new biotechnological processes, such as enzyme-assisted flavor production, are very likely to receive more attention in the near future. The application of enzyme technology is expanding very fast as new enzymes enter the market. It is already possible to produce hydrolyzed proteins for use as flavors and flavor enhancers.

9.4.1.2 Plant Tissue Culture Technology and Plant Breeding

Plant tissue culture (PTC) is considered simply as a method of producing large numbers of identical plants without facing traditional problems of climate, season, and disease [102]. Traditionally, plant tissue cultures have been viewed as biocatalysts for the production of secondary metabolites including naturally derived food ingredients. The use of plant tissue cultures has been proven successful for the production of specific metabolites, higher metabolic yields in cultured plant cells as compared to the intact plants [60].

Many types of plant cells from a variety of species can be cultivated *in vitro* by supplying them with nutrients and growth substances under strict aseptic conditions. Cultures of individual cells, grown in a fermenter, can be used in preference to whole plants for producing flavorings. A recent development with considerable promise is the cultivation of "hairy root cultures." From undifferentiated cells (or callus) grown *in vitro*, whole plants can be generated by adjusting the proportions of various growth factors in the surrounding medium [103].

Natural vanilla flavor and natural vanillin are in high demand and alternative production routes such as using PTC or microbial systems are being sought [60]. "*De novo*" synthesis of vanillin and vanilla plant cells and plant callus tissue are important commercial products biosynthesized by PTC. Alternatively, hair roots of vanilla plant have been used as a biocatalyst to convert the precursor ferulic acid into vanillin [104].

For flavor and fragrance materials, PTC offers a means of rapid multiplication of stock selected for superior product quality, yield, disease resistance, and thereby, improving the overall economic performance of the industry. Plant tissue culture enables a rapid build up of plant material; multiplication rates of 1:4 per month can be routinely achieved. Thus, beginning with only one explant producing four shoots in the first month, 50,000 plantlets can be produced in 7 months. As the multiplication rates can be forecast with great accuracy, production can be planned so that the plantlets can be transferred to conventional growing systems at the most suitable time of year [102]. The disadvantage of this technology is that plant cells grow slowly (compared to the growth of microorganisms), are prone to contamination, and have a low volumetric productivity [104].

Another approach could be to breed cultivates with the aim of enriching the flavor content significantly. It is certainly valuable exploring the opinion of flavor enriching fruits by specific breeding. This could lead to cost-effective production of all natural from-the-name-material (FTNM) derived flavors. In this way, the rather laborious precursor or *in vivo* approach using microbes and enzymes can be substituted by an even more environmentally friendly *in vivo* or *in planta* approach [104].

9.4.1.3 Recombinant DNA Technology

Recombinant DNA technology allows the efficient production of enzymes or flavor compounds [5]. Sensory characteristics such as color or flavor can also be manipulated by recombinant DNA techniques. Two intensively sweet proteins, thaumatin and monellin, are produced in the fruits of the African plants, katemfe (*Thaumatococcus danieli* Benth.) and serendipity (*Diooscoreophyllum cumminsii* Dield.), respectively. The thaumatin gene has been introduced into potato as a model system, while the monellin gene was expressed in tomato and lettuce to enhance the sweet taste [105].

Another example is the production of green note compounds using lipoxygenase pathway enzymes. Lipoxygenase and hydroperoxide lyase are the determinant enzymes for the conversion of fatty acids into natural food flavor components. However, these compounds present in natural sources at very low levels and because of their instable nature it has been difficult to purify these compounds. As a result, considerable efforts have been made to clone these two enzymes for commercial uses in the production of natural flavor components. Many plant lipoxygenase from different plants have been cloned and expressed in *E. coli* or yeast [5].

The future generation of bioflavors could involve reprogramming of metabolic pathways in microbes and plants to arrive at novel or improved flavor top notes or building blocks. In most cases the pathways and the enzymes involved in flavor biogenesis are not known and hence, it is currently impossible to over-express such pathways by genetic tools [104].

9.4.2 Smoked Flavor Production by Pyrolysis

The process of curing and smoking food has been carried out to preserve both meat and fish, although the emphasis has gradually changed from preservation to the imparting of a very pleasant flavor to the food product. Initially, the smoking process was carried out in a kiln with little or no control over the smoking process. However, today smoke generators are designed to ensure the burning of hardwood sawdust under controlled conditions of forced air recycling and temperature. The direct use of wood smoke is relatively inefficient and that is why the use of natural liquid smoke flavors has gained wide acceptance [35].

Processing of meat and fish, such as curing and/or smoking, creates a characteristic flavor in the products especially in sausages or bacon [106]. Commercial smoke flavorings used in the food industry have different organoleptic features, due to their different compositions. Many factors have an important influence in the composition of these mixtures such as the processes followed in their manufacture [107].

There are several studies investigating the flavor composition of smoked products that gives the products their characteristic flavors. The main components responsible for the smoke flavor are phenol derivatives. In addition to the flavoring compounds arising from wood pyrolysis, flavoring compounds derived from plants are also present. The soluble fraction mostly contains fatty acids and fatty esters, in addition to acids, alcohols, carbonyls, esters, furans, lactones, furans, and many miscellaneous compounds [108]. The phenolic fraction of traditional kiln-smoked Bacon contains phenols, furfuryl alcohol, and cyclotene while different compounds, such as 2,4,5-trimethyl-3(2*H*)-furanone, have been isolated from cooked Oscar Mayer bacon [109].

9.4.3 Meat-Like Flavors

The economical value and high popularity of meat lead to the production of meat-like flavors through process chemistry. Several heat-induced reactions lead to the formation of meat flavors. These reactions are the pyrolysis of peptides and amino acids, the degradation of sugars, the oxidation, dehydration, and decarboxylation of lipids, the degradation of thiamin and ribonucleotides, and interactions involving sugars, amino acids, fats, H_2S, and NH_3 [110].

Meat flavor is due to a great number of volatiles from different chemical classes. However, most of the odorants described as meaty aroma contain sulfur. The two most important reactions which generate meaty aroma compounds are the reactions between sulfur containing amino acids and reducing sugars (Maillard reaction) and the thermal degradation of thiamin [35]. Sulfur-containing furans are the basic chemicals responsible for the aroma of thermally treated meat.

The developments of commercial meat flavorings, in which authentic constituents from meat are replaced by cheaper precursor sources, have started with the analysis of natural meat flavors and model systems. These so-called processed meat flavors are normally produced by thermal treatment of a reaction mixture consisting of an amino acid source, e.g., hydrolyzed vegetable proteins (HVPs) or autolyzed yeast extracts (AYE), a reducing sugar source, and an additional sulfur source (e.g., cysteine, methionine, or thiamin), which facilitates the formation of sulfur-containing aroma compounds. Varying the ingredients or the reaction conditions (e.g., pH, temperature, pressure, or processing time) and adding special aroma essences (e.g., essential oils, smoke essences, or synthetic aroma chemicals) yield different species-specific flavorings (e.g., pork, beef, or lamb) or products [111].

The beginning of thermal processes to produce meat-like flavorings is associated to the introduction of HVPs that are widely used as seasoning and ingredients in processed food products. These thermally processed products bring both a characteristic flavor of their own (primarily from processing) and an abundance of flavor enhancers [35]. Hydrolyzed vegetable or plant protein can be defined as mixtures containing amino acids and frequently other substances such as salt and peptides, obtained by the hydrolysis of vegetable proteins. The HVP bouillon does not posses a sharp flavor but rather is described as a warm, salty, and spicy sensation that is associated with enhanced meat extracts [112].

The three available industrial methods for the production of HVPs are [35] (1) enzymatic reactions in which bitter peptides are formed; (2) alkaline hydrolysis, which typically results in an unacceptable flavor profile and unbalanced amino acid content; and (3) acid hydrolysis, which is the most preferred method that is cost effective and yields a range of good flavors.

The secondary meat-like flavors can be obtained using autolyzed yeast extracts. When yeast cells die, they automatically break up, a process called autolysis, and a collection of protein, fats, vitamins, minerals, and, most importantly, MSG, a flavor enhancer, remain. MSG is commonly used in prepared foods like lunch meats, broths, and potato chips due to its ability to enhance the flavor of many foods and is also found naturally in many food sources such as tomatoes, mushrooms, and carrageenan derived from algae. During fermentation of soy sauce the degree of autolysis depends on the temperature/time conditions and the pH of the medium. The process is terminated by raising the temperature in order to inactivate the enzyme system. The degree of protein and carbohydrate breakdown can be varied to satisfy particular flavor characteristics. The flavor of the end product is determined, to a large extent, by the choice of yeast and the nature of the substrate. Aerobically grown *S. cerevisiae* gives a much stronger, more beef-like flavor than anaerobically grown *Saccharomyces carlsbergensis* [35].

REFERENCES

1. Jelinek, G., *Sensory Evaluation of Food Theory and Practice*, Ellis Horwood, England, U.K., p. 17, 1985.
2. Belitz, H.D. and Grosch, W., *Food Chemistry*, D. Hadziyev (Trans.) (English), Springer Werlag, Berlin, Germany, p. 319, 1999.
3. Meillgaard, M.C., Civille, G.V., and Carr, B.T., *Sensory Evaluation Technique*, 3rd edn., CRC Press, Boca Raton, FL, chap. 2, 1999.
4. Giese, J., Modern alchemy—Use of flavors in food, *Food Tech. Chicago*, 48, 106, 1994.
5. Sun Pan, B., Kuo, J.M., and Wu, C.M., Flavor compounds in foods, in *Chemical and Functional Properties of Food Components*, 3rd edn., Sikorski, Z., Ed., CRC Press, Boca Raton, FL, p. 295, 2007.
6. Smith Palmer, A., Stewart, J., and Fyfe, L., Antimicrobial properties of plant essential oils and essences against five important food-borne pathogens, *Lett. Appl. Microbiol.*, 26, 118, 1998.
7. Tassou, C., Koutsoumanis, K., and Nychas, G.J.E., Inhibition of *Salmonella enteritidis* and *Staphylococcus aureus* in nutrient broth by mint essential oil, *Food Res. Int.*, 33, 273, 2000.
8. Inouye, S., Takizawa, T., and Yamaguchi, H., Antibacterial activity of essential oils and their major constituents against respiratory tract pathogens by gaseous contact, *J. Antimicrob. Chemother.*, 47, 565, 2001.

9. Burt, S., Essential oils: Their antibacterial properties and potential applications in foods: a review, *Int. J. Food Microbiol.*, 94, 223, 2004.

10. Chryssavgi, G., Vassiliki, P., Athanasios, M., Kibouris, T., and Komaitis, M., Essential oil composition of *Pistacia lentiscus* L. and *Myrtus communis* L.: Evaluation of antioxidant capacity of methanolic extracts, *Food Chem.*, 107, 1120, 2008.

11. Matiucci-Pereira, C.A. and Ferreira-Maia, J., Study of the antioxidant activity and essential oil from wild basil (*Ocimum gratissimum* L.) leaf, *Cienc. Tecnol. Aliment.*, 27, 624, 2007.

12. Barra, A., Coroneo, V., Dessi, S., Cabras, P., and Angioni, A., Characterization of the volatile constituents in the essential oil of *Pistacia lentiscus* L. from different origins and its antifungal and antioxidant activity, *J. Agric. Food Chem.*, 55, 7093, 2007.

13. Bozin, B., Mimica Dukic, N., Samojlik, I., and Jovin, E., Antimicrobial and antioxidant properties of rosemary and sage (*Rosmarinus officinalis* L. and *Salvia officinalis* L., Lamiaceae) essential oils, *J. Agric. Food Chem.*, 55, 7879, 2007.

14. Hu, X.-L., Han, Z.-X., Liu, Y.-F., Zhao, H., and Yin, F.-J., Study on the antioxidant activity and the chemical components of essential oil obtained from the lavender, *Food Sci. Technol.*, 13, 115, 2006.

15. Fasseas, M.K., Mountzouris, K.C., Tarantilis, P.A., Polissiou, M., and Zervas, G., Antioxidant activity in meat treated with oregano and sage essential oils, *Food Chem.*, 106, 1188, 2008.

16. El Ghorab, A., Shibamoto, T., and Ozcan, M., Chemical composition and antioxidant activities of buds and leaves of capers (Capparis ovata Desf. var. canescens) cultivated in Turkey, *J. Essent. Oil Res.*, 19, 72, 2007.

17. El Ghorab, A.H., Fadel, H.M., and El Massry, K.F., The Egyptian Eucalyptus camaldulensis var. brevirostris: Chemical compositions of the fruit volatile oil and antioxidant activity, *Flavor Fragr. J.*, 17, 306, 2002.

18. Matook, S.M. and Hashinaga, F., Evaluation of the antioxidant activity of extracts from buntan (*Citrus grandis* Osbeck) fruit tissues, *Food Chem.*, 94, 529, 2006.

19. Giamperi, L., Fraternale, D., and Ricci, D., Essential oil composition and antioxidant activity of peels of Citrus sinensis blood and blond Citrus aurantium, *Rivista Italiana EPPOS*, 12, 21, 2002.

20. Stauffer, C.E., Eye and taste appeal, *Baking Snack*, 15, 31, 1993.

21. Farrell, K.T., *Spices, Condiments, and Seasonings*, Avi Publishing, Westport, CT, p. 258, 1985.

22. Lewis, Y.S., *Spices and Herbs for the Food Industry*, Food Trade Press, Orpington, U.K., p. 12, 1984.

23. Eskin, N.A.M., *Biochemistry of Foods*, Second edition, Academic Press, Inc., San Diego, CA, p.134, 1990.

24. Buttery, R.G., Vegetable and fruit flavors, in *Flavor Research Recent Advances*, Teranishi, R., Flath, R., and Sugisawo, H., Eds., Marcel Dekker Inc., New York, p. 175, 1981.

25. Lewinsohn, E., Schalechet, F., Wilkinson, J., Matsui, K., Tadmor, Y., Nam, K.H., Amar, O., Lastochkin, E., Larkov, O., Ravid, U., Hiatt, W. et al., Enhanced levels of the aroma and flavor compound *S*-linalool by metabolic engineering of the terpenoid pathway in tomato fruits, *Plant Physiol.*, 127, 1256, 2001.

26. Galliard, T. and Matthew, J.A., Lipoxygenase-mediated cleavage of fatty acids to carbonyl fragments in tomato fruits, *Phytochemistry*, 16, 339, 1977.

27. Croteau, R. and Karp, F., Origin of natural odorants, in *Perfumes: Art, Science and Technology*, Muller P.M. and Lamparsky, D., Eds., Elsevier Applied Science, London, U.K., p. 101, 1991.

28. Pichersky, E., Lewinsohn, E., and Croteau, R., Purification and characterization of *S*-linalool synthase, an enzyme involved in the production of floral scent in *Clarkia breweri*, *Arch. Biochem. Biophys.*, 316, 803, 1995.

29. Buttery, R.G., Importance of lipid derived volatiles to vegetable and fruit flavor, in *Flavor Chemistry of Lipid Foods*, Min D.B. and Smouse, T.H., Eds., American Oil Chemists Society, Washington, DC, p. 156, 1989.

30. Sarry, J.E. and Gunata, Z., Plant and microbial glycoside hydrolases: Volatiles release from glycosidic aroma precursors, *Food Chem.*, 87, 509, 2004.

31. Eskin, N.A.M., *Plant Pigments, Flavors, and Textures: The Chemistry and Biochemistry of Selected Compounds*, Academic Press, New York, p. 219, 1979.

32. Mitiku, S.B., Ukeda, H., and Sawamura, M., Enantiomeric distribution of alpha-pinene, beta-pinene, sabinene, and limonene in various citrus essential oils, in *Food Flavors and Chemistry: Advances of the New Millennium*, Spanien, M., Shahidi, F., Parliament, T.H., Mussian, C., Ho, C.T., and Contis, E.T., Eds., The Royal Society of Chemistry, Cambridge, U.K., p. 216, 2001.

33. Paillard, N.M.M., The flavour of apples, pears and quinces, in *Food Flavours*, Morton, D. and MacLeod, A.J., Eds., Elsevier Science Publishing Company Inc., Amsterdam, The Netherlands, p. 1, 1990.

34. Crouzet, J., Etievant, P., and Bayonove, C., The flavor of fruits, in *Food Flavors*, Morton, I.D. and Macleod, A.J., Eds., Elsevier Science Publishing Company Inc, Amsterdam, The Netherlands, p. 43, 1990.

35. Reineccius, M., Changes in food flavor due to processing, in *Flavor Chemistry and Technology*, 2nd edn., CRC Press, Boca Raton, FL, p. 103, 2006.

36. Flament, I., Coffee, cocoa and tea, *Food Rev. Int.*, 5, 317, 1989.

37. Clarke, R.J., The flavor of beverages, in *Food Flavors*, Morton, I.D. and Macleod, A.J., Eds., Elsevier Science Publishing Company Inc, Amsterdam, The Netherlands, p. 61, 1990.

38. Bokuchava, M.A. and Skobeleva, N.I., The flavor of beverages, in *Food Flavors*, Morton, I.D. and Macleod, A.J., Eds., Elsevier Science Publishing Company Inc., Amsterdam, The Netherlands, p. 49, 1990.

39. Belitz, H.D. and Grosch, W., *Food Chemistry*, 2nd edn., D. Hadviyev (Trans.), Springer Verlag, Berlin, Germany, p. 697, 1986.

40. Ziegleder, G., Composition of flavor extracts of raw and roasted cocoas, *Z. für Lebensmittel. und Forsch. A.*, 192, 521, 1991.

41. Uhl, S.R., *Handbook of Spices, Seasonings and Flavorings*, CRC Press, Boca Raton, FL, p. 217, 2000.

42. Röling, I., Kerler, J., Braster, M., Apriyantono, A., Stam, H., van Verseveld, H.W., Microorganisms with a taste for vanilla: Microbial ecology of traditional Indonesian vanilla curing, *Appl. Environ. Microbiol.*, 67, 1995, 2001.

43. De Faveri, D., Torre, P., Aliakbarian, B., Dominguez, J.M., Perego, P., and Converti, A., Response surface modeling of vanillin production by *Escherichia coli* JM109pBB1, *Biochem. Eng. J.*, 36, 268, 2007.

44. Barghini, P., Di Gioia, D., Fava, F., and Ruzzi, M., Vanillin production using metabolically engineered *Escherichia coli* under non-growing conditions, *Microb. Cell Fact.*, 6, 13, 2007.

45. Chang, S.S. and Huang, A.S., Food flavors, *Food Technol.*, 43, 99, 1989.

46. Nagodawithana, T., Flavor enhancers: Their probable mode of action, *Food Technol.*, 48, 79, 1994.

47. Farmer, L.J., The role of nutrients in meat flavour formation, *P. Nutr. Soc.*, 53, 327, 1994.

48. Whitfield, F.B., Volatiles from interactions of maillard reactions and lipids, *Crit. Rev. Food Sci. Nutr.*, 31, 1, 1992.

49. Izzo, H.V. and Ho, C.T., Peptide-specific Maillard reaction products: A new pathway for flavor chemistry, *Trends Food Sci. Technol.*, 3, 253, 1992.

50. Ho, C.T. and Hartman, G.J., Formation of oxazolines and oxazoles in Strecker degradation of D l-alanine and L-cysteine with 2,3-butanedione, *J. Agric. Food Chem.*, 30, 793, 1982.

51. Werkhoff, P., Güntert, M., and Hopp, R., Dihydro-1,3,5-dithiazines-unusual flavor compounds with remarkable organoleptic properties, *Food Rev. Int.*, 8, 391, 1992.

52. Melton, S.L., Jafar, S., Sykes, D., and Trigiano, M.K., Review of stability measurements for frying oils and fried food flavor, *J. Am. Oil Chem. Soc.*, 71, 1301, 1994.

53. Battcock, M. and Azam-Ali, S., Fermented fruits and vegetables. A global perspective, FAO Agricultural Services Bulletin, No. 134, 1998.

54. Smit, G., Wouters, J.T.M., and Meijer, W.C., Quality and flavor of fermented products, in *Cultured Products: History, Principles and Applications*, Marcel Dekker, Inc., New York, chap. 5. 2004.

55. Nout, M.J.R., Upgrading traditional biotechnological processes future, in *Applications of Biotechnology in Traditional Fermented Foods*, Shakow, A., Ed., The National Academies Press, Inc., Washington, DC, p. 9, 1992.

56. Wampler, D.J., Flavor encapsulation: A method for providing maximum stability for dry flavor systems, *Cereal Chem.*, 37, 817, 1992.

57. Smit, G., van Hylckama Vlieg, J.E.T., Smit, B.A., and Wim, J.M., Fermentative compounds of flavor compounds by lactic acid bacteria, *Aust. J. Dairy Technol.*, 57, 61, 2002.

58. Feron, G. and Wache, Y., *Microbial Biotechnology of Food Flavor Production*, Taylor & Francis Group, New York, 2006.

59. Kroger, M., Kurman, J.A., and Rasic, J.L., Fermented milks-past, present, and future in applications of biotechnology, in *Traditional Fermented Foods*, Shakow, A., Ed., The National Academies Press, Inc., New York, p. 59, 1992.

60. Knorr, D., Caster, C., Dörneburg, H., Dorn, R., Graff, S., Havkin-Frenkel, D., Podstolski, A., and Werrman, U., Biosynthesis and yield improvement of food ingredients from plant cell and tissue cultures, *Food Technol.*, 47, 57, 1993.

61. Wittkowski, R., Baltes, W., and Jennings, W.G., Analysis of liquid smoke and smoked meat volatiles by headspace gas chromatography, *Food Chem.*, 37, 135, 1990.

62. Reineccius, G.A., Off flavors in foods, *Crit. Rev. Food Sci. Nutr.*, 29, 381, 1991.

63. Merkle, J.A. and Larick, D.K., Conditions for extraction and concentration of beef fat volatiles with supercritical carbon dioxide, *J. Food Sci.*, 59, 478, 1994.

64. Moyler, D.A. and Heath, H.B., Liquid carbon dioxide extraction of essential oils, in *Developments in Food Science: Flavors and Fragrances: A World Perspective*, Lawrence B.W., Mookherjee, B.D., and Willis, B.J., Eds., Elsevier Publication, Amsterdam, The Netherlands, p. 41, 1988.

65. Hu, Y., Xia, W., and Ge, C., Characterization of fermented silver carp sausages inoculated with mixed starter culture, *Lebensm. Wiss. Technol.*, 41, 730, 2008.

66. Hammes, W.P., Haller, D., and Gänzle, M.G., Fermented meat, in *Handbook of Fermented Functional Foods*, Farnworth, E.R., Ed., CRC Press, Boca Raton, FL, p. 251, 2003.

67. Ebeler, S.E., Unlocking the secrets of wine flavor, *Food Rev. Int.*, 17, 45, 2001.

68. Koprivnjak, O., Conte, L., and Totis, N., Influence of olive fruit storage in bags on oil quality and composition of volatile compounds, *Food Technol. Biotechnol.*, 40, 129, 2002.

69. Kalua, C.M., Allen, M.S., Bedgood, D.R., Bishop, A.G., Prenzler, P.D., and Robards, K., Olive oil volatile compounds, flavor development and quality: A critical review, *Food Chem.*, 100, 273, 2007.

70. McFeeters, R.F., Fermentation microorganisms and flavor changes in fermented foods, *J. Food Sci.*, 69, 35, 2004.

71. Sabatini, N. and Marsilio, V., Volatile compounds in table olives (*Olea Europaea* L., *Nocellara del Belice* cultivar), *Food Chem.*, 107, 1522, 2008.

72. Barber, S., Torner, M.J., Martinez Anaya, M.A., and Benedito de Barber, C., Microflora of the sour dough of wheat flour bread: Biochemical characteristics and baking performance of wheat doughs elaborated with mixtures of pure microorganisms, *Z. Lebensm. Unters. For.*, 189, 6, 1989.

73. Hoffmann, T. and Schieberle, P., Formation of aroma-active Strecker aldehydes by a direct oxidation of Amadori compounds, *J. Agric. Food Chem.*, 48, 4301, 2000.

74. Hansen, A., Lund, B., and Lewis, M.J., Flavor of sourdough rye bread crumb, *Lebensm. Wiss. Technol.*, 22, 141, 1989.

75. Hatchwell, L.C., Overcoming flavor challenges in low-fat frozen desserts, *Food Technol.*, 48, 98, 1994.

76. Wang, L., Park, S., Chung, J., Baik, J., and Park, S., The compounds contributing to the greenness of green tea, *J. Food Sci.*, 69, 301, 2004.

77. Wang, L.F., Lee, L.-F., Chung, J.-D., Baik, J.-H., So, S., and Park S.-K., Discrimination of teas with different degrees of fermentation by SPME–GC analysis of the characteristic volatile flavor compounds, *Food Chem.*, 109, 196, 2008.

78. Owuor, P.O., Obanda, M., Nyirenda, H.E., and Mandala, W.L., Influence of region of production on clonal black tea chemical characteristics, *Food Chem.*, 108, 263, 2008.

79. Van Osnabrugge, W., How to flavor baked goods and snacks effectively, *Food Technol.*, 43, 74, 1989.

80. Farrell, H.M. Jr., Behe, M.J., and Enyeart, J.A., Binding of para-nitrophenyl phosphate and other aromatic compounds by beta-lactoglobulin, *J. Dairy Sci.*, 70, 252, 1987.

81. O'Neill, T.E. and Kinsella, J.E., Binding of alkanone flavors to beta-lactoglobulin: Effects of conformational and chemical modification, *J. Agric. Food Chem.*, 35, 770, 1987.

82. O'Keefe, S.F., Wilson, L.A., Resurreccion, A.P., and Murphy, P.A., Determination of the binding of hexanal to soy glycinin and beta-conglycinin in an aqueous model system using a headspace technique, *J. Agric. Food Chem.*, 39, 1022, 1991.

83. O'Keefe, S.F., Resurreccion, A.P., Wilson, L.A., and Murphy, P.A., Temperature effect on binding of volatile flavor compounds to soy protein in aqueous model systems, *J. Food Sci.*, 56, 802, 1991.

84. O'Neill, T.E. and Kinsella, J.E., Effect of heat treatment and modification on conformation and flavor binding by β-lactoglobulin, *J. Food Sci.*, 52, 98, 1987.

85. Godshall, M.A. and Solms, J., Flavor and sweetener interactions with starch, *J. Food Technol.*, 46, 140, 1992.

86. Gassenmeier, K. and Schieberle, P., Formation of the intense flavor compound trans-4,5-epoxy-(E)-2-decenal in thermally treated fats, *J. Am. Oil Chem. Soc.*, 71, 1315, 1994.

87. Shin, M.G., Yoon, S.H., Rhee, J.B., and Kwon, T.-W., Correlation between oxidative deterioration of unsaturated lipid and normal-hexanal during storage of brown rice, *J. Food Sci.*, 51, 460, 1986.

88. Drost, B.W., van den Berg, R., Freijee, F.J.M., van der Velde, E.G., and Hollemans, M., Flavor stability, *J. Am. Soc. Brewing Chem.*, 48, 124, 1990.

89. Collins, Y., McSweeney, P.L.H., and Wilkinson, M.G., Lipolysis and free fatty acid catabolism in cheese: A review of current knowledge, *Int. Dairy J.*, 13, 841, 2003.

90. Hsieh, R.J. and Kinsella, J.E., Lipoxygenase generation of specific volatile flavor carbonyl compounds in fish tissues, *J. Agric. Food Chem.*, 37, 279, 1989.

91. Theerakulkait, C., Barrett, D.M., and McDaniel, M.R., Sweet corn germ enzymes affect odor formation, *J. Food Sci.*, 60, 1034, 1995.

92. McKellar, R.C., Development of off-flavors in ultra-high temperature and pasteurized milk as a function of proteolysis, *J. Dairy. Sci.*, 64, 2138, 1981.

93. Naim, M., Striem, B.J., Kanner, J., and Peleg, H., Potential of ferulic acid as a precursor to off-flavors in stored orange juice, *J. Food Sci.*, 53, 500, 1988.

94. Harlander, S., Engineering the foods of the future, *Cereal Food World*, 35, 1106, 1990.

95. Hessing, M. and van der Lugt, J.P., Enzyme-assisted synthesis of flavors, *World Ing.*, 1, 14, 1995.

96. Havkin Frenkel, D. and Belanger, F., *Biotechnology in Flavor Production*, Wiley-Blackwell, Inc., New York, p. 232, 2008.

97. Gatfield, I.L., Production of flavor and aroma compounds by biotechnology, *Food Technol.*, 40, 110, 1988.

98. Pommer, K., New proteolytic enzymes for the production of savory ingredients, *Cereal Food World*, 40, 745, 1995.

99. Whitehead, I.M., Muller, B.L., and Dean, C., Industrial use of soybean lipoxygenase for the production of natural green note flavor compounds, *Cereal Food World*, 40, 193, 1995.

100. Sun, S.Y. and Xu, Y., Solid-state fermentation for "whole-cell synthetic lipase" production from *Rhizopus chinensis* and identification of the functional enzyme, *Process Biochem.*, 43, 219, 2008.

101. Fukushige, H. and Hildebrand, D.F., A simple and efficient system for green note compound biogenesis by use of certain lipoxygenase and hydroperoxide lyase sources, *J. Agric. Food Chem.*, 53, 6877, 2005.

102. Stones, T.W., *Flavors and Fragrances: A World Perspective*, Lawrence, B.W., Mookherjee, B.D., and Willis, B.J., Eds., Elsevier, Amsterdam, The Netherlands, p. 115, 1988.

103. Madden, D., *Food Biotechnology: An Introduction*, International Life Sciences Institute Press, Brussels, Belgium, p. 10, 1995.

104. Laane, C., Generation of taste through biotechnology, *Int. Food Ingred.*, 1, 23, 1996.

105. Olempska-Beer, Z.S., Kuznesof, P.M., DiNovi, M., and Smith, M.J., Plant biotechnology and food safety, *Food Technol.*, 47, 64, 1993.

106. Shahidi, F., Flavor of muscle foods-an overview, in *Flavor of Meat, Meat Products and Sea Foods*, Blackie Academic and Professional, Inc., London, U.K., p. 1, 1998.

107. Guillen, M.D. and Manzanos, M.J., Characterization of the components of a salty smoke flavoring preparation, *Food Chem.*, 58, 97, 1997.

108. Mottram, D.S., Koutsidis, G., Oruna-Concha, M.-J., Ntova, M., and Elmore, J.S., Analysis of important flavor precursors in meat, in *Handbook of Flavor Characterization. Sensory Analysis, Chemistry, and Physiology*, Deibler, K.D. and Delwiche, J., Eds., Marcel Dekker, Inc., New York, Chap. 33, 2004.

109. Ho, C., Lee, K.N., and Jin, O.Z., Isolation and identification of volatile flavor compounds in fried bacon, *J. Agric. Food Chem.*, 31, 336, 1983.

110. Lawrie, R.A. and Ledward, D.A., *Lawrie's Meat Science*, 7th edn., Woodhead Publishing Ltd., England, U.K., p. 279, 2006.

111. Ruther, J. and Baltes, W., Sulfur-containing furans in commercial meat flavorings, *J. Agric. Food Chem.*, 42, 2254, 1994.

112. Ashurst, P.R., *Food Flavorings*, 3rd edn., Aspen Publishers, Gaithersburg, MD, p. 134, 1999.

10 Food Acids: Organic Acids, Volatile Organic Acids, and Phenolic Acids

Y. Sedat Velioğlu

CONTENTS

10.1 ORGANIC ACIDS IN FOODS: INTRODUCTION

The term "organic acid" refers to organic compounds with acidic properties. The acidity of organic acids is associated with their carboxyl group ($C(=O)OH$), and therefore they are called carboxylic acids. Carboxylic acids undergo a substitution reaction in which the –OH groups are replaced. Organic acids can be classified according to the type of carbon chain (aliphatic, alicyclic, aromatic, and heterocyclic), their extent of saturation and substitution, and the number of carboxyl groups (mono-, di-, tri-carboxylic). Monocarboxylic acids, such as formic (1C), acetic (2C), propionic (3C), and butyric (4C) acids, are highly volatile liquids with a pungent taste. Butyric acids and others that have an aliphatic chain with more than four atoms are oily compounds, and hence considered as fatty acids. Lactic acid (3C) is a monocarboxylic acid; however, it contains a hydroxyl group and is considered as an alpha hydroxyl acid. Succinic acid (4C) is a dicarboxylic acid. Lactic and succinic acids have low volatility and thus are not considered as volatile acids. Oxalic (2C), malic (4C), and tartaric (4C) acids are also dicarboxylic acids; however, the last two contain one and two hydroxyl groups, respectively. Citric acid (6C) is the best-known tricarboxylic acid with one hydroxyl group that is found in foods. Its isomer, isocitric acid, is used for authenticity determination in several juices. Pyruvic acid is a keto acid, produced in biochemical processes and is found in small amounts in microorganism-involved products, such as in yogurt and wine. Fumaric acid, which is the isomer of maleic acid, is an intermediate in the Krebs cycle and is found naturally in several plants and mushrooms. Some basic properties of organic acids found in foods are given in Table 10.1 and their chemical structures are shown in Figure 10.1. Organic acids that are present in foods come are either intrinsic, or added to foods to carry out some function (acidifier, chelator, antioxidant synergist, and antimicrobial), or produced by microorganisms. Organic acids are not considered as nutrients but they give a characteristic taste to food, increase palatability, and promote digestion. Therefore, they

TABLE 10.1
Some Properties of Organic Acids in Foods

Acid	Number of COOH Groups	Number of OH Groups	Chemical Formula
Acetic	1	—	$C_2H_4O_2$
Propionic	1	—	$C_3H_6O_2$
Butyric	1	—	$C_4H_8O_2$
Lactic	1	1	$C_3H_6O_3$
Oxalic	2	—	$C_2H_2O_4$
Fumaric	2	—	$C_4H_4O_4$
Succinic	2	—	$C_4H_6O_4$
Malic	2	1	$C_4H_6O_5$
Tartaric	2	2	$C_4H_6O_6$
Citric	3	1	$C_6H_8O_4$

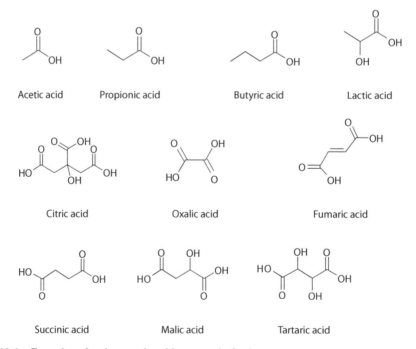

FIGURE 10.1   Formulae of main organic acids present in foods.

are one of the major contributors to the flavor, besides sugars and volatile compounds. A significant number of organic acids are involved in acid metabolism. Some organic acids decrease during ripening. In plants, a small part of them bind to cations to form salts. The amount of these salts can vary in the range of 3% in lemons or 20%–30% in pears [1–6].

Phenolic acids are described in detail in this chapter; amino acids and fatty acids are mentioned in Chapters 3 and 5, respectively.

10.1.1 FRUITS AND VEGETABLES

While some of the organic acids (e.g., malic, pyruvic, and oxalacetic) are involved in the sugar metabolism in fruits, some others (e.g., propionic acid) are converted into ethylene during ripening or storage. In fruits, the levels of acids generally decline during ripening, presumably due to their utilization as respiratory substrates. However, in banana pulp, an increase in acidity occurs during ripening. The astringent taste of unripe banana is at least partly attributed to its oxalic acid content, which undergoes significant decarboxylation during ripening, probably by the action of oxalate oxidase [7].

The most widely occurring and abundant acids in fruits are citric and malic acids. Malic acid is predominant in pomme and stone fruits, consisting of 50%–90% of all acids, while citric acid is the most abundant in berries, citrus fruits, pomegranate, pears, and tropical fruits, Table 10.2.

Besides grapes, several fruits such as avocados, cherries, lemons, and bananas contain tartaric acid. The tartaric acid content in 11 different Turkish white grape varieties was found in the range of 4.98–7.48 g/L. This was followed by citric (31–181 mg/L) and malic (1.36–3.47 g/L) acids [8]. The content of tartaric acid is quite high in rosehip (700–2600 mg/100 g). Many of the fruits and vegetables contain salicylic acid. Oxalic acid is quite common in vegetables and occurs in small quantities in some fruits [9,10].

Other acids such as succinic, oxalacetic, isocitric, lactic, fumaric, and pyruvic acids occur in small quantities as metabolic intermediates (e.g., in the tricarboxylic acid cycle, glyoxylate cycle, or the shikimic acid pathway) [11]. There are a few notable exceptions: unlike other berries, bilberry produces primarily malic acid; the avocado is deficient in the major plant acids: citric and malic. Malic acid is the predominant acid in apples, plums, cherries, and apricots. Isocitric and succinic acids are also present in many fruits. Oxalic acid is mainly found in vegetables; however, blackberries, grapefruit, grapes, lemons, oranges, pears, and plums also contain this acid. Other acids such as quinic, shikimic, lactic, glyceric, glycolic, glyoxylic, oxaloacetic, benzoic, fumaric, citramalic, alpha-ketoglutaric, pyruvic, nicotinic, mucic, and lacto-isocitric acids can be present in fruits and vegetables in small or trace amounts.

Citric and malic acids are also the predominant acids in vegetables and their relative abundance varies with the vegetable. For example, in potato, sweet potato, leguminous seeds, many leafy vegetables, tomato, and beetroot, citric acid is the main acid. However, malic acid dominates in cucurbits, lettuce, artichoke, broccoli, cauliflower, okra, onion, celery, carrot, parsnip, turnip, and green beans [12] (Tables 10.3 and 10.4).

Besides other factors such as color and firmness, soluble solids to acids ratio is a good index to determine harvest time. For prunes, this value has to be 12–15 [13]. A maximum limit was set for citric acid to isocitric acid ratio to reveal the dilution factor of juices: 130 for apricot puree and orange juice, 80 for grapefruit juice, and 185 for raspberry juice [14]. The total acid content varies widely from <0.2% in papaya and avocado [3] to 8.3% in lemon juice [15].

As in fruits, the major organic acids in vegetables are malic and citric acids, followed by oxalic acid (Table 10.5). The content of free titratable acids in vegetables is 0.2–0.4 g/100 g of fresh tissue, an amount that is low in comparison to fruits. Oxalic acid occurs in larger amounts in some vegetables.

TABLE 10.2
Malic, Citric, and Salicylic Acid Contents (mg/100 g)
in Fruits

Fruits	Acids		
	Malic	Citric	Salicylic
Apple	270–790	9–30	0.31
Apricot	700–1300	140–700	2.6
Banana	240–500	80–390	
Blackberry	860–950	15–21	
Blackcurrant	220–440	2350–3110	3.1
Blueberry	850	523	
Cherry, morello	1800		
Cherry, sweet	730–1140	10–15	0.85
Date	1260		4.5
Gooseberry	720	720	
Grape	220–650	23	1.4
Grapefruit	50–310	1000–1460	0.68
Kiwi fruit	470–530	980–1010	0.32
Lemon	200	3500–7200	0.18
Mango	74	200–327	0.11
Orange	40–190	600–1880	2.4
Peach	280–370	160–320	0.58
Pear	100–240	80–200	
Pineapple	87–100	580–670	2.1
Plum	820–1990	23–55	0.14
Pomegranate	100	500	0.07
Quince	930		
Red currant	240–640	1690–2300	5.1
Rose hip	1900–4000		
Strawberry	90–340	670–940	1.4

Source: Adapted and calculated from Souci, S.W. et al., *Food Composition and Nutrition Tables*, 6th rev, Medpharm Scientific
Publishers, Stuttgart, Germany, 2000, 1182.

Note: Blackberry: isocitric 0.81; grape: tartaric 0.39–0.67; rose hip:
tartaric 0.7–2.6; banana: acetic 6–9.6; mango: tartaric 81.
Salicylic: hazelnut 0.14; peanut 1.1; almond 3; pistachio 0.55.

10.1.2 WINE AND BEER

Tartaric acid is the main organic acid in grapes at the levels of 5–10 g/L, followed by malic acid
(2–4 g/L) and a small amount of citric acid. Other acids such as succinic, acetic, lactic, and propionic acids are formed in limited content in wines during fermentation. An increased content of
these acids indicates improper fermentation. The content of organic acids in commercial Austrian
and Spanish red and white wines is shown in Table 10.6 [16,17].

Potassium bitartrate and calcium tartrate are responsible for the physical stability of wines. A
portion of tartaric acid slowly esterifies with ethanol to form ethyl bitartrate. Malic acid is converted
to lactic acid (0–2.5 g/L) during malolactic fermentation, and the taste of wine becomes weaker.
Succinic and acetic acids are also formed during fermentation [18,19]. The content of organic acids
in vinegar and wines is shown in Table 10.7.

TABLE 10.3
Malic, Citric, and Salicylic Acid Contents
(mg/100 g) in Vegetables

Vegetables	Acids		
	Malic	Citric	Salicylic
Artichoke	170	100	
Asparagus	70–120	30–90	0.14
Aubergine	170	10	0.3
Beetroot	37	195	0.18
Broccoli	120	210	0.65
Brussel sprouts	200	240–450	0.07
Carrot	240–309	10–55	0.23
Cauliflower	201	20	0.16
Cucumber	240	20	0.78
French beans	167–188	14–32	0.11
Green peppers	60	262	1,2
Horseradish	680		
Kale	50–380	90–350	32
Kohlrabi		149–160	
Lettuce	13–130	13	
Mushroom	124	8.5	0.24
Onion	170–190	20	0.16
Parsnip	350	130	0.45
Peas	6	85	
Potato	80–104	142–650	0.12
Pumpkin	199	6,5	0.12
Radishes		100	1,2
Spinach	42	24	0.58
Sweet corn	29	21	0.13
Tomato	20–230	130–680	
White cabbage	60–600	50–150	

Source: Adapted from Souci, S.W. et al., *Food Composition and Nutrition Tables*, 6th rev, Medpharum Scientific Publishers, Stuttgart, Germany, 2000, 1182.

Note: French beans: isocitric 1.5; tomato: lactic 6, acetic 8.

The acids mentioned above and oxalic acid were found in lower amounts in beer. Their contents were as follows (mg/L): citric acid: 74–202; lactic acid: 559–631; malic acid: 40–68; oxalic acid: 12–25. Tartaric acid was not detected in three of the four samples, with the only one containing 24 mg/L [20].

10.1.3 NUTS, SEEDS, AND GRAINS

Nuts, seeds, and grains are not considered as sources of organic acids. Other than amino acids and fatty acids, they contain phenolic acids and phytic acid. Phytic acid and its salts, phytats, are regarded as the primary storage form of both phosphate and inositol in seeds, mostly within the hulls. The phytic acid content of cereals (whole grain) varies from 0.5% to 2.0% [21].

10.1.4 MILK AND YOGURT

The organic acid content of milk varies in the range of 0.12%–0.21% (av. 0.17%) (w/w), which is around 1.2% in dry matter. Distribution of organic acids in milk serum alongside with yogurt and white cheese is shown in Table 10.8 [22–24]. Citric acid (1600 mg/L) is the predominant organic acid in milk. During storage, it disappears rapidly as a result of bacterial growth. Lactic and acetic acids are degradation products of lactose. Other acids shown in the table are produced from lactose, citric acid, and the hydrolysis of fat. Milk also contains nitrogenous acidic compounds, namely orotic acid (4–30 mg/kg) and hippuric acid (4 mg/kg). Orotic acid as well as total creatinine and uric acid are suitable indicators used for the determination of the proportion of milk in foods. The average values for the whole milk and skimmed milk powders are 50.6 and 66.4 mg/100 g solids, respectively [9,22].

Generally, the production of some organic acids, such as lactic, formic, acetic, propionic, and succinic acids, in yogurt is the result of metabolic activities of starter cultures. These acids contribute to the aroma of yogurt, especially lactic acid is very important in the formation of the typical yogurt flavor. This acid gives a sharp, acidic and refreshing taste to yogurt. During the manufacture of yogurt, there is an appreciable increase in the level of some organic acids such as citric and orotic acids. The level of organic acids in yogurt is dependent on several variables such as the starter bacteria, type of milk, and incubation temperature [25,26].

TABLE 10.4
Succinic Acid Contents (mg/100 g) in Vegetables

Vegetables	Succinic Acid
Asparagus	4–13
Carrot	5.0
Mushroom	10
Potato	3.7
Sweet corn	7.7
White cabbage	1.1–130

Source: Adapted from Souci, S.W. et al., *Food Composition and Nutrition Tables*, 6th rev, Medpharm Scientific Publishers, Stuttgart, Germany, 2000, 1182.

10.1.5 COFFEE AND COCOA

Some major organic acids were determined in four different coffee samples. The content (mg/g) of organic acids in cherry coffee beans were as follows: citric, 11.5; quinic, 6.7; malic, 4.7; succinic,

TABLE 10.5
Total Oxalic Acid Content (mg/100 g) in Fruits and Vegetables

Vegetables	Total Oxalic Acid	Fruits	Total Oxalic Acid
Artichoke	8.8	Apple	0.3–0.7
Aubergine	9.5	Apricot	6.8
Beetroot	181	Blackberry	12
Brussel sprouts	6.1	Cherry (sweet)	7.2
Carrot	6.1	Cherry (morello)	4.7
French beans	44	Gooseberry	19
Green peppers	16	Grape	8
Kale	7.5	Mango	36
Onion	5.5	Pear	6.2
Parsley	0–10	Plum	12
Red cabbage	7.4	Raspberry	16
Spinach	422	Red currant	9.9
White cabbage	0–13	Strawberry	16

Source: Adapted from Souci, S.W. et al., *Food Composition and Nutrition Tables*, 6th rev, Medpharm Scientific Publishers, Stuttgart, Germany, 2000, 1182.

TABLE 10.6
Organic Acids in Wine[a] (mg/L)

Acids	White Wine		Red Wine	
Acetic	210–250	271	530–650	709
Citric	290–1170	272	890–1560	677
Lactic	410–560	4753	2660–3370	1083
Malic	2370–3170	484	0.0–30	410
Succinic	400–550	250	450–580	236
Tartaric	1190–1680	1450	1060–1310	2918

Sources: Adapted from Vonach, R. et al., *J. Chromatogr. A*, 824, 159, 1998; Mato, I. et al., *Anal. Chim. Acta*, 565, 190, 2006.

[a] For each type of wine, the data in the first column is adapted from Vonach et al., 1998 [16] and the data in the second column is adapted from Mato et al., 2006 [17].

TABLE 10.7
Concentrations (g/L) of Organic Acids in Several Vinegars

Acids	Balsamic Vinegar		Wine Vinegar	Apple Vinegar
	Not Aged	Aged		
Acetic	62.5	54.9	70.9	50.9
Citric a	0.35	1.88	0.09	0.02
Formic	0.04	0.53	0.01	0.28
Lactic	0.49	1.24	0.76	0.38
Malic	3.60	11.6	0.66	3.56
Succinic	0.40	0.78	0.51	0.27
Tartaric	1.58	2.55	0.91	Not det.

Source: Adapted from Caligiani, A. et al., *Anal. Chim. Acta*, 585, 110, 2007.

3.3, and fumaric, 0.021 [27]. Besides these acids, the presence of acetic (trace) and oxalic (trace–0.7) acids were reported [28]. Formic and acetic acids predominate among the volatile acids in coffee. Malonic, succinic, glutaric, and malic acids are only minor constituents. Itaconic, citraconic, and mesaconic acids are degradation products of citric acid, while fumaric and maleic acids are degradation products of malic acid. Organic acids present in cocoa (1.2%–1.6%) are formed during cocoa fermentation, and consist of mostly acetic, citric (0.45%–0.75%), and oxalic (0.32%–0.50%) acids. Eight brands of cocoa were found to contain 1.22%–1.64% total acids, 0.79%–1.25% volatile acids, and 0.19%–0.71% acetic acid [9].

10.1.6 Honey

Honey acidity is mainly due to organic acids whose quantity is lower than 0.5%. Acidity contributes to the honey flavor, stability against microorganisms, enhancement of chemical reactions, and antibacterial and antioxidant activities. Gluconic acid, resulting from the action of honey's glucose oxidase on glucose, provides the major contribution to acidity and is in equilibrium with gluconolactone. Other organic acids together with inorganic anions also contribute to the acidity of honey [29]. The acid level is mostly dependent on the time elapsed between the nectar collection by bees and

TABLE 10.8
Organic Acid Content (mg/100 g) of Milk,
Yogurt, and White Cheese

Acids	Milk	Yogurt	White Cheese
Orotic	0.4–3	8.92	—
Citric	160	213.3	124
Pyruvic	—	1.9	—
Lactic	2	964.9	270
Uric	—	3.58	—
Formic	4	71.8	95
Acetic	3	6.2	126
Propionic	—	6.8	54
Oxalic	2	—	—
Others	1	—	—

Sources: Walstra P., *Dairy Technology: Principles of Milk Properties and Processes,* Marcel Dekker Inc., New York, 1999, 752; Fernandez-Garcia, E. and McGregor, J.U., *J. Dairy Sci.,* 77, 2934, 1994; Akalin, A.S. et al., *Milchwissenschaft,* 54, 260, 1997.

the final honey density in the honeycomb cells. Other acids such as acetic, butyric, lactic, citric, succinic, formic, malic, maleic, and oxalic acids are also present in small amounts [9]. The organic acid contents can be used as indicators of deterioration caused by storage or aging, or even to measure purity and authenticity [29]. Among the 12 organic acids detected in the 24 broom honey samples, gluconic (16.2 g/kg), isobutyric (5.1 g/kg), and butyric (4.2 g/kg) acids were identified in the majority of samples at high concentrations. Other acids such as oxalic, pyruvic, malic, citric, succinic, fumaric, propionic, 2-oxopentanoic, and glutaric acids were also determined in some of the honeys in minor quantities [9]. The content of gluconic acid in honeys of various botanical origins was found to be 1.6–4.93 g/kg [30]. The authors have indicated that citric, pyruvic, gluconic, malic, and quinic acids can be used to define the botanical origin of honey. Among these acids, gluconic acid is produced during the transformation of nectar to honey. Formic, malic, tartaric, and citric acids were found in the nectars. The other acids are the fermentation products of honey.

10.1.7 MEAT

The predominant acid in muscle tissue is lactic acid (0.2%–0.8% on fresh meat weight basis) that is formed by glycolysis, followed by glycolic (0.1%) and succinic acids (0.05%). The pyruvate that is a generated as the end product of glycolysis is converted to lactic acid with lactic dehydrogenase, and, since the metabolic waste products cannot be removed without a blood stream, the lactic acid accumulates in the muscle. Other acids of the Krebs cycle are present in negligible amounts [9,31,32]. The aerobic mechanism in muscle produces energy from glycogen, which normally comprises about 1% of the muscle weight. When the muscle is contracting rapidly, its oxygen supply becomes inadequate for the support of ATP resynthesis via an aerobic metabolism. Under these conditions, the aerobic metabolism supplies energy for a short time, converting glycogen to lactic acid, especially after slaughtering [33]. In beef muscle, the glycogen level drops rapidly from the initial value of 56 to 10 μmol/g at 48 h, postmortem. In the same period, the lactic acid level increases from 13.1 to 82.4 μmol/g, i.e., from 0.117% to 0.735% [34]. The main constituent of organic acids in fish meat is also lactic acid, with a content of 0.2%–1.0% [35]. During the storage of fish, formic, acetic, propionic, *n*-butyric and isobutyric, *n*-valeric and isovaleric acids are formed. A study on oxidized sardine oil revealed that propionic acid followed by acetic acid were the dominant acids. Although the concentrations of butyric and valeric acids are much lower, their odor thresholds make them more important than other acids [36].

Lactic and acetic acids are used in the beef industry for the decontamination of carcasses. The effectiveness of these acids is dependent on the concentration and temperature of the acid solution, the exposure time and application pressure, the application stage in the slaughtering process, the tissue type, the type of microorganisms, and the level of initial concentration [37]. Therefore, a higher lactic and/or acetic acid concentration might be expected in meats treated with these acids.

10.1.8 Dough and Bread

In bread fermentation, yeast converts more than 95% of the glucose into carbon dioxide and ethanol. During this conversion, a small amount of pyruvic acid is produced. *Lactobacilli* present in yeast and flour may be associated with flavor development even in yeast leavened breads [38]. Besides, the other flavor compounds in breads, or mostly in sourdough breads, higher amount of lactic and acetic acids and lower amounts of citric and malic acids, are produced as a result of the activity of lactic acid bacteria. The ratio of lactic to acetic acid is important for the final flavor. The addition of sucrose to dough increases bacterial production of lactic and acetic acids [39]; however, during baking, a significant amount of these acids will be evaporated.

10.1.9 Organic Acids as Food Additives

Organic acids can be found naturally, or produced during fermentation, or intentionally added to foods during processing. Organic acids are mainly used in foods as acidifiers. They are responsible for the fresh and tart taste of fresh and processed foods, and therefore are often used to balance taste. The color of fruits is also related to the type of acids present in the product because of the reducing enzymatic browning or the increasing stability of natural pigments, especially anthocyanins and myoglobin, but not chlorophylls and carotenes. Acids stabilize color, reduce turbidity, change melt characteristics, prevent splattering, or enhance gelling. They are also indirectly effective on sterilization because of lowering pH and increasing sensitivity of bacteria or their spores to heat. Another indirect effect of acids is chelation of metals such as iron and copper, to avoid rancidity in oils and non-enzymatic browning reactions in fruit and vegetable products. Acids also act as antioxidant synergists [40,41]. Some of them act as fungicides or are more effective at inhibiting bacterial growth. The mode of action of organic acids is related to the pH reduction of the substrate, the acidification of internal components of cell membranes by ionization of the undissociated acid molecule, or the disruption of substrate transport by the alternation of cell membrane permeability. The undissociated portion of the acid molecule is primarily responsible for antimicrobial activity; therefore, the effectiveness depends on the dissociation constants (pK_a) of the acid [42]. The limits for acidifiers in related food groups are given in detail in the Codex Alimentarius documents under the title "GSFA provisions," which is available online. The information about a specific organic acid can be searched by using names, or the INS number or the CAS number. A database of the properties of additives is also available on CD-ROM [43]. Detailed information about individual acidifiers is given in [41]. The sourness of organic acids differs from each other. The relative amounts of selected organic acids that give the same sourness and their taste thresholds are shown in Table 10.9 [44,45]. Brief information about acids that can be used as additives in foods is given below:

Citric acid: International Numbering System (INS) No: 330; FW: 192.13; Chem. name: 2-hydroxy-1,2,3-propenetricarboxylic acid. Citric acid is a premier acid in the food industry. The functional uses of citric acid are acidulant, sequestrant, and antioxidant synergist. It is a flavor enhancer [46] and a leavening agent [43]. It may be produced from sources such as lemon or pineapple juice or

TABLE 10.9
Median Taste Thresholds (N) and Relative
(Anhydrous Citric Acid = 100) Amounts to
Give Same Sourness for Selected Organic Acids

Acids	Thresholds (N)	Amounts to Give Same Sourness
Acetic	0.0018	
Citric	0.0023	109[a]
Tartaric	0.0012	85 ± 5
Formic	0.0018	
Lactic	0.0016	130 ± 10
Malic	0.0016	90 ± 5
Oxalic	0.0026	
Succinic	0.0032	
Fumaric		75 ± 5

Sources: Adapted from Amerine, M.A. et al., *Principles of Sensory Evaluation of Food*, Academic Press, New York, 1965, 30–144; Lange, H.J., *Untersuchungs-methoden in der Konservenindustrie*, Verlag Paul Parey, Berlin und Hamburg, 1972, 298.

[a] Monohydrate; taste threshold for hydrochloric acid = 0.0009.

by fermentation using *Candida* spp. or non-toxicogenic strains of *Aspergillus niger*. The acceptable daily intake (ADI) is not limited for citric acid and for its calcium, potassium, sodium, and ammonium salts. It has been demonstrated that citric acid is more effective than acetic and lactic acids for inhibiting the growth of thermophilic bacteria [42]. Citric acid has been used in a broad range of foods including fruit products, dairy industry, confectionary, etc.

Malic acid: (INS No: 296; FW: 134.09) Chem. name: DL malic acid; 2-hydroxybutanedioic acid, hydroxy succinic acid. Malic acid is the second most popular general purpose food acid, although only less than one-tenth of the quantity of citric acid is used in food. The functional uses include acidity regulator and as a flavoring agent [46]. The ADI is not limited. The usage of malic acid is similar to that of citric acid.

Tartaric acid: INS No: 334, FW: 150.09; Chem. name: L-tartaric acid, L-2,3-dihydroxybutonedioic acid, L-2,3-dihydroxybutonedioic acid, L-2,3-dihydroxysuccinic acid. Even though tartaric acid can be used in a broad range of foods, its usage is limited to grape-flavored foods. Tartaric acid is extracted from the residues of the wine industry.

The functional uses include antioxidant, acid, sequestrant, and flavoring agent. An ADI of 0–30 mg/kg dw was established [46].

Acetic acid: INS No: 260, FW: 60.05; Chem. name: ethanoic acid. The preservation of foods with acetic acid is an old technique that dates back to centuries. It is the main acid of vinegar and constitutes approximately 4%. It can be produced by fermentation or also by the oxidation of acetaldehyde, methanol, and carbon dioxide. The functional uses are as an acidifier and flavoring agent. The ADI is not limited for acetic acid and for its potassium and sodium salts [46]. Potassium, sodium, and calcium salts of acetic acid are commonly used.

Lactic acid: INS No: 270, FW: 90.08, Chem. name: 2-hydroxypropanoic acid, 2-hydroxypropionic acid. Lactic acid is primarily found in milk products, mainly in yogurt, kefir, etc. Lactic acid is also produced in muscle. It is commercially obtained by lactic fermentation of sugars or prepared synthetically. The functional use is as an acidifier and can be used in food in desired amounts (*quantum satis*) directly or as its salts [46].

Fumaric acid: INS No: 270, FW: 116.07, Chem. name: *trans*-butanedioic acid, *trans*-1,2-ethylene-dicarboxylic acid. Fumaric acid is naturally found in rice, sugar cane, wine, plant leaves, bean sprouts, and edible mushrooms in small amounts (is an intermediate in the tricarboxylic acid (Krebs) cycle), and is also produced in sun-exposed skin. The functional uses are as an acidity regulator and a flavoring agent. The ADI is not specified for fumaric acid. The ferrous salt of fumaric acid can be used as iron fortifier. The common application areas include bakery, confectionary, and drinks. It can also be used by mixing with other acidifiers.

Propionic acid: INS No: 280, FW: 74.08, Chem. name: propanoic acid, ethylformic acid. It occurs in food as a result of natural processing. It can be found in some cheeses up to 1%, produced by several bacteria or fungi [46]. The functional uses include preservative, antimold, antirope agent, flavoring agent. The ADI is not limited for propionic acid. Its activity is pH dependent; and it is commonly used in packaged bread in the quantity of 1–3 g/kg [47].

Adipic acid: INS No: 355, FW: 146.14, Chem. names: hexanedioic acid, 1,4-butanedicarboxylic acid. Adipic acid is not commonly used in the food industry. Functional uses include acidity regulator, flavoring agent, firming agent, raising agent, and stabilizer. An ADI of 0–5 mg/kg dw was established. It is used mostly in bakery in the range of 1–2 g/kg. Ammonium (E 359), potassium (E 357), and sodium (E 356) salts are also used as additives [47].

Phosphoric acid: INS No: 338, FW: 98; other chemical name: orthophosphoric acid. Phosphoric acid is an inorganic acid, the second most common acidifier (after citric acid—60%) used in the food industry—around 25% with its salts. All other acidifiers together account for 15% [9]. Phosphoric acid can be used in foods as an acidulant, sequestrant, and antioxidant synergist. The quantity used varies in a broad range of 1–20 g/kg (up to 2 g/L in beverages). No ADI has been established for phosphoric acid, however the maximum tolerable daily intake (MTDI) from all sources is limited to 70 mg/kg dw.

10.2 VOLATILE ACIDS

The term volatile acids refer to steam-distillable acids. The primary volatile acid in foods is acetic acid, and volatile acidity is often expressed as acetic acid. Propionic, formic, and butyric acids are also volatile acids found in some foods. Lactic and succinic acids are slightly steam distillable and therefore may not be considered as a part of the volatile acids.

The most significant food, which contains volatile acids, is vinegar. It contains high amounts (4%–8%) of volatile acids, mainly acetic acid, as compared to other foods. Organic acid composition of several vinegars is shown in Table 10.7 [48]. The volatile acid content of sound fruits and vegetables is negligible. Volatile acids are by-products of microbial metabolism. Acetic acid bacteria (e.g., *Acetobacter aceti*) can convert glucose to ethanol and acetic acid. Several yeasts (*Klockera, Hansenula*, and *Saccharomyces*) and most lactic acid bacteria can produce acetic acid. Formic and acetic acids can be formed in considerable amounts during mashing in fruit juice production. The amount can be 30–70 mg/L as acetic acid in juices produced from sound fruits. In grape juice produced from sound grapes, the formic and acetic acid contents were found to be 18 and 23 g/L, respectively. In apple juice the formic acid content was 16–26 mg/L [3].

The quantity of acetic acid formed during alcoholic fermentation usually does not exceed 0.3 g/L in wine. The U.S. limits for volatile acids in wine are 1.2 and 1.1 g/L for red and white table wines, respectively. The aroma threshold for acetic acid in red wine varies from 0.6 to 0.9 g/L. Elimination of air and the use of sulfur dioxide will limit the increased amount of acetic acid in wine. Formic acid is usually found in diseased wines, propionic acid is usually found in traces in old wines. On the contrary, the production of acetic acid is desired in vinegar production. The acetic acid bacteria convert the alcohol into acetic acid by the process of oxidation. The alcohol content of the solution used in vinegar production should be in the range 10%–13% [30]. In addition to acetic acid, pickled vegetables contain several volatile acids. Acetic acid is commonly added to cucumber fermentation brines to suppress the growth of undesirable microorganisms. Propionic and butyric acids can be produced from lactic acid during anaerobic spoilage of salt stock cucumbers. Succinic acid is one of the less volatile acids, which may be formed from malic acid by lactic acid bacteria, and lactic acid is formed from sugar fermentation and malolactic degradation of malic acid [49]. In spoiled fermented cucumber, the acids formed during spoilage were (in the order of concentration) acetic acid > butyric acid > propionic acid. The lactic acid does not remain [50]. An upper limit was set for lactic and volatile acids (as acetic acid) as 0.5 and 0.4 g/L, respectively in apple, grape, grapefruit, orange, and pear juices. The excess content of these acids is considered as an indicator of fermentation [14]. Additionally, volatile acids have been mentioned already for some of the above foods.

10.3 PHENOLIC ACIDS

10.3.1 INTRODUCTION

Polyphenolic compounds are widely distributed in all plant organs, and therefore are an integral part of the human diet. The term phenolics encompass over 8000 naturally occurring compounds [51]. Phenolic compounds possess at least one aromatic ring with one or more hydroxyl groups attached to it. It has been [52] indicated that polyphenols can be divided in at least 10 different classes depending on their basic chemical structure, including low (phenolic acids), intermediate (flavonoids) or high (hydrolysable or condensed tannins) molecular weight compounds [53]. Public interest to food phenolics has increased because of the antioxidant and free radical scavenging abilities associated with some phenolics and their potential effects on human health [54].

The presence of polyphenols in plant foods was largely influenced by genetic factors.

The biosynthesis and accumulation of these metabolites are endogenously controlled by the process during the developmental differentiation, or regulated by exogenous factors [55]. Other factors, such as germination, degree of ripeness, temperature, processing, and storage also influence the content of phenolics in plants [54,56,57]. For instance, UV light increased the phenolic acid content in tomato [58]. Different washing treatments had no effect on the phenolic content in iceberg lettuce, but wounding induced the synthesis of phenylalanine ammonia lyase to produce a phenolic acid (caffeic acid) [59].

Phenolic acids constitute one of the major parts of phenolics. They account for about one-third of dietary phenolics. Phenolic acids are the phenols possessing one carboxylic acid. They are secondary metabolites that were commonly found in plant-derived foods. These compounds are effective on sensory and nutritional properties of plant foods [54]. Like other phenolic compounds, phenolic acids also contribute to the off-flavor and undesirable taste such as sour, bitter, astringent, and phenolic-like. Taste thresholds of phenolic acids in foods can vary in a wide range (i.e., 300 ppm for protocatechuic acid and 240 ppm for syringic acid). Phenolic acids are partly responsible for discoloration because some of these compounds are polyphenoloxidase substrates and involved in enzymatic browning in the presence of oxygen and turn to o-quinones, which later constitute the brown-colored pigments. Also they contribute to darkening by forming metal complexes with ions like iron and copper [60,61]. Procyanidins are the most significant phenolic fraction with sensory characteristics. Studies reported in the literature show that the contribution of phenolic acids to astringency is small [62]. Besides acting as a pro-oxidant in certain circumstances, phenolic compounds also interact with proteins, vitamins, and minerals [63,64]. The simultaneous consumption of ≥1 cup of coffee, a major source of caffeic acid, inhibits the absorption of iron in the meal at the rate of 40% [65].

Phenolic acids can be in a free or bound form in plant cell, but are mostly in the bounded form. Esters and amides are the most frequently reported forms, whereas glycosides rarely occur [66], in which the hydroxycinnamic acids being the more common. For example, in brown rice and millet rice, bound phenolic acids comprise 80%–90% and 53%–74% of the total phenolic acids, respectively [67]. Hydroxybenzoic acids are the main components of complex structures like hydrolysable tannins and lignin. They are usually covalently linked to the cell wall polysaccharides or they directly consist of the cell wall fraction, like lignin [54,68].

Two classes of phenolic acids can be distinguished: derivatives of benzoic acid and derivatives of cinnamic acid. Hydroxybenzoic and hydroxycinnamic acid derivatives contain C_6–C_1 and C_6–C_3 structures, respectively. Hydroxybenzoic acids and their aldehydes (e.g., vanillin, syringaldehyde, p-hydroxybenzaldehyde) are fairly common in higher plants. These compounds can be found free, although their corresponding methyl and ethyl esters and glycosides occur very commonly in free and/or bound forms [54]. Hydroxycinnamic acids are the most important group in phenylpropanoids. Representative structures of hydroxybenzoic and hydroxycinnamic acids are shown in Figure 10.2 [69]. Hydroxybenzoic acids are also components of gallotannins or ellagitannins [70]. Hydroxybenzoic acids are the simplest structures within the polyphenolic compounds. Besides their health effects, these compounds have significant effects on taxonomic studies [53].

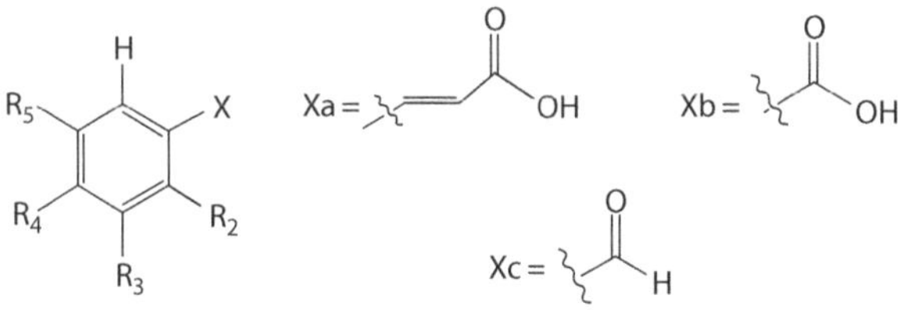

Common Name	R_2	R_3	R_4	R_5	X
Cinnamic acids					
Cinnamic acid	–H	–H	–H	–H	a
o-Coumaric acid	–OH	–H	–H	–H	a
p-Coumaric acid	–H	–H	–OH	–H	a
m-Coumaric acid	–H	–OH	–H	–H	a
Ferulic acid	–H	–OCH$_3$	–OH	–H	a
Sinapic acid	–H	–OCH$_3$	–OH	–OCH$_3$	a
Caffeic acid	–H	–OH	–OH	–H	a
Benzoic acids					
Benzoic acid	–H	–H	–H	–H	b
Salicylic acid	–OH	–H	–H	–H	b
p-Hydroxybenzoic acid	–H	–H	–OH	–H	b
Vanillic acid	–H	–OCH$_3$	–OH	–H	b
Syringic acid	–H	–OCH$_3$	–OH	–OCH$_3$	b
Protocatechuic acid	–H	–OH	–OH	–H	b
Gentisic acid	–OH	–H	–H	–OH	b
Gallic acid	–OH	–OH	–OH	–OH	b
Veratric acid	–H	–OCH$_3$	–OCH$_3$	–H	b
Aldehydes					
Syringaldehyde	–H	–OCH$_3$	–OH	–OCH$_3$	c
Vanillin	–H	–OCH$_3$	–OH	–H	c

FIGURE 10.2 Structures of prominent phenolic acids. (From Robbins, R.J., *J. Agric. Food Chem.*, 51, 2866, 2003.)

Hydroxycinnamic acids are the most abundant members of dietary hydroxycinnamates including ferulic, caffeic, and p-coumaric acids which are the most widespread ones. Of these acids, caffeic acid is ester linked to quinic acid (Figure 10.3) to form chlorogenic acids and found at high levels in coffee [66,71]. The most common esters are caffeic, coumaric and ferulic acids with D-quinic and, in addition, with D-glucose. Since quinic acid have four OH groups, four bonding possibilities exist [9]. After the invention that there are many hydroxycinnamic acid esters with quinic acid, the term *chlorogenic acids*, is being used as a general term to describe esters formed between certain trans cinnamic acids and quinic acid. The neochlorogenic (3-o-caffeoylyquinic acid) and cryptochlorogenic acids (4-o-caffeoylyquinic acid) are the most common isomers of chlorogenic acid (5-o-caffeoylyquinic acid); they are shown in Figure 10.3 [71]. Cinnamic acids may also be conjugated to other molecules including certain hydroxy acids, amino compounds, polysaccharides, and lipids [79,81]. Esterification of hydroxycinnamic acids with organic acids is also quite common. Tartaric acid esters of several hydroxycinnamic acids are shown in Figure 10.4 [72]. Derivatives of phenolic acids differ in patterns of the hydroxylations and methoxylations of their aromatic rings [73–75]. As a general rule, hydroxybenzoic acids are present as conjugates and the content of individual hydroxybenzoic acids rarely exceeds 0.1 mg/L in juices [76]. Unlike hydroxycinnamates, hydroxybenzoic acid derivatives are mainly present in foods in the form of glycosides; p-hydroxybenzoic, vanillic, and protocatechuic acids are the most common forms [70,73–76]. The role of phenolic

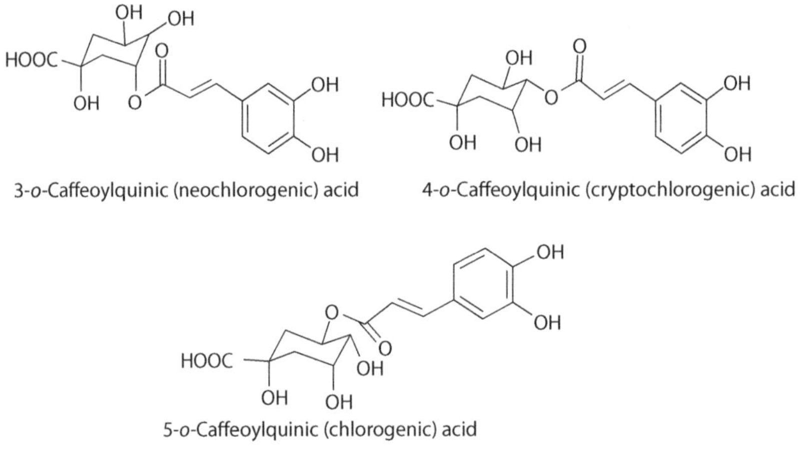

3-o-Caffeoylquinic (neochlorogenic) acid   4-o-Caffeoylquinic (cryptochlorogenic) acid

5-o-Caffeoylquinic (chlorogenic) acid

FIGURE 10.3   Chlorogenic acid isomers.

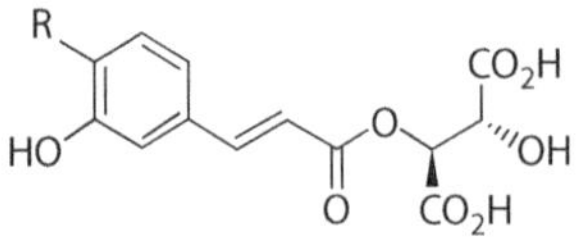

Hydroxycinnamic Acid Ester	R
trans-Caffeoyltartaric acid (caftaric acid)	OH
trans-p-Coumaroyltartaric acid (cutaric acid)	H
trans-Feruloyltartaric acid (fertaric acid)	OCH_3

FIGURE 10.4   Tartaric acid esters of several hydroxycinnamic acids.

acids in plants is not well known, they have been connected with diverse functions, including nutrient uptake, protein synthesis, enzyme activity, photosynthesis, structural components, and allelopathy. Cinnamic and benzoic acid derivatives exist in all plant foods and dispersed throughout all parts of the plant [69].

The biosynthetic pathway (Figure 10.5) of polyphenols, including phenolic acids is well known. Phenylalanine formed in plants via the shikimate pathway is a common precursor for most of the phenolic compounds. Forming hydroxycinnamic acids from phenylalanine requires hydroxylation and methylation steps. The formation of hydroxybenzoic acids is simple and they can directly be formed from the corresponding hydroxycinnamic acids with the loss of acetate or with an alternate path stemming from an intermediate in the phenylpropanoid pathway [69,77,78].

It is quite difficult to estimate daily phenolic acid consumption because of different nutritional habits of populations or individual preferences. The total intake of polyphenols was estimated at approximately 1g/day and phenolic acids account for about one third (ca. 330 mg/day) of this intake. Another independent estimation for phenolic acids was at 200 mg/day. Regularly consumed beverages such as wine, coffee, beer, and tea provide a major part of phenolic acids in the diet [61,79,80]. Fruits, fruit juices, and chocolate also constitute a significant part in the diet. Clifford estimates

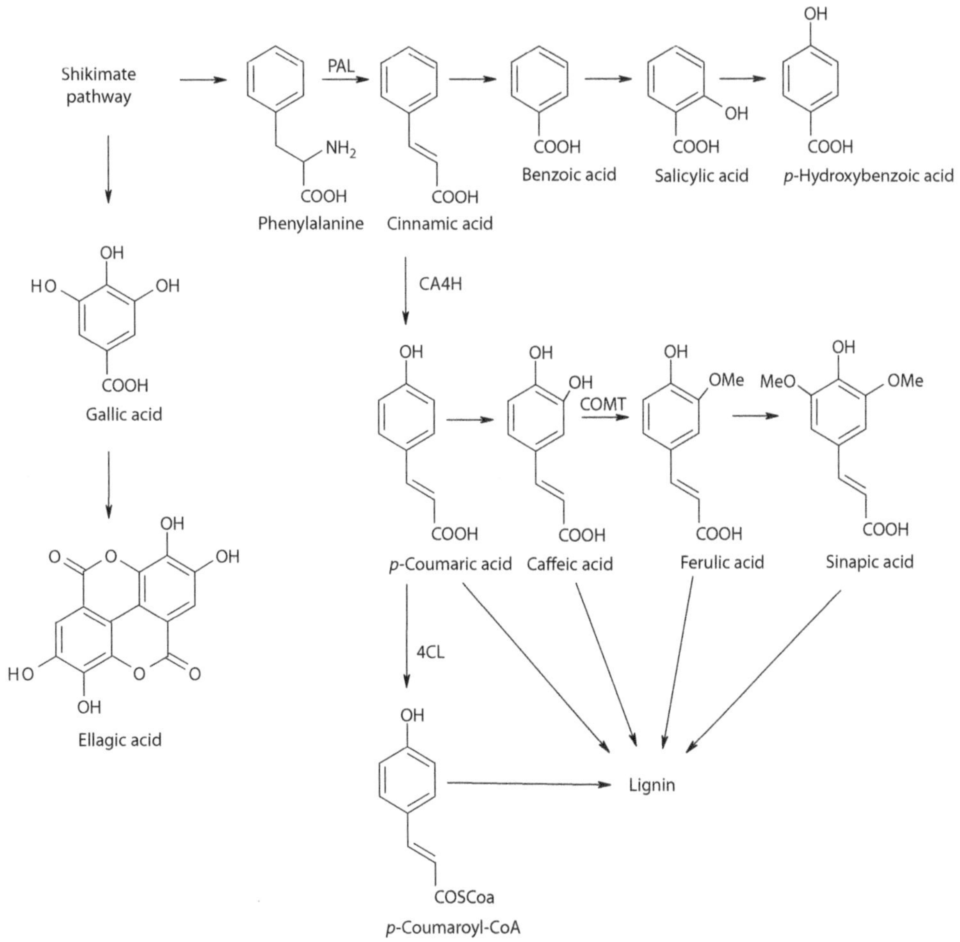

FIGURE 10.5 Biosynthesis of phenolic acids. (Adapted from Duthie, G. and Crozier, A., *Curr. Opin. Lipid*, 11, 43, 2000; Hakkinen, S., Flavonols and phenolic acids in berries and berry products, Doctoral dissertation, Kuopio University Publications, Kuopio, Finland, *D. Med. Sci.*, 221, 92, 2000.)

that the daily cinnamate intake could be as high as 1 g/day, especially in people consuming certain amount of coffee, brans, and citrus fruits [81]. The total consumption of phenolics only from fruit and vegetables was estimated at 473 mg/day in United States [82,83]. Chen et al. calculated this amount as 111.7 mg/day [84]. The daily hydroxycinnamic and hydroxybenzoic acid consumptions have been estimated at 211 mg/day (206 of which was caffeic acid from coffee only) and 11 mg/day, respectively in Germany [85]. It can be concluded from these data that the daily consumption of phenolic acids varies in a broad range.

10.3.2 ANALYTICAL TECHNIQUES FOR THE DETERMINATION OF PHENOLIC ACIDS IN FOODS

Several methods were implemented for determination of phenolic acids in foods, including thin layer chromatography, gas chromatography, and high performance liquid chromatography (HPLC). HPLC is currently a major choice of researchers because of its versatility, sensitivity, and precision. Phenolic compounds consist of a phenol ring and show intense absorption in the UV region; therefore HPLC with diode array detector becomes the most useful technique and has almost swept away all other techniques within a decade. A HPLC-MS spectrum of several phenolic acids is shown in Figure 10.6 [86]. Structures and determinations of all chlorogenic acids, including dicaffeoylquinic and caffeoylferuloylquinic acids by HPLC-MS were given by [81]. A sample spectrum for several chlorogenic acids is given in Figure 10.7 [81].

The analytical chemistry of phenolic compounds was summarized in detail by several authors including [53,69,87,88]. From these articles it can be concluded that the extraction of phenolic acids can be achieved using methanol and ethanol (mostly with different proportions of water), ethyl acetate, acetone, chloroform, pressurized low-polarity water [89] or, rarely, with water itself. In some samples, such as in wine or clear juices, direct HPLC injection can be applied after several minor alterations [90]. But in a majority of the studies, aqueous methanol or ethyl acetate was used as extraction solvent. As a general treatment, free phenolics were extracted with diethylether from the extract and the water phase was treated with HCl or in common with NaOH under nitrogen to liberate the insoluble-bound esterified phenolic acids. The use of solid-phase cartridges is also another application to remove undesired compounds. To avoid losses in

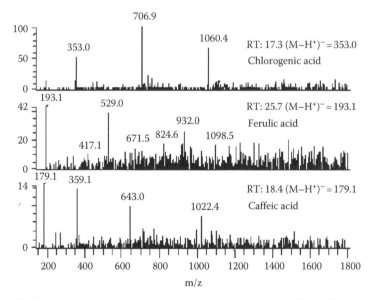

FIGURE 10.6 HPLC-MS spectra of chlorogenic, ferulic, and caffeic acids. (From Simonovska, B. et al., *J. Chromatogr. A*, 1016, 89, 2003. With permission.)

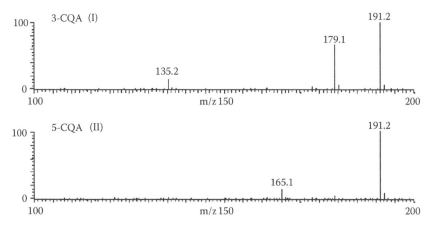

FIGURE 10.7 HPLC-MS spectra of 3-*o*-caffeoylquinic (3-CQA) and 5-*o*-caffeoylquinic (5-CQA) acids. (Adapted from Clifford, M.N., *J. Sci. Food Agric.*, 79, 362, 1999. With permission.)

hydroxycinnamic acid derivatives during alkaline hydrolysis, the addition of 1% ascorbic acid or EDTA is recommended [91] and 2 mM NaF is used to inactivate the polyphenoloxidases and prevent phenolic degradation due to browning. Enzymatic hydrolysis with β-glucosidase can also be applied.

The use of C_{18} RP columns and the gradient technique using solvents such as diluted acetic or formic acid along with acidified methanol were usually preferred by most workers [53,78,88,92], who clearly separated nine phenolic acid standards in less than 40 min using 11:89 acetonitrile–water (Figure 10.8). Others [84,93,94] also achieved resolution on separation of the phenolic acids using HPLC. With the help of solid-phase extraction cartridges [94], successfully determined 10 phenolic acids in apple and pear with HPLC along with some other phenolics in less than 60 min using 70% acetone as an extraction solvent. Detailed data about separation and determination of several chlorogenic acids and a tartaric acid derivative is available through [71,81,95–97].

Besides comparing with authentic standards, the UV data acquired from diode array detector [87,93] is frequently used for the determination of individual phenolic acids and derivatives. Most of the benzoic acid derivatives display their absorption maxima at 246–262 nm with a shoulder at 290–315 nm, except gallic and syringic acid, which have absorption maxima at 271 and 275 nm, respectively. The cinnamic acids, due to the additional conjugation, absorb in two regions, one in the range 225–235 nm, and the other between 290 and 330 nm [87,69].

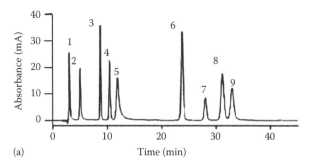

(a)

FIGURE 10.8 HPLC chromatogram of several phenolic acids. (Adapted from Mas, S. et al., *Talanta*, 71, 1455, 2007. With permission.)

Mass spectrometric data to identify phenolic acids is getting more prevalent. Carini et al. [96] and Kammerer et al. [98] published valuable data about the determination of several phenolic acids and derivatives by HPLC/MS, in mate and black carrot, respectively. The data about the determination of phenolic compounds in wine was reviewed by Monagas et al. [72]. Positive ion thermospray LC/MS using a discharge electrode to effect solvent-mediated chemical ionization yields $[M + H]^+$ ions and some fragmentation to principal losses of H_2O, CO, and CO_2, when applied to the analysis of simple phenolic acids [99]. Electrospray LC/MS is a more promising method and used successfully to determine phenolic acids and conjugates. Mass spectral data of several phenolic acids in flow injection analysis particle beam electron impact (FIA-PB-EI) MS is shown in Table 10.10 [100].

TABLE 10.10
Absorption Maxima and MS Data of Phenolic Acids and Related Compounds

Phenolic Acids	Absorbance Max (nm)		MS Data (From Bocchi et al. [100])	
	From Lee [87]	From Ho et al. [93]	[M]⁺	[M − OH]⁺
Benzoic Acids				
4-Hydroxybenzoic (*p*-hydroxybenzoic)	256	252	138	121
3-Hydroxybenzoic (*m*-hydroxybenzoic)	237, 296			
2-Hydroxybenzoic (salicylic)	242, 301	300	138	—
2,6-Dihydroxybenzoic (γ-resorcylic)	247, 306			
2,6-Dimethoxybenzoic	243, 280			
2,6-Dihydroxybenzoic (gentisic)	236, 327	328	154	137
β-Resorcylic (2,4-dihydroxybenzoic)	256, 294			
2,4-Dimethoxybenzoic (veratric acid)	256, 292		182	165
2,3-Dihydroxybenzoic	245, 314			
3,4-Dihydroxybenzoic (protocatechuic)	258, 294	258	154	137
4-Hydroxy-3-methoxybenzoic (vanillic)	260, 292	258, 290	168	151
3,4-Dimethoxybenzoic	260, 292			
3,5-Dihydroxybenzoic (α-resorcylic)	249, 306			
3,5-Dimethoxybenzoic	249, 305			
3,4,5-Trihydroxybenzoic (gallic)	271	270	170	153
4-Hydroxy-3,5-dimethoxybenzoic (syringic)	275		198	181
2,4,6-Trihydroxybenzoic	256, 293			
Ellagic acid		290, 366		
Cinnamic Acids				
3,4-Dihydroxycinnamic (caffeic)	234, 290	294, 322	180	163
4-Hydroxycinnamic (*p*-coumaric)	232, 298	310	164	147
4-Hydroxy-3,5-dimethoxycinnamic (sinapic)	238, 323		224	207
4-Hydroxyl-3-methoxycinnamic (ferulic)	236, 292	296, 342	194	177
3-Hydroxycinnamic (*m*-coumaric)	235, 279			
2-Hydroxycinnamic (*o*-coumaric)	234, 276		164	147
Derivatives				
p-Hydroxybenzaldehyde		283		
Chlorogenic acid (5-caffeoylquinic)		298, 324		
Vanillin (4-hydroxy-3-methoxybenzaldehyde)		278, 308		
Syringaldehyde		306		

10.3.3 Distribution of Phenolic Acids in Foods

The distribution of phenolic acids in plants is not homogeneous. They are mainly localized in the cell wall. Phenolic acid distribution changes during ripening and storage. The processing can result in an increase in some phenolic compounds; especially in foods rich in phenolic compounds. For example, raspberry juice obtained by the diffusion extraction method and stored at 65°C for a long period resulted in increased ellagic acid content. Jam making also increases the amount of this compound because of the release of ellagic acid from the cell wall by the hydrolysis of ellagitannins [101].

The predominant hydroxycinnamic acid in fruits is caffeic acid constituting more than 75% of the total hydroxycinnamic acids in apples, apricots, blackberries, plums, and tomatoes. However, p-coumaric acid is the major hydroxycinnamic acid in citrus fruits and pineapple [64]. Probably the most well-known, bound hydroxycinnamic acid is chlorogenic acid, which is combined from caffeic and quinic acids. Chlorogenic acids are generally the most abundant phenolic acids and constitute 75%–100% of the total hydroxycinnamic acid content of most fruits. Also, a single cup of coffee may contain up to 350 mg chlorogenic acid [70]. Therefore, it can be concluded that the most important phenolic acid is caffeic acid in human diet. Another major source of phenolic acids in diet is ferulic acid, which is covalently linked to plant cell walls and is especially abundant in an insoluble form in cereal brans [66]. Excluding several fruits such as raspberry, blackberry, red currant, strawberry, and blackcurrant, and vegetables such as horseradish and onions, the amount of hydroxybenzoic acids is usually very low. These fruits and vegetables mainly contain gallic, p-hydroxybenzoic, and protocatechuic acids. The ratio of chlorogenic acid and isomers, namely, neochlorogenic and cryptochlorogenic acid was found as 3.9:78.7:18.4 in prune [71]. Concentrations of neochlorogenic and 3-p-coumaroylquinic acid in sweet cherries ranged from 19.5 to 53.0 and from 7.5 to 50.6 mg/100 g fw, respectively [95]. The p-coumaroylquinic acid content in cherries and plums was found as 40–450 and 4–40 mg/kg, respectively. The content of glucosides of caffeic and p-coumaric acid in tomato was 15–18 and 19–68 mg/kg, respectively [74]. Total hydroxycinnamate content (mg/kg fresh weight) of several fruits summarized by [102] is as follows: apple 263–308, bilberry 170–347, blueberry 226–315, blackcurrant 104–167, cherry 100–1900, cranberry 191, red grapes 5–19, red wine 4–13, white grapes 5.5, white wine 1–34, orange 136–163, peach 54–148, plum 500–900, red raspberry 3–35, and strawberry 14–69. European cranberrybush (*Viburnum opulus* L.) berries are the major source of chlorogenic acid, containing 2037 mg/kg chlorogenic acid which consists 54% of the total phenolics. The astringency that is mostly caused by this compound can easily be removed from the juice using activated carbon [103]. Cinnamates and hydroxybenzoic acid derivatives in foods were well documented by [81,104]. These articles might be beneficial for those who need to be informed about the nature, occurrence, and dietary burden of these compounds. The total phenolic acid content (the sum of p-coumaric, ferulic, and sinapic acids) varies with maturation. In rye caryopses, phenolic acid content was the highest (44.24 µg/g dw) at the initial stage of development and the lowest (6.52 µg/g dw) in matured grains [105]. The phenolic acid content in various foods is summarized in Table 10.11.

10.3.4 Structure–Antioxidant Activity Relationship in Phenolic Acids

The antioxidant activity of phenolic compounds exists from their ability to donate a hydrogen atom to a free radical from an aromatic hydroxy (OH) group and the ability of an aromatic compound to support an unpaired electron as a result of the delocalization around the electron system [77]. The antioxidant activity of phenolic acids can be considered as moderate within all phenolic compounds, while that of tannins was the strongest and that of most quinones, isoflavones and lignans tested were the weakest. The radical scavenging activity of phenolic acids depends on the number and position of the hydroxy groups and methoxy substitutions in the molecules. Benzoic acid and cinnamic acid, without the hydroxy group had no activity. The radical scavenging activity of hydroxybenzoic acids

TABLE 10.11
Phenolic Acid Contents in Foods (mg/kg or mg/L)

Fruit/Vegetable	p-Coumaric	Caffeic	Ferulic	Chlorogenic	Sinapic	Neo-Chlorogenic
Fruits						
Apple					50–500[3]	
Apple pulp-juice	25[2]		<0.5[2]	81[2], 16.6[6]	9.0[2]	
Apricot				37–123[5]		26–132[5]
Avocado	4[4]		3.2[4]	5.6[4]	2.2[4]	
Blueberry	6–20[1]	1860–2090[1]	13–14[1]			
Cherries				180–1150[3]		200–1300[2]
Currant, black	20–44[1]	68–84[1]	18–24[1]			
Currant, red	5–15[1]	3–8[1]	1–3[1]			
Currant, black						
Grape juice	0.3[2]	0.4–2.3[2], 9.5[6]	0.2[2]			
Kiwi		600–1000[3]				
Pear juice (commercial)	0.08–0.12[2]	0.01–0.07[2]		1.14–7.88[2]		
Pear					15–600[3]	
Peach				43–282[5]		33–142[5]
Plum	140–1150[3]			15–142[5]	15–142[5]	88–771[5]
Raspberry, red	6–25[1]	4–10[1]	3–17[1]			
Raspberries				60–100[3]		
Strawberry	7–27[1]	Max 7[1]	2[1]			
Orange juice, commercial			0.54[2]	2.0[2]		
Vegetables						
Artichoke				433[5]	450[3]	
Eggplant			600–660[3]	575–632[5]	310[4]	
Carrot		1[4]		23–121[5]	100[4]	
Cauliflower	3.1[4]			1.4[4]	1.5[4]	
Green bean					2.9[4]	
Jerusalem artichoke					76[4]	
Potato	30[4]	260[4]	94[4]	22–71[5]	100–190[3]	8.6[4]
Spinach, frozen					1.4[4]	

Others

p-Hydroxybenzoic: blueberry 5–6[1]; currant-red 9–13[1]; strawberry 10–36.[1] Gallic: blueberry 3–7[1]; currant-black-red 3–13[1]; grape juice 0.7,[2] 12.5[6]; raspberry-red 15–59[1]; strawberry 1–44.[1] Ellagic: apple pulp-juice 0.6[6]; currant-black 4–11[1]; grape juice 2.5[6]; red raspberry 19–38[1]; strawberry 90–402.[1] Protocatechuic: carrot 4–6[4]; green bean 2.6[4]; spinach-frozen 1.1.[4]

Parsnip: sinapic 11[4]; soy bean: sinapic 20[4]; tomatoes: chlorogenic 12–71[2,5]; turnip: ferulic 4.2[4]; sinapic 2.1[4]; beet-red: sinapic 1.9[4]; broccoli: neochlorogenic 58[5]; cabbage-Chinese: chlorogenic 1.4[4]; lemon: chlorogenic 5.8.[2]

Sources: Hakkinen, S., Flavonols and phenolic acids in berries and berry products, Doctoral dissertation, Kuopio University Publications, Kuopio, Finland, *D. Med. Sci.*, 221, 92 pp, 2000; Escarpa, A. and Gonzalez, M.C., *Crit. Rev. Anal. Chem.*, 31, 57, 2001; Manach, C. et al., *Am. J. Clin. Nutr.*, 79, 727, 2004; Mattila, P. and Hellstrom, J., *J. Food Comp. Anal.*, 20, 152, 2007; Herrmann, K., *Crit. Rev. Food Sci. Nutr.*, 28, 315, 1989; Harborne, J.B., *The Flavonoids: Advances in Research since 1986*, Chapman & Hall, London, 1993, 688 pp.

Note: The data were adapted from the more comprehensive tables provided by [51,53,70,74,75,78]. Please see the references to find out original data.

decreased in the following order: gallic acid (3,4,5-trihydroxy, 3.52 mM) > dihydroxybenzoates (2,4-dihydroxybenzoic acid, 1.22 mM; protocatechuic acid, 1.15 mM) > monohydroxybenzoates (*o*-, *m*-, *p*-hydroxy, 0.03–0.04; except for syringic acid 1.39 mM, having *p*-hydroxy and methoxy). When considering dihydroxy cinnamates such as caffeic acid (1.31 mM) and chlorogenic acid (1.56 mM) they had a stronger activity when compared to monohydroxycinnamates (*m*- and *o*-coumaric acids (0.82 and 0.93 mM, respectively), but they were less effective than other monohydroxycinnamates (*p*-coumaric and ferulic or isoferulic acids: 1.53–1.96 mM). This was due to the influence of the hydroxyl position in hydroxycinnamic acids. The substitution of the 3- or 4-hydroxyl group of caffeic acid (1.31 mM) by a methoxy group as in ferulic acid (1.92 mM) and isoferulic acid (1.53 mM) obviously raised the activity [106]. In general, cinnamic acids have stronger antioxidant activity than their benzoic counterparts. As for the influence of aromatic substitution in the kinetic test the antioxidant activity increased in the sequence *p*-hydroxy < *p*-hydroxymethoxy < dihydroxy < *p*-dihydroxydimethoxy. However when experiments were made in the LDL system the sequence changed and the antioxidant capacity of the dihydroxy acids was equal to or higher than that of the *p*-hydroxydimethoxy acids [107].

The antioxidant activity of the aromatic ring is enhanced by the presence of the propenoid side group (as in the case of hydroxycinnamic acid derivatives), instead of the carboxylic group of benzoic acid derivatives, where the conjugated double bond in the side chain has a stabilizing resonance effect on the phenoxyl radical. The stability of this radical is further enhanced by the two methoxy groups [108,109] and this also indicated that the two major factors effecting antioxidant activity of the phenolic acids are: the substitutions on the aromatic ring and the structure of the side chain. From the structure–antioxidant activity relationship studies performed, it may be concluded that the number of phenolic groups and the type of alkyl spacers between the carboxylic acid and the aromatic ring strongly influence the antioxidant activity [110,111] indicate that the phenols with two vicinal hydroxy groups on an aromatic residue are better free radical scavengers than the polyphenols with a single hydroxyl group per aromatic residue.

Monohydroxy benzoic acids show no antioxidant activity in the *o*- and *p*-positions in terms of the hydrogen donating capacity against the radicals generated in the aqueous phase, but the *m*-hydroxy acids have an antioxidant activity. Antioxidant activities (TEAC value, mM, relative to trolox) of several monohydroxybenzoic acids were as follows: 2-hydroxy (*o*-) 0.04, 3-hydroxy (*m*-) 0.84, and 4-hydroxy (*p*-) 0.08. This is consistent with the electron withdrawing potential of the single carboxyl function group on the phenol ring affecting the *o*- and *p*-positions. The monohydroxy benzoates are, however, effective hydroxyl radical scavengers due to their propensity to the hydroxylation and the high reactivity of the hydroxyl radical [112,113].

The phenolic acids have lower antioxidant activities against peroxy radicals (ORAC) than the flavonoids. The ORAC capacities of several hydroxy benzoic and hydroxycinnamic acids were found to be as follows [114]: *Hydroxybenzoic acids*; 4-hydroxy (0.17), protocatechuic acid (2.06), gallic acid (1.74), 2,5-dihydroxybenzoic acid (1.20), vanillic acid (1.11) and syringic acid (1.27); *Hydroxycinnamic acids*; caffeic acid (2.23), *p*-coumaric acid (1.09), ferulic acid (1.33), and sinapic acid (1.66). It is well known that polyphenols with catechol groups (aromatic rings with two hydroxyl groups in the ortho position) have greater antioxidant activity [115]. As seen in the above list, protocatechuic and caffeic acids, both having two hydroxy groups have higher antioxidant activity

When considering whole food, the hydrolysis method was another factor in antioxidant activity. In wheat bran, the alkaline hydrolysable fraction, mostly ferulic acids, had greater antioxidant activity as compared to the acid hydrolysable fractions [116].

10.3.5 HEALTH EFFECTS OF PHENOLIC ACIDS

Metabolic energy production involves the oxidation of food materials in the body. Reactive oxygen and nitrogen species (called free radicals) are continuously produced as a by-product during the electron transport chain. Environmental factors, such as pollution, radiation, xenobiotics can also generate free radicals. Free radicals can also be produced in the immune system's cells as a part of an immune response [117]. Production of these species may thus result in damage to the proteins, lipids, and DNA in the respiring mitochondria, leading to mutations in the mitochondrial DNA that have been associated with a wide range of human diseases [118]. Body has its own defense mechanism, called endogenous antioxidants consisting of mainly superoxide dismutase (SOD), glutathione peroxidase (GPx) and catalase enzymes to avoid the harmful effects of these radicals. Several other enzymes such as lipases, proteases and DNA repair enzymes also contribute to the antioxidant defense system. Under ideal conditions the body would be in a steady state with the free radicals produced and quenched by endogenous antioxidants. But in most cases this balance is not perfect and cells need exogenous antioxidants which have to be provided by diet consisting of vitamin C, vitamin E, carotenoids, phenolics, and antioxidant enzyme related minerals such as zinc, selenium, manganese, and copper. Phenolic compounds mainly act as a chain breakers or chain terminators, causing blockage of the lipid radical starting chain oxidation [118], and therefore dietary polyphenols receive considerable interest for their presumed role in the prevention of various degenerative diseases such as cancers, cardiovascular diseases, and osteoporosis which are associated with aging [115,119]. Among the variety of phenolic compounds, phenolic acids have also attracted considerable interest in the past few years [73,75,77]. Their role is based on numerous animal studies and some clinical and epidemiological studies. The chemical structure of polyphenols is effective on their antioxidant activity potential, antioxidant activity and intestinal absorption. The major parameters on these properties are molecular weight, glycosylation, and esterification [119]. Current evidence strongly supports a contribution of polyphenols to the prevention of cardiovascular diseases, cancers, and osteoporosis and suggests a role in the prevention of neurodegenerative diseases and diabetes mellitus. However it is still impossible to evaluate the individual and social benefits that increases in polyphenol intake could have for the general population containing particular groups at specific disease risk [115]. Currently we do not have enough data to make a clear interpretation about the health effects of phenolics. As we know; there is a relationship between polyphenol ingestion and plasma antioxidant level, moreover epidemiological studies have shown a correlation between an increased consumption of phenolic antioxidants and a reduced risk of cardiovascular disease and certain types of cancer. Phenolic antioxidants function as terminators of free radicals and chelators of metal ions that are capable of catalyzing lipid peroxidation. The data on health effects of polyphenols is mostly derived from *in vitro* or animal experiments, which are often performed with much higher doses than a person normally consumes [115,119]. The role of polyphenolic antioxidants *in vivo* is not clear yet. A WHO report indicated that there is enough evidence on the possible protective effects of flavonoids in cardiovascular disease but not for cancers [120].

Phenolic acids can be directly absorbed in the small intestine. The structure of phenolics has a major impact on intestinal absorption. The major parameters are molecular weight, glycosilation, and esterification [115]. Low-molecular weight phenols, including phenolic acids, are unable to precipitate protein. It has been shown that chlorogenic acid in coffee reduced iron bioavailability. Chlorogenic and caffeic acids have been shown to reduce zinc absorption in rats [54]. Phenolic acids significantly contribute to the flavor in food products. A summary of this contribution is given by [64]. Phenolic acids from beer are present in blood after being largely metabolized to the form of glucuronide and sulfate conjugates. The extent of conjugation is related to the chemical structure; monohydroxy derivatives had lowest conjugation degree while dihydroxy derivatives had the highest [80]. The metabolic transformations of benzoic acids include conjugation with sulfate, glucuronate and glycine. Methylation may also occur. Some of the dietary hydroxybenzoic acids such as protocatechuic acid can be converted during metabolism to vanillic acid [104]. Most of the metabolism is in the form of glucuronidation or the formation of glucuronide conjugates [121].

The absorption efficiency of polyphenols in humans is generally low. Unlike flavonoids, phenolic acids have not been extensively studied and are not considered to be of great nutritional interest [122]. In a limited number of papers it is revealed that, the absorption rate of caffeic acid was 95%, while it is 33% in chlorogenic acid, which is an ester of caffeic acid with quinic acid [123]. It has been also indicated that chlorogenic acid is not well absorbed from the digestive tract in rats unlike caffeic acid, and it is subject to almost no structural change to the easily absorbed forms [124].

Finally, even though beneficial properties of phenolic compounds are well known, it has to be also kept in mind that further research regarding the potential toxicities associated with flavonoids and other phenolics is required if these plant-derived products are to be used for therapy.

REFERENCES

1. McMurry, J., *Organic Chemistry*. Brooks/Cole Publ. Company, Belmont, CA, pp. 738–758, 1984.
2. Wolfe, D.H., *General, Organic and Biological Chemistry*. McGraw-Hill Books, New York, 884 pp., 1986.
3. Schobinger, U., *Handbuch der Lebensmitteltechnologie, Frucht- und Gemuseseafte*. (Translated by J. Acar). Hacettepe University Press, Turkey, 602 pp., 1988.
4. Meislich, H., Carboxylic acids and their derivatives. In: *Organic Chemistry*. McGraw Hill Professional Book Group, Columbus, OH, pp. 91–99, 1999, Chapter 11.
5. Gomis, D.B., HPLC analysis of organic acids. In: *Food Analysis by HPLC*. L.M.L. Nollet (Ed.), Marcel Dekker, New York, Basel, Switzerland, pp. 477–492, 2000, Chapter 11.
6. IUPAC, Compendium of Chemical Terminology, Electronic version, http://goldbook.iupac.org/ (accessed March 31, 2009-04-04), 2009.
7. Seymour, G.B., Taylor, J.E., and Tucker, G.A., *Biochemistry of Fruit Ripening*. Chapman and Hall, London, U.K., 454 pp., 1993.
8. Soyer, Y., Koca, N., and Karadeniz, F., Organic acid profile of Turkish white grapes and grape juices. *J. Food Comp. Anal.*, 16, 629–636, 2003.
9. Belitz, H.D. and Grosch, W., Fruits and fruit products. In: *Food Chemistry* (2nd edition). Springer, Berlin, Germany, pp. 748–800, 1999.
10. Souci, S.W., Fachmann, W., and Kraut, H., *Food Composition and Nutrition Tables* (6th rev.). Medpharum Scientific Publishers, Stuttgart, Germany, 1182 pp., 2000.
11. Haard, N.F. and Chism, G.W., Characteristics of edible plant tissues. In: *Food Chemistry*. O.R. Fennema (Ed.), Marcel Dekker Inc., New York, pp. 943–1011, 1996.
12. Salunkhe, D.K., Bolin, H.R., and Reddy, N.R., *Storage, Processing, and Nutritional Quality of Fruits and Vegetables*. Vol. I, *Fresh Fruits and Vegetables*. CRC Press, Boca Raton, FL, 323 pp., 1991.
13. Westwood, M.N., *Temperate-Zone Pomology: Physiology and Culture* (3rd edition). Timber Press, Portland, OR, 523 pp., 1993.
14. Bielig, H.J., Faethe, W., Fuchs, G., Koch, J., Wallrauch, S., and Wucherpfennig, K., *RSK-Values, The Complete Manual*. Verlag Flüssiges Obst Gmbh, Schönborn, Germany, 197 pp., 1987.
15. Nelson, P.E. and Tressler, D.K., *Fruit and Vegetable Juice Processing Technology*. Avi Publ. Company Inc., Westport, CT, 163 pp., 1980.
16. Vonach, R., Lendl, B., and Kellner, R. High-performance liquid chromatography with real-time Fourier-transform infrared detection for the determination of carbohydrates, alcohols and organic acids in wines. *J. Chromatogr. A*, 824, 159–167, 1998.
17. Mato, I., Huidobro, J.F., Simal-Lozano, J., and Sancho, M.T., Simultaneous determination of organic acids in beverages by capillary zone electrophoresis. *Anal. Chim. Acta*, 565, 190–197, 2006.
18. Edwards, T.L., Singleton, V.L., and Boulton, R.B., Formation of ethyl esters of tartaric acid during wine aging: Chemical and sensory effects. *Am. J. Enol. Vitic.*, 36, 118–124, 1985.
19. Boulton, R.B., Singleton, V.L., Bisson, L.F., and Kunkee, R.E., Juice and wine acidity. In: *Principles and Practices of Wine Making*. Chapman and Hall, New York, pp. 521–538, 1996.
20. Pérez-Ruiz, T., Martínez-Lozano, Tomás, V., and Martín, J., High-performance liquid chromatographic separation and quantification of citric, lactic, malic, oxalic and tartaric acids using a post-column photo-chemical reaction and chemiluminescence detection. *J. Chromatogr. A*, 1026, 57–64, 2004.
21. Lasztity, R. and Lasztity, L., Phytic acid in cereal technology. In: *Advances in Cereal Science and Technology*, Vol. 10. Y. Pomeranz (Ed.), American Association of Cereal Chemists, Inc., St. Paul, MN, pp. 309–371, 1990.
22. Walstra P., *Dairy Technology: Principles of Milk Properties and Processes*. Marcel Dekker Inc., New York, 752 pp., 1999.
23. Fernandez-Garcia, E. and McGregor, J.U., Determination of organic acids during the fermentation and cold storage of yogurt. *J. Dairy Sci.*, 77, 2934–2939, 1994.

24. Akalin, A.S., Kinik, O., and Gonc, S., Determination of organic acids in commercial cheeses by high performance liquid chromatography. *Milchwissenschaft*, 54, 260–262, 1997.
25. Rasic, J.L. and Kurmann, J.A., *Yoghurt: Scientific Grounds, Technology, Manufacture and Preparations*. Tech. Dairy Publ. House, Copenhagen, DK, 427 pp., 1978.
26. Tamime, A.Y. and Robinson, R.K. *Yoghurt Science and Technology* (2nd edition). Woodhead Publishing Limited, Cambridge, U.K., 619 pp., 1999.
27. Jham, G.L., Fernandes, S.A., Garcia, C.F., and Da Silva, A.A., Comparison of GC and HPLC for the quantification of organic acids in coffee. *Phytochem. Anal.*, 13, 99–104, 2002.
28. Van der Stegen, G.H.D. and Van Duijn, J., Analysis of normal organic acids in coffee. *Proceedings of 12th ASIC Colloquium*. Montreux, ASIC, Paris, pp. 238–246, 1987.
29. Cavia, M.M., Fernández-Muiño, M.A., Alonso-Torre, S.R., Huidobro, J.F., and Sancho, M.T. Evolution of acidity of honeys from continental climates: Influence of induced granulation. *Food Chem.*, 100, 1728–1733, 2007.
30. Lee, F.A., *Basic Food Chemistry* (2nd edition). The Avi Publ. Comp, Westport, CT, 564 pp., 1983.
31. Kauffman, R.G., Meat composition. In: *Meat Science and Applications*. Y.H. Hui (Ed.), Marcel Dekker Inc., New York, pp. 1–19, 2001.
32. Greaser, M.L., Postmortem muscle chemistry. In: *Meat Science and Applications*. Y.H. Hui (Ed.), Marcel Dekker Inc., New York, pp. 21–38, 2001.
33. Forrest, J.C., Aberle E.D., Hedrick, H.B., Judge, M.D., and Merkel, R.A., *Principles of Meat Science*. W.H. Freeman & Company, San Francisco, CA, 417 pp., 1975.
34. Pearson, A.M. and Young R.B., *Muscle and Meat Biochemistry*. Academic Press Inc., San Diego, CA, 457 pp., 1989.
35. Osako, K., Hossain, M.A., Kuwahara, K., Okamoto, A., Yamaguchi, A., and Nozaki, Y., Quality aspect of fish sauce prepared from underutilized fatty Japanese anchovy and rabbit fish. *Fish. Sci.*, 71, 1347–1355, 2005.
36. Durnford, E. and Shahidi, F., Flavour of fish meat. In: *Flavor of Meat, Meat Products and Seafoods*. F. Shahidi (Ed.), Blackie Academic and Professional, London, U.K., pp. 131–158, 1998.
37. Sofos, J.N., *Improving the Safety of Fresh Meat*. Woodhead Publ. Ltd., Cambridge, U.K., 354 pp., 2005.
38. Cauvain, S., Bread Making: Improving Quality. Woodhead Publ. Ltd., Cambridge, U.K., 608 pp., 2003.
39. Rehman, S.U., Paterson, A., and Piggott, J.R., Flavour in sourdough breads: A review. *Trends Food Sci. Techol.*, 17, 557–566, 2006.
40. Dauthy, M.E., Fruit and Vegetable Processing. FAO Food and Agricultural Services Bulletin No 119, Rome, http://www.fao.org (accessed 06.04.2009), 1995.
41. Doores, S., pH control agents and acidulants. In: *Food Additives* (2nd edition). A.L. Branen (Ed.), Marcel Dekker Inc., New York, pp. 621–660, 2001.
42. Barbosa-Canovas, G.V., Fernandez-Molina, J.J., Alzamora, S.M., Tapia, M.S., Lopez-Malo, A., and Chanes, J.W., General considerations for preservation of fruits and vegetables by combined methods for rural areas. In: *Handling and Preservation of Fruits and Vegetables by Combined Methods for Rural Areas*. FAO Agricultural Services Bulletins No. 149, pp. 39–53, 2003.
43. Clydesdale, F.M., *Food Additives: Toxicology, Regulation and Properties*. Database on CD-ROM. CRC Press, Boca-Raton, FL, 1997.
44. Amerine, M.A., Pangborn, R.M., and Roessler, E.B., The sense of taste. In: *Principles of Sensory Evaluation of Food*. Academic Press, New York and London, pp. 30–144, 1965.
45. Lange, H.J., *Untersuchungs-Methoden in der Konservenindustrie*. Verlag Paul Parey, Berlin und Hamburg, 298 pp., 1972.
46. FAO, General Standard for Food Additives. (GSFA-online) www.codecsalimentarius.net/gsfaonline/additives/index.html (accessed 06.04.2009), 2007.
47. Norfad, Nordic Food Additive Database. www.norfad.dk (accessed on 9.4.2009), 2009.
48. Caligiani, A., Acquotti, D., Palla, G., and Bocchi, V., Identification and quantification of the main organic components of vinegars by high resolution ^{1}H NMR spectroscopy. *Anal. Chim. Acta*, 585, 110–119, 2007.
49. McFeeters, R.F., Single injection HPLC analysis of acids, sugars and alcohols in cucumber fermentations. *J. Agric. Food Chem.*, 41, 1439–1443, 1993.
50. Fleming, H.P., Daeschel, M.A., McFeeters, R.F., and Pierson, M.D., Butyric acid spoilage of fermented cucumbers. *J. Food Sci.*, 54, 636–639, 1989.
51. Harborne, J.B., *The Flavonoids: Advances in Research Since 1986*. Chapman and Hall, London and New York, 688 pp., 1993.
52. Harborne, J.B., *Plant Phenolics* (Vol. 1.) In: *Methods in Plant Biochemistry*. P.M. Dey and J.B. Harborne (Eds.), Academic Press, London, U.K., 1989.

53. Escarpa, A. and Gonzalez, M.C., An overview of analytical chemistry of phenolic compounds in foods. *Crit. Rev. Anal. Chem.*, 31, 57–139, 2001.
54. Bravo, L., Polyphenols: Chemistry, dietary sources, metabolism, and nutritional significance. *Nutr. Revs.*, 56, 317–333, 1998.
55. Howard, L.R., Clark, J.R., and Brownmiller, C., Antioxidant capacity and phenolic content in blueberries as affected by genotype and growing season. *J. Sci. Food Agric.*, 83, 1238–1247, 2003.
56. Asami D.K., Hong, Y.J., Barrett, D.M., and Mitchell A.E., Processing-induced changes in total phenolics and procyanidins in clingstone peaches. *J. Sci. Food Agric.*, 83, 56–63, 2003.
57. Caldwell, C.R., Britz, S.J., and Mirecki, R.M., Effect of temperature, elevated carbon dioxide, and drought during seed development on the isoflavone content of dwarf soybean [*Glycine max* (L.,) Merrill] grown in controlled environments. *J. Agric. Food Chem.*, 53, 1125–1129, 2005.
58. Luthria, D.L., Mukhopadhyay, S., and Krizek, D.T., Content of total phenolics and phenolic acids in tomato (*Lycopersicon esculentum* Mill.) fruits as influenced by cultivar and solar UV radiation. *J. Food Comp. Anal.*, 19, 771–777, 2006.
59. Beltran, D., Selma, M.V., Marin, A., and Gil, M.I., Ozonated water extends the shelf life of fresh-cut lettuce. *J. Agric. Food Chem.*, 53, 5654–5663, 2005.
60. Haard, N.F. and Chism, G.W., Characteristics of edible plant tissues. In: *Food Chemistry.* O.R. Fennema (Ed.), Marcel Dekker, Basel, New York, Hong Kong, pp. 943–1011, 1996.
61. Scalbert, A. and Williamson, G., Dietary intake and bioavailability of polyphenols. *J. Nutr.*, 130, 2073S-2085S, 2000.
62. Spanos, G.A. and Wrolstad, R.E., Phenolics of apple, pear, and white grape juices and their changes with processing and storage. *J. Agric. Food Chem.*, 40, 1478–1487, 1992.
63. Galati, G. and O'Brien, P.J., Potential toxicity of flavonoids and other dietary phenolics: Significance for their chemopreventive and anticancer properties. *Free Rad. Biol. Med.*, 37, 287–303, 2004.
64. Lule, S.U. and Xia, W., Food phenolics, pros and cons: A review. *Food Revs. Int.*, 21, 367–388, 2005.
65. Zijp, I.M., Korver, O., and Tijburg, L.B.M., Effect of tea and other dietary factors on iron absorption. *Crit. Revs. Food Sci. Nutr.*, 40, 371–39, 2000.
66. Karakaya, S., Bioavailability of phenolic compounds. *Crit. Revs. Food Sci. Nutr.*, 44, 453–464, 2004.
67. Zhou, Z., Robards, K., Helliwell, S., and Blanchard, C., The distribution of phenolic acids in rice. *Food Chem.*, 87, 401–406, 2004.
68. Schuster, B. and Herrmann, K., Hydroxybenzoic and hydroxycinnamic acid derivatives in soft fruits. *Phytochemistry*, 24, 2761–2764, 1985.
69. Robbins, R.J., Phenolic acids in foods: An overview of analytical methodology. *J. Agric. Food Chem.*, 51, 2866–2887, 2003.
70. Manach, C., Scalbert, A., Morand, C., Remesy, C., and Jimenez, L., Polyphenols: Food sources and bio-availability. *Am. J. Clin. Nutr.*, 79, 727–747, 2004.
71. Nakatani N., Kayano S., Kikuzaki, H, Sumino, K, Katagiri, K., and Mitani, T., Identification, quantitative determination and antioxidative activities of chlorogenic acid isomers in prune (*Prunus domestica* L.). *J. Agric. Food Chem.*, 48, 5512–5516, 2000.
72. Monagas, M., Bartolome, B., and Gomez-Cordovez C., Updated knowledge about the presence of phenolic compounds in wine. *Crit. Rev. Food Sci. Nutr.*, 45, 85–118, 2005.
73. Shahidi, F. and Naczk, M., *Food Phenolics: Sources, Chemistry, Effects, Applications.* Technomic Publishing Company Inc., Lancaster, U.K., 331 pp., 1995.
74. Herrmann, K., Occurrence and content of hydroxycinnamic and hydroxybenzoic acid compounds in foods. *Crit. Rev. Food Sci. Nutr.*, 28, 315–347, 1989.
75. Mattila, P. and Hellstrom, J., Phenolic acids in potatoes, vegetables and some of their products. *J. Food Comp. Anal.*, 20, 152–160, 2007.
76. Li, P., Wang, X.Q., Wang, H.Z., and Wu, Y.N., High performance liquid chromatographic determination of phenolic acids in fruits and vegetables. *Biomed. Environ. Sci.*, 6, 389–398, 1993.
77. Duthie, G. and Crozier, A., Plant-derived phenolic antioxidants. *Curr. Opin. Lipid*, 11, 43–47, 2000.
78. Hakkinen, S., Flavonols and phenolic acids in berries and berry products. Doctoral dissertation. Kuopio University Publications, Kuopio, Finland, *D. Med. Sci.* 221, 92 pp., 2000.
79. Clifford, M.N., Wu, W., and Kuhnert, N., The chlorogenic acids of *Hemerocallis. Food Chem.*, 95, 574–578, 2006.
80. Nardini, M., Natella, F., Scaccini, C., and Ghiselli, A., Phenolic acids from beer and absorbed and extensively metabolized in humans. *J. Nutr. Biochem.*, 17, 14–22, 2006.
81. Clifford, M.N., Chlorogenic acids and other cinnamates-nature, occurrence and dietary burden. *J. Sci. Food Agric.*, 79, 362–372, 1999.

82. Vinson, J.A., Hao, Y., Su, X., and Zubik, L., Phenolic antioxidant quantity and quality in foods: Vegetables. *J. Agric. Food Chem.*, 46, 3630–3634, 1999.

83. Vinson, J.A., Su, X., Zubik, L., and Bose, P., Phenol antioxidant quantity and quality in foods: Fruits. *J. Agric. Food Chem.*, 49, 5315–5321, 2001.

84. Chen, H., Zuo, Y., and Deng, Y., Separation and determination of flavonoids and other phenolic compounds in cranberry juice by high-performance liquid chromatography. *J. Chromatogr. A*, 913, 387–395, 2001.

85. Radtke, J., Linseisen, J., and Wolfram, G., Phenolic acid intake of adults in a Bavarian subgroup of the national food composition survey. *Z. Ernahungswiss*, 37, 190–197, 1998.

86. Simonovska, B., Vovk, I., Andrensek, S., Valentova, K., and Ulrichova, J., Investigation of phenolic acids in yacon (*Smallanthus sonchifolius*) leaves and tubers. *J. Chromatogr. A*, 1016, 89–98, 2003.

87. Lee, H.S., HPLC analysis of phenolic compounds. In: *Food Analysis by HPLC*, L.M.L. Nollet (Ed.), Marcel Dekker, Basel, New York, pp. 775–824, 2000, Chapter 19.

88. Naczk, M. and Shahidi, F., Extraction and analysis of phenolics in food. *J. Chromatogr. A*, 1054, 95–111, 2004.

89. Kim, J.W. and Mazza, G., Optimization of extraction of phenolic compounds from flax shives by pressurized low-polarity water. *J. Agric. Food Chem.*, 54, 7575–7584, 2006.

90. Betes-Saura, C., Andres-Lacueva, C., and Lamuela-Raventos, R.M., Phenolics in white free run juices and wines from Penedes by high-performance liquid chromatography: Changes during vinification. *J. Agric. Food Chem.*, 44, 3040–3046, 1996.

91. Tomas-Barberan, F.A., Gil, M.I., Cremin, P., Waterhouse, A.L., Hess-Pierce, B., and Kader, A.A., HPLC-DAD-ESIMS analysis of phenolic compounds in nectarines, peaches, and plums. *J. Agric. Food Chem.*, 49, 4748–4760, 2001.

92. Mas, S., Fonrodona, G., Tauler, R., and Barbosa, J., Determination of phenolic acids in strawberry samples by means of fast liquid chromatography and multivariate curve resolution methods. *Talanta*, 71, 1455–1463, 2007.

93. Ho, P., Hogg, T.A., and Silva M.C.M., Application of a liquid chromatographic method for the determination of phenolic compounds and furans in fortified wines. *Food Chem.*, 64, 115–122, 1999.

94. Schieber, A., Keller, P., and Carle, R., Determination of phenolic acids and flavonoids of apple and pear by high-performance liquid chromatography. *J. Chromatogr. A*, 910, 265–273, 2001.

95. Mozetic, B., Trebse, P., and Hribar, J., Determination and quantitation of anthocyanins and hydroxycinnamic acids in different cultivars of sweet cherries (*Prunus avium* L.) from Nova Gorica region (Slovenia). *Food Technol. Biotechnol.*, 40, 207–212, 2002.

96. Carini, M., Facino, R.M., Aldini, G., Calloni, M., and Colombo, L., Characterization of phenolic antioxidants from mate (*Ilex paraguayensis*) by liquid chromatography/tandem mass spectrometry. *Rapid Commun. Mass. Spectrom.*, 12, 1813–1819, 1998.

97. Pellati, F., Benvenuti, S., Melegari, M., and Lasseigne, T., Variability in the composition of anti-oxidant compounds in *Echinacea* species by HPLC. *Phytochem. Anal.*, 16, 77–85, 2005.

98. Kammerer, D., Carle, R., and Schieber, A., Characterization of phenolic acids in black carrots (*Daucus carota* ssp. *sativus* var. *atrorubens* Alef.) by high-performance liquid chromatography/electrospray ionization mass spectrometry. *Rapid Commun. Mass Spectrom.*, 18, 1331–1340, 2004.

99. Mellon, F.A., Self, R., and Startin J.R., Mass spectrometry of natural substances in food, in *Bioactive Non-Nutrients in Foods*. Royal Society of Chemistry, Cambridge, UK, 2000, pp. 93–169.

100. Bocchi, C., Careri, M., Groppi, F., Mangia, A., Manini, P., and Mori, G., Comparative investigation of UV, electrochemical and particle beam mass spectrometric detection for the high-performance liquid chromatographic determination of benzoic and cinnamic acids and of their corresponding phenolic acids. *J. Chromatogr. A*, 753, 157–170, 1996.

101. Linsday, D.G., Maximizing the functional benefits of plant foods. In: *Functional Foods*. G.R. Gibson and C.M. Williams (Eds.), CRC Press, Boca Raton, FL, pp. 183–208, 2001.

102. Heinonen, I.M. and Meyer, A.S., Antioxidants in fruits, berries and vegetables. In: *Fruit and Vegetable Processing*. W. Jongen (Ed.), CRC Press, Boca-Raton, FL, pp. 23–51, 2002.

103. Velioglu, Y.S., Ekici, L., and Poyrazoglu, E.S., Phenolic composition of European cranberrybush (*Viburnum opulus* L.) berries and astringency removal of its commercial juice. *Int. J. Food Sci. Technol.*, 41, 1011–1015, 2006.

104. Tomas-Barberan, F.A. and Clifford, M.N., Dietary hydroxybenzoic acid derivatives-nature, occurrence and dietary burden. *J. Sci. Food Agric.*, 80, 1024–1032, 2000.

105. Weidner, S., Amarowicz, R., Karamac, M., and Fraczek, E., Changes in endogenous phenolic acids during development of Secale cereale caryopses and after dehydration treatment unripe rye grains. *Plant Physiol. Biochem.*, 38, 595–602, 2000.
106. Cai, Y.Z., Sun, M., Xing, J., Luo, Q., and Corke, H., Structure-radical scavenging activity relationships of phenolic compounds from traditional Chinese medicinal plants. *Life Sciences*, 78, 2782–2788, 2006.
107. Natella, F., Nardini, M., Felice, M.D., and Scaccini, C., Benzoic and cinnamic acid derivatives as antioxidants: Structure-activity relation. *J. Agric. Food Chem.*, 47, 1453–1459, 1999.
108. Virgili, F., Scaccini, C., Hoppe, P.P., Kramer, K., and Packer L., Plant phenols and cardiovascular disease: Antioxidants and modulators of cell response. In: *Nutraceuticals in Health and Disease Prevention*. K. Kramer (Ed.), Marcel Dekker, New York, pp. 187–214, 2001.
109. Shahidi, F. and Wanasundara, P.K.J., Phenolic antioxidants. *Crit. Rev. Food Sci. Nutr.*, 32, 67–103, 1992.
110. Siquet, C., Paiva-Martins, F., Lima, J.L.F.C, Reis, S., and Borges, F., Antioxidant profile of dihydroxy-and trihydroxyphenolic acids-A structure-activity relationship study. *Free Rad. Res.*, 40, 433–442, 2006.
111. Scalbert, A., Johnson, I.T., and Saltmarsh, M., Polyphenols: Antioxidants and beyond. *Am. J. Clin. Nutr.*, 81 (Suppl.), 215S–217S, 2005.
112. Rice-Evans, C.A., Miller, N.J., and Paganga, G., Structure–antioxidant activity relationships of flavonoids and phenolic acids. *Free Rad. Biol. Med.*, 20, 933–956, 1996.
113. Rice-Evans, C.A., Miller, N.J., and Paganga, G., Antioxidant properties of phenolic compounds. *Trends Plant Sci.*, 2, 152–159, 1997.
114. Guo, C., Cao, G., Sofic, E., and Prior, R.L., High performance liquid chromatography coupled with coulometric array detection of electroactive components in fruits and vegetables: relationship to oxygen radical absorbance capacity. *J. Agric. Food Chem.*, 45, 1787–1796, 1997.
115. Scalbert, A., Manach, C., Morand, C., Remesy, C., and Jimenez, L., Dietary polyphenols and the prevention of diseases. *Crit. Rev. Food Sci. Nutr.*, 45, 287–306, 2005.
116. Kim, K.H., Tsao, R., Yang, R., and Cui, S.W., Phenolic acid profiles and antioxidant activities of wheat bran extracts and the effect of hydrolysis conditions. *Food Chem.*, 95, 466–473, 2006.
117. Kaur, C. and Kapoor, H.C., Antioxidants in fruits and vegetables-the millenium's health. *Int. J. Food Sci. Technol.*, 36, 703–725, 2001.
118. Willcox, J.K., Ash, S.L., and Catignani, G.L., Antioxidants and prevention of chronic disease. *Crit. Rev. Food Sci. Nutr.*, 44, 295–295, 2004.
119. Scalbert, A., Morand, C., Manach, C., and Remesy, C., Absorption and metabolism of polyphenols in the gut and impact on health. *Biomed. Pharmacother.*, 56, 276–282, 2002.
120. WHO, Diet, nutrition and the prevention chronic diseases. WHO Technical Report Series No 916. WHO, Geneva, Switzerland, 149 pp., 2003.
121. Reddy, L., Odhav, B., and Bhoola, K.D., Natural products for cancer prevention: A global perspective. *Pharmacol. Ther.*, 99, 1–13, 2003.
122. Watanabe, H., Yashiro, T., Tohjo, Y., and Konishi, Y., Non-involvement of the human monocarboxylic acid transporter 1 (MCT1) in the transport of phenolic acid. *Biosci. Biotechnol. Biochem.*, 70, 1928–1933, 2006.
123. Olthof, M.R., Hollman, P.C.H., and Katan, M.B., Chlorogenic acid and caffeic acid are absorbed in humans. *J. Nutr.*, 131, 66–71, 2001.
124. Azuma, K., Ippoushi, K., Nakayama, M., Ito, H., Higashio, H., and Terao, J., Absorption of chlorogenic acid and caffeic acid in rats after oral administration. *J. Agric. Food Chem.*, 48, 5496–5500, 2000.

11 Biological Oxidations: Enzymatic and Nonenzymatic Browning Reactions and Control Mechanisms

Fahrettin Göğüş, Sibel Fadıloğlu, and Çiğdem Soysal

CONTENTS

11.1 INTRODUCTION

Browning reactions that take place during processing and storage of foods are widespread phenomena and are important in terms of the alteration of appearance, flavor, and nutritive value. Browning is considered to be desirable if it enhances the appearance and flavor of a food product in terms of tradition and consumer acceptance like in the cases of coffee, tea, snack foods, beer, and in toasting of bread. However, in many other instances, such as fruits, vegetables, and frozen and dehydrated foods, browning is undesirable as it results in off-flavors and colors. Another significant adverse effect of browning is the lowering of the nutritive value of the food substance. This problem is of considerable importance to the food industry as it reduces consumer acceptability, and therefore causes significant economic impact, both to primary food producers and the food processing industry.

There are two main mechanisms of browning reactions:

- Nonenzymatic browning (the Maillard reaction, caramelization, and ascorbic acid oxidation)
- Enzymatic browning

The Maillard reaction occurs between carbonyl compounds, especially reducing sugars, and compounds with free amino groups, such as amines, amino acids, and proteins [1]. This reaction may have either beneficial or detrimental effects [2]. Since the Maillard reaction affects the sensory characteristics of cooked foods such as flavor, aroma, and color [3], it may be desirable in baked, fried, or roasted foods [4] as in the manufacture of coffee, tea, beer, bread, and cake [5]. However, the reaction that occurs upon heating or prolonged storage has been associated with the formation of compounds with a strong radical scavenging activity and is one of the deteriorative processes that take places in stored foods [6–8]. The Maillard reaction is a complex reaction and is influenced by many factors, such as temperature, pH, time, water activity, types and concentrations of amino acids, and sugars [9,10]. Changing any of these factors will alter the reaction rate, reaction pathways, and reaction end-products.

Caramelization of sugar is another type of nonenzymatic browning reaction. These reactions are responsible for the characteristic flavors, aromas, and the brown color of many cooked foods.

It involves the heat-induced decomposition of sugars, normally monosaccharides. They undergo initial enolization and progress to subsequent complex reactions, such as dehydration, dicarboxylic cleaving, and aldol condensation [11]. Caramelization products vary in chemical and physical properties and in their constituents depending on the temperature, the pH, and the duration of heating [12,13].

Browning of foods is also caused by ascorbic acid, which is easily oxidized and decomposed under food processing and storage conditions. The formation of dehydroascorbic acid and diketo-gluconic acids from ascorbic acid occur during the reaction and are capable of interacting with the free amino acids nonenzymatically, producing a red-to-brown discoloration [7]. Some factors such as pH, oxygen, ascorbic acid concentration, and temperature affect the ascorbic acid oxidation.

Enzymatic browning of fruits and vegetables due to mechanical injury during postharvest storage or processing is caused by the catalytic action of the enzyme polyphenol oxidase (PPO) [14,15]. PPO is a very important enzyme for food chemists and processors because its action leads to major economic losses in fresh fruits and vegetables, such as potatoes, apples, grapes, bananas, many tropical fruits, lettuce, and other leafy vegetables [16]. This browning occurs due to the oxidation and dehydrogenation of colorless polyphenols present in the plants. PPOs catalyze the hydroxylation of monophenols to o-diphenols and the oxidation of o-diphenols to o-quinones. Then these quinones may condense and react nonenzymatically with other phenolic compounds, amino acids, proteins, and other cellular constituents to produce brown melanoidin pigments [17,18]. The rate of enzymatic browning in fruit and vegetables is governed by the pH, temperature, active PPO content, phenolic content, and oxygen availability within the tissue. Heat treatments or the application of sulfites are commonly used methods of inactivating the enzyme.

11.2 ENZYMATIC BROWNING AND ITS PREVENTION

Browning of fruits and vegetables due to mechanical or physiological injury during postharvest storage or processing is believed to be one of the main causes of quality loss. It is caused by the enzymatic oxidation of phenols to quinones by enzymes, typically PPOs, in the presence of oxygen. These quinones are highly reactive substances that normally react further with other quinones, amino acids, or proteins, leading to the formation of dark-colored pigments [19–23]. These products are high molecular weight polymers, called melanins and melanoproteins, which are produced as a result of enzymatic browning [24]. Although browning reactions, in some food products, result in good appearance in terms of color, these kinds of reactions, in general, lead to undesirable results with respect to texture, sweetness, and overall flavor that cause food quality deteriorations and create economic losses [7,25–27]. It is estimated that up to 50% losses in some tropical fruits occur as a result of enzymatic browning [18]. Therefore, inhibition studies have gained more importance for these types of reactions in food and vegetable processing technology [28]. PPO has been given more attention in food technology in this regard. Peroxidase, which performs single-electron oxidation in the presence of hydrogen peroxide, is the other enzyme considered to be involved in enzymatic browning [29–32]. Peroxidase catalyzes the oxidation of phenols by the H_2O_2 produced during respiration or as a consequence of the wounding of tissue [33]. The role of peroxidase in enzymatic browning is still under discussion, because the internal level of hydrogen peroxide in plants limits peroxidase activity [34]. It has been proposed that PPO could act as a promoter of peroxidase activity by generating hydrogen peroxide and quinine, which are used by the peroxidase as substrate, although peroxidase alone has no oxidative activity [27].

11.2.1 POLYPHENOL OXIDASE

PPOs are ubiquitous, copper-containing metalloproteins that are found almost universally in animals, plants, fungi, and bacteria. Probably there is no common PPO; the enzymes found in animals, plants, and fungi are different with respect to their sequences, size, glycosylation, and activation [35]. PPOs also known as tyrosinase, catechol oxidase, catecholase, phenolase, monophenol oxidase, and cresolase were first discovered in 1856 in mushrooms [18]. Based on the substrate specificity, the International Enzyme Nomenclature has designated monophenol monooxygenase or tyrosinase as EC 1.14.18.1, which hydroxylates monophenols to o-diphenols; diphenol oxidase or catechol oxidase or diphenol oxygen oxidoreductase as EC 1.10.3.1, which dehydrogenates o-dihydroxyphenols to o-quinones; and laccase or p-diphenol oxygen oxidoreductase as EC 1.10.3.2. In this chapter, PPO will be referred to as monophenol and diphenoloxidases to distinguish them from laccase. For PPOs isolated from a great number of fruits and vegetables, there exist different substrate specificities and degrees of inhibition while its level in plants is dependent on the species, cultivar, maturity, and age. In general, PPO activity is very low in young plants, and is often undetectable [36].

Monophenol oxidase catalyzes the hydroxylation of monophenols to o-diphenols (Figure 11.1). The enzyme is referred to as tyrosinase in animals, since L-tyrosine is the major monophenolic substrate [37]. In mammals, L-tyrosine is the initial substrate in the pathway leading to the final products of black-brown eumelanins, red-yellow pheomelanins, or a mixture of pheomelanins and eumelanins. In plants, the enzyme is sometimes referred to as cresolase owing to the ability of the enzyme to utilize the monophenolic substrate, cresol. In microorganisms and plants, a large number of structurally different monophenols, diphenols, and polyphenols serve as substrates for tyrosinase. As many plants are rich in polyphenols, the name PPO has been frequently used for this enzyme [38].

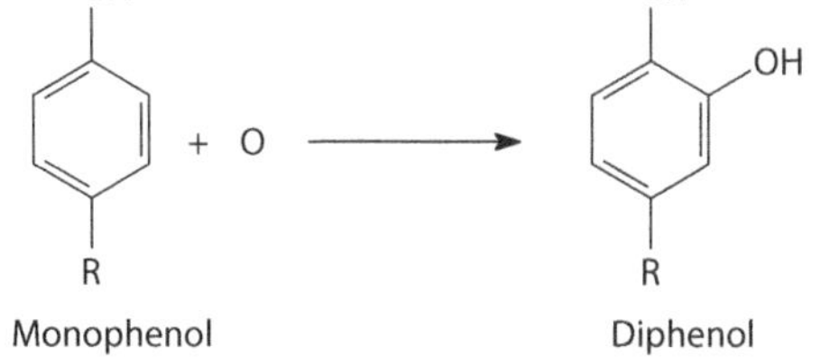

FIGURE 11.1 Oxidation of monophenol producing the diphenol.

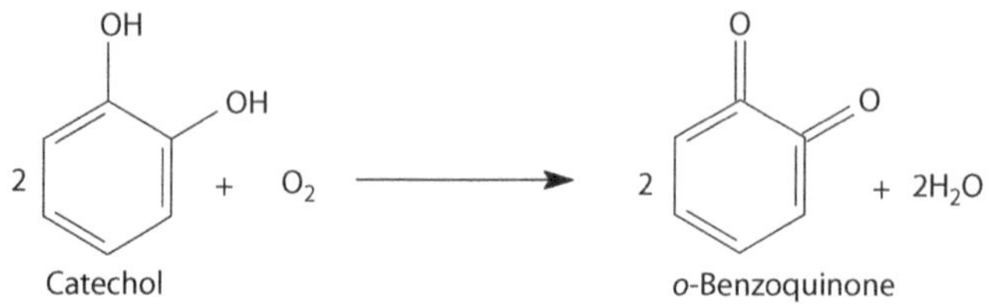

OH
|
OH

2   + O$_2$ $\longrightarrow$ 2   + 2H$_2$O

Catechol   o-Benzoquinone

FIGURE 11.2   Oxidation of catechol (*o*-diphenol) to quinones.

The oxidation of diphenols to quinones in the presence of oxygen is catalyzed by diphenol oxidase. The reaction is best illustrated by the oxidation of catechol (Figure 11.2). They are moderately glycosylated copper-binding proteins using molecular oxygen as a cofactor and exhibiting two copper-binding domains [39]. The catechol oxidases contain copper as the prosthetic group and for the enzyme to act on its substrate phenols, the Cu^{2+} must first be reduced to Cu$^+$ in which state the enzyme can bind O$_2$. The structure of the active site of the enzyme, in which copper is bound by six or seven histidine residues and a single cysteine residue, is highly conserved [40]. The active site of *Neurospora crassa* PPO contains two copper atoms, each of which is liganded to three histidine residues [41,42].

Laccase was first described by Yoshida [43], and was characterized as a metal-containing oxidase by Bertrand [44]. It has the unique ability of oxidizing *p*-diphenols, thus allowing it to be distinguished from *o*-diphenol oxidases, such as catechol oxidase (Figure 11.3) [45]. Basically any substrate with characteristics similar to a *p*-diphenol will be oxidized by laccases. In addition, at least some of the fungal laccases can also oxidize monophenols such as cresol and some are able to oxidize ascorbic acid. Laccases can be roughly divided into two major groups, which show clear differences, i.e., those from higher plants and those from fungi [46]. The occurrence of laccases in higher plants appears to be far more limited than in fungi. The entire family of the *Anacardiaceae*, of which the lacquer tree is a member, appears to contain laccase in the resin ducts and in the secreted resin [47]. Reports on the presence of laccase in other plant species are more limited. Cell cultures of *Acer pseudoplatanus* have been shown to produce and secrete laccase [48,49], and *Pinus taeda* tissue has been shown to contain eight laccases, all expressed predominantly in the xylem tissue [50]. Five distinct laccases have been shown to be present in the xylem tissue of *Populus*

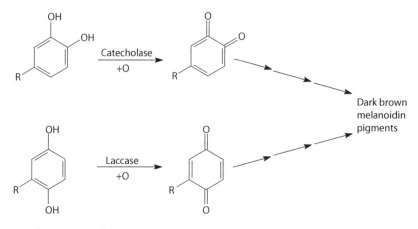

FIGURE 11.3   Comparison of reactions catalyzed by catecholase (*o*-DPO) and laccase (*p*-DPO). (From Walker, J.R.L., Enzymatic browning in fruits: Its biochemistry and control, in *Enzymatic Browning and Its Prevention*, Lee, C.Y. and Whitaker, J.R., Eds., American Chemical Society, Washington, DC, 1995. With permission.)

euramericana [51]. Other higher plant species also appear to contain laccases, although their characterization is less convincing [52]. The presence of laccase has been documented in virtually every fungus examined for it [46].

11.2.1.1 Physiological Significance of Polyphenol Oxidase

PPO has been ascribed many potential functions in plants and has long been of interest because of its many undesirable side effects especially in the food industry. The functions of PPOs in higher plants are not fully known but they are thought to play a key part in the plant's defense mechanism against disease causing microorganisms and insect attack [45]. When microbial infection occurs, the cell integrity is broken and the enzymatic reaction takes place. An impervious scab of melanin forms that acts as a physical antimicrobial barrier. The quinones formed during the reaction are known to denature proteins in the invading microorganisms and the polymeric phenol complexes can act as inhibitors of microbial growth [36]. Whether due to mechanical injury or to cellular disruption from disease, the quinones produced by the resulting interaction of the PPO-phenolic compounds are very reactive, making them good candidates for their involvement in protection from other organisms. Thus, the disruption of the plastid results in the activation of latent PPO, which reacts with the phenolics released from the vacuole [53]. PPO is a relatively stable enzyme and the release of latent activity can take place over long time periods [54], resulting in the continuous production of quinones, despite the inhibitory properties of the end products of PPO [53]. Quinones can also behave as direct toxic compounds against pathogens [55,56].

The possible function of a PPO that has been speculated is in the modulation of the photosystem I (PSI) reduction of molecular oxygen (the Mehler reaction or pseudocyclic photophosphorylation) [57]. Normally, photosynthetic energy is utilized in the reduction of intermediates necessary for biosynthetic processes ($NADP^+$, nitrite, etc.), but it has been suggested that under circumstances that limit carbon dioxide availability, destructive, high oxygen concentrations can occur temporarily. PPO acts to oxidize the reduced quinones (internal to the membrane) generated by PSI activity, thereby preventing molecular oxygen accumulation and regenerating substrate for the continued operation of the cycle [58].

In addition to its general occurrence in plants, PPO is also found in microorganisms, especially in fungi, and in some animal organs. PPO plays an important role in the sclerization of insects and crustaceans during the molting cycle [59]. Phenoloxidase in the cuticle of the spiny lobsters is found in a latent form, pro-phenoloxidase (pro-PO). Activation of pro-PO *in vivo* appears to be related to the molting cycle and may be important in the regulation of pigment formation. In order to increase in size, crustaceans must replace their confining exoskeletons with larger ones. This periodic shedding is termed molting. The pro-PO activating system was reportedly initiated by microbial products, possibly through limited proteolysis. In crayfish, β-1,3-glucans from fungi and algae specifically activate pro-PO and hemocytes. The pro-PO activating system has been proposed to comprise a recognition system, which is important in host defense in arthropods [60].

Although it is very likely from present information that laccases are present throughout the plant kingdom, it is difficult to determine their activity in crude extracts against a background of PPO and peroxidase activity. The occurrence, development, and function of higher plant laccases are still under investigation. Fungal laccases have been studied far more extensively. Their function in fungi is quite varied and includes the regulation of morphology, the control of virulence and nutrition, and their ability to delignify woody tissues [46].

11.2.1.2 Characteristics of Polyphenol Oxidase

Plant PPOs, which are encoded by nuclear genes, are localized in plastids [61], where they seem to be associated with the internal thylakoid membranes, thus remaining physically separated from their phenolic substrates stored in the vacuole [22,39,62]. Therefore, the oxidation reactions are only initiated after senescence or an environmental stress (such as pathogen attack or injury) where the biological barriers between the enzymes and the substrates are destructed, allowing PPO and phenolic substrates to mix. The enzyme is located in a variety of cell fractions, both in the organelles where it may be tightly bound to membranes and in the soluble fraction. Inactive or latent PPO forms have been frequently reported in plants and require activation. Many plant PPO preparations from different tissues display latency, assessed by an increase in enzyme activity at pH 6.0 or higher when treated with activating agents such as proteases (trypsin or proteinase K) or low millimolar concentrations of SDS. Examples include leaves, fruits, and roots [22]. In fruits, PPO latency has been reported in the loquat fruit [22], mangoes [63], avocados [64], and grapes [65]. The conversion of particulate forms of the enzyme to soluble forms occurs in fruits when they are exposed to stress conditions or during ripening and storage.

PPOs isolated from food sources are oligomers and contain one copper prosthetic group per subunit [66]. There are two copper ions at the active center but it is not known if they complex with oxygen, or the substrates, or both, or whether one copper complexes with oxygen and the other with the substrates [67]. The enzyme shows both oxidase (oxidize o-diphenols) and monooxygenase (oxidize monophenols) activity yielding o-quinone in both cases. In the initial step, oxygen would complex with the active-site of the cuprous ion, and this complex undergoes attack by the bound catechol. The loss of water from such an initial adduct could yield a molecule with the vinylogous ozone moiety (Figure 11.4) [68].

The affinity of PPOs for their phenolic substrates depends on the stereochemistry of the substrate together with the position and nature of its substitutions (e.g., hydroxyl, methyl, and glycosyl) [39]. The substrate specificity of PPO changes with the type and the source of the enzyme. Whereas PPOs from animal tissues are relatively specific for tyrosine and DOPA, those from fungi and higher plants act on a wide range of monophenol and o-diphenols [14].

Monophenols are more slowly acting substrates as they have to be hydroxylated prior to their oxidation to the corresponding o-quinones [7]. The commonest natural substrates for monophenol oxidase are probably tyrosine and p-coumaric acid or their derivatives. All o-diphenol oxidases require the basic o-dihydroxyphenol structure for oxidase activity so that catechol is the simplest possible, but not necessarily the best, substrate; 4-methyl catechol is usually the fastest [45]. The rate of oxidation of o-diphenols by PPO increases with increasing electron withdrawing power of substituents in the *para* position. o-Diphenol substitution ($-CH_3$) at one of the positions adjacent to the $-OH$ groups prevents oxidation. These positions should remain free for oxidation to take place [14].

Most PPO preparations from potato, apple, mushroom, and bean possess both monophenol and diphenol oxidase activities, whereas those from tea leaf, tobacco, mango, banana, pear, peach, and sweet cherry have been reported not to act on monohydroxyphenols [28]. Whether a single enzyme system exhibits both mono- and diphenol oxidase activities is still unclear. It was suggested that both cresolase and catecholase functions are catalyzed by a single site. Verdedoncella apple PPO showed both monophenol and diphenol oxidase activities with a reaction mechanism involving one

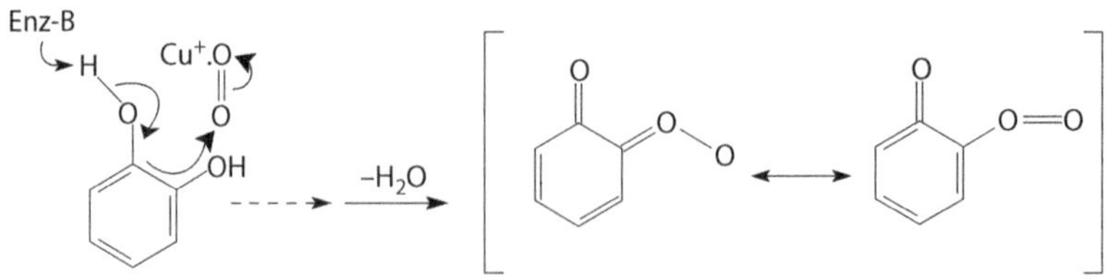

FIGURE 11.4 Role of copper in the oxidase activity of the enzyme. (Adapted from Walsh, C., *Enzymatic Reaction Mechanism*, W.H. Freeman & Company, San Francisco, CA, 1979, 461. With permission.)

single active site with two enzyme forms [69]. When both monophenol and diphenol oxidases are present in plants, the ratio of the monophenol to diphenol oxidase activity is usually 1:10 or as low as 1:40 [70].

The specificity of PPO toward its phenolic substrates has been extensively studied and many discrepancies in the apparent K_m values appear between species, cultivars, and experimental conditions [71]. The affinity of plant PPO for the phenolic substrates is generally low. The K_m is high, 2–6 mM [70], which is higher than most of the fungi and bacteria, around 0.1 mM. The affinity to oxygen is also relatively low, ranging from 0.1 to 0.5 mM [14]. Only a few studies have been devoted to the effect of the second substrate, i.e., oxygen. With apple PPO, it was shown that the reaction is an ordered mechanism whereby oxygen is the first substrate to bind to the enzyme [70]. In some fruits, the best substrate of PPO is a compound not occurring as a phenolic constituent. Simple phenols such as catechol are commonly used in phenolase assays in spite of not always being found in association with the enzyme. However, the presence of a potential substrate does not ensure its role as an actual substrate in phenolase action leading to enzymatic browning [28]. Laccases demonstrate a higher affinity for molecules with a high steric environment than catechol oxidases do. Plant laccases generally have a low specificity with regard to the reducing substrate [39].

The optimum pH for the activity of an enzyme preparation from any one source usually varies with different substrates and is characteristic of the substrate as well as the enzyme preparation. The strawberry PPO showed maximum activity at pH 5.5 with catechol, and at pH 4.5 with 4-methylcatechol as substrate [72]. The type of buffer and the purity of the enzyme affect the optimum pH. Isoenzymes have also different pH optima. Enzyme preparations obtained from the same fruit or vegetable at various stages of maturity have been reported to differ in optimum pH of activity as well [14]. Very different pH optima have been reported for PPO activity ranging from acidic pH (3.5) in grape [73] to basic pH (9.0) in Amasya apple [74]. Two pH optima for PPO have also been reported in literature. The pH optima for the Jonagored apple PPO are 5.0 and 7.5 [75] and for the eggplant PPO, 4.0 and 6.5–7.5 [71].

Similar to the other properties of PPO, there is also inter- and intraspecies diversity in the number of PPO isoforms and their molecular masses. Depending on the source of the enzyme, the molecular weights vary between 29,000 and 200,000 with subunit molecular weight from 29,000 to 67,000. This variation in molecular mass was supposed to be the result of the modification of PPO proteins during purification, both by proteolytic cleavage and by chemical modification with reactive quinones [22]. The plantain PPO is present in multiple forms in which the molecular weights range between 30,000 and 70,000 [76]. Kiwifruit contained four catecholase isoenzymes with molecular weights of 15,000, 20,000, 25,000, and 45,000 Da. The dominant isoenzyme was 25,000 as determined by gel electrophoresis [77]. The banana PPO has four isoenzymes with a peptide molecular weight of 31,000 in SDS-Page electrophoresis and 62,000 in sucrose gradient ultracentrifugation [78]. The PPO in wheat grains showed three to five isoenzymes for different varieties [79]. For lobster PPO, it was reported that PPO from the Western Australian lobster had two isoforms while the Florida spiny lobster PPO had three isoforms [59]. Also, PPO is found in multiple forms in pear [80], strawberry [72], avocado [64], and mushroom [81].

Properties discussed indicate high heterogeneity in the expression of PPO concerning enzymatic activity (optimum pH, latency, specificity, etc.), number of isoforms, and apparent molecular mass. This heterogeneity may be due to differential genomic expression of the species, the physiological state or the nature of the tissues studied [71].

11.2.1.3 Mechanism of Oxidation

The overall mechanism of plant PPOs is still incompletely understood although the mechanism for fungal PPO (*N. crassa*) was presented by Lerch [82]. A model was proposed in which the site

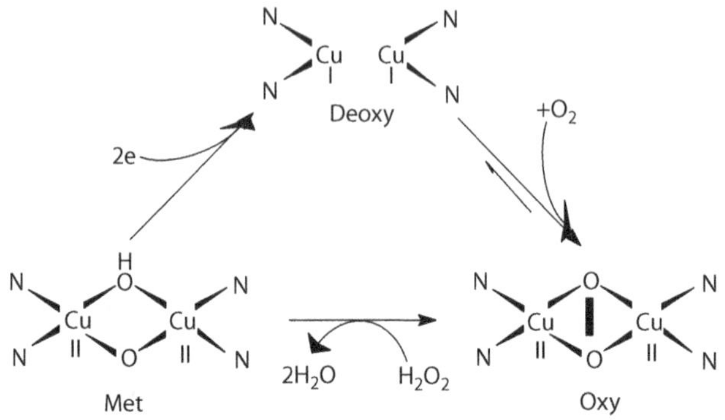

FIGURE 11.5 Interrelation of the three functional states of tyrosinase. (From Lerch, K., Tyrosinase: Molecular and active site structure, in *Enzymatic Browning and Its Prevention*, Lee, C.Y. and Whitaker, J.R., Eds., American Chemical Society, Washington, DC, 1995, 64. With permission.)

of interaction with the phenolic substrate, whether mono- or diphenol, was based on the binuclear center of copper. The basic functional molecular unit for the fungal enzyme appeared to be a single-chain protein with two copper atoms per molecule, liganded in part by histidine [82,83]. The active site of copper appeared to be binuclear and occurred in different functional states: *met*, *oxy*, and *deoxy* [7]. Interrelation of these three functional states of the active site of PPO is shown in Figure 11.5. The binuclear copper site of *met* PPO consists of two tetragonal Cu(II) ions bridged by one or two ligands. The *oxy* form can be obtained by treatment of *met* PPO with reducing agents in the presence of molecular oxygen. *Deoxy* PPO was assigned a bicuprous Cu(I)–Cu(I) structure. Besides reacting with molecular oxygen reversibly, *deoxy* PPO also binds one carbon monoxide per molecule [38]. These three forms of PPO account for both the mono and diphenolase activities [84].

Based on the information on the binuclear copper complex and on the results of how different anions and organic ligands interact with the site, pathways of hydroxylation and oxidation reaction mechanisms of tyrosinase were proposed as shown in Figure 11.6 [18]. The proposed mechanism of action for *N. crassa* PPO appears to fit the data for most PPOs. The proposed mechanisms for hydroxylation and dehydrogenation reactions with phenols probably occur by separate pathways but are linked by a common *deoxy* PPO intermediate (*deoxy* in Figure 11.6). Proposed intermediates in the *o*-diphenol oxidation pathway are shown in Figure 11.6a.

The diphenol oxidase activity involves the oxidation of two molecules of *o*-diphenol to two molecules of *o*-quinone with the concomitant 4e⁻ reduction of molecular oxygen to two molecules of water. For this reaction, the binuclear copper site in *met* and *oxy* PPO is geometrically correct for axial coordination of both orthophenolic oxygens to the Cu(II) ions with a Cu–Cu distance of 3.4–3.6 Å. This would allow for an efficient 2e⁻ transfer from the *o*-diphenol substrate to the binuclear site [38]. In this proposed pathway, O_2 is bound first to the two Cu(I) groups of *deoxy* PPO to give *oxy* PPO in which the O_2 has the characteristics of peroxide [85]. The two Cu(II) groups of *oxy* PPO then bind to two hydroxyl groups of catechol, replacing the two water molecules or OH groups, to form the O_2·catechol·enzyme complex. The catechol is oxidized to benzoquinone, leaving the enzyme as *met* PPO. Another molecule of catechol binds to *met* PPO, and is oxidized to benzoquinone, in the process reducing the enzyme to *deoxy* PPO, thereby completing the cycle. Two catechol molecules are oxidized in a complete cycle, consuming two atoms of O_2 [86].

The mechanism of monophenol hydroxylation by PPO is shown in Figure 14.6b. *In vitro*, the reaction begins with *met* PPO of cycle (a). PPO in its "resting" state is mainly the *met* form, which is unable to act on monophenols. In most cases, a diphenol is necessary as the reducing agent to obtain the *deoxy* form, the only one capable of reacting with molecular oxygen and continuing in the

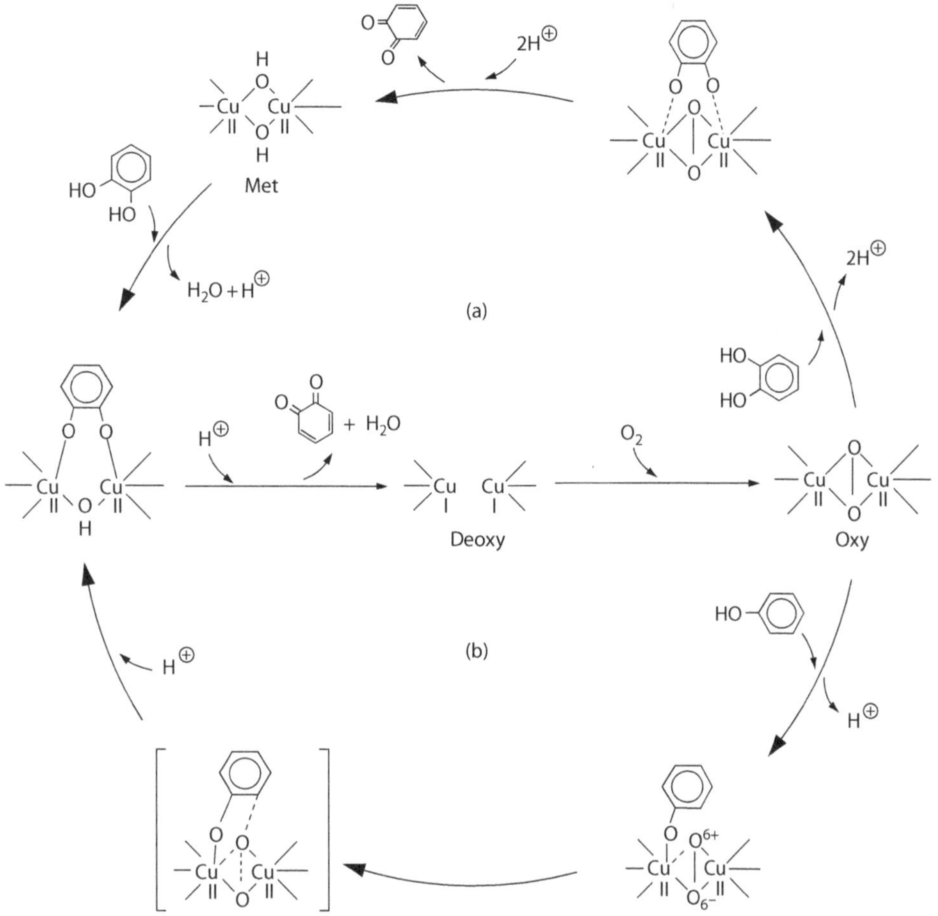

FIGURE 11.6 Proposed kinetic scheme depicting the mechanisms of oxidation of *o*-diphenol (catechol; top a) and monophenol (phenol; bottom b) for the *N. crassa* polyphenol oxidase. (From Whitaker, J.R. and Lee, C.Y., Recent advances in chemistry of enzymatic browning, in *Enzymatic Browning and Its Prevention*, Lee, C.Y. and Whitaker, J.R., Eds., American Chemical Society, Washington, DC, 1995, 3. With permission.)

catalytic action [87]. For this reason the monophenolase activity presents a characteristic lag time which exists until a sufficient amount of catechol (needed to reduce the *met* form to the *deoxy* one) is produced by the small amount of the *oxy* form generally present in the resting enzyme preparations. Obviously, this lag time cannot be observed (or is proportionally reduced) in the presence of small amounts of catechols, reducing agents, or traces of transition metal ions such as Fe^{2+} [84]. These reducing compounds convert the *met* PPO to the *deoxy* form, which can then bind O_2 to give the *oxy* PPO. The *oxy* form then binds the monophenol *via* one of the Cu(II)s, displacing the water molecule. Subsequently, the steric orientation of the bound monophenol and O_2 are altered to place the *o*-position of the monophenol adjacent to the second Cu(II) in the *oxy* PPO, so that the *o*-position is hydroxylated by the bound –O–O– moiety. The initial product of the monophenol is catechol bound to both Cu(II)s. In the last step, the bound catechol is two-electron oxidized and released as quinone, reforming the Cu(I)–Cu(I) deoxy PPO, to complete the cycle. The *deoxy* PPO can again bind O_2 without cycling through the *met* form [86]. Only the first cycle of the hydroxylation on a monophenol requires starting at the *met* PPO; all subsequent cycles begin with the *deoxy* PPO [18]. Importantly, the proposed catalytic cycle directly leads to the quinone product and no catechol is released in between [88].

The mechanisms for monophenol and *o*-diphenol oxidation by PPO indicate that *o*-quinone is the final product of the enzyme. The quinone formation is both enzyme and oxygen dependent. Once this has taken place, the subsequent reactions occur spontaneously and no longer depend on the presence of PPO or oxygen. The first reaction is thought to be a secondary hydroxylation of the *o*-quinone or of excess *o*-diphenol. The resultant compound (triphenolic trihydroxybenzene) interacts with *o*-quinone to form hydroxyquinones which can undergo polymerization and are progressively converted to red and red-brown polymers, and finally to the brown melanins which appear at the site of plant tissue injury [7].

The mechanism of laccase has been reported as a two-step reaction. The first step in the oxidation of quinol by laccase was the formation of the semiquinone, with the transfer of an electron from the substrate to the copper in the enzyme. The second step was a nonenzymatic disproportionation reaction between two semiquinone molecules to give one molecule of quinone and one of quinol. The function of copper and electron transfer in the reaction mechanism has been studied by using inorganic ions, electron paramagnetic resonance, and spectrophotometric methods [14].

11.2.1.4 Phenolic Substrates

Phenolic compounds are secondary metabolites that are derivatives of the pentose phosphate, shikimate, and phenylpropanoid pathways in plants [89]. Structurally, the phenolic compounds comprise an aromatic ring, bearing one or more hydroxyl substituents, and range from simple phenolic molecules to highly polymerized compounds [90]. Despite this structural diversity, the group of compounds is often referred to as polyphenols. These compounds play an important role in growth and reproduction, providing protection against pathogens and predators [90], besides contributing toward the color and sensory characteristics of fruits and vegetables [91].

Generally speaking, human and animal PPO tend to be relatively specific for L-tyrosine and L-DOPA, whereas a wider substrate range (regardless of stereochemistry) is known for plant and fungal enzymes [56]. Phenolic acids and flavonoids are the main dietary phenolic compounds which participate in browning reactions. Phenolic acids consist of two subgroups, i.e., the hydroxybenzoic and hydroxycinnamic acids. Hydroxybenzoic acids include gallic, *p*-hydroxybenzoic, protocatechuic, vanillic, and syringic acids. Hydroxycinnamic acids, on the other hand, are aromatic compounds with caffeic, ferulic, *p*-coumaric, and sinapic acids being the most common. The most important member of this group of compounds in food material is chlorogenic acid, which is the key substrate for enzymatic browning, particularly in apples and pears [7]. Flavonoids constitute the largest group of plant phenolics, accounting for over half of the 8000 naturally occurring phenolic compounds [92]. These polyphenolic compounds are characterized by two aromatic cycles (A and B rings) linked by a heterocycle (C ring). They are classified according to the oxidation degree of the C ring, and include flavonols, anthocyanins, and flavon-3-ols [39]. In food material, the important flavonoids are the catechins, leucoanthocyanidins (sometimes called the food tannins), anthocyanins, and the flavonols [7]. The common flavonoids are usually poor substrates of *o*-diphenol oxidases but they may be oxidized by diphenol oxidases *via* coupled reactions. For example anthocyanins and procyanidins are poor substrates for *o*-diphenol oxidases but undergo significant breakdown in the presence of favorable substrates such as chlorogenic acid. These reactions could lead to serious losses of pigment in some fruits and have been demonstrated in grapes [26] and strawberry [93]. The flavonols also participate in browning reactions and are widely distributed in plant tissues. The most commonly occurring flavonols are kaempferol, quercetin, and myricetin [7].

In principle, any simple monophenol and the corresponding diphenol could behave as a PPO substrate, taking into due account that diphenols are quinonized much faster than monophenols, owing to K_s as well as k_{cat} differences. As a general rule, bulky and/or crowded substituents too close to the phenolic hydroxyl, prevent or at least render more difficult any enzyme/substrate interaction [84]. In the majority of cases chlorogenic acid, caffeic acid, catechin, 3,4-dihydroxyphenylpropionic (hydrocaffeic) acid, and 4-methyl catechol were all readily oxidized but the rate of oxidation was much reduced with 3,4-dihydroxyphenylacetic acid and 3,4-dihydroxybenzoic (protocatechuic) acid while the isomeric 2,3-dihydroxybenzoic acid was not oxidized to *o*-diphenol oxidases from apples or other fruits. These support the view that the nature of the sidechain is critical and therefore may play a part in the enzyme/substrate interaction [45]. Structures of some phenolic substrates are given in Figure 11.7. Aromatic amines and *o*-aminophenols which are structural analogues of monophenols and *o*-diphenols are also oxidized by PPO. They were found to undergo the same catalytic reactions as ortho hydroxylation and oxidation. In the presence of tyrosinase, arylamines are converted to an *o*-aminophenols (hydroxylation) and subsequently oxidized to *o*-quinoneimines [38]. 2-Amino-3-hydroxybenzoic (3-hydroxyanthranilic) acid was studied and it was concluded that, besides being a poor PPO substrate, it behaves as a catechol both in stimulating the hydroxylase activity of the enzyme toward monophenolic substrates and in yielding the corresponding *o*-quinoneimine carboxylic acid as the product [84].

When a PPO substrate, either mono- or diphenolic, contains a sidechain (usually in the 4-position relative to the phenolic hydroxyl) bearing a nucleophilic moiety such as $-NH_2$, $-SH$, $-COO^-$, etc., the formed *o*-quinone tends to undergo an intramolecular attack leading to a bicyclic catechol, which is again a PPO substrate, and in addition can give rise to a redox exchange with the still noncyclized quinone. The extent of such cyclization reactions is pH-dependent, so for example in the case of tyrosine and DOPA it is severely reduced at low pH values, where the amino group exists as the conjugate acid and is therefore unable to carry out the nucleophilic attack. The ability attitude to intramolecular cyclization following quinonization is one of the most outstanding features of several PPO substrates and is of the highest importance with respect to the metabolism of phenolics [84].

In some fruits, PPOs use other phenolic substrates; for example a relative of DOPA, 3,4-dihydroxyphenylethylamine (dopamine), is the major substrate in bananas and DOPA is the natural substrate in the leaves of broad beans. The grape PPO acts on *p*-coumaryl and caffeoyl-tartaric (caftaric) acids while dates contain an unusual combination of diphenol oxidase substrates including a range of caffeoyl-shikimic acids; these are analogous to the ubiquitous isomers of chlorogenic acid [45].

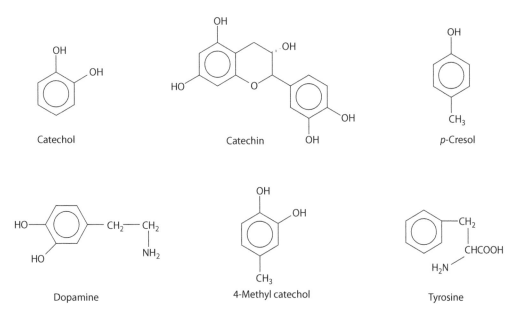

FIGURE 11.7 Structures of some PPO substrates.

11.2.2 Control of Enzymatic Browning

The browning of foods during processing and storage decreases their sensory qualities due to associated changes in the color and flavor, besides increased softening and altered nutritional properties [94]. Therefore, its control is essential for preserving the quality of a food. Various techniques and mechanisms have been developed over the years and most of them try to eliminate one or more of the essential components (oxygen, enzyme, copper, or substrate) from the reaction. The exclusion of oxygen from the cut surface of fruits and vegetables prevents the browning reaction since it occurs rapidly upon exposure to oxygen. The removal of copper by the chelating agents is another way to prevent browning because PPO requires the copper prosthetic group in order to act in enzymatic browning. Chemical modification of the phenolic substrates and inhibition of the enzyme by certain chemicals are other alternatives to prevent browning. On the other hand, inactivation of the enzyme by heat treatment is also commonly used for the control of browning in canned or frozen fruits and vegetables, but not applicable in the storage of fresh foods.

Heat treatment is the most widely utilized method for stabilizing foods because of its capacity to destroy microorganisms and to inactivate enzymes. However, several problems may arise through the use of heat and this leads to unfavorable texture changes and the development of off-flavors. Such problems may occur possibly in the processing of pre-peeled potatoes, apples, pears, and peaches. It is therefore essential to control the heating time very carefully at high temperatures, so that the enzymes are inactivated while avoiding significant changes in flavor and texture. In general, exposure of PPO to temperatures of 70°C–90°C destroys their catalytic activity [28]. However, the time taken for the complete heat inactivation of PPO varies considerably with the food product.

Chutintrasri and Noomhorm [95] studied the thermal inactivation of PPO in pineapple puree. Thermal treatment between 40°C and 90°C in relation to exposure time showed that the inactivation of PPO followed first-order kinetics and high activation energy reflects a greater sensitivity of PPO to temperature change. The PPO activity reduced by approximately 60% after exposure to 40°C–60°C for 30 min. Denaturation increased rapidly above 75°C and residual activity was only about 7% at 85°C and 1.2% at 90°C for 5 min. In another study, heat inactivation of crude PPO from six apple cultivars showed that the time required for 50% reduction of initial activity at 68°C and 78°C was in the range of 25.6–91.2 min and 2.4–4.3 min, respectively [96]. They also reported that PPO in apples was relatively more heat stable than PPO in most fruits. In a recent study, PPO activity in Algerian dates was found to be more heat stable than peroxidase activity [97].

A combined effect of heat treatment (70°C for 2 min) and chemicals (ascorbic acid, citric acid, EDTA, sodium metabisulfite) were studied by Almeida and Nagueira [98] in avocado, banana, apple, pear, peach, potato, eggplant, mushroom, and hearts-of-palm. They reported that, with the exception of PPO from avocado, the most adequate method in the control of PPO was a combination of ascorbic acid, citric acid, and heat treatment.

Lee et al. [99] studied the effect of blanching on catalase, lipoxygenase, and PPO activities in green beans. They found that the enzymes studied were inactivated completely at 82°C and above for 3.5 min. Furthermore, Ma et al. [100] reported that water blanching at 100°C for 3 min or 94°C for 5 min prevents darkening in frozen sweet potatoes by significantly decreasing the PPO activity.

Adverse effects of blanching in terms of losses in vitamins, flavors, colors, texture, carbohydrates, and other water-soluble components have led to the application of combined microwave heating and blanching for a short time at moderate temperatures and it has been reported to be most efficient method for enzyme inactivation. Vina et al. [101] compared the quality of Brussels sprouts treated at 50°C by microwave then blanched in boiling water with samples blanched directly in water at 100°C. Microwave pre-blanching showed no deleterious effect on total chlorophyll, radical scavenging activity, total flavonoids, and ascorbic acid content of Brussels sprouts. A similar study has been presented by Devece et al. [102]. A combined microwave and hot-water bath treatment of mushrooms achieved complete PPO inactivation in a short time. Mushroom weight loss and shrinkage were also prevented by short processing time.

Ultrafiltration, an alternative method for the control of enzymatic browning, has been studied by many workers [103–105]. This technique is used for the separation of the enzyme from liquid systems based on differences in molecular weights of its contents [106–108]. Tanada-Palmu et al. [109] suggested that polysulfone membranes with a cut-off of 20 kDa were sufficient for the removal of PPO from banana extract at 600 kPa transmembrane pressures. In a similar study, ultrafiltration of banana juice through a 10,000 molecular weight cut-off membrane removed all PPO and reduced juice browning [110]. Goodwin and Morris [105] studied the effect of ultrafiltration on wine quality and browning. They found that the efficiency of ultrafiltration depends on the cultivar type. Hence, ultrafiltration with 50,000 molecular weight cut-off membrane before fermentation slowed oxidation in the Seyval wine, but not in the Vidal wine.

High hydrostatic pressure (HHP) processing is another method for the preservation of foods. It is a nonthermal preservation technique and causes little or no change in the organoleptic and nutritional quality of product unlike most conventional heat treatments. Enzymatic reactions may be enhanced or inhibited by pressure and pressure-induced changes in the enzyme–substrate interaction, changes in the reaction mechanism, the effect on a particular rate-limiting step, or the overall catalytic rate [111].

Microorganisms and deteriorative enzymes can be inhibited or inactivated depending on the amount of pressure and time applied to the product. Microorganisms (other than spores) can be inactivated at mild pressures (<300 MPa), but most of the food quality-related enzymes are rather pressure-stable [112], and their inactivation is strongly dependent on the type of enzyme, pH, medium composition, temperature, and so on [113].

Ballestra et al. [114] studied the effects of high pressure treatments (100–500 MPa at 4°C for 10 min) on freshly sliced mushrooms and on liquid extract. They have reported that pressure above 200 MPa results in significant enzymatic browning and PPO activation in treated mushrooms. Furthermore, PPO activation enhances if pressure is increased from 200 to 500 MPa both in whole tissue and in liquid extract. Therefore, it has been concluded that stabilization of mushroom by high pressure alone could not be possible and a combination of HHP with thermal or chemical treatments is necessary. The combined treatments, like mild heating and chemicals have also been suggested by other workers in order to increase the efficiency of HHP. For example, Phunchaisri and Apichartsrangkoon [115] studied the combined effect of heat treatment and HHP on PPO in fresh lychees and lychees preserved in syrup. A pressure of 600 MPa, at 60°C for 20 min caused extensive inactivation of PPO (90%) in fresh lychee but not in lychee preserved in syrup due to baroprotection by the syrup.

In a separate study, the combined treatment of high pressure and mild heat at 400 MPa and 70°C for 10 min results in 95% inactivation of food quality-related enzymes in carrot juice [116]. At the optimum condition, 97.0% of lipoxygenase, 95.8% of PPO, and 89.4% of pectinmethylesterase were inactivated. It has been reported that combined treatment was very efficient for the production of high-quality carrot juice. Hernandez and Cano [117] also studied the combined effect of HHP (50–500 MPa) and heat treatment (20°C–60°C) on pectinmethylesterase, peroxidase, and PPO activities in tomato puree. The reduction in enzyme activities was reported as 32.5% (at 150 MPa for 30°C), 25% (at 350 MPa for 20°C), and 10% (at 200 MPa for 20°C) for pectinmethylesterase, peroxidase, and PPO, respectively.

Combined treatments of HHP with some chemicals were also reported to prevent enzymatic browning during the storage of foods. Prestamo et al. [118] revealed that treatment of apple fruit after HHP treatment (400 MPa at 5°C for 30 min) with 20 mM ascorbic acid prevents the formation of brown color after 2 months of storage at 5°C.

The HHP treatment on peach puree with and without antibrowning agents has been studied by Guerrero-Beltran et al. [119]. The PPO activation was observed at 103, 207, and 310 MPa in the presence of ascorbic acid. However, increase in pressure to 517 MPa caused inhibition of PPO in peach purees containing ascorbic acid. Addition of cysteine only caused 50% inactivation of PPO but 99% inactivation was obtained after HHP treatment at 517 MPa. They concluded that treatments at high pressure for long times either in the presence or absence of antibrowning agents prevent peach puree from discoloration during storage at low temperatures.

Manothermosonication (MTS), a combined treatment of heat and ultrasound under moderate pressure, is another alternative to conventional heat treatment in order to inactivate enzymes and microorganisms [120–122]. The ultrasound generates the cavitation or bubble implosion in the media. This implosion can cause inactivation of the enzyme and destruction of microorganisms. The simultaneous pressure treatment maximizes the intensity of the explosion, which increases the level of inactivation. The MTS technique avoids the adverse effects of elevated temperatures on quality and also results in reduced energy requirements and therefore reduced costs.

Lopez et al. [120] studied the inactivation of three food quality enzymes—peroxidase, PPO, and lipoxygenase—by MTS. They revealed that a synergistic effect of temperature, pressure, and ultrasound results in reduced enzyme resistance and heat treatment for inactivation. Ultrasonic wave amplitude was found to be more effective on enzyme destruction than static pressure at constant temperature. This combined treatment was reported to be very useful in solving problems caused by thermostable enzymes.

Supercritical carbon dioxide (SCCO$_2$) was reported to inactivate peroxidase [123], pectinesterase [124], and PPO [125]. Chen et al. [126] studied the inactivation of PPO from Florida spiny lobster, broken shrimp, and potato. They reported that SCCO$_2$ causes a change in secondary structures of the PPO and the activity of the enzyme declines. A similar effect of SCCO$_2$ was observed for the PPO in apple juice [127]. The residual activity of PPO was 38.5% at 30 MPa at 55°C for 60 min indicating that SCCO$_2$ treatment could effectively prevent the browning of cloudy apple juice. However, the restoration of PPO residual activity during storage at 4°C was observed. Similarly, the potato PPO treated by high pressure CO$_2$ regained 28% of the original activity during the first 2 weeks of frozen storage [126].

Chemical agents are applied during processing and storage of foods in order to prevent the deteriorative effect of enzymatic browning. The use of sulfiting agents, such as sulfur dioxide and sodium bisulfite, to inactivate enzymes is well known in the industry. Indeed, sulfite is believed to be the most effective chemical additive used to prevent enzymatic discoloration in most produce. However, the FDA has limited their use to only a few applications because of adverse health effects in certain populations [128]. Representative examples of other chemical preservation agents that are alternatives to sulfites include acidulants (citric, malic, and phosphoric acids), chelators (EDTA, phosphates, and organic acids), reducing agents (ascorbic acid and analogs, cysteine, and glutathione), complexing agents (cyclodextrins), and enzyme inhibitors (aromatic carboxylic acids and substituted resorcinols).

Acidulants as preservation agents are widely used for controlling enzymatic browning. The most commonly employed are citric, malic, and phosphoric acids that occur naturally in tissues. In general their action is to lower tissue pH and thus to decrease the rate of enzymatic browning. Citric acid has long been used to prevent enzymatic browning due to its double inhibitory action on lowering the pH of the medium and chelating with the copper moiety of the enzyme.

Ascorbic acid, the best known alternative to sulfite, is commonly used to prevent enzyme discoloration of fruits by reducing the colorless *o*-quinones to diphenols [129]. In the study of Rocha and Morais [130], the inhibitory effect of three chemical dips using ascorbic acid (AA), citric acid (CA), and calcium chloride (CC) on the PPO activity of minimally processed apple during cold storage showed that an ascorbic acid dip (42.6 mM) of 5 min duration was the most efficient chemical treatment in reducing the PPO activity. A 92% inhibition was achieved after 7 days of storage at 4°C. On the other hand, a similar combination of these chemicals (ascorbic acid, citric acid, and sodium chloride) was found to be very effective on controlling browning in the puree of cherimoya fruit during freezing, storage, and thawing [131].

Cysteine is an alternative reducing agent and reduces PPO activity due to the formation of the colorless Cys-quinone-adducts [132]. İyidoğan and Bayındırlı [133] studied the effect of anti-browning agents (L-cysteine, kojic acid, and 4-hexylresorcinol) on the enzymatic browning in the cloudy apple juice and found that L-cysteine was the most significant one. They also reported that 89.2% inhibition of PPO in cloudy apple juice was obtained with the combination of 3.96 mM L-cysteine, 2.78 mM kojic acid, and 2.34 mM 4-hexylresorcinol at the end of 24 h storage period.

Chelators have been used in various food processing applications for enzyme inactivation [129]. EDTA is a chelating agent permitted for use in the food industry as a chemical preservative. Maximum chelating efficiency occurs at the higher pH values where carboxyl groups exist in a dissociated state [134]. EDTA is generally used in combination with other chemical treatments for the prevention of enzymatic browning in foods. A typical combination of anti-browning agents might consist of a chemical reducing agent (ascorbic acid), an acidulant (citric acid), and a chelating agent (EDTA) [37]. Ascorbic acid also has a chelating effect on the prosthetic group of PPO. Plantain-PPO is inhibited by copper complexing agents and reducing agents [76].

Substituted resorcinols and aromatic carboxylic acids were reported as potent inhibitors of PPO. 4-Hexylresorcinol exhibits potential use in the food industry due to its nontoxic, nonmutagenic, and noncarcinogenic properties. This substance interacts with PPO and forms an inactive complex incapable of catalyzing the browning reaction [135]. It has been reported that the mixtures of 4-hexylresorcinol and ascorbic acid prevent the browning of vacuum-packaged apple slices [136]. A study on the inhibition of PPO in mango puree with 4-hexylresorcinol, cysteine, and ascorbic acid has shown that the synergistic effect of higher concentrations of cysteine or ascorbic acid with 4-Hexylresorcinol results in reduced PPO activity and less darkening [137].

The inhibitory character of aromatic carboxylic acids is related to the presence of the benzene ring. The inhibition of PPO by benzoic and cinnamic acid series has been reported [138–140] and it was suggested that better inhibition was obtained when the −COOH group was directly attached to the benzene ring, while hydroxylation, methylation, or esterification considerably decreased the inhibitory effects of these acids. Robert et al. [141] studied inhibitory effects of various carboxylic acids on two forms of palmito (*Acanthophoenix rubra*) PPO. Cinnamic acid was found to have the greatest inhibitory effect and it was concluded that acids with a double bond between the benzene ring and the carboxylic group showed the highest inhibitory effect. This inhibitory effect was decreased by substitutions on the benzene ring.

A number of natural antibrowning agents have been identified by researchers. The inhibitory effect of honey on enzymatic browning of fresh-cut apples was studied either by simply immersing apple slices or vacuum impregnating in a honey solution. Vacuum impregnation with honey was found to be more effective in controlling browning than simple immersion treatment. So honey treatment in combination with the vacuum impregnating operation may have a great potential for developing high quality fresh cut fruits [142]. In another study, the effects of the onion extract on pear PPO and browning of pear have been investigated [143]. The PPO of pear was inhibited by the onion extract and the inhibitory effect of the extract increased with increasing temperature and time. It has been suggested that the retardation of pear juice browning by onion extract seems to be caused by the inhibition of pear PPO. A similar study on the inhibitory effect of onion extract on peach PPO showed that heated onion extract exhibits stronger inhibitory effect and prevents browning in peach juice [144]. Alternatively, pineapple juice was found to be an effective browning inhibitor for both fresh apple rings kept in air or vacuum packed and for dried apples [145].

11.3 NONENZYMATIC BROWNING REACTIONS

Nonenzymatic browning (NEB) reactions occur during the thermal processing and storage of foods. These reactions cause some desirable and undesirable chemical and structural changes in foods. The mostly known changes are formation of brown color, production of flavors or off-flavors, nutritional loss followed by the reduction of ascorbic acid, amino acids, and invert sugars, and formation of some toxic and mutagenic compounds such as imidazols, HMF, acrylamide, advanced glycation end-products (pentosidine and argpyrimidine), and melanoidines. There are many different substrates and pathways for NEB. Oxidation of ascorbic acid in citrus fruits may cause loss of vitamin C with subsequent darkening of the fruit. Sugars in fruit juices can caramelize when exposed to excessively high temperatures during concentration. The Maillard-type of browning of reducing sugars and free amines in dehydrated potatoes results in off-flavors. In dried powdery food products, similar Maillard browning results in decreased solubility and brown color. The reaction is common for foods containing both reducing sugar and protein and enough moisture to act as a reaction phase.

In general, browning reactions are deleterious to the nutritional value of the food concerned, and can occur during processing as well as during storage of food products. It is therefore imperative to arrest these reactions, thereby not only preventing any nutritional changes, but also other changes which might render the food unacceptable to the consumer.

11.3.1 The Maillard Reaction and Its Prevention

The Maillard reaction, also referred to as nonenzymatic glycation or browning, was characterized in 1912. He observed that yellow–brown colors were produced when glucose was heated with glycine. He repeated the experiment with many different reducing sugars and amino acids [1].

The Maillard reaction in fact is a series of nonenzymatic reactions involving the reaction between the carbonyl groups of reducing-sugars with the amino groups of amino acids, polypeptides, proteins, enzymes, nucleic acids, or phospholipids, forming Schiff bases, and followed by their Amadori rearrangement. The end product of the reaction is a brown pigment known as melanoidin. In spite of all the work that has been done and the great progress that has been achieved, the mechanism of the Maillard reaction is still not wholly understood. The series of the reactions, some in parallel, produce various numbers of Maillard reaction products (MRPs).

The importance of MRPs is due to their role as flavors, colors, and antioxidants. The reactions between the single amino acid/glucose combinations resulted in the production of aroma which were thought to be responsible for bread, crusty biscuits, or cake [146]. Maillard reactions were found to be responsible for aroma volatiles formed during the short intense heating step of manufacture of white chocolate [147]. Formation of flavor compounds resulted from Maillard reactions were reviewed by van Boekel [148]. The occurrence of brown color caused by the Maillard reaction in processed foods is important in acceptability. The formation of brown color has been studied in different articles [149,150]. Pasta browning was summarized by Feillet et al. [151]. Since the studies of Franzke and Iwainsky in 1954 [152], there is accumulating evidence for the ability of MRP to decrease the oxidation rate of lipids. Various mechanisms are involved in the antioxidant capacity of MRPs, like radical chain-breaking activity [153], scavenging of reactive oxygen species, and metal chelation [154].

Some negative consequences of the Maillard reaction have been discussed in various studies [155,156]. One of the most obvious negative consequences of the Maillard reaction in food is the loss of nutritive value of proteins involved, with a loss of quality and a possible decrease of food safety. In various milk products (milk powder, UHT milks, sterilized milks), the Maillard reaction leads to the loss of nutritional value. A review of the MRPs used as markers for the loss of nutritional value of proteins was discussed by Erbersdobler and Somoza [157]. Nakamura et al. [156] reported that R-chymotryptic digestibility of squid tropomyosin have been altered at the early stage of the Maillard reaction. High temperature cooking such as grilling or frying leads to heterocyclic amines that are potent mutagens. Nagao et al. [158] identified mutagenic compounds in instant and caffeine-free coffee. They consisted of dicarbonyl compounds, methylglyoxal, diacetyl, and glyoxal, from which the methylglyoxal presented highest mutagenic activity. Also in brewed and instant coffee, strong chemiluminescence due to singlet oxygen and excited carbonyls originated by the Maillard reaction of sugars and amino acids were identified [159].

Swedish scientists announced that certain foods that are processed or cooked at high temperatures contain relatively high amounts of a known carcinogen, acrylamide [160]. Several months after the Swedish announcement, a number of research groups simultaneously discovered that acrylamide is formed during the Maillard reaction, and the major reactants leading to the formation of acrylamide

TABLE 11.1
Some Desirable and Undesirable Effects of the Maillard Reaction in Foods

Desirable
    Formation of antioxidants
    Formation of desirable food flavors and
       aromas
    Formation of desirable colors
Undesirable
    Loss in nutritional value of protein
    Formation of undesirable colors and flavors
    Formation of reactive oxygen species
    Formation of carcinogenic products
    Formation of mutagenic products
    Formation of advanced glycation end-products

are sugars and the amino acid asparagine [161–163]. Following these findings, significant progress has been made on acrylamide formed during the Maillard reaction. Studies on acrylamide formation during the Maillard reaction have been summarized in different review articles [164,165].

The Maillard reaction also leads to the formation of advanced glycation end-products (AGEs). By oxidation, the sugars, the Schiff base, and/or the Amadori compound can form AGE precursors and radical species accelerating the glycation process. The formation of AGEs involves structural and functional changes in proteins. Several studies have shown that changes in protein structure by the effect of the formation of AGEs are behind the development of various pathologies associated with hyperglycemia; e.g., Alzheimer's disease [166], cataracts [167], and kidney disorders [168]. Desirable and undesirable effects of the Maillard reaction are listed in Table 11.1.

Clearly, the Maillard reaction pathway leads to some desirable and some undesirable effects in foods. The task of the food technologist is then to optimize by finding the best balance between the favorable and unfavorable effects of the reaction in a given process. It could be a question of minimizing the formation of acrylamide while obtaining an optimal flavor and color production when grilling meats; of maximizing the antioxidant production while minimizing flavor and color production in milk drying, etc. Understanding the Maillard reaction is therefore an added value for the food processing techniques, like grilling, roasting, baking, and drying.

11.3.1.1 Mechanism of the Maillard Reaction

The Maillard reaction has been named after the French chemist Louis Camile Maillard who first described it. Hodge [169] developed the first coherent scheme in 1953. This first scheme has been modified by Martins et al. [5] (Figure 11.8). The mechanism of the reaction is taught to occur at seven steps (A–G) with three main steps as initial, intermediate, and final stage:

  I. Initial stage (colorless, no absorption in the near-UV)
    A. Sugar-amine condensation
    B. Amadori rearrangement
 II. Intermediate stage (colorless, or yellow with strong absorption in the near-UV)
    C. Sugar dehydration
    D. Sugar fragmentation
    E. Amino acid degradation

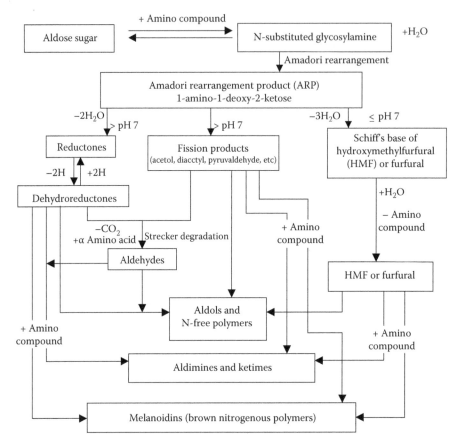

FIGURE 11.8 General pathway of the Maillard reaction. (From Martins, S.I.F.S. et al., *Trends Food Sci. Technol.*, 11, 364, 2001. With permission.)

III. Final stage (highly colored)
   F. Aldol condensation
   G. Aldehyde-amine polymerization; formation of heterocyclic nitrogen compounds (melanoidines)

The initial stage of the Maillard browning is a condensation reaction between the carbonyl group of an aldose and the free amino group of an amino acid, to give an N-substituted aldosylamine. This reaction, which is initiated by an attack of the nucleophilic amino compound on the carbonyl carbon, is reversible and requires an acidic catalyst. The Schiff base is in equilibrium with the cyclic aldosylamine structure as shown in Figure 11.9 [170].

The next step, the Amadori rearrangement, is essential to the continuation of the Maillard reaction. It is an acid-base catalyzed conversion of the N-substituted aldosylamine to an N-substituted 1-amino-1-deoxy-2-ketose (Figure 11.10) [171]. The Amadori reaction is catalyzed by weak acids, where the protonation of the Schiff base and the subsequent proton shift constitute the critical steps; the amino acid serve as their own acid catalysts so the reaction is rapid even in the absence of added acid. If a ketose is allowed to react with the amino compound, a ketosylamine is formed (Figure 11.11). This leads to a 2-amino-2-deoxyaldose through Heyn's rearrangement, which is similar to the Amadori rearrangement. Both rearrangements result in a conversion of an α-hydroxyamino compound to an α-aminocarbonyl compound.

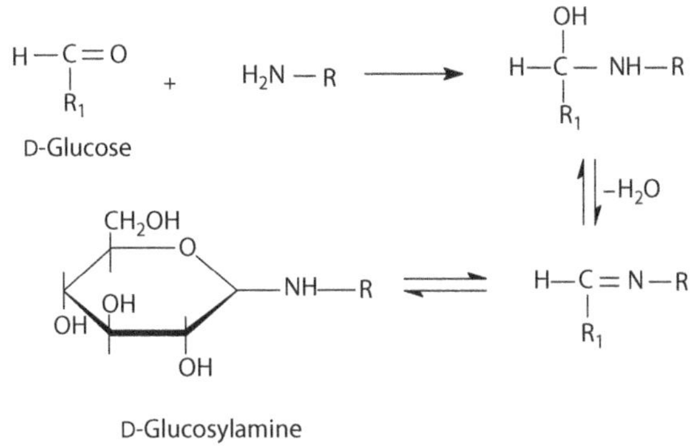

FIGURE 11.9 Reaction of aldose with amino compound to give an aldosylamine ($R_1 = (CHOH)_4CH_2OH$). (From Hurst, D.T., Recent developments in the study of non-enzymic browning and its inhibition by sulphur dioxide, *B.F.M.I.R.A. Scientific and Technical Surveys*, No. 75, Leatherhead, England, U.K., 1972. With permission.)

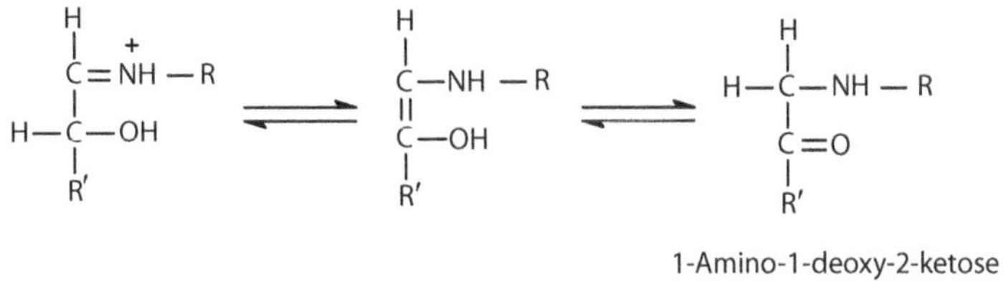

1-Amino-1-deoxy-2-ketose

FIGURE 11.10 The Amadori rearrangement. (From Kort, M.J., *Adv. Carbohydr. Chem. Biochem.*, 25, 311, 1970. With permission.)

2-Amino-2-deoxyaldose

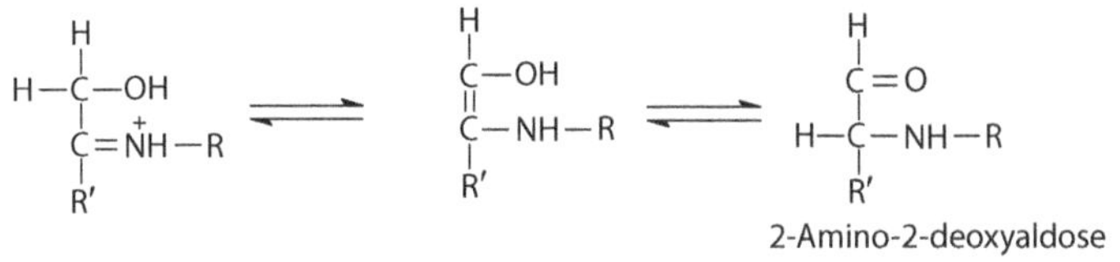

FIGURE 11.11 Heyn's rearrangement. (From Kort, M.J., *Adv. Carbohydr. Chem. Biochem.*, 25, 311, 1970. With permission.)

The formation reactions of the Amadori rearrangement products proceed slowly at room temperature because a tautomeric shift to the open chain form of the reducing sugar is required for the initial reaction to occur. In the reaction of aldoses with the amino acids blocking the Amadori rearrangement step the formation of brown pigment is stopped entirely. Once the Amadori rearrangement products are formed they are more stable than the glycosylamines in most acidic environments, although they are heat labile. Upon heating, the Amadori rearrangement products undergo fission and colorless reductones and fluorescent substances are formed.

In the intermediate stage, three degradation pathways exist; sugar dehydration, sugar fragmentation, and Strecker degradation:

Sugar dehydration: There are two types of sugar dehydration both of which depend on pH.

In acidic conditions, at pH 7 or below, it undergoes mainly 1,2-enolization with the formation of furfural (when pentoses are involved) or HMF (when hexoses are involved). In alkaline medium, at pH > 7, the degradation of the Amadori compound is thought to involve mainly 2,3-enolization, where reductones, such as 4-hydroxy-5-methyl-2,3-dihydrofuran-3-one, and a variety of fission products, including acetol, pyruvaldehyde, and diacetyl are formed.

Sugar fragmentation: The accepted mechanism is dealdolization, the reverse of aldol condensation. The products are aldols, amino-free polymers, and free amino compounds.

Strecker degradation: All fission products are highly reactive and take part in further reactions. Carbonyl groups can condense with free amino groups, which results in the incorporation of nitrogen into the reaction products. It is actually the reaction of a reductone and an α-amino acid. Carbon dioxide is liberated to form an aldehyde with one fewer carbon and the original amino acid. The aldehydes formed condense with themselves, with sugar fragments, with furfurals, and with the other degradation products to form brown pigments.

The possible reactions involved in the final stage are aldol condensation, aldehyde-amine polymerization, and formation of heterocyclic nitrogen compounds (pyrroles, imidazoles, pyridines, and pyrazines). The polymerization of products from the second step and copolymerization with amino compounds yield the colored products. Finally both water-soluble and water-insoluble pigments, called melanoidins, are produced, and their structure is dependent on the types of amino acid, sugar, and the sugar:amino acid ratio. The melanoidins vary widely in molecular weight and contain several discrete chromophores [172]. Melanoidins may be distinguished from one to another on the basis of their molecular weight and solubility. Melanoidins with a molecular weight <500 Da represent low molecular weight compounds and are soluble in water or organic solvents. Soluble melanoidins with molecular weights >12,000 Da can easily be prepared by dialysis of aldose-amino acid reaction mixtures, as the browning reactions proceed the melanoidins eventually become insoluble in water and precipitate. A general formula $C_{67}H_{76}O_{32}$, "(sugar + amino acid) -2~(3H$_2$O)", has been proposed for melanoidin based on the consideration that the main pathway of melanoidin formation involves the reactions of amino compounds with deoxyosones, furfural, and other fragmentation products which accompany the dehydration reaction [173]. Rubinsztain et al. [174] found that basic amino acids and sugars form melanoidins that have higher molecular weights than those formed by neutral or acidic amino acids. This indicates that cross-linking is probable.

The first critical step for the formation of acrylamide is the amino-carbonyl reaction between asparagine and a carbonyl substance, preferably α-hydroxycarbonyls such as reducing sugars, resulting in the corresponding *N*-glycosyl conjugation and forming the Schiff base as a key intermediate after dehydration under elevated temperatures. Under low moisture conditions, both the *N*-glycosyl conjugation and the Schiff base are relatively stable. In an aqueous system, however, the Schiff base may hydrolyze to the precursors or rearrange to the Amadori compound, which is not an efficient precursor in acrylamide formation. Even under low moisture conditions, this reaction is most likely the main pathway initiating the early Maillard reaction stage that leads to 1- and 3-deoxyosones, which further decompose, generating color and flavor. This is in accordance with the relatively low transformation yield of asparagine to acrylamide [175]. Alternatively, the Schiff base may be apt to make decarboxylation either directly *via* the Schiff betaine or *via* the intermediary oxazolidine-5-one to generate the azomethine ylide I, which furnishes the decarboxylated Amadori product after tautomerization [176]. In summary, acrylamide may be released *via* the following pathways: (a) directly from azomethine ylide I [177]; (b) β-elimination reaction from the Maillard intermediate, i.e., decarboxylated Amadori product [176]; and (c) loss of ammonia from 3-aminopropionamide deriving from the azomethine ylide II. Such a reaction has been shown to preferentially proceed under aqueous conditions in the absence of sugars [178]. Besides the main reaction precursors (reducing sugars and amino acids) and key intermediates, it is suggested that fat or oils can also play an important role in the acrylamide formation pathway.

11.3.1.2 Variables Affecting the Maillard Reaction

In view of the large numbers of stages and the sensitivity of each stage in the reaction to changes in the conditions, it is very complicated to investigate and control the whole Maillard reaction. The extent of, and the pathway for, change are controlled by many factors. These factors may be distinguished as those relating to the reactant composition, e.g., type of amino acid, sugar type, sugar:amino acid ratio, and external factors such as moisture level, temperature, and pH. These will now be considered in detail.

11.3.1.2.1 Amino Acid Type

Amino acid type which is used or present in the medium is very important in the Maillard reaction. Some amino acids have two reactive groups therefore these types of amino acids rapidly react with the sugars to produce brown pigment. Ashoor and Zent [4] have classified the amino acids according to their reactivity with the sugars at pH 9.0. They classified lysine, glycine, trytophan, and tyrosine as the most reactive amino acids. Kwak and Lim [179] have measured the color intensity of 60 model systems heated for 3 h at pH 6.5 and 100°C. They found that the color intensity of the MRPs from the lysine was the highest. It was followed by basic and nonpolar amino acids such as arginine, phenylalanine, leucine, isoleucine, and valine. The next group was the acidic amino acids such as aspartic acid and glutamic acid. The lowest value was found for cysteine. Lysine is often hypothesized as being the most reactive amino acid due to its two reactive α- and ε-amino groups [180]. Despite this logical deduction, numerous studies have provided contradictory results. Wolfrom et al. [181] studied the effect of amino acid type on the extent of browning. L-Arginine and 4-aminobutyric acid were found as the most intense and rapid color developing amino acids. They have been followed by glycine, alanine, serine, and L-proline. Aspartic acid, L-glutamic acid, and L-glutamine were found to behave similarly to glycine. Lysine was not included in this study.

Göğüş et al. [182] studied the effects of temperature and amino acid type on the extent of both HMF content and browning in model systems at pH 3.5. They compared the browning rates of three major amino acids of grape juice (proline, arginine, and glutamine) in the presence of glucose and fructose. They found that in the model systems with glutamine, the accumulation of HMF and browning were higher than the model systems containing arginine or proline. Figure 11.12 shows the brown pigment formation measured at 420 nm for the model systems containing proline, arginine, or glutamine together with glucose and fructose. They also reported that the presence of more than one amino acid in combination in model systems caused an additive increase in HMF accumulation and browning. Massaro and Labuza [183] studied total parental nutrition (TPN) solution at pH 5.5–5.7. They found that tryptophan and cysteine produced high brown pigment than lysine on a molar ratio basis.

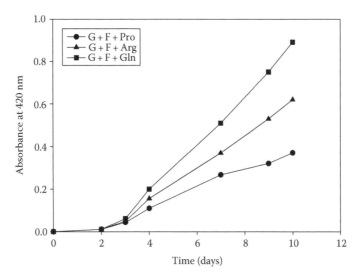

FIGURE 11.12 Production of brown pigments by different amino acids at 55°C and pH 3.5 as a function of time (G, glucose; F, fructose; Pro, proline; Arg, arginine; Gln, glutamine).

The type of amino acids and sugars were found as the main factors in flavor compound formation [184]. For example; meat-related flavor compounds are mainly sulfur containing compounds, derived from cysteine and ribose (coming from the nucleotides), while the amino acid proline gives rise to typical bread, rice, and popcorn flavors.

van Boekel [148] pointed out that in the case of proteins or peptides the reactive amino group is the ε-amino group of lysine, because the α-amino groups are tied up in the peptide bond and are not available for the Maillard reaction or the Strecker reaction and concluded that this results in a different behavior of amino acids compared to proteins and peptides. In the case of proteins, the Maillard reaction often leads to cross-link formation, while brown pigments are for a large part covalently attached to the proteins [185].

11.3.1.2.2 Type of Sugar

Reducing sugars are essential ingredients in these reactions, providing the carbonyl groups for interaction with the free amino groups of amino acids, peptides, and proteins [7]. Low molecular weight compounds tend to be more reactive than the high molecular weight compounds as a result of greater steric hindrance in the latter. Accordingly, aldopentoses are generally more reactive than aldohexoses and monosaccharides are more reactive than di- or oligosaccharides. Aldoses in general appear to be more reactive than ketoses, apparently a consequence of the more sterically hindered carbonyl group of ketoses [172].

The concentration of the acyclic or straight-chain form determines reactivity since it is only the acyclic form with free aldose or hexose group that can react. When a crystalline sugar is dissolved in a suitable solvent, it undergoes mutarotation which is the equilibrium reaction between the α-pyranose, α-furanose, β-pyranose, β-furanose (cyclic), and straight-chain (acyclic) forms. It was reported by Labuza and Baisier [186] that the browning is faster if the concentrations of the acyclic state in solution are higher.

The MRPs from glucose showed a higher degree of browning than those from fructose [187], this being ascribed to the keto sugar proceeding through imine intermediates that favor the formation of the Heyns products, while the aldose sugars proceeded through the Amadori products [169]. The rate of browning of the Heyns products is known to be slower than that of the Amadori products [188]. Hayashi and Namiki [187] and Kwak and Lim [179] found that xylose, which is a pentose showed much higher reactivity than glucose and fructose that are hexoses.

Maillard [1] found that the increasing order of brown pigment formation is: D-xylose > L-arabinose > hexoses > disaccharides. Fructose did not condense with amino acids in dilute solution although scientists have since confirmed that a definite interaction does take place. D-Fructose has also been reported to brown at a much faster rate than glucose during the initial stages of the browning reaction, but it then falls behind. This was confirmed using model systems containing glucose-glycine and fructose-glycine [189]. Sucrose, as a nonreducing sugar, will only participate when the glycosidic bond is hydrolyzed and the reducing monosaccharide constituents released. Hydrolysis of the glycosidic bond in sucrose is facilitated by a low pH, resulting in an increase in the Maillard reaction rate in protein–sucrose systems.

Claeys et al. [190] studied the kinetics of acrylamide by heating a model system consisting of asparagine and glucose, fructose, or sucrose (pH 6) at temperatures between 140°C and 200°C. Acrylamide formation appeared to proceed faster and to be more temperature sensitive in the asparagine glucose than in the asparagine-fructose model system. Significantly less acrylamide was formed in the asparagine-sucrose model system as compared to the model systems with glucose or fructose.

11.3.1.2.3 Sugar:Amine Ratio

The extent of browning seems to vary to the initial concentrations and ratio of the reactants. O'Brian and Morrisey [172] stated that, an excess of reducing sugar over amino compound promotes the rate of Maillard browning, since there are mechanistic differences in the destruction of sugar compared to the amino acid. However, Warmbier et al. [191] found that, the browning rate increased to a maximum at glucose:lysine ratio of 1:3 (present in casein). Since, the initial step of formation of the Schiff base is dependent on the concentration of both, sugar and amino acid, the Schiff base formation increases with decreasing sugar:amino acid ratio.

Labuza and Baisier [186] showed that the ratio of amino acid to reducing sugar can affect the rate of browning. The effect of increasing the amino acid concentration shows a greater increase in browning than that of increasing the sugar content on a molar basis and the increase for both is greater than the relative concentration increase.

Renn and Sathe [192] investigated the effects of pH, temperature, and reactant molar ratios on the L-leucine and D-glucose Maillard browning reaction in an aqueous system. At 100°C and pH 9 and 10, samples with excess leucine had higher mean values for color than the mean color values for samples using glucose/leucine ratios of 1:1 or 2:1. At 122.5°C and pH 9 and 10, samples with glucose/leucine ratios of 1:2 had a significantly higher rate of glucose loss than samples containing glucose/leucine ratios of 2:1 and 1:1 at the corresponding pH.

Kato et al. [193] observed that, at low concentrations of glycine, fructose browned faster than glucose, whereas, at high amino acid concentration, the reverse occurred. However, Martins and van Boekel [194] pointed out that the rate of browning and the rate constant of the step in the reaction network that results in color formation, are two different things. They showed that increasing the initial reactant concentrations should not influence the reaction rate constants since, for the reaction of sugar (S) with the amino acid (A), it follows that

$$-\frac{d[S]}{dt} = -\frac{d[A]}{dt} = k_1[S][A]$$

The overall rate of loss of the S and the A is equal to the rate constant times their concentration. So, the overall rate depends on the concentration.

11.3.1.2.4 Effect of pH

Both the initial pH of the product and the buffering capacity of the system, influence the rate and direction of the Maillard reaction. Borsook and Wasteneys [195] were the first to investigate systematically the influence of pH on the extent of interaction of glucose and free amino-nitrogen. They made quantitative observations on the interaction of glucose with glycine, and of glucose with various enzymatic digests of protein material. Determinations of pH and Van Slyke amino nitrogen were made with time; in each case the initial pH values were set by phosphate buffers and the solutions contained 1% glycine and 13.2% glucose. An appreciable loss of amino-nitrogen occurred in 48 h; the loss increased with increasing pH. It was shown that the optimum pH for the whole reaction is in the range pH 6–9 [4]. With regard to the influence of pH on the Maillard reaction, Labuza and Baisier [186] observed that the substrate loss increased with increasing pH, up to a pH of about 10, with little, if any, browning occurring below pH 6.

The rate of the reaction is also dependent on the concentration of the acyclic sugar present. The amount of acyclic sugar increases with increasing pH and so increases the rate of reaction. The formation of the enediol anion is also considered a key reaction in both reversible and irreversible sugar reactions. Reversible sugar reactions are ionization, mutarotation, enolization, and isomerization, and these reactions imply that the sugar moiety remains intact. Irreversible sugar reactions imply that the sugar moiety is eventually degraded in organic acids. Since ionization is a rate-determining step, pH obviously has an influence on sugar reactions and hence on the Maillard reaction [194]. Furthermore, the main degradation pathways of the Amadori compound, namely enolization and retro-aldolization, were shown to be strongly dependent on the reaction pH [196]. The pH-dependence of the Maillard reaction for the amino acid reagent can, at least qualitatively, be described by the effect of protonation of the amino acid. The amount of unprotonated amino group, which is considered to be the reactive species, increases obviously with increasing pH.

The nonenzymatic browning reactions of fructose and fructose-lysine aqueous model systems were investigated at 100°C between pH 4.0 and pH 12.0 by measuring the loss of reactants and monitoring the pattern of UV-absorbance and brown color development [2]. At all the pH values tested, the loss of fructose was lower in the presence than in the absence of lysine. The promoting effect of pH was clear on the browning development and was in agreement with the earlier studies [4,197].

Lertittikul et al. [198] monitored the browning products by heating the solution containing 2% porcine plasma protein (PPP) and 2% glucose adjusted to various pHs (8, 9, 10, 11, and 12) at 100°C for different times (0, 2, 4, 6, and 8 h). Among all the MRPs tested, those derived from the PPP-glucose system at pH 12 rendered the highest browning and intermediate products. However, an electrophoretic study revealed that cross-linked proteins with high molecular weight were formed in the PPP-glucose model system to a greater extent at pHs 8 and 9, than at pHs 10–12.

The pH-dependence of acrylamide formation exhibited a maximum around pH 8; lower pH enhanced elimination and decelerated the formation of acrylamide in heated foods [199].

11.3.1.2.5　Temperature Effect

The most important influence on the kinetics of the Maillard reaction is temperature. Increasing temperature results in a rapidly increasing rate of browning. Not only the rate of browning but also the character of the reaction is determined by temperature. In model systems, the rate of browning increases 2–3 times for each 10° rise in temperature. In foods containing fructose, the increase may be 5–10 times for each 10° rise. Temperature also affects the composition of the MRPs formed. If the color intensity is measured, it may also be increased with increasing temperature because of the changing composition and increasing carbon content of the pigment.

Benzing-Purdie et al. [200] worked with equimolecular amounts of D-xylose and glycine in aqueous solution at temperatures of 22°C, 68°C, and 100°C. They reported that an increase in temperature leads to an increase in aromatic character in both high and low molecular weight products. The structure of the melanoidins synthesized at room temperature differs considerably from those synthesized at higher temperatures in that they have different types of aliphatic carbons and fewer unsaturated carbons.

As with other chemical reactions, the Arrhenius relationship provides a good description of the temperature dependence of the Maillard browning reaction:

$$k = k_0 \exp(-E_a/RT)$$

where
　　k is the reaction rate constant of nonenzymatic browning
　　k_0 a frequency constant (independent of temperature)
　　E_a the activation energy
　　R the universal gas constant
　　T the absolute temperature

Linear, exponential, and hyperbolic functions have also been employed to correlate browning but were found to be valid only over a limited range.

Quantitative knowledge is very important in order to control the Maillard reaction in food processing operations. However, it should be noted that the various reaction steps have different temperature sensitivities. Therefore, it is a necessary first to specify what has to be measured for the determination of the activation energy of the Maillard reaction. In most cases, the reported activation energies will reflect more elementary reaction steps, and will be the resultant of one or more rate-controlling steps. The variation in activation energies reported in the literature is vast. Unfortunately, activation energies reported for the same reaction step can easily differ by a factor of 4.

This probably reflects the importance of the experimental conditions: mechanisms may change depending on these conditions, and this will be reflected in the activation energies [201].

Activation energies can be determined for browning reactions in various food systems as a function of water content, from which the temperature dependence of the browning reaction at different water contents can be elucidated. Miao and Roos [202] studied the effects of water contents on nonenzymatic browning (NEB) rates of amorphous, carbohydrate-based food model systems containing L-lysine and D-xylose as reactants at different temperatures (40°C, 50°C, 60°C, 70°C, 80°C, and 90°C) of spray drying. They modeled the data by using the Brunauer-Emmett-Teller and Guggenheim-Anderson-deBoer equations. The rate of browning increased with water content and temperature, but a lower $(T - T_g)$ was needed for browning at decreasing water content. Water content seemed to affect the activation energy of NEB, and higher water contents decreased the temperature dependence of the NEB. At higher temperatures, the NEB became less water content dependent and enhanced the browning in spray-drying. The temperature dependence of nonenzymatic browning could also be modeled using the Williams-Landel-Ferry (WLF) equation, but the WLF constants were dependent on the water content.

Bozkurt et al. [203] evaluated the reaction orders, rate constants, and activation energies for HMF accumulation and brown pigment formation (BPF) in boiled grape juice and its model systems at 55°C, 65°C, and 75°C over 10 days at pH 4.0. The calculated activation energies for HMF accumulation and brown pigment formation were in the range of 49.7–103 kJ/mol and 116–132 kJ/mol, respectively. Göğüş et al. [182] studied the effect of temperature in various model systems containing amino acids in single or in combination, in the presence of glucose and fructose. They found that the rate of browning increased 3.20 times from 55°C to 65°C and 3.50 times from 65°C to 75°C in a glutamine (0.015 mol/L), glucose (2.0 mol/L), and fructose (2.0 mol/L) model at pH 3.5 (Figure 11.13).

Pedreschi et al. [204] investigated acrylamide formation and changes in color of fried potato strips in relation to frying temperature. Acrylamide formation decreased dramatically as the frying temperature decreased from 190°C to 150°C. Color showed high correlations with French fry acrylamide content.

A paper summarized the characterization of parameters that influence the formation and degradation of acrylamide in heated foods [199]. It has been found that the higher temperature (200°C)

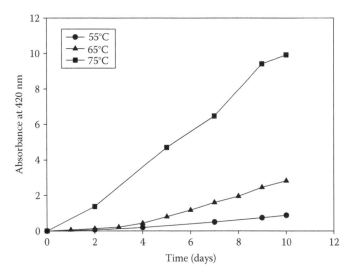

FIGURE 11.13 Effect of temperature on brown pigment formation in a "glucose + fructose + glutamine" model system (G, glucose; F, fructose; Gln, glutamine).

combined with prolonged heating led to reduced levels of acrylamide, due to the elimination and degradation processes. At certain concentrations, the presence of asparagine or monosaccharides (in particular fructose, glucose, and glyceraldehyde) was found to increase the net content of acrylamide. Addition of other free amino acids or a protein-rich food component strongly reduced the acrylamide content, probably by promoting competing reactions and/or covalently binding the formed acrylamide.

11.3.1.2.6 Water Activity

Nonenzymatic browning reactions may occur as a result of heating, dehydrating, or concentrating food constituents. The role of bound and unbound water in browning reactions has been investigated by several authors. Almost all of the results showed that a maximum browning rate occurs at water activities between 0.4 and 0.6 depending on the type of the food substance. At lower water activities the reaction rate decreases as a result of the increasing diffusion resistance due to high viscosity [205], whereas at higher water activities the reaction rate again slows down, due to the dilution of the reactants [206]. Furthermore at higher water activities, water behaves as a reaction product and blocks the formation of reaction intermediates that are produced together with water.

Labuza et al. [206] explained the properties based on the water that plays a role in the chemical reactions: (1) water as a solvent in which reactants are able to dissolve, be transported, and react; and (2) water as a reactant which participates in specific reactions.

According to their investigations, even at low water activities sucrose may be hydrolyzed to form reducing sugars which have a potential for browning. They have found that the formation of browning products of dehydrated skim milk at 54°C is the maximum at 75% relative humidity, and the rate decreases with decreasing relative humidity. They proposed that water not only accelerates the reaction, but also shortens the induction period. The acceleration of the rate may be due to the increased availability and mobility of the reactants. The effect of induction time may indicate that the formation of pigment has a different pathway at the lower humidity, which requires more time for color development, or that sufficient intermediates must be built up before they can dissolve and react to form the pigment [206]. They have also examined browning products formation in pea-soup at a wide range of water activities from 0.0 to 0.9 again the browning rate was the maximum at a water activity of 0.75. At higher water activities the rate of reaction decreased sharply. The major reason is that at this point the maximum amount of reactants can be dissolved in solution; the problem is no longer diffusion, but rather that as water content increases the reactants are diluted. Since by the law of mass action, the rate of a reaction is proportional to concentration, a decrease in concentration by the dilution with water will decrease the rate.

Warmbier et al. [191] have studied the effect of water activity on BPF by using a humectant, glycerol. They prepared a model system containing K-sorbate, glucose, casein, aprezon-B oil, microcrystalline cellulose, and variable amounts of water. They measured the browning rate at 420 nm and found that there is a linear relationship between time and BPF. The maximum rate of nonenzymatic browning occurred at water activities 0.45–0.55. They also found that when the temperature increases the browning maximum shifts to higher water activities. They proposed that liquid humectant (glycerol) can cause the water activity for the maximum browning to be shifted downward.

Another important study on brown pigment formation and water activity was carried out by Eichner and Karel [205] on model systems containing variable amounts of microcrystalline cellulose MCC, methylcellulose, glycerol, and gum arabic together with the reducing sugar (glucose) and amino acid (glycine). They found that addition of the MCC to the original system containing sugar, amino acid, glycerol, and water had no effect on the browning rate, whereas the system with gum arabic showed the lowest rate. Gum arabic also shifted the water activity for maximum browning upward. At high water activities ($a_w > 0.75$) all of the hydrocolloids showed similar behavior. Finally they concluded that browning rate in a sugar-amino system is not simply related to the water activity. Optimum browning conditions are determined by either the amount of water, or the state of water binding in a distinct system.

Eichner [207] have studied the effect of water activity in the different steps of the Maillard mechanism by determining the reaction intermediates in a model system containing glucose-lysine (1:1 mol) and MCC (14 g/g glucose) at 40°C, at water activities between 0.23 and 0.82. He found that (1) the decrease in the free amino group of lysine was minimum at a_w 0.23, and maximum around water activities 0.62–0.75; (2) not only the rate of formation of reducing browning intermediates, but also their rate of decomposition appeared to increase at higher water activities and the nature of the reducing intermediates changed due to a change in water activity; and (3) at higher water activities, browning was proceeding directly *via* 1,2 eneaminol intermediate of the Amadori rearrangement, without using the Amadori rearrangement products (ARPs).

The effect of water activity in a range between 0.11 and 0.80 was observed by measuring both BPF and HMF accumulation in concentrated orange juice with and without xanthan gum and microcrystalline cellulose [208]. They reported that the samples containing 1.5% xanthan gum and 0.5% microcrystalline cellulose showed the lowest BPF and HMF accumulation, whereas the browning rate and HMF accumulation in pure orange juice concentrate were the highest at all temperatures studied (45°C, 60°C, and 75°C). Figure 11.14 shows the effect of water activity on both HMF accumulation and browning of pure orange juice concentrate after 48 days of storage at 45°C. It is seen that both HMF accumulation and browning show a maximum at a_w 0.50. However, HMF accumulation has a wider range of maximum compared to the sharp maximum of browning.

Ameur et al. [209] determined the critic a_w value where HMF began to form, to describe better the relationship between the water activity and the formation of HMF. This HMF level was defined as the analytical limit of quantification of 3.60 mg/kg. They found that the water activity allowing HMF formation was significantly higher at 300°C (critical a_w of 0.7) than at 250°C or 200°C (mean critical a_w of 0.51).

Mundt and Wedzicha [210] also explored the effect of a_w on the rate of browning in biscuit doughs at 105°C–135°C. They found that, contrary to common expectation, the value of a_w in the range 0.04–0.4 did not affect the measured rate of browning. They concluded that this result has significant implications to the understanding of water relations at high temperatures, where the thermodynamic parameters governing water sorption are generally inaccessible.

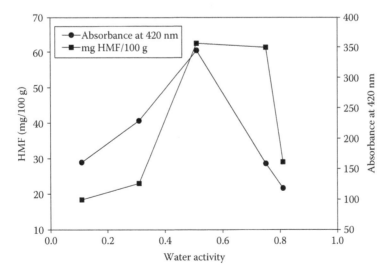

FIGURE 11.14 Influence of water activity on HMF accumulation and brown pigment formation in orange juice concentrate at 45°C for 48 days.

11.3.1.3 Control Mechanisms of the Maillard Reaction

The Maillard reaction is desirable in terms of the formation of color and flavor. However, in many other instances, such as fruits, vegetables, frozen, and dehydrated foods, browning is undesirable as it results in off-flavors and colors with the loss of nutritive value. It can also cause the formation of some toxic and mutagenic compounds in some cases. Therefore, it is important to know how to control or minimize the Maillard reaction in food processing. A variety of methods have been proposed for controlling the Maillard reaction. The Maillard reaction can be controlled or inhibited by controlling temperature, time, water activity, and pH; reduction of reducing sugar and/or amino nitrogen content; packaging with gas; application of high hydrostatic pressure, and the use of chemical inhibitors such as sulfites, flavonoids, and cations.

11.3.1.3.1 Temperature

The rate of browning increases with increasing temperature or time. Since these reactions have been shown to have a high temperature coefficient, lowering of the temperature during the processing or the storage of food products can help to minimize these reactions.

11.3.1.3.2 Water Activity

The Maillard reaction being moisture dependent for optimum activity, it can be inhibited by reducing the moisture content through dehydrating procedures. Sherwin and Labuza [211] showed the role of moisture in the Maillard browning reaction rate in intermediate moisture foods. They found that the addition of humectants to a model formulation caused significant plasticization and a lowering of the T_g curve. However, with the addition of the sorbitol, a solid-state humectant, there was no change in reaction rate at equal moisture content, in comparison with the control. For the glycerol formulation, a liquid phase humectant, there were high rates of browning, between $a_w = 0.11$ to 0.75, that were greater than for the control formulation. They concluded that a solvency-based molecular mechanism describes the effect of increased moisture on reaction rate in a semi-moist food.

11.3.1.3.3 pH

The pH of the Maillard browning reaction is an important parameter for both the reaction rate and the characteristics of the products. Since the Maillard reaction is generally favored at the more alkaline conditions, lowering of the pH might provide a good method of control.

11.3.1.3.4 Biochemical Agents

Removal or conversion of one of the reactants of the Maillard reaction controls the browning. For instance, glucose or other fermentable sugars in liquid egg whites, egg yolks, or whole eggs can be removed by yeast or bacterial fermentations. The use of commercial glucose oxidase-catalase preparations is a widely used alternate commercial method for removing glucose. Gluconic acid produced from glucose is acceptable in small concentrations. This enzyme has been used to remove glucose from egg prior to spray drying [7]. Another study has been performed to control the Maillard browning in fried potato chips by yeast fermentation [212]. They studied the effect of different combinations of yeast concentration and fermentation time. They observed that browning of chips decreased 60% and the yield of acceptable chips increased considerably after the yeast treatment at the optimum yeast concentrations.

11.3.1.3.5 Modified Atmosphere Packaging

Modified atmosphere packaging is useful in excluding oxygen by using an inert gas. This reduces the possibility of lipid oxidation, which in turn could give rise to reducing substances capable of interacting with amino acids. While this reaction does not appear to influence the initial carbonylamino reaction, exclusion of oxygen is thought to effect other reactions involved in the browning process [7]. Maillard reactions are the main causes of brown color formation in glucose syrup.

Effects of modified atmosphere packaging (MAP) on browning in glucose syrups stored at 25°C and 45°C were studied [213]. They examined different atmospheres such as air, 100% N_2, 90% N_2/10% O_2, 25% CO_2/75% N_2, 75% CO_2/25% N_2, and vacuum. The oxygen content and pressure affected the Maillard reaction. They found that in the absence of oxygen, the browning rate in the glucose syrup samples was low while it increased with the increase in oxygen content.

11.3.1.3.6 High Pressure

Another parameter controlling the Maillard reaction is high (hydrostatic) pressure. Moreno et al. [214] studied glucose-lysine model systems over a range of pH values (5–10) in unbuffered and buffered media at 60°C either under atmospheric pressure or at 400 MPa. They reported that in the buffered media, at pH values less than or equal to 8.0, the pressure slowed the Maillard reaction from the initial stages. These effects were attributed to the pH drop caused by the pressure-induced dissociation of the acid groups. However, they also noted that, in the unbuffered and buffered media at an initial pH = 10.2, the high pressure accelerated the formation and subsequent degradation of ARP, leading to increased levels of intermediate and advanced reaction products. Komthong et al. [215] also studied the effect of high hydrostatic pressure (100 MPa) combined with pH (6.0, 7.0, and 8.0) and temperatures (80°C and 90°C) on the Maillard reaction of the sugar (glucose or fructose)-amino acid (leucine, lysine, or glutamate) solution models. They found that both the formation of browning products and HMF content decreased by the high pressure treatment.

11.3.1.3.7 Chemical Inhibitors

In view of the wide occurrence of the Maillard reaction products during the production and storage of various different food products, it would be of great interest to limit this reaction in the undesirable cases. A variety of chemical inhibitors have been used for that purpose. Sulfites are the most widely used inhibitors. However, the restrictions to the use of sulfite agents in foods promoted the scientists to develop alternatives to sulfites. Therefore; calcium salts, thiols, aspartic and glutamic acids, phenolic acids, and various flavonoids have been studied as alternatives to sulfites. On the other hand, scientists have made efforts for finding the chemical way to reduce the amount of acrylamide in the Maillard reaction during the last few years. It has been reported that cations effectively prevented the formation of the Schiff base, which is the key intermediate leading to acrylamide, and mainly changed the reaction path toward the dehydration of glucose leading to HMF and furfural [216]. It has also been found that aminoguanidine could inhibit the Maillard reaction both *in vitro* and *in vivo*. The action of aminoguanidine is probably due to the trapping of intermediates of the advanced Maillard reaction, such as 3-DG, leading to the inhibition of further progress of the Maillard reaction [217].

Sulfites/Sulfur dioxides are commonly used as food additives. They are known as food preservatives but also have an important role as inhibitors of enzymatic and nonenzymatic browning. However, they are subject to regulatory restrictions because of their adverse effects on health.

In the context of a food additive, "sulfur dioxide" refers to a mixture of oxospecies of sulfur in an oxidation state of (+4), i.e., SO_2, HSO_3^-, SO_3^{2-}, $S_2O_5^{2-}$. Sulfite and hydrogen sulfite ions are formed by the ionization of the species H_2SO_3, which is more correctly written as $SO_2 \cdot H_2O$. In practice it is very difficult to provide a quantitative description of these ionic and nonionic forms in foods. To avoid ambiguity the term S(IV) is used to describe the mixture of sulfur species when it is not necessary or possible to identify the detailed composition.

Sulfur dioxide/Sulfites can be used to control nonenzymatic browning because of their ability to react with the carbonyl intermediates. A variety of carbonyl intermediates can be formed during the nonenzymatic browning process, including simple carbonyls, dicarbonyls, and α, β-unsaturated carbonyls. Sulfites can react with all of these intermediates and thus block the formation of the brown pigments. Sulfites inhibit the Maillard reaction by binding with the intermediates formed during the early stage of the reaction to form sulfonates. The important sulfonates are formed as a result of the replacement of the hydroxyl group at position 4 of the

3-deoxyhexosulose intermediate. In the Maillard reaction of glucose, 3-deoxyhexosulose (DH) is converted to 3,4-dideoxy-4-sulfo-D-erythrohexosulose (DSH) [218]. The formation of DSH involves the conversion of DH to 3,4-dideoxyhexosulose-3-ene (DDH) which is probably also the most reactive known intermediate. Wedzicha and Vakalis [219] developed a kinetic model for the sulfite inhibition of the browning in the homogeneous aqueous "glucose + glycine + S(IV)" system. It was found that the first determining step of this conversion, in fact, catalyzed by S(IV) in reactions leading to monofructoseglycine (MFG) or in the decomposition of MFG to DH. The formation of MFG is subject to general acid- or base-catalysis. It has been shown that sulfite and disulfite ions are capable of being effective acid-base catalysts.

Flavonoids are known for their ability to control the Maillard browning. Colahan-Sederstrom and Peterson [220] studied the effect of Epicatechin (EC) to inhibit the thermal development of aromatic MRPs formed during ultrahigh-temperature (UHT) processing of bovine milk. They reported that the addition of EC to raw fluid milk prior to UHT processing reduced the overall thermal formation of key aroma-active compounds in comparison to the traditional UHT milk sample. Epicatechin (EC) and epigallocatechin gallate (EGCG) have also been used in another study to investigate their effect on Maillard browning in the UHT milk and glucose/glycine model system [221]. They found that, EC and EGCG reduced Maillard fluorescence at the 0.1 mmol/L level, while fluorescence was negligible with added flavonoids at 1.0 mmol/L in the model glucose/glycine system. When these flavonoids were added to milk, they reduced the production of the Maillard associated fluorescence with UHT processing. They also reported that EC and EGCG reduced the total color difference during thermal processing.

The prevention of acrylamide formation has been widely studied following the discovery of the formation of acrylamide during the Maillard reaction [163,222,223]. Acrylamide has been classified as "probably carcinogenic to humans" (class 2A) by the International Agency for Research on Cancer [224]. Becalski et al. [163] reported a decreased acrylamide formation when adding rosemary herb to the oil used for frying potato slices. Relatively lower amounts of acrylamide after the addition of a flavonoid spice mix have also been reported by Fernandez et al. [223]. Lindsay and Jang [222] found the suppression of acrylamide formation by polyvalent cations and polyanionic compounds in fried potato products. Effect of NaCl in reducing acrylamide formation also has been evaluated in potato chips during frying [225]. The soaking of potato slices in NaCl solution before frying dramatically reduced acrylamide formation in potato chips in 90% of the chips in comparison to the control chips. Gokmen and Senyuva [226] recently reported that dipping potatoes into calcium chloride solution inhibited the formation of acrylamide by up to 95% during frying. They found that the cations prevented the formation of the Schiff base of asparagines and so, the formation of acrylamide.

Kwak and Lim [227] investigated the effects of antibrowning agents (cysteine, glutathione, sodium sulfite, pentasodium tripolyphosphate, citric acid, and oxalic acid) and phenolic acids (ferulic, hydroxybenzoic, syringic, and vanillic acids) on the inhibition of browning in a glucose-lysine model system. They found that citric acid was the most efficient antibrowning agent during storage in air at 30°C and inhibited browning to 36% after 4 weeks. They also reported that its antibrowning capacity was increased by 8%–15% in the presence of any of the phenolic acids.

11.3.2 CARAMELIZATION

The degradation of sugars in the absence of amino acids and proteins by heating them over their melting point and thereby causing color and flavor changes is called caramelization. If this reaction is not carefully controlled it could lead to the production of unpleasant, burnt, and bitter products. If the reactions are carried out under controlled conditions, pleasant qualities of caramel are obtained.

During caramelization several flavor components as well as polymeric caramels are produced. Caramels are a complex mixture of various high molecular weight components. They can be classified into three groups: caramelans ($C_{24}H_{36}O_{18}$), caramelens ($C_{36}H_{50}O_{25}$), and caramelins ($C_{125}H_{188}O_{80}$). These polymers are often used as colors in commercial food products, from soft drinks to soy sauce, confectionary, and ice-cream. They are labeled as E150. Commercial caramels are produced directly by heating sugar, or by heating sugar in the presence of co-factors, such as ammonia or sulfite. This results in caramels with different colors or charged caramels. These aspects are very important for the use of different caramels in foods. Caramels used to color soft drinks should be negatively charged to prevent reactions with phosphates that cause precipitation and loss of color. On the other hand, caramels used for bakery goods should be positively charged. Caramelization starts at relatively high temperatures as compared to the other browning reactions, and depends on the type of sugar. In foods often several different carbohydrates and other components are present; all these may influence the caramelization temperature as well as the different steps and reactions, and thus the final flavors and colors that are produced. Diacetyl is an important flavor compound, produced during the first stages of caramelization. Besides diacetyl, hundreds of other flavor compounds are produced, for instance furans like HMF and hydroxyacetylfuran (HAF), furanones such as hydroxydimethylfuranone (HDF), dihydroxydimethylfuranone (DDF), and maltol from disaccharides, and hydroxymaltol from monosaccharides.

It has been found that the pyrolysis of sucrose, glucose, and starch all produced caramels of similar composition [228]. However, it has been established that the fragmentation of sugars occurs to a significant extent at pH values below neutrality [229] and increases considerably at high pH values and temperatures, yielding colored N-free polymers [230]. The degradation of sugars has been proposed in two ways, as acidic and alkaline.

In acidic degradation, 1,2-enediol forms from the aldose or ketose after a series of dehydration reactions. If the initial sugar is a hexose, 1,2-enediol is converted to HMF. If it is a pentose, 1,2-enediol is converted to 2-furaldehyde. 3-Deoxyaldose-2-ene, 3-deoxyosulose, and osulos-3-ene are intermediates in the acidic degradation of fructose. The last series of reactions include both fragmentation reactions (flavor production) and polymerization reactions (color production).

In alkaline degradation, the initial reaction is the transformation *via* the 1,2- and 2,3-enediol. Under strong alkaline conditions, continuous enolization progresses along the carbon chain, resulting in a complex mixture of cleavage products such as saccharinic acid, lactic acid, and 2,4-dihydroxybutyric acid. The development of color is very complex and involves a series of polymerization reactions.

Ajandouz et al. [2] studied the caramelization of fructose in both acidic and alkaline conditions. They found that at initial pH values ranging from 4.0 to 7.0, a progressive accumulation of the intermediate degradation products occurred as a function of time and no lag time was observed, whereas a high level of fructose degradation occurring during the initial stages of the heating period was reported at pH 8.0 and pH 9.0. They also found that the rate of the caramelization reaction of glucose increased exponentially from pH 7.0 up to pH 12.0, while a linear increase in the rate of the caramelization reaction of fructose occurred in the same pH range.

The contribution of the caramelization reaction to the overall browning in model systems of glucose and fructose in the presence and absence of amino acids have been investigated by Bozkurt [231] (Figure 11.15). He found that the contribution of the caramelization of glucose and fructose to the overall browning of glucose, fructose, and glutamine changed between 25% and 50% according to the studied temperature (55°C–75°C). It was reported that the increase in temperature increased the contribution of caramelization.

Caramelization of sucrose requires a temperature of about 200°C. Reactions involved in the caramelization of sucrose include mutarotation, enolization and isomerization, dehydration and fragmentation, anhydride formation, and polymerization. The extent to which the reaction occurs depends upon pH, temperature, and heating time. Sucrose, held at 160°C as a melt, will hydrolyze to glucose and fructose anhydride. The production of water and organic acids such as acetic, formic, and pyruvic during sucrose caramelization will enhance the hydrolysis. Hydrolytic products, glucose and fructose, are reactants in the formation of caramel and volatile flavor compounds

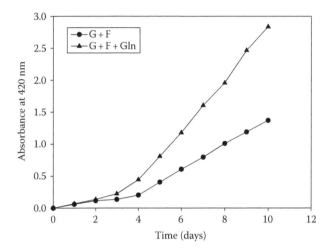

FIGURE 11.15   Contribution of caramelization of glucose and fructose on brown pigment formation at 65°C in a "glucose + fructose + glutamine" model system (G, glucose; F, fructose; Gln, glutamine).

[232]. Heating sucrose to 300°C in a stream of nitrogen produces volatile furan compounds; namely 2-methyl furan, furan, 2-hydroxyacetyl furan, and other volatile reaction products (methanol, acetone, acrolein, propanol, and acetaldehyde) [230]. A major compound (2-furfural) and volatile thermal degradation products (2-furyl methyl ketone, 2-furyl propyl ketone, methyl-benzo(*b*)furan, and 5-methyl-3-hydro-furan-2-one) were identified by heating sucrose in an open evaporation dish at 180°C for 90 min [233].

11.3.3   ASCORBIC ACID OXIDATION

Chemically, L-ascorbic acid belongs to the family of carbohydrates, functionally it is classified as an organic acid and a reducing reagent, and physiologically the compound is attributed to the class of essential vitamins [234]. Ascorbic acid is an antioxidant but it contributes to browning of foods because it is easily oxidized and decomposed under common storage and processing conditions. For this reason, ascorbic acid is responsible for the browning of most of the fruit juices and concentrates. It has a special importance in the citrus industry.

The factors affecting ascorbic acid degradation are pH, oxygen, ascorbic acid concentration, temperature, light, metals, citric acid, and so on. L-Ascorbic acid in its reduced form is of utmost importance in nutrition and food processing. However, once it has degraded, it results in the production of reactive carbonyl compounds that act as intermediates in nonenzymatic browning in foods. These reactive carbonyl compounds undergo further reactions leading to the formation of brown pigments. Studies showed that L-ascorbic acid decomposition resulted in furfural and HMF formation [235].

The reaction of ascorbic acid in fruit juices and concentrates is very much dependent on pH, as the browning process is inversely proportional to pH over a range of 2.0–3.5. Juices with a higher pH are much less susceptible to browning, for example, orange juice at a pH of 3.4. Below pH 4.0, browning is due primarily to the decomposition of ascorbic acid to furfural [235]. Some investigators reported that during the degradation of L-ascorbic acid, a decrease in pH due to the formation of carbonyl compounds was noted. The measurement of sample pH after storage showed a constant pH of 3.65 [181].

Monsalve et al. [236] studied the degradation of dehydroascorbic acid and chlorogenic acid, at pH 6, and at variable water activities and temperatures in model systems containing cellulose, dehydroascorbic acid and/or chlorogenic acid. They found that at 45°C nonenzymatic browning of dehydroascorbic acid-cellulose model system increased linearly through a certain time and then it

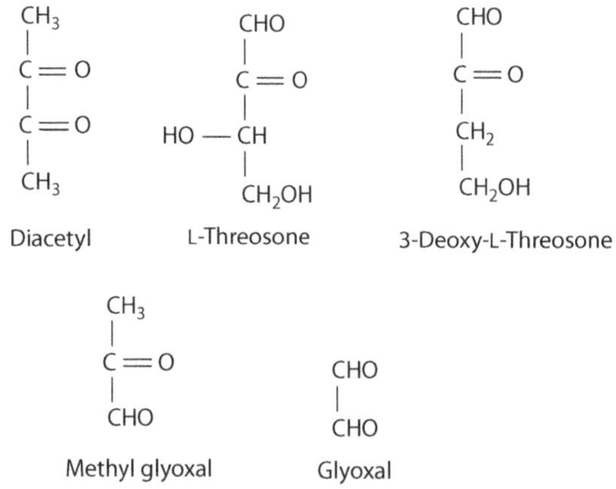

FIGURE 11.16   L-Ascorbic acid degradation products (α-dicarbonyl compounds).

remains nearly constant. The data showed that there is an inverse relationship between the brown product formation and water activity, and between the temperature and ascorbic acid loss.

It is accepted that there are two possible pathways, namely oxidative and nonoxidative, for the nonenzymatic browning reaction of L-ascorbic acid. In the course of the nonenzymatic browning reaction of L-ascorbic acid, a wide range of degradation products have been reported in the literature [234,235,237]. In the presence of oxygen, ascorbic acid is degraded primarily to dehydroascorbic acid *via* a monoanion. The lactone of dehydroascorbic acid is hydrolyzed to form 2,3-diketogulonic acid, which does not show vitamin C activity. Decarboxylation of 2,3-diketogulonic acid produces xylosone that is further degraded to reductones and furan compounds. Dehydroascorbic acid cannot be formed in nonoxidative conditions, where ascorbic acid can undergo hydrolysis. Furfural can be formed from ascorbic acid in both oxidative and nonoxidative conditions. Shinoda et al. [238] have proposed pathways for the formation of the furan precursors furfural and 2-furoic acid. They found that the formation of furfural from ascorbic acid in model orange juice was repressed by the presence of ethanol and mannitol acting as free radical scavengers.

Schulz et al. [234] used dehydro-Shinoda-ascorbic acid, which is the first characteristic intermediate of L-ascorbic acid degradation *via* the oxidative route, as a starting material of the reaction to distinguish it from the nonoxidative pathway. They found that dehydro-L-ascorbic acid yielded five α-dicarbonyl compounds, namely glyoxal, methylglyoxal, diacetyl, L-threosone, and 3-deoxy-L-threosone (Figure 11.16). They concluded that these α-dicarbonyl compounds are formed from L-ascorbic acid on the oxidative pathway. But, they pointed out that these products can also be produced *via* the nonoxidative route. They also found that 3-deoxy-L-pentosone is exclusively formed *via* the nonoxidative route.

REFERENCES

1. Maillard, L.C. Action of amino acids on sugars. Formation of melanoidins in a methodical way, *Compt. Rend.*, 154, 66, 1912.
2. Ajandouz, E.H. et al. Effect of pH on caramelization and Maillard reaction kinetics in fructose–lysine model systems, *J. Food Sci.*, 66, 926, 2001.
3. Gazzani, G. et al. Mutagenic activity of the Maillard reaction products of ribose with different amino acids, *J. Food Sci.*, 52, 757, 1987.
4. Ashoor, S.H. and Zent, J.B. Maillard browning of common amino acids and sugars, *J. Food Sci.*, 49, 206, 1984.
5. Martins, S.I.F.S., Jongen, W.M.F., and van Boekel M.A.J.S. A review of Maillard reaction in food and implications to kinetic modeling, *Trends Food Sci. Technol.*, 11, 364, 2001.

6. Davidek, J., Velsek, J., and Pokorny, J. *Development in Food Science: Chemical Changes during Food Processing*, vol. 21, Elsevier Science Publishers, Amsterdam, The Netherlands, 1990, p. 117.
7. Eskin, N.A.M. *Biochemistry of Foods*, 2nd ed., Academic Press, New York, 1990, Chap. 5, p. 9.
8. Macrane, R., Robinson, R.K., and Saadler, M.J. *Encyclopedia of Food Science, Food Technology and Nutrition*, vol. 1, Academic Press Limited, London, U.K., 1993, p. 146.
9. Ames, J.M. Control of the Maillard reaction in food systems, *Trends Food Sci. Technol.*, 1, 150, 1990.
10. Wijewickreme, A.N., Krejpcio, Z., and Kitts, D.D. Hydroxyl scavenging activity of glucose, fructose, and ribose–lysine model Maillard products, *J. Food Sci.*, 64, 457, 1999.
11. Kroh, L.W. Caramelization in food and beverages, *Food Chem.*, 51, 373, 1994.
12. del Buera, M.P. et al. Nonenzymatic browning in liquid model systems of high water activity: Kinetics of color changes due to caramelization of various single sugars, *J. Food Sci.*, 52, 1059, 1987.
13. Shallenberger, R.S. and Birch, G.C. *Sugar Chemistry*, AVI Publishing Co., Inc., Westport, CT, 1975, p. 169.
14. Lee, C.Y. Browning reaction, enzymatic, in *Wiley Encyclopedia of Food Science and Technology*, Francis, F.J., Ed., John Wiley & Sons, New York, 1999, p. 208.
15. Mayer, A.M. and Harel, E. Polyphenoloxidases and their significance in fruits and vegetables, in *Food Enzymology*, Fox, P.F., Ed., Elsevier Applied Science, London, U.K., 1991, p. 373.
16. Whitaker, J.R. Enzymes, in *Food Chemistry*, Fennema, O.R., Ed., Marcel Dekker, New York, 1996, p. 431.
17. Rouet-Mayer, M.A., Ralambosa, J., and Philippon, J. Roles of o-quinones and their polymers in the enzymic browning of apples, *Phytochemistry*, 29, 435, 1990.
18. Whitaker, J.R. and Lee, C. Y. Recent advances in chemistry of enzymatic browning, in *Enzymatic Browning and Its Prevention*, Lee, C.Y. and Whitaker, J.R., Eds., American Chemical Society, Washington, DC, 1995, p. 3.
19. Richard-Forget, F.C. et al. Oxidation of chlorogenic acid, catechins, and 4-methylcatechol in model solutions by apple polyphenoloxidase, *J. Agric. Food Chem.*, 40, 2114, 1992.
20. Weemaes, C.A. et al. Activity, electrophoretic characteristics and heat inactivation of polyphenoloxidases from apples, avocados, grapes, pears and plums, *Lebensm.-Wiss-u-Technol.*, 31, 44, 1998.
21. Özoğlu, H. and Bayindirli, A. Inhibition of enzymic browning in cloudy apple juice with selected anti-browning agents, *Food Control*, 13, 213, 2002.
22. Selles-Marchart, S., Casado-Vela, J., and Bru-Martinez, R. Isolation of a latent polyphenoloxidase from loquat fruit (*Eriobotrya japonica Lindl.*): Kinetic characterization and comparison with the active form, *Arch. Biochem. Biophys.*, 446, 175, 2006.
23. Nunez-Delicado, E. et al. Characterization of polyphenoloxidase from Napoleon grape, *Food Chem.*, 100, 108, 2007.
24. Wong, M.K., Dimick, P.S., and Hammerstedt, R.H. Extraction and high performance liquid chromatographic enrichment of polyphenoloxidase from theobroma cacao seeds, *J. Food Sci.*, 55, 1108, 1990.
25. Erat, M., Sakiroglu, H., and Kufrevioglu, O.I. Purification and characterization of polyphenol oxidase from Ferula sp., *Food Chem.*, 95, 503, 2006.
26. Cheynier, V., Osse, C., and Rigaud, J. Oxidation of grape juice phenolic compounds in model solutions, *J. Food Sci.*, 53, 1729, 1988.
27. Richard-Forget, F.C. and Gauillard, F.A. Oxidation of chlorogenic acid, catechins, and 4-methylcatechol in model solutions by combinations of pear (*Pyrus communis Cv. Williams*) polyphenol oxidase and peroxidase: A possible involvement of peroxidase in enzymatic browning, *J. Agric. Food Chem.*, 45, 2472, 1997.
28. Vamos-Vigyazo, L. Polyphenol oxidase and peroxidase in fruits and vegetables, *Crit. Rev. Food Sci.*, 15, 49, 1981.
29. Degl'Innocenti, E. et al. Biochemical study of leaf browning in minimally processed leaves of lettuce (*Lactuca sativa L. var. acephala*), *J. Agric. Food Chem.*, 53, 9980, 2005.
30. Degl'Innocenti, E. et al. Physiological basis of sensitivity to enzymatic browning in "lettuce", "escarole" and "rocket salad" when stored as fresh-cut products, *Food Chem.*, 104, 209, 2007.
31. Nicolas, J. et al. Polyphenols and enzymatic browning, in *Polyphenolic Phenomena*, INRA Ed., Paris, 1993, p. 165.
32. Robinson, D.S. Peroxidase and their significance in fruits and vegetables, in *Food Enzymology*, Fox, P.F., Ed., Elsevier, London, U.K., 1991, p. 399.
33. Amiot, J.M. et al. Phenolic compounds and oxidative mechanism in fruit and vegetables, in *Phytochemistry of Fruit and Vegetables*, Tomas-Barberan, F.A. and Robins, R.J., Eds., Claredon Press, Oxford, U.K., 1997, p. 51.

34. Tomas-Barberan, F.A. and Espin, J.C. Phenolic compounds and related enzymes as determinants of quality in fruits and vegetables, *J. Sci. Food Agric.*, 81, 853, 2001.
35. Jaenicke, E. and Decker, H. Tyrosinases from crustaceans form hexamers, *Biochem. J.*, 371, 515, 2003.
36. Zawistowski, J., Biliaderis, C.G., and Eskin, N.A.M. Polyphenoloxidase, in *Oxidative Enzymes in Food*, Robinson, D.S. and Eskin, N.A.M., Eds., Elsevier Science, New York, 1991, p. 217.
37. Marshall, M.R., Kim, J., and Wei, C.-I. *Enzymatic Browning in Fruits, Vegetables and Seafoods*, FAO, Rome, Italy, 2000, http://www.fao.org
38. Lerch, K. Tyrosinase: Molecular and active site structure, in *Enzymatic Browning and Its Prevention*, Lee, C.Y. and Whitaker, J.R., Eds., American Chemical Society, Washington, DC, 1995, p. 64.
39. Pourcel, L. et al. Flavonoid oxidation in plants: From biochemical properties to physiological functions, *Trends Plant Sci.*, 12, 29, 2007.
40. Mayer, A.M. Polyphenol oxidases in plants and fungi: Going places? A review, *Phytochemistry*, 67, 2318, 2006.
41. Lerch, K. *Neurospora* tyrosinase: Structural, spectroscopic and catalytic activity, *Mol. Cell. Biochem.*, 52, 125, 1983.
42. Huber, M., Hintermann, G., and Lerch, K. Primary structure of tyrosinase from *Streptomyces glaucescens*, *Biochemistry*, 24, 6038, 1985.
43. Yoshida, H. Chemistry of lacquer (*urushi*), *J. Chem. Soc.*, 43, 472, 1883.
44. Bertrand, G. Sur la laccase et sur le pouvoir oxydant de cette diastase, *CR Acad. Sci. (Paris)*, 120, 266, 1985.
45. Walker, J.R.L. Enzymatic browning in fruits: Its biochemistry and control, in *Enzymatic Browning and Its Prevention*, Lee, C.Y. and Whitaker, J.R., Eds., American Chemical Society, Washington, DC, 1995.
46. Mayer, A.M. and Staples, R.C. Laccase: New functions for an old enzyme, *Phytochemistry*, 60, 551, 2002.
47. Huttermann, A., Mai, C., and Kharazipour, A. Modification of lignin for the production of new compounded materials, *Appl. Microbiol. Biotechnol.*, 55, 387, 2001.
48. Bligny, R. and Douce, R. Excretion of laccase by sycamore (*Acer pseudoplatanus* L.) cells. Purification and properties of the enzyme, *Biochem. J.*, 209, 489, 1983.
49. Tezuka, K. et al. Occurrence of heterogeneity on N-linked oligosaccharides attached to sycamore (*Acer pseudoplatanus* L.) laccase of excretion, *Biochem. Mol. Biol. Int.*, 29, 395, 1993.
50. Sato, Y. et al. Molecular cloning and expression of eight cDNAs in loblolly pine (*Pinus taeda*), *J. Plant Res.*, 114, 147, 2001.
51. Ranocha, P. et al. Biochemical characterization, molecular cloning and expression of laccases a divergent gene family in poplar, *Eur. J. Biochem.*, 259, 485, 1999.
52. Dean, J.F.D. and Eriksson, K.-E.L. Laccase and the deposition of lignin in vascular plants, *Holzforschung*, 48(Suppl.), 21, 1994.
53. Vaughn, K.C. and Duke, S.O. Function of polyphenol oxidase in higher plants, *Physiol. Plant.*, 60, 106, 1984.
54. Montalbini, P., Buchanan, B.B., and Hutchesan, S.W. Effect of rust infection on rates of photochemical polychemical oxidation of latent polyphenol oxidase of Vicia faba chloroplast membranes, *Physiol. Plant Pathol.*, 18, 51, 1981.
55. Takahama, U. Oxidation of vacuolar and apoplastic phenolic substrates by peroxidase: Physiological significance of the oxidation reactions, *Phytochemistry*, 3, 207, 2004.
56. Mayer, H.A. and Harel, E. Polyphenol oxidases in plants, *Phytochemistry*, 18, 193, 1979.
57. Tolbert, N.E. Activation of polyphenoloxidase of chloroplast, *Plant Physiol.*, 51, 234, 1973.
58. Sherman, T.D., Gardeur, T.L., and Lax, A.R. Implications of the phylogenetic distribution of polyphenol oxidase in plants, in *Enzymatic Browning and Its Prevention*, Lee, C.Y. and Whitaker, J.R., Eds., American Chemical Society, Washington, DC, 1995, p. 103.
59. Chen, J.S. et al. Comparison of phenoloxidase activity from Florida spiny lobster and western Australian lobster, *J. Food Sci.*, 56, 154, 1991.
60. Ali, M.T. et al. Activation mechanisms of pro-phenoloxidase on melanosis development in Florida spiny lobster and Western Australian lobster (*Panulirus argus*), *J. Food Sci.*, 59, 1024, 1994.
61. Chevalier, T. et al. Molecular cloning and characterization of apricot fruit polyphenol oxidase, *Plant Physiol.*, 119, 1261, 1999.
62. Yoruk, R. and Marshall, M.R. Physicochemical properties and function of plant polyphenol oxidase: A review, *J. Food Chem.*, 27, 361, 2003.
63. Robinson, S.P., Loveys, B.R., and Chacko, E.K. Polyphenoloxidase enzymes in the sap and skin of mango fruit, *Aust. J. Plant Physiol.*, 20, 99, 1993.

64. Kahn, V. Some biochemical properties of polyphenol oxidase from two avocado varieties differing in their browning rates, *J. Food Sci.*, 42, 38, 1977.

65. Sanchez-Ferrer, A., Bru, R., and Garcia-Carmona, F. Novel procedure for extraction of a latent grape polyphenoloxidase using temperature-induced phase separation in Triton X-114, *Plant Physiol.*, 91, 1481, 1989.

66. Richardson, T. and Hyslop, D.B. Enzymes, in *Food Chemistry*, 2nd ed., Fennema, O.R., Ed., Marcel Dekker, New York, 1985, p. 446.

67. Leydon Webb, J. *Enzyme and Metabolic Inhibitors*, vol. 2, Academic Press, New York, 1966, p. 301.

68. Walsh, C. *Enzymatic Reaction Mechanism*, W.H. Freeman & Company, San Francisco, CA, 1979, p. 461.

69. Espin, J.C. et al. Monophenolase activity of polyphenol oxidase from verdedoncella apple, *J. Agric. Food Chem.*, 43, 2807, 1995.

70. Nicolas, J.J. et al. Enzymatic browning reactions in apple and apple products, *Crit. Rev. Food Sci.*, 34, 109, 1994.

71. Marques, L., Fleuriet, A., and Macheix, J.J. Fruit polyphenol oxidases, new data on an old problem, in *Enzymatic Browning and Its Prevention*, Lee, C.Y. and Whitaker, J.R., Eds., American Chemical Society, Washington, DC, 1995, p. 90.

72. Wesche-Ebeling, P. and Montgomery, M.W. Strawberry polyphenol oxidase: Purification and characterization, *J. Food Sci.*, 55, 1315, 1990.

73. Valero, E., Varon, R., and Garcia-Carmona, F. Characterization of polyphenoloxidase from Airen grapes, *J. Food Sci.*, 53, 1482, 1988.

74. Oktay, M. et al. Polyphenol oxidase from Amasya apple, *J. Food Sci.*, 60, 495, 1995.

75. Rocha, A.M.C.N. and Morais, A.M.M.B. Characterization of polyphenoloxidase (PPO) from "Jonagored" apple, *Food Control*, 12, 85, 2001.

76. Ngalani, J.A., Signoret, A., and Crouzet, J. Partial purification and properties of plantain polyphenoloxidase, *Food Chem.*, 48, 341, 1993.

77. Park, E.Y. and Luh, B.S. Polyphenol oxidase of kiwifruit, *J. Food Sci.*, 50, 678, 1985.

78. Galeazzi, M.A.M., Sgarbieri, V.C., and Constantinides, S.M. Isolation, purification and physicochemical characterization of polyphenoloxidases from a drawf variety of banana (*musa cavendishii* L.), *J. Food Sci.*, 46, 150, 1981.

79. Tikoo, S. et al. Studies on polyphenol oxidase in wheat grains, *Cereal Chem.*, 50, 520, 1973.

80. Halim, D.H. and Montgomery, M.W. Polyphenol oxidase of d'Anjou pears (*pyrus communis* L.), *J. Food Sci.*, 43, 603, 1978.

81. Ratcliffee, B. et al. Tyrosinase, laccase and peroxidase in mushrooms (Agaricus, crimini, oyster, and shiitake), *J. Food Sci.*, 59, 824, 1994.

82. Lerch, K. Copper monooxygenases, tyrosinases and dopamine-monooxygenase, in *Metal Ions in Biological Systems. Copper Proteins*, Siegel, H., Ed., Dekker, New York, 1981, p. 143.

83. Solomon, E.I. Copper proteins, in *Metal Ions in Biology*, Spiro, T.G., Ed., John Wiley & Sons, New York, 1981, p. 183.

84. Rescigno, A. et al. Tyrosinase inhibition: General and applied aspects, *J. Enzyme Inhib. Med. Chem.*, 17, 207, 2002.

85. Solomon, E.I., Baldwin, M.J., and Lowrey, M.D. Electronic structures of active sites in copper proteins: Contributions to reactivity, *Chem. Rev.*, 92, 521, 1992.

86. Whitaker, J.R. Polyphenol oxidase, in *Food Enzymes Structure and Mechanism*, Wong, D.W.S., Ed., Chapman & Hall, New York, 1995, p. 271.

87. Sánchez-Ferrer, A. et al. Tyrosinase: A comprehensive review of its mechanism, *Biochim. Biophys. Acta*, 1247, 1, 1995.

88. Decker, H. et al. Similar enzyme activation and catalysis in hemocyanins and tyrosinases, *Gene*, 398, 183, 2007.

89. Randhir, R., Lin, Y.-T., and Shetty, K. Phenolics, their antioxidant and antimicrobial activity in dark germinated fenugreek sprouts in response to peptide and phytochemical elicitors, *Asia Pacific J. Clin. Nutr.*, 13, 295, 2004.

90. Bravo, L. Polyphenols: Chemistry, dietary sources, metabolism, and nutritional significance, *Nutr. Rev.*, 56, 317, 1998.

91. Alasalvar, C. et al. Comparison of volatiles, phenolics, sugars, antioxidant vitamins, and sensory quality of different colored carrot varieties, *J. Agric. Food Chem.*, 49, 1410, 2001.

92. Balasundram, N., Sundram, K., and Samman, S. Phenolic compounds in plants and agri-industrial by-products: Antioxidant activity, occurrence, and potential uses, *Food Chem.*, 99, 191, 2006.

93. Wesche-Ebeling, P. and Montgomery, M.W. Strawberry polyphenoloxidase: Its role in anthocyanin degradation, *J. Food Sci.*, 55, 731, 1990.
94. Martinez, M.V. and Whitaker, J.R. The biochemistry and control of enzymatic browning, *Trends Food Sci. Technol.*, 6, 195, 1995.
95. Chutintrasri, B. and Noomhorm, A. Thermal inactivation of polyphenoloxidase in pineapple puree, *Lebensm-Wiss-u-Technol.*, 39, 492, 2006.
96. Yemenicioğlu, A., Özkan, M., and Cemeroğlu, B. Heat inactivation kinetics of apple polyphenoloxidase and activation of its latent form, *J. Food Sci.*, 62, 508, 1997.
97. Khali, M. and Selselet-Attou, G. Effect of heat treatment on polyphenol oxidase and peroxidase activities in Algerian stored dates, *Afr. J. Biotechnol.*, 6, 790, 2007.
98. Almeida, M.E. and Nogueira, J.N. The control of polyphenol oxidase activity in fruits and vegetables—A study of the interactions between the chemical compounds used and heat treatment, *Plant Foods Hum. Nutr.*, 43, 245, 1995.
99. Lee, C.Y., Smith, N.L., and Hawbecker, D.E. Enzyme activity and quality of frozen green beans as affected by blanching and storage, *J. Food Qual.*, 11, 279, 1988.
100. Ma, S.X. et al. Prevention of enzymatic darkening in frozen sweet potatoes [*Ipomoea batatas (L.) Lam.*] by water blanching: Relationship among darkening, phenols, and polyphenol oxidase activity, *J. Agric. Food Chem.*, 40, 864, 1992.
101. Vina, S.Z. et al. Quality of Brussels sprouts (*Brassica oleracea L. Gemmifera DC*) as affected by blanching method, *J. Food Eng.*, 80, 218, 2007.
102. Devece, C. et al. Enzyme inactivation analysis for industrial blanching applications: Comparison of microwave, conventional, and combination heat treatments on mushroom polyphenol oxidase activity, *J. Agric. Food Chem.*, 47, 4506, 1999.
103. Flores, J.H. et al. Ultrafiltration of White Riesling juice: Effect of oxidation and pre-UF juice treatment on flux, composition, and stability, *Am. J. Enol. Vitic.*, 39, 180, 1988.
104. Sims, C.A., Johnson, R.P., and Bates, R.P. Quality of a nonsulfited *Vitis rotundifolia* and a *Euvitis* hybrid white wine produced from ultrafiltered juice, *Am. J. Enol. Vitic.*, 40, 272, 1989.
105. Goodwin, C.O. and Morris, J.R. Effect of ultrafiltration on wine quality and browning, *Am. J. Enol. Vilic.*, 42, 347, 1991.
106. Drake, S.R. and Nelson, J.W. Apple juice quality as influenced by ultrafiltration, *J. Food Qual.*, 9, 399, 1986.
107. Hsu, J.C. et al. Heat unstable proteins in grapejuice and wine. II. Characterization and removal by ultrafiltration, *Am. J. Enol. Vitic.*, 38, 17, 1987.
108. Padilla, O.I. and McClellan M.R. Molecular weight cut-off of ultrafiltration membranes and the quality and stability of apple juice, *J. Food Sci.*, 54, 1250, 1989.
109. Tanada-Palmu, P., Jardine, J., and Matta, V. Production of a banana (*Musa cavendishii*) extract containing no polyphenol oxidase by ultrafiltration, *J. Sci. Food Agric.*, 79, 643, 1999.
110. Sims, C.A., Bates, R.P., and Arreola, A.G. Color, polyphenol oxidase, and sensory changes in banana juice as affected by heat and ultrafiltration, *J. Food Qual.*, 17, 371, 1994.
111. Ludikhuyze, L.R. and Hendrickx, M.E. Effects of high pressure on chemical reactions related to food quality, in *Ultra High Pressure Treatments of Foods*, Hendrickx, M.E. and Knorr, D., Eds., Kluwer Academic/Plenum Publishers, New York, 2001, p. 167.
112. Seyderhelm, I. et al. Pressure induced inactivation of selected food enzymes, *J. Food Sci.*, 61, 308, 1996.
113. Hendrickx, M. et al. Effects of high pressure on enzymes related to food quality, *Trends Food Sci. Technol.*, 9, 197, 1998.
114. Ballestra, P. et al. Effect of high-pressure treatment on polyphenol oxidase activity of the *Agaricus bisporus* mushroom, *High Press. Res.*, 22, 677, 2002.
115. Phunchaisri, C. and Apichartsrangkoon, A. Effects of ultra-high pressure on biochemical and physical modification of lychee (*Litchi chinensis Sonn.*), *Food Chem.*, 93, 57, 2005.
116. Kim, Y.-S. et al. Effects of combined treatment of high hydrostatic pressure and mild heat on the quality of carrot juice, *J. Food Sci.*, 66, 1355, 2001.
117. Hernandez, A. and Cano, M.P. High-pressure and temperature effects on enzyme inactivation in tomato puree, *Agric. Food Chem.*, 46, 266, 1998.
118. Prestamo, G., Sanz, P.D., and Arroyo, G. Fruit preservation under high hydrostatic pressure, *High Press. Res.*, 19, 535, 2000.
119. Guerrero-Beltran, J.A., Barbosa-Canovas, G.V., and Swanson, B.G. High hydrostatic pressure processing of peach puree with and without antibrowning agents, *J. Food Process. Preserv.*, 28, 69, 2004.

120. Lopez, P. et al. Inactivation of peroxidase, lipoxygenase, and polyphenol oxidase by manothermosonication, *J. Agric. Food Chem.*, 42, 252, 1994.

121. Vercet, A., Lopez, P., and Burgos, J. Inactivation of heat-resistant lipase and protease from *Pseudomonas fluorescens* by manothermosonication, *J. Dairy Sci.*, 80, 29, 1997.

122. Vercet, A. et al. The effects of manothermosonication on tomato pectic enzymes and tomato paste rheological properties, *J. Food Eng.*, 53, 273, 2002.

123. Christianson, D.D. et al. Supercritical fluid extraction of dry-milled corn germ with carbon dioxide, *J. Food Sci.*, 49, 229, 1984.

124. Arreola, A.G. Effects of supercritical CO_2 on some quality attributes of single strength orange juice, Masters thesis, University of Florida, Gainesville, FL, 1990.

125. Zemel, G.P. Low pH inactivation of polyphenoloxidase, Masters thesis, University of Florida, Gainesville, FL, 1989.

126. Chen, J.S. et al. Inactivation of polyphenol oxidase by high-pressure carbon dioxide, *J. Agric. Food Chem.*, 40, 2345, 1992.

127. Gui, F. et al. Change of polyphenol oxidase activity, color, and browning degree during storage of cloudy apple juice treated by supercritical carbon dioxide, *Eur. Food Res. Technol.*, 223, 427, 2006.

128. Sapers, G.M. Browning of foods: Control by sulfites, antioxidants, and other means, *Food Technol.*, 47, 75, 1993.

129. McEvily, A.J., Iyengar, R., and Otwell, W.S. Inhibition of enzymatic browning in foods and beverages, *Crit. Rev. Food Sci.*, 32, 253, 1992.

130. Rocha, A.M.C.N. and Morais, A.M.M.B. Polyphenol oxidase activity of minimally processed "Jonagord" apples (*Malus domestica*), *J. Food Process. Preserv.*, 29, 8, 2005.

131. Manzocco, L., Mastrocola, D., and Poiana, M. Control of browning in frozen puree of cherimoya (*Annona cherimola, Mill.*) fruit, *Sci. Des Aliments*, 18, 101, 1998.

132. Dudley, E.D. and Hotchkiss, J.H. Cysteine as an inhibitor of polyphenol oxidase, *J. Food Biochem.*, 13, 65, 1989.

133. İyidoğan, N.F. and Bayındırlı, A. Effect of L-cysteine, kojic acid and 4-hexylresorcinol combination on inhibition of enzymatic browning in Amasya apple juice, *J. Food Eng.*, 62, 299, 2004.

134. Dziezak, J.D. Preservatives: Antioxidants—The ultimate answer to oxidation, *Food Technol.*, 40, 94, 1986.

135. Lambrecht, H.S. Sulfite substitutes for the prevention of enzymatic browning in foods, in *Enzymatic Browning and Its Prevention*, Lee, C.Y. and Whitaker, J.R., Eds., American Chemical Society, Washington, DC, 1995, p. 313.

136. Luo, Y. and Barbosa-Canovas, G.V. Enzymatic browning and its inhibition in new apple cultivars slices using 4-hexylresorcinol in combination with ascorbic acid, *Food Sci. Technol. Int.*, 3, 195, 1997.

137. Guerrero-Beltran, J.A., Swanson, B.G., and Barbosa-Canovas, G.V. Inhibition of polyphenol oxidase in mango puree with 4-hexylresorcinol, cysteine and ascorbic acid, *Lebensm-Wiss-u-Technol.*, 38, 625, 2005.

138. Walker, J.R.L. The control of enzymic browning in fruit juices by cinnamic acids, *J. Food Technol.*, 11, 341, 1976.

139. Gunata, Y.Z., Sapis, J.C., and Moutonet, M. Substrates and aromatic carboxylic acid inhibitors of grape polyphenoloxidases, *Phytochemistry*, 26, 1573, 1987.

140. Kermasha, S., Goetghebeur, M., and Monfette, A. Studies on inhibition of mushroom polyphenol oxidase using chlorogenic acid as substrate, *J. Agric. Food Chem.*, 41, 526, 1993.

141. Robert, C., Rouch, C., and Cadet, F. Inhibition of palmito (*Acanthophoenix rubra*) polyphenol oxidase by carboxylic acids, *Food Chem.*, 59, 355,1997.

142. Jeon, M. and Zhao, Y. Honey in combination with vacuum impregnation to prevent enzymatic browning of fresh-cut apples, *Int. J. Food Sci. Nutr.*, 56, 165, 2005.

143. Kim, M.-J., Kim, C.Y., and Park, I. Prevention of enzymatic browning of pear by onion extract, *Food Chem.*, 89, 181, 2005.

144. Kim, C.Y. et al. Inhibition of polyphenol oxidase and peach juice browning by onion extract, *Food Sci. Biotechnol.*, 16, 421, 2007.

145. Lozano-De-Gonzalez, P.G. et al. Enzymatic browning inhibited in fresh and dried apple rings by pinapple juice, *J. Food Sci.*, 58, 399, 1993.

146. Lane, M.J. and Nursten, H.E. Variety of odors produced in Maillard model systems and how they are influenced by reaction conditions, in *The Maillard Reaction in Foods and Nutrition*, Waller, G.R. and Feather, M.S., Eds., *ACS Symposium Series*, vol. 215, American Chemical Society, Washington, DC, 1983, p. 141.

147. Muresan, S. et al. Aroma profile development of intermediate chocolate products. I. Volatile constituents of block-milk, *Food Chem.*, 68, 167, 2000.
148. van Boekel, M.A.J.S. Formation of flavour compounds in the Maillard reaction, *Biotech. Adv.*, 24, 230, 2006.
149. Matiacevich, S.B. and Buera, M.P. A critical evaluation of fluorescence as a potential marker for the Maillard reaction, *Food Chem.*, 95, 423, 2006.
150. Knerr, T. et al. Formation of a novel colored product during the Maillard reaction of D-glucose, *J. Agric. Food Chem.*, 49, 1966, 2001.
151. Feillet, P., Autran, J.C., and Icard-Verniere, C. Pasta brownness: An assessment, *J. Cereal Sci.*, 32, 215, 2000.
152. Franzke, C. and Iwainsky, H. Antioxidant action of melanoidins, *Deutsche Lebensmittel-Rundschau*, 580, 251, 1954.
153. Morales, F.J. and Babbel, M.B. Antiradical efficiency of Maillard reaction mixtures in a hydrophilic media, *J. Agric. Food Chem.*, 50, 2788, 2002.
154. Bersuder, P., Hole, M., and Smith, G. Antioxidants from a heated histidine-glucose model system. Investigation of the copper (II) binding ability, *JAOCS*, 78, 1079, 2001.
155. Ferrer, E. et al. Review: Indicators of damage of protein quality and nutritional value of milk, *Food Sci. Technol. Int.*, 5, 447, 1999.
156. Nakamura, A. et al. Changes in allergenicity and digestibility of squid tropomyosin during the Maillard reaction with ribose, *J. Agric. Food Chem.*, 54, 9529, 2006.
157. Erbersdobler, H.F. and Somoza, V. Forty years of furosine—Forty years of using Maillard reaction products as indicators of the nutritional quality of foods, *Mol. Nutr. Food Res.*, 51, 423, 2007.
158. Nagao, M. et al. Mutagens in coffee and tea, *Mutat. Res.*, 68, 101, 1979.
159. Kato, T., Hiramoto, K., and Kikugawa, K. Possible occurrence of new mutagens with the DNA breaking activity in coffee, *Mutat. Res.*, 306, 9, 1994.
160. Swedish National Food Administration. Information about Acrylamide in Food, April 24, 2002. www.slv. se (accessed March 30, 2009).
161. Mottram, D.S., Wedzicha, B.I., and Dodson, A.T. Acrylamide is formed in the Maillard reaction, *Nature*, 419, 448, 2002.
162. Stadler, R.H. et al. Acrylamide from Maillard reaction products, *Nature*, 419, 449, 2002.
163. Becalski, A. et al. Acrylamide in foods: Occurrence, sources and modeling, *J. Agric. Food Chem.*, 51, 802, 2003.
164. Taeymans, D. et al. A review of acrylamide: An industry perspective on research, analysis, formation and control, *Crit. Rev. Food Sci.*, 44, 323, 2004.
165. Zhang, Y. and Zhang, Y. Formation and reduction of acrylamide in Maillard reaction: A review based on the current state of knowledge, *Crit. Rev. Food Sci.*, 47, 521, 2007.
166. Prakash Reddy, V. et al. Involvement of Maillard reactions in Alzheimer disease, *Neurotox. Res.*, 4, 191, 2002.
167. Tessier, F., Obrenovich, M., and Monnier, V.M. Structure and mechanism of formation of human lens fluorophore LM-1. Relationship to vesperlysine A and the advanced Maillard reaction in aging, diabetes, and cataractogenesis, *J. Biol. Chem.*, 274, 20796, 1999.
168. Waanders, F. et al. Renal accumulation of pentosidine in non-diabetic proteinuria-induced renal damage in rats, *Nephrol. Dial. Transpl.*, 20, 2060, 2005.
169. Hodge, J.E. Chemistry of browning reactions in model systems, *J. Agric. Food Chem.*, 1, 928, 1953.
170. Hurst, D.T. Recent developments in the study of non-enzymic browning and its inhibition by sulphur dioxide, B.F.M.I.R.A. Scientific and Technical Surveys, No. 75, Leatherhead, England, U.K., 1972.
171. Kort, M.J. Reactions of free sugars with aqueous ammonia, *Adv. Carbohydr. Chem. Biochem.*, 25, 311, 1970.
172. O'Brian, J. and Morrisey, P.A. Nutritional and toxicological aspects of the Maillard browning reaction in foods, *Crit. Rev. Food Sci.*, 28, 211, 1989.
173. Kato, H. and Tsuchida, H. Estimation of melanoidin structure by pyrolysis and oxidation, *Prog. Food Nutr. Sci.*, 5, 147, 1981.
174. Rubinsztain, Y. et al. Investigation on the structural units of melanoidins, *Org. Geochem.*, 6, 797, 1984.
175. Blank, I. Current status of acrylamide research in food: Measurement, safety assessment, and formation, *Ann. N. Y. Acad. Sci.*, 1043, 30, 2005.
176. Yaylayan, V.A., Wnorowski, A., and Locas, C.P. Why asparagine needs carbohydrates to generate acrylamide, *J. Agric. Food Chem.*, 51, 1753, 2003.

177. Zyzak, D.V. et al. Acrylamide formation mechanism in heated foods, *J. Agric. Food Chem.*, 51, 4782, 2003.

178. Granvogl, M. et al. Quantitation of 3-aminopropionamide in potatoes—A minor but potent precursor in acrylamide formation, *J. Agric. Food Chem.*, 52, 4751, 2004.

179. Kwak, E.J. and Lim, S.I. The effect of sugar, amino acid, metal ion, and NaCl on model Maillard reaction under pH control, *Amino Acids*, 27, 85, 2004.

180. Miller, R., Olsson, K., and Pernemalm, P. Formation of aromatic compounds from carbohydrates. IX. Reaction of D-glucose and lysine in slightly acidic, aqueous solution, *Acta Chem. Scand. B*, 38, 689, 1984.

181. Wolfrom, M.L., Naoki, K., and Derek, H. Factors affecting the Maillard reaction between the sugars and amino acids studies on the nonenzymic browning of dehydrated orange juice, *J. Agric. Food Chem.*, 22, 796, 1974.

182. Göğüş, F., Bozkurt, H., and Eren, S. Nonenzymic browning reactions in multi sugar and amino acid systems, *J. Food Process. Preserv.*, 22, 81, 1998.

183. Massaro, S.A. and Labuza, T.P. Browning and amino acid loss in model total parental nutrition solutions, *J. Food Sci.*, 55, 821, 1990.

184. Jousse, F. et al. Simplified kinetic scheme of flavour formation by the Maillard reaction, *J. Food Sci.*, 67, 2534, 2002.

185. Brands, C.M.J., Wedzicha, B.L., and van Boekel, M.A.J.S. Quantification of melanoidin concentration in sugar–casein systems, *J. Agric. Food Chem.*, 50, 1178, 2002.

186. Labuza, T.P. and Baisier, W.M. The kinetics of nonenzymatic browning, in *Physical Chemistry of Foods*, Schwarzberg, H.G. and Hartel, R.W., Eds., Marcel Dekker, Inc., New York, 1992, p. 595.

187. Hayashi, T. and Namiki, M. Role of sugar fragmentation in an early stage browning of amino-carbonyl reaction of sugar with amino acid, *Agric. Biol. Chem.*, 50, 1965, 1986.

188. Pilkova, L., Pokorny, J., and Davidek, J. Browning reactions of Heyns rearrangement products, *Nahrung*, 34, 759, 1990.

189. Kato, K., Watanabe, K., and Sato, Y. Effect of the Maillard reaction on the attributes of ovalbumin, *J. Agric. Food Chem.*, 29, 540, 1978.

190. Claeys, W.L., De Vleeschouwer, K., and Hendrickx, M.E. Kinetics of acrylamide formation and elimination during heating of an asparagine-sugar model system, *J. Agric. Food Chem.*, 53, 9999, 2005.

191. Warmbier, H.C., Schnickels, R.A., and Labuza, T.P. Effect of glycerol on nonenzymatic browning in a solid intermediate moisture model food system, *J. Food Sci.*, 20, 218, 1976.

192. Renn, P.T. and Sathe, S.K. Effects of pH, temperature, and reactant molar ratio on L-leucine and D-glucose Maillard browning reaction in an aqueous system, *J. Agric. Food Chem.*, 45, 3782, 1997.

193. Kato, H., Yamamoto, M., and Fujimaki, M. Mechanisms of browning degradation of D-fructose in special comparison with D-glucose-glycine reaction, *Agric. Biol. Chem.*, 33, 939, 1969.

194. Martins, S.I.F.S. and van Boekel, M.A.J.S. Kinetics of the glucose/glycine Maillard reaction pathways: Influences of pH and reactant initial concentrations, *Food Chem.*, 92, 437, 2005.

195. Borsook, H. and Wasteneys, H. The interaction of free amino-nitrogen and glucose, *Biochem. J.*, 19, 1028, 1925.

196. Huyghues-Despoints, A. and Yaylayan, V.A. Retro-aldol and redox reactions of Amadori compounds: Mechanistic studies with variously labelled D-[13 C] glucose, *J. Agric. Food Chem.*, 44, 672, 1996.

197. Baxter, J.H. Free amino acid stability in reducing sugar systems, *J. Food Sci.*, 60, 405, 1995.

198. Lertittikul, W., Benjakul, S., and Tanaka, M. Characteristics and antioxidative activity of Maillard reaction products from a porcine plasma protein-glucose model system as influenced by pH, *Food Chem.*, 100, 669, 2007.

199. Rydberg, P. et al. Factors that influence the acrylamide content of heated foods, *Chemistry and Safety of Acrylamide in Food Advances in Experimental Medicine and Biology*, vol. 561, Springer Science + Business Media, Inc., New York, 2005, p. 317.

200. Benzing-Purdie, L.M., Ripmeester, J.A., and Ratcliffe, C.I. Effects of temperature on Maillard reaction products, *J. Agric. Food Chem.*, 33, 31, 1985.

201. van Boekel, M.A.J.S. Kinetic aspects of the Maillard reaction: A critical review, *Nahrung*, 45, 150, 2001.

202. Miao, S. and Roos, Y.H. Nonenzymatic browning kinetics of a carbohydrate-based low-moisture food system at temperatures applicable to spray drying, *J. Agric. Food Chem.*, 52, 5250, 2004.

203. Bozkurt, H., Göğüş, F., and Eren, S. Nonenzymic browning reactions in boiled grape juice and its models during storage, *Food Chem.*, 64, 89, 1999.

204. Pedreschi, F., Kaack, K., and Granby, K. Acrylamide content and color development in fried potato strips, *Food Res. Int.*, 39, 40, 2006.

205. Eichner, K. and Karel, M. The influence of water content and water activity on brown pigment formation, *J. Agric. Food Chem.*, 20, 218, 1972.

206. Labuza, T.P., Tannenbaum, S.R., and Karel, M. Water content and stability of low moisture and intermediate moisture foods, *Food Technol.*, 24, 543, 1970.

207. Eichner, K. The influence of water content on brown pigment formation in dehydrated foods and model systems and the inhibition of fat oxidation by browning intermediates, in *Water Relations of Foods*, Duckworth, R.B., Ed., Academic Press, New York, 1975, p. 417.

208. Göğüş, F., Düzdemir, C., and Eren, S. Effects of some hydrocolloids and water activity on nonenzymic browning of concentrated orange juice, *Nahrung*, 44, 438, 2000.

209. Ameur, L.A., Mathieu, O., and Lalanne, V. Comparison of the effects of sucrose and hexose on furfural formation and browning in cookies baked at different temperatures, *Food Chem.*, 101, 1407, 2007.

210. Mundt, S. and Wedzicha, B.L. A kinetic model for browning in the baking of biscuits: Effects of water activity and temperature, *Lebensm-Wiss-u-Technol.*, 40, 1078, 2007.

211. Sherwin, C.P. and Labuza, T.P. Role of moisture in Maillard browning reaction rate in intermediate moisture foods: Comparing solvent phase and matrix properties, *J. Food Sci.*, 68, 588, 2003.

212. Jaiswal, A.K., Shukla, R., and Gupta, G. Control of Maillard browning in potato chips by yeast prefermentation, *J. Food Sci. Technol. Mysore*, 40, 546, 2003.

213. Rais, A. and Aroujalian, A. Reduction of the glucose syrup browning rate by the use of modified atmosphere packaging, *J. Food Eng.*, 80, 370, 2007.

214. Moreno, F.J. et al. High-pressure effects on Maillard reaction between glucose and lysine, *J. Agric. Food Chem.*, 51, 394, 2003.

215. Komthong, P. et al. Effect of high hydrostatic pressure combined with pH and temperature on glucose/fructose-leucine/lysine/glutamate browning reactions, *J. Fac. Agric. Kyushu Univ.*, 48, 135, 2003.

216. Gokmen, V. and Senyuva, H.Z. Acrylamide formation is prevented by divalent cations during the Maillard reaction, *Food Chem.*, 103, 196, 2007.

217. Edelstein, D. and Brownlee, M. Mechanistic studies of advanced glycosylation end product inhibition by aminoguanidine, *Diabetes*, 41, 26, 1992.

218. Ingles, D.L. The formation of sulphonic acids from the reaction of reducing sugars with sulphite, *Aust. J. Chem.*, 15, 342, 1962.

219. Wedzicha, B.L. and Vakalis N. Kinetics of the sulphite-inhibited Maillard reaction: The effect of sulphite ion, *Food Chem.*, 27, 259, 1988.

220. Colahan-Sederstrom, P.M. and Peterson, D.G. Inhibition of key aroma compound generated during ultrahigh-temperature processing of bovine milk *via* epicatechin addition, *J. Agric. Food Chem.*, 53, 398, 2005.

221. Schamberger, G.P. and Labuza, T.P. Effect of green tea flavonoids on Maillard browning in UHT milk, *Lebensm-Wiss-u-Technol.*, 40, 1410, 2007.

222. Lindsay, R.C. and Jang, S. Model systems for evaluating factors affecting acrylamide formation in deep fried foods, *Chemistry and Safety of Acrylamide in Food Advances in Experimental Medicine and Biology*, vol. 561, Springer Science + Business Media, Inc., New York, 2005, p. 329.

223. Fernandez, S., Kurppa, L., and Hyvonen, L. Content of acrylamide decreased in potato chips with addition of a proprietary flavoniod spice mix (Flavomare) in frying, *Inn. Food Technol.*, 18, 24, 2003.

224. International Agency for Research on Cancer, *IARC Monographs on the Evaluation of Carcinogenic Risks to Humans, Some Industrial Chemicals, Acrylamide*, vol. 60, Lyon, France: IARC, 1994, p. 389.

225. Pedreschi, F. et al. Color kinetics and acrylamide formation in NaCl soaked potato chips, *J. Food Eng.*, 79, 989, 2007.

226. Gokmen, V. and Senyuva, H.Z. Effects of some cations on the formation of acrylamide and furfurals in glucose-asparagine model system, *Eur. Food Res. Technol.*, 225, 815, 2007.

227. Kwak, E.J. and Lim, S.I. Inhibition of browning by antibrowning agents and phenolic acids or cinnamic acid in the glucose-lysine model, *J. Sci. Food Agric.*, 85, 1337, 2005.

228. Bryce, D.J. and Greenwood, C.T. Thermal degradation of starch, *Staerke*, 15, 166, 1963.

229. O'Beirne, D. Effect of pH on nonenzymatic browning during storage in apple juice concentrate prepared from bramley's seedling apples, *J. Food Sci.*, 51, 1073, 1986.

230. Myers, D.V. and Howell, J.C. Characterization and specification of caramel colours: An overview, *Food Chem. Toxicol.*, 30, 356, 1992.

231. Bozkurt, H. Kinetics of color changes due to Maillard Reactions in model systems, Masters thesis, University of Gaziantep, Gaziantep, Turkey, 1996.

232. Fagerson, I.S. Thermal degradation of carbohydrates—A review, *J. Agric. Food Chem.*, 17, 747, 1969.

233. Kitts, D.D. et al. Chemistry and genotoxicity of caramelized sucrose, *Mol. Nutr. Food Res.*, 50, 1180, 2006.

234. Schulz, A. et al. Electrospray ionization mass spectrometric investigations of α-dicarbonyl compounds—Probing intermediates formed in the course of the nonenzymatic browning reaction of L-ascorbic acid, *Int. J. Mass Spectrom.*, 262, 169, 2007.

235. Huelin, P.E. Studies on the anaerobic decomposition of ascorbic acid, *Food Res.*, 15, 78, 1953.

236. Monsalve, G.A., Powers, J.A., and Leung, H.K. Browning of dehydro ascorbic acid and chlorogenic acid as a function of water activity, *J. Food Sci.*, 55, 1425, 1990.

237. Yuan, J.P. and Chen, F. Degradation of ascorbic acid in aqueous solution, *J. Agric. Food Chem.*, 46, 5078, 1998.

238. Shinoda, Y. et al. Browning of model orange juice solution: Factors affecting the formation of decomposition products, *Biosci. Biotech. Biochem.*, 69, 2129, 2005.

12 Lipid Oxidation and Control of Oxidation

Sotirios Kiokias, Theodoros H. Varzakas, Ioannis S. Arvanitoyannis, and Athanasios E. Labropoulos

CONTENTS

12.1 TYPES OF OXIDATION IN OIL MODEL SYSTEMS

12.1.1 INTRODUCTION

Free radicals (R•, ROO•, etc.) can be defined as any chemical species having one or more unpaired electrons, a wide definition that covers hydrogen atoms, transitions metals, the oxygen molecule itself, etc.[1] One of the major areas in which carbon free radicals and oxygen free radicals are involved is lipid oxidation that determines the quality and shelf life of foods.[2] Oxidative rancidity leads to the production of spoiled off-flavor products and is commonly associated with the organoleptic appraisal of food products.[3]

The oxidation of edible oils rich in polyunsaturated fatty acids is a major concern of the food industry because it is directly related to economic, nutritional, flavor, safety, and storage problems.[4] Edible oils are susceptible to oxidation that produces undesirable volatile compounds and causes detrimental flavor effects in oil-based foods.[5] The oxidation process is an important reaction occurring between unsaturated lipids and atmospheric oxygen and it is usually accelerated by the presence of metals, light, heat, and several initiators.[6] Though lipid oxidation has been extensively studied in bulk oils, the relevant mechanisms have not been elucidated yet in lipid based emulsions.[7] A better understanding of the factors monitoring the oxidative deterioration in these systems would offer strategies to improve the organoleptic and nutritional value of relevant food products.

12.1.2 AUTOXIDATION

Autoxidation is the oxidative deterioration of unsaturated fatty acids via an autocatalytic process consisting of a free radical mechanism. This indicates that the intermediates are radicals (R•, odd electron species) and that the reaction involves an initiation step and a propagation sequence, which continues until the operation of one or more termination steps.[8] Autoxidation of lipid molecules is briefly described in the following reactions[9]:

(I) Initiation

$$RH \Rightarrow R^{\bullet} + H^{\bullet} \qquad \text{(Reaction 12.1)}$$

$$\text{or } X^{\bullet} + RH \Rightarrow R^{\bullet} + XH \qquad \text{(Reaction 12.2)}$$

(II) Propagation

$$R^{\bullet} + O_2 \Rightarrow ROO^{\bullet} \qquad \text{(Reaction 12.3)}$$

$$ROO^{\bullet} + RH \Rightarrow ROOH + R^{\bullet} \qquad \text{(Reaction 12.4)}$$

(III) Termination

$$R^{\bullet} + R^{\bullet} \Rightarrow R - R \qquad \text{(Reaction 12.5)}$$

$$R^{\bullet} + ROO^{\bullet} \Rightarrow ROOR \qquad \text{(Reaction 12.6)}$$

$$ROO^{\bullet} + ROO^{\bullet} \Rightarrow ROOR + O_2 \qquad \text{(Reaction 12.7)}$$

In the *initiation step* (Reactions 12.1 and 12.2), hydrogen is abstracted from an olefinic acid molecule (RH) to form alkyl radicals (R•), usually in the presence of a catalyst, such as metal ions, light, heat, or irradiation, at a relatively slow rate. The duration of the initiation stage varies for different lipids and depends on the degree of unsaturation and on the presence of natural antioxidants.[10]

In the *propagation sequence* (Reactions 12.3 and 12.4), given an adequate supply of oxygen, the reaction between alkyl radicals and molecular oxygen is very fast and peroxyl radicals are formed (ROO•). These react with another fatty acid molecule producing hydroperoxides (ROOH) and new free radicals that contribute to the chain by reacting with another oxygen molecule. Hydroperoxide molecules can decompose in the presence of metals to produce alkoxyl radicals (RO•), which cleave into a complex mixture of aldehydes and other products, i.e., secondary oxidation products.[11]

The mutual annihilation of free radicals is known as the *termination stage* (Reactions 12.5 through 12.7), when the free radicals R• and ROO• interact to form stable, non-radical products. The rate of oxidation of fatty acids increases with their degree of unsaturation. The relative rate of autoxidation of oleate, linoleate, and linolenate is in the order of 1:40:100 on the basis of oxygen uptake and 1:12:25 on the basis of peroxide formation.[12]

12.1.3 AZO-INITIATED OXIDATION

An intrinsic problem during the investigation of lipid oxidative deterioration is the uncertainty about the rate of initiative reactions. One possible way of overcoming this problem is to introduce into the reaction mixture a compound that decomposes at a constant rate to free radicals (X•) capable of extracting a hydrogen atom from the fatty acid (RH) and consequently initiating the autoxidation process.

The compounds most frequently used for this are the so-called azo-initiators[13] (X−N=N=X) which thermally decompose to highly reactive carbon-centered radicals.

Therefore, azo-initiators are useful for *in vitro* studies of lipid peroxidation generating free radicals spontaneously as following[14]:

$$X - N = N - X \Rightarrow 2X^• + N_2 \qquad \text{(Reaction 12.8)}$$

The water-soluble azo-initiator AAPH [2,2-azo-bis (2-amidinopropane) dihydrochloride] can be used to produce radicals in the aqueous phase, whereas the lipid-soluble AMVN [2,2'-azo-bis-(2,4-dimethylvaleronitrile)] is commonly used to produce radicals in the lipid phase.[15]

12.1.4 PHOTOSENSITIZED OXIDATION

Photo-oxidation involves direct reaction of light-activated, singlet oxygen ($^1O_2$) with unsaturated fatty acids and the subsequent formation of hydroperoxides.[16] In the most stable triplet state (two unpaired electrons in a magnetic field), oxygen is not very reactive with unsaturated compounds. Photosensitized oxidation involves reaction between a double bond and highly reactive singlet oxygen (paired electrons and no magnetic moment) produced from ordinary triplet oxygen by light in the presence of a sensitizer, such as chlorophyll, erythrosine, or methylene blue[17] (Reactions 12.9 and 12.10)

$$^1\text{sens} + h\nu \Rightarrow {}^1\text{sens}^* \Rightarrow {}^3\text{sens}^* + {}^3O_2 \qquad \text{(Reaction 12.9)}$$

$$^3\text{sens}^* + {}^3O_2 \Rightarrow {}^1\text{sens} + {}^1O_2 \qquad \text{(Reaction 12.10)}$$

Singlet oxygen oxidation differs from autoxidation in several important respects: (1) It is an ene and not a radical chain reaction, (2) it gives products which are similar in type but not identical in structure to those obtained by autoxidation, and (3) it is a quicker reaction and its rate is related to the number of double bonds rather than the number of doubly activated allylic groups.[10]

12.1.5 METAL CATALYZED OXIDATION

Many natural oils contain metals such as cobalt, iron, magnesium, and copper, possessing two or more valence states with a suitable oxidation–reduction potential and serving thereby as excellent prooxidants in lipid oxidation reactions.[18] Contamination of oils with specific metals (copper, iron, etc.) can also occur during the refining procedure.

Metals can initiate fatty acid oxidation by reaction with oxygen. The anion thus produced can either lose an electron to give singlet oxygen or react with a proton to form a peroxyl radical, which serves as a good chain initiator[19] (Reaction 12.11).

$$M^{n+} + O_2 \Rightarrow M^{(n+1)+} + O^{\bullet -2} \begin{cases} {}^1O_2 \\ \\ OH^{\bullet} \end{cases} \qquad \text{(Reaction 12.11)}$$

Many oxygenated complexes of transition metals have now been isolated and used as catalysts for oxidation of olefins, whereas recent evidence supports the initiation of autoxidation through the formation of a metal hydroperoxide catalyst complex.[20]

Once a small amount of hydroperoxides is formed, the transition metals can promote decomposition of the pre-formed hydroperoxides due to their unpaired electrons in 3d and 4d orbitals.[21] A metal, capable of existing in two valence states typically acts as follows (Reactions 12.12 through 12.14):

$$M^{n+} + ROOH \Rightarrow RO^{\bullet} + OH^{-} + M^{(n+1)+} \qquad \text{(Reaction 12.12)}$$

$$M^{(n+1)+} + ROOH \Rightarrow ROO^{\bullet} + H^{+} + M^{n+} \qquad \text{(Reaction 12.13)}$$

$$2ROOH \Rightarrow RO^{*} + ROO^{\bullet} + H_2O \,(\text{Net reaction}) \qquad \text{(Reaction 12.14)}$$

In a system containing multivalent metal ions, such as Cu^{+}—Cu^{2+} or Fe^{2+}—Fe^{3+}, the hydroperoxides can readily decompose to produce both $RO^{\bullet}$ and $ROO^{\bullet}$ as the metal ions undergo oxidation–reduction.[22] Fukuzaka and Fujii[23] reported that ferrous ions could catalyze the formation of alkoxyl radicals from linoleic acid hydroperoxides during oxidation of food emulsions.

Chelation of metals by certain compounds decreases their prooxidant effect by reducing their redox potential and stabilizing the oxidizing form of the metal.[4] A few natural acids (citric, phosphoric, tartaric, oxalic, etc.) and ethylenediamintetraacetic acid (EDTA) can chelate metals and thereby increase oxidation stability in oil model systems.[9]

12.1.6 ENZYME-CATALYZED OXIDATION

The basic chemistry of enzyme-catalyzed oxidation of food lipids such as in cereal products, or in many fruits and vegetables is the same as for autoxidation, but the enzyme lipoxygenase (LPX) is very specific for the substrate and for the method of oxidation.[24] Lipoxygenases are globulins with molecular weights ranging from $0.6–1 \times 10^5$ Da, containing one iron atom per molecule at the active site.

LPX type-1, can be found in many natural sources (e.g., potato, tomato, and soybean prefer polyunsaturated fatty acids), having as the best substrate linoleic acid (18:2), which is oxidized to 9 and 13 hydroperoxides.[8]

These hydroperoxides suffer fragmentation to give short-chain compounds (hexanal, 9-oxononanoic acid, 2-nonenal) some of which have marked and characteristic odors. LPX type-2 (present in gooseberry, soybean, and legumes) catalyzes oxidation of acylglycerols, whereas cooxidation of other plant components, e.g. carotenoids, may also occur.[25] In that case, hydroperoxides may suffer enzyme-catalyzed reactions to give a mixture of products of oxidative deterioration as given in lipoxygenase enzymatic reaction (Reaction 12.15):

$$ROH, RCHO, RH + O_2 \longrightarrow ROOH \longrightarrow ROOR, RCO_2H \qquad \text{(Reaction 12.15)}$$

12.1.7   DECOMPOSITION OF LIPID HYDROPEROXIDES ("OFF-FLAVOR" OXIDATION PRODUCTS)

A large body of scientific evidence suggests that the progressive loss of food palatability during lipid oxidation is due to the production of short chain compounds from the decomposition of lipid hydroperoxides.[4,26] The volatile compounds produced from the oxidation of edible oils are influenced by the composition of the hydroperoxides and positions of oxidative cleavage of double bonds in the fatty acids.[27]

A variety of compounds such as hydrocarbons, alcohols, furans, aldehydes, ketones, and acid compounds are formed as secondary oxidation products and are responsible for the undesirable flavors and odors associated with rancid fat.[28] The "off-flavor" properties of these compounds depend on the structure, concentration, threshold values, and the tested system. Aliphatic aldehydes are the most important volatile breakdown products because they are major contributors to unpleasant odors and flavors in food products.[29] The peroxidation pathway from linoleic acid to various volatiles is determined in several researchs,[7,30] by using various techniques (Gas chromatography mass spectrometry, GC–MS, and electron spin resonance spectroscopy, ESR), identified the volatile aldehydes that are produced during the oxidation of sunflower oil. In both cases, hexanal was the major aldehyde product of hydroperoxide decomposition, whereas pentanal, 2-heptenal, 2-octenal, 2-nonenal, 2,4-nonadienal, and 2,4-decadienal were also identified.

12.1.8   LIPID OXIDATION IN FOOD EMULSIONS

Because many common foods are emulsified materials (mayonnaise, coffee creamers, salad dressing, etc.), a better understanding of lipid oxidation mechanisms in emulsions is crucial for the formulation, production, and storage of these products.[31,32] Moreover, apart from their technological importance, emulsion systems generally mimic the amphiphilic nature and the basic structural characteristics of important biological membranes (e.g., phospholipids), which are also prone to *in vivo* oxidative degradation when attacked by singlet oxygen and free radicals.[33] In that aspect, *in vitro* research on the oxidative stability and antioxidation of model emulsions could provide with useful information of nutritional interest and thereby serve as pilot studies for *in vivo* clinical trials.[34]

Emulsions are thermodynamically unstable systems because of the positive energy required to increase the surface area between the oil and water phases.[35] Generally, the stability of food emulsions is complex because it covers a large number of phenomena, including flocculation, coalescence, creaming, and final phase separation.[36,37] Oil-in-water emulsions consist of three different components: water (the continuous phase), oil (the dispersed phase), and surface-active agents (emulsifiers at the interface).

In such a system, the rate of oxidation is influenced by the emulsion composition (relative concentrations of substrate and emulsifier) and especially by the partition of the emulsifier between the interface and the water phase.[38] Other factors influencing lipid oxidation in emulsions are particle size of the oil droplets, the ratio of oxidizable to non-oxidizable compounds in the emulsion droplets, and the packing properties of the surface-active molecules.[39] In addition, the amount and composition of the oil phase in an emulsion are important factors that influence oxidative stability, formation of volatiles, and partition of the decomposition products, between the oil and water phase.[40]

A certain body of recent research has focused on the microstructural stability of protein stabilized oil-in-water emulsions that are structurally similar to recently developed foodstuffs (e.g., dairy alternative or "fresh cheese type" products, etc.).[41–43] The image of such an emulsion has been visualized by the use of Confocal Laser Scanning Microscopy (CSLM).[42] However, not much research has been done yet on the oxidative destabilization of these emulsion systems. A better understanding of the factors monitoring the oxidative deterioration of emulsions would offer antioxidant strategies to improve the organoleptic and nutritional value of the related products.

12.2   LIPID STABILITY MEASUREMENTS

12.2.1   Evaluation of Primary Oxidation Products ("Lipid Hydroperoxides")

12.2.1.1   Measurement of Peroxide Value (POV)

One of the most commonly used methods for measuring rancidity is the peroxide value (POV), which is expressed in milliequivalents of oxygen per kilogram of fat or oil.[44] This test is performed by iodometry, based on the reduction of hydroperoxide group (ROOH) with the iodide ion (I^-). The concentration of the present peroxide is proportional to the amount of the released iodine (I_2), which is assessed by titration against a standardized solution of sodium thiosulphate ($Na_2S_2O_3$), using a starch indicator.[45] (Reactions 12.16 and 12.17):

$$2ROOH + 2H^+ + 2KI \rightarrow I_2 + 2ROH + H_2O + K_2O \qquad \text{(Reaction 12.16)}$$

$$I_2 + 2Na_2S_2O_3 \rightarrow Na_2S_4O_6 + 2NaI \qquad \text{(Reaction 12.17)}$$

The iodometric method is rapid and applicable to all normal fats and oils. It is highly empirical however and results may be affected by the structure and reactivity of the peroxides as well as by the reaction temperature and time.[46]

Freshly refined fats and oils should have a POV value of <1 with levels >3 being the absolute maximum of the acceptable value, whereas it can be as high as 10 before any off-flavors that can be detected sensorically.[9]

12.2.1.2   Determination of Lipid Hydroperoxides with the Ferric Thiocyanate Method

As an alternative approach to the determination of peroxide values, a colorimetric method has been reported in the literature.[47] This method is based on the principle that the reduction of hydroperoxides can be accompanied by oxidation of Fe^{2+} to Fe^{3+} and determination of Fe^{3+} as ferric thiocyanate (Reactions 12.18 through 12.20). In more details, in the presence of hydroperoxides, the reaction of ferrous chloride with ammonium thiocyanate leads to the production of the red ferric thiocyanate chromophore that absorbs at 500 nm.[8] Therefore, the higher the intensity of the chromophore the higher the amount of hydroperoxides in the sample. This method was found to be the most sensitive when compared with eight other photometric methods for evaluating fat deterioration.[48] During the last years, there is an increasing application of the ferric thiocyanate method, in particular in experiments evaluating the lipid oxidation in emulsion systems.[49]

$$ROOH + Fe^{2+} \rightarrow RO^\bullet + Fe^{3+} \qquad \text{(Reaction 12.18)}$$

$$RO^\bullet + Fe^{2+} + H^+ \rightarrow ROH + Fe^{3+} \qquad \text{(Reaction 12.19)}$$

$$Fe^{3+} + 5SCN^- \rightarrow Fe(SCN)_5^{2-} \qquad \text{(Reaction 12.20)}$$

12.2.1.3   Measurement of Conjugated Dienes (CD)

Oxidation of polyunsaturated fatty acids is accompanied by an increase in the ultraviolet absorption of the product. The spectra of the oxidized lipids are characterized by an intense absorption near 233 nm, with a lesser secondary absorption maximum due to ketone dienes in the region 260–280 nm.[50] The appearance of conjugated dienes in oxidized lipids is due to the double bond shift following free radical attack on hydrogens of methylene groups separating the double bonds in these compounds.[51] Obviously, the determination of the conjugated dienes is a more sensitive oxidative indicator in linoleate (18:2) rich-substrates containing conjugated double bonds. In these lipid systems, the amount of conjugated dienes can be quantified on the basis of absorbance at 232 nm and the molar absorptivity of linoleic acid, according to the following equation.[52]

$$\text{Conjugated dienes (g/100 g of oil)} = \frac{1.0769 * \text{Abs}232}{\text{Oil concentration in the sample (g/L)}}$$

Many authors have concluded that the conjugated diene method might be used as an index of stability of lipids in place of, or in addition to PV. It is faster than iodometric PV determination, much simpler, does not depend on chemical reactions or color development, and requires a smaller sample size.[53] However, the presence of compounds absorbing in the region of the conjugated diene formation may interfere with such determinations. It has been proposed that interference in complex systems can be minimized or eliminated by derivative spectroscopy, using photodiode detection of spectra and computer analysis of the data.[54]

12.2.2 Determination of Secondary Oxidation Products ("Off-Flavor" Volatiles)

12.2.2.1 *p*-Anisidine Value Test

The extent of oxidation in fats and oils can be determined by the measurement of the formation of carbonyl compounds. The *para*-anisidine test determines the amount of aldehydes (principally 2-alkenals and 2,4-alkadienals) in vegetable and animal fats. Aldehydes in oil react with the *p*-anisidine reagent under acidic conditions to give yellowish products that absorb at 350 nm.[55]

The *para*-anisidine value (PAV) is defined as 100 times the absorbance of a solution resulting from the reaction of 1 g of fat or oil in 100 mL of a mixture of solvent and *p*-anisidine, measured at 350 nm in a 10 mm cell.[56] A drawback of the method is that nonvolatile aldehydes also, for example 2,5 oxo-glycerides, can contribute to the absorption. As a rule of thumb, it can be said that good quality oil should have a PAV of less than 10.[57] A good correlation between the *p*-anisidine value of salad oils and their organoleptic scores has been reported in the literature.[58]

The so-called TOTOX value has been used as an oxidative indicator,[59] combining peroxide (POV) and *p*-anisidine values (PAV) as follows:

$$\text{TOTOX value} = 2(\text{POV}) + \text{PAV} \qquad \text{(Reaction 12.21)}$$

Although the totox value gives, somewhat a better, indication of the quality and rancidity of the oil, still it does always correlate well with sensorial observations, as it holds for both the POV and PAV values.[60]

12.2.2.2 Determination of Thiobarbituric Acid Related Substances

The thiobarbituric acid (TBA) test is one of the most frequently used methods to assess lipid peroxidation, basically based on the determination of malonaldehyde which is assumed to be an important lipid oxidation product in food and biological systems.[61]

The reaction of malonaldehyde with the TBA reagent (Reaction 12.22) produces a pink complex with an absorption maximum at 532 nm.[62] Therefore the increase in the amount of the producing red pigment as oxidative rancidity advances has been applied as a reliable oxidative indicator to a wide variety of foods. Several attempts have been made to establish a relationship between TBA values and the development of undesirable flavors in fats and oils.[63] It has been shown that flavor threshold values correlate well with TBA results of vegetable oils such as those of soybean, corn, and safflower.[64] However the chemical complexity of food and biological samples imposes certain limitations on the use of the TBA test for evaluating their oxidative state.[44]

Production of TBAR chromophore (532 nm) during lipid oxidation

12.2.2.3 Determination of Volatile "Off-Flavor Products" with Gas Chromatography

12.2.2.3.1 Headspace Techniques (SPME Extraction)

Various gas chromatographic (GC) methods, such as direct injection, dynamic headspace, and static headspace, have been used for the analysis of volatile products, resulting from the oxidative deterioration of vegetable oils.[65] Though advantages and disadvantages are apparent with each GC method, for routine analyses, static headspace is the method of choice because it is rapid and requires no cleaning between samples.[66]

Solid-phase microextraction (SPME) is a versatile new sample preparation technique that has been used to measure the volatile products in lipid oxidation.[67,68]

In headspace SPME, there are two processes involved: the release of analytes from their matrix and the adsorption of analytes by the fiber coating.[69] The volatile organic analytes are extracted, concentrated in the coating and transferred to the analytical instrument for desorption and analysis.[70] In comparison to well-established techniques, SPME is inexpensive, solvent free, and convenient.[71] In addition, because relatively mild conditions can be used, i.e., systems at equilibrium and temperatures less than 50°C, SPME gives a better quantitative estimate of the flavor profile.[72]

During the last years, appropriate SPME extraction techniques have been developed and successfully applied during oxidation of food related o/w emulsions in order to describe their volatile profile during oxidative deteroration.[73] It has been found that the use of natural antioxidant mixtures has effectively inhibited the production of volatile aldehydes[7,29] and thereby protect the final products from their organoleptic and nutritional deterioration. Kiokias and Oreopoulou[7] described the profile of certain aldehydes, extracted with SPME during oxidation of sunflower o/w emulsions.

12.2.2.3.2 Non-Headspace Techniques

In these techniques, the samples to be analyzed are directly introduced into the gas chromatic column.[74] To maintain the quality of the column it is necessary to use a pre-column, which should be replaced regularly. However, a disadvantage of these techniques can be that primary oxidation products can react on the column to form additional values.

Moreover, GC-sniffing techniques are applied to determine "off-flavor" compounds.[75] With this method, a portion of the gas stream which leaves the column is led to the flame ionization detector while another part is used for sensorial determination.

12.2.3 Other Methods for the Evaluation of Oxidative Rancidity

12.2.3.1 Measurement of Induction Time of Oxidation (Rancimat, Swift Test)

Though it is scientifically accepted that lipid oxidation generally proceeds slowly in the initial stages, after a certain time (widely known as induction time) the oxidation rate starts increasing exponentially due to the fact that more unstable hydroperoxides are produced and subsequently are more easily broken down.[4,27] Induction time is an important parameter for the quality of the lipid substrate, dependent on many factors, such as type of oil, fatty acid composition, and presence of catalysts (e.g., metals, light, etc.).[60]

Several methods have been developed for measuring the induction period.

The active oxygen test[45] involves POV measurements of oil samples after bubbling of air in their interior and heating at 98°C. The Rancimat[60,75] is an automated type of Swift Test, in which effluent gases after bubbling through the oil are led into a tube containing distilled water. During oxidation reactions various acids (e.g., formic, acetic, and propionic) are formed increasing the conductivity of the solution, which is recorded between two platinum electrodes.

Drawbacks of these methods[57] are that only bulk oils can be tested whereas volatile components may be lost during the oxidation process giving incorrect results for induction times.

12.2.3.2 Other Recent Oxidation Techniques

Quite a few spectrophotometric techniques have been developed during the last years to estimate oxidative rancidity.[76] Apart from the determination of conjugated diene hydroperoxides at 232 nm, the UV absorption at 268 nm has been commonly applied as a measure of volatiles due to the presence of unsaturated aldehydes absorbed in the region.[77] Infrared spectroscopy has been used to follow the formation of trans double bonds. Moreover, chemi- and bioluminescence have also been used, based on the fact that the breakdown of the peroxides during oxidation is accompanied by the emission of light.[4] Proton NMR has been used as a method to assess rancidity and to follow oxidation; in tested edible oils a good correlation between relative changes and TOTOX values has been established.[53] Polarographic methods have been used, in which the oxidizing compounds are reduced at a dripping mercury electrode.[78]

Moreover with recently developed HPLC systems, it is in principle possible to identify various products of lipid oxidation.[9] By using different types of columns and varying detection wavelengths a range of oxidizing products that cannot be analyzed by other techniques (such as alcohols and 2,5 glycerides, etc.) have been identified.[79]

For instance, Thin Layer Chromatography (TLC) has been used to rapidly assess the overall quality of oil.

12.3 CONTROL OF LIPID OXIDATION VITAMINS AS NATURAL ANTIOXIDANTS

12.3.1 General Information about Lipid Antioxidants

Lipid antioxidants can be broadly defined as any compounds serving to inhibit oxidative processes that cause deterioration of food lipids thereby improving the quality and extending the shelf life of the food products.[80,81] It has been widely accepted that antioxidants for use in food systems must satisfy the following criteria[82,83]:

1. Being inexpensive, nontoxic, and effective at low concentrations
2. Having high stability and capability of surviving processing
3. Having no odor, taste, or color of its own but easy to incorporate and having a good solubility in the product

Antioxidants may be added directly to the food system or as a solution in the food's oil phase, in a food grade solvent or in an emulsified form that may be sprayed onto the food product.[84]

Based on their function, food antioxidants are classified as chain breaking or primary antioxidants, and synergists or secondary antioxidants.[85]

The primary antioxidants are free radical scavengers (FRS) that delay or inhibit the initiation step or interrupt the propagation step of autoxidation.[86]

They mainly donate hydrogen atoms to the lipid radicals (ROO* or RO*) to produce lipid derivatives and antioxidant radicals (FRS*) that are more stable and less readily available to promote autoxidation[87] (Reaction 12.23). Chemical properties, including hydrogen bond energies, resonance delocalization, and susceptibility to oxidation, as well as the energy of the resulting free radical will influence the antioxidant effectiveness of the FRS.[88] The antioxidant radical is stabilized by delocalization of the unpaired electron around a phenol ring to form stable resonance hybrids and finally participates in termination reactions resulting in non-radical dimers[88] (Reaction 12.24). This means that each FRS is capable of inactivating at least two free radicals, the first being inactivated when the FRS interacts with the peroxyl radicals and the second when the FRS* enters a termination reaction with another peroxyl radical. This procedure can be pictured in (Reaction 12.25)

$$ROO^\bullet \text{ or } RO^\bullet + FRS \Rightarrow ROOH \text{ or } ROH + FRS^\bullet \qquad \text{(Reaction 12.23)}$$

$$FRS^\bullet + FRS^\bullet \Rightarrow FRS - FRS \qquad \text{(Reaction 12.24)}$$

(Reaction 12.25)

Primary antioxidants are mono- or poly-hydroxy phenols with various ring substitutions including synthetic (BHA, BHT, PG), and natural compounds (tocopherols, carotenoids).[89] Their antioxidant potency is determined by several factors, including chemical reactivity of the antioxidant toward the radical, concentration and mobility of the antioxidant in the microenvironment, and interaction with other antioxidants.[90,91]

Compounds that retard the rate of lipid oxidation by processes other than that of free radical scavenging are termed secondary antioxidants.[92] Secondary antioxidants usually show antioxidant activity if a second minor component is present in the sample.[93] Some of them are often called synergists because they promote the action of the primary antioxidants; citric acid, ascorbic acid, ascorbyl palmitate, lecithin, and tartaric acid are good examples of synergists.[93] In the category of secondary antioxidants are also included compounds that bind metal ions (EDTA), reducing agents (e.g., ascorbic acid), and singlet oxygen quenchers (e.g., carotenoids).[94]

Regarding their origin, antioxidants are divided into synthetic and natural ones. The synthetic antioxidants mainly contain phenolic groups (such as butylated hydroxyanisole-BHA, propyl gallates-PG, etc.). Despite their superior efficacy and their high stability, there is increasing concern about the safety of synthetic antioxidants, including potential toxicity, allergenicity, etc.[95,96]

Thus the replacement of synthetic antioxidants by the "safer natural antioxidant" has been increasingly advocated, nowadays. Indeed, the food industry has a strong preference for the use of natural antioxidants, some of which may exist inherently in foods or be added intentionally during their processing.[97,98]

As shown by recent studies, a number of natural extracts from selected herbs (such as rosemary, sage, oregano, and thyme), which were found to be rich in polyphenols, flavonoids, and other compounds, have been well proved to be effective in retarding the development of rancidity in oils and fatty acids.[99,100] Indeed, the activity of the natural extracts has been found to be dependent, among other factors, on the type and polarity of the extraction solvent, the isolation procedures, and the active components from the raw materials.[101,102]

The following sections focus on the antioxidant potential of several natural vitamins, such as carotenoids (provitamin-A), tocopherols (vitamin-E), and ascorbic acid (vitamin-C) against the oxidative deterioration of oil based systems. For each specific compound, the mechanisms of action as well as the results of the relevant studies are reported, whereas some information on synergistic interactions between these vitamins is also recorded.

12.3.2 CAROTENOIDS AS RADICAL SCAVENGERS AND OXYGEN QUENCHERS

Carotenoids are a class of natural pigments, familiar to all through the orange-red to yellow colors of many fruits and vegetables as well as for the provitamin A activity that some of them possess.[103] Most carotenoids are 40-carbon terpenoids having isoprene as their basic structural unit. A general subdivision is into "carotenes" which are strictly hydrocarbons (α- and β-carotene, lycopene) and "xanthophylls" which contain polar end groups reflecting an oxidative step in their formation (lutein, bixin, capsanthin, etc.).[104]

A large body of scientific evidence suggests that carotenoids scavenge and deactivate free radicals both *in vitro* and *in vivo*.[105] It has been reported that their antioxidant action is determined by (1) electron transfer reactions and the stability of the antioxidant free radical; (2) the interplay with other antioxidants; and (3) their structure and the oxygen pressure of the microenvironment.[103,104] Moreover, the antioxidant activity of carotenoids is characterized by literature data for (1) their relative rate of oxidation by a range of free radicals, or (2) their capacity to inhibit lipid peroxidation in multilamellar liposomes.[106–109]

According to Mordi[110] the antioxidant activity of carotenoids is a direct consequence of the chemistry of their long polyene chain: a highly reactive, electron-rich system of conjugated double bonds susceptible to attack by electrophilic reagents, and forming stabilized radicals. Therefore, this structural feature is mainly responsible for the chemical reactivity of carotenoids toward oxidizing agents and free radicals, and consequently, for any antioxidant role.[111]

β-Carotene has received considerable attention in recent times as a putative chain breaking antioxidant, although it does not have the characteristic structural features associated with conventional primary antioxidants, but its ability to interact with free radicals including peroxyl radicals is well documented.[112] Burton and Ingold[113] were the first researchers who investigated the mechanisms by which β-carotene acts as a chain breaking antioxidant. In fact, the extensive system of conjugated double bonds makes carotenoids very susceptible to radical addition, a procedure which eventually leads to the free radical form of the carotenoid molecule.

According to this particular mechanism, β-carotene is capable of scavenging peroxyl radicals (Reaction 12.26). The resulting carbon centered radical (ROO-β-CAR•) reacts rapidly and reversibly with oxygen to form a new, chain-carrying peroxyl radical (ROO-β-CAR-OO•). The carbon centered radical is resonance stabilized to such an extent that when the oxygen pressure is lowered the equilibrium of the Reaction 12.27 shifts sufficiently to the left, to effectively lower the concentration of the peroxyl radicals and hence reduces the amount of autoxidation in the system.[114] Furthermore, the β-carotene radical adduct can also undergo termination by reaction with another peroxyl radical (Reactions 12.27 and 12.28).

$$\beta\text{-CAR} + ROO^{\bullet} \rightarrow ROO\text{-}\beta\text{-CAR}^{\bullet} \qquad \text{(Reaction 12.26)}$$

$$ROO\text{-}\beta\text{-CAR}^{\bullet} + O_2 \leftrightarrow ROO\text{-}\beta\text{-CAR-OO}^{\bullet} \qquad \text{(Reaction 12.27)}$$

$$ROO\text{-}\beta\text{-CAR}^{\bullet} + ROO^{\bullet} \rightarrow \text{inactive products} \qquad \text{(Reaction 12.28)}$$

To understand the mechanism of antioxidant activity of the carotenoids it is also important to analyze the oxidation products that are formed during their action as antioxidants. A relationship between product-forming oxidation reactions to carotenoid antioxidant effects has been additionally proposed.[115,116]

In fact, in living organisms the partial pressure of oxygen in the capillaries of active muscle is only about 20 Torr, while in tissue it must be considerably lower.[117] Certainly, this is consistent with the theory that carotenoids can operate as biological antioxidants under certain conditions, a theory that has been proved in many clinical trials.[118,119]

The carotenoid activity during oxidation is strongly influenced by the oxygen pressure (pO_2) of the experimental conditions. Kiokias and Oreopoulou[7] have shown that certain natural carotenoid mixtures (paprika, bixin and tomato, and palm-oil preparations) inhibited the azo-initiated oxidation of sunflower oil-in-water emulsions (operated rapidly under low pO_2) in terms of both primary and secondary oxidation products. However, other studies[120,121] concluded that carotenoids not only did not inhibit aerial lipid autoxidation (high pO_2) but even exerted a prooxidant character, a phenomenon also observed at high carotenoid concentrations[122] that could be due mainly to a more increased formation of carotene-peroxyl radicals, promoting the propagation of autoxidation.

Interestingly, Kiokias[8] observed that during aerial autoxidation of bulk and emulsified sunflower oil, various carotenoids strongly inhibited the production of volatile aldehydes, though they exerted no effect against the formation of hydroperoxides in the earlier stage of autoxidation, a finding that has been also reported by Warner and Frankel[123] during autoxidation of soybean oil.

The discovery that carotenoids deactivate singlet molecular oxygen was an important advance in understanding their technological and biological effects.[124] According to Min and Bradley[28] the addition of various carotenoids to foods containing unsaturated oils improves their shelf life. The mechanism by which carotenoids, and especially β-carotene which has been widely studied, act as oxygen quenchers can be summarized as follows[94]:

In the presence of β-carotene, singlet oxygen will preferentially transfer exchange energy to produce the triplet state carotene, while oxygen comes back to its ground energy state to be inactivated (Reaction 12.29). Triplet state β-carotene releases energy in the form of heat, and the carotenoid is returned to its normal energy state (Reaction 12.30). In this way carotenoids act so effectively that one carotenoid molecule is able to quench up ~1000 molecules of singlet oxygen.

$$^1O_2 + \beta\text{-carotene} \rightarrow {}^3\beta\text{-carotene}^* + {}^3O_2 \qquad \text{(Reaction 12.29)}$$

$$^3\beta\text{-carotene}^* \rightarrow \beta\text{-carotene} + \text{heat} \qquad \text{(Reaction 12.30)}$$

During chlorophyll sensitized photoxidation of edible oils, the carotenoid antioxidant effect was enhanced with an increasing concentration and number of double bonds.[124–126] It has been reported that capsanthin which contains 11 conjugated double bonds, a conjugated keto-group, and a cyclopentane ring had higher antiphotooxidative activity than β-carotene, which has the same number of double bonds but neither of the functional groups.[125,126] In a recent review paper by Kiokias and Gordon,[127] the antioxidant activity of various carotenoids both *in vitro* and *in vivo* were summarized along with a reference of any reported prooxidant effects.

12.3.3 TOCOPHEROLS (VITAMIN-E) AS LIPID ANTIOXIDANTS

Tocopherols and tocotrienols comprise a group of eight chromanol homologs that possess vitamin E activity in the diet. They are natural monophenolic compounds that can be found in plant tissues (e.g., nuts, vegetable oils, fruits, and vegetables).[128]

The tocopherols consist of a 6-chromanol group and an apolar phytyl chain, with the different forms (α-, β-, γ-, δ-) varying in the number and location of the methyl groups.[129] The four tocotrienols are similar to corresponding tocopherols with the only difference being the unsaturation in the side chain at position 3′, 7′, and 11′. In vegetable oils, the tocopherol content depends very much on the growing conditions of the plant from which the oil is extracted as well as on the processing and storage conditions.[4]

In literature, due to differences in the used and tested systems there is still contradiction about the *in vitro* and *in vivo* antioxidant effects of tocopherol and tocotrienol isomers.[130,131] Several factors have been reported to influence the tocopherol against lipid oxidation, such as

1. The rate by which tocopherols can transfer hydrogen atoms from the 6-hydroxyl group to the lipoperoxyl radical, which is ranked as following, between the different isomers: alpha > beta > gamma > delta.
2. Reactivity of the tocopheryl radical/stoichiometry that governs its conversion to the non-radical species. For instance, the sterically hindered α-tocopheryl radical is converted to a non-radical species slower than other isomers with δ-tocopherol being the fastest.
3. Stability of the tocopherols at elevated temperatures is in the following order: delta > gamma > beta > alpha.

Taking into account that different orders of isomer activity have been previously reported, it can be concluded that antioxidant potential of tocopherols is mainly dependent on the tested lipid system. Indeed, the antioxidant activities of tocopherols have been investigated in various lipid substrates, including vegetable oils, animal fats, emulsions, PUFAs, etc.[132] Huang et al.[133] proposed that the relative antioxidant activity of different tocopherols depends on temperature, lipid composition, physical state (bulk phase, emulsion), and tocopherol concentration.

In general, tocopherols behave as chain breaking antioxidants by competing with the fatty acids of their lipid substrate (RH) for the chain peroxyl radicals (ROO[*]).[3] α-Tocopherols, for instance, donate a hydrogen to a peroxyl radical resulting in a α-tocopherol semiquinone radical (Reaction 12.31). This may further donate another hydrogen to produce methyl tocopherol quinone or react with another α-tocopheryl semiquinone radical (Reaction 12.32) to produce α-tocopherol dimer (Reaction 12.33) which also possesses antioxidant activity[134]:

$$\text{ROO}^{\bullet} + \alpha - \text{toc} \Rightarrow \text{ROOH} + \alpha\text{-toc}-\text{semiq}^{\bullet} \quad \text{(Reaction 12.31)}$$

$$\alpha\text{-toc}-\text{semiq}^{\bullet} + \text{ROO}^{\bullet} \Rightarrow \text{ROOH and methyl}-\text{tocqin} \quad \text{(Reaction 12.32)}$$

$$\alpha\text{-toc}-\text{semiq}^{\bullet} + \alpha\text{-toc}-\text{semiq}^{\bullet} \Rightarrow \alpha\text{-toc}-\text{dimer} \quad \text{(Reaction 12.33)}$$

Several authors[9,130] claimed that adding tocopherols to vegetable oils (even to the refined oils when a small tocopherol fraction was removed during deodorization) hardly shows any improvement in the oxidative stability due to the fact that an optimum concentration of these compounds is still present.

In food manufacturing practice, it is recommended to keep the amount of the total α-tocopherol (natural or added) at levels between 50 and 500 ppm, depending on the kind of food product, because at sufficiently higher levels of addition the effect may become prooxidant.[10] On the basis of hydroperoxide formation, α-tocopherol was found to exert an antioxidant activity at low concentrations but this changed to prooxidant effect at high concentrations (e.g. >1000 ppm).[51] However, α-tocopherol still acted as an antioxidant with respect to the formation of volatiles, which is the determining factor in flavor keep-ability of oil- and fat-containing foods.[135] α-Tocopherol is considered as the major antioxidant of olive oil with an activity dependent on both concentration and temperature.[136] In aqueous linoleic acid micelles solution, α-tocopherol was a more effective antioxidant than γ–tocopherol.[137] In another study, γ-tocopherol retained antioxidant activity at higher concentrations than α-tocopherol, though its effectiveness was not increased with concentration in soybean oil.[138] Generally, it has been found that under common test conditions in oils, the antioxidant activity decreases from δ-tocopherol to α-tocopherol.[139,140] However, several authors found that δ-tocopherol acted as a better antioxidant than α-tocopherol, during autoxidation of linolate-rich substrates.[129,141] It can be hypothesized that that the oxidation rate of samples containing tocopherols can be related to the number of methyl substituents on the chromanol ring, that would affect the stability of tocopherol and the tocopheroxyl radical, as well as the kinetics of hydrogen donation to a lipid radical.

12.3.4 ACTIVITY OF ASCORBIC ACID (VITAMIN-C) AGAINST LIPID OXIDATION

L-ascorbic acid, or vitamin C, is very widespread in nature, and it is increasingly gaining importance as a versatile natural food additive due both to its vitamin activity and its ability to improve the quality and extend the shelf life of many food products.[142] Ascorbic acid is attractive as an antioxidant because as a food additive it has no usage limits and is highly recognized by the consumer as a nature-identical product.[94] Vegetable foodstuffs (oranges, blackcurrants, parsley, green peppers, etc.) are the richest sources of vitamin C, but also animal products contain relatively smaller amounts. L-ascorbic acid is a six carbon weak acid with a pK_a of 4.2, which is reversibly oxidized due to its enediol structure[143] with the loss of an electron to form the free radical semihydroascorbic acid according to the Reaction 12.34. Activity of ascorbic acid as an antioxidant is shown in the ascorbic acid reaction (Reaction 12.34).

Compared with other species, this radical is relatively stable and its further oxidation results in dehydroascorbic acid (DHASc) which probably exists *in vivo* in multiple

L-Ascorbic anion   Semi-dehydroascorbic radical   Dehydroascorbic acid

(Reaction 12.34)

forms[86], and can be reduced back to ascorbic acid by the same intermediate radical.

In food systems, ascorbic acid is mainly a secondary antioxidant that can scavenge oxygen, act synergistically with chelators, and regenerate primary antioxidants. Indeed, depending on the conditions ascorbic acid can act through several mechanisms[130,145,146]: (1) hydrogen donation to regenerate the stable antioxidant radical; (2) metal inactivation to reduce the initiation of the metals; (3) hydroperoxide reduction to produce stable alcohols by non-radical processes; and (4) oxygen scavenging.[99]

In principle, ascorbic acid and its salts (sodium or calcium ascorbate) are water soluble antioxidants, not widely applicable for lipid systems but extensively used in beverages.[128] In aqueous systems containing metals, ascorbic acid may also act as a prooxidant by reducing the metals that become active catalysts of oxidation in their lower valences. However, in the absence of added metals, ascorbic acid is an effective antioxidant at high concentrations.[78] The action of ascorbic acid in lipid autoxidation is dependent on concentration, the presence of metal ions, and other antioxidants.[47] It has been shown that ascorbates can protect plasma and LDL lipids from peroxidative damage, and it may inhibit the binding of copper ions to LDL.[146] In several countries, ascorbic palmitate is used in fat containing foods due to its lipid solubility. However whether ascorbic palmitate exerts a better activity than ascorbic acid is fully dependent on the tested system. Indeed, Cort[147] observed that during oxidation of soya bean oil at 45°C, ascorbyl palmitate was much more effective than ascorbic acid, whereas exactly the opposite trend was found by Kiokias and Gordon[29] in auto-oxidation of olive oil in water emulsions. Interestingly, in the latest research ascorbic acid was a superior antioxidant than its lipid homologue against the oxidative deterioration of olive oil.

12.3.5 ACTIVITY OF NATURAL ANTIOXIDANT MIXTURES AGAINST LIPID OXIDATION

Recent studies[99,148,149] have shown that some natural antioxidants, such as vitamin E, vitamin C, and carotenoids can exhibit synergistic interactions, with the consequence that a combination of these compounds has a better antioxidant activity than the sum of the individual components.

Mixtures of tocopherols and ascorbic acid have been reported to exhibit a synergistic effect during lipid oxidation. In a system containing both of these compounds, tocopherols are the primary free radical scavengers (FRS) because they are present in the lipid phase.[150] The water soluble ascorbic acid can donate a hydrogen to tocopherols, so that tocopheroxyl radicals (TOC$^\bullet$), are reduced back to tocopherols (TOC), whereas ascorbic acid is converted to dehydroascorbic acid according to the Reaction 12.35:

$$TOC^\bullet + Asc \Rightarrow TOC + DHAsc \qquad \text{(Reaction 12.35)}$$

By this synergistic mechanism, tocopherols and ascorbic acid can mutually reinforce one another by regenerating their oxidized forms.[151] Radical exchange reactions between lipid radicals, tocopherols, and ascorbic acid are the basis of numerous approaches for stabilizing oil and foods with their mixtures.[97]

Carotenoids were also reported to play a role in recycling phenolic antioxidants, e.g., tocopherols after one-electron oxidation.[152] A recent laser photolysis study[153] has shown that carotenoid radicals are reduced by α- or β-tocopherols via an electron transfer mechanism (Reactions 12.36 and 12.37).

$$CAR^\bullet + TOH \rightarrow CAR + TO^\bullet \qquad \text{(Reaction 12.36)}$$

$$CAR + TO^\bullet \rightarrow CAR^\bullet + TOH \qquad \text{(Reaction 12.37)}$$

According to Bohm et al.[154] a synergistic effect, observed in cell protection by β-carotene and vitamin E, may occur because it is not only quenching oxy-radicals but also repairing the α-tocopheroxyl radicals that are produced when α-tocopherol scavenges an oxy-radical. Such a synergistic mechanism requires that the CAR$^\bullet$ is reconverted to CAR. However it has been also suggested that α-tocopherol protects β-carotene from being oxidized and not the other way around.[114] Henry et al.[121] found that a combination of β-carotene with α-tocopherol exhibited a stronger antioxidant effect than the individual antioxidants. Similarly, Lievonen[155] found that a mixture of lutein with γ-tocopherol was strongly antioxidative while lutein with α-carotene had no effect on the oxidation of a purified triglyceride fraction from rapeseed oil. Kiokias and Gordon[29] observed synergistic effects between norbixin and α-, δ-tocopherols against the production of "off-flavor" volatiles during the autoxidation of olive oil-in-water emulsions.

Several *in vitro* studies have investigated interactions between carotenoids and vitamin C. It has been claimed[154] that ascorbic acid can reduce the carotenoid radical cations in methanol according to Reaction 12.38:

$$CAR^{\bullet+} + AscH_2 \rightarrow CAR + AscH^\bullet + H^+ \qquad \text{(Reaction 12.38)}$$

According to Kanner[156] the presence of an ascorbic acid–cupric ion couple at relatively high concentration inhibits carotene degradation in a β-carotene linoleate model system and thereby reinforces the antioxidant activity of the carotenoid. A tendency of increased carotenoid antioxidant activity in the presence of ascorbic acid has been observed during azo-initiated oxidation of sunflower oil-in-water emulsions.[8]

In the same oxidation system, an enhanced antioxidant activity of carotenoid mixtures (lutein, lycopene, paprika, bixin, etc.) have been reported[7] as compared to each separate compound. Moreover, Kiokias and Gordon[29] found that mixtures of olive oil phenolics with various carotenoids exhibited a strong activity against the autoxidation of bulk and emulsified olive oil, whereas individual carotenoids presented no inhibitory effect.

12.4 LEGISLATION RELATED TO OIL

12.4.1 E.U. LEGISLATION FOR OILS AND FATS

According to E.U. Directive 96/3/EC (entry into force 17/2/1996) the bulk transport in sea-going vessels of liquid oils or fats that are to be processed, and that are intended for or likely to be used for human consumption, is permitted in tanks that are not exclusively reserved for the transport of foodstuffs, subject to the following conditions:

  I. That, where the oil or fat is transported in a stainless steel tank, or tank lined with epoxy resin or technical equivalent, the immediately previous cargo transported in the tank shall have been a foodstuff, or a cargo from the list of acceptable previous cargoes set out in the Annex,
 II. That, where the oil or fat is transported in a tank of materials other than those in point I, the three previous cargoes transported in the tanks shall have been foodstuffs, or from the list of acceptable previous cargoes set out in the Annex.

The captain of the sea-going vessel transporting, in tanks, bulk liquid oils and fats intended for or likely to be used for human consumption shall keep accurate documentary evidence relating to the three previous cargoes carried in the tanks concerned, and the effectiveness of the cleaning process applied between these cargoes.

Following Regulation (EEC) No. 136/66 (entry into force 1/10/1966) the Council shall fix a single production target price, a single market target price, a single intervention price, and a single threshold price for olive oil for the Community. Olive oil bought in by intervention agencies shall not be sold by them on the Community market on terms which might impede price formation at the level of the market target price. When olive oil is exported to third countries: (1) the difference between prices within the Community and prices on the world market may be covered by a refund where the former are higher than the latter and (2) a levy equal at most to the difference between prices on the world market and prices within the Community may be charged where the former are higher than the latter. Virgin olive oil produced by mechanical processes and free from any admixtures of other types of oil or of olive oil extracted in a different manner. Virgin olive oil is classified as follows: (a) Extra : olive oil of absolutely perfect flavor, with a free fatty acid content expressed as oleic acid of not more than 1 g per 100 g; (b) Fine : olive oil with the same characteristics as "Extra" but with a free fatty acid content expressed as oleic acid of not more than 1.75 g per 100 g; (c) Ordinary: olive oil of good flavor with a free fatty acid content expressed as oleic acid of not more than 3.73 g per 100 g; (d) Lampante: off-flavor olive oil or olive oil with a free fatty acid content expressed as oleic acid of more than 3.73 g per 100 g.

Regulation (EEC) No 2568/91(entry into force 8/9/1991) makes clear that the characteristics of the oils shall be determined in accordance with the methods of analysis set out below: (a) for the determination of the free fatty acids, expressed as the percentage of oleic acid, (b) for the determination of the peroxide index, (c) for the determination of aliphatic alcohols, (d) for the determination of the sterol content, (e) for the determination of erythrodiol and uvaol, (f) for the determination of the saturated fatty acids in position 2 of the triglyceride, (g) for the determination of the trilinolein content, (h) for spectrophotometric analysis, (i) for the determination of the fatty acid composition, (j) for the determination of the volatile halogenated solvents, (k) for the evaluation of the organoleptic characteristics of virgin olive oil, and (l) for proof that refining has taken place.

Regulation (EC) No.2991/94 (entry into force 1/1/1996) laid down standards for: milk fats, fats, and fats composed of plant and/or animal products with a fat content of at least 10% but less than 90% by weight, intended for human consumption. The fat content excluding salt must be at least two-thirds of the dry matter. The Regulation applies also to products which remain solid at a temperature of 20°C, and which are suitable for use as spreads. The products may not be supplied or transferred without processing to the ultimate consumer either directly or through mass caterers, unless they meet the requirements set out. It has no application to the designation of products the exact nature of which is clear from traditional usage and/or when the designations are clearly used to describe a characteristic quality of the product and to concentrated products (butter, margarine, blends) with a fat content of 90% or more. The following information must be indicated in the labeling and presentation of the products: the sales description, the total percentage fat content by weight at the time of production for products, the vegetable, milk, or other animal fat content in decreasing order of weighted importance as a percentage by total weight at the time of production for compound fats and the percentage salt content must be indicated in a particularly legible manner in the list of ingredients.

According to Regulation (EC) No.2815/98 (entry into force 31/10/2001) the designation of origin shall relate to a geographical area and may mention only: (a) a geographical area whose name has been registered as a protected designation of origin or protected geographical indication and/ or (b) for the purposes of this Regulation: a Member State, the European Community, a third country. The designation of origin, where this indicates the European Community or a Member State shall correspond to the geographical area in which the "extra virgin olive oil" or "virgin olive oil" was obtained. However, in the case of blends of "extra virgin olive oils" or "virgin olive oils" in which more than 75% originates in the same Member State or in the Community, the main origin may be designated provided that it is followed by the indication "selection of (extra) virgin olive oils more than 75% of which was obtained in…(designation of origin)." An extra virgin or virgin olive oil shall be deemed to have been obtained in a geographical area for the purposes of this paragraph only if that oil has been extracted from olives in a mill located within that area.

Regulation (EC) No.1019/2002 (entry into force 1/11/2002) laid down specific standards for retail-stage marketing of the olive oils and olive-pomace oils. Oils shall be presented to the final consumer in packaging of a maximum capacity of 5 L. Such packaging shall be fitted with an opening system that can no longer be sealed after the first time it is opened and shall be labeled. However, in the case of oils intended for consumption in mass caterers, the Member States may set a maximum capacity exceeding 5 L. Certain categories of oil are (a) extra virgin olive oil, (b) virgin olive oil, (c) olive oil composed of refined olive oils and virgin olive oils, and (d) olive-pomace oil. Only extra virgin and virgin olive oil may bear a designation of origin on the labeling.

Some representative points and comments (repeals, modifications, amendments) of the E.U. Directive/ Regulations for oils–fats are given in Table 12.1.

12.4.2   U.S. LEGISLATION FOR OIL

According to International agreement on olive oil and table olives (Table 12.2) aims to the modernization of olive cultivation, olive oil extraction, and table olive processing:

TABLE 12.1
E.U. Directive and Regulations (Main Points and Comments) with Regard to Oils—Fats

Directive—Title	Main Points	Comments
E.U. 96/3/EC (entry into force 17/2/1996)—Hygiene of foodstuffs as regards the transport of bulk liquid oils and fats by sea	Definition of equivalent conditions to ensure the protection of public health and the safety and wholesomeness of the foodstuffs concerned. The bulk transport in sea-going vessels of liquid oils or fats is permitted in tanks that are not exclusively reserved for the transport of foodstuffs. Register of the three previous cargoes carried in the tanks concerned, and the effectiveness of the cleaning process applied between these cargoes.	
Regulation No. 136/66/EEC (entry into force 1/10/1966)—Establishment of a common organization of the market in oils and fats	An export refund may be paid on exports of olive oil and rapeseed oil to cover the difference between the world and Community market prices.	Amendments—Regulation (EEC) No.1253/70 (entry into force 1/7/1970) No.1547/72 (entry into force 24/7/1972) No.1707/73 (entry into force 1/9/1973) No.2560/77 (entry into force 1/1/1978) No.1562/78 (entry into force 1/11/1978) No.1585/80(entry into force 27/6/1980) No. 3454/80 (entry into force 1/1/1980) No.1413/82 (entry into force 21/5/1982) No.2260/84 (entry into force 3/8/1984) No.231/85 (entry into force 3/2/1985) No.1454/86 (entry into force 24/5/1986) No.1915/87 (entry into force 3/7/1987) No.3994/87 (entry into force 1/1/1988) No.1098/88 (entry into force 29/4/1988) No.2210/88 (entry into force 26/7/1988) No.3499/90 (entry into force 5/12/1990) No.3577/90 (entry into force 1/1/1991) No.1720/91 (entry into force 26/6/1991) No.356/92 (entry into force 1/11/1992) No.2046/92 (entry into force 30/7/1992) No.3179/93 (entry into force 23/11/1993) No.3290/94 (entry into force 1/7/1995) No.1581/96 (entry into force 19/8/1996) No. 1638/98 (entry into force 4/8/1998) No.2702/1999 (entry into force 1/1/2000) No.2826/2000 (entry into force 1/1/2001)

TABLE 12.1 (continued)
E.U. Directive and Regulations (Main Points and Comments) with
Regard to Oils—Fats

Directive—Title	Main Points	Comments
	Marketing standards, in particular covering labeling and quality grading, may be laid down for olive oil.	Replacement and correction of articles
	Olive oil used for humanitarian aid purposes is purchased on the Community market or comes from intervention stocks.	
Regulation (EEC) No 2568/91 (entry into force 8/9/1991)— The characteristics of olive oil and olive-residue oil and on the relevant methods of analysis	The characteristics of the oils shall be determined in accordance with the methods of analysis.	
	The content of free fatty acids is expressed as acidity calculated conventionally.	
	All the equipment used shall be free from reducing or oxidizing substances	
Regulation (EC) No.2991/94 (entry into force 1/1/1996)— Marketing standards for olive oil	Application to milk fats and fats composed of plant and/or animal products.	
	Not applicable to concentrated products.	
	The products are supplied to the consumer only after processing.	
	Products imported into the Community must comply with the Directive.	
Regulation (EC) No.2815/98 (entry into force 31/10/2001)— Marketing standards for olive oil	The designation of origin shall relate to a geographical area.	
	The "extra virgin olive oil" and "virgin olive oil" shall be packaged in an establishment approved for that purpose.	
	Designation of origin checks in the packaging plants.	
Regulation (EC) No.1019/2002 (entry into force 1/11/2002)— Standards for spreadable fats	Specific standards for retail-stage marketing of the olive oils and olive-pomace oils.	Amendment
	Oils shall be presented to the final consumer in packaging of a maximum capacity of 5 L.	Regulation (EC) No.1964/2002 (entry into force 1/7/2002)
	Specific way of labeling	The products which were legally produced, were labeled and got into circulation before 1/1/2003 will be marketed until the consumption of the stocks.

1. To encourage research and development to elaborate techniques that could
   a. Modernize olive husbandry and the olive-products industry through technical and scientific planning
   b. Improve the quality of the products obtained therefrom
   c. Reduce the cost of production of the products obtained, particularly that olive oil, with a view to improving the position of that oil in the overall market for fluid edible vegetable oils
   d. Improve the situation of the olive products industry as regards the environment to abate any harmful effects
2. To encourage the transfer of technology and training in the olive sector

"Olive oil" shall be restricted to oil obtained solely from the olive, to the exclusion of oil obtained by solvent or re-esterification processes and of any mixture with oils of other kinds. "Virgin olive oil" is oil which is obtained from the fruit of the olive tree solely by mechanical or other physical means under conditions, and particularly thermal conditions, that do not lead to deterioration of the oil, and which has not undergone any treatment other than washing, decantation, centrifugation, and filtration, to the exclusion of oil obtained by solvent or re-esterification processes and of any mixture with oils of other kinds. The main points of this agreement are given in Table 12.2.

12.4.3 Canadian Legislation for Oil

For Edible oil products Act (Table 12.3) "edible oil product" means a food substance of whatever origin, source, or composition that is manufactured for human consumption wholly or in part from a fat or oil other than that of milk. No person shall manufacture or sell an edible oil product, other than oleomargarine, manufactured by any process by which fat or oil other than that of milk has been added to or mixed or blended with a dairy product in such manner that the resultant edible oil product is an imitation of or resembles a dairy product. The Lieutenant Governor in Council may make regulations (a) designating the edible oil products or classes of edible oil products to which this Act applies; (b) providing for the issue of licenses to manufacturers and wholesalers of any edible oil product and prescribing the form, terms, and conditions thereof and the fees to be paid therefor, and providing for the renewal, suspension, and cancellation thereof, (c) prescribing standards for the operation and maintenance of premises and facilities in which any edible oil product is manufactured, packed, or stored, (d) prescribing the standards of quality for and the composition of any edible oil product or class of edible oil product, (e) providing for the detention and confiscation of any edible oil product that does not comply with this Act and the regulations, (f) respecting the advertising of any edible oil product or class of edible oil product, (g) requiring and providing for the identification by labeling or otherwise of any edible oil product or class of edible oil product sold

---

TABLE 12.2
U.S. Agreement for Oil

Title	Main Points	Comments
International agreement on olive oil and table olives, 1986	The objectives of this agreement were the international cooperation, the modernization of olive cultivation, olive oil extraction, and table olive processing.	Amendments
	Definitions (olive oil crop year, olive products, etc.)	1993 (replacements and extension of the previous agreement)
		2000 (prolongation of olive oil agreement)

TABLE 12.3
Canadian Act for Oil

Title	Main Points	Comments
Edible oil products Act, 1990	Definitions (edible oil product, etc.) This Act applies to every edible oil product and class of edible oil product designated in the regulations.	Amendments This Act is amended in 1994, 1999, and 2001.

or offered for sale, (h) prescribing the powers and duties of inspectors and analysts, (i) prescribing the records to be kept by manufacturers and wholesalers of any edible oil product, (j) exempting any manufacturer, wholesaler, or retailer of any edible oil product from this Act and the regulations, and prescribing terms and conditions therefor, (k) respecting any matter necessary or advisable to carry out effectively the intent and purpose of this Act. The main points of this Act are given in Table 12.3.

REFERENCES

1. Hamilton, R.J., Kalu, C., Prisk, E., Padley, B.F., and Pierce, H., Chemistry of free radicals in lipids, *Food Chem.*, 60, 193, 1997.
2. Rice-Evans, A.C., Measurement of total antioxidant action as a marker of antioxidant status *in vivo*. Proceedings and limitations, *Free Radical Res.*, 33, 59, 2000.
3. Frankel, E.N., Antioxidants in lipid foods and their impact on food quality, *Food Chem.*, 57, 51–56, 1996.
4. Akoh, C.C. and Min, B.D., *Food Lipid Chemistry, Nutrition and Biotechnology*, Marcel Dekker, New York, 1997.
5. Gordon, M.H., Dietary antioxidants in disease, *Nat. Product Rep.*, 13, 265, 1996.
6. Kiritsakis, A., *Olive Oil-from Tree to the Table*, 2nd ed., Food and Nutrition Press, Trumbull, CT, 1998.
7. Kiokias, S. and Oreopoulou, V., Antioxidant properties of natural carotenoid preparations against the AAPH-oxidation of food emulsions, *Innov. Food Sci. Emerg. Tech.*, 7, 132–139, 2006.
8. Kiokias, S., In *vitro* and in *vivo* antioxidant properties of natural carotenoid mixtures. PhD thesis, School of Food Biosciences, The University of Reading, Reading, U.K., 2002.
9. Allen, J.C. and Hamilton, R.J., *Rancidity in Foods*, 3rd ed., Chapman & Hall, London, U.K., pp. 1–22, 1994.
10. Gunstone, F.D., *Fatty Acid and Lipid Chemistry*, Chapman & Hall, Glasgow, U.K., 1996.
11. Allen, J.C. and Hamilton, R.J., *Rancidity in Foods*, 3rd ed., Chapman & Hall, London, U.K., 1994.
12. Hsieh R.J. and Kinsella, E.J., Oxidation of PUFAs, mechanisms products and inhibition with emphasis on fish oil, *Adv. Food. Nutr. Res.*, 33, 233, 1985.
13. Liegeois, C., Lernseau, G., and Collins, S., Measuring antioxidant efficiency of wort, malt, and hops against the 2,2,-azobis (2-qmidinopropane) dihydrochloride-induced oxidation of an aqueous dispersion of linoleic acid, *J. Agric. Food. Chem.*, 48, 1129, 2000.
14. Niki, E., Free radical initiator as sources of water or lipid soluble peroxyl radicals, *Methods Enzymol.*, 186, 100, 1990.
15. Esterbauer, H., The chemistry of oxidation of lipoproteins, in *Oxidative Stress, Lipoproteins, and Cardiovascular Dysfunction*, Rice-Evans, C. and Bruckdorfer, K.R. (Eds.), Portland Press, London, U.K., 1980.
16. Fakourelis, E.C. and Min, B.D., Effects of chlorophyll and β-carotene on the oxidation stability of olive oil, *J. Food Sci.*, 52, 234, 1987.
17. David, W.R. and Dorr, A.L., *Food Lipid Chemistry, Nutrition and Biotechnology*, Akoh, C.C. and Min, B.D. (Eds.), Marcel Dekker, New York, 1997.
18. Dunford, H.B., Free radicals in iron-containing systems, *Free Radical Biol. Med.*, 3, 405, 1987.

19. Mistry, B.S. and Min, B.D., Oxidised flavor compounds in edible oils. Developments in food science, in *Off Flavours in Foods and Beverages*, Charalambous, G. (Ed.), Elsevier, Amsterdam, the Netherlands, 1992.

20. Villiere, A., Vian, M., Brownec, I., Moreau, N., and Genot, C.J., Oxidative stability of bovine serum albumin- and sodium caseinate stabilised emulsions depends on metal availability, *J. Agric. Food Chem.*, 53, 1514–1520, 2005.

21. Chen, J.H., Lee, T.C., and Ho, C.T. Antioxidant effect and kinetics study of capsanthin on the chlorophyll-sensitized photooxidation of soybean oil and selected flavor compounds, *ACS Symp. Ser.*, 660, 188, 1997.

22. Mei, L. et al. Iron-catalysed lipid oxidation in emulsions as affected by surfactant, pH and NaCl, *Food Chem.*, 61, 407, 1998.

23. Fukuzaka, K. and Fujii, T., Peroxide dependent and independent lipid peroxidation: Site specific mechanisms of initiation by chelated iron and inhibition by α-tocopherol, *Lipids*, 7, 227, 1998.

24 Yamamoto. S., Enzymatic lipid peroxidation, reactions of mammalians lipoxyganases, *Free Radical Biol. Med.*, 10, 149, 1991.

25. Canfield, L.M., Bulux, J., and Serrano, J., Plasma response to oral β-carotene in Guatemala schoolchildren, *Am. J. Clin. Nutr.*, 54, 539, 1991.

26. Aparicio, R. and Morales, M.T., Relationship between volatile compounds and sensory attributes of olive oil, *J. Am. Oil. Chem. Soc.*, 73, 1253, 1996.

27. Frankel, E.N., Recent advances in lipid oxidation, *J. Sci. Food. Agric.*, 54, 495, 1991.

28. Min, B.D. and Bradley, D.G., Fat and oil flavours, in *Encyclopedia of Food Science & Technology*, Wiley, New York, 1992.

29. Kiokias, S. and Gordon, M., Antioxidant properties of annatto carotenoids, *Food Chem.*, 83, 523, 2003.

30. Keszler, A., Krisky, T., and Nenesh, A., Mechanisms of volatile compounds production during storage of sunflower oil, *J. Agric. Food. Chem.*, 48, 1567, 2000.

31. Ponginelbi, I., Nawar, W.W., and Chinchoti, P., Oxidation of linoleic acid in emulsions. Effects of substrate, emulsifiers, and sugar concentration, *J. Am. Oil Chem. Soc.*, 76, 131, 1999.

32. Kiokias, S., Lampa, K., Tsimogiannis, D., and Oreopoulou, V., Inhibition of oxidative deterioration in food emulsions, in *Proccedings of INTRAFOOD-EFFoST Conference*, Valencia, Spain, Vol. 2, p. 1237, 2005.

33. Halliwell, B. and Gutteridge, J., *Free Radicals in Biology and Medicine*, 2nd ed., Clarendon Press, Oxford, U.K., 1995.

34. Kiokias, S. and Gordon, M., Dietary supplementation with a natural carotenoid mixture decreases oxidative stress, *Eur. J. Clin. Nutr.*, 57, 1135, 2003.

35. Dickinson, E., *Introduction to Food Colloids*, Oxford University Press, Oxford, U.K., 1997.

36. Friberg, E. and Larsson K., *Food Emulsions*, 3rd ed., Marcel Dekker, New York, 1996.

37. Kiokias, S., Reiffers-Magnani, C., and Bot, A., Stability of whey protein stabilized oil in water emulsions during chilled storage and temperature cycling. *J. Agric. Food. Chem.*, 52, 3823, 2004.

38. Coupland, I.N. and Mc Clements, D.J., Lipid oxidation in food emulsions, *Trends Food. Sci. Technol.*, 7, 83, 1996.

39. Labuza, T.P., Kinetics of lipid oxidation in foods, *Crit. Rev. Food. Technol.*, 2, 355, 1971.

40. McClements, D.J. and Decker, E.A., *J. Food Sci.*, 65, 1270, 2000.

41. Kiokias, S. and Bot, A., Temperature cycling stability of pre-heated acidified whey protein-stabilised o/w emulsion gels in relation to the internal surface area of the emulsion, *Food Hydrocolloids*, 20, 246, 2006.

42. Kiokias, S. and Bot, A., Effect of protein denaturation on temperature cycling stability of heat-treated acidified protein-stabilised o/w emulsions, *Food Hydrocolloids*, 19, 493, 2005.

43. Dickinson, E., Milk protein interfacial layers and the relationship to emulsion stability and rheology, *Colloids Surf. B: Biointerfaces*, 20, 197, 2001.

44. Moore, K. and Roberts, L.S., Measurement of lipid peroxidation, *Radical Res.*, 28, 659, 1998.

45. Warner, K. and Eskin, N.M., *Methods to Assess Quality and Stability of Oils and Fat Containing Foods*, AOCS Press, Champaign, IL, 1995.

46. Gray, I.J., Measurement of lipid oxidation, *J. Am. Oil Chem. Soc.*, 55, 539, 1978.

47. Ueda, S., Hayashi, T., and Namiki, M., Effect of ascorbic acid on lipid autoxidation in a model food system, *Agric. Biol. Chem.*, 50, 1, 1986.

48. Tsoukalas, V. and Grosch, W., Analysis of food deterioration-Comparison of some photometric tests. *J. Am. Oil Chem. Soc.*, 54, 490, 1977.

49. Hu, M., McClements, J., and Decker, E.A., Impact of whey protein emulsifiers on the oxidative stability of salmon oil-in-water emulsions, *J. Agric. Food Chem.,* 51, 1435, 2003.
50. Logani, K.M. and Davis, R.E., Lipid oxidation: Biological effects and antioxidants, *Lipids*, 12, 485, 1980.
51. Huang, W., Frankel, E.N., Schwarz, K., and German, B., Effect of pH on antioxidant activity of α-tocopherol and trolox in oil-in-water emulsions, *J. Agric. Food Chem.,* 42, 2496, 1996.
52. IUPAC, *Standard Methods of Analysis of Oils Fats and Derivatives*, 7th ed., Paquot, C. and Hautfenne, H. (Eds.), Blackwell Scientific Publications, Oxford, U.K., 1987.
53. Wanasundara, N.U., Shahidi, F., and Jablonski, C.R., Comparison of standard and NMR methodologies for assessment of oxidative stability of soybean oils, *Food Chem.*, 52, 249, 1995.
54. Wheatley, A.R., Some recent trends in the analytical chemistry of lipid peroxidation, *Trends Anal. Chem.*, 19, 620, 2000.
55. White, P.J., Conjugated diene, anisidine value and carbonyl value analysis, in *Methods to Assess Quality and Stability of Oil and Fat Containing Foods,* Warner, K. and Eskin, M. (Eds.), AOCS Publisher, Champaign, IL, p. 159, 1995.
56. Kasuga A., Aoyagi, Y., and Sugahara, T., *J. Food Sci.*, 60, 1113, 1995.
57. Rossel, J.B., *Rancidity in Foods*, 3rd ed., Allen, J.C. and Hamilton, R.J. (Eds.), Blackie Academie & Professional, Glasgow, U.K., 1996.
58. List, G.R. et al., Oxidation and quality of soybean oil: A preliminary study of the anisidine test, *J. Am. Oil Chem. Soc.,* 51, 17, 1974.
59. Osborn, H.T. and Akoh, C.C., Effect of emulsifier type, droplet size, and oil concentration on lipid oxidation in structured lipid-based oil-in-water emulsions, *Food Chem.*, 84, 451, 2004.
60. Shaidi, F. and Wanasadura, U.N., Methods for measuring oxidative stability in fats and oils. In: *Food Lipid Chemistry, Nutrition and Biotechnology*, Akoh, C.C, and Min, B.D. (Eds.), Marcel Dekker, New York, 1997.
61. Coupland, J.N., Zhu, Z., Wan, H., McClements, D.J., Nawar, W.W., and Chinachoti, P., Droplet composition affects the rate of oxidation of emulsified ethyl linoleate, *J. Am. Oil Chem. Soc.,* 73, 795, 1996.
62. Tarladgis, B.G., Pearson, A.M., and Dugan, R.L., Chemistry of the 2-TBA test for determination of oxidative stability in foods, *J. Sci. Food. Agric.*, 15, 602, 1964.
63. McDonald, R.E. and Hultin, H.O., *J. Food Sci.*, 52, 15, 1987.
64. Hawrysh, J.Z., Stability of conola oil, in *Canola and Rapeseed Oil. Production, Chemistry, Nutrition, and Processing,* Shahidi, F. (Ed.), Van Nortland Reinhlod, New York, p. 99, 1990.
65. Schieberle, P., *Characteristion of Food: Emerging Methods*, Gaonkar, A.G. (Ed.), Elsevier, London, U.K., 1995.
66. Snyder, J.M., Frankel, E.N., and Selke, E., Capillary gas chromatographic analysis of headspace volatiles from vegetable oils, *J. Am. Oil Chem. Soc.*, 62, 1675, 1985.
67. Zabaras, D. and Wyllie, S.G., Re-arrangement of p-menthane terpenes by Carboxen during HS-SPME, *J. Sep. Sci.,* 25, 685, 2002.
68. Steffen, A. and Pawliszyn, J., Analysis of flavor volatiles using headspace SPME, *J. Agric. Food Chem.,* 92, 401, 1996.
69. Jimenez, A., Beltran, G., and Aguilera, M.P., Application of SPME to the analysis of volatile compounds in virgin olive oils, *J. Agric. Food Chem.,* 46, 2744, 1998.
70. Brunton, P.N., Cronin, A.D., and Durkan, R., A comparison of SPME fibres for measurement of hexanal, and pentanal in cooked turkey, *Food Chem.,* 68, 339, 2000.
71. Steenson, D.F., Lee, J.H., and Min, B.D., SPME microextraction of volatile soybean oil and corn oil compounds, *J. Food Sci.,* 67, 71, 2002.
72. Yang, X. and Reppard, T., Solid phase microextraction for flavors analysis. *J. Agric. Food Chem.*, 41, 1925, 1994.
73. Beltran, G., Aguilera, M.P., and Gordon, M.H., Solid phase microextraction of volatile oxidation compounds in oil-in-water emulsions, *Food Chem.*, 92, 401, 2005.
74. Zegarska, Z. et al., *Milchwissenchaft,* 51, 195, 1996.
75. Karahadian, C. and Lindsay, R.C., *Flavor Chemistry: Trends and Developments*, Teranishi, R. et al. (Eds.), ACS, Washington, DC, 1989.
76. Slater, F.T., Overview of methods used for detecting lipid peroxidation, *Methods Enzymol.,* 300, 283, 1995.
77. Oreopoulou, V., Extraction of natural antioxidants, in *Extraction Optimization in Food Engineering*, Tzia, C. and Liadiakis, G. (Eds.), Marcel Dekker, New York, pp. 329–346, 2003.
78. Porter, L.W., Black, D.E., and Drolet, M.A., Use of polyamide oxidative fluorescence test on lipid emulsions: Contrast in relative effectiveness of antioxidants in bulk versus dispersed systems, *J. Agric. Food Chem.,* 37, 615, 1989.

79. Facino, R.M., Carini, M., Aldini, G., and Colombo, L., Characterization of the intermediate products of lipid peroxidation in phosphatidylcholine liposomes by fast-atom bombardment mass spectrometry and tandem mass spectrometry techniques, *Rapid Commun. Mass Spec.,* 10, 1148, 1996.
80. Krinsky, N.I., Mechanism of actions of biological antioxidants, *Proc. Soc. Exp. Biol. Med.,* 200, 298, 1992.
81. Loliger, J., The use of antioxidants in foods, in *Free Radicals and Food Additives,* Aruoma, I.O. and Haliwell, B. (Eds.), Taylor & Francis, London, U.K., 1990.
82. Fennema, O.R., *Food Chemistry*, Marcel Dekker, New York, 1996.
83. Kumpulainen, J.K. and Salonen, T.J., *Natural Antioxidants and Food Quality in Atherosclerosis and Cancer Prevention*, Royal Society of Chemistry, Cambridge, U.K., 1996.
84. Larson, R.A., Plant defenses against oxidative stress, *Arch. Insect Biochem. Physiol.,* 29, 175, 2003.
85. Logani, M.K. and Davis, R.E., Lipid oxidation and effect of antioxidants, *Lipids,* 15, 485, 1981.
86. Buettner, G.R. and Moseley, P.L., EPR spin trapping of free radicals produced by bleomycin and ascorbate, *Free Radical Res. Commun.,* 19, 589, 1993.
87. Cao, G. and Prior, R., Measurement of oxygen radical absorbance capacity in biological samples, *Methods Enzymol.,* 299, 350, 1999.
88. Nawar, W.W., Lipids, in *Food Chemistry*, 3rd ed., Fennema, O. (Ed.), Dekker, New York, 1996.
89. Papadopulos, G. and Boskou, D., Antioxidant effect of natural phenols in olive oil, *J. Am. Oil Chem. Soc.,* 68, 669, 1991.
90. Papas, M.A., Diet and antioxidant status, in *Antioxidants in Human Health and Disease*, Basu, K.T., Temple, J.N., and Gerg, L.M. (Eds.), CABI Publishing, New York, 1999.
91. Tsusihasci, H., Kigosi, M., and Niki, E., Action of β-carotene as an antioxidant against lipid peroxidation, *Arch. Biochem. Biophys.,* 323, 137, 1995.
92. Gordon, M.H., The mechanism of antioxidant action *in vitro*, in *Food Antioxidants*, Hudson, F.J.B. (Eds.), Elsevier Applied Science, London, U.K., 1990.
93. Minotti, G., Sources and role of iron in lipid peroxidation, *Chem. Res. Toxicol.,* 6, 134, 1993.
94. Reisch, W.D., Dorris, A., and Eitenmiller, R., Antioxidants, in *Food Lipid Chemistry, Nutrition and Biotechnology*, Akoh, C.C. and Min, B.D. (Eds.), Marcel Dekker, New York, 1997.
95. Hudson, B.J. and Lewis, J.I., Polyhydroxy flavonoid antioxidants for edible oils: Structural criteria for activity, *Food Chem.,* 10, 47, 1983.
96. Thompson, M. and Moldens, P., Cytoxicity of BHA and BHT in rat hepatocytes, *Biochem. Pharmacol.,* 37, 2201, 1983.
97. Loliger, J., Natural antioxidants, *Lipid Technol.,* 3, 58, 1993.
98. Ramarathnam, N. et al., The contribution of plant food antioxidants to human health, *Trends Food Sci. Technol.,* 6, 75, 1995.
99. Frankel, E.N. et al., Antioxidant activity of a rosemary extract and its constituents, carnosic acid, carnosol, and rosemarin acid, in bulk oil and oil-in-water emulsion, *J. Agric. Food. Chem.,* 44, 131, 1996.
100. Takacsova, M., Pribela, A., and Faktorova, M., Study of the antioxidative effect of thyme, sage, juniper, and oregano, *Die Nahung*, 39, 241, 1995.
101. Abdalla, E.A. and Roozen, P.J., Effect of plant extracts on the oxidative stability of sunflower oil and emulsion, *Food Chem.,* 64, 323, 1999.
102. Vekiari, S. et al., Oregano flavonoids as lipid antioxidants, *J. Am. Oil Chemists Soc.,* 70, 483, 1993.
103. Packer, L., Antioxidant action of carotenoids-in vitro and in vivo protection against oxidation of human LDL, *Ann. N. Y. Acad. Sci.,* 691, 48, 1993.
104. Khachik, F., Beecher, G.R., and Smith, J.C., Lutein, lycopene, and their oxidative metabolites in chemoprevention of cancer, *J. Cell. Biochem.,* 22, 236, 1997.
105. Tessa. J. et al., Interactions between carotenoids and the CCl_3O_2 radical, *Am. Chem. Soc.,* 117, 8322, 1995.
106. Bast, A., Haanen, G.R., and VandenBerg, H., Antioxidant effects of Carotenoids, *Int. J. Vit. Nutr. Res.,* 68, 399, 1998.
107. Mortensen, A. and Skibsted, L.H., Importance of carotenoid structure in radical-scavenging reactions, *J. Agric. Food. Chem.,* 45, 2970, 1997.
108. Mortensen, A. and Skibsted, L.H., Reactivity of β-carotene towards peroxyl radicals studied by later flash state photolysis, *FEBS Lett.,* 426, 392, 1998.
109. Soffers, A.M.F., Boerma, M.G., Laanre, C., and Rietjens, I.M., Antioxidant activities of carotenoids: Quantitative relationships between theoretical calculations and experimental, *Free Radical Res.,* 30, 233, 1999.
110. Mordi, R.C., Carotenoids-function and degradation, *Chem. Industry,* 110, 79, 1993.

111. Jorgensen, K. and Skibsted, L.H., Carotenoid scavenging of radicals—Effect of carotenoid structure and oxygen partial pressure on antioxidative activity, *Zeitshrift furLebensmitteln,* 196, 423, 1993.

112. Burton, W.G., Antioxidant action of carotenoids, *Br. J. Nutr.,* 119, 109, 1988.

113. Burton, W.G and Ingold, K.U., Beta carotene: An unusual type of lipid antioxidant, *Science,* 224, 569, 1984.

114. Palozza, P. and Krinsky, N.I., Antioxidant effects of carotenoids in vitro and in vivo. An overview, *Methods Enzymol.,* 213, 403, 1992.

115. Palozza, P., Calvello, G., and Bartoli, G.M., Prooxidant activity of β-carotene under 100% oxygen pressure in rat liver microsomes, *Free Radical Biol. Med.,* 19, 887, 1995.

116. Liebler, D.C., Antioxidant reaction of carotenoids, *Ann. N. Y. Acad. Sci.,* 691, 20, 1993.

117. Kasaikina, O.T. et al., Effect of environmental factors on the β-carotene reactivity toward oxygen and free radicals, *Biol. Membr.,* 15, 168, 1998.

118. Olmedella, B., Granado, F., and Blanco, I., Supplementation with lutein and α-tocopherol in separate or combined oral doses in controlled men, *Cancer Lett.,* 114, 179, 1997.

119. Dugas, T.R., Morel, D.W., and Harrison, E.H., Dietary supplementation with beta-carotene, but not with lycopene, inhibits endothelial cell mediated oxidation of LDL lipoprotein, *Free Radical Biol. Med.,* 26, 1238, 1999.

120. Heinonen, M. et al., Inhibition of oxidation in 10% oil in water emulsions by β-carotene with α-, γ-, and δ-tocopherols, *J. Am. Oil Chem. Soc.,* 74, 1047, 1997.

121. Henry, L.K., Gatignani, G.L., and Scwharz, S., The influence of carotenoids and tocopherols on the stability of safflower seed oil during heat-catalysed oxidation, *J. Am. Oil Chem. Soc.,* 75, 1399, 1998.

122. Lominsky, S., Grossman, S., and Bergaman, H., In vitro and in vivo effects of β-carotene in rat epidermal lipoxygenase, *Int. J. Vit. Res.,* 67, 407, 1997.

123. Warner, K. and Frankel, E.N., Effects of β-carotene on light stability of soyabean oil, *J. Am. Oil Chem. Soc.,* 64, 213, 1987; Kiritsakis, A. and Dugan, L.R., Studies in photooxidation of olive oil, *J. Am. Oil Chem. Soc.,* 62, 892, 1985.

124. Lee, H.S. and Min, B.D., Effects, quenching mechanisms, and kinetics of carotenoid in chlorophyll-sensitized photo-oxidation of soyabean, *J. Agric. Food Chem.,* 38, 1630, 1990.

125. Chen, J.H., Lee, T.C., and Ho, C.T., Antioxidant activities of caffeic acid and its related hydrocynamic acid compounds, *J. Agric. Food Chem.,* 45, 2374, 1997.

126. Nielsen, B.R. et al., Singlet versus triplet reactivity in photodegradation of C40 carotenoids, *J. Agric. Food Chem.,* 44, 2106, 1997.

127. Kiokias, S. and Gordon M., Properties of carotenoids in vitro and in vivo, *Food Rev. Int.,* 20, 99, 2004.

128. Eitenmiller, R. and Laden, W.O., Vitamins, in *Analyzing Food for Nutrition Labeling and Hazardous Contaminants,* Jern, I.J. and Ikins, W.C. (Eds.), Dekker, New York, 1995.

129. Cillard, J. and Cillard, P., Behaviour of α-, γ-, and δ-tocopherols with linoleic acid in aqueous media, *J. Am. Oil Chem. Soc.,* 57, 39, 1980.

130. Madhavi, L.D., Singhal, S.R., and Kulkavni, R.P., Technological aspects of foods antioxidants, in *Food Antioxidants: Technological Toxicological and Health Perspectives,* Madhavi, D.L. et al. (Eds.), Dekker, New York, pp. 159–266, 1996.

131. Schuler, P., Natural antioxidants exploited commercially, in *Food Antioxidants,* Hudson, F.B. (Ed.), Elsevier Applied Sciences, London, U.K., 1990.

132. Gottstein, T. and Gross, W., Model study of different antioxidant properties of α- and γ-tocopherols in fats, *Food Sci. Technol.,* 92, 139, 1990.

133. Huang, W., Schwarz, K., German, B., Frankel E.N., and Hopia, A.I., Antioxidant activity of α-tocopherol and trolox in different lipids substrates: Bulk oil-in-water emulsions. *J. Agric. Food Chem.,* 44, 444, 1996.

134. Kamal-Eldin, A. and Appelwist, L.A., The chemistry and antioxidant properties of tocopherols and tocotrienols, *Lipids,* 31, 671, 1996.

135. Frankel, E.N., Natural and biological antioxidants in food and biological systems. Their mechanisms of action, application and implications, *Lipid Technol.,* 49, 77, 1995.

136. Marinova, E.M. and Yanislieva, N.V., Effect of temperature on antioxidant action of inhibitors in lipid autoxidation, *J. Sci. Food. Agric.,* 60, 313, 1992.

137. Pryor, W.A., Cornicall, J.A., Dorall, L.I., and Tait, B., A rapid screening test to determine the antioxidant activity of natural and synthetic antioxidants, *J. Org. Chem.,* 58, 3521, 1993.

138. Jung, M.Y. and Min, D.B., Effects of α-, γ-, and δ-tocopherols on oxidative stability of soybean oils, *J. Food Sci.,* 55, 1464, 1992.

139. Lee, H.S. and Montag, A., Antioxidative wirksamkeit von tocochromanolen, *Fett. Techn.*, 94, 213, 1992.
140. Hopia, A.I., Huang, S.W., Schwarz, K., German, B., and Frankel E.N., Effect of different lipid systems on antioxidant activity of rosemary constituents. Carnosol and carnosoic acid with and without α-tocopherol, *J. Agric. Food Chem.*, 44, 2030, 1996.
141. Lee, H.S. and Wander, R.J., Relative activities of several tocopherols, *J. Sci. Food Agric.*, 10, 537, 1959.
142. Coultate, T.P., *Food—The Chemistry of Its Components*, 2nd ed., The Royal Society of Chemistry, Cambridge, U.K., 1996.
143. Johnson, E.L., Food technology of the antioxidant nutrients, *Crit. Rev. Food Sci. Nutr.*, 35, 149, 1995.
144. Counsel, J.N. and Horning D.H., *Vitamin C, Ascorbic Acid*, Applied Science, London, U.K., 1981.
145. Liao, M.T. and Seib, P.A., Selected reactions of L-ascorbic related to foods, *Food Technol.*, 41, 104, 1987.
146. Stait, E.S. and Leake, S.D., Ascorbic acid can either increase or decrease LDL modification, *FEBS Lett.*, 34, 263, 1994.
147. Cort, W.M., Antioxidant properties of ascorbic acid in foods, *Adv. Chem. J.*, 200, 533, 1982.
148. Frankel, E.N., Huang, W., Kanner, J., and German, B., Interfacial phenomena in the evaluation of antioxidants: Bulk oils vs emulsions, *J. Agric. Food Chem.*, 42, 1054, 1994.
149. Niki, E., Antioxidants in relation to lipid peroxidation, *Chem. Phys. Lipids*, 44, 227, 1987.
150. McCay, P.B., Vitamin E: Interactions with free radicals and ascorbate, *Ann. Rev. Nutr.*, 5, 323, 1985.
151. Englard, S. and Seifter, S., The biochemical functions of ascorbic acid, *Ann. Rev. Nutr.*, 6, 365, 1986.
152. Tan, B. and Saley, M.H., Antioxidant activities of tocopherols and tocotrienols on plant color carotenes, *Am. Chem. Soc.*, 202, 39, 1991.
153. Mortensen, A. and Skibsted, L.H., Reactivity of β-carotene towards peroxyl radicals studied by later flash state photolysis, *FEBS Lett.*, 426, 396, 1998.
154. Bohm, F., Edge, R., Lange, L., McGarvey, J., and Truscott, T.G., Carotenoids enhance vitamin E antioxidant activity, *Am. Chem. Soc.*, 119, 621, 1997.
155. Lievonen, S., The effects of carotenoids on lipid oxidation. PhD Thesis. Department of Food Science, University of Helsinki, Helsinki, Finland, 1996.
156. Kanner, J., Pro-oxidant and antioxidant effects of ascorbic acid and metal salts in a β-carotene linoleate model system, *J. Food Sci.*, 42, 60, 1977.

REFERENCES (IT)

E.U. Directive and Regulations with regard to oils and fats

E.U. 96/3/EC (http://europa.eu.int/smartapi/cgi/sga_doc?smartapi!celexplus!prod!DocNumber&1g=en&type_doc=Directive&an_doc=96&nu_doc=3) (accessed 20/03/2009).

Regulation No. 136/66/EEC (http://europa.eu.int/smartapi/cgi/sga_doc?smartapi!celexplus!prod!DocNumber&1g=en&type_doc=Regulation&an_doc=66&nu_doc=136) (accessed 20/03/2009).

Regulation (EEC) No 2568/91 (http://europa.eu.int/smartapi/cgi/sga_doc?smartapi!celexplus!prod!DocNumber&1g=en&type_doc=Regulation&an_doc=91&nu_doc=2568) (accessed 20/03/2009).

Regulation (EC) No.2991/94 (http://europa.eu.int/smartapi/cgi/sga_doc?smartapi!celexplus!prod!DocNumber&1g=en&type_doc=Regulation&an_doc=94&nu_doc=2991) (accessed 20/03/2009).

Regulation (EC) No.2815/98 (http://europa.eu.int/smartapi/cgi/sga_doc?smartapi!celexplus!prod!DocNumber&1g=en&type_doc=Regulation&an_doc=98&nu_doc=2815) (accessed 20/03/2009).

Regulation (EC) No.1019/2002 (http://europa.eu.int/smartapi/cgi/sga_doc?smartapi!celexplus!prod!DocNumber&1g=en&type_doc=Regulation&an_doc=2002&nu_doc=1019) (accessed 20/03/2009).

U.S. legislation related to oil

http://r0.unctad.org/commodities/agreements/oliveen.pdf (accessed 20/03/2008).

http://www.amazon.com/gp/product/9211122791/103-4320927-0003049?v=glance&n=283155 (accessed 20/03/2009).

Canadian legislation for oil

http://www.e-laws.gov.on.ca/DBLaws/RepealedStatutes/English/90e01_e.htm (accessed 20/03/2008).

http://www.canlii.org/on/laws/sta/e-1/20041201/whole.html (accessed 20/03/2009).

13 Food Additives and Contaminants

Theodoros H. Varzakas, Ioannis S. Arvanitoyannis,
and Athanasios E. Labropoulos

CONTENTS

13.1 INTRODUCTION

Food additives have been used for centuries to improve food quality. Smoke, alcohols, and spices have been extensively used for the last 10,000 years as additives for food preservation. The above-mentioned additives as well as a restricted number of additives comprised the main food additives until the Industrial Revolution. The Industrial Revolution brought so many changes in foods and asked for improved quality as well as quantity of the manufactured foods. For this reason many chemical substances were developed either for preservation or for color and/or odor enhancement. In the 1960s, over 2500 different chemical substances were used toward food manufacturing. In the United States over 2500 different additives were used to manufacture over 15,000 different foods. The desire for nutritional, functional, and tasty foods is an ongoing process. An additive is used to improve the shape, color, aroma, and extend the shelf life of a food. The following categories of additives are described:

- Coloring agents (Tables 13.1 through 13.3)
- Acids (Table 13.5)
- Aromatic substances (Table 13.6)
- Toxic metals (Tables 13.9 and 13.10)
- Biocides
- Preservatives (Table 13.11)
- Emulsifiers (Table 13.13)
- Leavening agents (Table 13.14)

There has been an intense skepticism regarding the safe use of additives in foods. In the 1960s and 1970s the increase of toxicological information caused an increase in the knowledge of possible risks derived from the consumption of foods containing additives. It was observed that the use of food additives has toxicological effects in humans. It is for this reason that in this chapter the limits of food additives use as well as pesticide residues' control are mentioned. It is well known that there is a plethora of risks derived from additives, but also there are benefits from their use in food production. Additives will continue to play a significant role in food production since consumers continue to desire healthier, tastier, and occasionally functional foods and as the population of earth continues to increase.

13.2 CODEX COMMITTEE ON FOOD ADDITIVES AND CONTAMINANTS

In the *Codex Procedure Manual* 13 edition (2003), it is clearly written that "All provisions in respect of food additives (including processing aids) and contaminants contained in Codex commodity standards should be referred to the Codex Committee on Food Additives and Contaminants preferably after the standards have been advanced to step 5 of the Procedure for the Elaboration of Codex Standards or before they are considered by the Commodity Committee concerned at Step 7, though such reference should not be allowed to delay the progress of the Standard to the subsequent Steps of the Procedure."

All provisions in respect of food additives will require to be endorsed by the Codex Committee on Food Additives and Contaminants, on the basis of technological justification submitted by the commodity committees and of the recommendations of the Joint FAO/WHO Expert Committee on Food Additives concerning the "safety-in-use (acceptable daily intake (ADI) and other restrictions) and an estimate of the potential and, where possible, the actual intake of the food additives, ensuring conformity with the General Principles for the Use of Food Additives".

13.3 SWEETENERS—GENERAL PHILOSOPHY OF DIRECTIVE 94/35/EC ON SWEETENERS FOR USE IN FOODSTUFFS

For sweeteners to be included in this Directive, they first have to comply with the general criteria set out in Annex II of the Food Additives Framework Directive 89/107/EEC (OJL 40, 11,2,89, pp, 27–33). Under these criteria, food additives may only be approved if it has been demonstrated that they perform a useful purpose, are safe, and do not mislead the consumer, The recitals of the Directive 94/35/EC on sweeteners for use in foodstuffs further explain that the use of sweeteners to replace sugar is justified for the production of

1. Energy-reduced foods
2. Noncarcinogenic foods (i.e., foods that are unlikely to cause tooth decay)
3. Foods without added sugars, for the extension of shelf life through the replacement of sugar and for the production of dietetic products

13.3.1 DEFINITION OF SWEETENER (FSA, (REGULATION 2(1))

For the purposes of these regulations, a sweetener is defined as a food additive that is used or intended to be used either to impart a sweet taste to food or as a tabletop sweetener. Tabletop sweeteners are products that consist of, or include, any permitted sweeteners and are intended for sale to the ultimate consumer, normally for use as an alternative to sugar. Foods with sweetening properties, such as sugar and honey, are not additives and are excluded from the scope of this legislation. The Sweeteners in Food Regulations 1995 do not apply where a substance listed as a permitted sweetener is used for purposes other than sweetening, for example where sorbitol is used as a humectant in accordance with the Miscellaneous Food Additives Regulations 1995 and parallel Northern Ireland legislation.

13.3.2   Permitted Sweeteners

(Regulations 2(1), 3(1), 4(a), and Schedule 1)

The only sweeteners that are permitted for sale to the ultimate consumer or for use in or on food are those listed in Schedule 1 of the regulations whose specific purity criteria are in compliance with that stated in the annex to Directive 95/31/EC.

13.3.3   Foods Allowed to Contain Permitted Sweeteners

(Regulations 3(2), 3(3), and Schedule 1)

Permitted sweeteners are only allowed to be used in or on foods that fall within one of the categories listed in Schedule 1 of the regulations. A maximum usable dose for each permitted sweetener, varying according to the food category, is also specified within Schedule 1 and this must be respected. The use of two or more sweeteners in a single food is permitted, provided suitable categories exist and the maximum level for each individual sweetener is observed. The sale of foods that do not comply with these provisions is illegal.

13.3.4   Sweeteners in Compound Foods—Carry-Over

(Regulation 2(1) (amended 1997) and Regulation 5A)

The regulations have been amended to include provisions on carry-over (Regulation 5A) to bring them into line with the GB Regulations on Colors and Miscellaneous Food Additives. These provisions allow the presence of a permitted sweetener in a compound food, to the extent that the sweetener is allowed by the regulations in one of the ingredients of the compound food. However, the definition of 'compound foods' in Regulation 2(1) means that permitted sweeteners are only allowed in the following compound foods:

1. Those with no added sugar or that are energy-reduced
2. Dietary foods intended for a low-calorie diet (excluding those specifically prepared for infants and young children)
3. Those with a long shelf life

The regulations also provide for what is commonly known as "reverse carry-over." This means permitted sweeteners can be present in foods (such as intermediary products) in which they would not otherwise be permitted, provided that these foods are to be used solely in the preparation of a compound food that will conform to the regulations.

13.3.5   Foods Not Allowed to Contain Sweeteners

(Regulation 3(4) (amended 1997) and Regulation 5)

The use of sweeteners in any foods for infants and young children is prohibited. This is specified in the Council Directive 89/398/EEC on the approximation of the laws of the member states relating to foodstuffs intended for particular nutritional uses (OJL 186, 30.6.89, pp. 27–32) and this prohibition now includes foods for infants and young children not in good health. The sale of such products containing sweeteners is also prohibited. Foods for infants and young children, generally known as "baby foods," include foods specially prepared for infants and young children who are in good health; or whose digestive processes or metabolism is disturbed; or who have a special physiological condition where they would be able to obtain benefit from controlled consumption of certain substances in foods. For the purposes of this prohibition, Regulation 2(1) defines "infants" as children under the age of 12 months and "young children" as children aged between 1 and 3 years. These definitions reflect those given in Article 1(2) of the Directive 91/321/EEC on infant formulae and follow-on formulae (OJL 175, 4.7.91, pp. 35–49) that are made under Directive 89/398/EEC.

The terms "maximum usable dose" and *"quantum satis"*

(Regulations 2(3)(c) and 2(3)(d))

These expressions are explained in the General Guidance Notes, Section 1, paragraphs 19 and 15 respectively.

The terms "with no added sugar" and "energy-reduced"

(Regulations 2(3)(a) and 2(3)(b))

Many of the categories listed in Schedule 1 of the regulations are described as "with no added sugar" or "energy-reduced." The final product must comply with the definitions of these terms and the effect is to further restrict the type of foods in which sweeteners may be used. However, the actual terms "with no added sugar" or "energy-reduced" are not required by these Regulations to be used in the labeling of such products. Whatever description is used for those products must be in accordance with the Food Labeling Regulations 1996.

"Energy-reduced" foods are foods with an energy value reduced by at least 30% compared with the original or a similar food. The legislation does not define the precise basis for this comparison, but wherever possible it should be by reference to one or more products that are currently on the market. If it is not possible to identify a comparable product that is currently on the market, the comparison could be made on the basis of previously marketed products. In an extreme case where it is not possible to identify an actual product, the comparison might be made with a hypothetically equivalent product, the composition of which is based on the use of sucrose rather than permitted sweeteners.

13.3.6 Additional Labeling Requirements for Tabletop Sweeteners

(Regulation 4(b))

The regulations include labeling requirements that apply to *tabletop sweeteners* only, In addition to the requirements contained within existing U.K. labeling legislation, tabletop sweeteners must include on their labels the phrase:

"[*Name of sweetener(s)*]-based tabletop sweetener"

Furthermore, where tabletop sweeteners contain polyols and/or aspartame, the following phrases must also be included on their labels:

for polyols—"excessive consumption may induce laxative effects"

for aspartame—"contains a source of phenylalanine"

For the purposes of these Regulations, polyols are considered to be sorbitol and sorbitol syrup (E420 (i) and (ii)), mannitol (E421), isomalt (E953), maltitol and maltitol syrup (E965 (i) and (ii)), lactitol (E966), and xylitol (E967).

The Additive and Food Contaminants (AFC) Panel of the European Food Safety Authority[1] has evaluated the new long-term study on the carcinogenicity of aspartame conducted by the European Ramazzini Foundation (European Foundation of Oncology and Environmental Sciences) in Bologna, Italy. In its opinion, the Panel concluded, on the basis of all the evidence currently available, that there is neither need to further review the safety of aspartame nor to revise the previously established Acceptable Daily Intake (ADI) for aspartame (40 mg/kg body weight).

The Panel also noted that intakes of aspartame in Europe, with levels up to 10 mg/kg body weight/day, are well below the ADI.

Aspartame, an intense sweetener, has been authorized for use in foods and as a tabletop sweetener for more than 20 years in many countries throughout the world. Extensive investigations have been carried out on aspartame and its breakdown products through experimental animal and human studies, intake studies, and post-marketing surveillance.

In addition to a number of safety evaluations conducted in the past, the Scientific Committee on Food (SCF) carried out a review of all original and more recent studies on aspartame in 2002 and reconfirmed that aspartame is safe for human consumption.

13.4 ADDITIVES IN FOOD PACKAGING

13.4.1 FDA REQUIREMENTS

Food packaging additives in the United States include slips, antistats, antioxidants, colorants, anti-fogs, antimicrobials, and oxygen scavengers.

To obtain direct food contact approval, according to FDA requirements, materials must meet extractability requirements. The packaging material must not adulterate the food.

A significant change in the FDA approval procedure was instituted in January 2000 with the new Food Contact Notification (FCN) system. To get approval for a new food-contact substance (FCS), the producer submits information including composition; intended use including additive level, usage temperature, and type of food the substance will contact; and data on migration of the substance into food.

Migration studies can use food-simulating solvents, such as 10% ethanol to represent aqueous, acidic and low-alcohol foods, or food oil or 50% or 95% ethanol to represent fatty foods. Experimental temperature and duration are set at the most extreme anticipated conditions. For example, the most extreme condition requires heating at 121°C for 2 h, followed by 10 days at 40°C, in a special cell designed to withstand the extremes of temperature and pressure. The FDA uses migration data to estimate consumer exposure to the substance. In the previous system, an application could take several years to get approval. With the FCN system, the FDA has 120 days to review the application and object based on safety grounds, or the substance may be marketed. Expectations were that the new system would result in many more applications for new products from companies that otherwise would not have tried marketing in food applications.[2]

Although food companies have been driving a trend toward irradiation to prolong shelf life, general FDA guidance on package irradiation has not yet been published. Only a few polymers are approved for gamma irradiation of prepackaged food and these were approved with additive packages prevalent in the 1960s.

13.4.2 SYNTHETIC/VEGETABLE-DERIVED ADDITIVES

The trend continues toward use of synthetic or vegetable-oil-based rather than animal-fat-derived additives, agree industry experts. In Europe, this trend is due to the concern of bovine spongiform encephalopathy (BSE). Since many companies have customers globally, the concern has spread to the United States as well. Some companies continue to be interested in vegetable-based additives for kosher-certified applications.[2]

13.4.3 SLIPS AND ANTISTATS

Slip and antistat additives, which function at the surface of the plastic part, are traditionally migratory. The additives are difficult to predict and control because migration occurs over time and depends on part/film thickness and polymer crystallinity.

Ampacet is a U.S. company that has introduced nonmigratory, surface-functional slip and antistat products that fit a need for controllable and predictable performance in premium films. Other advantages of the slip are that it can be used at higher temperatures than conventional slips and has no adverse effect on sealing. The nonmigratory antistat does not interact with adhesives, has no effect on sealing or printing, and has high thermal stability. Its antistatic properties last longer than those of conventional migrating antistats and its surface resistivity is similar at 50% and 12% relative humidity. The nonmigratory additives are used at much higher levels than traditional additives and so are more expensive. They are extensively used in coextruded structures, such as a film that has slip on the inside but not on the outside. One trend for antistatic additives is the use of longer chain materials which provide higher temperature processing stability that is in increasing demand, as extruders are pushed to faster rates. A trend for slip additives is to use higher purity slips with reduced short chain (four- to nine-carbon) impurities.[2]

These higher purity slips are characterized by lower organoleptics, or taste and odor components. A company recently introduced a low organoleptic erucamide slip product called Incroslip that has advantages for the bottled water and beer industries.

Plastic screw-type bottle closures contain high amounts of slip to enable torque release. Irradiation or sterilization of bottles by UV or ozone can degrade the trace amounts of by-products inherent in erucamide and produce off-tastes and odors.[2]

13.4.4 ANTIOXIDANTS

High performance antioxidants are manufactured by chemical companies which perform well under harsh conditions such as gamma irradiation of food packaging. These usually have an improved solid phosphate or they are amine oxides derived from vegetable oils. Other products are phenolic antioxidants with reduced water extractability thus pertaining an advantage to liquid food products and to those applications requiring excellent post thermo-oxidative stability at 150°C after 14 days hot boiling water extraction. The antioxidant should not discolor, as typical phenolics do in either a quinone methide dimerization reaction in the dark that causes pinking or in a gas-staining reaction with prompt oxides of nitrogen that causes yellowing (Table 13.12).[2]

Vitamin E antioxidants are also used for improved organoleptic properties for sensitive applications such as plastic milk and beverage bottles. These are powerful stabilizers, effective at very low concentrations, because they react with carbon-centered radicals.

13.4.5 COLORANTS

A variety of organic and inorganic colorants are allowed by FDA for indirect food contact; other colorants are exempted from FDA regulation based on migration testing in a specific polymer for a specific application (Table 13.1). However, dyes are not allowed by FDA.

The industry trend toward thinner parts (membranes, laminated films) creates a need for a higher colorant loading to maintain color intensity and opacity (Table 13.2). This has driven demand for

---

TABLE 13.1
Quantity of Coloring Agents Used in Certain Foods

Food Categories	Dye Concentration (mg/kg)	Average Quantity (mg/kg)
Caramels-sweets	10–400	100
Drinks	5–200	75
Cereals	200–500	350
Maraschino cherry	100–400	200
Pet foods	100–400	200
Ice creams	10–200	30
Sausage	40–250	125
Snack	25–500	200

Source: CCIC, Guidelines for good manufacturing practices: Use of certified FD&C colors in food, Certified Color Industry Committee, *Food Technol.*, 22(8), 14, 1968.

TABLE 13.2
Representative Examples of Tint Arising from Mixing of Basic Dyes

Tint	FD&C Blue No. 1	FD&C Blue No. 2	FD&C Red No. 3	FD&C Red No. 40	FD&C Yellow No. 5	FD&C Yellow No. 6
Strawberry			5	95		
Black	36			22		42
Egg white					85	15
Cinnamon	5			35	60	
Green	3				97	
Green mint	25				75	
Orange						100
Grape	20			80		
Black cherry	5			95		
Chocolate	10			45	45	
	8			52	40	
Caramel	6		21		64	9
Peach				60		40
Blackberry	5		75			20

Source: Adapted from CCIC, *Food Technol.*, 22(8), 14, 1968.

higher pigment levels in "super concentrates" to maintain cost effectiveness. "High-efficiency" concentrates can contain pigment loadings of 75% or greater, compared to levels of 20%–50% for conventional concentrates (Table 13.3).[2]

The presence of flavored colorants (peach and raspberry), flavors (caramel, citric acid, and vanilla), and food preservatives (sodium nitrite, sodium nitrate, sodium benzoate, benzoic acid, potassium sorbate, and sodium chloride) in *Escherichia coli* suspension during exposure to sunlight was evaluated.[3] He reported that these additives did not change the extent of cell survival. No effect on viability and mutation induction (kanamycin resistant) was also recorded when cells were in contact with any of the additives for 80 min in the dark. However, when the relevant additive was present in cell suspension during sunlight exposure the number of induced mutations was increased to varying extents over that seen with sunlight alone. Raspberry and peach increased the number of mutations in a dose-dependent manner, while vanilla produced mutations in an additive fashion. Nitrite, nitrate, benzoate, sorbate, and benzoic acid increased mutation somewhat additively over that of sunlight. Sodium chloride and citric acid were not effective. The impact of this investigation reflects the significance of the combination of sunlight and chemical food additives as potential risk, which requires special attention, and necessitates further investigations to evaluate the risk.

13.4.6 Antifogs

In fresh-cut produce packaging, antifogs prevent the film from fogging so that the consumer can see the product clearly. The use of antifogs in fresh food packaging is on the increase and will continue into the future as new applications as well as new polymer entrants into the fresh food packaging industry continue to evolve.[2]

Antifogs act as a surfactant, so that moisture given off by produce forms a transparent, continuous film on the package surface rather than forming beads of water. Antifogs can be impregnated into the film as an additive or applied as a liquid coating. A new trend toward microwaving of fresh-cut produce packages, such as spinach products, has led to challenges of meeting performance requirements and regulatory requirements, which are stricter at elevated temperatures, producing films with

TABLE 13.3

Dyes Used for Food; Name, Origin, Functionality, Adverse Effects, ADI, and Applications

E	Name	Origin	Functionality	Effects	ADI[a]	Product Uses
E102	Tartrazine	Synthetic dye	Yellow color	Headaches in children, allergic reactions in adults Forbidden in Norway and Austria	0–7.5	Fruit juices, cake mixtures, soups, ice creams, sauces, jams, yoghurt, sweets, gums, lollipops
E104	Cinolin yellow	Synthetic dye of coal tar	Cloudy yellow color	Low absorption from gastroenteric system Forbidden in Norway, United States, Australia, Japan	0–0.5	Smoked cod, ice creams
E107	Yellow 2G	Synthetic dye of coal tar	Yellow color	Allergic reactions Forbidden in Norway, United States, Switzerland, Japan, Sweden	—	
E120	Carmine	Natural color of extraction of dried insects	Red	Cancer	0–2.5	Ice creams, alcoholic drinks
E122	Azorubin	Synthetic color	Red	Allergic reactions to asthmatic people and those who are sensitive in aspirin, edema, retention of gastric juices	0–4.0	
E128	Red 2G	Synthetic dye	Red	In the intestine it gets transformed to aniline, which causes methemoglobinemia	0–0.1	Sausages, cooked meat products, jams, drinks
E150 δ	Ammonium sulfite, caramel color	Synthetic dye derived from carbohydrates	Brown	Gastroenteric problems, reduction in white blood cells of patients with B6 vitamin in low levels	—	Glucose tablets, ice cream, baking flour, total milled bread
E151	Bright black BN (boron nitride) or PN (brilliant black tetrasodium 4-acetamido-5-hydroxy-6-[7-sulfonato-4-(4-sulfonato-phenylazo)-1-naphthylazo] naphthalene-1,7-disulfonate)	Synthetic color of coal tar and nitrogen dye	Black	Intestine cysts in mice	0–1	Black sauces, chocolate mouse

Source: CFR, Code of Federal Regulations Title 21, Office of the Federal Register, U.S. Government Printing Office, Washington, DC, 1988.

[a] ADI, acceptable daily intake mg/kg body weight.

an anti-fog coating. The coating also aids in improving the shelf life of the vegetables. The extended shelf life is due in part to the permeability of the package, but also to the "synergistic effect" of the antifog coating that reduces moisture, which can encourage the growth of spoilage bacteria.[2]

13.4.7 ANTIMICROBIALS

The use of antimicrobials, or biocides, in packaging is a growing trend in the global food packaging industry. In the United States many of the antimicrobials in use protect the packaging or the packaging raw materials, although recent interest has been in antimicrobials to protect the packaged food. Antimicrobials that are incorporated into food packaging are regulated by the FDA under the Federal Food, Drug and Cosmetic Act (FFDCA).

Under the FFDCA, the FDA ensures that such antimicrobial uses are safe with respect to any potential human dietary intake. Unrelated to federal requirements under the FFCDA, antimicrobial products used in food packaging that have no intended antimicrobial effect on the processed food in the package are subject to EPA registration as pesticides under the Federal Insecticide, Fungicide and Rodenticide Act (FIFRA). Antimicrobials added to food or delivered to food via the packaging are treated as direct food additives and are not subject to FIFRA. Research at Clemson University has looked at coating food packages with nisin, particularly for hot dog packaging. Nisin can be compounded into the packaging polymer or applied as a powder or a coating. It is widely used in Europe but not extensively in the United States, although it has FDA approval.[2]

Silver compounds are also used in Europe and have FDA approval for some applications. Other research is looking at additives that produce the antimicrobial chlorine dioxide under certain relative humidity or UV light conditions. The advantage of these systems is that the antimicrobial could protect any product within the package, not just what comes in contact with a protective coating. Chlorine dioxide is also less expensive and effective for a broader range of microorganisms than nisin. The additives are currently being used in a sachet inside the package, but can be compounded into the packaging polymer.[2]

13.4.8 FRESHNESS INDICATORS

A major trend in the food industry over the past several years has been the consumer's desire for freshness in taste and appearance and freshness indicators are an upcoming technology. Modified atmosphere packaging controls the flow of carbon dioxide through the food package to extend shelf life, but is dependent on storage conditions in the store or home. Indicators to show when a food has begun to decay are currently being used in bulk packaging, such as adhesive labels. Current research is focused on making indicators cost-effective for individual packaging. Indicator dyes work by either changing as a function of time and temperature or by reacting with a food degradation product. For example, an indicator in a sensor or in the packaging film could react with an amine given off by fish at the beginning of decay.[2]

13.4.9 OXYGEN SCAVENGERS

The use of oxygen absorbers is a relatively new additive trend in food packaging. Oxygen scavengers are especially important in the trend toward single-serve packages due to the smaller packages' increased surface volume and exposure to oxygen. Commercial oxygen scavengers include iron oxide powders enclosed in sachets, additives incorporated into the packaging polymer or a polymer layer extruded as part of the package to maintain freshness by absorbing headspace oxygen and oxygen that enters the package. They can be incorporated directly into the walls of the package either as an existing layer within the package or as a distinct scavenging layer.

The oxygen scavenging polymer system consists of an oxidizable resin, ethylene methyl acrylate cyclohexene methyl acrylate (EMCM) and a masterbatch containing a photoinitiator and a cobalt salt catalyst. Other oxygen scavenging polymers based on nylon, polypropylene, polybutadiene, and polyisoprene degrade on oxidation into by-products that can migrate into packaged food and cause off-taste or odor.

EMCM does not degrade into compounds responsible for off-taste or odor and the photoinitiator allows the inactive polymer to be stored and then activated by UV light during the package filling process.[2]

Cryovac introduced an improved oxygen scavenging film that reportedly removes oxygen 20% faster than before. The film slowed down microbial growth and oxidative deterioration of flavors, color, and nutrients. A Nestle fresh pasta package using the film recently won an award for technical innovation from the Flexible Packaging Association. The oxygen scavenging process increases the shelf life of refrigerated pasta.[2]

13.5 ADDITIVE CATEGORIES

13.5.1 ACIDS

13.5.1.1 Lactic Acid

Dubos[4] concluded that lactic acid has a bacteriostatic effect on *Mycobacterium tuberculosis*, which increased as pH decreased. Experiments carried out with *Bacillus coagulans* in tomato paste showed that lactic acid was four times more effective regarding the inhibition of bacterial growth compared to malic, citric, propionic, and acetic acid.[5]

Lactic acid is used in the jams, sweets, and drinks industries. This acid is the best acid to control the acidity and assure the transparency of brine in pickles.[6] Calcium lactate can be used as a taste enhancer, as a baking processing aid, to proof the dough, as an inhibitor of decolorization of fruits and vegetables, as a gelatinization factor during pectins dehydration, or as an improver of the properties of milk powder or condensed milk.[7] The ethyl esters of lactic acid can be used to enhance taste. Calcium lactate can be used in dietetic foods as well as nutrition supplements (Table 13.4).

Lactic acid is an intermediate product of human metabolism. In the cases of pneumonia, tuberculosis, and heart failure, a non-physiological quantity of acid was detected in human blood. A growth problem was detected when acid was injected into water at a quantity of 40 mg/100 mL or in food at 45.6 mg/100 mL in hamsters. Lactic acid proved to be lethal in newborn fed milk with an unknown quantity of acid.[8] The poisonous effect of acid has to do with its isomeric form. Babies fed with milk that has been acidified with D(−) or DL form suffered acidosis, lost weight, and dehydrated. Hence, only the L(+) form can be used in premature newborn babies. Acid is a food ingredient and an intermediate metabolite of human beings, therefore, there are no established limits of day consumption for humans.[9]

13.5.1.2 Succinic Acid

FDA allowed the use of succinic acid as a taste enhancer and as a pH regulator.[7] Succinic acid reacts with proteins and is used to modify the dough plasticity.[6] Succinic acid derivatives can be used as taste enhancing agents or in combination with paraffins as a protective layer for fruits and vegetables. They can be used in pills production at a percentage not greater than 4.5%–5.5% of gelatin percentage and 15% of the total weight of capsule. Many derivatives of succinic acid are used as ingredients of paper and paperboard in food packaging.

The succinic acid anhydrite is ideal for baking powder production of the Allied Chemical Co.[6,10] The low level of acid hydrolysis is important during mixing of dough, since it is required that the additional acid should not react with soda during mixing until the product swells.

In an acute toxicity study in rats with a daily subcutaneous injection of 0.5 mg for a period of 60 days, and with a dose increase of 2 mg/day during the fourth week, no detrimental effect was observed. Acid is produced in some fruits and constitutes an intermediate product of the cycle of Krebs. Hence, no established limits can be determined for humans.

TABLE 13.4
Maximum Amount of Organic Acids in Various Foods

Foods	Acetic Acid	Calcium Acetate	Sodium Acetate	Sodium Diacetate	Adipic Acid	Caprilic Acid	Malic Acid	Succinic Acid
Baked foods	0.25	0.2		0.4	0.05	0.013		
Drinks. nonalcoholic drinks					0.005		3.4	
Breakfast cereals			0.007					
Cheese	0.8	0.02				0.04		
Chewing gum	0.5						3.0	
Spices	9.0				5.0			0.084
Dairy products	0.8				0.45			
Oils-fats	0.5		0.5	0.1	0.3	0.005		
Frozen dairy desserts					0.004	0.005		
Gelatin		0.2			0.55	0.005	0.8	
Sauces	3.0			0.25		0.1		
Hard candies			0.15				6.9	
Jam			0.12					2.6
Meat products	0.6		0.12	0.1	0.3	0.005		0.0061
Fruit juices							3.5	
Snacks			0.6	0.05	1.3	0.016		
Soft candies			0.2	0.1		0.005	3.0	

Source: CFR, Code of Federal Regulations Title 21, Office of the Federal Register, U.S. Government Printing Office, Washington, DC, 1988.

13.5.1.3 Fumaric Acid

Fumaric acid is responsible for a sour taste in foodstuffs. It is widely used in fruit juices, in desserts, in the frozen dough of biscuits, wild cherry liqueur, and in wines. It can also be used as a coating agent in caramels and bread. Fumaric acid contributes to the extension of the shelf life of baking powder due to its restricted solubility and the low humidity absorption. It attains very good antioxidant properties and is used to avoid the rancidity of pork fat, butter, milk powder, sausages, bacon, walnuts, and chips.[6,11] It could also be used as a preservative in green foodstuffs and fishes in the same way as sodium benzoate does. CFR[7] allowed the use of fumaric acid and its salts as dietetic products and nutritional additives. It can also be used as a source of available iron in the human organism if fumaric acid is combined with iron. Many fumaric acid derivatives have been approved for use in foodstuffs.

13.5.1.4 Malic Acid

CFR[7] allowed the use of malic acid in foodstuffs as an acidifier, aroma enhancer, and pH regulator. Malic acid also contributes to the non-browning of fruits and acts synergistically with antioxidant substances.[6]

Malic acid is used in the production of iced fruits, marmalades, nonalcoholic carbonated drinks as well as drinks originating from fruits. The limits of use have been defined accordingly: 3.4% nonalcoholic drinks, 35% chewing gums, 0.8% gelatins and puddings, 6.9% hard candies, 3.5% in processed fruits and juices, 3% in soft candies, and 0.7% in all other foodstuffs.

13.5.1.5 Tartaric Acid

CFR[7] allowed the use of tartaric acid in foodstuffs as an acidifier, solidifying agent, taste-enhancing agent, as a material for maintenance of humidity, and pH regulator. Tartaric acid can be used in drinks manufacturing as an enhancing agent of the red color of wine.[6] It can also be used in tarts, marmalades, and candies. Mixing of tartaric acid and citric acid can be used in the production of hard candies with special flavors such as apple and wild cherry. This acid reacts with other antioxidant substances to avoid rancidity and discoloration in cheeses.[6] Tartaric acid and its salts with potassium constitute ingredients of baking powders, L(+) isomer and its salts can be consumed up to 30 mg/kg body weight.[9]

13.5.1.6 Citric Acid

Citric acid is a common metabolite of plants and animals. It is widely used in foodstuffs and pharmaceutical industry as well as the chemical industry for isolation of ions and neutralization of bases. Citric acid esters can be used as plasticizers in the manufacture of polymers and as adhesives. Citric acid can be used as an acidifier, taste enhancer, and acts synergistically with other antioxidant substances. Citric acid and its salts are used in ice creams, drinks, fruits, jams, and as an acidifier in the manufacturing of canned vegetables. Calcium citrate is used as a stabilizer in pepper, potatoes, tomato, and beans during their processing. It has also been widely applied in dairy products manufacturing.[6] Moreover, it can be used in creamy sauces and soft cheeses. Sodium citrate can be used as an emulsifier. This proves to be the main acidifying factor in carbonated drinks entailing a rancid taste.[6,10] It can be used in line with other substances as an antioxidant and as a decelerator of fruits' browning. It is more hygroscopic than fumaric acid and could cause problems during storage of powdered products. Citric acid can be found in animal tissues and constitutes an intermediate product of the Krebs cycle, hence, the daily consumption levels for the acid and its salts have not been determined yet.[12]

13.5.1.7 Ascorbic Acid

Ascorbic acid is being used as an antimicrobial and antioxidative factor and enhances the uniformity and color stability. Ascorbic acid and its salts with Na and Ca are used as food additives.[7] The D isomer of ascorbic acid compared to the L isomer has no biological value and gets oxidized faster than ascorbic acid with the result of the protection of vitamin C from oxidation.[10] It is used as a pH regulator to avoid the enzymatic browning of fruits and vegetables. Plants and all mammals except man, apes, and Guinea pigs synthesize ascorbic acid, hence for this reason these three need to consume alternative sources of acid such as vegetables. High doses of ascorbic acid are recommended for cancer therapy and common cold. Ascorbic acid does not cause problems if it is consumed in high doses compared to vitamins A and D that might cause problems.

13.5.1.8 Acetic Acid

Acetic acid is used as a pH regulator, solvent, and as a pharmaceutical ingredient.[7] It is safe when consumed in combination with the proper processing conditions. It can also be used as an additive in mustard, ketchup, mayonnaise, and sauces.

Acetic acid is well known for the various unpleasant effects on humans such as allergic symptoms, ulcer,[13] anaesthesia,[14] and epidermic reactions[15] until death.[16] Acid is made up on plant and animal tissues. There is no restriction on the daily consumption of acid from humans.

13.5.1.9 Sorbic Acid

Sorbic acid and its salts are widely being used as fungicides and preservatives in pickles, mayonnaise, salads, spices, fruit and vegetables pulping, jams, frozen salads, syrups, beer, wines, sweets, cheeses, yoghurt, fishes, meat, poultry, and in various bakery products.[6,17] Sorbic acid is the least harmful preservative. According to a subchronic study of 2 months, carried out in 25 female and 25 male mice that consumed 40 mg acid/kg body weight, this did not lead to severe effects on the

TABLE 13.5
Uses and Indicated Quantities from Sorbic Acid Use

Dairy products (cheese, sour cream, yoghurt)	0.05–0.30
Bakery (cakes, dough, sugar crust, garnish)	0.03–0.30
Vegetables (fresh salads, boiled vegetables, pickles, olives, starters)	0.02–0.20
Fruits (dry fruits, fruit juices, fruit salads, jam, syrup)	0.02–0.25
Drinks (wine, carbonated drinks, fruit drinks)	0.02–0.10
Miscellaneous (smoked and salted fish, mayonnaise, margarine, sweets)	0.05–0.20

guinea pigs. According to studies carried out in rats fed food containing 1% or 2% sorbic acid for 80 days, no histologic abnormalities or growth problems were observed (Table 13.5).[18]

13.5.1.10 Propionic Acid

Propionic acid is extensively used in cheeses, sweets, gelatins, puddings, jams, drinks, soft candies, Swedish cheese; at a concentration of 1%. Calcium propionate can be used as an antimicrobial agent and acts as an inhibitor of the mould formation in bread dough. No daily consumption limits were determined due to the fact that propionic acid is an intermediate metabolite of human metabolism.[9]

13.5.1.11 Benzoic Acid

CFR[7] approved benzoic acid for use at a concentration of 0.1%. The acid and its sodium salt can be used in processed foods, as food and drink preservatives with pH less than 4.5. It has an inhibitory action on mould growth and bacteria belonging to the following species: *Bacillaceae, Enterobacteriaceae,* and *Micrococcaceae*. Benzoic acid and its sodium salt can be used in the preservation of carbonated drinks and noncarbonated drinks, fruit juices, jams, mayonnaise, mustard, pickles, bakery products, and ketchup.[19,20]

13.5.2 Coloring Agents

The synthetic dyes used nowadays are divided into the following three categories:

1. FD&C dyes: Certified for food use, medicines, and cosmetics.
2. D&C dyes: Dyes considered as safe for medicinal and cosmetics use when coming into contact with muciferous tissue or when they get absorbed.
3. External D&C dyes: Dyes, which due to their toxicity do not get certified for product use due to be consumed, however, they are considered as safe for use in products of exterior applications.

Moreover, coloring agents used in foodstuffs can be divided into the following two classes:

1. Dyes requiring certification
2. Dyes getting exempted from certification

13.5.2.1 Dyes Requiring Certification for Their Use

13.5.2.1.1   FD&C Red No. 2 (Amaranth)

This dye was one of the first seven allowed dyes for use in 1906. Amaranth is a red brown powder easily dissolved in water and producing a deep purple or a sea red solution.

In experiments where amaranth was used at a percentage over 5% of the total dye, no pathological findings were reported with the exception of mutations, and tumor development in rats. Long-term experiments for 7 years in dogs injected with the above-mentioned dyes at 2% did not cause any pathological problems. Similar experiments in mice, rats, rabbits, hamsters, cats, and dogs showed neither reproduction problems nor teratogenic problems or other unpleasant effects.[21] Hence, it was decided that it can be included in the list of the approved dyes. However, two experiments carried out in the ex-Soviet Union reported that the use of this dye caused carcinogenesis as well as embryotoxic reactions in mice that consumed 0.8%–1.6% of amaranth.

13.5.2.1.2 FD&C Red No. 4

This dye is allowed in foods since 1929. It is a red colored powder easily dissolved in water producing an orange-yellow solution. It was originally used as an additive in butter and margarine. It does not cause carcinogenesis in mice when contained in food at 5% for a period of 2 years. It was concluded that this dye was toxic when consumed at 1% of the food for a period of 7 years. In another experiment carried out in dogs that consumed dye at 2% for a period of 6 months, three out of four died. Due to the above-mentioned experiments the use of this dye was forbidden in 1976.[22]

13.5.2.1.3 Citrus Red No. 2

The use of this dye is restricted, and used mainly in leather dying. Nutritional studies showed that its use causes carcinogenesis in cats and dogs. Further experiments showed that this dye causes carcinogenesis.

13.5.2.1.4 FD&C Red No. 40

This dye is allowed in the United States and Canada following thorough experiments. However, it is not allowed in the United Kingdom, Sweden, and Europe.

13.5.2.1.5 FD&C Yellow No. 3, FD&C Yellow No. 4

These dyes are orange colored and fat soluble. They were used in 1918 in the United States as additives in margarine production. They were reported to be liver-toxic in dogs and cats.[23] These dyes caused weight loss in guinea pigs even at 0.05%.[24] Mice consuming a quantity of dye with their food higher than 0.05% for a period greater than 2 years showed heart atrophy or hypertrophy. Dyes are metabolized under the effect of gastric acid conditions[25,26] and their metabolites caused liver cancer in rats as well as urinary bladder cancer in dogs.[27] Due to these reported experimental results these dyes were rejected from the list of dyes that are allowed to be used.

13.5.2.2 Dyes Not Requiring Certification for Their Use

13.5.2.2.1 Annatto Extracts

One of the older dyes used in foods, textiles, and cosmetics. The extraction of these dyes is carried out from the pericarp of the seeds of the plant Annatto tree (*Bixa orellana* L.). Annatto oil is produced with the distillation of the coloring agents from the seeds of Annatto tree (*Bixa orellana* L.) using an edible vegetable oil. The main coloring agent is norbixin (E 160b). It is supplied as an oil-soluble extract which contains mainly bixin (a mono methyl ester) or as a water-soluble product that is produced by extraction with aqueous alkali and therefore contains mainly norbixin, the free acid.

Processing is primarily carried out by abrading away the pigment in a suspending agent. Abrasion may be followed by aqueous alkaline hydrolysis with simultaneous production of norbixin. Traditionally, water or vegetable oil is used as a suspending agent, although solvent processing is now also employed to produce more purified annatto extracts. Annatto is usually marketed as an extract of the annatto seed, containing amounts of the active pigments bixin or norbixin that can vary from less than 1% to over 85%.

In 2002, the Secretariat of the Joint FAO/WHO Expert Committee on Food Additives (JECFA) requested information relating to the toxicity, intake, and specifications of annatto extracts. Previous intake estimates for annatto had provided ambiguous results because the bixin/norbixin content of the annatto extract was unclear. As a consequence, many of the stated use levels, such as some of those appearing in the draft CCFAC General Standard for Food Additives (GSFA)[28] overestimated true use levels by more than an order of magnitude. This resulted in estimates of intake of annatto extracts that appeared to exceed the JECFA ADI of 0.065 mg/kg body weight/day (based on pure bixin). Previous intake estimates for annatto provided ambiguous results because the bixin/norbixin content of annatto extract was unclear. European annatto producers consulted with the food industry to determine use levels of specific annatto extracts. These data were combined with the levels of bixin/norbixin in particular extracts to estimate the concentration of bixin/norbixin in foods. Concentrations in food were combined with data about food consumption using various methods to estimate consumer intakes, which ranged from less than 1%–163% of the acceptable daily intake (0.065 mg/kg body weight/day). Higher intake estimates are conservative because they assume that a consumer always chooses a food that is colored with annatto extracts. In practice this is extremely unlikely, since annatto is associated only with certain product/flavor combinations.[29]

Food colors differ from many other food additives in that the level required to meet a technological need can vary considerably from food to food. This is because colors are usually associated with particular flavors and so the need for a coloring agent, for example an orange color such as annatto, will vary greatly depending on whether the flavor is vanilla, lemon, orange, or mango. Another source of variability is the opacity of foods. The more opaque a food is the higher the amount of color required to achieve a given shade. As a consequence, typical use levels (i.e. use in the majority of colored food products) do not necessarily reflect the maximum use level required to achieve the color density required for certain color/flavor/food combinations. It is therefore necessary to take this potential source of variability carefully into account when estimating consumer intakes.

13.5.2.2.2   Betalains (Dehydrated Sugar Beets)

These dyes belong to the plants of the family Centrospermae, and more specifically plants such as beetroots and bougainvillea. Beetroot contains both dyes, the red (betacyanins) and the yellow (betaxanthins). Betalains are sensitive to light, pH, and heat.[30–32] They are dissolved easily in water giving solutions of blackberry or cherry coloration. Beetroot dyes are used in foods with low storage time and do not require high and extended heat during their processing. In heat processing cases dye addition could be carried out just before or after the end of processing.[30] Betalains are used at 0.1%–1% of the total dye in hard candies, yoghurt, sauces, cakes, meat substitutes, mild drinks, and gelatinous desserts.

13.5.2.2.3   Anthocyanins

Anthocyanins are used as pH indicators. The dye is red at an acidic environment whereas it becomes blue as pH increases. Anthocyanins have more intense color at pH = 3.5,[31] hence, for this reason these dyes are used in foods with low pH. They get discolored in the presence of amino acids[33] while they get oxidized in the presence of ascorbic acid.

13.5.2.2.4   Saffron

Saffron is an expensive dye since it is required to get 165,000 flower stains for the production of 50 g pure dye. The most important Saffron dyes are crocin and crocetin. Crocin has a yellow-orange color, dissolves in hot water whereas it dissolves less in alcohol. Crocetin dissolves little in water and more in organic solvents. Saffron is used for its aroma and its color is resistant to sunlight, moulds, and pH and has a high dying capacity. It is used at a concentration of 1–260 ppm in cooked food, soups, and confectionery products.

13.5.2.2.5   Cochineal Extracts

They are derived from the body of female insects of species *Coccus cacti*. This insect can be found in the Canary Islands and the dye costs a great deal. It can be used in the production of candies, sweets, alcoholic and nonalcoholic drinks, jams, eye shadows, rouge, and as a coating for pills at a percentage of 0.04%–0.2%.

13.5.2.2.6   Chlorophyll

Chlorophyll is the most abundant natural dye of the plants. However, it is not widely used in foods.[33] Its green color is decomposed easily even at mild processing conditions. Due to its structure, during processing, magnesium gets replaced by hydrogen destroying the purple ring giving a dark brown color.[34] Chlorophyll stabilizes with the replacement of magnesium ions with copper ions. Chlorophyll extracts are forbidden for use in foods in the United States. The use of complex chlorophyll-copper is allowed in the EU and Canada.

13.5.2.2.7   Carotenoids

Carotenoids are sensitive to alkali and very sensitive to air, light, and high temperature. They are undissolved in water, ethyl alcohol, and glycerol. β-Carotene is one of the fewest dyes with nutritive value because it gets transformed into provitamin A in the human organism and can be used without quantitative limitations. It takes a yellow-orange color in foods when used at concentrations of 2–50 ppm. β-Carotene constitutes one of the basic ingredients of butter, margarine, cheeses, ice creams, yoghurt, and pasta products.

β-Carotene (E 160 a (i)) is obtained by solvent extraction of edible plants (e.g. lucerne) or from algae (*Dunaliella salina*), and can be naturally accompanied by minor carotenoids (e.g. α-carotene). Nearly pure β-carotene (E 160 a (ii)) is available from the fungus *Blaskeslea trispora*. The coloring principle of red pepper extracts (E 160 c) is capsanthin, accompanied with capsorubin and various minor carotenoids. Lycopene (E 160 d) is found in several fruits (e.g., papaya) and represents the main carotenoid of red tomatoes. Lutein (E 161 b) is found in several plants (e.g., alfalfa) or flowers (e.g., marigold; *Tagetes erecta* L.). In marigold, lutein is acylated with various fatty acids.

A sensitive HPLC multimethod was developed by Breithaupt[35] for the determination of the carotenoid food additives (CFA) norbixin, bixin, capsanthin, lutein, canthaxanthin, b-apo-80-carotenal, b-apo-80-carotenoic acid ethyl ester, b-carotene, and lycopene in processed food using an RP C30 column. For unequivocal identification, the mass spectra of all analytes were recorded using LC–(APcI) MS. For extraction, a manual process as well as accelerated solvent extraction (ASE) was applied. Important ASE parameters were optimized. ASE was used for the first time to extract CFA from various food matrices. Average recoveries for all analytes ranged from 88.7% to 103.3% (manual extraction) and from 91.0% to 99.6% (ASE), with the exception of norbixin using ASE (67.4%). Limits of quantitation (LOQ) ranged from 0.53 to 0.79 mg/L. The presented ASE method can be used to monitor both, forbidden application of CFA or the compliance of food with legal limits.

13.5.2.2.8   Titanium Dioxide (No. 77891)

This dye exists in nature in three crystalline forms—anatase, brookite, and rutile—however, only one is used as an additive). It possesses an intense white color, is resistant to sunlight, oxidation, pH, and the presence of microbes. Only the synthetically produced titanium dioxide can be used as a food additive.[30] This dye cannot be dissolved in all solvents. The allowed quantity of dye in foods is up to 1% and can be used in confectionery products in the formation of white parts as well as a background. Finally, it can be used in the production of pills and cosmetics.

13.5.2.2.9   Iron Oxides and Hydroxides

These synthetic dyes impart a large range of colors varying from red, yellow, and black with exceptional stability in light and temperature. Natural dyes are not acceptable as food additives due to the difficulties encountered to isolate the pure substance. They are insoluble in most solvents; however, they can be dissolved in hydrogen chloride. The maximum allowed quantity of these dyes is lower than 0.25% w/w in fish paste and in pet food.

13.5.3 AROMATIC SUBSTANCES

Aroma is a food and drink property causing stimulation of the olfactory centers (Table 13.6). The role of the aromatic substances in foods is summarized below:

1. Impart a characteristic aroma, e.g. vanillin gives a vanillin aroma to ice creams
2. Enhance, complement, or modify the already existing aroma, e.g. vanillin addition to moderate chocolate or cocoa's smell
3. Cover a non-desirable odor

TABLE 13.6
Aroma and Taste Enhancers in Food; Name, Origin, Functionality, Adverse Effects, ADI, and Applications

E	Name	Origin	Functionality	Effects	ADI[a]	Product Uses
E256	Malic acid	Natural (potatoes, apples, etc.) or artificial (commercial use)	Aromatic agent/ acidifying agent	Avoid the use of artificially made malic acid in baby foods		Canned tomatoes, peas, orange puree with low calories, concentrated tomato juice, fruit juices
E620	L-Glutamate	Natural (amino acid of plants and animals) and artificial (commercial use)	Aroma and taste enhancer	Its accumulation could cause damage to brain cells of guinea pigs	0–120	Sauces, soups, aromatic lasagna and processes cheeses
E621	Acidic monosodium glutamate	Artificial	Taste enhancer	Its accumulation could cause brain tissue disorders	0–120	Instant soups and sauces
E622	Acidic potassium glutamate	Artificial	Taste enhancer	High doses could cause phenomena related to high potassium	—	Vegetable soups
E627	Disodium Guanylate	Artificial	Taste enhancer	It should be avoided by patients with high uric acid due to the fact of being raw material of the synthesis of uric acid	—	Canned meat products, crackers, vegetable soups and sauces
E631	Sodium inosinate	Natural or artificial	Taste enhancer	Similar to E 627		Processed meta products, sauces and soups
E635	Ribonucleic sodium	Artificial	Taste enhancer	Similar to E 627		Similar to E627

TABLE 13.6 (continued)
Aroma and Taste Enhancers in Food; Name, Origin, Functionality, Adverse Effects, ADI, and Applications

E	Name	Origin	Functionality	Effects	ADI[a]	Product Uses
E421	Mannitol	Natural or artificial	Sweetener and noncaking	Some people may develop nausea, vomiting, diarrhea		Sweets, ice creams, gums
E905	Microcrystalline wax	Artificial	Antifoaming agent and glazing agent	High doses cause diarrhea and irritation		Dried fruits, citrus fruits, gums
E951	Aspartame	Artificial	Sweetener	Phenylketonuria patients should avoid it		Soft drinks, milk drinks, fruit juices, dry nuts, nutrition supplements, mustard.
E952	Cyclamic acid and its sodium and calcium salts	Artificial	Sweetener	Cyclamics get excreted unalterable by the human body except some others that could disrupt it and cause undesirable side effects		Low calorie products

Source: Adapted from Van den Ouweland, G.A.M. and Swaine, J.R.L., *Perfumer Flavorist*, 5, 15, 1980.

[a] ADI, acceptable daily intake mg/kg body weight.

The odor of a chemical substance has a direct relationship with its chemical formula. Compounds such as nitrogen, oxygen, and sulfur play a significant role. More specifically, the addition of a hydroxyl group into the molecule of a compound reduces or restricts its odor, whereas replacement of a hydroxyl group with a ketone group enhances the odor (Table 13.7).

13.5.3.1 Polycyclic Aromatic Hydrocarbons

Polycyclic aromatic hydrocarbons (PAHs) comprise the largest class of chemical compounds known to be cancer causing agents. Some, while not carcinogenic, may also act synergically with others. PAHs are being found in water, air, soil, and, therefore also in food. They originate from diverse sources such as tobacco smoke, engine exhausts, petroleum distillates, and coal-derived products, with combustion sources predominating. However, PAHs may also form directly in food as a result of several heat processes such as grilling,[36] roasting, smoke drying, and smoking.

PAHs are compounds consisting of two or more condensed aromatic rings, lineated together, either *cata*-annellated (linearly-, or angularly), or *peri*-condensed.

Formation, factors affecting concentrations, legal limits and occurrence of polycyclic aromatic hydrocarbons in smoked meat products and smoke flavor additives are briefly reviewed by Simko.[37] The most widely employed techniques such as thin-layer chromatography (TLC), gas chromatography (GC), and high-performance liquid chromatography (HPLC) are evaluated. Moreover, sample preparation, pre-separation procedures, separation and detection systems being used for the determination have been evaluated with emphasis on the latest developments in applied food analysis and

TABLE 13.7
Saturated Organic Acids and Characteristics with Regard to Aroma

Acid	Characteristics
Acetic	Strong aroma, fine, carmine
Propionic	Rancid, fatty, milky
Butyric	Fatty, milky, rancid
Valeric	Rancid, sweet, chocolate
Caprylic	Fatty, fruit smell, soapy
Myristic	Soapy
Palmitic	Soapy, waxy
Stearic	Soapy, waxy

Sources: WHO, *Environmental Health Criteria 3: Lead*, World Health Organization, Geneva, 1976; Reilly, C., *Metal Contamination of Food*, 2nd ed., Elsevier Applied Science, London, 1991.

TABLE 13.8
Aldehydes Found in Bovine, Chicken, and Pork Fat

Aldehydes	Bovine	Chicken	Pork
C5	+	−	−
C6	+	+	+
C7	+	+	−
C7 2t	+	+	+
C7 2t 4C	−	+	+
C8	+	−	+
C8 2t	+	+	+
C9	+	+	+
C9 2t	+	+	+
C9 2t 4C	−	+	+
C10	−	+	+

Sources: Dabeka, R.W. et al., *Food Add. Contam.*, 89, 1987. WHO, *International Standards for Drinking Waters*, 2nd ed., World Health Organization, Geneva, 1963. Reilly, C., *Metal Contamination of Food*, 2nd ed., Elsevier Applied Science, London, 1991.

the chosen data regarding the concentration of polycyclic aromatic hydrocarbons in smoked meat products and smoke flavor additives are summarized (Table 13.8).

13.5.3.2 Flavoring Substances

Munro and Danielewska-Nikiel[38] conducted a study to determine the margins of safety between non-observed-effect levels (NOELs) and estimates of daily intake for 809 flavoring substances evaluated by the Joint FAO/WHO Expert Committee on Food Additives (JECFA) between 2000 and 2004. Estimates of daily intake were calculated by means of two methods, the maximized survey-derived daily intake (MSDI) and the possible average daily intake (PADI). The MSDI estimates were based on the production volume of flavoring agents as reported by industry, whereas the higher and more conservative PADI estimates were derived by multiplying the anticipated average use level of a flavoring substance in each of the 33 food categories by the average amount of food consumed daily from that food category and summing the overall intake of the 33 food categories. These intake estimates were used to calculate the margins of safety for the flavoring agents to determine whether adequate margins of safety would still exist in the event that the MSDIs used by JECFA to evaluate the safety of flavoring substances underestimated daily intakes. Based on the calculation of the margins of safety using the MSDI values, 99.9% of the 809 flavoring substances evaluated by JECFA have margins of safety of greater than 100. In comparison, 98% of the flavoring substances have margins of safety of greater than 100 when the margins of safety were calculated according to the PADI values. The results indicate that if the MSDI estimates used by JECFA for the evaluation of the safety of flavoring substances were underestimated, a wide margin of safety exists for all but a few of the flavoring substances even when intakes were estimated from the PADI values.

Although the Procedure for the Safety Evaluation of Flavoring Agents does not require toxicological data in every instance, but rather relies primarily on structure–activity relationships to assess safety, this study confirmed that on the basis of a more traditional toxicological process of evaluating chemical safety, the procedure showed that the evaluated substances did not present a concern for safety at current estimates of intake.[38]

13.5.4 Toxic Substances

It is difficult to distinguish clearly between the essential metals (micronutrients) and the toxic substances in human diet. Nearly all metals are toxic for humans if they get absorbed by the organism at high concentrations. However, it is possible to differentiate between those elements that are known definitely as essential and those that cause toxicosis at extremely low concentrations and those that do not possess known therapeutic normal functions. The most dangerous heavy metals, which occur nearly everywhere and particularly in agricultural products are mercury, lead, and cadmium. Other toxic metals are arsenic, boron, beryllium, selenium, and others. The most probable sources of food contamination with toxic substances are soil, agricultural chemicals and fertilizers, contaminated water from industrial wastes, the various food-processing stages, plants that can absorb toxic substances from the contaminated land and store them in their tissues, as well as the muddy depositions of pit wastes.

The commonest metals contaminating foods are detailed below:

13.5.4.1 Lead

The presence of lead in the human food chain constitutes a major hygienic problem in the world. It exists in every organ or tissue of the human body and varies between 100 and 400 mg or approximately 1.7 μg/g tissue.[39] Over 90% lead is present in the bones. It constitutes a physiological ingredient of human diet (Table 13.9). The daily food uptake of lead is estimated to be approximately equal to 100–300 μg, with particularly high percentages during intense environmental pollution.[40] Lead absorption contained in foods is estimated to be approximately equal to 10% for adults and 40% for children.[41] Many dietetic factors seem to affect the uptake levels of lead. Hence, low calcium levels, iron deficiency, and diets rich in carbohydrates but low in proteins as well as diets containing high levels of vitamin D lead to an increased absorption of lead. In the healthy adult approximately 90% of the absorbed lead gets excreted by urine and feces.

Lead infection can be distinguished as acute and chronic. Symptoms of the former include blood, nervous, gastroenteric, and hepatic failures.[41] In general, phenomena such as anorexia, indigestion, and constipation occur followed by colic with intense paroxysmal abdominal pain. Although lead encephalopathy has been observed in little children[42,43] very little is known regarding the symptoms

TABLE 13.9
Lead Concentration in Food and Drinks

Food	Range (mg/kg)	Average (mg/kg)
Cereals	<0.01–0.81	0.17
Meat and fish	<0.01–0.70	0.17
Fresh fruits	<0.01–0.76	0.12
Canned fruits	004–10.0	0.40
Fresh vegetables	<0.01–1.5	0.22
Canned vegetables	0.01–1.5	0.24
Milk	<0.01–0.08	0.03
Water	1–50[a]	5[a]
Alcoholic drinks	50–100[a]	

Sources: WHO, *Environmental Health Criteria 3: Lead*, World
Health Organization, Geneva, Switzerland, 1976; Reilly,
C., *Metal Contamination of Food*, 2nd edn, Elsevier
Applied Science, London, U.K., 1991.

[a] μg/L.

of chronic encephalopathy. Some of the known typical clinical symptoms are mild anemia, overactivity, aggressive behavior, mental disorder, peripheral neuropathy, paralysis, and renal failure.[40]

13.5.4.2 Mercury (Hg)

Mercury is considered as the most hazardous heavy metal met in the food chain due to its continuous presence in the environment, its bioaccumulation and transportation into the water chain and its high levels in a large variety of foods. It exists in three different forms: elementary mercury, mercuric mercury, and acetyl mercury. The chemical form affects largely the absorption, its distribution in the body tissues, and its biological half life. The standard human diet contains less than 50 μg mercury/kg food,[44] whereas seafood stands for the main source of mercury.

Mercury is a cumulative poison stored mainly in the liver and the kidneys. The concentration levels depend on the type of organism and the chemical form. Mercury, in its metallic form, is less absorbed, easily excreted from the organism, and hence it is improbable to cause poisoning. On the contrary, both organic and inorganic substances of mercury are particularly toxic for humans. Hence, methyl mercury is considered as one of the six most dangerous chemicals existing in the environment.[45] It is absorbed by the intestine, enters the bloodstream rapidly, and binds to the plasma proteins. Methyl mercury could prove to be neurotoxic for adults and embryos since it accumulates in the brain.[46] It can cause irreversible damage to the central nervous system leading to disorder, terror, restriction of the optical field, blindness, loss of hearing, and finally death.[41] Selenium seems to react to mercury poisoning in many animal species.[47]

The most important source of mercury or methyl mercury uptake by man is fishery products (fish and seafood) and mainly hunted fish, found on the top of the food chain, such as swordfish, tuna, bass, mackerel fish, and shark. Big hunted fish with the highest percentage of methyl mercury are often migratory and can be found in waters of high or low infection in mercury, during their life time.

The European Union has implemented Regulation 466/2001 of the Commission, according to which the maximum permitted tolerance values in total mercury in foodstuffs amounts to 0.5 mg/kg fresh weight product in fishery products, except for certain fishes where the maximum level is 1 mg/kg fresh weight product.

At an international level, in June 2003, the temporary permitted weekly uptake percentage was decreased from 3.3 μg/kg body weight and determined at 1.6 μg/kg body weight.

Mercury uptake greatly depends on the quantity and type of fish consumed and varies significantly between member states. According to data published by SCOOP (Scientific Co-operation on questions relating to food) the average rate of exposure to total mercury through nutrition is estimated to be 109 μg/kg food. In different member states the average day consumption of fishery products per person varies between 10 and 80 g (weekly consumption varies between 70 and 560 g). Hence, the weekly exposure to total mercury via nutrition varies from 7 to 61 μg/person and in a person weighing 60 kg the weekly uptake varies from 0.1 to 1 μg. People and children consuming large quantities of fishery products may take methyl mercury near or exceeding the determined temporary percentage of the weekly uptake.

Inclusion of the determination of heavy metals (mercury, cadmium, and lead) in fresh and frozen fish in the Official Food Control programs by member states is of outmost importance under the current circumstances.

The EFSA published an opinion in 2004 regarding the danger to human health from the consumption of food that may contain mercury. It is estimated that the European consumers' uptake of methyl mercury is close to the international uptake level. EFSA suggested the realization of additional studies, mainly in particular groups such as pregnant women, or women due to be pregnant, women suckling, and children. Until completion of these studies, EFSA recommended that consumers and sensitive population groups should get informed about mercury. According to EFSA's opinion these groups should not consume more than a small portion (<100 g) of a big hunting fish per week, such as shark or swordfish. Along the same line, tuna fish should not be consumed more than twice a week.

13.5.4.3   Cadmium (Cd)

Cadmium poisoning is very frequent due to its increased solubility in organic acids. It is considered as one of the most hazardous micronutrients in foods and the environment due to its high toxicity.[48] Levels of cadmium in food are normally very low if contamination has not occurred (Table 13.10). The total range varies between 0.095 and 0.987 mg/kg. Increased cadmium levels are mainly observed in meat and seafood. The daily consumption of cadmium has been estimated in many countries to be 10–80 μg/day.[49] According to WHO, cadmium level in drinkable water is defined to be 10 μg/L.[50] The most reported food poisoning cases due to cadmium come from Japan and Australia.[51]

It has been estimated that under normal conditions approximately 6% of cadmium existing in consumed foods, is absorbed by the human organism.[41] It has also been reported that high calcium and

TABLE 13.10
Cadmium Concentration in Selected Foods

Food	Cadmium Concentration (μg/kg)
Bread	<2–43
Potatoes	<2–51
Cabbage	<2–26
Apples	<2–19
Poultry	<2–69
Minced beef	<2–28
Kidney (sheep)	13–2000
Prawns	17–913
Seafood	50–3660
Drinkable water	<1–21 μg/L

Source: Reilly, C., *Metal Contamination of Food*, 2nd edn, Elsevier Applied Science, London, U.K., 1991; Dabeka, R.W. et al., *Food Addit. Contam.*, 4, 89, 1987; WHO, *International Standards for Drinking Waters*, 2nd edn, World Health Organization, Geneva, Switzerland, 1963.

protein levels in the diet tend to considerably increase cadmium absorption. The highest percentage of cadmium that gets absorbed is withheld in the kidneys. Hence, long-term cadmium absorption means serious kidney failure, as well as bone attack leading to brittleness and possible breakdown of the skeleton.[52] Cadmium toxicity constitutes the main cause of the disease itai-itai observed in some population groups in Japan.[51] The high levels of cadmium in the diet are considered responsible for the increase of the rate of different cancers in humans.[53] Cadmium toxicity is significantly inhibited by the presence of selenium, zinc, and cobalt.

13.5.4.4 Arsenic (As)

Arsenic is traditionally, closely linked to homicides and suicides. Its toxicity has to do with the chemical form of this element. Hence, inorganic compounds of arsenic are more toxic, followed by organic compounds and finally arsenic in gas form.[54] During the past, products based on arsenic such as parasiticides, insecticides, fungicides, wood preservatives, and other similar products were widely used. Nowadays, all the above are forbidden in many countries due to the proved toxicity of arsenic.

Arsenic currently exists in most foods because of its wide dispersion into the environment and its past use as a chemical in agriculture. With the exemption of seafood, it is usually present in foodstuffs at a concentration lower than 0.5 mg/kg. According to FAO/WHO the maximum permissible day consumption uptake level is prescribed at a concentration of 2 μg/kg body weight.[28] Arsenic also exists in all drinkable waters at a concentration varying between 0 and 0.2 mg/L. The maximum concentration in drinkable water is 0.01 mg arsenic/L.[55]

Arsenic with three or five valences contained in foods can be easily absorbed by the gastroenteric tube. Then it can be easily transported into all tissues and organs. It is mainly accumulated on the skin, hair, nails, and to a certain degree on the bones and the muscles. The total arsenic level in the human organism has been estimated to be 14–20 mg.[56]

Arsenic in general is a poison with its five valences form less toxic than that of the three valences form. It can bind to organic sulfhydryl groups so that it inhibits the action of many enzymes, especially those involved in cellular metabolism and respiration.[41] The clinical symptoms are associated with dilation and increased permeability of capillary tubes, especially in the intestine. Chronic arsenic poisoning means appetite loss leading to weight loss, gastrointestinal disorders, peripheral neuritis, conjunctivitis, hyperkeratosis, and skin melanosis. Moreover, it is a cancer suspect agent.[57]

13.5.4.5 Selenium (Se)

Despite being one of the essential micronutrients in human and animal diet, its uptake in particular in high concentrations often leads to the appearance of the toxic syndrome. Seleniosis of farm animals has been widely reported in many parts of the world.[41]

Selenium presence in the human food chain is mainly affected by its levels in agricultural soils. Hence, its daily uptake varies depending heavily on the geographical area. 50–200 μg/day were proposed as normal uptake selenium levels, with the exception of infants and children under 6 years of age where it is suggested as 10–40 μg/day and 20–120 μg/day, respectively.

Selenium is present in foodstuffs as the compounds selenium cysteine and selenium methionine. Approximately 80% of the organic selenium present in foodstuffs seems to be absorbed by the organism. Absorption is greater in foods of plant origin compared to foods of animal origin.[8]

The safety limit between selenium, as an important micronutrient in human diet, and the appearance of toxic symptoms is quite small. The main symptoms from the uptake of an increased selenium quantity are dermatitis, vertigo, fragile nails, gastric disorders, hair loss, and garlic odor during breathing.

13.5.4.6 Antimonium (Sb)

Antimonium is toxic and constitutes a major ingredient of many foodstuffs. High levels of antimonium in foods arise from contamination due to containers glazed with enamel containing antimonium. These containers might be used for food storage or as cooking utensils.

Very little is known about the nutritional uptake of antimonium. The daily uptake is about 0.25–1.25 mg for American children.[58] The allowed permissible concentration of antimonium in drinkable water is 0.1 mg/L. Antimonium is stored mainly in liver, kidneys, and skin. Poisoning symptoms include colic, nausea, weakness, and collapse with slow or irregular respiration and reduced body temperature.

13.5.4.7 Aluminum (Al)

Aluminum is widely used in many industrial applications. Aluminum compounds are used in the food industries both as additives and as different utensils (cooking utensils or storage containers). Moreover, they can be used in pharmaceutical and cosmetic industries. Its wide use is due to the low cost, it does not undergo oxidation, it is recyclable, flexible, and easy to handle and is extracted from bauxite, which is abundant in nature.

The concentration of aluminum in different foodstuffs varies from 0.05 to 129 mg/kg with an average concentration of approximately 12.6 mg/kg.[59] With the exception of certain spices and tea leaves, the normal aluminum levels in foodstuffs are very low. On the contrary, the increased uptake of antimonium by humans comes from its use as a food additive. Contamination could also occur from the use of aluminum containers and aluminum cans in the food industries. Water does not constitute a significant source of aluminum uptake.

The chemical form of aluminum affects aluminum absorption. Furthermore, the parathyroid hormones, vitamin D and iron seem to affect aluminum absorption.[41] In the human bloodstream aluminum is stored mainly in the liver, kidneys, spleen, bones, and heart and brain tissues.

13.5.4.8 Tin (Sn)

Tin is widely distributed in low quantities in most lands. It exists in lower than 1.0 mg/kg concentrations in all main food groups, except canned vegetables (9–80 mg/kg) and fruits (12–129 mg/kg).[60] However, these values today are considered too high since the industry uses polyacrylic resin as a coating agent. The use of tinned cans in the canning industry could be a source of contamination with tin.

Tin found in foodstuffs seems to be less absorbed by the human organism and is excreted mainly by stool.[21] Low quantities of the absorbed tin could be withheld by kidneys, liver, and bones. High tin levels in foodstuffs could cause acute poisoning. The lethal toxic dose for humans is 5–7 mg/kg body weight. Chronic poisoning leads to delay in growth, anemia, and histopathologic lesions in the liver. Moreover, tin affects iron absorption and the formation of haemoglobin.[41]

13.5.5 Medicinal Residues

The use of medicines as additives in animal nutrition is approved since the 1950s. These non-nutritional additives contain hormones, antibiotics, sulfonamides, nitrofurans, and arsenic compounds. These additives have a very significant effect on the increase in yield of dairy products.

Antibiotics and other medicines are given to the productive animals in therapeutic doses (200–1000 g medicine/ton food, 220–1100 mg/kg) for disease treatment, in preventative doses (100–400 g medicine/ton food, 110–440 mg/kg) for the prevention of diseases caused by bacteria or protozoa and finally at doses higher or equal than 200 g (2.2–220 mg/kg) medicine/ton food for 2 weeks or more with the aim to increase the food yield and the acceleration of animal growth.[61,62]

Despite the fact that the mechanisms by which medicines enrich the animal food and accelerate their growth are hardly tangible, the following have been reported concerning their use[61,63]:

- They suppress the microorganisms responsible for mild but non-recognizable infections.
- They reduce the microbial production of toxins that decelerate growth.
- Antimicrobial factors reduce the destruction of essential nutrients from microbes in the gastroenteric system or increase the biosynthesis of vitamins or other growth factors.
- They improve the absorption capacity and use of nutrients.

However, the use of these antimicrobial medicines could result in the presence of residues in foods of animal origin. Some representative examples of possible sources of contamination of dairy and meat products from medicinal residues is the extended use or the high dose of approved medicines, frequent mammary gland therapy with antibiotics, misuse of these medicines (e.g., the wrong mixing of contaminated milk with uncontaminated milk), twofold increase of the dose, use of dry period therapy in milking cows, and improper application of the directions of use of these medicines.[64–66]

Medicinal residues in foods should be avoided for the following reasons[67]:

- Some residues could cause temperamental reactions and in particular in sensitive groups of the population that could be very serious.
- In general, the presence of medicinal residues exceeding the determined allowed levels is illegal.
- Some medicinal residues in liquid dairy products have the ability to intervene in starter cultures that are used in processed dairy products.
- The presence of residues shows that the food could be coming from an animal that had suffered a severe infection.
- The concern of public regarding food safety, hygiene, and food quality has greatly increased nowadays.
- The most important argument is that the presence of residues leads to the formation of microorganisms resistant to medicines, which are human pathogens.

13.5.6 Germicides

The term germicides refers to a group of chemicals used at a world level in agricultural products with the aim to control, destruct, or inhibit weeds, insects, fungi, and other harmful plants or animals. Approximately 320 active ingredients of parasiticides are available in thousands of different combinations.[68]

Out of the 500,000 estimated plant or animal species and microorganisms, the well known harmful plants and animals are less than 1%. However, this small percentage is enough to cause large scale economic destruction. Hence, it is reported that insects, pathogenic plants, and weeds destruct approximately 37% of the American agricultural production, with losses reaching 50%–60% in the developing world. The total cost involved in challenging harmful plants and animals is too high since it is required to use huge amounts of germicides and becomes even higher if one includes the indirect cost incurred by the destruction of beneficial organisms, the perturbation of ecological systems, and the appearance of infections and diseases in humans.[69]

Germicides consist of herbicides, insecticides, and fungicides. There exist three main chemical groups of insecticides: organochlorine compounds, organophosphorous, and carbamides. A fourth group constitutes the synthetic pyrethroids (synthetic chemicals associated with natural pyrethroids abundant in the chrysanthemum). These compounds generally have a low toxicity in mammals, including humans, and most of them are biodegradable. The most common herbicides are organochlorine compounds, whereas the most common fungicides are pentachlorophenol, cadmium chloride, and commercially known as Benomyl, Captan, Maneb, Thiram, Zineb, etc. Most of them are based on genetic mutations and have heavy metals as their active ingredients.

Most germicides have a carcinogenic, teratogenic, and embryotoxic action at a high degree whereas some of them affect even the central nervous system. The organochlorine germicides bioaccumulate in human tissues from the liver enzymes without getting detoxified and excreted. The result is the production of epoxides and peroxides that cause membrane damage and lead to the formation of free radicals. These free radicals interact with DNA acting as a mutation material.[70] Moreover; organochlorine compounds restrict the transfer of inorganic material through the cell membranes and inhibit cell respiration. Organophosphoric compounds inhibit acetylcholinesterase, a key enzyme involved in the process of transfer of nervous impulses. These germicides are often particularly toxic and because of their volatile character, they have an immediate effect upon inhaling. They are more biodegradable compared to the organochlorine compounds; however, they are still suspects for chronic toxic effects. Moreover, carbamides are poisons of the nervous system with varying toxicity.

When germicides are not used in the right way, they result in the presence of residues in food and may cause severe problems in the consumers' health. Their use in agricultural products presents three relevant but distinct hazards that determine the possibility a failure or injury that might occur as described below[68]:

1. Environmental hazards, associated with unpleasant effects on non-target organisms and contamination of groundwater.
2. Occupational hazards associated with the farmers working in the fields, who breathe the germicides. These are the most significant ones, because human exposure to germicides is direct and hence, health damage is more probable.
3. The presence of residues in the foodstuffs.

Occupational hazards could be considerably reduced with strict controls and application of the required preventive technology. Environmental effects are associated with certain germicides, and more specifically heavy metal compounds and organochlorine compounds. These are particularly toxic and resistant to biodegradation.

Over the last few years, the public worry about the presence of germicide residues in foodstuffs has increased. The discovery that certain germicides, considered safe some years ago, cause hazardous effects in human health has increased anxiety even more. This pushed the U.S. National Research Council (NRC) in 1985 to start the study of methods to determine the residue limits of these germicides in foods. Hence, according to NRC[71] a very important parameter was used for expressing the oncogenic strength of a germicide. This is expressed as Q^* (tumors/mg germicide/kg body weight/day) and shows the possibility of a germicide to cause tumors in guinea pigs. Hence, a high value of Q^* shows a strong oncogenic reaction (i.e., formation of additional tumors) at a given dose. These estimations for different germicides are average values derived from many positive oncogenic studies in animals. The nutritional oncogenic hazard is estimated by multiplying Q^* with the exposure, e.g., food consumption × germicide residue.

Finally, it should be mentioned that there are tables comprising the limits that FAO/WHO have determined regarding the acceptable daily intake (ADI) and the maximum residue limits (MRLs).

13.5.7 Preservatives

13.5.7.1 Antimicrobial Substances

It is well known that the growth of microorganisms in foods is undesirable with the exception of the useful microbial fermentations in certain products that are essential for the product to acquire its organoleptic characteristics and its final structure. The development of undesirable microorganisms in foods leads to the change of appearance, structure, consistency, taste, color, and nutritional product value. Moreover, certain microorganisms are toxic for humans, causing infections or endotoxin formation following their growth in foods. Thus, it is required to inhibit the microorganisms' growth in foods with the aim of food preservation and enhancement and assurance of their quality (Table 13.11).

Food preservation has been accomplished through the application of physical or chemical methods.[17] The natural methods of inhibition or destruction of microorganisms consist of the change of temperature through the application of high (e.g., pasteurization, sterilization) or low temperatures (refrigeration, freezing), the exposure to ionizing radiation and water removal (e.g., drying). The chemical methods of food preservation include the use of desirable microorganisms

TABLE 13.11
Preservatives in Food; Name, Origin, Functionality, Adverse Effects, ADI, and Applications

E	Name	Origin	Functionality	Effects	ADI[a]	Product Uses
E200	Sorbic acid	Natural, from fruits or synthetic from ketene	Preservative	Skin irritant in direct contact	0–25	Yoghurt, soft drinks, sweets, wine
E210	Benzoic acid	Natural substance, synthetically manufactured	Preservative	Allergic reactions Not recommended for hyperactive children	0–5	Jams, beer, fruit juices, yoghurts, soft drinks
E211	Sodium benzoate	From benzoic acid	Preservative	Allergic reactions	0–5	caviar, sweets, sauces (soya, takos, barbeque), soft drinks
E212	Potassium benzoate	From benzoic acid	Preservative	Allergic reactions	0–5	Margarine, concentrated fruit juices
E213	Calcium benzoate	From benzoic acid	Preservative	Allergic reactions	0–5	Fruit condensed juices
E214	p-Hydroxybenzoic ethyl	From benzoic acid	Preservative	Allergic reactions Similar to E215, E216, E217, E218, E219	0–10	Beer, fruit juices, aromatic substances, syrups, yams
E216	p-Hydroxybenzoic propyl	From benzoic acid	Preservative	Similar to E 214		Similar to E 214
E218	p-Hydroxybenzoic methyl	Synthetic	Preservative	Causes allergic and skin reactions	0–10	Cooked beets, sauces chocolate fillings
E220	Sulfur dioxide	Chemically produced	Stabilizer, preservative, aromatic compound	Destroys vitamin B, possible gastric problems	0–0.7	Fruit salads, dry fruits, beer, gelatin, fruit juices, sausages
E221	Sodium thiosulphate	From sulfurous acid	Equipment sterilization, antioxidant	Allergic reactions Not recommended to those having problems with liver	0–0.7	Frozen prawns, potatoes, beer, and wine
E222	Sodium hydrogen sulfite	From sulfurous acid	Preservative	Allergic reactions in patients suffering from asthma	0–0.7	Beer, wine, milk, dairy products, fruit juices.
E223	Sodium bisulfite	Sulfurous acid salt	Antioxidant, preservative	Allergic and skin reactions	0–0.7	Frozen prawns, potatoes, alcoholic drinks, pickles, orange juice
E224	Potassium bisulfite	Synthetic	Preservative	Allergic reactions	0–0.7	Frozen potatoes, seafood, wines, fruit juices
E226	Calcium sulfite	Synthetic	Preservative	Allergic reactions	—	

E-number	Name	Source	Function	Effects	ADI[a]	Foods
E227	Calcium hydrogen sulfite	Synthetic	Preservative	Allergic reactions	—	Beer, jams, gel
E230	Diphenyl	Synthetic produced from benzene	Preservative	Exposure to diphenyl causes nausea, vomiting, eye and nose irritations	0–0.5	Processing of fruit skin
E236	Formic acid	Synthetically produced	Preservative	No consequences Forbidden in United Kingdom	0–3	No technological applications
E239	Hexamethyl enetetramine	Derived from formaldehyde and ammonia.	Preservative	Gastroenteric perturbation Possible carcinogenesis	0–0.15	Provolone cheese, marinated foods, and red herrings
E249	Potassium nitrite	Synthetic salt	Meat preservative	It enters the bloodstream and affects hemoglobin Headaches, methemoglobinemia If it reacts with amines it gives carcinogenic nitrosamines	0–0.2	Cooked meat, sausages, smoked fish
E250	Sodium nitrite	Synthetic salt	Preservative	Similar to E249	0–0.2	Meat products, sausages, bacon, frozen pizza
E251	Sodium nitrate	Natural salt	Preservative, color stabilizer	Nitric could be converted into nitrates Similar to E249	0–5	Bacon, ham, cheeses, frozen pizza
E252	Potassium nitrate	Natural salt or synthetic	Preservative, color stabilizer	Anemia, kidneys infection. Similar to E249	—	Sausages, cooked pork meats, Dutch cheeses, canned meat
E261	Potassium acetate	Synthetic	Natural colors preservative	Low toxicity, gets excreted in urine	—	Provolone cheese, marinated foods, and red herrings
E270	Lactic acid	Natural substance	Preservative	Gastroenteric disorders in infants	—	Margarine, carbonated drinks, infant milk
E281	Sodium propionate	Natural propionic acid salt	Preservative	Possible correlation with migraines	—	Processed cheese, bakery and dairy products
E282	Calcium propionate	Propionic acid salt	Preservative	Skin irritant in pure form in bakery workers	—	Processed cheese, dairy and baking products, frozen pizza.
E290	Carbon dioxide	Natural gas	Preservative packaging gas	In the stomach, it increases gastric juice secretions	—	Juices, soft drinks

[a] ADI, acceptable daily intake mg/kg body weight.

for fermentations or the direct use of chemical additives that act as antimicrobial factors. Moreover, some chemical additives such as salt, phosphoric salts, and antioxidants sometimes exert a direct or indirect antimicrobial action. In this chapter we will only refer to the chemicals used exclusively as antimicrobial substances in food, pharmaceuticals, or other materials.[17,72]

Nowadays, the needs of the modern market as well as the requirements of the modern consumer have led to the wide use of chemical antimicrobial substances in foods. In general, the modern marketing system is largely based on the use of antimicrobial substances.

There are limits for all antimicrobial substances that are used. Antimicrobial action depends on the following factors:

- The physical and chemical properties (e.g. solubility, toxicity)
- The types of microorganisms involved
- The type and the properties of the preserved product

The use of combinations of inhibitory factors is well known[73] as the principle of microbial hurdles. Finally, it seems that pH affects the range of the antimicrobial action of a chemical antimicrobial substance.

The use of antimicrobial substances is regulated and approved by competent authorities of each country while at an international level the FAO (Food and Agriculture Organization) and the WHO (World Health Organization) deal with such matters. Some of these antimicrobial substances are characterized as GRAS (Generally Recognized as Safe) by the FDA (Food and Drug Administration) and are acquitted from the food additives regulations. These substances should be approved for a specific use. Moreover, they must obey good manufacturing practices and be properly labeled.

Burdock and Carabin[74] reported on the history and description of GRAS, a system for review and approval of ingredients for addition to food. The GRAS approval process for a food ingredient relies on the judgment of "… experts qualified by scientific training and experience to evaluate its safety…" the end product of which is no better or worse than that by the FDA, but often more expeditious. The process and requirements for a successful GRAS determination are discussed and compared with that of the food additive petition (FAP) process. The future of the GRAS process is assured by its history of successful performance, bringing safe food ingredients to the consumer in a timely manner.

GRAS is probably now more useful than before in examining, for example: (1) macronutrients, that cannot be tested by conventional means (at more than 5% of the diet of test animals), (2) biotechnologically produced or novel foods and responding to the concept "as safe as" conventional food, and (3) high doses of nutrients (e.g., vitamins or specific foods such as meat in "Atkins-type" diets) for which no animal models exist.

The most important approved chemicals acting as antimicrobial substances are described below.

13.5.7.1.1 Sorbic Acid and Its Salts

Sorbic acid is a compound existing both in nature and produced synthetically. Nowadays, it is widely used; due to its low cost. Sorbic acid and its salts constitute very effective antimicrobial substances against many yeasts, moulds, and bacteria. Certain microorganisms such as lactic acid bacteria as well as specific yeasts and moulds are resistant to the inhibition by sorbic acid. The activity of sorbic acid against microorganisms is an operation derived from synergic or antagonistic effects with product composition, pH, water activity, chemical additives, storage temperature, microbial flora, atmospheric gases, and packaging.[75] A significant factor affecting the antimicrobial action of sorbic acid is the pH of the substrate. The maximum pH for activity is 6.5 whereas the measureable inhibition of microbial growth has been reported even at pH 7.0. Sorbic acid concentrations (<0.3%) can only inhibit and not inactivate microorganisms in foods. Higher concentrations can inactivate microorganisms; however, they cannot be used or allowed in foods, due to their unpleasant effect on product taste. Finally, it is well known that sorbic acid is one of the less harmful preservatives used in foods.

13.5.7.1.2   Benzoic Acid and Its Salts

Benzoic acid can be widely used as a preservative in the food, pharmaceutical, and cosmetics industries. They possess very powerful antimicrobial and antioxidant properties. In the food industries, they are widely used as preservatives in foods with pH less than 4.5 due to their low cost and easy incorporation into products. The narrow pH range in which they act, their undesirable taste, and their toxicological profile which is less desirable than other antimicrobial substances, constitute limitations in their use.[76] Concentrations of preservatives usually vary from 0.05% up to 0.1%. However, either lower concentrations are employed or they can be used in combination with other antimicrobial substances when taste problems are evident.[72,77] As antimicrobial substances, benzoic acid and its salts are more effective against yeasts and bacteria compared to moulds.[19,72] Their antimicrobial action varies depending on the food, its pH and water activity, as well as on the type and microorganism species. Some yeasts, which destroy food of medium humidity, seem to be resistant in the inhibition caused by benzoic acid and its salts.[76–81] Moreover, there are certain microorganisms capable of metabolizing benzoic acid. Reports on its use as an antimicrobial substance showed that these compounds do not cause unpleasant effects on human health.[19,72,76] It is generally considered as a safe food preservative.

13.5.7.1.3   Propionic Acid and Its Salts

They exist naturally in Swedish type cheeses at a percentage of 1%, derived from Propionibacterium that are involved in ripening of these cheeses. Besides the mould inhibition in cheeses, they can be used as preservatives in baked products, where they cause inhibition of fungi as well as some bacterial species. Moreover, they could be used as antimicrobial factors at levels between 0.1% and 0.38%. Their antimicrobial action depends on the type of microbe and the product pH (highest activity at pH = 6.0). Propionic acid and its salts can be categorized as GRAS when used in concentrations that do not surpass the normal amount required for the accomplishment of the desired effect.

13.5.7.1.4   Esters of Parahydroxybenzoic Acid

They are used as antimicrobial substances in cosmetics, medicines, and in foods. Compared to benzoic acid their solubility in water is higher and decreases with the increase of the number of carbon atoms. These compounds are stable in air and resistant to cold and heat, and steam sterilization. They are often used in concentrations varying between 0.05% and 0.1%. They exert antimicrobial action against yeasts, moulds, and bacteria. The microbial inhibition increases when the carbon chain of the ester is not branched. The use of these compounds is favored in high pH products where other antimicrobial factors are ineffective. They could also be effectively used in conjunction with benzoic acid depending on the cost, pH, and taste, and more particularly in products with slightly acidic pH.[72]

Methyl paraben (CAS No. 99-76-3) is a methyl ester of p-hydroxybenzoic acid. It is a stable, nonvolatile compound used as an antimicrobial preservative in foods, drugs, and cosmetics for over 50 years. Methyl paraben is readily and completely absorbed through the skin and from the gastrointestinal tract. It is hydrolyzed to p-hydroxybenzoic acid, conjugated, and the conjugates are rapidly excreted in the urine. There is no solid evidence of accumulation. Acute toxicity studies in animals indicate that methyl paraben is practically nontoxic by both oral and parenteral routes. In a population with normal skin, methyl paraben is practically nonirritating and non-sensitizing. In chronic administration studies, no-observed-effect levels (NOEL) as high as 1050 mg/kg were reported and a no-observed-adverse-effect level (NOAEL) in the rat of 5700 mg/kg is posited. Methyl paraben is not carcinogenic or mutagenic. It is not teratogenic or embryotoxic and is negative in the uterotrophic assay. The mechanism of cytotoxic action of parabens may be linked to mitochondrial failure dependent on induction of membrane permeability transition accompanied by the mitochondrial depolarization and depletion of cellular ATP through uncoupling of oxidative phosphorylation.

Parabens are reported to cause contact dermatitis reactions in some individuals on cutaneous exposure. Parabens were implicated in numerous cases of contact sensitivity associated with cutaneous exposure but the mechanism of this sensitivity is still unknown.[82] Sensitization occurred when medications containing parabens were applied to damaged or broken skin. Allergic reactions to ingested parabens were reported, although rigorous evidence of the allergenicity of ingested paraben is lacking.

13.5.7.1.5 Sulfur Dioxide and Salts of Sulfuric Acid

Sulfuric acid is one of the most traditional antimicrobial substances, widely employed in products such as wines.[72,83] It is soluble in water and gives sulfurous acid and its ions when it comes into contact with water in foods. The ions percentage increases with the reduction of pH value. Sulfuric acid salts can be used as preservatives and are more useful since they are available in dry forms. The antimicrobial action against yeasts and moulds and bacteria is selective, with certain species to be more sensitive in inhibition compared to other species.[72] Bacteria are generally more sensitive in inhibition.[19] Sulfur dioxide and sulfuric acid salts are GRAS substances; however, their level of use in wines is restricted to 0.035%. Its presence at higher levels lead to undesired taste and is generally forbidden in foodstuffs that are considered as sources of thiamine due to the fact that they inactivate it.[19,72,84–86]

13.5.7.1.6 Carbon Dioxide

It is a gas that solidifies at −78.5°C, thus forming dry ice. It is used in the maintenance of carbonated drinks, vegetables, fruits, meat, fish, and wines. The dry ice form can be used to store or transport products at low temperatures. As an ingredient that modifies temperatures in order to store fruits and vegetables, it delays respiration and ripening, and inhibits the growth of yeasts and moulds. It can also be used in carbonated drinks as a fermentation factor and as an inhibitor of microbial growth. It inhibits oxidation changes in beer.[72] The antimicrobial activity depends on its concentration, the type of microorganisms, water activity, and storage temperature. Since these factors vary, carbon dioxide could not exert any action, stimulate or inhibit growth, or be lethal for microorganisms.[72,87–90] From the toxicological point of view, it is well known that the respiration of 30%–60% carbon dioxide in the presence of 20% oxygen could cause death in animals, whereas man's exposure in atmospheres containing more than 10% could lead to the loss of senses.[72] Moreover, respiration of smaller quantities for an extended period of time could be deleterious.

13.5.7.1.7 Epoxides

Ethylene and propylene oxides are cyclic ethers with only one oxygen atom connected with two neighboring carbon atoms of the same chain.[72,89] Ethylene oxide gas, could go through most of the organic materials without causing any damage, and hence it very useful for the sterilization of materials sensitive to temperature. Epoxides are drastic against yeasts, moulds and insects, but less severe against bacteria. The antimicrobial action of these two epoxides depends on the humidity and their ease of penetration into organic materials.

13.5.7.1.8 Hydrogen Peroxide

Hydrogen peroxide is an oxidative as well as discolorative agent with antimicrobial properties. It is an unstable compound and dissociates easily to form water and oxygen. The dissociation and the antimicrobial activity of this compound increases at higher temperatures. The antimicrobial action arises from its oxidative properties and depends on the concentration, pH, temperature, and exposure time.[90–92] Besides its use as an additive, hydrogen peroxide is produced by lactobacilli in foods. It is allowed to be used in cheese manufacturing, whey powder processing, and in other applications.[11,89,93,94]

13.5.7.2 Antioxidants

Antioxidants are substances used in food preservation together with antimicrobial substances. It is important to realize that besides microbial destruction oxidative degradation of polyunsaturated fatty acids contributes significantly at the time of food preservation. Fatty acid oxidation is a complex process of chemical and biological reactions leading to the formation of a large number of products. Fat oxidation results in changes in taste and aroma of foods with a high concentration in oils and fats, changes in their structure due to the reaction of products derived from oxidation of fats with proteins and loss of their nutritional value due to destruction of vitamins, amino acids, and essential fatty acids.[95] Furthermore, fat oxidation products are directly related to the development of a number of diseases such as arteriosclerosis, coronary heart disease, and cancer, as well as the process of aging of cells.

Antioxidants play an important role in the deceleration of lipid oxidation reactions in foodstuffs. According to FDA they are defined as "substances used as preservatives, with the aim to reduce spoilage, rancidity or food discoloration, which are derived from oxidations." Addition of antioxidants in foodstuffs is either intentional (direct addition into product) or symptomatic (migration of antioxidants from packaging material into product). The right and effective use of antioxidants depends on the understanding of (a) the chemistry of oils and fats, and (b) the mechanism of oxidation and their operation as substances, which result in food oxidation (Table 13.12).[96]

The approval of an antioxidant for food use requires extended toxicological studies that consist of studies for possible transformation, carcinogenic, and teratogenic effects. Finally, there are some antioxidant substances that have been recognized as safe (GRAS).

Twelve food additives and six secondary plant products were analyzed with regard to their antioxidant activity by using three different test systems (Trolox equivalent antioxidant capacity, photochemiluminescence, ferric reducing antioxidant power) as reported by Kranl et al.[97] The results differed depending on the assay. All the food additives showed antioxidant activities comparable to the calibration substance Trolox. In contrast, the secondary plant products had up to 16 times higher antioxidant potential. This might be a good reason for the food industry to opt for natural antioxidants instead of synthetic ones in order to get storage stability for the processed food items—which, according to recent surveys, is in the interest of the consumers.

Lipid peroxidation is a reaction occurring in food products that leads to off-flavor and other quality losses (e.g., changes in color and texture). To stabilize products, the food industry uses food additives endowed with antioxidant activity. Typical antioxidants are butylated hydroxytoluene (BHT) and butylated hydroxyanisole (BHA) as well as ascorbic acid and their derivatives. During the last decade, natural antioxidants have been gaining ground at the expense of synthetic ones. Secondary plant products, widely known for their health promoting effects, have been evaluated for their use as food ingredients with stabilizing effects (Figure 13.1).[98]

An overview on most of the analytical methods existing to determine the antioxidant activity was compiled by Bohm and Schlesier.[99]

The three recent EU directives that fixed the maximum permitted levels (MPL) for food additives for all member states[100] also include the general obligation to establish national systems for monitoring the intake of these substances in order to evaluate their safe use. Leclercq et al.[101] considered additives with primary antioxidant technological function or those with an acceptable daily intake (ADI) that was established by the Scientific Committee for Food (SCF): gallates, butylated hydroxyanisole (BHA), butylated hydroxytoluene (BHT), and erythorbic acid. The potential intake of these additives in Italy was estimated by means of a hierarchical approach using either step by step or more refined methods. The likelihood of the current ADI to be exceeded was very low for erythorbic acid, BHA, and gallates. On the other hand, the theoretical maximum daily intake (TMDI) of BHT was above the current ADI. The three food categories found to be the main potential sources of BHT were "pastry, cake, and biscuits;" "chewing gums;" and "vegetables oils and margarine,". They overall contributed 74% of the

TABLE 13.12
Antioxidants in Food; Name, Origin, Functionality, Adverse Effects, ADI, and Applications

E	Name	Origin	Functionality	Effects	ADI[a]	Product Uses
E302	L-Ascorbic calcium	Synthetic salt	Preservative, antioxidant	Low possibility for stone formation in the kidneys	—	Madrilene
E310	Galic propyl ester	Synthetic substance	Antioxidant	Gastric and skin irritations. Not allowed in baby foods	0–0.5	Vegetable oils, margarine, snacks, gums
E311	Galic oktylester	Similar to E310	Similar to E310	Similar to E310	0–0.5	Similar to E310
E312	Galic dodecyl ester	Similar to E310	Similar to E310	Similar to E310	0–0.5	Similar to E310
E320	Butylic hydroxyanisole (BHA)	Synthetic derived from p-methoxynenol and isobutane	Antioxidant	Not allowed in baby foods. Forbidden in Japan	0–5	Biscuits, sweets, sauces, chips, soft drinks, margarine
E321	Butylic hydroxytoluene (BHT)	Synthetic derived from cresol and isobutylene	Antioxidant	In pigs, 1 g causes death in 2 weeks. Not allowed in baby foods	0–0.5	Gums, chips, margarine, peanuts, sauce cubes, mash
E325	Lactic sodium	Lactic acid salt	Humectant. It maintains pH levels	None	—	Cheese, gel, jam, ice creams, margarine
E326	Lactic potassium	Similar to E311	Similar to E311	Similar to E311	0–0.5	Similar to E325

[a] ADI, acceptable daily intake mg/kg body weight.

TMDI. The actual use of BHT in these food categories is discussed, together with other aspects such as losses of this substance in the technological process and percentage of ingestion in the case of chewing gums.

13.5.7.3 Antibiotics in Animal Feed

Animals and humans constitute overlapping reservoirs of resistance, and consequently the use of antimicrobials in animals can have an impact on public health. For example, the occurrence of vancomycin-resistant enterococci in food-animals is associated with the use of avoparcin, a glycopeptide antibiotic used as a feed additive for the growth promotion of animals. Vancomycin resistant enterococci and vancomycin resistance determinants can therefore spread from animals to humans. The bans on avoparcin and other antibiotics as growth promoters in the EU have provided scientists with a unique opportunity to investigate the effects of the withdrawal of a major antimicrobial with respect to the selective pressure on the occurrence and spread of antimicrobial resistance. Data shows that although the levels of resistance in animals and food, and consequently in humans, were markedly reduced after the termination of use, the effects on animal health and productivity were very minor.[102]

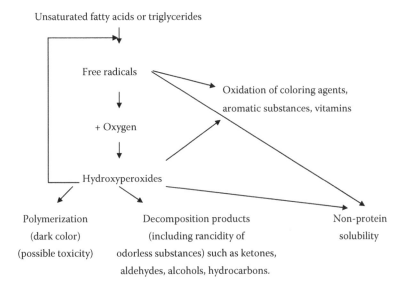

Unsaturated fatty acids or triglycerides

Free radicals

Oxidation of coloring agents,
aromatic substances, vitamins

+ Oxygen

Hydroxyperoxides

Polymerization	Decomposition products	Non-protein
(dark color)	(including rancidity of	solubility
(possible toxicity)	odorless substances) such as ketones,	
	aldehydes, alcohols, hydrocarbons.	

FIGURE 13.1 Mechanisms of fatty acid oxidation.

Food animal production in intensive systems and, within those, various types of chemical additives included in the compound diets fed has received much adverse attention from the media, consumers, politicians, legislators, and advisory groups in recent years according to Pugh.[103] No additive class received more adverse comment than the antibiotics used for the purpose of enhancing the efficiency of animal production. Pugh[103] considered the safety of the antibiotic feed additives (AFAs) against the background of the regulatory measures in place, defined their role, and described the relevant concerns. Regarding the health risk that underpinned the AFA bans he made a warning over the precedent created by the use of the precautionary principle in the recent banning of six of their number.

McEvoy[104] reviewed the legislative framework controlling the use of veterinary medicines and zootechnical feed additives in the EU. From a contamination perspective, "problem" compounds include sulfonamides, tetracyclines, nitroimidazoles, nitrofurans, ionophore coccidiostats, and nicarbazin. The literature on each of these was reviewed and examples of interventions to minimize contamination were given. Examples of contaminants include naturally occurring and synthetic toxic environmental compounds (e.g. mycotoxins and dioxins) that may contaminate feed raw materials. Zootechnical feed additives and veterinary medicines may also contaminate unmedicated feeding stuffs due to carry over during feed production. Contaminated feed can cause deleterious health effects in the animals and, through "secondary exposure" of consumers to products derived from these animals, may be harmful to people.

13.5.8 PACKAGING MATERIALS

Materials widely used in food packaging consist of paper, cellulose products, cellophane and metals (aluminum, tin, stainless steel, etc.), glass, plastic, and various materials such as wood and textiles. Most of these materials have been used for years and have caused minor problems. On the contrary, plastics bring into contact with food a large number of chemical ingredients, which have not been used before by the food industry; however, they offer new advantages as new packaging materials. Migration of these substances, being used in plastics processing, into food causes the biggest worry and concern regarding food safety problems and dangers.

All plastics, besides the basic polymer derived from the oil industry, consist of a number of other substances either added intentionally during their preparation or processing or found inside them unavoidably as residues from polymerization reactions. Polymers are high molecular weight compounds, inert and with restricted solubility in aqueous and oily systems, hence it is improbable that they migrate into food at a high concentration.[105] Even if packaging fragments get consumed by mistake, they will not react with body fluids present in the gastroenteric system. The concern on safety due to the use of plastics as packaging materials mainly arises from the possible toxicity of other lower molecular weight compounds that may appear in the final product and get filtered in food during its storage.

The two main reasons of food adulteration from plastic packaging materials are the following:

- Polymerization residues that consist of monomers, oligomers (with molecular weight higher than 200), catalysts (mainly metallic salts and organic peroxides), solvents, emulsifiers and water proof materials, impurities, infectious plant materials, inhibitors, and degradation products (Table 13.13).
- The various auxiliary processing materials, such as antioxidants, plasticizers, light and heat stabilizers, coloring agents, lubricants, fungicides, etc.

TABLE 13.13
Emulsifiers in Food; Name, Origin, Functionality, Adverse Effects, ADI, and Applications

E	Name	Origin	Functionality	Effects	ADI[a]	Product Uses
E400	Alginic acid	Natural extract	Insoluble in cold water	None	0–5	Flans, yoghurt, jams
E407	Carageenan	Extract from algae	Stabilizer, leavening agent	Maybe not safe in big quantities	—	Condensed milk, chocolate milk, ice cream, yoghurt, biscuits, and toothpastes
E412	Guar gum	Φυτικό εκχύλισμα	Stabilizer, leavening agent	Big quantities cause nausea	—	Παγωτά, σάλτσες, γιαούρτι, τυρί κρέμα
E414	Acacia Gum	Natural substance	Stabilizer, leavening agent	None	—	Wine, beer, canned vegetables
E420	Sorbitol	Natural substance	Stabilizer, sweetener	Big quantities cause flatulence, abdominal pains, diarrhea	—	Chocolate, sweets, ice creams, jams for diabetics, ready cakes, gums
E422	Glycerin (Glycerol)	Natural substance Also synthetically manufactured	Aromatic	None	—	Jams, gel, ice creams, desserts, drinks
E461	Methylcellulose	From cellulose	Emulsifier, stabilizer, leavening agent	Big quantities might cause nausea	—	Drinks, dietetic products, sauces, soft drinks

[a] ADI, acceptable daily intake mg/kg body weight.

The more volatile the monomers, the lower their concentration with time, however, low levels of these monomers could continue to be found in the final product.[105,106] Styrene and acrylonitrile residues are generally removed with higher difficulties.

13.6 TOXICOLOGICAL CONCERNS

The Threshold of Toxicological Concern (TTC) refers to the possibility of establishing a human exposure threshold value for all chemicals, below which there is no significant risk to human health. The concept that exposure thresholds can be identified for individual chemicals in the diet, below which no appreciable harm to health is likely to occur, is already widely embodied in the practice of many regulatory bodies in setting acceptable daily intakes (ADIs) for chemicals for which the toxicological profile is known. However, the TTC concept goes further than this in proposing that a *de minimis* value can be identified for any chemical, including those of unknown toxicity, taking the chemical structure into consideration.

This concept forms the scientific basis of the U.S. Food and Drug Administration (FDA) 1995 Threshold of Regulation for indirect food additives. The TTC principle has also been adopted by the Joint FAO/WHO Expert Committee on Food Additives (Commission of the European Communities, 1997; JECFA, 1997)[107] in its evaluations of flavoring substances.[108]

Hattan and Kahl[109] reported on the current developments in food additive toxicology in the United States. They mentioned that a recently published proposal (Fed. Reg. 66 (2001) for mandatory submission of information on all plant-derived bioengineered foods fed to humans or animals will be reviewed. Under this proposal, information such as data on identity, level and function of the introduced substance(s); an estimate of dietary exposure; allergenic potential of the protein; data relevant to other safety issues that may be associated with the substance; selection of a comparable food; historic uses of comparable food; composition and characteristics of bioengineered food versus those of the comparable food should be provided. In addition, characterization of the parent plant; construction of the transformation vector and introduced genetic material along with number of insertion sites and genes; data on the genetic material and any newly inserted genes for antibiotic resistance should be submitted with the notification. The Interagency Coordinating Committee for Validation of Alternative Methods (ICCVAM) was identified by the U.S. Congress as the organization to review and validate new alternative toxicological test methods for 14 U.S. government agencies. The validated and accepted alternative toxicity tests will be incorporated into toxicity testing recommendations for regulatory agencies (Table 13.14).

13.7 DO ADDITIVES CAUSE HEALTH PROBLEMS?

Crohn's disease is a chronic granulomatous inflammation of the gastrointestinal tract that was first described in the beginning of the twentieth century. The histological similarity with intestinal tuberculosis has led to the assumption of an involvement of mycobacteria and mycobacterial antigens, respectively, in the etiology. A major defense mechanism against mycobacterial lipid antigens is the CD1 system that includes the CD1 molecules for antigen presentation and the natural killer T cells for recognition and subsequent production of cytokines like interferone-gamma and tumor necrosis factor-alpha. These cytokines promote granulomatous transformation. Various food additives, especially emulsifiants, thickeners, surface-finishing agents, and contaminants like plasticizers share structural domains with mycobacterial lipids. It was therefore hypothesized by Traunmuller[110] that these compounds are able to stimulate by molecular mimicry the CD1 system in the gastrointestinal mucosa and to trigger the pro-inflammatory cytokine cascade (Table 13.15).

These are diglycerides of long-chained fatty acids esterified with acetic, lactic, citric, or tartaric acid (E472a–d) and fatty acid esters of propylene glycol (E477), polyglycerol (E476), or polyoxyethylene sorbitan (E432–E436). They are frequently used in the industrialized production of sweets, cream desserts, sauces, spreads, and margarine. The increased consumption of the latter

TABLE 13.14

Leavening Agents in Food; Name, Origin, Functionality, Adverse Effects, ADI, and Applications

E	Name	Origin	Functionality	Effects	ADI[a]	Product Uses
E503	Ammonium carbonate	Synthetic	Leavening agent	Stimulation of muciferous glands	—	Cocoa, baby foods
E510	Ammonium chloride	Synthetic	Flour improver	Should be avoided by patients suffering hepatic failure and liver cirrhosis due to diuretic properties	—	Bakery products
E514	Sodium sulfate	Natural (from minerals)	Flour improver	Due to sodium increase in the body it should be avoided by patients suffering cardiac and kidney failure	—	Brewing
E518	Magnesium sulfate	Natural (seawater)	Hardening and noncaking properties	It could cause toxic phenomena in patients suffering kidney failure if it does not get excreted	—	Table salt and pharmaceuticals
E540	Calcium pyrophosphate	Natural (minerals) or artificial	Nutrition supplement	Danger in perturbation of the balance between phosphorous-calcium	0–70	Processed cheeses and healthy diet products
E541	Aluminum sodium phosphate (acidic)	Artificial	Flour improver	Accumulates in brain nervous tissues of patients with Alzheimer	0–6	Cakes and fries
E541	Aluminum sodium phosphate (basic)	Συνθετικό	Emulsifier and leavening agent	Similar to E541 (acidic)		Similar to E541 (acidic)

E-number	Name	Origin	Function	Effect / Danger	ADI[a]	Food products
E544	Polyphosphoric calcium	Artificial	Emulsifier and leavening agent	Danger in perturbation of the balance between phosphorous-calcium	0–70	Dairy products
E545	Polyphosphoric ammonium	Artificial ammonium	Complex properties	Similar to E544	—	Imported salt, garlic and onion powder, sugar fine, and chewing gums
E553	B-talk (French chalk)	Natural	Anticaking and leavening properties	Wheezing and cough due to chemical inflammation of small bronchi		
E554	Aluminum sodium silicate	Natural or/and artificial	Non-agglomerating agent	Neurotoxic, causing Alzheimer's disease	—	Dried food in powder form, salt, hard cheeses in pieces, rice
E556	Calcium-aluminum silicate	Natural (mineral)	Non-agglomerating agent	Similar to E554		Rice, chewing gums, hard cheeses in pieces, salt

[a] ADI, acceptable daily intake mg/kg body weight.

TABLE 13.15
Suggested Additives to Be Banned from School Meals

Colorings
E102—Tartrazine
E104—Quinoline Yellow
E107—Yellow 2G
E110—Sunset Yellow
E120—Cochineal
E122—Carmoisine
E123—Amaranth
E124—Ponceau 4R
128—Red 2G
E131—Patent Blue V
E132—Indigo Carmine
133—Brilliant Blue FCF
E151—Black PN
154—Brown FK
155—Brown HT

Preservatives
E211—Sodium benzoate
E212—Potassium benzoate
E213—Calcium benzoate
E214—Ethyl-4-hydroxybenzoate
E215—Ethyl-4-hydroxybenzoate sodium salt
E216—Propyl 4-hydroxybenzoate
E217—Propyl 4-hydroxybenzoate sodium salt
E218—Methyl 4-hydroxybenzoate
E219—Methyl 4-hydroxybenzoate sodium salt
E220—Sulfur dioxide
E221—Sodium sulfite
E222—Sodium hydrogen sulfite
E223—Sodium metabisulfite
E224—Potassium metabisulfite
E226—Calcium sulfite
E227—Calcium hydrogen sulfite
E250—Sodium nitrite
E251—Sodium nitrate

Flavorings/Enhancers
E621—Monosodium glutamate
E635—Sodium 5-ribonucleotide

Sweeteners
Aspartame
Acesulfame K
Sodium saccharine

was already suspected to play a role in the etiopathogenesis of CD.[111] Furthermore, waxes of animal or vegetable origin like beeswax, carnauba wax, and candelilla wax (E901–E914) used as surface-finishing agents for sweets and citrus fruits contain long-chained fatty acid esters.[112]

Other candidates are dialkyl phthalate esters (di[2-ethylhexyl] phthalate, DEHP; diisononyl phthalate, DINP), used as plasticizers for polyvinyl chloride (PVC) since the 1930s.[113,114] They were found in concentrations up to 45% in children's toys and teething rings made of soft PVC and migrate into the saliva during chewing and mouthing activities. Ready-to-serve meals are heavily contaminated with plasticizers when PVC gloves were worn during cooking and packaging.[115]

The understanding of Crohn's disease as a CD1-mediated delayed-type hypersensitivity to certain food additives would lead to strong emphasis on a dietary treatment.

Van den Brandt et al.[116] concluded that epidemiology can contribute significantly to hazard identification and characterization, and exposure assessment. Epidemiologic studies directly contribute data on risk (or benefit) in humans as the investigated species, and in the full food intake range normally encountered by humans. Areas of contribution of epidemiology to the risk assessment process are identified, and ideas for tailoring epidemiologic studies to the risk assessment procedures are suggested, dealing with data collection, analyses, and reporting of both existing and new epidemiologic studies. The paper by Van den Brandt et al.[116] described a scheme to classify epidemiologic studies for use in risk assessment, and deals with combining evidence from multiple studies. Using a matrix approach, the potential contribution to each of the steps in the risk assessment process is evaluated for categories of food substances. The contribution to risk assessment of specific food substances depends on the quality of the exposure information.

Sartkaya and Cakır[117] evaluated four food preservatives (sodium nitrate, sodium nitrite, potassium nitrate, and potassium nitrite) and then five combinations at a concentration of 25 mM for genotoxicity in the somatic mutation and recombination test (SMART) of *Drosophila melanogaster*. Three-day-old larvae, trans-heterozygous including two linked recessive wing hair mutations (multiple wing hairs and flare) were fed at different concentrations of the test compounds (25, 50, 75, and 100 mM) in standard Drosophila Instant Medium. The wings of the emerging adult flies were scored for the presence of spots of mutant cells, which can result from either somatic mutation or mitotic recombination. Moreover, lethal doses of food preservatives used were determined in the experiments. A positive correlation was observed between total mutations and the number of wings having undergone mutation. For the evaluation of genotoxic effects, the frequencies of spots per wing in the treated series were compared to the control group, which is distilled water. Chemicals used were ranked as sodium nitrite, potassium nitrite, sodium nitrate, and potassium nitrate according to their genotoxic and toxic effects. Moreover, the genotoxic and toxic effects produced by the combined treatments considerably increased, especially when the four chemicals were mixed. This study has revealed that the correct administration of food preservatives/additives may have a significant effect on human health.

Although potassium sorbate (PS), ascorbic acid, and ferric or ferrous salts (Fe-salts) are used widely in combination as food additives, the strong reactivity of PS and oxidative potency of ascorbic acid in the presence of Fe-salts might form toxic compounds in food during its deposit and distribution. Kitano et al.[118] evaluated the reaction mixture of PS, ascorbic acid, and Fe-salts for mutagenicity and DNA-damaging activity by means of the Ames test and rec-assay. Effective lethality was observed in the rec-assay. No mutagenicity was induced in either *Salmonella typhimurium* strains TA98 (with or without S-9 mix) or TA100 (with S-9 mix). In contrast, a dose-dependent mutagenic effect was obtained when applied to strain TA100 without S-9 mix. The mutagenic activity became stronger, increasing with the reaction period. Furthermore, the reaction products obtained in a nitrogen blanket did not display any mutagenic and DNA-damaging activity. PS, ascorbic acid, and Fe-salts were inactive when they were used separately. Omission of one component from the mixture of PS, ascorbic acid, and Fe-salt turned the reaction system inactive. These results demonstrate that ascorbic acid and Fe-salt oxidized PS and the oxidative products caused mutagenicity and DNA-damaging activity.

Nowadays there is a strong shift toward natural compounds. The interest in the possible use of natural compounds to prevent microbial growth has notably increased in response to the consumer pressure to reduce or eliminate chemically synthesized additives in foods. Minimally processed fruits are an important area of potential growth in rapidly expanding fresh cut produce. However, the degree of safety obtained with the currently applied preservation methods seems to be insufficient. Lanciotti et al.[119] gave an overview of the application of natural compounds, such as hexanal, 2-(E)-hexenal, hexyl acetate and citrus essential oils, to improve the shelf life and the safety of minimally processed fruits as well as their mechanisms of action.

Plants and plant products can represent a source of natural alternatives to prolong the shelf life and the safety of food. Some of these volatile compounds were found to play a key role in the defense systems of fresh produce against decay causing microorganisms. Some of these compounds are produced throughout the lipoxygenase pathway that catalyzes the oxygenation of unsaturated fatty acids, forming fatty acid hydroperoxides.

Aldehydes and the related alcohols are produced by the action of hydroperoxide lyases, isomerases, and dehydrogenases. Many of the natural aromas of fruits and vegetables responsible for their "green notes", such as hexanal, hexanol, 2-(E)-hexenal, and 3-(Z)-hexenol, are carbon compounds formed through this pathway.

These compounds are also important constituents of the aroma of tomatoes, tea, strawberry, olive oil, grape, apples, and pear. Moreover, plant essential oils, composed mainly by terpenoids, were extensively studied for their antimicrobial activity against many microorganisms including several pathogens.[120] In particular, the activity of oils from Labiatae and citrus fruits were investigated.

According to the Hyperactive Children's Support Group and Organics Research the additives shown in Table 13.15 should be banned from school meals.

13.8   RISK ASSESSMENT OF FOOD ADDITIVES

Edler et al.[121] presented a review on the mathematical methods and statistical techniques presently available for hazard assessment. Existing practices of the JECFA, FDA, EPA, etc. were examined for their similarities and differences. A framework is established for the development of new and improved quantitative methodologies. They concluded that mathematical modeling of the dose–response relationship would substantially improve the risk assessment process. An adequate characterization of the dose–response relationship by mathematical modeling clearly requires the use of a sufficient number of dose groups to achieve a range of different response levels. This need not necessarily lead to an increase in the total number of animals in the study if an appropriate design is used. Chemical-specific data relating to the mode or mechanism of action and/or the toxicokinetics of the chemical should be used for dose–response characterization whenever possible. It is concluded that a single method of hazard characterization would not be suitable for all kinds of risk assessments, and that a range of different approaches is necessary so that the method used is the most appropriate both for the data available and for the risk characterization. Future refinements to dose–response characterization should incorporate more clearly the extent of uncertainty and variability in the resulting output.

FAO/WHO encourages member countries to develop national food control measures based on risk assessment in order to assure a proper protection level to consumers and facilitate fair trade. This is particularly important for developing countries as WTO members because it is clearly stated in the Sanitary and Phytosanitary Measures (SPS) Agreement that: (a) SPS measures should be based on risk assessment techniques developed by relevant international organizations; and (b) Codex standards which are based on risk assessment are regarded as the international norm in trade dispute settlement. When conducting risk assessment on food chemicals (including additives and contaminants) in developing countries, in most cases it is not necessary to conduct their own hazard characterization because the ADIs or PTWIs of food chemicals developed by international expert groups (e.g., JECFA) are universally applicable and also developing countries do not have the resources to repeat those expensive toxicological studies. On the other hand, it is necessary to conduct exposure assessment in developing countries because exposure to food chemicals varies from country to country.[122] This is not only crucial in setting national standards, but also very important for developing countries to participate in the process of developing the Codex standards. In addition to food standard development, risk assessment is equally useful in setting up priorities in imported food inspection and evaluating the success of various food safety control measures.

Renwick et al.[123] presented a review of risk characterization, the final step in risk assessment of exposures to food chemicals. The report is the second publication of the project "Food Safety in Europe: Risk Assessment of Chemicals in the Food and Diet (FOSIE)."[124] The science underpinning the hazard identification, hazard characterization, and exposure assessment steps has been published in a previous report.[125] Risk characterization is the stage of risk assessment that integrates information from exposure assessment and hazard characterization into advice suitable for use in decision making. The focus of this review was primarily on risk characterization of low molecular weight chemicals, but consideration was also given to micronutrients and nutritional supplements, macronutrients, and whole foods. Problem formulation, as discussed here, is a preliminary step in risk assessment that considers whether an assessment is needed, who should be involved in the process and further risk management, and how the information will provide the necessary support for risk management. The report described good evaluation practice as an organizational process and the necessary condition under which risk assessment of chemicals should be planned, performed, scrutinized, and reported. The outcome of risk characterization may be quantitative estimates of risks, if any, associated with different levels of exposure, or advice on particular levels of exposure that would be without appreciable risk to health, e.g. a guidance value such as the acceptable daily intake (ADI). It should be recognized that risk characterization is often an iterative and evolving process.

ADI is derived from NOAEL or other starting point, such as the benchmark dose (BMD), by the use of an uncertainty or adjustment factor. In contrast, for non-threshold effects a quantitative hazard estimate can be calculated by extrapolation, usually in a linear fashion, from an observed incidence within the experimental dose–response range to a given low incidence at a low dose. This traditional approach is based on the assumption that there may not be a threshold dose for effects involving genotoxicity. Alternatively, for compounds that are genotoxic, advice may be given that the exposure should be reduced to the lowest possible level.

A case-by-case consideration and evaluation is needed for hazard and risk characterization of whole foods. The initial approach to novel foods requires consideration of the extent to which the novel food differs from any traditional counterparts, or other related products, and hence whether it can be considered as safe as traditional counterparts/related products (the principle of substantial equivalence). As for macronutrients, epidemiological data identifying adverse effects, including allergic reactions, may also exist. Human trials on whole foods, including novel foods, will only be carried out when no serious adverse effects are expected.

REFERENCES

1. EFSA, EFSA assesses new aspartame study and reconfirms its safety. Press Release, Parma, May 5, 2006.
2. Markarian, J. Additives in food packaging, *Plast. Addit. Compound.*, 4(4), 16–21, 2002.
3. Salih, F.M., Risk assessment of combined photogenotoxic effects of sunlight and food additives, *Sci. Total Environ.*, 362(1–3): 68–73, 2006.
4. Dubos, R.J., The effect of organic acids on mammalian tubercle bacilli, *J. Exp. Med.*, 92, 319, 1950.
5. Rice, A.C. and Pederson, C.S., Factors influencing growth of *Bacillus coagulans* in canned tomato juice and specific organic acids, *Food Res.*, 19, 124, 1954.
6. Gardner, W.H., Acidulants in food processing, In: *Handbook of Food Additives*, 2nd edn, Vol. 1, Furia, T.E., Ed. CFC Press, Cleveland, OH, pp. 225–270, 1972.
7. CFR, Code of Federal Regulations Title 21, Office of the Federal Register. U.S. Government Printing Office, Washington, DC, 1988.

8. Young, V, R., Nahapetian, A., and Janghorbani, M., Selenium bioavailability with reference to human nutrition, *Am. J. Clin. Nutr.*, 35, 1076, 1982.

9. FAO/WHO, Toxological evaluation of certain food additives with a review of general principles and of Specifications, 17th Report of the Joint Food and Agriculture Organization of the United Nations/ World Health Organization Expert Committee on Food Additives, WHO Tech. Report Ser. No. 539, FAO Nutrition Meetings Report Ser. No. 53, 1973.

10. Gardner, W.H., *Food Acidulants*. Allied Chemical Corporation, New York, 1966.

11. Lewis, R.J. Sr., *Food Additives Handbook*. Van Nostrand Reinhold, New York, 1989.

12. FAO/WHO, Specifications for the identity and purity of food additives and their toxicological evaluation: Emulsifiers, stabilizers, bleaching and maturing agents, 7th Report of the Joint Food and Agriculture Organization of the United Nations/World Health Organization Expert Committee on Food Additives, WHO Tech, Report Ser. No. 281, FAO Nutrition Meetings Report Ser. No. 35, 1963.

13. Tuft, L. and Ettelson, L.N., Canker sores from allergy to weak organic acid (citric and acetic), *J. Allergy*, 27, 536, 1956.

14. Wiseman, R.D. and Adler, D.K., Acetic acid sensitivity as a cause of cold urticatia, *J. Allergy*, 27, 50, 1956.

15. Weil, A.J. and Rogers, H.E., Allergic reactivity to simple aliphatic acids in man, *J. Invest. Dermatol.*, 17, 227, 1951.

16. Palmer, A.A., Two fatal cases of poisoning by acetic acid, *Med. J. Aust.*, 1, 687, 1932.

17. Sofos, J.N. and Busta, F.F., Sorbic acid and sorbates. In: *Antimicrobials in Foods*, 2nd edn, Davidson P.M. and Branen A.L., Eds. Marcel Dekker, New York, pp, 49–94, 1993.

18. FAO/WHO, Evaluation of certain good additives, 18th Report of the Joint Food and Agriculture Organization of the United Nations/World Health Organization Expert Committee on Food Additives, WHO Tech. Report Ser. No. 557, FAO Nutrition Meetings Report Ser. No. 54, 1974.

19. Chichester, D.F. and Tanner, F.W. Jr., Antimicrobial food additives. In: *Handbook of Food Additives*, 2nd edn, Vol. 1, Furia, T.E., Ed. CRC Press, Cleveland, OH, pp. 115–184, 1972.

20. Lück, E., *Antimicrobial Food Additives, Characteristics, Uses, Effects*. Springer-Verlag, Berlin, Germany, 1980.

21. WHO, Trace elements in human nutrition, Tech. Rep. Ser. No 532. World Health Organization, Geneva, Switzerland, 1973.

22. U.S. FDA, Termination of provisional listing of FD&C Red No. 1. Title 21, Code of Federal Regulation, Part 81, Sect. 81.10, Office of the Federal Register. General Services Administration, Washington, DC, 1983b.

23. Allamark, M.G., Grice, H.C., and Lu, F.C., Chronic toxicity studies on food colors. Part I. Observations on the toxicity of FD&C Yellow No. 3 (Oil Yellow AB) and FD&C Yellow No. 4 (Oil Yellow OB) in rats, *J. Pharm. Pharmacol.*, 7, 591, 1955.

24. Hansen, W.H. et al., Chronic oral toxicity of ponceau 3R, *Toxicol. Appl. Pharmacol.*, 5, 105, 1963.

25. Harrow, L.S. and Jones, J.H. The decomposition of azo colors in acid solution, *J. Assoc. Off. Agric. Chem.*, 37, 1012, 1954.

26. Radomaski, J.L. and Harrow, L.S., The metabolism of 1-(o-talylazo)-2-naphthy-lamine (Yellow OB) in rats, *Ind. Med. Surg.*, 35, 882, 1966.

27. Clayson, D.B. and Gardner, R.K., Carcinogenic aromatic amine and related compounds. In: *Chemical Carcinogens*, Searle, C.E., Ed., pp. 366–461. American Chemical Society, Washington, DC, 1976.

28. CAC, Contaminants, Joint FAO/WHO Food Standards Program, Codex Alimentarius, Vol, XVII. World Health Organization, Geneva, Switzerland, 1984.

29. Tennant, D.R. and O'Callaghan, M., Survey of usage and estimated intakes of annatto extracts, *Food Res. Int.*, 38, 911, 2005.

30. Marmion, D.M., *Handbook of U.S. Colorants for Foods, Drugs and Cosmetics*. Wiley, New York, 1984.

31. Newsome, R.L., Natural and synthetic coloring agents. In: *Food Additives*, Branen, A.L., Davidson, P.M., and Salminen, S., Eds. Marcel Dekker, New York, 1990.

32. Rayner, P., Colors, In: *Food Additive User's Handbook*, Smith, J., Ed. Blackie, London, U.K., 1991.

33. Sankaranaranan, R, Food colors, *Ind. Food Packer*, 35, 25, 1981.

34. Francis, F.J., Pigments and other contaminants, In: *Food Chemistry*, Fennema, O.R., Ed. Marcel Dekker, Inc., New York., 1985.

35. Breithaupt, D.E., Simultaneous HPLC determination of carotenoids used as food coloring additives: Applicability of accelerated solvent extraction, *Food Chem.*, 86, 449, 2004.

36. Panek, J., Davıdek, J., and Jehlickova, Z., *Natural Toxic Compounds of Foods. Formation and Change during Food Processing and Storage*, Davıdek J., Ed. CRC Press, Boca Raton, FL, 1995, p. 195.

37. Simko, P., Determination of polycyclic aromatic hydrocarbons in smoked meat products and smoke flavoring food additives, *J. Chromatogr. B*, 770, 3, 2002.
38. Munro, I.C. and Danielewska-Nikiel, B., Comparison of estimated daily intakes of flavoring substances with no-observed-effect levels, *Food Chem. Toxicol.*, 44, 758, 2005.
39. Barry, P.S., Lead concentration in human tissues, *Br. J. Industr. Med.*, 32, 119, 1975.
40. WHO, *Environmental Health Criteria 3: Lead*. World Health Organization, Geneva, Switzerland, 1976.
41. Reilly, C., *Metal Contamination of Food*, 2nd edn. Elsevier Applied Science, London, U.K., 1991.
42. NAS, *Airborne Lead in Perspective*. National Academy of Sciences, Washington, DC, 1972.
43. Reilly, A. and Reilly, C., Patterns of lead pollution in the Zambian environment, *Med. J. Zambia*, 6, 125, 1972.
44. Bouquiaux, J., *CEC European Symposium on the Problems of Contamination of Man and His Environment by Mercury and Cadmium*, CID, Luxemborg, 1974.
45. Bennet, B.G., Six most dangerous chemicals named, Monitoring and Assessment Research Center, London, U.K., on behalf of UNEP/ILO/WHO International Program on Chemical Safety, 1984.
46. Berlin, M.H., Clarkson, T.W., and Frieberg, L.T. Maximum allowable concentrations of mercury compounds, *Arch. Environ. Health*, 6, 27, 1963.
47. Stoewsand, G.S., Bache, C.A., and Lisk, D.J., Dietary selenium protection of methylmercury intoxication of Japanese quail, *Bull. Environ. Contam. Toxicol.*, 11, 152, 1974.
48. Vos, G., Hovens, J.P.C., and Delft, W.V., Arsenic, cadmium, lead and mercury in meat, livers and kidneys of cattle slaughtered in The Netherlands during 1980–1985, *Food Addit. Contam.*, 4, 73, 1987.
49. Dabeka, R.W., McKenzie, A.D., and Lacroix, G.M.A. Dietary intakes of lead, cadmium, arsenic and fluoride by Canadian adults, a 24-hours duplicate diet study, *Food Addit. Contam.*, 4, 89, 1987.
50. WHO, *International Standards for Drinking Waters*, 2nd edn. World Health Organization, Geneva, Switzerland, 1963.
51. Rayment, G.E., Best, E.K., and Hamilton, D.J., Cadmium in fertilizers and soil amendments, *Chemistry International Conference*, Aug 28–Sept 2, 1989. Royal Australian Chemical Institute, Brisbane, Australia, 1989.
52. Frieberg, L. et al., *Cadmium in the Environment*, 2nd edn. CRC Press, Boca Raton, FL, 1974.
53. Browning, E., *Toxicity of Industrial Metals*. Butterworths, London, U.K., 1969.
54. Buck, W.B., Toxicity of inorganic and aliphatic organic arsenicals. In: *Toxicity of Heavy Metals in the Environment*, Ohm, F.W., Ed. Marcel Dekker, New York, pp. 357–369, 1978.
55. Drinking Water Standards, Public Health Service Pub. No. 956. U.S. Government Printers, Washington, DC, 1962.
56. Schroeder, H.A. and Balassa, J.J., Abnormal trace metals in man: arsenic, *J. Chron. Dis.*, 19, 85, 1966.
57. IARC, *Evaluation of Carcinogenic Risk of Chemicals to Man, Some Inorganic and Organometallic Compounds*, Vol. 2. International Agency for Research on Cancer, Lyon, France, 1973.
58. Murthy, G.K., Rhea, U., and Peeler, J.R., Antimony in the diet of children, *Environ. Sci. Technol.*, 5, 436, 1971.
59. Pennington, J.A.T. and Jones, J.W., Aluminum in American diets. In: *Aluminum in Health, A Critical Review*, Gitelman, H.J., Ed. Marcel Dekker, New York, pp. 67–100, 1988.
60. Sherlock, J.C. and Smart, G.A., Tin in foods and the diet, *Food Addit. Contam.*, 1, 277, 1984.
61. Moorman, M.A. and Koenig, E., Antibiotic residues and their implications in foods, *Scope*, 7, 4, 1992.
62. NAS, *The Effects on Human Health of Subtherapeutic Use of Antimicrobials in Animal Feeds*. National Academy of Sciences, Washington, DC, 1980.
63. Franco, D.A., Webb, J., and Taylor, C.E., Antibiotic and sulfonamide residues in meat: Implications for human health, *J. Food Protect.*, 53, 178, 1990.
64. Booth, J.M., and Harding, F., Testing for antibiotic residues in milk, *Vet. Rec.*, 119, 565, 1986.
65. Jones, G.M. and Seymour, E.H., Cowside antibiotic residue testing, *J. Dairy Sci.*, 71, 1691–1699, 1988.
66. McEwen, S.A., Black, W.D., and Meek, A.H. Antibiotic residue prevention methods, farm management, and occurrence of antibiotic residues in milk, *J. Dairy Sci.*, 74, 2128–2137, 1991.
67. Brady, M. S. and Katz, S. E., Antibiotic/antimicrobial residues in milk, *J. Food Protect.*, 51, 8, 1988.
68. Hotchkiss, J.H., Pesticide residue controls to ensure food safety, *Crit. Rev. Food Sci. Nutr.*, 31, 191, 1992.
69. Pimentel, D., The dimensions of the pesticide question. In: *Ecology, Economics, Ethics: The Broken Circle*, Birmann, F.H. and Kellert, S.R., Eds. Yale University Press, New Haven, CT, pp. 59–69, 1991.
70. Pryor, W.A., *Free Radicals in Biology*, Vol. IV. Academic Press, New York, 1980.
71. NRC, *Regulating Pesticides in Food, The Delaney Paradox*. National Academy Press, Washington, DC, 1987.

72. Sofos, J.N. and Busta, F.F., Chemical food preservatives. In: *Principles and Practice of Disinfection, Preservation and Sterilization*, 2nd edn, Russell, A.D., Hugo, W.B., and Ayliffe G.A.J., Eds. Blackwell Scientific Publications, London, U.K., pp, 351–397, 1992.

73. Leistner, L., Hurdle technology applied to meat products of the shelf stable product and intermediate moisture food types. In: *Properties of Water in Foods*, Sinators D. and Multon J.C., Eds. Martinus Nijhoff Publishers, Dordrecht, the Netherlands, pp. 309–329, 1985.

74. Burdock, G.A. and Carabin, I.G., Generally recognized as safe (GRAS): History and description, *Toxicol. Lett.* 150, 3, 2004.

75. Sofos, J.N. and Busta, F.F., Antimicrobial activity of sorbate, *J. Food Protect.*, 44, 614, 1981.

76. Chipley, J.R., Sodium benzoate and benzoic acid. In: *Antimicrobials in Foods*, 2nd edn, Davinsonand, P.M. and Branen A.L., Eds. Marcel Dekker, New York, pp. 11–48, 1993.

77. Jermini, M.F.G. and Schmidt-Lorenz, W., Activity of Na-benzoate and ethyl-paraben against osmotolerant yeasts at different water activity values, *J. Food Protect.*, 50, 920, 1987.

78. Warth, A.D., Mechanism of resistance of Saccharomyces Bailii to benzoic, sorbic and other weak acids used as food preservatives, *J. Appl. Bacteriol.*, 43, 215–230, 1977.

79. Warth, A.D., Resistance of yeast species to benzoic and sorbic acids and to sulfur dioxide, *J. Food Protect.*, 48, 564, 1985.

80. Warth, A.D., Effect of benzoic acid on growth yield of yeasts differing in their resistance to preservatives, *Appl., Environ. Microbiol.*, 54, 2091, 1988.

81. Warth, A.D., Relationships among cell size, membrane permeability, and preservative resistance in yeast species, *Appl., Environ. Microbiol.*, 55, 2995, 1989.

82. Soni, M.G. et al., Evaluation of the health aspects of methyl paraben: A review of the published literature, *Food Chem. Toxicol.*, 40, 1335, 2002.

83. Banks, J.G., Nychas, G.J., and Board, R.G. Sulphite preservation of meat products. In: *Preservatives in the Food, Pharmaceutical and Environmental Industries,* Board, R.B., Allwood, M.C., and Banks J.G., Eds. Society for Applied Bacteriology Technical Series No. 22, Blackwell Scientific Publications, Oxford, pp. 17–33, 1987.

84. Daniel, J.W., Preservatives. In: *Food Toxicology-Real or Imaginary Problem?* Gimson G.G. and Walker R., Eds. Taylor & Francis, London, U.K., pp. 229–237, 1985.

85. Walker, R., Toxicological aspects of food preservatives. In: *Nutritional and Toxicological Aspects of Food Processing*, Walker R. and Quarttrucci, E., Eds. Taylor & Francis, London, U.K., pp. 25–49, 1988.

86. Ough, C.S., Sulfur dioxide and sulfites. In: *Antimicrobials in Foods*, 2nd edn, Davidson P.M. and Branen A.L, Eds. Marcel Dekker, New York, pp. 137–190, 1993a.

87. Enfors, S.O. and Molin, G. The influence of high concentrations of carbon dioxide on the germination of bacteria spores, *J. Appl. Bacteriol.*, 45, 279, 1978.

88. Clark, D.S. and Takacs, J., Gases as preservatives. In: *Microbial Ecology of Foods*, Vol. I, *Factors Affecting Life and Death of Microorganisms.* International Commission on Microbiological Specifications of Foods, Academic Press, New York, pp. 170–192, 1980.

89. Davidson, P.M. et al., Naturally occurring and miscellaneous food antimicrobials. In: *Antimicrobials in Foods*, Branen, A.L. and Davinson, P.M., Eds. Marcel Dekker, New York, pp. 371–419, 1983.

90. Foegeding, P.M. and Busta, F.F. Effect of carbon dioxide, nitrogen, and hydrogen gases on germination of Clostridium botulinum spores, *J. Food Protect.*, 46, 987, 1983.

91. Smith, Q.J. and Brown, K.L., The resistance of dry spore of Baccilus subtilis var, globibii (NCIB 80958) to solutions of hydrogen peroxide in relation to aseptic packaging, *J. Food Technol.*, 15, 169, 1980.

92. El-Gendy, S.M. et al., Survival and growth of Clostridium species in the presence of hydrogen peroxide, *J. Food Protect.*, 43, 431, 1980.

93. Stevenson, K.E. and Shafer, B.D., Bacterial spore resistance to hydrogen peroxide, *Food Technol.*, 37(11), 111, 1983.

94. Cords, B.R. and Dychdala, G.R., Sanitizers: Halogens, surface-active agents, and peroxides. In: *Antimicrobials in Foods*, 2nd edn, Davinson P.M., and Branen A. L., Eds. Marcel Dekker, New York, pp. 469–538, 1993.

95. Dziezak, J.D., Antioxidants, the ultimate answer to oxidation, *Food Technol.*, 40(9), 94, 1986.

96. Stuckey, B.N., Antioxidants as food stabilizers. In: *CRC Handbook of Food Additives*, Vol. I, Furia, T.E., Ed. CRC Press, Boca Raton, FL, pp. 185–223, 1972.

97. Kranl, K. et al. Comparing antioxidative food additives and secondary plant products—Use of different assays, *Food Chem.*, 93, 171, 2005.

98. Britt, C. et al., Influence of cherry tissue on lipid oxidation and heterocyclic aromatic amine formation in ground beef patties, *J. Agric. Food Chem.*, 46, 4891, 1998.

99. Bohm, V. and Schlesier, K., Methods to evaluate the antioxidant activity. In: *Production Practices and Quality Assessment of Food Crops*, Dris, R. and Jain, S.M., Eds., *Quality Handling and Evaluation*, Vol. 3. Kluwer Academic Publishers, Dordrecht, the Netherlands, pp. 55–71, 2004 (1-4020-1700-6).

100. European Commission, European Parliament and Council Directive No. 95/2/EC of 20 February 1995 on additives other than colours and sweeteners for use in foodstuffs. Official Journal No. L61, 18.3.95.

101. Leclercq, C., Arcella, D., and Turrini, A. Estimates of the theoretical maximum daily intake of erythorbic acid, gallates, butylated hydroxyanisole (BHA) and butylated hydroxytoluene (BHT) in Italy: A stepwise approach, *Food Chem. Toxicol.* 38, 1075, 2000.

102. Wegener, H.C., Antibiotics in animal feed and their role in resistance development, *Curr. Opin. Microbiol.*, 6, 439, 2003.

103. Pugh, D.M., The EU precautionary bans of animal feed additive antibiotics, *Toxicol. Lett.*, 128, 35, 2002.

104. McEvoy, J.D.G., Contamination of animal feedingstuffs as a cause of residues in food: A review of regulatory aspects, incidence and control, *Anal. Chim. Acta*, 473, 3, 2002.

105. Crosby, N.T., *Food Packaging Materials*. Applied Science Publications., London, U.K.., 1981.

106. Crompton, T.R., *Additive Migration from Plastics into Food*. Pergamon Press, Oxford, 1979.

107. Joint FAO/WHO Expert Committee on Food Additives, List of substances scheduled for evaluation and request for data. Rome, 2002.

108. Barlow, S.M. et al., Threshold of toxicological concern for chemical substances present in the diet, Report of a workshop held in Paris, France, 1999, Organized by the ILSI Europe Threshold of Toxicological Concern Task Force, *Food Chem. Toxicol.*, 39, 893, 2001.

109. Hattan, D.G. and Kahl, L.S., Current developments in food additive toxicology in the USA, *Toxicology*, 181, 417, 2002.

110. Traunmuller, F., Etiology of Crohn's disease: Do certain food additives cause intestinal inflammation by molecular mimicry of mycobacterial lipids? *Med. Hypotheses*, 65, 859, 2005.

111. Guthy, E., Morbus Crohn und Nahrungsfette, *Dtsch Med. Wochenschr.*, 107, 71, 1982.

112. Hamilton J.K., Ed., *Waxes: Chemistry, Molecular Biology and Functions*. The Oily Press, Dundee, Scotland, pp. 1–90, 257–310, 1995.

113. Bouma K. and Schakel D.J., Migration of phthalates from PVC into saliva simulant by dynamic extraction, *Food Addit. Contam.*, 19, 602, 2002.

114. Latini, G., De Felice, C., and Verrotti, A., Plasticizers, infant nutrition and reproductive health, *Reprod. Toxicol.*, 19, 27, 2004.

115. Tsumura, Y. et al., Eleven phthalate esters and di(2-ethylhexyl) adipate in one-week duplicate diet samples obtained from hospitals and their estimated daily intake, *Food Addit. Contam*, 18, 449, 2001.

116. Van den Brandt, P. et al., Evaluation of national assessments of intake of annatto extract (bixin). Evaluation of certain food additives and contaminants (44th report of the Joint FAO/WHO Expert Committee on Food Additives). WHO Technical Report Series No. 859, pp. 485–492, 1995a.

117. Sarıkaya, R. and Cakır, S., Genotoxicity testing of four food preservatives and their combinations in the *Drosophila* wing spot test, *Environ. Toxicol. Pharmacol.*, 20, 424, 2005.

118. Kitano, K. et al., Mutagenicity and DNA-damaging activity caused by decomposed products of potassium sorbate reacting with ascorbic acid in the presence of Fe salt, *Food Chem. Toxicol.*, 40, 1589, 2002.

119. Lanciotti, R. et al., Use of natural aroma compounds to improve shelflife and safety of minimally processed fruits, *Trends Food Sci. Technol.*, 15, 201, 2004.

120. Delaquis, P.J., Stanich, K., Girard, B., and Mazza, G., Antimicrobial activity of individual and mixed fractions of dill, cilantro, coriander and eucalyptus essential oils, *Int. J. Food Microbiol.*, 74, 101, 2002.

121. Edler, L. et al., Mathematical modelling and quantitative methods, *Food Chem. Toxicol.*, 40, 283, 2002.

122. Chen, J., Challenges to developing countries after joining WTO: Risk assessment of chemicals in food, *Toxicology*, 198, 3, 2004.

123. Renwick, A. et al., Risk characterisation of chemicals in food and diet, *Food Chem. Toxicol.*, 41, 1211, 2003.

124. Barlow, S., Dybing, E., Edler, L., Eisenbrand, G., Kroes, R., Van den Brandt, P. Eds., Food safety in Europe (FOSIE): Risk assessment of chemicals in food and diet, *Food Chem. Toxicol.*, 40(2/3), 137, 2002.

125. CCIC, Guidelines for good manufacturing practices: Use of certified FD&C colors in food, Certified Color Industry Committee, *Food Technol.*, 22(8), 14, 1968.

126. Van den Ouweland, G.A.M. and Swaine, J.R.L., Investigation of the species specific flavor of meat, *Perfumer Flavorist*, 5, 15, 1980.

Legislation

Council Directive 1996, (96/51/EC of 23 July 1996) amending Directive 70/524/EEC concerning additives in feeding stuffs. *Off. J. Eur. Comm.*, L 235, 39–58.

Directive 89/107/EEC (OJ L 40, 11.2.89, pp. 27–33.

Directive 89/398/EEC (OJ L 186, 30.6.89, pp. 27–32).

Directive 91/321/EEC on infant formulae and follow-on formulae (OJ L 175, 4.7.91, pp. 35–49).

Internet Addresses

1. http://europa.eu.int/comm./food/food/animalnutrition/feedadditives/authoadditivesen.htm (accessed 20/03/2009).
2. http://europa.eu.int/comm./food/food/animalnutrition/feedadditives/update.pdf (accessed 20/03/2009).
3. http://europa.eu.int/comm./consumers/cons_safe/prodsafe/gpsd/rapex weekly/2004week 30.htm (accessed 20/03/2009).
4. http://www.cspinet.org/reports/chemcuisine.htm (accessed 20/03/2009).
5. http://www.cfsan.fda.gov/1 (accessed 20/03/2009).
6. http://www.x-sitez.com/allergy/additives/colors 100-181.htm (accessed 20/03/2009).
7. http://www.x-sitez.com/allergy/additives/preservatives 200-290.htm (accessed 20/03/2009).
8. http://www.x-sitez.com/allergy/additives/296-385.htm (accessed 20/03/2009).
9. http://www.x-sitez.com/allergy/additives/vege400-495.htm (accessed 20/03/2009).
10. http://www.x-sitez.com/allergy/additives/misa500-579.htm (accessed 20/03/2009).

14 Nutrigenomics and Nutrigenetics

Abdullah Ekmekci and Meltem Yalinay Cirak

CONTENTS

14.1 RELATIONSHIP BETWEEN GENES, NUTRITION, AND HEALTH: GENE–NUTRIENT INTERACTIONS

The discipline of genetics endeavors to understand the functions and effects of genes whereas the term genomics deals with the functions and effects of all the genes in the genome of organisms. The science of genomics aims to understand the entire genome using experimental approaches and applies to common conditions, such as cancer, infectious diseases (e.g., human immunodeficiency), cardiovascular diseases, and neurological disorders such as Parkinson's disease, Alzheimer's, and bipolar disorder. Multiple genes and environmental factors can impact the development of these diseases [1].

Nutritional genomics (nutrigenomics) is a combination of molecular nutrition and genomics that provides a junction between health, diet, and genomics [2]. The science of nutrigenomics exhibits the influence of bioactive food components on the genetics and gene expression of the individuals by investigating the interaction between genes, nutrition, and life style [3] (Figure 14.1). By affecting the genes, bioactive components can influence the concentration of specific proteins that are directly or indirectly related to the disease. Essential (calcium, selenium, zinc, folate, and some vitamins) and nonessential bioactive food components are known to affect cellular processes related to health, such as hormonal balance, carcinogen metabolism, cell signaling pathways, control of cell cycle, apoptosis, and angiogenesis immensely [4]. Nutrigenomics is concerned with the effect of nutrients on the genome, the transcriptome, the proteome, and the metabolome.

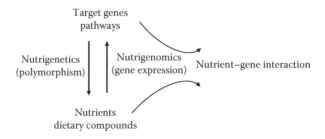

FIGURE 14.1 Nutrient–gene interactions.

Nutrigenetics can be considered as a subset of nutrigenomics and aims to clarify the different responses of the allelic variations in human DNA to the bioactive food components. While nutrigenomics helps for an optimal diet, nutrigenetics identifies the optimal diet for a given individual, i.e., a "personalized nutrition" [5].

The basic principles of nutrigenomics can fairly explain its range [6]. This is summarized here. Under certain circumstances, inappropriate diet can be a risk factor for some chronic diseases in some individuals. Dietary chemicals can alter either the gene expression and the gene itself, or the structure of the genome. The influence of diet on health depends on the genetic structure of an individual. Some genes and their common variants can be regulated by diet and play a role in chronic diseases. Eventually, dietary interventions based on the knowledge of nutritional requirement, nutritional status, and genotype can be used to optimize health and cure chronic diseases or improve individualized nutrition. An optimal diet can also influence the aging process [7].

Metabolites derived from lipids, carbohydrates, vitamins, hormones, and other components that are present in tissues, cells, and fluids at a set time point are the focus of study in metabolomics [8]. Metabolomics is a developing diagnostic tool for the metabolic classification of individuals, and has been employed in preclinical and clinical research for environmental, biomedical, and in toxicological applications [9]. Metabolomics explains the changes in the responses of cellular micronutrients to diet, due to the amount of bioactive food molecules and exposure time. Furthermore, human metabolic profiles can change in just few days due to smoking, alcohol, physical activities, and other environmental factors. The quantity of all the endogen and exogen nutrient-derived metabolites that exist in the body fluids such as urine and serum has not been specified yet [10].

14.1.1 Nutrients and Gene Expression: Nutritional Transcriptomics

A genome refers the entire set of genes. A transcriptome is the complete set of RNA produced by the genome, whereas a proteome is the set of proteins produced by the transcriptome. The study of the transcriptome is known as transcriptomics. The regulation of the rate of transcription of genes by food components represents an intriguing site for the regulation of an individual's phenotype [3]. There are approximately 33,000 genes in the human haploid genome. The functions of these genes can be controlled by activating or suppressing them at different times and situations in different cells, by means of covalent epigenetic changes, such as alternative splicing, exon shuffling, and methylation of the cytosines and histone proteins in the DNA; diverse proteins can be produced in four- to five fold numbers in these genes. Proteomics addresses three categories of biological interest: protein expression, protein structure, and protein function [11]. Miscellaneous ligand–receptor combinations such as peptide–protein, protein–protein, antigen–antibody, enzyme–substrate, membrane receptor–ligand or protein–DNA/RNA can be used for protein arrays.

A nutrient is defined in the postgenomic era as a fully characterized (physical, chemical, and physiological) constituent of a diet, natural or designed, that serves as a significant energy-yielding substrate for all the functions of the entire organism [12]. Approximately 40 micronutrients are required in the human diet. Suboptimal intakes of specific micronutrients have been associated with cardiovascular disease (B vitamins, vitamin E, and carotenoids), cancer (folate and carotenoids), neural tube defect (folate), and bone mass (vitamin D) [13]. As an example, the positive effect of ω-3 polyunsaturated fatty acids (PUFA) on cardiac arrhythmia, and how the nutritional effect of these fatty acids change according to the saturation rate have been demonstrated, whereas no effects of C 16–18 fatty acids like stearic acid and palmitic acid have been observed [14].

A nutrient–gene interaction pattern that influences gene expressions is bound to the gene control region, and this interaction is mediated by the nutrients that specifically bind to the transcription factors, which decrease or increase the transcription. The results of nutrient–gene interactions and dose–response researches are questionable considering the complexity of the chemical substances in the nutrients and the diversity of the genetic structure. Some other important factors such as age, gender, nutrition, and some unknown parameters can influence gene expression in human beings like in animals [15]. Nutrients alter gene expression by several mechanisms including DNA synthesis, DNA methylation, binding to transcription factors, alteration of the turnover of mRNA stability, and regulatory proteins [16].

Specific genes that are regulated by nutrients are susceptible genes that have a role directly or indirectly in the existence, progression, incidence, and phenotypic severity of some chronic diseases. Asparagine synthetase, which encodes the enzyme catalyzing the glutamine and ATP-dependent conversion of aspartic acid to asparagine, is an example of a gene regulation by the nutritional status of the cell. Asparagine synthetase gene expression is triggered by the depletion of amino acids in the extracellular environment and mRNA levels; the rate of transcription and mRNA stability increases consequently [17]. Dietary components can also modify the translation of RNA to proteins, as well as posttranslational modifications that can affect protein activity [18].

A dietary chemical may preferentially alter the expression of a susceptibility gene or its variant that in turn affects various gene–gene interactions. For example, the phytosterol genistein not only inhibits certain tyrosine kinases but also binds to estrogen receptors, and both reactions will affect the expression of different genes [19]. Some nutritional chemicals affect metabolism by influencing the signal transmission pathways. For example, green tea contains the polyphenol 11-epigallocatechin-3-gallate that inhibits the phosphorylation of tyrosine in Her-2/neu and EGF receptors [20]. From a molecular standpoint, nutrients are considered to be signaling molecules that, through appropriate cellular sensing mechanisms, result in the translation of these dietary signals into changes in gene, protein, and metabolite expressions [21].

Genetical genomics defines the studies that combine the gene expressions and marker genotypes in a population and targets genomic loci covering the gene-expression control differences [22]. Nutrigenomics is the study of the effects of bioactive compounds from food on gene expression. Nowadays, microarrays are popular for the assessment of the effect of a specific diet or nutrient on the expression of nearly the entire genome. The purposes of these gene expression profiling efforts can be summarized in three parts. First, the beneficial or adverse effects of many nutrients can be clarified by exploring the specific changes in gene expression. Second, gene expression profiling can help to identify "molecular biomarkers" that constitute important genes and proteins or metabolites that are altered in the pre-disease state. Third, it can also help to identify and characterize the basic molecular pathways of gene regulation by nutrients [21].

The control of gene expression is probably the first developed adaptation mechanism for survival. Organisms with single or multiple cells are sensitive to either the presence or absence of nutrient substances in the environment [23]. All the organisms respond to low and high levels of major nutrients such as glucose, fatty acids, and amino acids, and minor nutrients such as Ca^{2+}, Fe^{2+}, and vitamins by increasing and decreasing cellular protein production, which is necessary for cellular functions [24,25]. The use of functional foods (i.e., probiotics) as medicines is especially relevant to intrauterine and early childhood development during pregnancy, and hence nutrition can be thought of as functional because of its influences on prenatal development [26].

14.1.2 NUTRITIONAL EPIGENETICS

Epigenomics, the merged science of epigenetics and genomics, has arisen as a new discipline with the aim of understanding genetic regulation and its contribution to cellular growth and differentiation, disease, and aging [27]. Epigenetic changes involve not only gene expression like the methylation of cytosines in CG couples in DNA without changing the permutation of nucleotides of the gene and histone and non-histone proteins in nucleosomes but also the covalent alterations affecting the phenotype [28]. A current definition of epigenetics is, "the study of mitotically and/or meiotically heritable changes in gene function that do not involve changes in DNA sequence" [29]. Three systems, including DNA methylation, RNA-associated silencing, and histone modification are used to initiate and sustain epigenetic silencing. Disruption of one or other of these interacting systems can lead to inappropriate expression or silencing of genes, resulting in "epigenetic diseases" [30]. For example, the ICF (immunodeficiency, centromeric region instability, and facial anomalies) syndrome is caused by mutations in the DNMT3b gene, which is an essential enzyme required for the establishment of DNA methylation patterns [31].

Regulator molecules such as DNA methyltransferases, binding proteins recognizing the methylation sites, histone-modifying enzymes, chromatin remodeling factors, and their multimolecular complexes play a role in the epigenetic process. The activities of these regulator molecules can be changed by chemical compounds presenting in the environment and nutrients, like alcohol, polyphenols, zinc, selenium, folate, and some vitamins [3,4,32]. Hypomethylation could lead to the activation of oncogenes such as c-myc, c-fos, and c-ras [33]. Chronic administration of methionine- and choline-deficient diets result in the global hypomethylation of hepatic DNA and spontaneous tumor formation in rats [34]. In cancer cells, methylation of CpGs is known to contribute to gene silencing. Hypermethylation gives rise to inactivation of tumor suppressor genes by silencing the promoters and leading to cancer development; besides it causes the loss of genomic imprinting [35]. It is suggested that folate and alcohol intake may be associated with changes in promoter hypermethylation in sporadic colorectal cancer [36].

The best characterized epigenetic process is DNA methylation. During cell division, methylation patterns in the parental strand of DNA are maintained in the daughter strand by the action of DNA methyltransferase 1 (DNMT1) catalyzing the transfer of a methyl group from S-adenosyl methionine (SAM) to the 5′ position on cytosine residues [37]. The genome of the pre-implantation mammalian embryo undergoes extensive demethylation, and appropriate patterns of cytosine methylation are reestablished after implantation [38]. Genes traited from autosomal chromosomes with one allele from the mother and the other from the father are expressed almost equally. However, some of these genes are expressed preferably from either mother or father. Inhibition of the expression of one of these two autosomal gene copies by marking and allowing the other copy for expression is called genomic imprinting. All current data indicate that this marking occurs via allele-specific methylation of specific CpGs [38]. Promoter regions of many genes in the genome are rich in CpG dinucleotides. In non-embryonic cells, over 80% of the cytosine residues in CpGs of mammalian genomic DNA are methylated at position 5′ [39]. DNA methylation patterns might not be stable and irreversible, and could be responsive to dietary manipulation with methyl donor supplements [40]. Wolff et al. found that when pregnant females are fed a diet supplemented with methyl donors, a larger proportion of offspring have a wild-type coat color as compared with the offspring of mothers fed a standard diet [41]. This result argues that an environmental stimulus early in life can change the stable expression of genes and affect the phenotype of the adult [42]. Epigenetic deregulation of imprinted genes is associated with different human disease syndromes [43], and is frequently

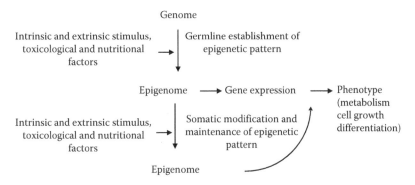

FIGURE 14.2 Interaction between the genome, the epigenome, and the phenotype.

observed in cancer as well [44]. Epigenetic mechanisms are susceptible to environmental influences and are reversible; nutrients such as folate, choline, and methionine can influence the epigenetic pattern (Figure 14.2). The epigenetic alterations and deregulations accumulate during aging [45].

Another class of epigenetic inheritance involves RNAs, which in the form of either noncoding RNA (Xist) or RNA interference (RNAi), can maintain the gene transcription state in a heritable manner [30].

14.2 HUMAN POPULATION VARIATIONS

Genetic epidemiology is a discipline that targets the identification of genetic polymorphisms involved in the development of diseases. While genetic polymorphisms in single gene disorders, which genetic epidemiology targets, can be studied more easily than complex diseases such as diabetes or cardiovascular disorders, the detection of the polymorphisms and their effects are more difficult in these complex diseases due to the absence of appropriate statistical methods. There may be some difficulties for obtaining results as (1) multiple genes and environmental factors can play a role in disease, (2) different genetic loci are independently associated with the same phenotype (genetic heterogeneity), (3) the presence of heterogeneity multidimensional interactions such as genotype profiles, and (4) the presence of low phenocopies that can be developed due to certain environmental factors. Nevertheless, new statistical methods and approaches are recommended for this problem [46].

14.2.1 NUTRIGENETICS AND INDIVIDUAL VARIATIONS

The individuals in a population (apart from some exceptions) have different genetic structures. Nutrigenetics explores the hereditary predisposition of individuals to a disease due to their genetic structures. Genetic structure of an individual can influence the equilibrium between health and disease. With the help of human genome research, human genetic material has been pointed out to have 99.9% similarities, and hereditary and polymorphic DNAs (nuclear and mitochondrial), or variations, existing as inside or outside genetic DNA sequences in the genome that differ from person to person and affect the individual's phenotype and health [47]. Despite the different approaches of the researchers, the most frequent polymorphic DNAs in the genome are usually single-nucleotide polymorphisms (SNPs), identified in every 300–1000 base pairs, and the variations in some double- or triple-nucleotide repeat numbers (variable number tandem repeats, VNTR) [48]. DNA sequence comparisons between the consensus sequence and publicly funded genome data have provided locations of 2.1 million SNPs. A random pair of human haploid genomes differed at a rate of 1 bp per 1250 on average, but heterogeneity was conspicuous in the level of polymorphism across the genome [49].

It is estimated that each of our genes contains approximately 10 deviations (or polymorphisms) in its code from the standard gene. However, not all polymorphisms have a functional impact, since SNPs occur in both the coding and more abundant noncoding regions of the human genome [50]. During the past decades, genetics has contributed to the discovery of many gene variants associated with human disease [51]. By May 2007, 69,734 of these variants had been annotated in the Cardiff Human Genome Mutation Database [52].

Mutations are the source of polymorphisms that occur at >1% frequency in the public genomes. These polymorphisms may rebound to the phenotype and may affect directly or indirectly the functions of the proteins and their interactions with the other proteins and substrates [53]. SNPs can be used not only as a marker in public gene researches but also for analyzing disparity of the personal responses to the nutrient complexes. An example of an SNP that has implications for nutrition practice is the one that occurs at base pair 677 of the gene that encodes methylenetetrahydrofolate reductase (MTHFR) [54], which catalyzes the reaction that produces 5-methyl tetrahydrofolate, a cofactor donating a methyl group to a reaction that converts homocysteine to methionine. The mutations concerning the MTHFR gene are the 677C $\rightarrow$ T or the 1298A $\rightarrow$ C, alone or in combination with C677T. These mutations are associated with reductions in the MTHFR enzyme activity. Reduction in MTHFR activity may cause increases in the plasma concentration of homocysteine, a risk factor for venous thromboembolic disease, ischemic arterial disease, and neural tube defects [55,56]. Two common polymorphisms of MTHFR, 677C $\rightarrow$ T and 1298A $\rightarrow$ C, may play a significant role in a number of other disease conditions, including cardiovascular disease, cerebrovascular disease, longevity, diabetes, psychiatric disease, renal failure, and replacement therapy [57]. MTHFR 677C $\rightarrow$ T polymorphism is also associated with a reduced risk of some forms of cancer. The protective effect of this folate-related polymorphism is dependent on adequate folate status [58]. Another example of a SNP is about a gene encoding the precursor of a polypeptide angiotensinogen having a role in the control of blood pressure. This SNP results in a guanine to arginine substitution in the promoter region of the gene. The AA genotype for the angiotensinogen G-6A polymorphism has been associated with higher levels of circulating angiotensinogen and essential hypertension [59].

Besides SNPs, haplotypes are also informative for linkage analysis among gene, disease, and nutrition in a population. A haplotype is a bonded SNP allele set existing on a single chromosome and transferred together to progeny (generations). Haplotype analysis is beneficial for personal or population-specific genetic mapping [60].

Nutrigenetics deals with the effect of genetic variation on the interaction between diet and disease or on nutrient requirements. Genetics has a virtual role in determining an individual's risk of developing a certain disease [21]. A prospective cohort study suggests that the functional catechol-O-methyltransferase (COMT) Val158Met polymorphism is associated with increased risk of acute coronary events, and it may interact with high, serum total homocysteine (tHcy) levels [61]. Another well-known example related to gene–nutrient interactions is lactose intolerance occurring in Northern Europe. A mutation that was formed in the lactase gene promoter of one or few individuals' years ago is transferred to next generations as a DNA variant for 1000 years, and today it is presenting as cis-acting variants affecting gene regulation in a great many number of persons [62]. Cis–trans acting variants: cis-acting variants in gene control sites are known as promoters and enhancers and influence transcript synthesis or stability specific to an allele, and they are close to the genes that are regulated. Trans-acting variants influence both alleles of a gene and are not close to the gene and generally exist on different chromosomes [63,64].

14.2.2 GENE–ENVIRONMENT INTERACTIONS

Food intake and nutrient exposure are key environmental factors that are involved in the pathogenesis and progression of the common polygenic, diet-related diseases. There is a bilateral dynamic interaction between human genome and nutrients that determines the gene expression and metabolic response. This interaction eventually influences the health condition or predisposition of a person to a disease [65]. If we always have the same genome and live in a constant environment at same situations, we will all certainly have similar health conditions and similar diseases. At least the expression of genetic information can be influenced by environmental factors such as the soil, water, and air pollution, oxygen pressure, allergens, stress, seasonal alterations, and circadian rhythm [66]. Statistically, the gene–environment interaction is defined as "a different effect of an environmental exposure on disease risk in persons with different genotypes," or, alternatively, "a different effect of a genotype on disease risk in persons with different environmental exposures" [67].

Systems biology approaches using methodologies and disciplines such as bioinformatics, microarray analysis, proteomics, and metabolomics are necessary for analyzing the differences of allele gene frequencies, and DNA haploid blocks and chemical complexity in nutrients, and, in addition, the interactions of gene–environment and gene–gene–environment among human populations.

14.3   NUTRIENT–DISEASE INTERACTIONS

Nutrition-related diseases can be considered as the outcome of very complex interactions between genetic and dietary factors, each one with only a modest effect. The analysis of a single dietary component with a single gene seems to be insufficient to provide adequate data [68]. Considering the cost-benefit evaluation of nutrient–disease interaction, there are a good many examples of which their positive contributions to health cannot be underestimated. Resveratrol, a polyphenol that occurs in abundance in grapes and red wine, has antioxidant and neuro-protective properties and therefore contribute to the beneficial effect of wine consumption on the neurodegenerative process such as Alzheimer's disease [69,70].

The first evidences related to gene, nutrient, and disease interaction are emerged from single gene mutation originated metabolic diseases like phenylketonuria. Phenylketonuria is an inborn error of metabolism resulting from a deficiency of phenylalanine hydroxylase and characterized by mental retardation and is treatable by a low phenylalanine diet [71].

In a comprehensive study in which hereditary and mitochondrial diseases had been excluded, 97 out of approximately 1000 genes were detected to possess single gene mutation related to metabolic diseases [72]. When the product categories of these abnormal phenotype related genes were analyzed, the most frequently detected ones were metabolism-related enzymes, 31.2% and modulatory proteins, 13.6% and the less frequently detected ones, <10%, were receptors, transcription factors, intracellular and extracellular matrix compounds, and the others.

Considering these nutrition-related diseases, obesity, cardiovascular diseases, and cancer are the most important and life-threatening ones for all the population.

14.3.1   Obesity

Obesity is the most frequent nutritional disorder in developed countries and has become a major health problem all over the world. Obesity and related metabolic disorders are involved as conspicuous risk factors in the development of chronic diseases like cardiovascular diseases and cancer [73,74].

Obesity is a condition resulting from a chronic imbalance between food intake and energy homeostasis [75,76]. The role of diet in the development of obesity is indisputable as an environmental factor. Besides environmental factors, genetic and behavioral factors are involved in this disease [77].

Obesity is also a matter of interest for nutrigenetics. To date, nearly 100 genes, and nowadays more are reported to be potentially involved in obesity and about more than 20 genes have been studied for diet response [10].

Among these potential genes, leptin and leptin receptor gene mutations have attracted attention for the prediction of obesity [78]. Adipose tissue has been described as a fat and energy depot and interestingly as an endocrine organ [79] secreting a number of hormones. Leptin is such a kind of hormone and one of the key regulators of fat metabolism and energy balance [76,80], in a way regulating appetite and metabolism. Studies concerning leptin and leptin receptor gene have shown that they have an association with obesity. Leptin is the product of the human homologue of mouse "obese" gene. A defective and unstable leptin molecule can be synthesized in consequence of homozygous mutation of this gene (ob/ob) and as a result hereditary obesity outcomes in mice [80,81]. Polymorphisms of the leptin receptor gene cause alternatively spliced variants and these have been shown to have associations between some of its variants and obesity, fat distribution, insulin levels, and blood pressure, and in addition with nutrient responses [76]. It has been demonstrated that overweight women carriers of the C polymorphisms of the leptin receptor gene lost more weight in response to a low-calorie diet than the non-carriers [1].

Leptin works by inhibiting the activity of neurons that contain neuropeptide Y (NPY), and agouti-related peptide (AgRP), which is also secreted by human adipocytes [82,83] and this protein is controlled by an obesity gene "agouti," implicated in energy homeostasis. The NPY neurons are key elements in the regulation of appetite. Experimentally, small doses of NPY injected into the brains of animals stimulate feeding, while selective destruction of the NPY neurons in mice causes them to become anorexic. The polymorphisms in noncoding regions of the gene for neuropeptide Y (NPYY5R) receptor gene have been shown to have strong correlation with the risk of obesity [84].

On the other hand leptin also works by increasing the activity of neurons expressing α-melanocyte-stimulating hormone (α-MSH). α-MSH is contrarily an important mediator of satiety and variations in the gene of the α-MSH receptor that cause obesity in humans [83]. In some studies, the association between obesity and mutations of the melanocortin 4-receptor and melanocortin 5-receptor genes has been established. While Dubern et al. pointed out the correlation between severe obesity in children and melanocortin pathway elements, particularly melanocortin receptor (MCR4) mutations; Chagnon et al. showed linkage or association with obesity phenotypes and MCR4 and especially MCR5 [85,86]. Melatonin has been shown to decrease circulating leptin levels in down-regulating leptin in the night time [87].

Moreover, studies about lipidogenesis and lipid metabolism have focused on a nuclear receptor, peroxisome proliferation-activated receptor γ (PPAR-γ). PPAR-γ gene is a good example for involvement in a gene–environment interaction. Activation of PPAR-γ leads to peroxisome proliferation and increased oxidation of fatty acids. PPAR-γ activation results in adipocyte differentiation as well as improved insulin signaling of mature adipocytes. PPAR-γ is associated with insulin resistance and blood pressure [5]. In individuals with a specific polymorphism in the PPAR-γ (Pro12Ala), a low polyunsaturated-to-saturated fat ratio has been demonstrated to be associated with an increase in body mass index and fasting insulin concentrations [88].

Furthermore, adipose tissues were found to have some possible links between obesity and insulin resistance [89]. Obesity often causes insulin resistance, a decline in the ability of insulin to stimulate glucose uptake in the body leads to compensatory oversecretion of this hormone by the pancreatic cells and eventually, to cell exhaustion and development of type-2 diabetes mellitus [11,75]. Likewise in type-1 diabetes mellitus, the role of genetic factors has been studied to highlight the relation between inflammation, oxidant stress, and insulin insensitivity. Nuclear factor kappa B (NFκB) is a crucial transcription factor for response to oxidative stress and inflammation. Investigations about NFκB gene have shown its possible role in the susceptibility to type-1 diabetes by means of allelic differentiations, individuals with the A10 allele may be more likely to develop diabetes compared with those with the A14 allele [90].

There are many studies about the interaction between obesity, inflammation, and immune response components. Adiponectin, angiotensinogen, adipsin acylation stimulating protein, tumor necrosis factor-alpha (TNF-α), interleukin 6 (IL-6), and plasminogen activator inhibitor-1 are proteins secreted by fat cells. These proteins regulate lipid metabolism, inflammation, cardiovascular functions, vascular haemostasis, and immunity [91]. There is growing evidence that low-level inflammation linked to the increased risk of developing cardiovascular disease and associated with obesity is mediated by pro-inflammatory adipokines [16]. Polymorphisms in adiponectin—another obesity-related gene—result in phenotypes such as insulin resistance, type-2 diabetes and increased adiposity. Interestingly, variations in adiponectin can result in altered circulating levels of its protein [92] like in the case of leptin molecule, and outlining the importance of taking these polymorphisms into account when tending to use these molecules as biomarkers [93].

Most recently the relation between immune response and obesity was analyzed and the interleukin-1 receptor antagonist gene polymorphisms were shown to associate with normal weight obese syndrome and obesity-related diseases [94].

The determination of the mutations of these kinds of genes in early life seems to be efficacious for handling the obesity problem in the near future, establishing the role of nutrigenetics in such issues.

14.3.2 Cardiovascular Diseases

In cardiovascular diseases lipoproteins such as high-density lipoprotein (HDL), low density lipoprotein (LDL), very low-density lipoprotein (VLDL), and chylomicrons are the primary targets to combat. In coronary heart diseases (CHD), genetic polymorphisms in several genes have been mentioned to have an effect on blood lipid levels. From a nutrigenetics approach, individual nutrient responses differ concerning these polymorphisms. Lipid lowering diets form a significant component in the prevention of CHD [95].

The apolipoprotein A1 (ApoA1) gene plays a role in lipid metabolism and CHD [1]. Apo A1 polymorphisms have associations with HDL levels. The G-to-A mutation in the promoter of ApoA1 gene causes increased HDL cholesterol concentrations and alters the response to polyunsaturated diet. Contrarily the A allele (variant) was shown to be associated with decreased serum HDL levels [96]. As a dietary recommendation women should avoid polyunsaturated fats in order to get rid of high serum HDL levels, and in men this type of fat's effect is important in the case of alcohol consumption and tobacco smoking [6].

Apolipoprotein E (ApoE) has a major role in the transport of chylomicrons, VLDL, and the variations in apolipoprotein E influence in the concentration of lipoproteins. ApoE gene polymorphisms consist of the three most concurred alleles ε2, ε3, and ε4). These alleles constitute three isomorphic proteins E2, E3, and E4, and six genotypes: ε2/ε2, ε2/ε3, ε2/ε4, ε3/ε3, ε3/ε4, and ε4/ε:4. The risk of cardiovascular disease and, nutrient and environmental responses are different in these three phenotypes. APOE3 is the most common phenotype and has a good response to low-fat diet and exercises. However APOE-ε4 variant was involved in 20% of the population. 1%–3% APOE protein mutations and its association with increased risk of cardiovascular diseases were detected in England [97]. As a dietary recommendation these individuals should avoided saturated fats, alcohol consumption, and tobacco smoking, and follow a low-fat diet. Nutrient responses to dietary fat intake regulation differ based on the genotypes. Individuals having APOE4 have good responses [5,98–100]. A study by Corella et al. showed that Apo E2 variants have lower LDL levels whereas E4 variants have correlation with plasma LDL levels and alcohol consumption [101].

Moreover several candidate genes and their variations such as cholesterol ester transfer protein (CETP), lipoprotein lipase (LPL), hepatic triglyceride lipase (HL), LDL-receptor, ATP binding cassette transporter A1 (ABCA1), and lecithin-cholesterol acyltransferase (LCAT) have been studied to demonstrate the individual's sensitivity toward developing CHD [102–108].

There are also some other studies concerning the effect of genetic variations to dietary cholesterol. Individuals with small, dense LDL particles (phenotype B) have an increased risk of coronary artery disease relative to those individuals exhibiting large, less dense LDL particles (phenotype A) [109]. An example of genotype environment interaction has been shown in these LDL patterns and low-fat diets in a study including 38 men with phenotype A. Phenotypic switching (A to B) was observed in 12 of these individuals when the fat content was lowered in the diet. Genetic factors seem to play an important role in dietary responses.

Additionally the role of environmental factors is significantly observed in fatty acids. In ancient times omega-6 to omega-3 essential fatty acids ratio in diets was approximately 1, whereas nowadays in Western diets the ratio is 15/1 to 16.7/1. A high omega-6/omega-3 ratio promotes the pathogenesis of many diseases, including cardiovascular disease, cancer, osteoporosis, and inflammatory, and autoimmune diseases. Contrarily increased levels of omega-3 PUFA (a lower omega-6/omega-3 ratio) exert suppressive effects through cytokines such as interleukin 1 beta (IL-l beta), TNF-α, and interleukin-6 (IL-6). The ratio of omega-6/omega-3 fatty acids is an important factor for the prevention and management of chronic diseases [110].

Genetic and environmental factors almost work together in CHD and dietary recommendations can be arranged based on nutrigenetic data of an individual.

14.3.3 Cancer Nutrigenomics

Human carcinogenesis is best characterized as an accumulation of alterations (genetic and epigenetic) in genes regulating cellular homeostasis, such as oncogenes, tumor suppressor genes, apoptosis regulating genes, and DNA repair genes [111]. Mutations in these three classes of genes can occur in the germline, resulting in hereditary predispositions to cancer, or in single somatic cells, resulting in sporadic tumors [112].

Carcinogenesis is known as a process composed of multiple stages (including tumor initiation, promotion, and progression) in which gene expression, and protein and metabolite function begin to operate aberrantly [113]. It is becoming increasingly apparent that the multiple changes in cancer cells, including chromosomal instability, activation of oncogenes (ras, myc) silencing of tumor suppressor genes (Rb, p53), and inactivation of DNA repair systems, are caused not only by genetic but also by epigenetic abnormalities [114].

The first alteration for multistep cancer development in normal tissue—single or more cells—is a mutation which is formed in an oncogene or a tumor suppressor gene spontaneously or by means of any carcinogen. This stage is very expeditious and irreversible. If these mutations had to occur simultaneously in a single cell, then the prevalence of cancer would be minimal [115]. The current doctrine is that these mutations occur over time, with each mutation engendering a clonal expansion resulting in a large number of cells that then form a substrate for subsequent mutations. If a mutation transfers to mitotic cells, it can cause clonal expansion initiating the neoplastic process. In these genetically altered cells, pre-neoplastic formation can occur via the promoting endogen and exogen factors (promotion and progression) [116]. In the final stage malignancy develops with the combination of additional genetic and epigenetic alterations (Figure 14.3). In a multistep carcinogenesis model, even if there is a point mutation or a major chromosomal defect, any of the significant alterations in DNA can constitute a clonal growth advantage to a cell. Invasion is merely the expansion of tumor cells into surrounding tissues as a result of continuous cell division. Metastasis is the spread of tumor cells to areas not directly adjacent to the primary tumor [117], and generally occurs in several stages: firstly a few cells separate from the primer pivotal, afterward tumor cells migration and lymph or blood vessels penetration occurs. These cells settle down temporarily to the vessels of the distant organs. From the blood vessel walls invasion is provided to the secondary pivotals. It begins to grow as a secondary tumor in the second zone [118].

Without a constant supply of nutrients and oxygen, the tumor cells are unable to multiply indefinitely. Continued growth requires development of new blood vessels (vascularization) in the tumor [119].

Dietary nutrients can influence cancer risk by inhibiting or enhancing carcinogenesis through diverse mechanisms of action [120]. Inhibitors of carcinogenesis can be clustered in two groups as natural and artificial: blocking and suppressing agents. In the first stage of cancer transformation, blocking agents inhibit not only the transformation of procarcinogens to carcinogens but also reaching the target molecules like/such as DNA, RNA, and proteins. However suppressing agents inhibit malignancy transition of cells in promotion and progression stages [121].

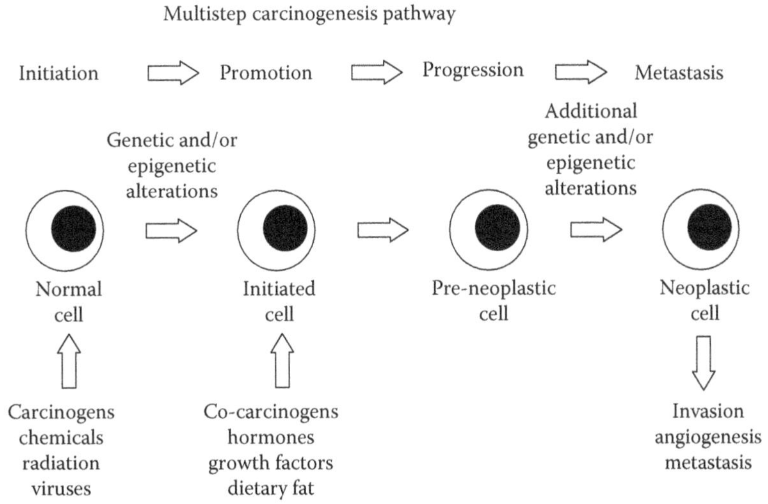

Multistep carcinogenesis pathway

FIGURE 14.3 Schematic presentation of the multistep carcinogenesis pathway.

Many dietary constituents can increase the risk of developing cancer (suspected human carcinogens), but there is also accumulating evidence from population as well as laboratory studies to support an inverse relationship between regular consumption of fruit and vegetables and the risk of specific cancers [122]. Theoretically dietary changes should prevent cancer development. When asked to suggest how best to do this, actually the answer is simple: "eat more fruit and vegetables" but the reasons are not so [123]. Evidence suggests that the cancer-protective effects of an individual's diet may reflect the combined effects of various vitamins, minerals, and other bioactive components such as flavonoids, isothiocyanates, and/or allium compounds rather than from the effect of a single ingredient [124]. However, bioactive nutrient compounds affect many processes such as carcinogen metabolism, DNA repair, hormonal regulation, cell division and cell differentiation, angiogenesis, apoptosis, and metastasis [122,125]. Besides the genetic profile of the individuals (nutrigenetic effect) influences the response to bioactive nutrient compounds by affecting the target zone, absorption, and metabolism. A well-known and fairly studied polymorphism example in the literature is Ala222Val polymorphism in the MTHFR gene in which folate metabolism changes, and in this situation it has been shown that in spite of a higher risk of neural tube defects [126] and cardiovascular disease [127] there is a lower risk for colon cancer [128].

Likewise epigenetic events such as methylation of DNA, methylation and phosphorylation of histone proteins, and modification of RNA can regulate the response to nutrient compounds by effecting the gene expression. Eventually synthesis and degradation of proteins and modification alterations in the end of the synthesis are all determinative in nutrient and compound responses [129].

Green tea is a widely consumed beverage all over the world [130]. Epigallocatechin gallate (EGCG), the main constituent of green tea extract has been shown to be a cancer preventive for a wide range of target organs such as esophagus, stomach, duodenum, colon, lung, liver, pancreas, breast, bladder, prostate, and skin [131]. In prostate cancer it exhibits these effects by inducing some mechanisms like apoptosis, inhibition of cell growth, and deregulation of cyclin kinase inhibitor WAF-1/p21-mediated cell cycle [132]. Resveratrol existing in fresh grape skin and red wine has been determined to have a striking effect on several molecular targets for preventing different stages of cancer development [120]. For example, it was found that resveratrol inhibits AP-1 activity induced by phorbol 12-myristate 13-acetate and ultraviolet C, and the MAPK pathway by interfering with c-Src protein tyrosine kinase and protein kinase C, which may provide a mechanism for the anticarcinogenic activity of resveratrol [133].

In many epidemiologic studies, meat consumption, particularly red meat, has been associated with higher risk of colorectal cancers or adenomas [134,135]. In several epidemiological, preclinical, and clinical studies related to colon cancer, the most significant factor determining the incidence and biologic properties of tumors has been pointed out to be nutrient bioactive compounds [136]. For example, fruits and vegetables are a main source of folate, and low folate intake has been associated with several types of cancer [128,137]. In most of the studies, although it has been demonstrated that there is a higher risk of colon cancer in folate deficiency and folate supplementation has been declared to reduce the risk [138]. In the folate mechanism, there are 50–100 genes involved directly or indirectly in related factors like receptors, binding proteins, enzymes, tissue-specific gene products, and folate derivative metabolites [139]. In these genes point mutations, deletion or insertion mutations, and rearrangements and genetic variations as polymorphisms, all influence the response of individuals to folate deficiency. Both mutation rate and gene expression may be influenced via DNA hypomethylation that is induced by folate deficiency [140].

Abnormal DNA methylation patterns are a hallmark of most cancers. Dietary factors may influence the methylation process in different ways. First, methyl groups available for the formation of S-adenosylmethionine (SAM) may be supplied from dietary factors. Second, dietary factors may modify the utilization of methyl groups by processes including shifts in DNA methyltransferase (Dnmt) activity. A third mechanism may relate to DNA demethylation activity. Finally, the DNA methylation patterns may influence the response to a bioactive food component [32].

14.3.4 Apoptosis

Apoptosis (programmed cell death) is a genetically controlled cell suicide process that has an essential role for the maintenance of homeostasis and prevention of cancer and some other diseases substantially in all living cells [141]. Apoptosis is an optional elimination method for irreversibly damaged cells. The two major pathways that initiate apoptosis are extrinsic (death receptor mediated) and intrinsic (mitochondrial mediated). In addition, mitogenic and stress-responsive pathways are involved in the regulation of apoptotic signaling [142].

Dietary bioactive components can regulate intracellular location of pro-apoptotic proteins (Bcl-2, Bcl-xL) or antiapoptotic proteins (Bax and Bak) promoting the release of cytochrome c from the mitochondria [143]. A lot of micronutrients like vitamins and minerals are necessary for DNA synthesis, DNA repair, and apoptosis. Micronutrient deficiencies can cause DNA damage, which may ultimately lead to cancer [137,144]. Folate deficiency causes chromosome breaks due to the massive incorporation of uracil in the DNA [145]. An anthocyanidin, delphinidin, a photochemopreventive existing in a lot of pigmented vegetable and fruits in abundant amounts inhibits UVB-mediated oxidative stress in human HaCaT keratinocytes and rat skin, and reduces DNA damage by an antioxidant effect and therefore it has been suggested to be efficacious in the treatment of UVB-mediated damage in human skin [146].

Lupeol, a triterpene is found in various edible plants such as olive, fig, mango, strawberry, and red grapes. It is a potent multi-target anticancer agent that induces apoptotic cell death of pancreatic cancer cells via modulation of Ras-induced proteinkinase Cα (pKCα)/ornithine decarboxylase (ODC), PI3K/Akt, MAPKs, and NFκB-signaling pathways [147]. Olive oil's bitter principle (i.e., oleuropein aglycone) is among the first examples of how selected nutrients from an Extra Virgin Olive Oil (EVOO)-rich "Mediterranean diet" directly regulate HER2-driven breast cancer disease [148].

Curcumin (diferuloylmethane) is a naturally occurring polyphenolic pigment, has antioxidant effect [149], and induces apoptosis in human colon cancer cells in a c-jun N-terminal kinase (JNK)-dependent manner [150], and has been found to suppress the activity of the antiapoptotic protein Bcl-2 in myeloma cells and colon cancer cells [151]. It is also demonstrated that curcumin inhibited human colon cancer cell growth by suppressing gene expression of EGFR through reducing the trans-activation activity of Egr-l [152]. Beta-carotene induces apoptosis by the activation of caspase-3 in human leukemia (HL-60), colon adenocarcinoma (HT-29) as well as melanoma (SK-MEL-2) cell lines [153]. Vitamin E induces apoptosis in many cancer cells, and has been found to selectively act as an anticancer drug, alone or in combination with chemotherapy and radiation [154]. It is suggested that, by targeting telomerase, supplemental vitamin E may be a useful adjuvant for the prevention of ovarian cancer and/or improved chemo-sensitivity, which could reduce the mortality of a disease that kills thousands of women annually [155].

14.3.5 ANGIOGENESIS

Development of blood vessels from *in situ* differentiating endothelial cells is called vasculogenesis, whereas sprouting of new blood vessels from the pre-existing ones is termed angiogenesis [156]. Angiogenesis does not initiate malignancy but angiogenesis is a prerequisite for the growth of solid tumors and their metastasis. The potential anti-angiogenic strategies targeted by chemopreventive agents could include the blocking of production and release of angiogenic factors, increase in anti-angiogenic factors, and inhibition of endothelial cell proliferation, migration, and survival, and disruption of the process of micro-vessel formation [157,158]. For example, grape seed extract possesses an anti-angiogenic potential, which is associated with its anti-proliferative, pro-apoptotic and inhibition of MMP-2 secretion in endothelial cells [159], and inhibits advanced human prostate tumor growth and angiogenesis [160]. Among the predominant isoflavones in soy, genistein is the most potent at inhibiting endothelial cell proliferation and *in vitro* angiogenesis, and inhibits carcinogenesis in animal models [161]. It has been reported that both green tea and one of its components, epigallocatechin-3-gallate (EGCG), significantly prevent angiogenesis [162]. Deguelin, a rotenoid isolated from several plant species, inhibits Akt and it is critical for angiogenesis, and therefore it is suggested that potential application in angioprevention and anti-angiogenic therapy [163]. Among the vitamin E group, tocotrienol has been shown to exhibit the novel anti-angiogenic effect in an *in vitro* system [164].

Finally prevalence of cancer development can be reduced with an appropriate diet and life style, physical activity, and weight control. There may be a lot of choices for an anti-cancer diet. For example such a low-calorie diet contains no refined sugar and salt, low total fat including essential fatty acids, balanced omega 3 and omega 6, no red meat, dark green vegetables rich in folic acid and chlorophyll, flax seed as a phytoestrogen source, at least ten vegetables like onion and radish rich in antioxidants containing vitamins C and E and phytochemicals, fruit service containing at least four kinds, useful probiotics, oral enzyme supplements, for vitamin D (Vit-D) adequate sunlight or 1000 IU/day Vit-D, ~200 μg/day selenium, and ~1000 μg/day vitamin B-12 [165].

14.4 FUTURE OF NUTRIGENOMICS

After completion of the human genome sequencing, a serial of new scientific concepts comprising omics have emerged. Metabolomics now promises to play a major role in the nutritional sciences [166,167]. From this point of view nutrigenomics and nutrigenetics are also developing quickly and provide an approach of "personalized medicine." The ultimate goal of nutrigenomics is to determine the dietary components that are most compatible with health for a specific individual. New developments allow identification of up to 500,000 SNPs in an individual, and with increasingly lowered pricing, these developments may explode the population-level potential for dietary optimization based on nutrigenomic approaches [98].

Genetic polymorphisms and epigenetic variations may change the response to diet compounds via influencing the absorption, metabolism, or molecular interaction zones. It is a well-known approach to give special diets to patients with congenital metabolism diseases like single gene mutation originated phenylketonuria and galactosemia in order to prevent the symptoms and phenotypic appearance.

Genetic information concerning single and multiple gene diseases are augmented day after day. Having knowledge about the target genes and their role in metabolic pathways and an advanced knowledge from nutrigenomic studies will give rise to solving a great many health problems. Nutritional diet requires a potent bioinformatics component. Shortly after birth, genetic tests for the determination of individual differences in DNA and analyzing all the information about the personal environment like life style can help predict potential risks related to diseases beforehand and facilitate the treatment procedure. Likewise, optimized diets can delay the aging process and provide a long-term aging by lightening some problems about age.

REFERENCES

1. Blum, K. et al., Nutrigenomics and pharmacogenomics: A scientific wonderland, *Soc. Sci. Inform.*, 45 (1), 35–52, 2006.
2. Afman, L. and Müller, M., Nutrigenomics: From molecular nutrition to prevention of disease, *J. Am. Diet. Assoc.*,106, 569–576, 2006.
3. Trujillo, E., Davis, C., and Milner, J., Nutrigenomics, proteomics, metabolomics, and the practice of dietetics, *J. Am. Diet. Assoc.*, 106, 403–413, 2006.
4. Davis, C. D. and Uthus, E. O., DNA methylation, cancer susceptibility, and nutrients interactions, *Exp. Biol. Med.*, 229, 988–995, 2004.
5. Ordovas, J. M. and Corella, D., Nutritional genomics, *Annu. Rev. Genom. Hum. Genet.*, 5, 71–118, 2004.
6. Kaput, J., Decoding the pyramid: A systems-biological approach to nutrigenomics, *Ann. N. Y. Acad. Sci.*, 1055, 64–79, 2005.
7. Kaput, J. and Rodriguez, R. L., Nutritional genomics: The next frontier in the postgenomic era, *Physiol. Genom.*, 16, 166–177, 2004.
8. German, J. B., Roberts, M. A., and Watkins, S. M., Personal metabolomics as a next generation nutritional assessment, *J. Nutr.*, 133 (12), 4260–4266, 2003.
9. Robertson, D. G., Metabonomics in toxicology: A review, *Toxicol. Sci.* 85, 809–822, 2005.
10. Kaput, J., Diet–disease gene interactions, *Nutrition,* 20, 26–31, 2004.
11. Kussmann, M., Raymond, F., and Affolter, M., OMICS-driven biomarker discovery in nutrition and health, *J. Biotechnol.*, 124 (4), 758–787, 2006.
12. Young, V. R., 2001 W.O. Atwater Memorial Lecture and the 2001 ASNS President's Lecture—Human nutrient requirements: The challenge of the post-genome era, *J. Nutr.*, 132 (4), 621–629, 2002.
13. Fairfield, M. and Fletcher, R. H.,Vitamins for chronic disease prevention in adults: Scientific review, *JAMA*, 287, 3116–3126, 2002.
14. Brouwer, I. A. et al., Association between n-3 fatty acid status in blood and electrocardiographic predictors of arrhythmia risk in healthy volunteers. *Am. J. Cardiol.,* 89, 629–631, 2002.
15. Whitney, A. R. et al., Individuality and variation in gene expression patterns in human blood, *Proc. Natl. Acad. Sci. U.S.A.*, 100, 1896–1901, 2003.
16. Evans, D. A., Hirsch, J. B., and Dushenkov, S., Phenolics, inflammation and nutrigenomics, *J. Sci. Food. Agric.*, 86, 2503–2509, 2006.
17. Jousse, C. et al., Evidence for multiple signaling pathways in the regulation of gene expression by amino acids in human cell lines, *J. Nutr.*, 130 (6), 1555–1560, 2000.
18. Davis, C.D. and Hord, N.G., Nutritional "omics" technologies for elucidating the role(s) of bioactive food components in colon cancer prevention, *J. Nutr.*, 135, 2694–2697, 2005.
19. Dixon, R. A. and Ferreira, D., Molecules of interest: Genistein, *Phytochemistry*, 60, 205–211, 2002.
20. Pianetti, S. et al., Green tea polyphenol epigallocatechin-3 gallate inhibits her-2/neu signaling, proliferation, and transformed phenotype of breast cancer cells, *Cancer Res.*, 62, 652–655, 2002.
21. Müller, M. and Kersten, S., Nutrigenomics: Goals and strategies. *Nat. Rev. Genet.*, 4, 315–322, 2003.
22. Jansen, R. C., Studying complex biological systems using multifactorial perturbation, *Nat. Rev. Genet.*, 4, 145–151, 2003.
23. Brosnan, M. P. et al., The stress response is repressed during fermentation in brewery strains of yeast, *J. Appl. Microbiol.*, 88, 746–755, 2000.
24. Levi, R. S. and Sanderson, I. R., Dietary regulation of gene expression, *Curr. Opin. Gastroenterol.*, 20 (2), 139–142, 2004.

25. Pegorier, J. P., Le May, C., and Girard, J., Control of gene expression by fatty acids, *J. Nutr.*, 134, 2444–2449, 2004.

26. Barker, J. and Meletis, C. D., Functional foods for childhood development, *Alt. Comp. Therap.*, 10, 131–134, 2004.

27. Callinan, P. A. and Feinberg, A. P., The emerging science of epigenomics, *Human. Mol. Genet.*, 15 (1), R95–R101, 2006.

28. Bird, A., DNA methylation patterns and epigenetic memory, *Genes. Dev.*, 16, 6–21, 2002.

29. Roloff, T. C. and Nuber, U.A., Chromatin, epigenetics and stem cells, *Eur. J. Cell Biol.*, 84, 123–135, 2005.

30. Egger, G. et al., Epigenetics in human disease and prospects for epigenetic therapy, *Nature*, 429, 457–463, 2004.

31. Okano, M. et al., DNA methyltransferases Dnmt3a and Dnmt3b are essential for *de novo* methylation and mammalian development, *Cell*, 99, 247–257, 1999.

32. Ross, S. A., Diet and DNA methylation interactions in cancer prevention, *Ann. N. Y. Acad. Sci.*, 983, 197–207, 2003.

33. Christman, J. K. et al., Reversibility of changes in nucleic acid methylation and gene expression induced in rat liver by severe dietary methyl deficiency, *Carcinogenesis*, 14, 551–557, 1993.

34. Poirier, L. A., The role of methionine in carcinogenesis in vivo, *Adv. Exp. Med. Biol.*, 206, 269–282, 1986.

35. Herman, J. G. and Baylin, S. B., Gene silencing in cancer in association with promoter hypermethylation, *N. Engl. J. Med.*, 349, 2042–5204, 2003.

36. Engeland, M. et al., Effects of dietary folate and alcohol intake on promoter methylation in sporadic colorectal cancer: The Netherlands Cohort Study on Diet and Cancer, *Cancer Res.*, 63, 3133–3137, 2003.

37. Bestor, T. H., The DNA methyltransferases of mammals, *Hum. Mol. Genet.*, 9, 2395–2402, 2000.

38. Reik, W. and Walter, J., Genomic imprinting: Parental influence on the genome, *Nat. Rev. Genet.*, 2, 21–32, 2001.

39. Ehrlich, M., Amount and distribution of 5-methyl cytosine in human DNA from different types of tissues or cells, *Nucleic Acids Res.*, 10, 2709–2721, 1982.

40. Bacino, C. A. et al., A double-blind therapeutic trial in Angelman syndrome using betaine and folic acid, *Am. J. Hum. Genet.*, 69, 679, 2001.

41. Wolff, G. L. et al., Maternal epigenetics and methyl supplements affect agouti gene expression in Avy/a mice, *FASEB J.*, 12, 949–957, 1998.

42. Jaenisch, R. and Bird, A., Epigenetic regulation of gene expression: How the genome integrates intrinsic and environmental signals, *Nat. Genet.*, 33, 245–254, 2003.

43. Arnaud, P. and Feil, R., Epigenetic deregulation of genomic imprinting in human disorders and following assisted reproduction, *Birth Defects Res. C. Embryo Today*, 75 (2), 81–97, 2005.

44. Feinberg, A. P. and Tycko, B., The history of cancer epigenetics, *Nat. Rev. Cancer*, 4, 143–153, 2004.

45. Feil, R., Environmental and nutritional effects on the epigenetic regulation of genes, *Mutat. Res.*, 600 (1–2), 46–57, 2006.

46. Heidema, A. G. et al., The challenge for genetic epidemiologists: How to analyze large numbers of SNPs in relation to complex diseases, *BMC Genet.*, 7, 23, 2006.

47. Livingston, R. J. et al., Pattern of sequence variation across 213 environmental response genes, *Genome Res.*, 14, 1821–1831, 2004.

48. Jiang, R. et al., Genome-wide evaluation of the public SNP databases, *Pharmacogenomics*, 4, 779–789, 2003.

49. Venter, J. C. et al., The sequence of the human genome, *Science*, 16 (291), 1304–1351, 2001.

50. Roche, H. M., Nutrigenomics—New approaches for human nutrition research, *J. Sci. Food Agric.*, 86 (8), 1156–1163, 2006.

51. Pampin, S. and Rodriguez-Rey, J. C., Functional analysis of regulatory single-nucleotide polymorphisms, *Curr. Opin. Lipidol.*, 18 (2), 194–198, 2007.

52. Cardiff Human Genome Mutation Database, http://archive.uwcm.ac.uk/uwcm/mg/docs/hohoho.html, 2007.

53. Online Mendelian Inheritance in Man, http://www3.ncbi.nlm.nih.gov/omim/, 2009.

54. Frosst, P. et al., A candidate genetic risk factor for vascular disease: A common mutation in methylenetetrahydrofolate reductase, *Nat. Genet.*, 10, 111–113, 1995.

55. Hanson, H. Q. et al., C677T and A1298C polymorphisms of methylene tetrahydrofolate reductase gene: Incidence and effect of combined genotypes on plasma fasting and post-methionine load homocysteine in vascular disease, *Clin. Chem.*, 47, 661–666, 2001.

56. Guieterrez Revilla, J. I. et al, C677T and A1298C MTHFR polymorphisms in the etiology of neural tube defects in Spanish population, *Med. Clin. (Barc.)*, 120, 441–445, 2003.

57. Fodinger, M., Horl, W. H., and Sunder-Plassmann, G., Molecular biology of 5,10-methylenetetrahydrofolate reductase, *J. Nephrol.*, 13 (1), 20–33, 2000.

58. Bailey, L. B., Folate, methyl-related nutrients, alcohol, and the MTHFR 677C→T polymorphism affect cancer risk: Intake recommendations, *J. Nutr.* 133, 3748S—3753S, 2003.

59. Svetkey, L. P. et al, Angiotensinogen genotype and blood pressure response in the dietary approaches to stop hypertension (DASH) study, *J. Hypertens.*, 19, 1949–1956, 2001.

60. Zhang, W. et al., Impact of population structure, effective bottleneck time, and allele frequency on linkage disequilibrium maps, *PNAS*, 101 (52), 18075–18080, 2004.

61. Voutilainen, S. et al., Functional COMT Val158Met polymorphism, risk of acute coronary events and serum homocysteine: The Kuopio Ischaemic Heart Disease Risk Factor Study, *PloS ONE*, 2 (1), e181, 2007.

62. Wang, Y. et al., The lactase persistence/non-persistence polymorphism is controlled by a cis-acting element, *Hum. Mol. Genet.*, 4, 657–662, 1995.

63. Rockman, M. V. and Wray, G. A., Abundant raw material for *cis*-regulatory evolution in humans, *Mol. Biol. Evol.*, 19, 1991–1992, 2002.

64. Pastinen, T., Ge, B., and Hudson, T. J., Influence of human genome polymorphism on gene expression, *Hum. Mol. Genet.*, 15, R9–R16, 2006.

65. Roche, H. M., Dietary lipids and gene expression, *Biochem. Soc. Trans.*, 32, 999–1002, 2004.

66. Kaput, J. et al., Horizons in Nutritional Science, The case for strategic international alliances to harness nutritional genomics for public and personal health, *Br. J. Nutr.*, 94, 623–632, 2005.

67. Ottmann, R., Gene-environment interaction: Definitions and study designs, *Prev. Med.*, 25 (6), 764–770, 1996.

68. Cocozza, S., Methodological aspects of the assessment of gene-nutrient interactions at the population level, *Nut. Metabol. Cardiovas. Dis.*, 17 (2), 82–88, 2007.

69. Savaskan, E. et al., Red wine ingredient resveratrol protects from beta-amyloid neurotoxicity, *Gerontology*, 49 (6), 380–383, 2003.

70. Jang, J. H. and Surh, Y. J., Protective effect of resveratrol on beta-amyloid-induced oxidative PC12 cell death, *Free Radic. Biol. Med.*, 34, 1100–1110, 2003.

71. Online Mendelian Inheritance in Man, http://www.ncbi.nlm.nih.gov/entrez/dispomim.cgi?id = 261600, 2007.

72. Jimenez-Sanchez, G., Human disease genes, *Nature*, 409, 853–855, 2001.

73. Sowers, J. R., Epstein, M., and Frohlich, E. D., Diabetes, hypertension, and cardiovascular disease–An update, *Hypertension*, 37, 1053–1059, 2001.

74. Bianchini, F., Kaaks, R., and Vainio, H., Overweight, obesity, and cancer risk, *Lancet Oncol.*, 3, 565–574, 2002.

75. Permana, P. A., DelParigi, A., and Tataranni, P. A., Microarray gene expression profiling in obesity and insulin resistance, *Nutrition*, 20, 134–138, 2004.

76. Loktionov, A., Common gene polymorphisms and nutrition: Emerging links with pathogenesis of multifactorial chronic diseases (review), *J. Nutr. Biochem.*, 14 (8), 426–51, 2003.

77. Bouchard, C. and Perusse, L., Heredity and body fat, *Ann. Rev. Nutr.*, 8, 259–277, 1988.

78. Subbiah, M. T., Nutrigenetics and nutraceuticals: The next wave riding on personalized medicine, *Transl. Res.*, 149 (2), 55–61, 2007.

79. Kim, S., Moustaid-Moussa, N., Secretory, endocrine and autocrine/paracrine function of the adipocyte, *J. Nutr.*, 130, 3110S–3115S, 2000.

80. Zhang, Y. et al., Positional cloning of the mouse obese gene and its human homologue, *Nature*, 372, 425–432, 1994.

81. Tartaglia, L. A. et al., Identification and expression cloning of a leptin receptor, OB-R, *Cell*, 83, 1263–1271, 1995.

82. Moussa, N. M. and Claycombe, K. J., The yellow mouse obesity syndrome and mechanisms of agouti-induced obesity, *Obes. Res.*, 7, 506–514, 1999.

83. Nahon, J. L., The melanocortins and melanin-concentrating hormone in the central regulation of feeding behavior and energy homeostasis, *C. R. Biol.*, 329 (8), 623–638, 2006.

84. Jenkinson, C. P. et al., Novel polymorphisms in the neuropeptide-Y Y5 receptor associated with obesity in Pima Indians, *Int. J. Obes.*, 24, 580–584, 2000.

85. Dubern, B. et al., Mutational analysis of melanocortin-4-receptor, agouti related protein and alpha–melanocyte-stimulating hormone genes in severely obese children, *J. Pediatr.*, 139, 204–209, 2001.

86. Chagnon, Y. C. et al., Linkage and association studies between the melanocortin receptors 4 and 5 genes and obesity related phenotypes in the Quebec Family Study, *Mol. Med.*, 3, 663–673, 1997.

87. Kus, I. et al., Pinealectomy increases and exogenous melatonin decreases leptin production in rat anterior pituitary cells: An immunohistochemical study, *Physiol. Res.*, 53 (4), 403–408, 2004.

88. Luan J. et al., Evidence for gene-nutrient interaction at the PPAR gamma locus, *Diabetes*, 50, 686–689, 2001.

89. Mohamed-Ali, V., Pinkney, J. H., and Coppack, S. W., Adipose tissue as an endocrine and paracrine organ, *Int. J. Obes. Relat. Metab. Disord.*, 22, 1145–1158, 1998.

90. Hegazy, D. et al., NFkappaB polymorphisms and susceptibility to type 1 diabetes, *Genes Immun.*, 2, 304–308, 2001.

91. Moustaid-Moussa, N. et al., Gene expression profiling in adipose tissue, in: *Genomics and Proteomics in Nutrition*, Berdanier, C.D., Moustaid-Moussa, N., Eds., Marcel Dekker Inc., New York, 2004, pp. 257–280.

92. Cancello, R.A., Poitou Tounian, Ch., and Clement, K., Adiposity signals, genetic and body weight regulation in humans, *Diabetes Metab.*, 30, 215–227, 2004.

93. Siest, G. et al., Pharmacogenomics and cardiovascular drugs: Need for integrated biological system with phenotypes and proteomic markers, *Eur. J. Pharmacol.*, 527 (1–3), 1–22, 2005.

94. Di Renzo, L. et al., Interleukin-1 (IL-1) receptor antagonist gene polymorphism in normal weight obese syndrome: Relationship to body composition and IL-1 alpha and beta plasma levels, *Pharmacol. Res.*, 55 (2), 131–138, 2007.

95. Mutch, D. M, Wahli, W., and Williamson, G., Nutrigenomics and nutrigenetics: The emerging faces of nutrition, *FASEB J.*, 19, 1602–1616, 2005.

96. Ordovas, J. M. et al., Polyunsaturated fatty acids modulate the effects of the APOA1 G-A polymorphism on HDL cholesterol concentrations in a sex-specific manner: The Framingham Study, *Am. J. Clin. Nutr.*, 75, 38–46, 2002.

97. Astley, S. B. and Elliott, R. M., The European Nutrigenomics Organisation – linking genomics, nutrition and health research, *Nutr. Bull.*, 29, 254–261, 2004.

98. Ferguson, L.R., Nutrigenomics: Integrating genomic approaches into nutrition, *Res. Mol. Diagn. Ther.*, 10 (2), 101–108, 2006.

99. Fogg-Johnson, N., and Kaput, J., Nutrigenomics: An emerging scientific discipline, *Food Technol.*, 57, 60–67, 2003.

100. Visvikis-Siest, S. and Marteau, J. B., Genetic variants predisposing to cardiovascular disease, *Curr. Opin. Lipidol.*, 17, 139–151, 2006.

101. Corella, D. et al., Alcohol drinking determines the effect of the APOE locus on LDL concentrations in men. The Framingham Offspring Study, *Am. J. Clin. Nutr.*, 73, 736–745, 2001.

102. Knoblauch, H. et al., Common haplotypes in five genes influence genetic variance of LDL and HDL cholesterol in the general population, *Hum. Mol. Genet.*, 11, 1477–1485, 2002.

103. Mar, R. et al., Association of the APOLIPOPROTEIN A1/C3/A4/A5 gene cluster with triglyceride levels and LDL particle size in familial combined hyperlipidemia, *Circ. Res.*, 94, 993–999, 2004.

104. Ordovas, J. M. and Schaefer, E. J. Genetic determinants of plasma lipid response to dietary intervention: The role of the APOA1/C3/A4 gene cluster and the APOE gene, *Br. J. Nutr.*, 83 (Suppl. 1), S127–S136, 2000.

105. Carmena-Ramon, R. F. et al., Influence of genetic variation at the apo A-I gene locus on lipid levels and response to diet in familial hypercholesterolemia, *Atherosclerosis*, 139, 107–113, 1998.

106. Zheng, K. Q. et al., Association between cholesteryl ester transfer protein gene polymorphisms and variations in lipid levels in patients with coronary heart disease, *Chin. Med. J. (Engl.)*, 117, 1288–1292, 2004.

107. Skoglund-Andersson, C. et al., Influence of common variants in the CETP, LPL, HL and APO E genes on LDL heterogeneity in healthy, middle-aged men, *Atherosclerosis*, 167, 311–317, 2003.

108. Evans, D. and Beil, F. U., The association of the R219K polymorphism in the ATP-binding cassette transporter 1 (ABCA1) gene with coronary heart disease and hyperlipidaemia, *J. Mol. Med.*, 81, 264–270, 2003.

109. Krauss, R. M., Dietary and genetic effects on LDL heterogeneity, *World Rev. Nutr. Diet*, 89, 12–22, 2001.

110. Simopoulos, A. P., Evolutionary aspects of diet, omega-6/omega-3 ratio and genetic variation: Nutritional implications for chronic diseases, *Biomed. Pharmacother.*, 60, 502–507, 2006.

111. Stanley, L. A., Molecular aspects of chemical carcinogenesis: The roles of oncogenes and tumour suppressor genes, *Toxicology,* 96 (3), 173–194, 1995.

112. Vogelstein, B. and Kinzler, K. W., Cancer genes and the pathways they control, *Nat. Med.*, 10, 789–799, 2004.

113. Go, V. L., Butrum, R. R., and Wong, D. A., Diet, nutrition, and cancer prevention: The postgenomic era, *J. Nutr.*, 133, 3830S–3836S, 2003.

114. Jones, P. A. and Baylin, S. B., The fundamental role of epigenetic events in cancer. *Nat. Rev. Genet.*, 3, 415–428, 2002.

115. Loeb, L. A, Loeb, K. R., and Anderson, J. P., Multiple mutations and cancer, *Proc. Natl. Acad. Sci. U. S. A.*, 100, 776–781, 2003.

116. Nowell, P. C., Tumor progression: A brief historical perspective, *Semin. Cancer Biol.,* 12, 261–266, 2002.

117. Oppenheimer, S. B., Cellular basis of cancer metastasis: A review of fundamentals and new advances, *Acta Histochem.*, 108 (5), 327–334, 2006.

118. Friedl, P. and Wolf, K., Tumour-cell invasion and migration: Diversity and escape mechanisms, *Nat. Rev. Cancer*, 3, 362–374, 2003.

119. Folkman, J., Role of angiogenesis in tumor growth and metastasis, *Semin. Oncol.*, 29, 515–518, 2002.

120. Milner, J. A. et al., Molecular targets for nutrients involved with cancer prevention, *Nutr. Cancer*, 41 (1–2), 1–16, 2001.

121. Wattenberg, L.W., Chemoprevention of cancer, *Cancer Res.,* 45, 1–8, 1985.

122. Surh, Y. J., Cancer chemoprevention with dietary phytochemicals, *Nat. Rev. Cancer,* 3, 768–780, 2003.

123. Collins, A. R. and Ferguson, L. R., Nutrition and carcinogenesis, *Mutat. Res.*, 551 (1–2), 1–8, 2004.

124. Davis, C. D., Nutritional interactions: Credentialing of molecular targets for cancer prevention, *Exp. Biol. Med.*, 232, 176–183, 2007.

125. Milner, J. A., Molecular targets for bioactive food components, *J. Nutr.*, 134, 2492S–2498S, 2004.

126. Czeizel, A. E. and Dudas, I., Prevention of the first occurrence of neural-tube defects by periconceptional vitamin supplementation, *N. Engl. J. Med.*, 327, 1832–1835, 1992.

127. Voutilainen, S. et al., Low dietary folate intakes associated with an excess incidence of acute coronary events: The Kuopio Ischemic eart Disease Risk Factor Study, *Circulation,* 103, 2674–2680, 2001.

128. Ma, J. et al., Methylenetetrahydrofolate reductase polymorphism, dietary interactions, and risk of colorectal cancer, *Cancer Res.*, 57 (6), 1098–1102, 1997.

129. Milner, J. A., Incorporing basic nutrition science into health interventions for cancer prevention, *J. Nutr.*, 133, 3820S–3826S, 2003.

130. Katiyar, S. K. and Mukhtar, H., Tea consumption and cancer, *World Rev. Nutr. Diet.,* 79, 154–184, 1996.

131. Fujiki, H. et al., Cancer prevention with green tea polyphenols for the general population, and for patients following cancer treatment, *Curr. Cancer Ther. Rev.*, 1, 109–114, 2005.

132. Adhami, V. M., Ahmad, N., and Mukhtar, H., Molecular targets for green tea in prostate cancer prevention, *J. Nutr.*, 133, 2417S–2424S, 2003.

133. Hu, R. et al., Resveratrol inhibits phorbol ester and UV-induced activator protein 1 activation by interfering with mitogen-activated protein kinase pathways. *Mol. Pharmacol.*, 60, 217, 2001.

134. Norat, T. et al., Meat consumption and colorectal cancer risk: Dose-response meta-analysis of epidemiological studies, *Int. J. Cancer.*, 98, 241–256, 2002.

135. Sandhu, M. S., White, I. R., and McPherson, K., Systematic review of the prospective cohort studies on meat consumption and colorectal cancer risk: A meta-analytical approach, *Cancer Epidemiol. Biomark. Prev.*, 10, 439–446, 2001.

136. Martinez, M. E., Primary prevention of colorectal cancer: Lifestyle, nutrition, exercise. Recent results, *Cancer Res.*, 166, 177–211, 2005.

137. Ames, B. N. and Wakimoto, P., Are vitamin and mineral deficiencies a major cancer risk? *Nat. Rev. Cancer*, 2, 694–704, 2002.

138. Kim, Y. I., Role of folate in colon cancer development and progression, *J. Nutr.*, 133, 37318–37319, 2003.

139. Molloy, A. M., Genetic variation and nutritional requirements, *World Rev. Nutr. Diet.*, 93, 153–163, 2004.

140. Jubb, A. M., Bell, S. M., and Quirke, P., Methylation and colorectal cancer, *J. Pathol.*, 195, 111–134, 2001.

141. Kerr, J. F., Wyllie, A.H., and Currie, A. R, Apoptosis: A basic biological phenomenon with wide-ranging implication in tissue kinetics, *Br. J. Cancer*, 26, 239–257, 1972.
142. Khan, N., Afaq, F., and Mukhtar, H., Apoptosis by dietary factors: The suicide solution for delaying cancer growth, *Carcinogenesis*, 28 (2), 233–239, 2007.
143. Chen, C. and King, A., Dietary cancer-chemopreventive compounds: From signaling and gene expression to pharmacological, *Trends Mol. Med.*, 26, 318–328, 2005.
144. Ames, B. N., Increasing longevity by tuning up metabolism, *EMBO Rep.*, 6, S1, S20–S24, 2005.
145. Blount, B. C. et al., Folate deficiency causes uracil misincorporation into human DNA and chromosome breakage: Implications for cancer and neuronal damage, *Proc. Natl. Acad. Sci. U.S.A.*, 94, 3290–3295, 1997.
146. Afaq, F. et al., Delphinidin, an anthocyanidin in pigmented fruits and vegetables, protects human HaCaT keratinocytes and mouse skin against UVB-mediated oxidative stress and apoptosis, *J. Invest. Dermatol.*, 127, 222–232, 2007.
147. Saleem, M. et al., Lupeol, a fruit and vegetable based triterpene, induces apoptotic death of human pancreatic adenocarcinoma cells via inhibition of Ras signaling pathway, *Carcinogenesis*, 26 (11), 1956–1964, 2005.
148. Menendez, J. A. et al., Olive oil's bitter principle reverses acquired autoresistance to trastuzumab (Herceptin™) in HER2-overexpressing breast cancer cells, *BMC Cancer*, 7, 80, 2007.
149. Oyama, Y. et al., Protective actions of 5′-n-alkylated curcumins on living cells suffering from oxidative stress, *Eur. J. Pharmacol.*, 360, 65–71, 1998.
150. Collett, G.P. and Campbell, F.C., Curcumin induces c-jun N-terminal kinase-dependent apoptosis in HCT116 human colon cancer cells, *Carcinogenesis*, 25, 11, 2183–2189, 2004.
151. Chauhan, D. P., Chemotherapeutic potential of curcumin for colorectal cancer, *Curr. Pharm. Des.*, 8 (19), 1695–1706, 2002.
152. Chen, A., Xu, J., and Johnson, A. C., Curcumin inhibits human colon cancer cell growth by suppressing gene expression of epidermal growth factor receptor through reducing the activity of the transcription factor Egr-1, *Oncogene*, 25, 278–287, 2006.
153. Palozza, P. et al., Mechanism of activation of caspase cascade during beta-carotene-induced apoptosis in human tumor cells, *Nutr. Cancer*, 47 (1), 76–87, 2003.
154. Borek, C., Dietary antioxidants and human cancer, *Integr. Cancer. Ther.*, 3 (4), 333–334, 2004.
155. Bermudez, Y. et al., Vitamin E suppresses telomerase activity in ovarian cancer cells, *Cancer Detect. Prev.*, 31 (2), 119–128, 2007.
156. Gupta, K. and Zhang, K., Angiogenesis: A curse or cure? *Postgrad. Med. J.*, 81, 236–242, 2005.
157. Kerbel, R. S., Tumor angiogenesis: Past, present and the near future, *Carcinogenesis*, 21, 505–515, 2000.
158. Sing, R. P. and Agarwal, R., Mechanisms of action of novel agents for prostate cancer chemoprevention, *Endocr. Relat. Cancer*, 13 (3), 751, 2006.
159. Agarwal, C. et al., Anti-angiogenic efficiency of grape seed extract in endothelial cells, *Oncol. Rep.*, 11, 681–685, 2004.
160. Sing, R. P. et al., Grape seed extract inhibits advanced human prostate tumor growth and angiogenesis and upregulates insulin-like growth factor binding protein-3, *Int. J. Cancer*, 108, 733–740, 2004.
161. Fotsis, T. et al., Genistein, a dietary-derived inhibitor of in vitro angiogenesis, *Proc. Natl. Acad. Sci. U.S.A.*, 90, 2690–2694, 1993.
162. Cao, Y. and Cao R., Angiogenesis inhibited by drinking tea, *Nature*, 398, 381, 1999.
163. Dell'Eva, R. et al., The Akt inhibitor deguelin, is an angiopreventive agent also acting on the NF-κB pathway, *Carcinogenesis*, 28 (2), 404–413, 2007.
164. Miyazawa, T. et al., Anti-angiogenic potential of tocotrienol in vitro, *Biochemistry (Moscow)*, 69 (1), 67–69, 2004.
165. Donaldson, M. S., Nutrition and cancer: A review of the evidence for anti-cancer diet, *Nutr. J.*, 3, 19–30, 2004.
166. Brazma, A. et al., Minimum information about a microarray experiment (MIAME)–toward standards for microarray data, *Nat. Genet.*, 29 (4), 365–371, 2001.
167. Ehrat, M. and Kresbach, G. M., DNA and protein microarrays and their contributions to proteomics and genomics, *Chimia*, 55, 35–39, 2001.

15 Pharmacogenomics and Toxicogenomics in Food Chemicals

Bensu Karahalil

CONTENTS

15.1 BIOMARKERS FOR CHEMICALS

Chemically induced toxicity at the level of the organism has been a subject of study over many years. Initially, changes in the overall physiology of the organism, lethal dose determination, body and organ weight, and gross pathology at the macroscopic level were studied. These parameters are still worth-while and part of routine toxicological examinations. The development of histological techniques made microscopic examination of tissues and determination of toxic effects possible at a more cellular level. After this, molecular techniques, including Southern, Northern, and Western blotting, became available for sensitive and early identification of specific molecular end points. Changes in the levels of particular proteins or metabolites in tissue, blood, or urine or in *in vitro* systems that correlate well with certain types of toxicity are now routinely assessed.[1]

The term "biomarkers" was defined by the National Academy of Sciences in the United States as follows: "A biomarker is a xenobiotically-induced variation in cellular or biochemical components or processes, structures, or functions that is measurable in a biological system or sample."[2] A biomarker is a measure of a biological response to a system change and can occur at the molecular, cellular, or whole organism level.[3] Biomarkers are quantifiable determinations of biological events that allow the differentiation of normal from abnormal biological conditions, e.g., unusual chemical metabolites *in vivo*, genetic changes, and alterations in gene expression, proteins and cell-based markers.[4] In this respect, it is necessary to have biomarkers to effect the determination and evaluation of exposure to xenobiotics. One reason biomarkers are necessary is that they also contribute to the knowledge about clinical pharmacology and toxicology and provide a basis for the designing of clinical trials that expeditiously and definitively evaluate safety and efficacy, provide information for guidance in dosing, and minimize interindividual variation in response.[5] They have the potential to have a significant impact on the measurement of biological response to exposures, such as drug, diet, and exercise. Biomarkers are routinely evaluated in a hypothesis driven manner, based on biological knowledge of the disease and the structure of the exposed compounds.

15.1.1 CLASSIFICATION OF BIOMARKERS

Biomonitoring is a useful tool for understanding the linkage between external chemical exposures and possible outcomes in humans.[6] Many of the conventional biomarkers used in preclinical drug safety studies are either insensitive, nonspecific, or they appear late in the pathogenesis of the lesion and, as such, qualify only as reporters or indicators of toxicity.

Biomarkers are classified based on their primary usage in population studies, into at least three categories:

1. Biomarkers of exposure
2. Biomarkers of effect
3. Biomarkers of susceptibility

Biomarkers of exposure are important in toxicology, because they are indicators of internal dose, or the amount of chemical exposure that has resulted in absorption into the body. Biomarkers of effect reflect changes following exposure at different levels and their reversibility. Biomarkers of susceptibility represent an inherent property of the individual, due to environmental or genetic factors, that can be determined without exposure. Biomarkers play a role in the use of pharmacogenetics, pharmacogenomics, and pharmacoproteomics for development of personalized medicine.

The use of genomics technologies capable of covering the gaps of conventional toxicology tests are cell and tissue-wide protein expression (proteomics) and metabolite profiling (metabolomics). Toxicogenomics together with proteomics, metabolomics, and transcriptomics can be combined with conventional toxicological approaches and high throughput molecular screening in an effort to verify the gene–environment interactions in disease and organ dysfunction. The combination of these technologies provides information about different cellular components such as mRNA, proteins, and metabolites, resulting from an action on the genome, and also about functional effects in cells, tissues, and biofluids (Figure 15.1).

Theoretically, a combination of a pair of markers would generate a combinatorial biomarker, but in practice the identification of combinatorial markers usually involves gene expression profiling or other highly multivariable assays. Combinatorial biomarkers enable better specificity and/or sensitivity than single markers.[3] Thus, markers in genomics technologies and in conventional toxicology

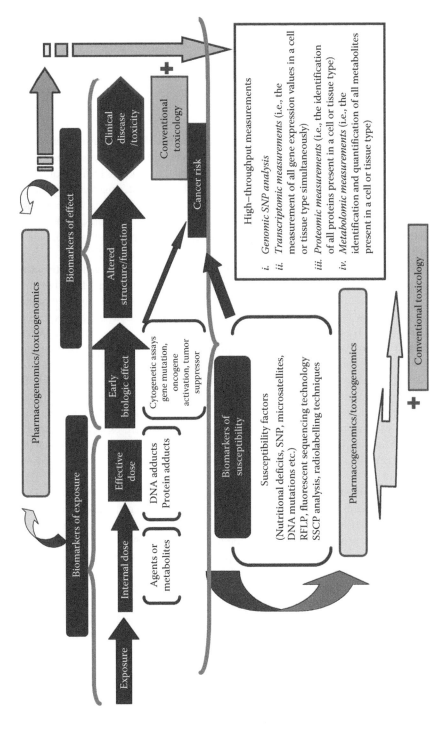

FIGURE 15. 1   Scheme of biomarkers including new technologies for genetic toxicology, susceptibility, and drug discovery investigations.

tests should combine to obtain more reliable and sensitive outcomes. Toxicogenomics studies are generally built on standard toxicology studies generating biological end point data, and as such, one goal of toxicogenomics is to detect relationships between changes in gene expression and in those biological parameters.

15.2 INTRODUCTION TO GENOMICS

First we will discuss genomics technologies and their usage in chemical safety. Then we will discuss how the transition from genetics to genomics took place. The key questions are: *What is genomics? What is in a genome? What is the history of genomics?*

Genomics is the study of the genome (all the genes of an organism). For example, humans have about 30,000 genes in their genome. Genetics is generally involved in the study of one or a few genes at a time. In contrast genomics attempts to obtain a global view of all the genes of an organism. To compare classical genetics with genomics, classical geneticists can find the location of the gene on a chromosome, then clone the gene, and sequence its DNA. By contrast, in genomics, scientists start by sequencing all the DNA of an organism and then they identify all the genes, and map the gene on the chromosome, and then try to determine the trait or functions of all the genes. Genomics provides a means of identifying all the genes through DNA sequencing, mapping the location, and figuring out all the functions of all the genes/gene products on all the chromosomes.[7–9]

Genomics is the study of an individual's gene structure including how the genes interact with each other and with the environment. Genomics came of age with the sequencing of the human genome along with the development of DNA microarrays and the computing power to analyze multiple variables. Together these allow for fully comprehensive and rapid investigations of gene expression.[10] The human genome has been the biggest project undertaken to date but there are many research projects around the world trying to map the gene sequences of other organisms. Researchers are using genomics to understand the effects of proteins on cell function, the association between food poisoning and bacteria-induced disease, the resistance of crops to environmental stress, the involvement of genes in fungal crop disease, the reaction of the body to tissue grafts and artificial prostheses, and so on. Thus, the use of genomics provides drug discovery, safer food, and crops adapted to environmental change.

The emergence of genomics has been exciting for anyone involved, especially in drug research and development. Early on, there was a great deal of excitement generated by the many promising ideas for the use of genomics and the benefits that genomics could bring in stimulating drug discovery and development.

Genomics is divided into two main groups: (1) structural genomics and (2) functional genomics. Structural genomics includes genome mapping, sequencing, gene and genome organization, genome manipulation, network structure, and protein structure characterization. Functional genomics includes transcriptomics, proteomics, and metabolomics. At the DNA level, genomics technologies used in all the studies mentioned above are, perhaps, the most important tools of technology being developed by mankind.

Genomics approaches have changed the way we do research in biology and medicine. Nowadays, we can measure the majority of mRNAs, proteins, metabolites, protein–protein interactions, genomic mutations, polymorphisms, epigenetic alterations, and micro RNAs in a single experiment. The data generated by these methods together with the knowledge derived by their analyses was unimaginable just a few years ago.[11] In drug safety assessment toxicogenomics, the discipline that uses genomics technologies was seen as a means to avoid late stage drug attrition by improving pharmaceutical companies' abilities to predict toxicity and to assess safety risks.[12–14]

15.2.1 GENOMICS TECHNOLOGIES

The benefits of genomics technologies are (1) provide better understanding of the molecular mechanisms of adverse biological processes, (2) identify novel expression profiles/biomarkers to aid prediction of toxicity of chemicals, (3) aid development of better *in vitro* models (based on mechanistic understanding that is more physiologically relevant), and (4) potential direct use as a screening tool.

15.2.2 STRUCTURAL GENOMICS

The first main group of genomics technology is structural genomics, which contains many applications and evaluation parameters. The variation in human genes was first realized thanks to pharmacogenetics studies.

15.2.2.1 Pharmacogenetics–Pharmacogenomics; Interchangeable Terms

The resemblance of identical twins in drug metabolism as compared with nonidentical twins established the importance of genetic inheritance in the disposal of many drugs. More recent developments have broadened pharmacogenetic approaches to include novel genomic scale techniques and the introduction of the term pharmacogenomics in the 1990s. Pharmacogenetics had its beginnings about 40 years ago when researchers realized that some adverse reactions were caused by genetically determined variations in enzyme activity. For example, prolonged muscle relaxation after suxamethonium was explained by an inherited deficiency of a plasma cholinesterase, and hemolysis caused by antimalarials was recognized as being associated with inherited variants of glucose-6-phosphate dehydrogenase.[15]

15.2.2.2 History of Pharmacogenetics and Pharmacogenomics

The history of pharmacogenetics stretches as far back as 510 BC, when Pythagoras noted that the ingestion of fava beans resulted in a potentially fatal reaction in some, but not all, individuals. Since then there have been numerous landmarks that have shaped this field of research, and have led to the current wave of interest.[16]

The origin and development of pharmacogenetics is traced to the early hints by Archibald Garrod, Haldane, and later by Williams. The concept was delineated by Motulsky in 1957 and described as pharmacogenetics by Vogel in 1959. Pharmacogenomics has been around in some form since the 1930s. In 1902, Archibald Garrod first asserted the hypothesis that genetic variations could cause adverse biological reactions when chemical substances were ingested. He also suggested that enzymes were responsible for detoxifying foreign substances, and that some people do not have the ability to eliminate certain foreign substances from the body because they lack enzymes required to break down these materials. For example, the term pharmacogenetics was to define inherited variability in response to drug treatment. Initially aimed at finding enzyme mutations that resulted in differences between individuals in drug metabolism, the field expanded to investigate genes controlling pharmacodynamic processes.[17,18]

After the study regarding fava beans, the first pharmacogenetic study took place in 1932, when the inability to taste a chemical compound known as phenylthiocarbamide was linked to an autosomal recessive trait. In the 1940s and 1950s, scientists first began to note "variable drug responses" in people taking various preventive medications. The science of pharmacogenomics, which aims to define the genetic determinants of drug effects, has evolved over the past 50 years. New technologies that became available in the 1950s, combined with a more genetic approach to investigation, disclosed new relationships between the genetic control of responses to exogenous substances and their metabolic fate.[19] This notion was strengthened by studies on family and twins in the 1960s and 1970s and extended by biochemical studies in the 1970s and 1980s.[20] Pharmacogenomics has an ever increasing impact on drug discovery and development. The Food and Drug Administration (FDA) is encouraging this effort and is putting significant emphasis and support into personalized medicine, promoting the translation of research findings into medical practice.

15.2.2.3 Pharmacogenomics Reflects the Evolution of Pharmacogenetics

Pharmacogenomics refers to the general study of all of the many different genes that determine drug behavior. Pharmacogenetics refers to the study of inherited differences (variation) in drug metabolism and response. The distinction between the two terms is considered arbitrary, however, and now the two terms are used interchangeably.[21] There are differences in terminology between pharmacogenetics and pharmacogenomics. Pharmacogenetics deals with differential effects of a drug (*in vivo*) in different patients dependent on the presence of inherited gene variants but pharmacogenomics with differential effects of compounds (*in vivo* or *in vitro*) on gene expression, among the entirety of the expressed genes. Expression profiling is essential in pharmacogenomics, Single Nucleotide Polymorphisms (SNP), and genomic expression in pharmacogenetics. Pharmacogenetics provides disease specific health care and it is a tool for compound selection and drug discovery. Pharmacogenetics focuses on patient variability instead of compound variability, in the question of one drug–many genomes (i.e. different patients).[22]

15.2.2.4 Some Differences between Pharmacogenomics and Pharmacogenetics

Pharmacogenomics is not just pharmacogenetics. There are three aspects of pharmacogenomics that make it different from classical pharmacogenetics:

1. Genetic xenobiotic response profile
2. The effect of xenobiotics on gene expression
3. Pharmacogenomics in drug discovery and drug development

15.2.2.4.1 Genetic Xenobiotic Response Profile

The response to drugs and xenobiotics varies with genotype. Some patients react well to drugs, while others may not benefit, or may even respond adversely. Individuals also experience different reactions to environmental agents. The sequencing of the human genome and the large-scale identification of genome polymorphisms have provided opportunities for understanding the genetic basis for individual differences in response to potential toxicants: an area of study that has come to be known as toxicogenetics/pharmacogenetics.[23] For SNPs to be used to predict a person's xenobiotic response, a person's DNA must be examined (sequenced) for the presence of specific SNPs.

15.2.2.4.2 Single Nucleotide Polymorphism

The mapping out of genotype–phenotype interactions is done mainly through SNP genotyping. SNPs (pronounced "snips") are naturally occurring variations of single nucleotides at set positions in a population's genome. SNPs are seen in 0.1% (approximately 3 million pairs) of all genomic DNA sequences and are the most common variation. SNPs occur once every 300–3000 base-pairs, if one compares the genomes of two unrelated individuals. On an average it can be said that one SNP occurs in the sequence of 1000 base-pairs. Thus in the entire genome an estimated 3.2 million SNPs would occur in 3.2 billion nucleotides. SNPs are responsible for most inherited variations.[15] Each SNP represents a difference in a single DNA building block, called a nucleotide. For example, a SNP may replace the nucleotide cytosine (C) with the nucleotide thymine (T) in a certain stretch of DNA.[24] Each variable base (SNP) results from a single error in DNA replication that occurred once in the history of mankind. Each SNP is characterized by only two bases. Common variants are more experimentally tractable than rare variants. SNPs result in functional differences by altering the quality and/or the quantity of cellular proteins.[25] Why has it been so difficult to identify genes determining disease or drug response? Most common human traits are not caused by a single variation, but probably by 20 or more genetic changes across the genome. SNPs will play a major role in associating sequence variations with heritable clinical phenotypes of drug or xenobiotic response. SNP discovery is of major interest in the post-genome era because SNPs have broad applications in biological fields, such as fine mapping, disease studies, population genetics, and molecular evolution. In collaboration with the National Human Genome Research Institute (NHGRI), the National Center for Biotechnology Information (NCBI) has established the dbSNP database to serve as a central repository for molecular variation.[26] Meanwhile, millions of SNPs have been available in the private domains, such as Celera's RefSNP database. Designed to serve as a general catalog of molecular variation to supplement the GenBank database, submissions can include a broad range of molecular polymorphisms: single base nucleotide substitutions, short deletion and insertion polymorphisms, microsatellite markers, and polymorphic insertion elements such as retrotransposons.[27] This provides us with an unprecedented opportunity to examine the local DNA sequence context of SNPs, and therefore to understand the molecular mechanisms of genome sequence evolution.[28]

15.2.2.4.3 What is the Human Genome Project?

The Human Genome Project (HGP) was the international collaborative research program whose goal was the complete mapping and understanding of all the genes of human beings. All our genes together are known as our "genome."[29] The HGP was officially launched in the United States on October 1, 1990 as a 15 year program to map and sequence the complete set of human chromosomes and those of several model organisms. The HGP is laying the groundwork for a revolution in medicine and biology. Its importance is underscored by the level of funding from the National Institutes of Health (NIH), the Department of Energy (DOE), the Wellcome Trust, and other governments and foundations around the world.[30–32]

15.2.2.4.4 Goals of the Human Genome Project

The HGP is a project that was undertaken in 1990. The Human Genome Project has successfully completed all the major goals in its current 5 year plan, covering the period 1993–1998. A new plan for 1998–2003 has been presented in which human DNA sequencing will be the major emphasis.[33] The goals of the HGP are

- To map our 30,000–50,000 genes and the sequencing of all of our DNA
- To determine the sequences of the 3 billion chemical base pairs that make up human DNA
- To investigate the nature, structure, function, and interaction of the genes, genomic elements, and genomes of humans and relevant pathogenic and model organisms
- To characterize the nature, distribution, and evolution of genetic variation in humans and other relevant organisms
- To study the relationship between genetic variation and the environment in the origins and characteristics of human populations and the causes, diagnoses, treatments, and prevention of disease
- To store this information in databases
- To improve tools for data analysis, transfer related technologies to the private sector, and address the ethical, legal, and social issues (ELSI) that may arise from the project[34]

15.2.2.4.5 Effect of Drugs on Gene Expression

Drugs affect gene expression. Variations in genes affect drug response. Gene expression studies before microarrays were dedicated to gene-finding and cloning. Many strategies were used to find mRNA transcripts which were differentially expressed between treated and untreated cell populations or tissues. These techniques were developed to find genes responsible for the differentiation of cell types in which specific gene expression is either present or absent between cell types or is switched on or off during the development of a cell lineage. Subsequently researchers applied the technology to looking for drug or chemical effects on increasing or decreasing the level of specific gene expression. It is worth noting that even with the microarray approach it is easier and more clear-cut to detect the on/off signals in gene expression observed between differentiated cell types than it is to detect the often transient and low magnitude changes in gene expression observed with drug and chemical treatment.[14]

15.2.2.4.6 Pharmacogenomics in Drug Discovery and Drug Development

One of the key factors in developing improved medicines lies in understanding the molecular basis of the complex diseases we treat. Genetic studies in conjunction with gene expression, proteomic, and metabolomic analyses provide a powerful tool to identify molecular subtypes of disease. Using these molecular data, pharmacogenomics has the potential to impact on the drug discovery and development process at many stages of the pipeline, contributing to both target identification and increased confidence in the therapeutic rationale.[35]

15.2.2.4.7 Toxicology to Toxicogenomics: Applications of Genomics in Toxicological Science

Toxicology, the study of poisons, focuses on substances and treatments that cause adverse effects in living things. A critical part of this study is the characterization of the adverse effects at the level of the organism, the tissue, the cell, and the molecular makeup of the cell. Thus, studies in toxicology measure effects on body weight and food consumption of an organism, on individual organ weights, on microscopic histopathology of tissues, and on cell viability, necrosis, and apoptosis. Recently added to the arsenal of end points that such toxicological studies can use is the measurement of levels of the thousands of proteins and mRNAs present in the cell.[36] The marriage of toxicology and genomics has created not only opportunities but also novel informatics challenges. The basic steps in generating a toxicogenomics knowledge base for the evaluation of novel compounds include the collection of data from conventional toxicology, establishing quality-checked and normalized expression data, construction of a reference database with a number of well-known and well-characterized compounds, and the selection of toxicological marker genes based on statistical methods.[37] If successfully implemented, this integration has the potential to serve as a powerful synergistic tool for understanding the relationship between gross toxicology and genome-level effects.[36]

15.2.2.4.8 Introduction of the Concept of Toxicogenomics

The concept of toxicogenomics was first introduced in 1999 and can be defined as the study of the relationship between the structure and activity of the genome (the cellular complement of genes) and the adverse biological effects of exogenous agents. The application of toxicogenomics provides an exceptional opportunity to identify the biological pathways and processes affected by exposure to pharmaceutical compounds and/or xenobiotics (exogenous agents). Toxicogenomics will provide a better understanding of the mechanism of toxicity and may facilitate the prediction of toxicity of unknown compounds. Mechanism-based markers of toxicity can be discovered and improved interspecies and *in vitro–in vivo* extrapolations which will drive model developments in toxicology. Toxicological assessment of chemical mixtures will benefit from the new molecular biological tools.[1,38] Toxicogenomics integrates functional genomics with classical toxicology and has great potential to revolutionize mechanistic research in general and in particular for chemical mixtures. Also, it is a scientific field that studies how the genome is involved in responses to environmental stressors and toxicants. Toxicogenomics combines studies of genetics, mRNA expression, cell and tissue-wide protein expression and metabolomics to understand the role of gene–environment interactions in disease. Transcriptomics, proteomics, and metabolomics are genomics technologies with great potential in toxicological sciences. Toxicogenomics involves the integration of conventional toxicological examinations with gene, protein, or metabolite expression profiles. An overview together with selected examples of the possibilities of genomics in toxicology is given. The expectations raised by toxicogenomics are earlier and more sensitive detection of toxicity. Furthermore, toxicogenomics will provide a better understanding of the mechanism of toxicity and may facilitate the prediction of toxicity of unknown compounds.[1,39,40] Gene-array technology that is central to the field of toxicogenomics is known as DNA micro-array. A technology that enables scientists to simultaneously monitor interactions among thousands of genes within the genome, it allows scientists to monitor the activity of thousands of genes simultaneously.[41]

15.2.2.4.9   The Need for Toxicogenomics

Toxicogenomics: Modern methods replace animal tests; Millions of animals are raised in the United States each year for routine toxicology tests, exposed to compounds in food additives, cosmetics, and industrial products, and then studied for ill effects. This is a time-honored way of identifying human health risks. It is also expensive and increasingly under attack by animal-rights activists as wasteful. Now the new approach, "toxicogenomics," has grown out of the human genome project. Rather than using animal pathology to identify illnesses, it probes human or animal genetic material printed on plates, called DNA arrays. Cancer researchers have already been using such arrays for several years to compare gene expression in healthy and diseased cells. Toxicologists are using the same technology to profile gene expression in cells exposed to test compounds.[42]

The drug development process is currently being hindered by non-optimal prediction of toxicity. Advances in molecular profiling approaches, such as transcriptomics, proteomics, and metabolomics, offer the potential to provide a more comprehensive insight into toxicological effects than hitherto possible.[43]

These new technologies present their own challenges, however, particularly in relation to standardization and assessment. The testing of new drugs, chemicals, food additives, and cosmetics may soon rely on toxicogenomics, an exciting branch of study evolving from the human genome project. The innovative approach goes straight to the source human DNA to sort out how cells react to certain chemicals, rather than relying on animal tests, which are notoriously poor predictors of toxicity in humans. Toxicogenomics may also allow researchers to rapidly evaluate new compounds for potential medicinal properties and let physicians tailor treatments and drugs to their patients' individual needs, based on their unique genetic makeup.

The GeneLib, ProteinLib, and PathwayLib components of ArrayTrack contain general but essential information for functional genomics research. These libraries also provide a basis for linking and integrating various omics data. For example, lists of genes, proteins, and metabolites derived from various omics platforms could be cross-linked based on their common identifiers through these three libraries. An additional library, ToxicantLib, is being developed for ArrayTrack and will similarly provide linkage between toxicological data and the different types of omics data. The ToxicantLib contains the chemical name and structure together with toxicological end points. Through the similarity comparison of the chemical structure of a toxicant with the structures of the metabolites in the PathwayLib, it provides the examination of the toxicity effect of a particular toxicant at the molecular level.[44]

15.2.3   FUNCTIONAL GENOMICS

Functional genomics is the understanding of the function of genes and other parts of the genome. Bioinformatics is that of gene sequences using new, high-throughput technologies; these technologies therefore must represent the new tools of medical discovery.[45]

1. Transcriptomics
2. Proteomics
3. Metabolomics

15.2.3.1   Transcriptomics

The composition of the genome determines many biological processes in an organism, thereby influencing the susceptibility toward genetic diseases or the response to xenobiotic compounds.[46] The transcriptome is the complete set of RNA products that can be produced from the genome, and transcriptomics is the study of the transcriptome. For example, microarrays and chips represent powerful tools for studies of diet–gene interactions. Identifying relevant diet–gene interactions benefits individuals seeking personalized dietary advice, and also improves public health recommendations by providing sound scientific evidence linking diet and health.[47,48] Unlike the genome, the transcriptome is extremely dynamic. Most of our cells contain the same genome regardless of the type of cell, stage of development or environmental conditions. Conversely, the transcriptome varies considerably in these differing circumstances due to different patterns of gene expression. Transcriptomics is therefore a global way of looking at gene expression patterns.[49]

15.2.3.2   Proteomics

Proteins are involved in all biological processes and can therefore be considered the functionally most important biological molecules. They are also particularly rich in biological information. The systematic identification and characterization of proteins, called proteomics, carries with it huge expectations, such as diagnostic and prognostic markers in blood serum and other body fluids; targets for pharmaceutical drugs; and improving the knowledge of fundamental biological processes.[50]

In Figure 15.2, transcription of DNA into RNA and further translation of RNA into proteins is shown. Correspondingly, genes are transcribed into RNA molecules, and after different processes these RNAs are translated into proteins.

15.2.3.3   Metabolomics

Concentrations, fluxes, and transport mechanisms of metabolites are known to be highly sensitive to disease and to drug intervention. Changes in recorded levels of, for example, glucose, cholesterol, and triglycerides not only have clinical implications but these metabolites can be used as outcome biomarkers. Metabolomics seeks to probe the mechanisms that contribute to changes in metabolic levels and fluxes and to understand the implications of such changes on disease, drug response, and nutritional intervention. A principal goal is to increase the range of available metabolite biomarkers, further understand their cellular and systemic distribution and facilitate their application in accelerating drug candidate progression. At the same time, the use of metabolomics makes possible the identification of new biomarkers for the ingestion of healthy food, something very important for public health.[51]

Proteomics and metabolomics detect only a small subset of the total proteome and metabolome, and fluxomics (metabolic flux analysis) tends to focus on discrete pathways determined in part by the substrates that are used for the analysis.[52] To date, gel-based proteomic analysis remains

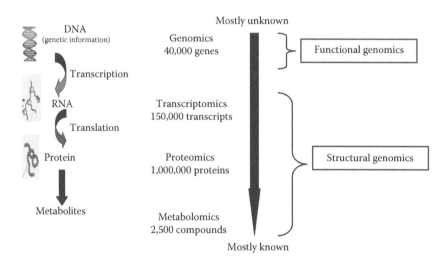

FIGURE 15.2   The relations between genomics technologies.

the main technique used in most of proteomic studies. Major advantages of gel-based proteomic analysis are its simplicity and availability in most of proteomic laboratories. It is probably the best "screening tool" to detect PTMs, although the specific identification is still required.[53]

15.3 GENOMICS IN MEDICINE AND FOOD CHEMICALS

Genomics is the study of the genome. Genetics originally is involved in the study of one or a few genes at a time. Genomics gets a global view of all the genes of an organism. Comparing classical genetics to genomics, classical geneticists can find the location of the gene on a chromosome, then clone the gene, and sequence its DNA. In genomics, researchers first sequence all the DNA of an organism, then identify the genes and map the gene on the chromosome, and then try to determine the trait or functions of all the genes. One of the important emerging points of genomics is to revolutionize the practice of medicine. That revolution, called "personalized medicine," includes the use of genomic information to improve the diagnosis of disease, as well as the prevention and treatment of disease. Personalized medicine tries to answer questions like: Why do some people get cancer and others don't? Why is cancer more aggressive in this person compared to that one? Why does this drug work for you and not me? Why does someone need twice the standard dose to be effectively treated? And why do others need only half of the standard dose? The goal of personalized medicine is to get the best medical outcomes by choosing treatments that work well with a person's genomic profile.[54] Each of these post-genomic approaches has already contributed to our understanding of specific aspects of the disease process and the development of diagnostic/prognostic clinical applications.[55]

On the other hand, food safety and food security are important global issues. "Omics" technologies also help to elucidate pathogen behavior at the molecular level and to develop better detection and typing systems. Omics-based tools enable researchers to explore complex biological processes in a quantitative and integrative manner via a systems biology approach. The application of omics technologies will play an important role in understanding how pathogens survive food safety barriers and interact with the host species, which will give rise to better tools for their control.[56]

15.3.1 DRUG DISCOVERY AND DEVELOPMENT

Personalized medicine is the treatment best suited for an individual patient. It is based on genomic as well as other factors that influence the response to drugs. Several "omics" technologies are being increasingly used for this purpose and personalized drug-discovery efforts are in progress in major therapeutic areas.[57] Individualized therapy may be also based on differential protein expression rather than a genetic polymorphism. Finally, proteomic technologies would enable discovery and development of drugs suitable for personalized therapy.[58]

Thanks to genomics technologies, there are more powerful medicines, better, safer drugs the first time, more accurate methods of determining appropriate drug dosages, advanced screening for disease, and a decrease in the overall cost of health care. The need to reduce adverse drug reactions and costs for the health care system may encourage many pharmacogenomic research programs by both private enterprises and government. The reduction of drug development time and the need to get more new drugs on the market are also causing pharmaceutical companies to increase pharmacogenomics use in their clinical research programs. Gene expression studies are highly applicable to drug development given the fundamental nature of gene transcription responses to xenobiotics. Specific gene expression responses have been established for several toxicities and compound classes.

15.3.2 FOOD CHEMICALS

Nutrition research has a history dating back to 1785, and the emergence of the "chemical revolution" when fundamental processes of metabolism and respiration were discerned. The advent of biochemical genetics in the mid-twentieth century, together with the recognition of person-to-person and population differences in response to food and foreign chemicals over the past 50 years, the focus of nutrition research evolved to mechanisms of individual variability in nutritional responses, and nutrigenomics more recently. Because nutrition impacts people on a daily basis in both health and disease, as well as in prevention and treatment of some common multifactorial diseases, nutrigenomics offers a promise for applications in public health and nutritional interventions guided by human genomic variation.[59]

We are exposed to thousands of chemical substances in our daily lives. Genomics is also entering food production and processing and has an important role in food safety. The quality and quantity of diet is a key determinant of human health and disease. Many dietary nutrients and components can act as signals and change the patterns of gene expression and protein expression and switch metabolic pathways to form distinctive dietary signatures.

The processes involved in the absorption, distribution, metabolism, and excretion of food additives, are dependent on their chemical structure. Some compounds such as glycerol esters, carbohydrates, proteins, or simple peptides (such as aspartame) are digested in the gastrointestinal tract, and the products are absorbed and utilized in normal intermediary metabolism. The enzymes of intermediary metabolism usually show both high specificity and high capacity. In consequence, there would be no difference between substrates which are food additives or produced from food additives, and other substrates absorbed from the diet. The possibility that structurally unrelated food additives could show either joint actions or interactions has been assessed based on their potential to share common sites and mechanisms of action or common pathways of elimination. The (International Life Sciences Institute) ILSI Europe Acceptable Daily Intake Task Force established an Expert Group of independent scientists to undertake an evaluation of the possibility of interactions occurring between the 350 food additives currently approved in the EU, as laid down in directives 94/35/EC, 94/36/EC, and 95/2/EC.[60]

In the past, the effects and mechanisms of dietary components on health and disease were investigated using functional assays or studies based on single genes or single physiological outcomes of measure. However, it is increasingly obvious that to understand the effects of nutrition on health and disease, molecular studies are indispensable.[61]

15.4 GENOMICS IN TOXICOLOGY; JOINT REPORTS (COT, COM, COC)

A joint symposium by the Committee on Toxicity (COT), Committee on Carcinogenicity (COC), and Committee on Mutagenicity (COM) was held in October 2001 on the use of genomics and proteomics in toxicology. The Committee reviewed the conclusions two times and the main conclusions were reached at the February 2004 meeting. The main conclusions at the February meeting are (1) To evaluate the functional significance of changes in gene expression and the distinction between adverse and adaptive changes, (2) The need for better toxicogenomic studies on dose-response assessment and the reversal of toxicological effects. At present toxicogenomic data could use part of the overall toxicological data, but not use it in the absence of prior knowledge about the toxicity of the chemical derived from conventional toxicological approaches, (3) To confirm key gene changes independently, for example by quantitative PCR analysis of mRNA (transcriptomic methods), (4) Both two dimensional gel methods and solid phase techniques were essential as complementary techniques (proteomics), (5) Metabolomics has potential for the development of biomarkers, (6) Transcriptomics, proteomics, and metabolomics needed to be considered for toxicological risk assessment, and (7) Epidemiological studies, showing exposure to chemicals, were needed using toxicogenomics.[62] There are many published studies with different chemicals using omics technologies.[63–65] Kier et al. have generated a database containing gene expression, histopathology, and clinical chemistry findings using the microarray with rat-specific toxicology related genes for 89 compounds. Analysis of the database indicates that treatment with toxic compounds induces specific gene expression patterns, dose- and time-dependent response, relationships in gene expression and gene expression–organ toxicity correlation. Their data also provided insight into toxic mechanism and gene regulation pathways. All results with both, rats *in vivo* and human hepatocytes *in vitro* suggest that microarrays with toxicologically relevant genes can be used routinely for the evaluation of chemical toxicity.[66] After discussions held at its February and May 2004 meetings, the COM reached a few conclusions. The COM agreed there were considerable difficulties in developing *in vitro* mutagenicity screening assays using toxicogenomic approaches and reviewed a study which had used HepG2 cells and considered that the data in the study were able to distinguish between genotoxic and non-genotoxic carcinogenomics in toxicology. On the other hand, the COC reached additional conclusions after discussions held at its June 2004 meeting. These conclusions were toxicogenomic studies in experimental animals regarding dose-response evaluation, investigations of reversibility, statistical handling of data and bioinformatics developments. These conclusions of COC are consistent with those reached by COT.[67] The COT committee established a working group to consider the risk assessment of multiple residues of pesticides and veterinary medicines in food, and of multiple sources of exposure to these substances on the request of the Food Standards Agency and consumer concerns.[68]

Toxicogenomics enables safety for humans and the environment without animal testing (ASAT). Due to legislation and requirements for chemical and food safety that recommend no animal tests there is an urgent need for reliable, quick and low-cost test methods that are not based on animal models. Toxicogenomics is expected to enable the development of such methods within a new concept of safety and risk assessment (ASAT).

15.5 SOME PUBLISHED STUDY EXAMPLES IN FOOD CHEMICALS USING GENOMICS TECHNOLOGIES

Echinacea, often used for the treatment of cold and viral infection, is another one of the top-five selling products in the United States. *In vitro* liver microsomal studies showed that echinacea may affect the activity of metabolizing enzymes (CYP3A, CYP1A2, CYP2D6, and CYP2C9). After 8 days of administration of 400 mg four times a day of an echinacea product, CYP1A2 and intestinal CYP3A activities were inhibited while hepatic CYP3A was induced. The findings show that the effect of echinacea on various CYP3A substrates may vary depending on the relative contribution of intestinal CYP3A versus the hepatic CYP3A in the individual substrate's clearance pathway to a given drug's overall clearance. *Ginkgo biloba* is often used for memory improvement. *In vitro* liver microsomal studies showed that ginkgo products affect the activity of metabolizing enzymes.[69]

Groten et al. dosed rats for 28 days with the four different food additives at four dose levels around the reported NOAEL. The samples chosen for transcript profiling were exposed to concentrations below the doses where obvious toxicity was observed, e.g. liver weight differences or histopathologically identified aberrances. Pooled RNA samples of rat livers were monitored for the expression of about 3500 genes using cDNA microarrays. Their data indicate that changes in hepatic transcription can be predictive for actual physiological changes that may occur when rats are exposed to food additives. These studies show the value of transcriptomics /proteomics in terms of predicting changes that occur at the physiological level. In a broader sense, these data demonstrated that a transcriptomics approach allows predictions of mechanisms of hepatotoxicity of food additives and of possible interactions.[40]

Kojic acid is a natural product and normally used as a food additive and preservative, a skin-whitening agent in cosmetics, a plant growth regulator and a chemical intermediate. Cheng et al. examined the overall biological effects of kojic acid on the gene expression profiling of human skin A375 malignant melanoma cells using DNA microarray technology. After treatment with kojic acid, a total of 361 differentially expressed genes were distinctively changed with 136 up-regulated genes and 225 down-regulated genes. Seven down-regulated genes of APOBEC1, ARHGEF16, CD22, FGFR3, GALNT1, UNC5C, and ZNF146 were shown to be the tumor suppressor genes in melanoma cancer cells. Thus, microarray technology coupled with RT-qPCR offered a high throughput method to explore the number of differentially expressed genes responding to kojic acid and their biological functions, and led to greater understanding of kojic acid effects on skin cancer therapy and related side effects. Moreover, the differentially expressed genes may become useful markers of skin malignant melanoma for further diagnostic and therapeutic applications.[70]

Mitchell et al. conducted a randomized controlled feeding study using Matrix-assisted laser desorption/ionization ionization-time of flight mass spectrometry (MALDI-TOF MS) assay to show that MALDI-TOF can be a potentially powerful tool in biomarker discovery. A four-period crossover study design was used and participants (38 healthy, nonsmoking men and women, aged 20–40) were assigned to each of the four diets in random order. Diets were an extension of a basal diet devoid of fruits and vegetables. In addition to one period of the basal diet, participants were fed a basal diet supplemented with either 436 g cruciferous (radish, sprouts, cauliflower, broccoli, and cabbage) or 190 g allium (chives, leeks, garlic, and onion), or 270 g apiaceous (dill, celery, parsley, parsnips, and carrots) vegetables. They collected blood samples on the last day of each diet. The results of this study indicate that MALDI-TOF could be a potentially powerful tool in biomarker discovery. Not only were we able to classify participants based on diet using only two peaks, but also to identify one of those peaks as a molecule that is relevant to diet.[71]

15.6 POTENTIAL IMPLICATIONS OF GENOMICS FOR REGULATORY AND RISK ASSESSMENT APPLICATIONS

Advances in genomics will have significant implications for risk assessment practice and regulatory decision making. EPA and other regulatory agencies are beginning to address the use of genomics data for various risk assessment applications, including the need to establish a link between genomic alterations and adverse outcomes of regulatory concern.[72] In February 2002, the OECD member countries decided that exploratory workshops should be held to develop a strategy concerning the future application of toxicogenomics in regulatory assessment of chemical safety. The OECD agreed, along with the International Program on Chemical Safety (IPCS), to hold twin workshops. The first workshop, focusing on human health aspects, was held on November 17–19, 2003 in Berlin, with the IPCS in the lead. This workshop recommended that case-studies involving chemicals with known modes-of-action and available toxicogenomics data should be developed and toxicogenomic endpoints should be added to existing guideline protocols in the area of hazard identification.[73]

Genomics information will not fundamentally alter risk assessment practices, but genomics will provide powerful new tools and insights to use for risk assessment.[74] Toxicogenomics, proteomics, and metabolomics technologies are at various stages of regulatory development and acceptance. Microarray technology, along with proteomics and metabolomics experiments, will help the EPA better screen the vast number of chemicals it is mandated to regulate. Using traditional tests, it can easily take 3–4 years and $20 million to test the toxicity of a pesticide.[75]

15.6.1 FOOD ADDITIVES

Food safety and food security are important global issues. Genomics, proteomics, and metabolomics make it possible to elucidate pathogen behavior at the molecular level and to develop better detection and typing systems. Omics-based tools enable researchers to explore complex biological processes in a quantitative and integrative manner via a systems biology approach. These methods of analysis are helping to identify genes that are potential targets for interventions, genes that play a role in pathogenesis, and genes that are responsible for specific survival and virulence characteristics.[76]

We often think of food additives as complex chemical substances produced by our modern society. However, the use of food additives dates to ancient times. Early people used salt to preserve meat and fish, herbs and spices to season foods, sugar to preserve fruits, and vinegar to pickle cucumbers.[77] Today American food manufacturers use nearly 3000 direct food additives. Some of these additives sound familiar, for example salt, sugar, yeast, etc., and others are unfamiliar, like ascorbic acid, butylated hydroxyanisol (BHA), sodium benzoate, etc. All food additives serve a useful function and must be approved by the Food and Drug Administration (FDA) before they can be included in food. Food additives are regulated by the FDA under the authority of the Food, Drug and Cosmetic Act of 1938 and its amendments in 1958 and 1960, and the Food Quality Protection Act.[78] Food additives are authorized for use in the European Union (EU) on the basis that they present no hazard to the health of the consumer at the proposed level of use.

In 2000, the International Life Sciences Institute (ILSI)-Europe Acceptable Daily Intake Task Force performed a safety evaluation of the combinatorial intake of 350 approved food additives.[60] The International Program on Chemical Safety (IPCS) is a cooperative program of the World Health Organization (WHO), the International Labour Organization (ILO), and the United Nations Environment Programme (UNEP). The main objectives of its risk assessment work are to provide, on a global basis, consensus assessments of priority chemicals (including pesticides) and to promote the development, validation, use, and harmonization of sound methodologies for risk assessment. The IPCS provides a number of comprehensive risk assessment documents, the most prominent of which are the Environmental Health Criteria monographs.[79] IPCS has a range of activities to evaluate the safety of food components, toxic natural constituents and contaminants as well as food additives and residues of pesticides and veterinary drugs. These activities include providing the secretariats and scientific advice to the Joint Expert Committee on Food Additives (JECFA) and the Joint Meeting on Pesticide Residues (JMPR) and carrying out international risk assessments of chemicals of concern such as acrylamide, produced as a by-product of food processing and cooking. For the assessment of chemicals in food, as with other chemicals assessment work, the development, harmonization, and use of internationally accepted, scientifically sound and transparent principles and methods are vitally important.[80]

In relation to chemical hazard/risk assessment, toxicogenomics could provide tools for (1) improving the understanding of mechanisms of toxicity, (2) identifying biomarkers of toxicity and exposure, (3) reducing uncertainty in grouping of chemicals for assessments, Quantitative Structure Activity Relationships (Q)SARs, inter-species extrapolation, and effects on susceptible populations, and (4) providing alternative methods for chemical screening, hazard identification, and characterization.[81]

15.7 TOXICOGENOMICS IN RISK ASSESSMENT

Genomic technologies are valuable tools for assessing drug and chemical safety. In particular, toxicogenomics provides additional information to aid in the assessment of dose-response relationships, but it was important to interpret such data in collaboration with information from existing conventional toxicological methods such as microscopical examination of fixed, sectioned, and stained tissues. Regarding drug development, one major application of toxicogenomics is the identification and characterization of early predictive biomarkers of toxicity. More sensitive and specific biomarkers will contribute not only to the understanding of the mechanisms involved in drug-induced toxic responses but also to improving human risk assessment that is fundamental to the drug approval process by the FDA or other regulatory agencies.[82] Microbial risk assessment is a critical discipline in the area of food safety. DNA microarray technology is highly applicable to the study of food safety.[83]

15.7.1 Applications and Needs

Toxicogenomics is changing the way toxicology is performed. It has significant contributions toward deciphering mechanisms of toxicity and aiding in risk assessment. Although most expression signatures still can't be interpreted or linked to biologic effects, officials at both the FDA and the EPA express optimism that the use of microarray data could help them better protect public health. First

TABLE 15.1
Toxicogenomics in Risk Assessment: Applications and Needs

Applications	Needs
Routine toxicogenomic analysis for compounds prior to innovation	Utility and impact will drive acceptance
Validation and improvements in technology	The statistical point of view
Molecular approaches to biomarker (specific and sensitive) identification and selection	Special statistical analysis
Biomarkers of exposure	Conserved genetic, protein, and metabolite annotations
Biomarkers of toxicity	Consistent analysis approaches
Investigating changes in thousands of genes, proteins or metabolites simultaneously	Provide the molecular basis of pharmacological and toxicological responses
Unraveling of mechanism of action	Effective and informative database
	Bioinformatics
Calibrating of instruments and standardization of assays	Availability of reference samples

of all, predictive nature needs to be established before widespread use in regulatory risk assessments. Reproducibility, reliability, and relevance must be implemented by validated methods. Applications of toxicogenomics are: (1) Assisting predevelopment toxicology by facilitating more rapid screens for compound toxicity (prospective prediction), (2) Allowing compound selection decisions to be based on safety as well as efficacy, (3) Retrospective explanation of toxicological liabilities, (4) An enhanced ability to extrapolate accurately between experimental animals and humans in the context of risk assessment, (5) Benchmark effects relative to competitor molecules, and (6) Verification of safety and validation of target-mediated toxicity.[12,84,85] Table 15.1 summarizes applications and needs of toxicogenomics in risk assessment.

15.8 FUTURE DIRECTIONS AND CHALLENGES

Current technology often generates more questions than answers for application in toxicogenomics and specifically in regulation. Integrating gene expression data with other omics data (proteomics and metabolomics research) as well as data from traditional toxicological studies and chemical structure information will provide more reliable conclusions and opportunity to develop a comprehensive and robust toxicogenomics software system to accommodate diverse data from various sources.[86] First of all, the reliability of technologies has to be considered. Sources of variability of both methodological and biological origin must be identified, quantified, and minimized to the extent possible. A major issue is what is needed to calibrate the instruments and standardize the assays? A problem which has been addressed is general accessibility of the databases. For instance, it is important to be able to correlate metabolic data with protein and gene expression. The question of how to validate a biomarker is an important issue. At present it is understood as confidence evaluation, raising and answering questions on reliability, field of application, and interpretation. In general terms, it can be stated that the extent of and the procedures for validation of a biomarker depend on the purpose of its use.[87] One of the most challenging aspects of implementing toxicogenomics in risk assessment involves establishing the appropriate supportive infrastructure to facilitate the effective management, integration, interpretation, and sharing of toxicogenomic data. Although databases provide effective data management solutions, the ability to integrate toxicogenomic data across chemical and biological space to develop mechanistic pathways and networks remains limited. With few exceptions, the lack of comprehensive reference databases hinders the ability to elucidate mechanisms of toxicity associated with changes in profiles regarding the required data.[88]

Research efforts are expended from various sides including government, academic organizations, and industry, realizing the importance of toxicogenomics and its down-stream effects.[89]

Toxicogenomic studies generate large quantities of raw data. Management of the full set of "raw" data collected from studies is a major challenge. No single method can currently be considered the most appropriate way to analyze toxicogenomic data.[90] No attempt should therefore be made at present to define specific statistical methods or software that should be used for the analysis of toxicogenomic studies. There have been rapid and extensive developments regarding the application of toxicogenomic methods to toxicological hazard characterization. It is possible that toxicogenomic approaches could potentially be applied in a limited number of situations but could not be applied routinely to toxicological risk assessment.

REFERENCES

1. Stierum, R. et al., Toxicogenomics concepts and applications to study hepatic effects of food additives and chemicals, Review, *Toxicol. Appl. Pharmacol.*, 207, 179, 2005.
2. Committee on Biological Markers of the National Research Council, Biological markers in environmental health research, *Environ. Health Perspect.*, 74, 3–9, 1987.
3. Koop, R., Combinatorial biomarkers: From early toxicology assays to patient population profiling, Review, *Drug Discov. Today*, 10, 781, 2005.
4. Au, W.W., Usefulness of biomarkers in population studies: From exposure to susceptibility and to prediction of cancer, *Int. J. Hyg. Environ. Health*, 210, 239, 2007.
5. Biomarkers Definitions Working Group, Biomarkers and surrogate endpoints, preferred definitions and conceptual framework, *Clin. Pharmacol. Ther.*, 69, 89, 2001.
6. Paustenbach, D. and Galbraith D., Biomonitoring and biomarkers: Exposure assessment will never be the same, *Environ. Health Perspect.*, 114, 1143, 2006.
7. Iradale, R. and Longley, M., Genetics, genicity and genomics—A new language for a new era, *Euroscreen, Ethics Genet. Screening*, 7, 5–8, 1997, ISSN 1355–7904.
8. Genomics and Its Impact on Science and Society, The Human Genome Project and Beyond U.S. Department of Energy Genome Research Programs, http://genomics.energy.gov (accessed April 30, 2009).
9. http://www.cdc.gov/genomics/activities/file/print/2009report/lingo.pdf.
10. Freeman, T., High throughput gene expression screening: Its emerging role in drug discovery, *Med. Res. Rev.* 20, 197, 2000.
11. Trevino, V., Falciani, F., and Barrera-Saldaña, H.A., DNA microarrays: A powerful genomic tool for biomedical and clinical research, *Mol. Med.*, 1, 2007, Epub ahead of print.
12. Pennie, W. et al., The principles and practices of toxicogenomics: Applications and opportunities, *Toxicol. Sci.*, 54, 277, 2000.
13. Ulrich, R. and Friend, S.H., Toxicogenomics and drug discovery: Will new technologies help us produce better drugs? *Nat. Rev. Drug Discov.*, 1, 84, 2002.
14. Lord, P.G., Nie, A., and McMillian, M., The evolution of gene expression studies in drug safety assessment, Johnson & Johnson Pharmaceutical Research & Development, *Toxicol. Mech. Meth.*, 16, 51, 2006.
15. Neelam, C., Shubha, R., and Padh, H., Pharmacogenetics: Genetic basis for rational drug therapy, Review, *Indian J. Pharm. Sci.*, 69, 180, 2007.
16. Pirmohamed, M., Pharmacogenetics and pharmacogenomics, *Br. J. Clin. Pharmacol.*, 52, 345, 2001.
17. Arranz, M.J. and De Leon, J., Nature Publishing Group, pharmacogenetics and pharmacogenomics of schizophrenia: A review of last decade of research, feature review, *Mol. Psychiatry*, 1, 2007.
18. Goldman, B.R., Pharmacogenomics: Privacy in the era of personalized medicine, *Northwest. J. Tech. Intellect. Property*, 4, 84, 2005.
19. Weber, W.W. and Cronin, M.T., Pharmacogenetic testing, *Clin. Chem.*, 1–24; *Encyclopedia of Analytical Chemistry*, Meyers, R.A., Ed., John Wiley & Sons Ltd, Chichester, U.K., ISBN 0471 97670 9.
20. Evans, W.E. and Relling, M.V., Insight review articles, moving towards individualized medicine with pharmacogenomics, *Nature*, 429, 464, 2004.
21. http://www.ncbi.nlm.nih.gov (accessed May 5, 2009).
22. Lindpainther, K., The search or individualized therapies role of pharmacogenomics in drug discovery and therapeutics, in *Pharmacogenomic,* Licinio, J. and Mweinheim, W., Eds., Wiley-VCH Verlag GmbH, Weinheim, Germany, 2002.

23. Orphanides, G. and Kimber, I., Toxicogenetics: Applications and opportunities, *Toxicol. Sci.* 75, 1, 2003.
24. Collins, A., Lonjou, C., and Morton, N.E., Genetic epidemiology of single-nucleotide polymorphisms, *Med. Sci.*, 96, 15173, 1999.
25. Saito, H., Translation of the human genome into clinical allergy, Review, *Allergol. Int.,* 52, 65, 2003.
26. http://www.ncbi.nlm. nih.gov/SNP (accessed May 05, 2009).
27. Sherry, S.T., Ward, M., and Sirotkin, K., dbSNP-database for single nucleotide polymorphisms and other classes of minor genetic variation, *Genome Res.*, 9, 677, 1999.
28. Zhang, F. and Zhao, Z., SNPNB: Analyzing neighboring-nucleotide biases on single nucleotide polymorphisms (SNPs), *Bioinformatics*, 21, 2517, 2005.
29. Collins, F.S. et al., A vision for the future of genomics research, *Nature*, 422, 835, 2003.
30. Venter, J.C. et al., Genomics: Shotgun sequencing of the human genome, *Science*, 280, 1540, 1998.
31. Venter, J.C. et al., The sequence of the human genome, *Science*, 291, 1304, 2001.
32. The Genome International Sequencing Consortium, Initial sequencing and analysis of the human genome, *Nature*, 409, 860, 2001.
33. Collins, F.S. et al., The members of the DOE and NIH planning groups: New goals for the U.S. Human Genome Project: 1998–2003, Review, *Science*, 282, 682, 1998.
34. Van Ommen, G.J., The Human Genome Project and the future of diagnostics, treatment and prevention, *J. Inherit. Metab. Dis.*, 25, 183, 2002.
35. Penny, M.A. and McHale, D., Pharmacogenomics and the drug discovery pipeline: When should it be implemented? *Am. J. Pharmacogenom.*, 5, 53, 2005.
36. Mattes, W.B. et al., Database development in toxicogenomics: Issues and efforts genomics and risk assessment, mini monography, *Environ. Health Perspect.*, 112, 495, 2004.
37. Gmuender, H., Wendt, M., and Koenig, J., Toxicogenomics and biomarker discovery for the prediction of long term toxicity, P24-01, S293, Abstracts of the EUROTOX 2006/6 CTDC Congress-43rd Congress of the European Societies of Toxicology & 6th Congress of Toxicology in Developing Countries, *Toxicol. Lett.*, 164S, S1–S324, 2006.
38. Nuwaysir, E.F. et al., Microarrays and toxicology: The advent of toxicogenomics, *Mol. Carcinog.*, 24, 153, 1999.
39. http://iccvam.niehs.nih.gov/methods/acutetox/invidocs/NCTPaper.pdf (accessed May 12, 2009).
40. Groten, J.P. et al., Toxicology of chemical mixtures: A challenging quest along empirical sciences, *Environ. Toxicol. Pharmacol.*, 18, 185, 2004.
41. Evans, W.E. and McLeod, H.L., Pharmacogenomics: Drug disposition, drug targets, and side effects, *N. Engl. J. Med.,* 348, 538, 2003.
42. Lovett, R.A., News focus toxicogenomics: Toxicologists brace for genomics revolution, *Science*, 289, 536, 2000.
43. Gallagher, W.M., Tweats, D., and Koenig, J., Omic profiling for drug safety assessment: Current trends and public-private partnerships, *Drug Discov. Today*, 14, 337, 2009.
44. Tong, W. et al., ArrayTrack—Supporting Toxicogenomic Research at the U.S. Food and Drug Administration National Center for Toxicological Research, *Environ. Health Perspect.*, 111, 1819, 2003.
45. Schoulink, K.G., Functional and comparative genomics of pathogenic bacteria, *Curr. Opin. Microbiol.*, 5, 20, 2002.
46. Heijne, W.H.M. et al., Systems toxicology: Applications of toxicogenomics, transcriptomics, proteomics and metabolomics in toxicology, *Expert Rev. Proteomics*, 2, 767, 2005.
47. El-Sohemy, A., Nutrigenetics, *Forum Nutr.*, 60, 25, 2007.
48. http://www.nugo.org/facts/28383/24027 (accessed May 09, 2009).
49. Patton, G.W. et al., Transcriptomic response to differentiation induction, *BMC Bioinform.*, 7, 81, 2006.
50. Aebersold, R., Constellations in a cellular universe commentary, *Nature*, 422, 115, 2003.
51. Harrigan, G.G., Metabolomics a 'systems' contribution to pharmaceutical discovery and drug development metabolomics, *Drug Discov. World*, Spring, 39–46, 2006.
52. Griffin, J.L. and Rosiers, C.D., Applications of metabolomics and proteomics to the mdx mouse model of Duchenne muscular dystrophy: Lessons from downstream of the transcriptome, *Genome Med.*, 1, 32, 2009.
53. Thongboonkerd, V., Proteomics in nephrology: Current status and future direction, *Am. J. Nephrol.*, 24, 360, 2004.
54. http://www.fda.gov (accessed May 09, 2009).

55. Tang, J. et al., Integrating post-genomic approaches as a strategy to advance our understanding of health and disease, *Genome Med.*, 30, 35, 2009.
56. Fratamico, P. and Luchansky, J., Applications of omics for food safety and security, Meeting Abstract, International Association for Food Protection, 2007, S19.
57. Jain, K.K., Challenges of drug discovery for personalized medicine, *Curr. Opin. Mol. Ther.*, 8, 487, 2006.
58. Jain, K.K., Role of pharmacoproteomics in the development of personalized medicine, *Pharmacogenomics*, 5, 331, 2004.
59. Ozdemir, V. et al., Genome-environment interactions and prospective technology assessment: Evolution from pharmacogenomics to nutrigenomics and ecogenomics, *OMICS*, 13, 1, 2009.
60. Groten, J.P. et al., An analysis of the possibility for health implications of joint actions and interactions between food additives, *Regulat. Toxicol. Pharmacol.*, 31, 77, 2000.
61. Liu-Stratton, Y., Roy, S., and Sen, C.K., DNA microarray technology in nutraceutical and food safety, *Toxicol. Lett.*, 150, 29, 2004.
62. Committe on Toxicity of Chemicals in Food Consumer Products and the Environment (COT), Toxicogenomics/proteomics/metabolomics: Update on literature retrieved after February 2004, TOX/2004/26, 1–12, www.food.gov.uk/multimedia/pdfs/TOX-2004-26.pdf (accessed April 30, 2009).
63. Jiang, Y. et al., Changes in gene-expression in carbon tetrachloride induced hepatotoxicity in mice following a 4 week repeated administration schedule and investigation of reversal of effects, *Toxicol. Sci.*, 79, 404–410, 2004.
64. Coen, M. et al., An integrated transcriptomic/metabonomic approach to the acute toxicity of paracetamol in mice, *J. Pharm. Biomed. Anal.*, 35, 93–105, 2004.
65. Keun, H.C. et al., A novel approach to scaling and comparing metabonomic trajectory data between experiments, *Chem. Res. Toxicol.*, 17, 579–587, 2004.
66. Kier, L.D. et al., Applications of microarrays with toxicologically relevant genes (tox genes) for the evaluation of chemical toxicants in Sprague Dawley rats in vivo and human hepatocytes in vitro, *Mutat. Res.*, 549, 101, 2004.
67. Committees on Toxicity, Mutagenicity and Carcinogenicity of Chemicals in Food, Consumer Products and the Environment (COT, COM, COC), Joint working paper on the use of toxicogenomics in toxicology, Update on statement published in 2002, Tox/2004/38 Annex B.
68. Committee on Toxicity of Chemicals in Food, Consumer Products and the Environment, Exposure to multiple additives and chemical contaminants in food, Tox/2004/39, http://www.food.gov.uk/science/ouradvisors/toxicity/cotreports/cotwgreports/cocktailreport (accessed September 09, 2008).
69. Huang, S.M. and Lesko, L.J., Drug-drug, drug–dietary supplement, and drug–citrus fruit and other food interactions: What have we learned? *J. Clin. Pharmacol.*, 44, 559, 2004.
70. Cheng, S.L. et al., Toxicogenomics of Kojic acid on gene expression profiling of A375, human malignant melanoma cells, *Biol. Pharm. Bull.*, 29, 655, 2006.
71. Mitchell, B.L. et al., Evaluation of matrix-assisted laser desorption/ionization-time of flight mass spectrometry proteomic profiling: Identification of alpha 2-HS glycoprotein B-chain as a biomarker of diet, *Proteomics*, 5, 2238, 2005.
72. Klaper, R. et al., Current use and future needs of genomics in ecological and human health risk assessment. International Workshop Report, EPA/600/R-04/039.
73. http://www.oecd.org (accessed May, 12, 2009).
74. Luebke, W.R. et al., Immunotoxicogenomics: The potential of genomics technology in the immunotoxicity risk assessment process, *Toxicol. Sci.*, 94, 22, 2006.
75. Freeman, K., Toxicogenomics data: The road to acceptance, *Environ. Health Perspect.*, 112, A678, 2004.
76. Fratamico, P. and Luchansky, J., Applications of omics for food safety and security, Meeting Abstract, Microbial Food Safety Research Unit, 19.
77. Food additives, U.S. Food and Drug Administration, FDA/IFIC; Food and Drug Administration/International Food Information Council Brochure, January 1992.
78. Enhancing the regulatory decision-making approval process for direct food ingredient technologies, workshop summary, Food Forum, Food and Nutrition Board, Institute of Medicine, National Academy Press, Washington, DC, 1999.
79. Younes, M. and Sonic-Mullin, C., Concepts of the international programme on chemical safety in the assessment of risks to human health from exposure to chemicals, *Int. J. Toxicol.*, 16, 461, 1997.
80. http://www.who.int/ipcs/food/en (accessed November 15, 2008).
81. Tong, W. et al., Quantitative structure-activity relationships (QSARs) for estrogen binding to the estrogen receptor: Predictions across species, *Environ. Health Perspect.*, 105, 1116, 1997.

82. Rosenblum, L.Y., Toxicogenomic applications to drug risk assessment, *Environ. Health Perspect.*, 111, A804, 2003.
83. Roy, S. and Sen, C.K., cDNA microarray screening in food safety, *Toxicology*, 221, 128, 2006.
84. Gatzidou, E.T., Zira, A.N., and Theocharis, S.E., Toxicogenomics: A pivotal piece in the puzzle of toxicological research, Review, *J. Appl. Toxicol.*, 27, 302, 2007.
85. Khor, T.O., Ibrahim, S., and Kong, A.N., Toxicogenomics in drug discovery and drug development: Potential applications and future challenges, Review, *Pharm. Res.*, 23, 1659, 2006.
86. Tong, W. et al., Development of public toxicogenomics software for microarray, data management and analysis, *Mutat. Res.*, 549, 241, 2004.
87. Gundert-Remy, U. et al., Molecular approaches to the identification of biomarkers of exposure and effect—report of an expert meeting organized by COST Action B15, *Toxicol. Lett.*, 156, 227, 2005.
88. Boverhof, D.R. and Zacharewski, T.R., Toxicogenomics in risk assessment: Applications and needs, *Toxicol. Sci.*, 89, 352, 2006.
89. Chung, T.H. et al., Recent progress in toxicogenomics research in South Korea, *BMC Proc.*, 209, 3, 2009.
90. Smart, R.C. and Hodgson, E., *Molecular and Biochemical Toxicology*, 4th edn., John Wiley & Sons, Inc. Publication, Hoboken, NJ, 2008.

Index

A